SPRINGER PROTOCOLS HANDBOOKS

For further volumes:
http://www.springer.com/series/8623

Springer Protocols Handbooks collects a diverse range of step-by-step laboratory methods and protocols from across the life and biomedical sciences. Each protocol is provided in the Springer Protocol format: readily-reproducible in a step-by-step fashion. Each protocol opens with an introductory overview, a list of the materials and reagents needed to complete the experiment, and is followed by a detailed procedure supported by a helpful notes section offering tips and tricks of the trade as well as troubleshooting advice. With a focus on large comprehensive protocol collections and an international authorship, Springer Protocols Handbooks are a valuable addition to the laboratory.

Protocols in Advanced Genomics and Allied Techniques

Aruna Pal

Animal Genetics and Breeding, West Bengal University of Animal and Fishery Sciences, Kolkata, West Bengal, India

Springer

Aruna Pal
Animal Genetics and Breeding
West Bengal University of Animal and Fishery Sciences
Kolkata, West Bengal, India

ISSN 1949-2448 ISSN 1949-2456 (electronic)
Springer Protocols Handbooks
ISBN 978-1-0716-1817-2 ISBN 978-1-0716-1818-9 (eBook)
https://doi.org/10.1007/978-1-0716-1818-9

This Springer imprint is published by the registered company Springer Science+Business Media, LLC part of Springer Nature.
The registered company address is: 1 New York Plaza, New York, NY 10004, U.S.A.

This book is dedicated to my parents, grandmother, grandfather, and my teachers.

Preface

I aimed to write a book with a complete structure of genomics studies along with allied topics essential for data validation. The idea first came to my mind while I was working as a Scientist in the Department of Biotechnology, Ministry of Science and Technology, Government of India. I worked as convenor of Human Genetics and Genomics Task Force. While evaluating the projects, I felt the need of a comprehensive book covering genomic technique and other techniques, which will be beneficial for the researcher to perform lab work with supplementation of bioinformatics and other analysis.

This book is meant for early stage researchers and students in particular. For any good publication it is equally important to conceive a fact with the help of multiple genomic and immunological tools. Next generation sequencing forms one of the main bases for advanced genomic tools. In this book, an attempt has been made to explain advanced techniques in a simple way, so that new researchers can easily conceive and execute it. It will also be equally helpful for the students of biotechnology. The content of the book starts with basic genomic techniques and gradually advanced techniques are described, particularly, next generation sequencing in a simple language and brief protocol. Nowadays for any good publication, multiple techniques need to be employed for a rigid conclusion. Hence, genomic techniques along with allied biotechnological and immunological techniques are used for research purposes to address a problem. For attaining a suitable conclusion, analysis with suitable bioinformatics tools is of utmost importance for any proteomics study, evolutionary study, or NGS analysis with big data. The basic feature of this book is easily perceptible language; it includes a wide range of topics commonly dealt under genomics and allied science, and the topics were described with underlying basic principles, suitable figures, illustrations, brief methodology, and ready references. Currently, there is an urgent need for such kinds of books with a wide range of topics together.

This book provides a hands-on tool for all the scientific researchers from both academic and R&D institutions in both government and private sectors, particularly from the field of life science. Apart from students, this book will be very much useful for teachers who teach them, since it covers a wide range of topics in concise and easy language for them to assimilate.

A course on genomics under biotechnology is offered at both undergraduate and postgraduate levels; this book will be most suitable for those who learn and also those who teach.

While writing this book, I felt excited to cover newer research progress. One interesting fact I would like to share is that the moment I feel a chapter is complete, I would find a newly published article or a very newer technique and achievement. Immediately I would include the new finding. Thus the time of my submission to the publisher has been extended. I believe science is very fascinating and there is no end to learn newer concepts. I believe that it is an attempt to compile the techniques to enable students and new researchers to explore further advanced concepts of science to reach new heights. It is our concerted effort globally to raise science to a great horizon for the benefit of mankind.

I believe that it is not only my effort but also of my students, teachers, parents, children, husband, friends, and all my well-wishers who had equally motivated and helped me in achieving this goal. Being a mother, I felt perhaps I am depriving my children of quality time with them. But they would co-operate and observe me working. I feel that this work will equally motivate my children too when they will grow up as well as my students to work together for science.

Kolkata, West Bengal, India ***Aruna Pal***

Contents

About the Author

Aruna Pal is an Assistant Professor at West Bengal University of Animal Fishery Sciences, India. She has served as a Scientist and Convener of Human Genetics Task Force at the Department of Biotechnology, Ministry of Science & Technology, Government of India. She has also worked as a Veterinary Officer at the Government of West Bengal and headed the Quality Control Laboratory, Frozen Semen Bull Station, PBGSBS (under the Central Project for Cattle Buffalo Breeding Programme, Ministry of Agriculture, Government of India). She was awarded a Junior Research Fellowship (Indian Council of Agricultural Research) for pursuing her M.V.Sc, and a Senior Research Fellowship (Indian Veterinary Research Institute) for pursuing her Ph.D. She has published 21 research papers in highly reputed journals, 2 books, 6 book chapters, 66 gene bank accession numbers (NCBI), and 18 research reviews. She is a pioneer in the documentation of new breeds of sheep, buffalo, and duck from the State of West Bengal, India. She is also an active member of nine scientific societies. She has been the Principal Investigator and co-Principal Investigator for three and six research projects, respectively.

Check for updates

Chapter 1

Conventional and Basic Genomic Techniques

Abstract

In this chapter, certain very basic and primary technologies pertaining to genomics have been recapitulated. The topics covered include basic protocols for genomic DNA isolation through various techniques, bacterial DNA isolation, Plasmid DNA isolation, and other electrophoretic techniques for its checking and visualization as agarose gel electrophoresis and Polyacrylamide gel electrophoresis. This chapter also includes certain basic polymorphism techniques suitable for detection of polymorphism, such as Restriction fragment length polymorphism (RFLP), Single stranded conformation polymorphism (SSCP). High-quality DNA is extremely essential for next generation sequencing studies, will be discussed in later chapters. The basic principle involved in genomic DNA isolation is the destruction of lipid and protein respectively through SDS and proteinase. DNA is precipitated through isopropanol. Plasmid DNA isolation involves alkali lysis which destroys bacterial DNA. In the electrophoretic techniques, DNA fragments are separated based on their mobility which depends on their fragment size. RFLP involves cutting of DNA fragments based on restriction enzymes, whereas SSCP involves differences in conformation of single stranded DNA due to their nucleotide variation and their mobility through PAGE.

Keywords DNA isolation, Electrophoresis, RFLP, SSCP, PAGE

In this chapter, we shall be dealing with some basic genomic techniques refreshing which form the basic platform to start with. The following topics will be dealt with as listed below:

1. Genomic DNA isolation.
2. Plasmid DNA isolation from *Escherichia coli* culture.
3. Horizontal Agarose Gel electrophoresis.
4. Vertical Polyacrylamide Gel electrophoresis.
5. Quality checking of DNA.
6. (RFLP)Restriction fragment length Polymorphism.
7. (SSCP) Single stranded conformation Polymorphism.

Aruna Pal, *Protocols in Advanced Genomics and Allied Techniques*, Springer Protocols Handbooks,
https://doi.org/10.1007/978-1-0716-1818-9_1,

1 Genomic DNA Isolation

Points to Remember

1. DNA is present in the nucleus of the cell. Apart from DNA, cell contains protein and lipid.
2. The aim is to destroy protein with Proteinase-K and lipid with SDS and separate the cell debris by extraction through organic solvent.
3. DNA is precipitated by isopropanol.
4. Good quality DNA is the primary criteria for any molecular biology work including next generation sequencing.

1.1 Genomic DNA Isolation from Mammalian Blood

1.1.1 Basic Principle

Genomic DNA is present in the nucleus of the cell. DNA is packaged with histone protein and present in the nucleus as chromatin material or chromosomes. Animal cell has lipoprotein cell membrane, which forms the cell boundary. A schematic and basic diagram of a typical animal cell is described in Fig. 1. Besides cell membrane, it has another cell organelle. Mammalian blood contains RBC, WBC and platelets. Mammalian RBC lacks nucleus, except Camel. Since DNA is present in the nucleus of WBC, RBC needs to be lysed with hypotonic solution (RBC lysis buffer). A schematic representation of DNA organization within the cell has been depicted in Fig. 2. DNA extraction buffer which is basically a hypotonic solution, is added to WBC. The Lipid present in the cell will be neutralized by SDS and Protein present in the cells will be degraded by Proteinase-K at 55 °C for 8–9 h. Phenol-Chloroform-Isoamyl alcohol will remove the cell debris from the aqueous solution of DNA. Sodium acetate acts as salt and Isopropanol precipitates DNA. DNA samples were incubated at 60 °C for a period of 2 h for the inhibition of action of DNAse if present and for dissolving the DNA pellet properly into TE buffer.

1.1.2 Protocol

Step 1. Procurement of Samples (Blood or Tissue)

Blood was collected aseptically (10 mL) from jugular vein under sterile conditions from each animal in centrifuge tube with 0.5 mL of 2.7% EDTA solution (as anticoagulant) and mixed thoroughly with the anticoagulant. Then the blood samples were transported to the laboratory in an ice box. The samples were stored in the refrigerator at −20 °C for further analysis.

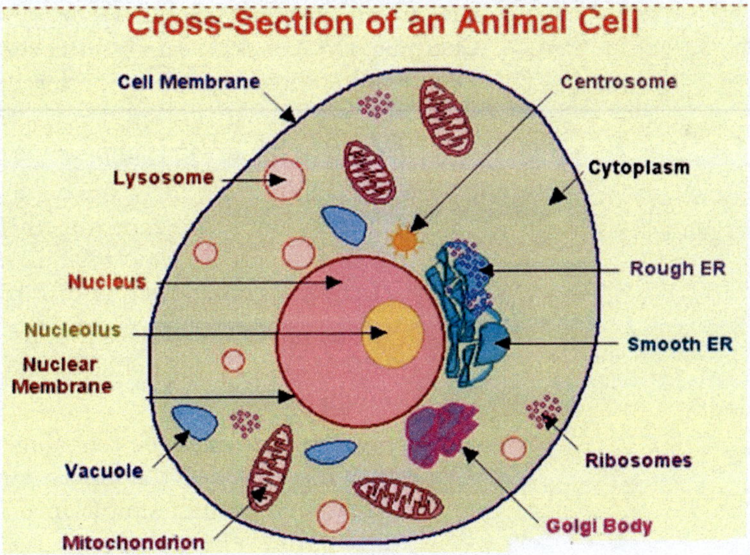

Fig. 1 A typical animal cell

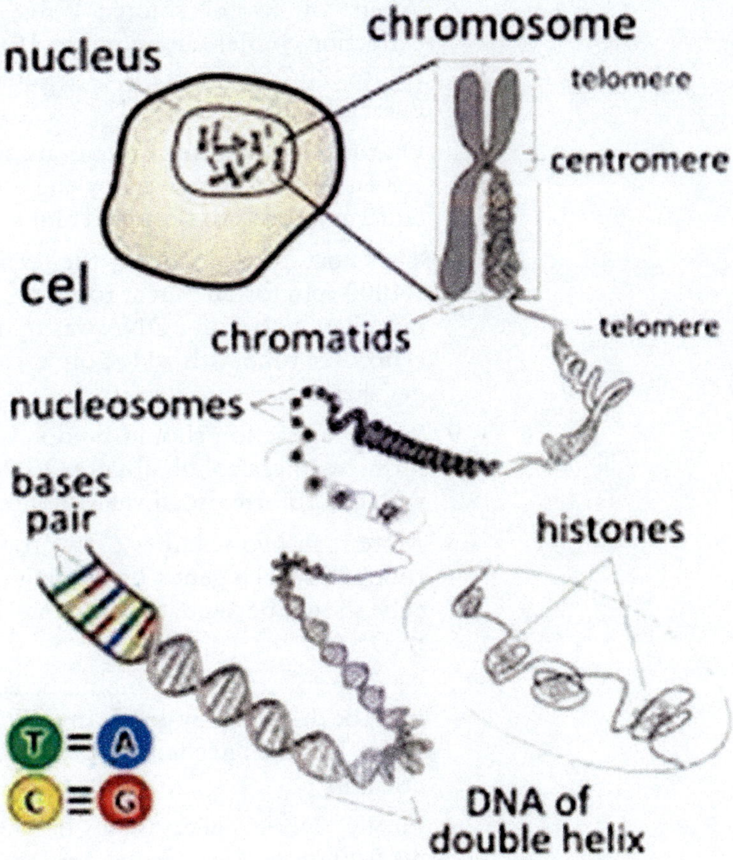

Fig. 2 Schematic representation of DNA organization in nucleus

Step 2. Extraction of Genomic DNA

Genomic DNA was isolated from the white blood cells using phenol-chloroform extraction method.

1. Thawing of the frozen samples was done in room temperature to get a uniform solution.
2. The tubes should be centrifuged @4000 rpm for 20 min at 4 °C. The supernatant possess plasma and lysed RBC need to be carefully pipetted out and discarded.
3. The pellet of WBC and RBC is to be mixed with double volume of RBC lysis buffer(pre-chilled) and kept in ice for 10 min by thorough mixing.
4. The tubes are again to be centrifuged @4000 rpm for a period of 20 min at 4 °C and the supernatant with the lysed RBC is to be discarded through simple inversion of the tubes without disturbing the WBC pellet.
5. The **steps 3** and **4** should be repeated until the WBC pellet is free of reddish tinge.
6. When "off white" colored WBC pellet was obtained, DNA extraction buffer along with 10% SDS and Proteinase K (20 mg/mL solution) is to be added and kept at 55 °C for overnight.
7. On the next day, the contents are to be transferred into clean, sterile, autoclaved polypropylene tubes. Equilibrated phenol of equal amount is to be added and mixed gently.
8. The next step involves the centrifugation of the tubes @4000 rpm for 20 min at room temperature. The upper aqueous phase possessing DNA was transferred into a fresh 15 mL centrifuge tube with wide bore micro tip, so that DNA do not get sheared.
9. Similar extraction should be done once with phenol: chloroform: isoamyl alcohol mixture (25:24:1) followed by extraction with chloroform: isoamyl alcohol mixture (24:1).
10. More than two volumes of isopropanol is to be mixed to the tube and shaken gently by swirling the tube once or twice. The tube should be kept at room temperature for the precipitation of DNA.
11. The precipitated DNA should be centrifuged and supernatant must be discarded by gentle inversion without disturbing DNA pellet. It is then needed to be washed twice with 500 μL of 70% ethanol.
12. Finally, DNA pellet is to be air dried and dissolved in 40–100 μL of TE buffer as per requirement.

13. The Eppendorf tubes containing DNA samples should be incubated at 60 °C for 2 h. DNA pellet will dissolve properly into TE buffer.
14. Samples are needed to be stored at 4 °C for a week so that DNA gets dissolved completely and subsequently at −20 °C till further use.

Note: Genomic DNA can be isolated from any body tissue. Each cell from any body tissue contains exactly the same genome. Since blood is the easily accessible tissue, it is routinely employed for DNA isolation. Semen can also be used for this purpose in case of male animals, with some modifications. Hair follicle is also a good source. In forensic science, blood stain and hair follicles are the main source.

Note: Since the genomic DNA is of high molecular weight, it is present just outside the well. Smear is present in case of shearing of genomic DNA or if contamination with protein is present.

Shearing of DNA occurs if rough handling is practiced after **step 8**, when the cell ruptures and DNA is exposed to media.

Protein contamination can occur if less quantity of proteinase K is added, if optimum temperature (55 °C) or time is not maintained.

1.1.3 Flow Chart

Flow chart:

Blood samples (10 mL) are collected in vials containing 0.5mL of 0.5M EDTA

↓ Stored at -20◦C

Samples are taken out, thawed, mixed with 3 volumes of RBC lysis buffer (prechiled)

↓ Centrifuged at 3000rpm for 20 min at 4◦C

Supernatant possessing plasma and lysed RBC in the tube need to be slanted with the intact pellet

↓

Pellet containing WBC and RBC was mixed with more than twice volumes of prechilled RBC lysis buffer

↓ Kept in ice for 10 min, mixed gently and centrifuged at 4000 rpm for 20 mins at 4◦C

The tarry colored supernatant with lysed RBC was discarded

↓

Step 4 and 5 is to be repeated thrice or four times

↓

10 to 15mL of NSS (8.5% NaCl) should be added and mixed with the WBC pellet

↓ Centrifuged at 4000rpm for 20min

The supernatant was discarded and then DNA extraction buffer should be added

↓ Vortexed

Subsequently 10% SDS and at 200 μL should be added

↓ Mixed gently

Proteinase K should be added in two pulse and is to be kept at 50◦C for overnight

↓

The next day phenol should be added in equal volume to the digested material

↓ Centrifuged at 4000rpm for 20 min at 4◦C

Upper aqueous phase should be collected and equal volume with phenol: chloroform: isoamyl alcohol (25:24:1)

↓ Centrifuged at 4000rpm for 20 min at 4◦C

The upper aqueous phase is to be collected and isopropanol should be added to precipitate the DNA

↓

Ethanol wash of the DNA ppt should be performed and supernatant should be discarded, the pellet should be intact

↓

The pellet is to be air dried and dissolved in TE buffer. The DNA is ready for use.

DNA is present in the nucleus. The basic principle is to extract DNA from nucleus by rupturing the cell membrane and nuclear membrane [1–6].

Application of Isolated Genomic DNA

1. Genomic DNA is helpful for characterization of complete gene sequence for a particular gene.
2. High-quality genomic DNA is necessary for studying marker assisted selection or genomic selection.
3. High-quality genomic DNA is equally a requirement for next generation sequencing studies as, for example, genome wide association studies or exome sequencing.

In the above context, we have described genomic DNA isolation from blood. For genomic DNA isolation from other tissue, similar protocol is used starting from step 6, when there is no need for lysis of RBCs.

1.2 Salt-Extraction Procedure for Extraction of High-Quality Genomic DNA for PCR-Based Techniques

1.2.1 Protocol

About 50–100 mg (1 cm^2) of tissue can be employed for DNA extraction. The fresh tissue should be homogenized for 10–15 s in 400 μL of sterile salt homogenizing buffer (0.4 M NaCl 10 mM Tris–HCl pH 8.0 and 2 mM EDTA pH 8.0), with the help of a Polytron Tissue Homogenizer. This is followed by 40 μL of 20% SDS (2% final concentration) and 8 μL of 20 mg/mL proteinase K (400 μg/mL final concentration). The samples are to be incubated at 55–65 °C for at least 1 h or overnight, followed by addition of 300 μL of 6 M NaCl (NaCl saturated H_2O) to each sample. Samples should be vortexed for 30 s at maximum speed, then centrifuged for 30 min at 10,000 × *g*. The supernatant should be transferred to fresh tubes. The next step is addition of isopropanol (equal volume) each sample mixed well followed by incubation at −20 °C for 1 h. Samples are then to be centrifuged for 20 min, 4 °C, at 10,000 × *g*. The pellet should be washed with 70% ethanol, dried and finally resuspended in 300–500 μL sterile dH_2O. Genomic DNA (5–15 ng) is dissolved in 10 μL of ddH_2O [7].

The purity of the DNA can be assessed from the A260/A280 ratio averaged as more than 1.77 for all organisms. RNA contamination or degradation of DNA will not be observed. The yield of DNA will range from 500 to 800 ng/mg fresh weight for all individuals sampled. The amount of tissue required for this method is less. 50 mg tissue was able to yield sufficient DNA so as to perform more than 3000 numbers of PCR reactions.

1.2.2 Bacterial Genomic DNA Isolation

Miniprep of Bacterial Genomic DNA

Bacterial DNA is to be isolated from a saturated liquid culture. The basic principle is lysis and removal of proteins by digestion with proteinase K. Cell wall debris, polysaccharides, and remaining proteins are to be removed by selective precipitation with CTAB, resulting in recovery of high-molecular-weight DNA from the resulting supernatant through isopropanol precipitation.

1.2.3 Materials

1. TE buffer.
2. 10% sodium dodecyl sulfate (SDS).
3. 20 mg/mL proteinase K (stored in small single-use aliquots at −20 °C).
4. 5 M NaCl CTAB/NaCl solution.
5. 24:1 chloroform/isoamyl alcohol.
6. 25:24:1 phenol/chloroform/isoamyl alcohol.
7. Isopropanol.
8. 70% ethanol.

1.2.4 Procedure

1. 5-mL liquid culture for the bacterial strain of interest is to be inoculated. It is to be grown in suitable conditions for that strain (i.e., appropriate medium, drug selection, temperature) till the culture is saturated. This process may require several hours to several days, based on the growth rate.
2. The culture is spinned in 1.5 mL microcentrifuge for 2 min, or till a compact pellet form. The supernatant is discarded.
3. The pellet is resuspended in 567 μL TE buffer through a process of repeated pipetting. 30 μL of 10% SDS and 3 μL of 20 mg/mL proteinase K should be added to attain a final concentration of 100 μg/mL proteinase K in 0.5% SDS, followed by thorough mixing and incubation for 1 h at 37 °C. The solution will appear viscous as soon as the detergent lyses the bacterial cell walls. It is not essential to predigest the bacterial cell wall with lysozyme.
4. 100 μL of 5 M NaCl is added and is to be mixed thoroughly. This is an important step since a CTAB–nucleic acid precipitate will occur if salt concentration drops below about 0.5 M at room temperature. The goal is the removal of cell wall debris, denatured protein, and polysaccharides complexed to CTAB, but keeping the nucleic acids intact in solution.
5. 80 μL of CTAB/NaCl solution is added and to be mixed thoroughly and incubated for 10 min at 65 °C.
6. An equal volume (0.7–0.8 mL) of chloroform/isoamyl alcohol is added, mixed thoroughly and centrifuged for 4–5 min in a microcentrifuge. This process removes CTAB–protein/polysaccharide complexes, leaving a white interface.

7. Aqueous, supernatant is to be transferred to a fresh microcentrifuge tube, leaving the interface behind. The next step is addition of an equal volume of phenol/chloroform/isoamyl alcohol. This is followed by centrifugation in a microcentrifuge for 5 min. A practical problem encountered is that with some bacterial strains the interface formed after chloroform extraction is not compact enough to allow easy removal of the supernatant. In such cases, most of the interface can be fished out with a sterile toothpick before removal of any supernatant. Remaining CTAB precipitate is then to be removed in the phenol/chloroform extraction.
8. The supernatant is transformed in a fresh tube. Added 0.6 volume of isopropanol for the precipitation of the nucleic acids. It is not essential to add salt since the NaCl concentration is already present. The tube is to be gently mixed till a stringy white DNA precipitate becomes clearly visible. The pellet is transferred to a fresh tube containing 70% ethanol as explained earlier. The precipitate may also be pelleted by centrifugation briefly at room temperature.

   Lack of stringy DNA precipitate in the above step indicates shearing of DNA into relatively low-molecular-weight pieces.
9. The DNA is washed with 70% ethanol for the removal of residual CTAB and again centrifuged at 5 min at room temperature to repellet it. The supernatant is removed and the pellet is briefly dried in a lyophilizer.
10. The pellet is dissolved in 100 μL TE buffer [6, 8, 9].

In Short:

Miniprep of Bacterial Genomic DNA

1. Bacterial strain is grown to saturation.
2. Spinned the 1.5 mL centrifuge tube for 2 min in microcentrifuge.
3. The content is resuspended in 567 μL TE buffer, 3 μL of 10% SDS, 3 μL of 20 mg/mL proteinase K. Mixed and incubated 1 h at 37 °C.
4. Added 100 μL of 5 M NaCl. Mix thoroughly.
5. Added 80 μL of CTAB/NaCl solution. Mix. Incubate 10 min at 65 °C.
6. Extracted with an equal volume of chloroform/isoamyl alcohol. Spinned the microcentrifuge tube for 5 min in microcentrifuge.
7. The aqueous phase is transferred to a fresh tube. The DNA is extracted with phenol/chloroform/isoamyl alcohol. Centrifuged for 5 min in microcentrifuge.

8. Transferred the aqueous phase to a fresh tube, followed by extraction of DNA with 0.6 vol isopropanol. Washed the precipitate with 70% ethanol. The supernatant is removed and pellet is dried in lyophilizer.
9. The pellet is resuspended in 100 μL TE buffer.

Large-Scale CsCl Prep of Bacterial Genomic DNA

There are certain cases when we need huge amount of exceptionally clean genomic DNA, particularly in Next generation sequencing (NGS) studies for the construction of genomic libraries, which will be discussed in details in later chapters. It involves purification of genomic DNA on a cesium chloride gradient on large scale.

Materials required: Cesium chloride 10 mg/mL, ethidium bromide, CsCl-saturated isopropanol or H_2O-saturated butanol 3 M sodium acetate, pH 5.2, centrifuge tubes. Wide-bored pipet 4-mL sealable centrifuge tubes, 3-mL plastic syringe with 15-G needle.

Preparation and Lysis of Cells

1. 100 mL culture of bacterial strain should be grown to saturation.
2. Cells are to be pelleted for 10 min at 6000 rpm). Supernatant is discarded. 50 mL Oakridge tube is sufficient for carrying out these operations.
3. Cells should be resuspended gently in 9.5 mL TE buffer. 0.5 mL of 10% SDS and 50 μL of 20 mg/mL proteinase K should be added and mix thoroughly and should be incubated for 1 h at 37 °C.

Precipitation and Purification of DNA

4. 1.8 mL of 5 M NaCl should be added and mix thoroughly.
5. 1.5 mL CTAB/NaCl solution should be added. This should be mixed thoroughly and incubate for 20 min at 65 °C.
6. An equal volume of chloroform/isoamyl alcohol is to be added and extracted thoroughly. Spinned for 10 min at 6000 × *g* (JA-20 rotor at 7000 rpm) at room temperature for the process of separation of the phases.
7. Aqueous supernatant should be transferred to a fresh tube with the help of a wide-bored pipette. In case the yield is high, the supernatant may be very viscous, an additional step for chloroform/isoamyl alcohol extraction, or a phenol/chloroform/isoamyl alcohol extraction, is optional but should not be necessary if the material is to be purified on a cesium chloride gradient.
8. 0.6 vol. isopropanol should be added and mixed until a stringy white DNA pellet precipitates out of solution and condenses into a tight mass. The precipitate should be transferred to 1 mL

of 70% ethanol in a fresh tube, by hooking it on the end of a Pasteur pipet that has been bent and sealed in a Bunsen flame.

9. The tube is centrifuged for 5 min at 10,000 × *g*. Supernatant is removed and the pellet should be redissolved in 4 mL TE buffer. This may take several hours to overnight— the DNA can be placed at 60 °C to hasten the process.
10. Then the DNA concentration should be measured on a spectrophotometer. The concentration is to be adjusted from 50 to 100 μg/mL, yielding upto 200–400 μg chromosomal DNA per 4 mL gradient. It is not recommended to spin larger quantities of chromosomal DNA on such a small gradient.
11. 4.3 g CsCl per 4 mL TE buffer should then be added and dissolve. 200 μL of 10 mg/mL ethidium bromide is then to be added and then should be transferred to 4-mL sealable centrifuge tubes. Volume adjustment should be done and balance tubes with CsCl in TE buffer (1.05 g/mL). Tubes must be sealed properly. Spin 4 h in a Beckman VTi80 rotor at 70,000 rpm, 15 °C, or overnight at 55,000 rpm, 15 °C.
12. Visualization of the gradient must be done under long wave UV lamp. A single band need to be visible. Removal of band should be done using a 15-g needle and a 3-mL plastic syringe. In case the DNA is intact, due to high-molecular-weight chromosomal DNA it will appear very viscous as the band is withdrawn from the gradient. As a result, it is advisable to use a wide-bore needle in order to avoid mechanical shearing of the DNA. As the band appears right at the top of the gradient, then the gradient is too dense. Amount of CsCl added in **step 11** is to be reduced.
13. Ethidium bromide should be removed by sequential extractions with CsCl-saturated isopropanol or water-saturated butanol.
14. Overnight dialysis should be done against 2 L TE buffer to remove CsCl.
15. DNA solution should be transferred to a fresh tube.

Note: Precipitation of DNA can also be carried out with 1/10th volume of 3 M Sodium acetate and 0.6 vol isopropanol, which can be further resuspended to required concentration.

Large-Scale CsCl Prep of Bacterial Genomic DNA (In Short)

1. 100 mL culture of bacterial strain is to be grown to saturation.
2. Spin 10 min at 4000 × *g*.
3. Resuspension of pellet should be done in 9.5 mL TE buffer, 0.5 mL of 10% SDS, and 50 μL of 20 mg/mL proteinase K. Mixing and incubation for 1 h at 37 °C is to be done.

4. 1.8 mL of 5 M NaCl is to be added and should be mixed thoroughly.
5. 1.5 mL CTAB/NaCl should be added and mixed. Incubation should be done for 20 min at 65 °C.
6. Extraction should be carried out with an equal volume of chloroform/isoamyl alcohol by spinning for 10 min at 6000 × *g*, room temperature.
7. Aqueous phase is to be transferred to a fresh tube. Extraction with phenol/chloroform/isoamyl alcohol is to be performed if necessary. Spin as in **step 6**.
8. Aqueous phase is to be transferred to a fresh tube. Precipitation of DNA should be done with 0.6 vol isopropanol. Precipitate is to be washed with 70% ethanol. Supernatant should be removed and the pellet should be resuspended in 4 mL TE buffer.
9. Measure DNA concentration. Concentration is to be adjusted to give 50–100 μg/mL. 4.3 g CsCl per 4 mL TE buffer should be added. Then 200 μL of 10 mg/mL ethidium bromide should be added. Transfer to sealable centrifuge tubes. Spinning should be done for 4 h at 70,000 rpm, 15 °C.
10. Visualize gradient with UV light. Bands are to be removed.
11. Ethidium bromide extraction with CsCl-saturated isopropanol should be done.
12. Dialyze overnight against 2 L TE buffer.

Reagents and Solutions

CTAB/NaCl solution (10% CTAB in 0.7 M NaCl): 4.1 g NaCl should be dissolved in 80 mL water and 10 g CTAB (hexadecyl trimethyl ammonium bromide) is to added slowly while heating and stirring. If necessary, heat to 65 °C to dissolve. Final volume should be adjusted to 100 mL.

1.3 DNA Isolation with Magnetic Beads

DNA isolation with magnetic bead is a rapid method of DNA isolation, where the basic principle is that, initially DNA is bound to the magnetic beads and they are separated from the solution rapidly using magnetic separation strand. In the next step DNA is eluted with appropriate elution buffer and then also the magnetic beads are separated using magnetic separation strand.

The basic principle for this Protocol is relied on the lysis and nuclease-inactivating properties of two chaotropic agents, regarded as guanidinium hydrochloride (GuHCl) and guanidinium thiocyanate (GuaSCN). These are used together with the use of magnetic beads as the affinity matrix. 4.5 mL blood gives optimum result and better DNA quantity and quality.

1. The leucocyte cell pellet obtained from buffy coat, should be suspended in 250 μL PBS.

2. Volumes of 250 μL lysis buffer A (50 mM Tris–HCl, 50 mM EDTA, 4 M GuHCl, 10 mM $CaCl_2$, 1% v/v Triton X-100, 2% *N*-Lauroyl-Sarcosine, pH = 7.5) and 50 μL of proteinase K (22.4 mg/mL, MERK) should be added, followed by homogenization of the lysate and incubation in 56 °C for 1.5 h.
3. A volume of 600 μL of lysis buffer B (50 mM Tris–HCl, 25 mM EDTA, 6 M GuaSCN, 3% v/v Triton X-100, 6% *N*-Lauroyl-Sarcosine, pH = 5.5) should be added and the suspension should be incubated at 70 °C for 10 min.
4. 500 μL of absolute isopropanol should be added to the lysate.
5. In the next step, magnetic silica particles 40 μL (@ 50 mg/mL) and PMSi-H1.0–5 is to be added for DNA binding.
6. The samples are to be homogenized and placed to MagnaRack (Invitrogen, Carlsbad, CA). In the next step, the magnetic beads must be washed three times starting with 700 μL wash buffer A (25 mM Tris–HCl, 4 M GuHCl, 30% v/v absolute isopropanol, pH 6.6), then twice with 500 μL wash buffer B (10 mM Tris–HCl, 100 mM NaCl, 80% v/v absolute isopropanol, pH 6.6).
7. In the final step, the magnetic beads are transferred to new 2 mL tubes followed by incubation at room temperature for 10 min. Caps should be open during the process of incubation. DNA is recovered in 1200 μL TE buffer sequentially, each time with 600 μL TE buffer (10 mM Tris–HCl, pH = 9) preheated in 65 °C [9].

2 Plasmid DNA Isolation from *Escherichia coli* Culture

Points to Remember

1. DNA is present in the nucleus of the cell. Apart from DNA, cell contains protein and lipid.
2. The aim is to destroy protein with Proteinase-K and lipid with SDS and separate the cell debris by extraction through organic solvent.
3. DNA is precipitated by isopropanol.
4. Good quality DNA is the primary criteria for any molecular biology work including next generation sequencing.

2.1 Basic Principle of the Process

The basic difference between Plasmid DNA isolation and Genomic DNA isolation is that Plasmid is single stranded DNA, whereas Genomic DNA is double stranded. The basic principle of DNA isolation is similar as Genomic DNA. Alkali is used for genomic DNA lysis and plasmid is extracted.

2.2 Protocol

From the overnight grown cultures, the plasmid DNA should be taken and extracted following the small scale alkali lysis method:

1. Bacterial culture (1.5 mL twice in the same tube) should be centrifuged @ 13,000 rpm at 4 °C for 1 min and the supernatant is discarded.
2. The pellet is dissolved in 100 μL of ice-cold Solution I, vortexed thoroughly, followed by incubation at room temperature for 5 min. In the next step, 200 μL of freshly prepared Solution II (RT) is to be added to the suspension and mixed gently by inversion and incubated on ice for 5 min. Next, 150 μL of ice-cold Solution III should be added and inverted gently for mixing.
3. The bacterial lysate should be centrifuged @13,000 rpm at 4 °C for 5 min. The supernatant is then to be transferred to a new microcentrifuge tube.
4. In the next step, phenol: chloroform: isoamyl alcohol (25:24:1) is added equal in volume to that of the supernatant and vortexed. Then the mixture is centrifuged @13,000 rpm at 4 °C for 2 min followed by the separation of the upper aqueous layer. This is to be transferred to a new microfuge tube.
5. In the next step, equal volume of chloroform: isoamyl alcohol (24:1) should be added and centrifuged for 2 min and the upper aqueous layer is to be transferred to a new microfuge tube.
6. The plasmid precipitation is done by adding twice the volumes of ethanol at room temperature and incubated for 2–5 min at room temperature, after brief vortexing. Then it is to be centrifuged @13,000 rpm at 4 °C for 5 min.
7. The supernatant should be discarded thoroughly by inverting the tube onto a paper towel and repeated washing with 70% ethanol, finally drying at room temperature for few min.
8. The plasmid should be dissolved in 25 μL of TE buffer (pH 8.0) containing 20 μg/mL DNase-free RNase and mixed gently and incubated at room temperature for 30 min.
9. The quality of the plasmid DNA is to be checked in 1% agarose gel and the concentration should be determined by densitometry method using gel documentation system. The plasmid obtained should be stored at −20 °C till further use [5, 10–12].

2.3 Application of Plasmid Isolation

1. One of the most important use of plasmid is its use as vector in recombinant DNA technology.

2. Plasmid isolation and its digestion with restriction enzyme is for its confirmation of inserted gene of interest.

3 Horizontal Agarose Gel Electrophoresis

3.1 Principle

Agarose is a linear polymer comprised of residues of D-galactose and L-galactose alternatively joined by -(13) and -(14) glycosidic linkages. Chains of agarose form helical fibers which aggregate into supercoiled structures with a radius of 2030 nm. Agarose forms gel like structure in a three-dimensional (3D) mesh of channels whose diameters range from 50 nm to >200 nm. Since DNA is negatively charged (due to phosphate group), DNA fragments will move towards anode. The technique separates double stranded DNA fragments in electric field. The smaller the DNA fragment, agarose percentage would be more.

3.2 Protocol

Horizontal submarine agarose gel electrophoresis is performed to check the quality of genomic DNA using 1% gel. One percent gel solution was prepared as follows:

Agarose: 0.3 g.

0.5× TBE buffer: 35 mL.

At first step, the gel casting tray should be prepared by sealing its end with adhesive tape. The comb was placed over it in so that a gap of 0.5 mm is present between the tips of the comb teeth and floor of the casting tray. As a result, the wells were completely sealed at bottom by agarose. A schematic representation of agarose gel electrophoresis has been depicted in Fig. 3a–d.

1% agarose suspended in 0.5× TBE buffer should be heated till the agarose is completely melted resulting in a clear transparent solution. Ethidium bromide (intercalating dye for binding with DNA) @ 5 μL/100 mL is added and mixed gently.

The agarose solution should be poured gently into the casting tray and cooled down at room temperature. Soon the gel solidifies, the comb and adhesive tapes were gently removed. Thus, the gel is ready for electrophoresis.

For loading the samples, 5 μL diluted DNA is mixed with 10 μL autoclaved double distilled water and 2 μL of 6× gel loading dye. The resulting mixed sample should be loaded into the well of agarose gel, submerged into 0.5× TBE buffer. 100 base pair ladder sequences should be used as marker. Electrophoresis is conducted at 100 V for 20–30 min. Gel is to be visualized under UV transilluminator followed by documentation through photography. DNA samples with intact band and those devoid of smearing should be used for further analysis [2–5, 10–12].

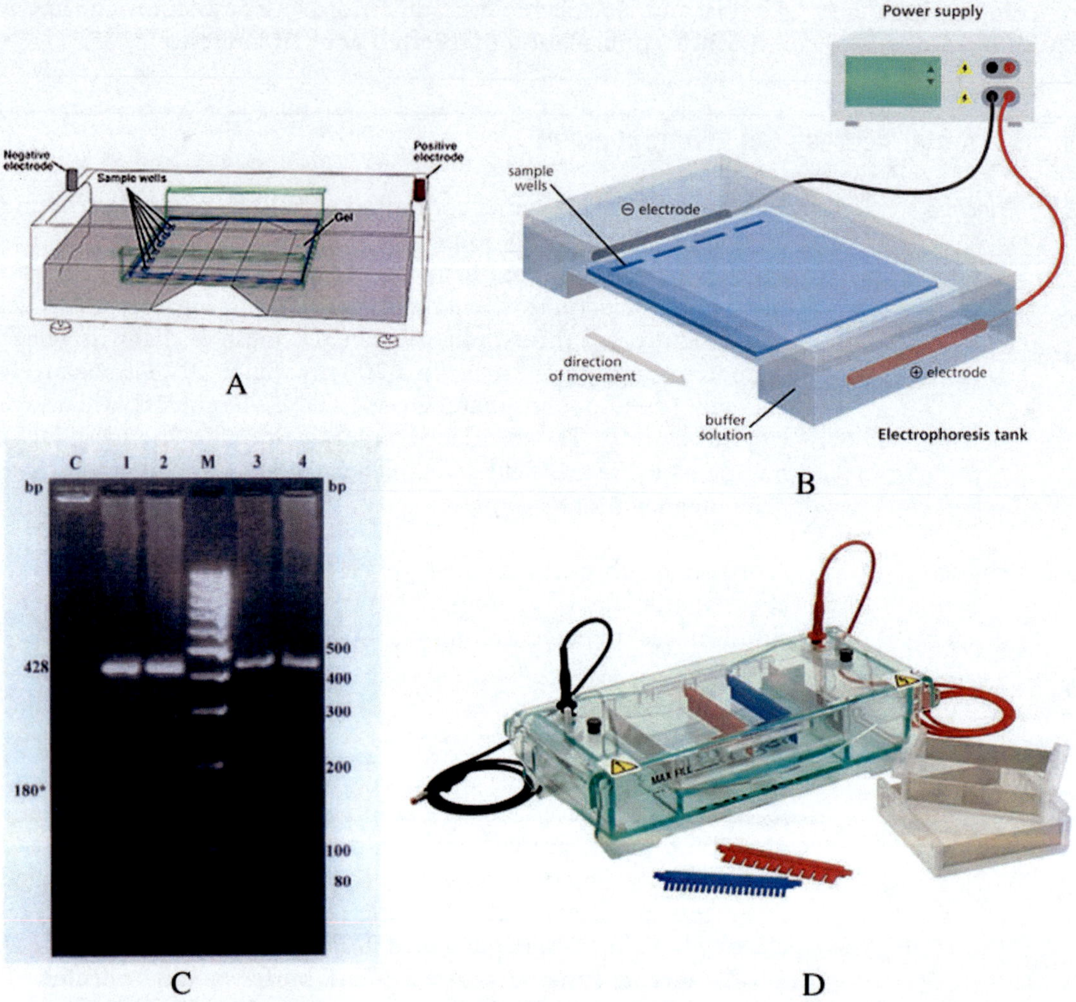

Fig. 3 (**a**, **b**) Horizontal agarose gel electrophoresis. (**c**) Diagrammatic representation of the photograph of agarose gel (3%) visualized through Gel documentation system under UV light. Lane C: Control. Lane 1, 2, 3 and 4: PCR product. Lane M: DNA marker. (**a**, **b**, **d**) Apparatus and schematic setup for horizontal agarose gel electrophoresis. (**d**) Apparatus for agarose gel electrophoresis

Note: In order to check DNA fragments or PCR products, the 3% agarose gel electrophoresis is normally used. If we need to separate smaller DNA fragments, the agarose percentage may be increased.

3.3 Application

1. Horizontal agarose gel electrophoresis is very much useful for quality checking of DNA.
2. It is equally useful for checking of PCR products and its molecular weight estimation.

3. Agarose gel electrophoresis with low melting agarose is helpful for purification of PCR product for cloning or sequencing operation.

4 Vertical Polyacrylamide Gel Electrophoresis

4.1 Principle

The scientists who worked on Cross-linked chains of polyacrylamide gel as matrix for electrophoresis were Raymond and Weintraub [13]. They used polyacrylamide, as electrically neutral gels for the separation of double-stranded DNA fragments based on the size and single-stranded DNAs as per the size and conformation.

The advantages of Polyacrylamide gel electrophoresis (PAGE) over agarose gel electrophoresis are as follows:

1. The resolving power is more in PAGE compared to that of agarose gel electrophoresis. Their efficacy is to separate or differentiate molecules of DNA whose lengths differ by as little as 0.1 (i.e., 1 bp in 1000 bp).
2. Larger quantities of DNA can be accommodated in the well than that of agarose gels. This technique can load up to 10 g of DNA to a single slot (1 cm 1 mm) of a typical polyacrylamide gel without losing resolution.
3. DNA recovered from polyacrylamide gels is of high pure grade. This can be utilized for the most demanding purposes (e.g., microinjection of mouse embryos).

   For detection and separation of DNA, PAGE is of two types:

   (a) Non-denaturing PAGE (used for double stranded DNA).

   (b) Denaturing PAGE (used for single stranded DNA).

PAGE is a more effective technique for DNA fragment separation compared to that of agarose gel electrophoresis with better resolving power.

(a) *Non-denaturing polyacrylamide gel electrophoresis.*

It is very much useful for separating large amounts of double stranded DNA fragments with smaller fragment size.

Double-stranded DNAs migrate through non denaturing polyacrylamide gels at rates inversely proportional to the $\log_{10}$ of their size. Electrophoretic mobility of DNA is influenced by their base composition and sequence, so that duplex DNAs of exactly the same size can differ in mobility by up to 10.

For casting the PAGE plate, the two glass plates are fixed in the casting apparatus and sealing the sides and bottom with the gasket. The composition of the polyacrylamide gel (8%) mix is as follows:

Acrylamide:Bisacrylamide (29:1)	5.3 mL
TBE (1×), pH 8.0	14.6 mL
Ammonium persulphate	66.6 μL
TEMED	13.3 μL
Total	20.0 mL

The stock solution of the acrylamide–biacrylamide is prepared by:

Acrylamide	29 g
Bisacrylamide	1 g
D.D.W.	Make up to 100 mL

TEMED aids in polymerization. Acrylamide gel mix is poured in between the plates and appropriate comb is inserted up to a certain depth as the volume to be loaded. The gel is set to polymerize at room temperature for at least 30 min. The lower reservoir of the gel tank is to be filled with 1× TBE buffer.

Soon, the polymerization of the gel is complete, combs should be removed and sample wells are rinsed with buffer. Then, the plates are attached to the gel electrophoresis apparatus. Proper precaution should be taken to avoid any air bubble in the bottom of the gel and the upper buffer chamber is filled with 1× TBE buffer, so that wells are covered.

Loading of the DNA samples are to be done with a gel loading dye (6×) with the help of gel loading tips.

DNA fragment (RE digested product): 10 μL.

6× gel loading dye: 2 μL.

The gel is to be run at 100 V for 5 h.

After the run is complete, the gel plates are detached from electrophoresis apparatus and the plates are separated carefully with the gel attached to one plate. A schematic representation of vertical PAGE is being depicted in Fig. 4a, b.

(b) *Denaturating PAGE.*

It is the technique used for single stranded DNA. Denaturing polyacrylamide gels are able to separate and purify single-stranded fragments of DNA. An agent (urea and/or, less frequently, formamide) is employed while gels are polymerized, in order to suppress base-pairing in nucleic acids. Denatured DNA migrates through these gels at a rate and is independent on its base composition and sequence.

The protocol is same as above, except formamide is used. The products were run in 12% polyacrylamide gel at 4 °C for 6–11 h at

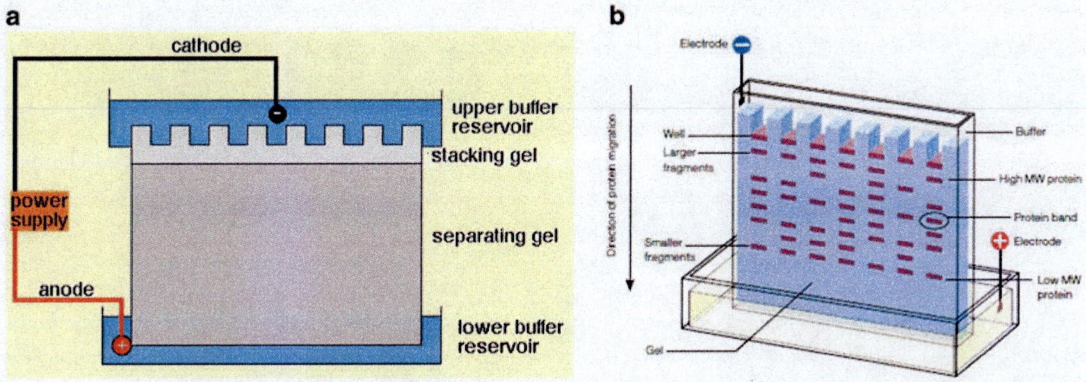

Fig. 4 (**a**, **b**) Vertical agarose gel electrophoresis

200 V depending upon the DNA fragment. DNA fragments are separated according to size and conformation.

Composition of polyacrylamide gel used for denaturing PAGE.

Reagents	Quantity in 30 mL mixture
30% acrylamide: bis-acrylamide (50:1)	12 mL
1× TBE	16.5 mL
10% Ammonium per-sulphate (APS)	100 μL
TEMED	20 μL
Glycerol (5%)	1.5 mL

4.2 Silver Staining in PAGE

4.2.1 Detection of DNA in PAGE

Ethidium bromide inhibits the polymerization of acrylamide; hence, it is not used for staining. However, staining with ethidium bromide can be employed for staining after electrophoresis. Since polyacrylamide quenches the fluorescence of the dye, the sensitivity for DNA detection may be somewhat diminished.

Most common and inexpensive technique for staining of DNA in PAGE is by silver staining.

Silver Staining of the Polyacrylamide Gel

Silver staining is a highly sensitive technique for post-electrophoretic detection of DNA bands in both polyacrylamide and agarose gels, as it can detect bands containing even less than 1.0 ng of DNA. Figure 5a depicts the silver-stained DNA fragments in PAGE, colored photograph in original. Figure 5b shows the silver-stained DNA fragments in black and white.

The procedure can be summarized as follows as described by Bassam et al. (1991) with slight modifications:

1. After electrophoresis is complete, the gel should be placed, still attached to one glass plate, in a plastic tray containing 10% glacial acetic acid. During rinsing, the gel will float free and the glass plate can then be removed.

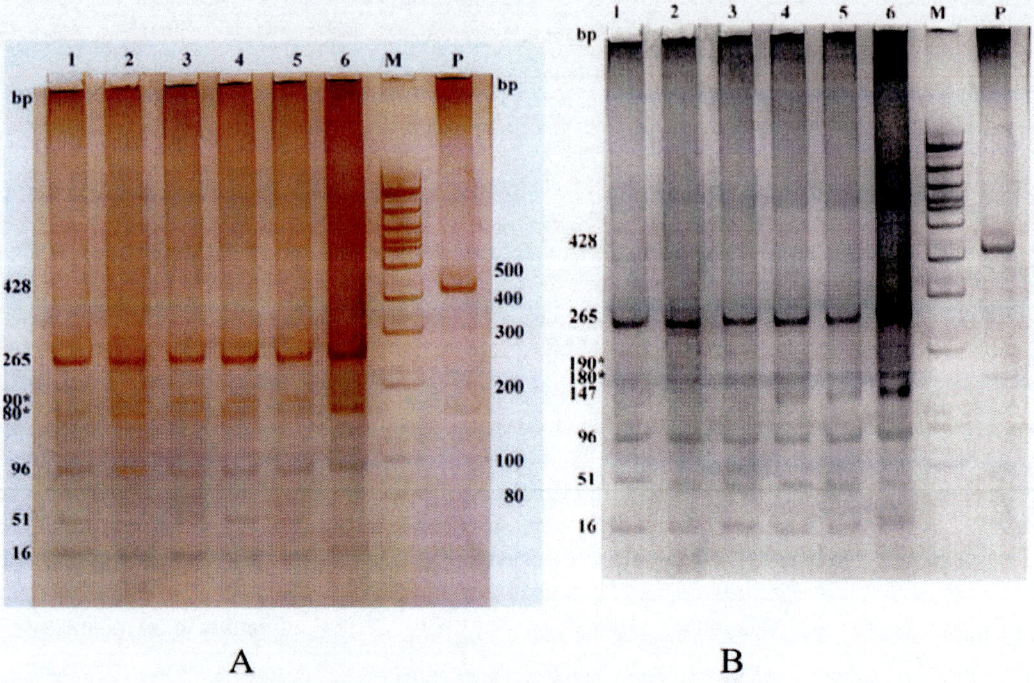

Fig. 5 Silver staining of PAGE, (**a**) Genotype 1, (**b**) Genotype 2

2. Gentle shaking of the plastic tray is to be continued for 30–45 min till the gel loading dye (containing bromophenol blue and xylene cyanol) disappears completely. This aids in fixing of the DNA band to the gel.
3. Then, the gel should be washed with distilled water thrice after decanting the glacial acetic acid to remove excess glacial acetic acid.
4. 0.1% silver nitrate (200 mL) in distilled water is to be poured in the plastic tray to cover the gel and gently shaken for 30 min for proper impregnation of the gel with $AgNO_3$ at weakly acidic pH.
5. After decanting the silver nitrate solution, the gel should be rinsed with DDW for 2–3 s to remove excess silver nitrate.
6. Three percent sodium carbonate containing 300 μL of formalin which act as developer is to be poured on the gel for the reduction of bound silver ions to metallic silver by alkaline formaldehyde.
7. The tray should be shaken gently under indirect light. Constant monitoring is to be done for the appearance of bands and background. When the ratio of signal to noise is at its maximum, the solution is removed.

8. 10% glacial acetic acid is to be added and shaken gently for 5 min.
9. Then, the gel should be transferred to 10% ethanol for long-term storage [2, 3, 5].

4.3 Application

1. PAGE with silver nitrate staining is very much helpful for detection and differentiation of DNA fragments or PCR products with slight difference in molecular weight.
2. PAGE is very useful for SSCP (single stranded conformation polymorphism) detection
3. SDS-PAGE is useful for protein detection.
4. It is useful for DGGE (differential gradient gel electrophoresis)

5 Quality Checking of DNA

Points to Remember

1. Spectrophotometry, fluorimetry, agarose gel electrophoresis and real-time PCR are the techniques for quality checking DNA.

5.1 Checking of Quality, Purity and Concentration of DNA

The genomic DNA isolated from the blood samples must be checked for its quality, purity and concentration. Good quality DNA can be used for further analysis.

It can be done by following methods:

In current days, scientific research has moved towards next generation sequencing and DNA microarray studies, which have revolutionized scientific research and are applicable to high-throughput analysis of biological systems. High quantities of pure, intact, double stranded, highly concentrated, genomic DNA without any contamination is the prime requisite. DNA banks need high qualities of pure DNA.

Hence, apart from conventional methods of DNA quality and quantity checking, many advanced methodologies have come up.

5.2 Evaluation of DNA Extraction

5.2.1 Spectrophotometer and Qubit Measurements

Purity of DNA extracted with a particular method should be assessed using an Eppendorf Biophotometer. The ratio of absorbance at 260 and 280 nm can be used to determine protein contamination, whereas the ratio of absorbance at 260 and 280 nm is to be calculated to determine guanidine contamination. Both the methods are efficient for DNA quality checking with higher values associated with better DNA purity.

5.3 Spectrophotometry

5.3.1 Purity of Genomic DNA

2 μL of DNA sample can be dissolved in 98 μL of DDW and reading should be taken on spectrophotometer at OD_{260} and OD_{280} against distilled water (100 μL) as a blank. Readings ranging from 1.7 to 1.9 are only used (in this case).

5.3.2 Concentration of Genomic DNA

$$\text{Amount of concentration of DNA } (\mu g/mL) = OD_{260} \times 50X$$

where, X = Dilution factor.

5.3.3 Purity of Genomic DNA

2 μL of DNA sample should be dissolved 98 μL of DDW and reading is to be taken on spectrophotometer at OD_{260} and OD_{280} against distilled water (100 μL) as a blank. Readings ranging from 1.7 to 1.9 are only used.

5.3.4 Concentration of Genomic DNA

$$\text{Amount of concentration of DNA } (\mu g/mL) = OD_{260} \times 50X$$

where, X = Dilution factor.

5.4 Qubit Fluorometer

Quantity of DNA extracted by the different methods can be efficiently assessed using Qubit 2.0 fluorometer. The basic principle is that the fluorescent dye binds to double stranded DNA. The Qubit fluorometer detects the fluorescence signal and finally DNA concentration is measured using DNA standards of known concentration. Qubit dsDNA BR Assay Kit may be efficiently employed for the DNA quantification.

On the basis on DNA concentration derived from the Qubit measurements and the volume of the DNA extract, total DNA yield is calculated with a simple multiplication step.

5.4.1 Gel Electrophoresis

The integrity of DNA extracted by each method can be efficiently checked by gel electrophoresis. 1 μL of each DNA extract is visualized in a 1.5% agarose gel with 0.5% ethidium bromide followed by visualization by U.V. illumination.

Quality of Genomic DNA

Horizontal submarine agarose gel electrophoresis is employed for checking the quality of genomic DNA with 1% gel. The composition of 1% gel solution is as follows:

Agarose: 0.3 g.

0.5× TBE buffer: 35 mL.

For the first step, the gel casting tray is prepared by sealing its end with adhesive tape. Second step involves placing of the comb over it maintaining a gap of 0.5 mm between the tips of the comb teeth and floor of the casting tray. Care should be taken so that the wells can be completely sealed at bottom by agarose.

The process for the preparation of 1% agarose gel includes agarose powder suspended in 0.5× TBE buffer, which need to be heated until the agarose gets completely melt and dissolves to give a clear transparent solution. Ethidium bromide (intercalating dye for visualization of DNA) is added soon the mixture cools to 60 °C @ 5 μL/100 mL.

The agarose solution should be poured into the casting tray and allowed to cool down at room temperature for the solidification of the gel, then the comb and adhesive tapes should gently be removed. Thus, the gel is made ready for electrophoresis.

For loading the samples, 10 μL autoclaved double distilled water is mixed with 5 μL diluted DNA and 2 μL of 6× gel loading dye and the mixed sample is to be loaded into the well of agarose gel which should be made submerged into 0.5× TBE buffer, 100 bp ladder sequences can be used as marker. Electrophoresis should be performed at 100 V for 20–30 min. Gel visualization is to be done by UV transilluminator followed by proper documentation through photography. The DNA samples with intact band and lack of smearing may be used for further analysis [2–6, 10–12].

5.4.2 *Real-Time PCR*

A real-time PCR targeting the gene is employed to detect the presence of amplifiable DNA in extracts of blood samples. The technique is similar to Ct values are used to determine the amount of amplifiable DNA obtained. Lesser Ct values are desirable in this regard.

In a real-time PCR assay a positive signal is detected through fluorescent signal, which is measured as Ct value. The Ct (cycle threshold) is explained as the number of cycles necessary for the fluorescent signal to cross the threshold, exceeding the background level. Ct levels are inversely proportional to the quantity of target nucleic acid in the sample. This implies that the lower the Ct level the greater the amount of target nucleic acid in the sample. Real time assays lead to 40 cycles of amplification. Ct values less than 29 indicates strong positive reactions, which indicate abundant target nucleic acid in the sample. Ct values ranging from 30 to 37 indicates positive reactions with moderate amounts of target nucleic acid. Ct values of 38–40 are weak reactions, which indicate minimal amounts of target nucleic acid which could represent an infection state or environmental contamination.

Quantitative PCR or real time PCR has been discussed in detail later in Chapter 9.

6 Restriction Fragment Length Polymorphism (RFLP)

Points to Remember

1. This technique identifies the variations at nucleotide level for various genes, which may be SNP, CNV using restriction sites.
2. Restriction enzymes recognize palindromic sites.

6.1 Basic Principle

It is the method of detecting SNP (single nucleotide polymorphism). Restriction enzyme (RE) cut at particular palindromic sequence of nucleotide. Nucleotide difference in one individual will lead to loss of RE cutting site.

6.1.1 Digestion Reaction Mixture

20 μL of each amplified product (sample) was digested with restriction enzyme. The composition of reaction mix was as follows:

Sl. No.	Reaction components	Amount (1×)
1.	PCR product	20 μL
2.	10× assay buffer for RE (buffer L)	2.5 μL
3.	Restriction enzyme (U/μL)	2.5 μL
	Total	25.00 μL

Level of restriction enzyme used for digestion and time for digestion is needed to be optimized for best result.

6.1.2 List of Restriction Enzymes Along with Their Cutting Site

Some of the most commonly used restriction enzymes are Alu-I, Eco-RI, BamH-I, Hinf-I, Hind-I etc.

If a researcher is interested for identifying the polymorphism in a particular gene or gene fragment of any species, the software for restriction mapper is very useful as described (Fig. 6).

Restriction enzymes are naturally observed in bacteria (and other prokaryotes). They are responsible for the recognition and binding to specific sequences of DNA, known as called **restriction sites**. The peculiarity of each restriction enzyme is that it recognizes just one or a few restriction sites, known as target sequence. As the RE identifies the target sequence, it will undergo a double-stranded cut in the DNA molecule. Normally, the cut is near to the restriction site and occurs in definite pattern.

6.1.3 Types of Restriction Enzyme

1. Type I enzymes cut at sites remote or far away from a recognition site. It needs ATP and *S*-adenosyl-L-methionine for its proper function. It is a multifunctional protein with multiple functions (both restriction and methylase activities).

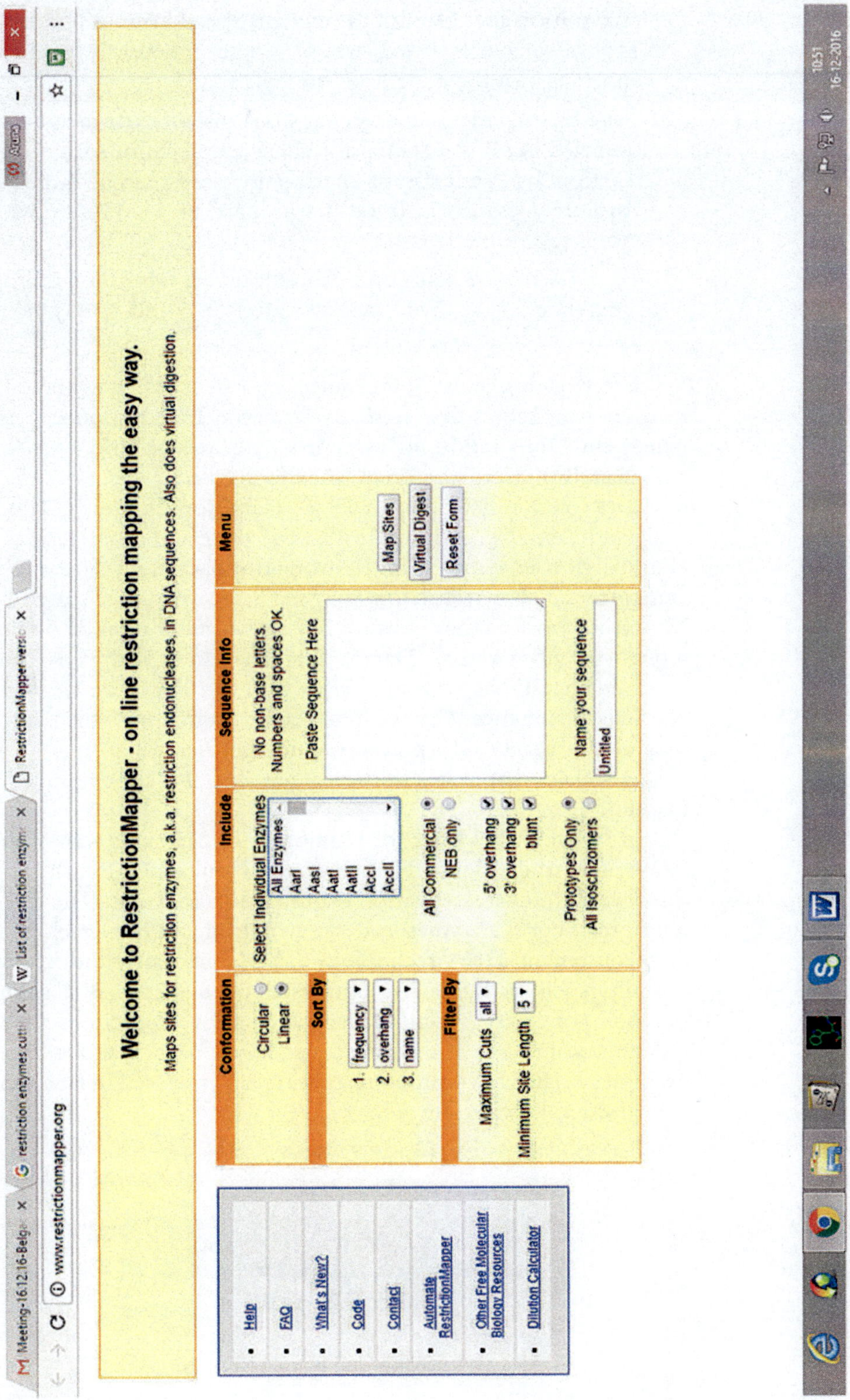

Fig. 6 Restriction Mapper: online restriction mapping

2. Type II enzymes cut within or at short specific distances from a recognition site. Most of the enzymes need magnesium. Unlike Type I enzyme, these are of single function (restriction) enzymes, not dependent on methylase.
3. Type III enzymes cut at sites at short distances from a recognition site. ATP is essential for their action. Stimulation of the reaction by *S*-adenosyl-L-methionine is essential, but is not absolutely essential. It occurs as part of a complex with a modification methylase.
4. Type IV enzymes act upon the modified DNA, for example, methylated, hydroxymethylated and glucosyl-hydroxymethylated DNA.

There is historical importance of restriction enzymes. The restriction enzymes first studied were type I restriction enzymes, which cut DNA randomly away from the recognition site. Arber and Meselson were the scientist who studied Type I restriction enzymes for the first time. In 1970, Hamilton O. Smith, Thomas Kelly and Kent Wilcox were the first to isolate and characterize type II restriction enzyme, *Hind*II, from the bacterium *Haemophilus influenzae*. These restriction enzymes were very much useful for laboratory work as they cleave DNA at the site of their recognition sequence. Afterwards, Daniel Nathans and Kathleen Danna detected that DNA of simian virus 40 (SV40) can be cleaved by restriction enzymes, that yields specific fragments which can be separated using PAGE (polyacrylamide gel electrophoresis). Thus, it was shown that restriction enzymes can also be employed for mapping DNA.

In 1978, Nobel Prize for Physiology or Medicine was awarded to Werner Arber, Daniel Nathans, and Hamilton O. Smith for the discovery and characterization of restriction enzymes .The discovery of restriction enzymes lead to revolution and the development of recombinant DNA technology. The large-scale production of proteins such as human insulin used by diabetics is a positive outcome of these recent advancements. A restriction enzyme recognizes and cuts at a DNA sequence at a particular site. As, for example, *Eco*RI is a common restriction enzyme used in labs. *Eco*RI cuts at the following site (Fig. 7):

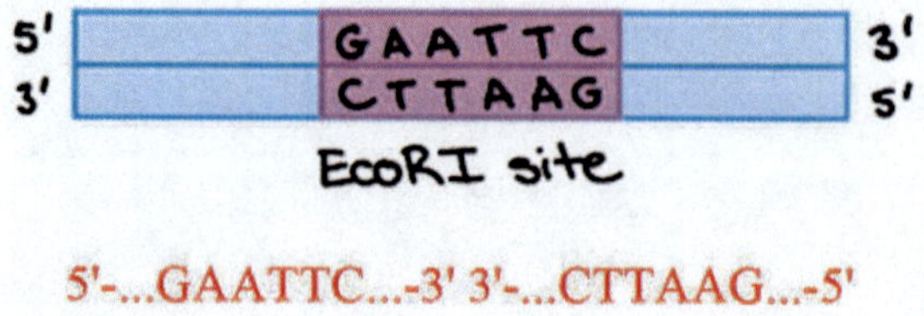

Fig. 7 EcoRIsite

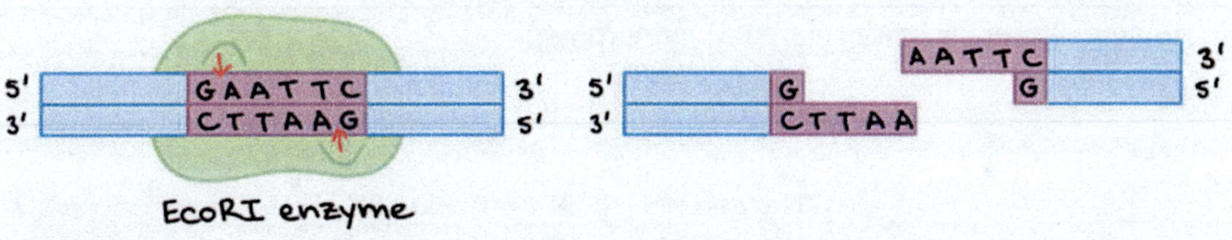

Fig. 8 Cutting site of EcoRI enzyme

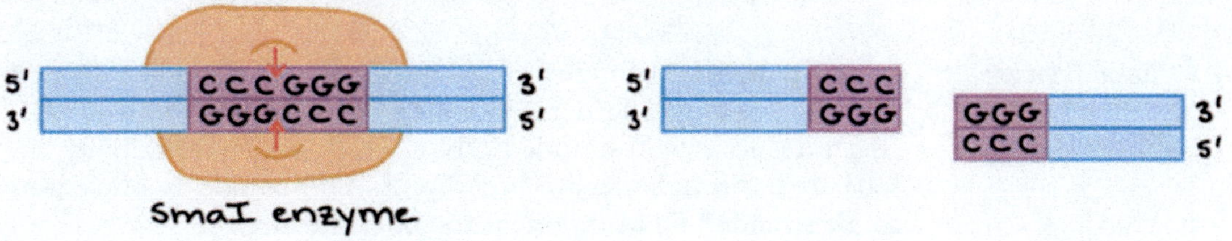

Fig. 9 Restriction site for SmaI enzyme

When EcoRI recognizes and cuts this site in a specific pattern at particular site. It gives rise to ends with single-stranded DNA "overhangs". The cutting site is being depicted in Fig. 8.

*Eco*RI enzyme binds to an *Eco*RI site and cut on both strands of the DNA. The pattern of the cut is:

5′-...G|AATTC...-3′ 3′-...CTTAA|G...-5′

Thus, it causes an overhang of 5′-AATT-3′ on each end of the cut DNA.

In case, another piece of DNA has matching overhangs (when that has also been cut by EcoRI), the overhangs will stick together by complementary base pairing. So, enzymes that leave single-stranded overhangs produce **sticky ends**. Sticky ends are useful in cloning since they hold two pieces of DNA together to be joined = by DNA ligase.

Some restriction enzymes are "blunt cutters," which cleaves DNA strands straight down the middle of a target sequence, leaving no overhang. A good example of a blunt cutter restriction enzyme is *Sma*I. The cutting site is being visualized in Fig. 9.

A *Sma*I enzyme binds to the *Sma*I restriction site, which is:

5′-...CCCGGG...-3′ 3′-...GGGCCC...5′

It leads to a cut right in the middle of this sequence on both strands, producing blunt ends. The cut sites are:

5′-...CCC|GGG...-3′ 3′-...GGG|CCC...5′

However restriction mapping can also be done by other related soft wares as DNASTAR, etc [2–4].

7 Single-Stranded Conformation Polymorphism

Points to Remember

1. The basic principle involved in SSCP is the Electrophoretic mobility of secondary structure of single stranded DNA.
2. It is a widely used technique for detecting SNPs.

7.1 Basic Principle

It is the method of detecting SNP (Single Nucleotide Polymorphism). SNPs will lead to different conformation of secondary structures for single stranded DNA, which differs in mobility and separated through denaturing PAGE. Formamide is preventing single stranded DNA fragment to anneal further.

Single stranded conformation polymorphism has been widely used for screening of various genomic variants amongst the samples, when the sample size is sufficiently high. It is equally applicable for a broad range of samples. Figure 10 depicts the PAGE representation for SSCP fragments.

Electrophoretic mobility difference is the main criteria for detection of variations of sequence . Mutation in DNA sequence (even a single point mutation) cause differences in mobility compared to that of wild type DNA after they are exposed to non-denaturing (or partly denaturing conditions). Formamide is essential for denaturing.

7.2 Protocol

1. 3 μL of amplified PCR-product is to be combined with 9 μL of formamide dye in PCR tubes (0.5 mL).
2. The next step involves denaturation of 12 μL of PCR-SSCP solution at 95 °C for a duration of 5 min and then immediate snapping in ice for 15 min.

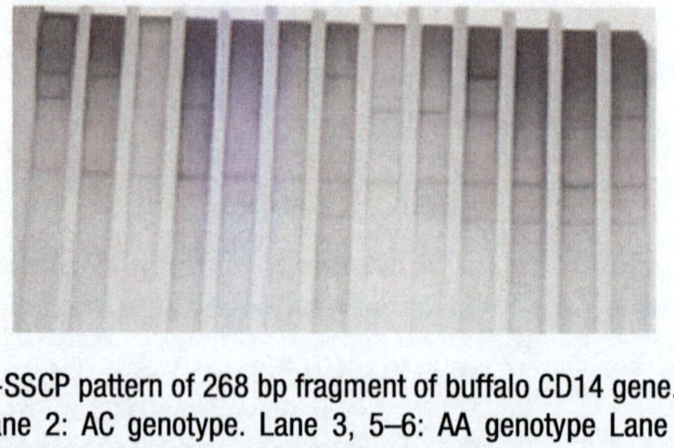

Fig. 10 PCR-SSCP pattern of 268 bp fragment of buffalo CD14 gene. Lane 1: AD genotype Lane 2: AC genotype. Lane 3, 5–6: AA genotype Lane 4, 8–9: AB genotype. Lane 7: AE genotype Lane 10: AH genotype. Lane 11–12: AF genotype Lane 13: AG genotype

3. The denatured products thus obtained need to be run in 12% polyacrylamide gel at 4 °C for 6–11 h at 200 V, which varies for different fragment with varying nucleotide.

Thus, eight different variants have been detected. Different variants correspond to differential size and mobility of secondary structures developed from DNA. Variability in nucleotide in the region is the cause for difference in secondary structure developed [10, 11].

References

1. Psifidi A, Dovas CI, Bramis G, Lazou T, Russel CL, Arsenos G, Banos G (2015) Comparison of eleven methods for genomic DNA extraction suitable for large-scale whole-genome genotyping and long-term DNA banking using blood samples. PLoS One 10(1): e0115960. https://doi.org/10.1371/journal.pone.0115960
2. Pal A, Chakravarty AK, Chatterjee PN (2013) Polymorphism of growth hormone gene and its association with seminal and sexual behavioral traits in crossbred cattle. Theriogenology 81:474–480
3. Pal A, Chakravarty AK, Bhattacharya TK, Sharma A (2005) Polymorphism of growth hormone gene and its association with expected milk production traits in dairy bulls. J Appl Anim Res 27(9):29–33
4. Pal A, Chakravarty AK, Bhattacharya TK, Joshi BK, Sharma A (2004) Detection of polymorphism of growth hormone gene for the analysis of relationship between allele type and growth traits in Karan Fries cattle. Asian Aust J Anim Sci 17(9):1334–1337
5. Pal A, Pal A, Chakravarty AK (2018) Mutations in growth hormone gene affect stability of protein structure leading to reduced growth, reproduction and milk production in cross bred cattle—an insight. Domest Anim Endocrinol 71:106405. https://doi.org/10.1016/j.domaniend.2019.106405
6. Sambrook J, Russell DW (2001) Molecular cloning: a laboratory manual, vol 1, 3rd edn. Cold Spring Harbor Laboratory Press, New York
7. Aljanabi SM, Martinez I (1997) Universal and rapid salt-extraction of high quality genomic DNA for PCR-based techniques. Nucleic Acids Res 25(22):4692–4693
8. Wilson K (1997) Preparation of genomic DNA from bacteria. In: Current protocols in molecular biology. Wiley. pp 2.4.1–2.4.5
9. Ling J, Wang H, Zhang S, Zhang D-d, Lai M-d, Zhu Y-m (2012) Optimization of genomic DNA extraction with magnetic bead-based semi-automatic system. Zhejiang Da Xue Xue Bao Yi Xue Ban 41(3):320–326
10. Pal A, Chatterjee PN, Sharma A (2014) Sequence characterization and polymorphism detection in bubaline CD14 gene. Buffalo Bull 32(2):138–156
11. Pal A, Sharma A, Bhattacharya TK, Chatterjee PN, Chakravarty AK (2011) Molecular characterization and SNP detection of CD14 gene of crossbred cattle. Mol Biol Int 2011:507346, 13 pages. https://doi.org/10.4061/2011/507346
12. Pal A, Pal A, Mallick AI, Biswas P, Chatterjee PN (2019) Molecular characterization of Bu-1 and TLR2 gene in Haringhata Black chicken. Genomics 112(1):472–483. https://doi.org/10.1016/j.ygeno.2019.03.010
13. Raymond S, Weintraub L (1959) Acrylamide gel as a supporting medium for zone electrophoresis. Science 130(3377):711. https://doi.org/10.1126/science.130.3377.711

Check for updates

Chapter 2

Recombinant DNA Technology

Abstract

This chapter discusses the protocols pertaining to the basic techniques for recombinant DNA technology such as polymerase chain reaction (PCR), extraction of total RNA, checking of quality, concentration and integrity of RNA, cDNA synthesis, and purification of cDNA. This chapter discusses the protocol pertaining to recombinant DNA technology and DNA cloning. The basic principle for PCR involves in vitro DNA replication with Taq DNA polymerase, forward and reverse primers, and dNTPs. The RNA extracted is very short lived; hence it is converted to cDNA using reverse transcriptase polymerase, which in turn can be stored for a longer period. For the purpose of recombinant DNA technology or DNA cloning, the desired gene of interest is amplified with PCR using the cDNA thus prepared as template. The amplified product is inserted in suitable vector (plasmid vector) and inserted in *Escherichia coli* culture. Positive colonies were identified through blue-white screening and sub-cultured in broth culture. As *E. coli* (DHα as preferred strain) multiplies, the plasmid vector containing the gene of interest also gets multiplied manifold.

Keywords Polymerase chain reaction, RNA extraction, Quality checking of RNA, cDNA synthesis, Recombinant DNA technology, Gene cloning

In this chapter, we discuss some basic techniques pertaining to recombinant DNA technology such as polymerase chain reaction, extraction of total RNA, checking of quality, concentration and integrity of RNA, cDNA synthesis, and purification of cDNA. This chapter discusses the protocol pertaining to recombinant technology and DNA cloning.

1 Polymerase Chain Reaction

Points to Remember

1. The basic principle of polymerase chain reaction is in vitro DNA amplification.

(continued)

Aruna Pal, *Protocols in Advanced Genomics and Allied Techniques*, Springer Protocols Handbooks, https://doi.org/10.1007/978-1-0716-1818-9_2,

2. PCR reaction conditions are maintained in such a manner that the ambient temperature is maintained for reaction of each step and for the enzymes to act.
3. The raw materials for DNA amplification are supplied externally viz. nucleotides (dNTPs), primers to bind to DNA strand, Taq DNA polymerase to carry on the reaction, and PCR buffer for reaction to carry out.
4. The T_m value is calculated as $4(G + C) + 2(A + T)$.
5. The number of DNA strands obtained is 2^n, where n refers to the number of cycles.

1.1 Principle

The basic principle for polymerase chain reaction, or PCR, resembles the process of DNA replication in vitro, resulting in multiple copies of the particular gene fragment. The primer binds to both the ends of the double-stranded DNA. The multiple copies of DNA obtained through PCR correspond to 2^n, where n refers to the number of cycles.

Reagents necessary for a standard PCR:

- *Primers*: The primers are the short series of nucleotides, usually of 20–30 nucleotides in length. These contain about similar numbers of four nucleotides where G and C residues are balanced. The design of primers is made in such a way to incorporate restriction sites. This leads to amplification of DNA fragments with restriction sites at both the ends, facilitating cloning.

  Automated DNA synthesizers are employed for the synthesis of oligonucleotide primers. They are useful in standard PCR procedures without the need for purification.
- *Template DNA*: The template refers to the DNA fragment which is to be amplified through PCR reaction. It may be genomic DNA, a recombinant plasmid, or other sample possessing DNA. Template DNA is usually dissolved in 10 mM Tris–Cl (pH 7.6) with a low concentration of EDTA (<0.1 mM).
- *Thermostable DNA polymerase.*

  DNA polymerase aids in the polymerization process. It is essential for it to be effective at higher temperatures, say 90–95 °C for denaturation. Hence, such a DNA polymerase is chosen, which is effective.
- *Taq* DNA polymerase is the most common form of enzyme employed for most forms of PCR. In cases when elongation from 3-mismatched primers occurs, a thermostable DNA polymerase with proofreading activity may be preferred.

- *Thermal cycler*: A thermal cycler is programmed to conduct the desired amplification program. Different companies are available in the market such as PerkinElmer, Applied Biosystem, and Biorad commercially marketing programmable thermal cyclers. It all depends on the choice of the investigator or the investigator's inclination, the available budget, and the range of uses of the machine. Figure 1 diagrammatically depicts the principle of PCR and a glimpse of PCR machine or thermocycler.

1.2 Protocol with Suitable Example

A region of 428 (bp) DNA fragment of growth hormone gene is amplified by using a pair of primers (forward and reverse).

1.2.1 Gene Amplification by Polymerase Chain Reaction: A Case Study

Forward primer : 5′CCGTGTCTATGAGAAGC3′.

Reverse primer : 5′GTTCTTGAGCAGCGCGT3′.

PCR Reaction Mixture

Standarization of the PCR cycling conditions are essential. Different combinations of a reaction mixture are set with different annealing temperatures ranging from 45 °C, 48 °C, 51 °C to 53 °C. Similarly, standardization is equally important for magnesium concentration for particular PCR reaction. Different concentrations of magnesium chloride (1.5–2 mM) may be standarized to find the optimum one.

PCR is carried out in a final volume of 25 μL using the following reaction components, which gave specific amplification:

Sl. No.	Reaction component	Final concentration
1.	dNTPS mix	200 mM each
2.	Primers Forward primer (F) Revers primer (R)	 20 pmol 20 pmol
3.	10× PCR assay buffer	1×
4.	$MgCl_2$ (15 mM)	2 mM
5.	Taq DNA polymerase (6 U/μL)	3 U
6.	Autoclaved DDW	Up to 25 μL

PCR Program

PCR-reactions were carried out in a thermocycler using the following program:

Sl. No.	Steps	Temperature (°C)	Time
1.	Initial denaturation	94	5 min
2.	Denaturation	94	1 min
3.	Annealing	46	45 s

(continued)

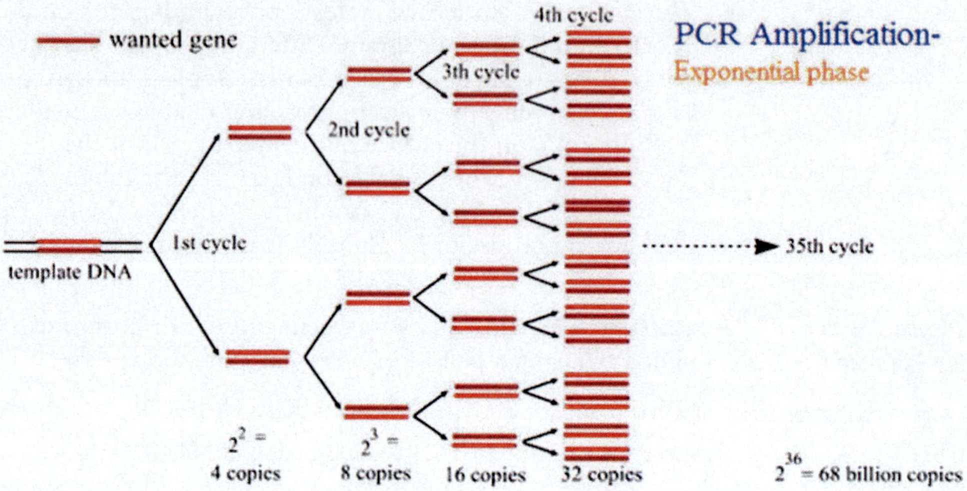

Fig. 1 Schematic representation of a polymerase chain reaction. Basic principle of Polymerase chain reaction (PCR)

Sl. No.	Steps	Temperature (°C)	Time
4.	Extension	72	50 s
5.	Final extension	72	10 s
6.	Go to Step (2) → 35 cycles		
7.	Storage	4	Till further analysis

Note: The annealing temperature varies from primer to primer. The annealing temperature for the primer is 1–2 °C lower than the T_m value of the primer. The T_m value (melting temperature) of the primer can be calculated as follows:

T_m value of the primer: $4(G + C) + 2\ (A + T)$.

G represents the total number of guanosine residues in the primer.

C represents the total number of cytosine residues in the primer.

A represents the total number of adenosine residues in the primer.

T represents the total number of thymine residues in the primer [1–8].

1.2.2 Types of PCR

1. Gradient PCR: It is very much useful for the optimization of PCR reactions, where varying annealing temperatures can be set.
2. Hot Start PCR: This technique involves initial denaturation at high temperatures and then adding primers. This is done to avoid mispairing of primers leading to non-specific PCR products detected as non-specific bands in electrophoresis. Mis-priming of primers usually happens at suboptimal annealing

temperatures. It equally avoids off-target amplification, when low concentration of complex templates are employed, for example, in the case of mammalian genomic DNA template. The chance for mispriming is increased by the increased ratio of primer:template DNA.

3. Inverse PCR: **Inverse PCR** is a technology employed where sequence information is present only on one side of the target region. Initially, a restriction enzyme is selected that cannot cut through the stretch of the known DNA. The optimum length of the recognition sequence needs to be six or more base pairs for the generation of reasonably long DNA segments for amplification by PCR.

   The target DNA is then cut with this restriction enzyme in order to generate a piece of DNA having compatible sticky ends, one upstream of the known sequence and the other downstream. The DNA fragment with two ends joins and forms a circle.

   The PCR primers are designed in such a way as to recognize the end regions of the known sequence. Each primer binds to a different strand of the circular DNA. Primers point "outward" into the unknown DNA. In the next step, PCR amplifies the unknown DNA to generate linear molecules possessing short stretches of known DNA at the ends, and in the middle, the restriction enzyme site. Degenerate primers are designed as per the amino acid sequences and possess various nucleotides at the wobble position.

   Inverse PCR is the process involving sequences DNA near a known sequence by observing a restriction enzyme recognition sequence away in the unknown region. It is followed by cutting out the template, and amplifying the entire piece with *Taq* polymerase.

4. Real-time PCR: It is also known as quantitative PCR. It is the technique developed when fluorescent dye as sybergreen or Taqman probe is used, which binds with the amplicon. More the copy number of mRNA or cDNA is produced, more will be the fluorescence detected, as it will bind with fluorescent dye. The details have been discussed in subsequent chapters.

1.2.3 Application

(a) It is a very useful technique for the detection of differential mRNA expression profiles in biological processes.

(b) It is useful for nutrigenomics study.

1.3 Primer Designing

It is an essential step for the initiation of any PCR reaction. There are certain software specially designed for primer synthesis (DNASIS, DNAstar). There are certain criteria for a set of good primers:

1. The forward and reverse primers should not be complementary; otherwise, they will bind among themselves instead of binding to the template.
2. GC content should not be very high; otherwise, they will lead to too high annealing temperatures.
3. Primers should not be too large with many nucleotides; otherwise, there will be difficulty in primer binding.

2 Extraction of Total RNA

Points to Remember

1. RNA is short lived and is the base material for cDNA synthesis.
2. The basic difference from DNA isolation is that guanidine isothiocyanate is added in the case of RNA isolation.
3. DNAase is added to remove traces of DNA, if any.
4. mRNA reveals only the transcribed portion, while DNA contains both intron (untranscribed) and exon (to be transcribed).

2.1 Basic Principle

RNA is short lived. It needs to be isolated immediately after collection.

2.2 Protocol

1. About 25 mg of animal tissue is to be triturated in pestle and mortar, and incubated with 1 mL trizol for 5 min at 25 °C (to allow complete dissociation of nucleoprotein complexes).
2. 1 mL triturate should be taken in chloroform/DEPC (an RNAse inhibitor)-treated Eppendorf tube (2 mL) and 1 mL chloroform is to be added to it, mixed thoroughly and left for incubation at 25 °C for 5 min.
3. The contents are to be centrifuged @10,000 rpm for 10 min at 4 °C. Soon after the process of centrifugation, the mixture is found to be separated into: (a) lower red phenol-chloroform phase, (b) white middle phase, and (c) colorless upper aqueous phase.
4. RNA is present in the upper aqueous phase, which should be pipetted to a fresh Eppendorf tube. The RNA is to be precipitated by adding an equal volume of 100% isopropanol. It is to be incubated at room temperature (20–25 °C) for 10 min and then centrifuged @ 10,000 rpm for another 10 min at 4 °C.

5. The supernatant should be discarded and 70% ethanol is to be added, left in freezer for 2 min in order to remove salts and again centrifuged @ 10,000 rpm for 10 min.
6. In the final stage, the supernatant is discarded, and the pellet should be air dried and dissolved in 20 μL of sterile 0.1% DEPC-treated distilled water or nuclease free water. The RNA obtained is to be stored at −70 °C until further use [1, 2, 4, 6, 8].

Trizol or a similar reagent is composed of a mono-phasic solution of phenol and guanidine isothiocyanate. The trizol method of RNA isolation is an improvement to the single-step RNA isolation method developed by Chomczynski and Sacchi1.

RNA later is a reagent that can be added to the sample from which RNA is to be isolated.

2.3 Other Methods for RNA Extraction

1. Directly from blood.

   RNA can be directly extracted from blood. In cases where very limited amount of blood is available, RNA can be directly isolated from blood (RiboPure™ Blood Kit). It involves the principle of column chromatography. The RNA later needs to be used for RNA stabilization. The RNA isolated is of highest purity and convenience, and includes both organic extraction and silica columns (no precipitation required). It is a very rapid method of RNA isolation [8].
2. Bacterial total RNA isolation.

   Reagents:
   (a) Stop solution comprising 5% buffer equilibrated phenol (pH 7.4) in ethanol.
   (b) RNAeasy mini kit and shredders.
   (c) DNase I double strength (dissolve the solid DNase I in 275 μL of RNase-free water).
   (d) DNase I.
   (e) RNA-guard.
   (f) OnePhor-All 10× Buffer.
   (g) DNase I Kit.
   (h) buffer ((RNAse-free/10 mM Tris/1 mM EDTA, pH 8.0) containing 1 mg/mL lysozyme, RNase-free water, ethanol, and ice).

2.4 Protocol

It involves harvesting of cells to grow bacterial cultures to the desired density. Before harvesting, 2 mL centrifuge tubes should be labeled and filled with 1/9th volume of the stop solution and kept on ice. As an example, for a 1.7-mL culture, 190 μL of stop solution is used. Transfer culture to tubes containing stop solution followed by capping and mixing by inversion is to be done. Samples can be kept on ice up to 20 min. Cells are to be pelleted at 4 °C at a maximum speed for 1 min in a microcentrifuge. The supernatant is discarded. The cell pellets are frozen in liquid nitrogen. The cells may be stored at −80 °C for future use or can be used immediately.

After thawing of the cells on ice, the RNAeasy protocol is followed for gram-negative bacteria. QIAgen RNA Isolation 1.

Important notes before starting:

(a) Do not overload the column.

(b) The use of Lysis Buffer RLT may lead to a precipitate upon storage. Warming redissolves it.

(c) Add 10 μL of 2 mercaptoethanol per 1 mL of Lysis Buffer RLT prior to use.

(d) The working solution of Wash Buffer RPE is prepared by mixing four volumes of ethanol (96–100%) to the supplied concentrate mixture.

1. The cell pellet should be loosened by flicking the bottom of the tube. The cells are resuspended in 100 μL of lysozyme-possessing TE buffer (by flicking the tube; incubate for 3–5 min on the bench top).
2. 350 μL of Lysis Buffer RLT is to be added to the sample and vortex vigorously, followed by centrifugation for 2 min at the maximum speed. The supernatant should be loaded to the QIAshredder spin column sitting in a 2-mL collection tube and then centrifuged for 2 min at the highest speed. The flow-through fraction is transferred from QIAshredder to another tube without affecting the cell-debris pellet in the collection tube.
3. 250 μL of ethanol (100%) is added to the lysate and mixed properly by pipetting. No need to centrifuge.
4. Sample (usually 700 μL) is added, including any precipitate which may have formed, to an RNeasy mini-spin column sitting in a 2-mL collection tube. Centrifugation for 15 s should be done at the maximum speed. Flow-through is discarded and the collection tube can be reused.
5. Pipetting of 350 μL of Wash Buffer RW1 into the RNeasy column should be done and centrifugation for 15 s at maximum speed is performed to wash. Flow-through is discarded; keep the collection tube.
6. 10 μL of Qiagen DNase I (double strength) stock solution is to be added to 70 μL of DNase I Buffer RDD (supplied with the RNase-Free DNase I Set) in a microcentrifuge tube. Mix the tube (DNase I is especially sensitive to physical denaturation) and centrifuge briefly to gather residual liquid from the sides of the tube.

7. DNase I incubation mix (80 μL) should be pipette and mixed directly onto the RNeasy silica membrane, and placed on the bench top for 30 min.
8. Pipetting of 350 μL of Wash Buffer RW1 into the RNeasy column is done and should be left on the bench top for 5 min.
9. The next step is centrifugation for 15 s at the maximum speed. Discard the flow-through and the collection tube.
10. Transfer of the RNeasy column is done into a new 2-mL collection tube. 500 μL of Wash Buffer RPE is pipetted onto the RNeasy column. Close the tube gently and centrifugation for 15 s is to be done at the maximum speed to wash the column. Discard flow-through and reuse the collection tube.
11. Another 500 μL of Wash Buffer is added with RPE to the RNeasy column. The tube is closed and centrifuge for 2 min at the maximum speed. Discard collection tube with flow-through.
12. The RNeasy column is placed in a new 2-mL collection tube and centrifuged for 1 min at the maximum speed.
13. For the elution process, transfer of the RNeasy column is to be done to a new 1.5-mL collection tube. Pipetting 30 μL of RNase-free water directly onto the silica-gel membrane is done, and it is kept on the bench top for 1 min. Then the tube is closed and centrifuged for 1 min at the maximum speed to elute.

2.4.1 Second DNase I Treatment

1. Eluates were pooled from four columns (care should be taken that the RNA concentration need not exceed the capacity of the column= 100 μg).
2. For each pooled RNA sample (110–120 μL), 10 μL of 10× One-Phor-All Buffer, 1.4 μL of RNA-guard, 5.0 μL (25 units) of Amersham DNase I based on 1.1 unit DNase I per microgram of cDNA fragment size 50–200 bases were added. It is incubated at 37 °C for 30 min.
3. 350 μL of Lysis Buffer RLT is added to the sample and mix.
4. 250 μL of 100% ethanol is added to the lysate and mixed well by pipetting. Do not centrifuge.
5. The sample (usually 700 μL) should be applied to an RNeasy mini spin column in a 2-mL collection tube. 15 s of centrifugation should be done at the maximum speed, followed by discarding flow-through and the collection tube.
6. The RNeasy column should be transferred to a new 2-mL collection tube. 500 μL of Wash Buffer was pipetted with RPE into the RNeasy column. Tube is gently closed and centrifugation is done for 15 s at the maximum speed to wash the

column. The flow-through is discarded and the collection tube can be reused.

7. Another 500 μL of Wash Buffer RPE is added to the RNeasy column. The tube was closed and centrifuged for 2 min at the maximum speed. Discard the collection tube with flow-through.
8. The RNeasy column can be placed in a new 2-mL collection tube and centrifuged for 1 min at the maximum speed.
9. For elution, the RNeasy column should be transferred to a new 1.5-mL collection tube. 50 μL of RNase-free water is pipetted directly onto the silica-gel membrane and is left on the bench top for 1 min. The tube is closed and centrifugation can be done for 1 min at the maximum speed to elute.

2.4.2 Application

RNA isolation and subsequent cDNA preparation is very important for studying the coding region of the gene.

3 Checking of Quality, Concentration, and Integrity of RNA

Points to Remember

1. Spectrophotometry, fluorimetry, agarose gel electrophoresis, and real-time PCR are the techniques for quality checking of DNA.

3.1 Basic Principle

The purity of the extracted RNA was checked in a UV spectrophotometer by measuring its absorbance at 260 nm (OD_{260}), followed by 280 nm (OD_{280}). It can also be assessed by electrophoresis by visualization of rRNA and mRNA.

3.2 Protocol

3.2.1 Using Spectrophotometry

The RNA samples showing the OD_{260}:OD_{280} value between 1.9 and 2.2 are considered of good quality and used further. At 260 nm, 1 OD unit equates to 40 μg/μL of RNA and therefore the concentration of the extracted RNA can be calculated as:

$$\text{RNA concentration } (\mu g/\mu L) = OD_{260} \times 40 \times \text{dilution factor.}$$

3.2.2 Using Electrophoresis

The integrity of the extracted RNA can be checked using 2.2 molar (M) formaldehyde denatured agarose gel electrophoresis. Agarose gel (1%) is prepared by melting 0.3 g of agarose in 30 mL of DEPC-treated water and cooled to 60 °C; formaldehyde is added to give the final concentration of 1× and 2.2 M, respectively. The gel is poured into the tray for solidification for 30 min at room

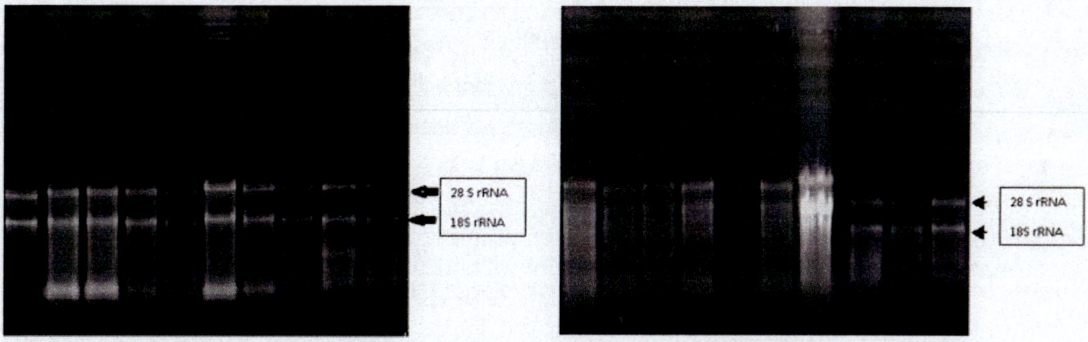

Fig. 2 Distinct band and smear of mRNA visualized in agarose gel electrophoresis

temperature. The RNA samples are loaded in the gel after mixing 2 μL RNA with 2 μL 10× gel running buffer, 4 μL formaldehyde, and 10 μL formamide. The mixture is kept at 65 °C for 15 min and chilled on ice at once and kept for 5 min. After a brief pre-run of the gel for 5 min at 5 v/cm, the sample can be loaded into the well and can be run in 1× gel running buffer for 2 h and viewed under a UV trans-illuminator. Good-quality RNA shows two intact bands (16s and 24s rRNA) with smearing between mRNA (Fig. 2) [1–5, 8].

4 cDNA Synthesis

Points to Remember
It involves the principle of reverse transcription of RNA to DNA.

4.1 Basic Principle

It involves the principle of reverse transcription from RNA to DNA.

4.2 Protocol

The total RNA has to be reverse transcribed with the help of the murine reverse transcriptase enzyme to synthesize single-strand cDNA. About 2 μg (2 μL) of total RNA should be taken in a 0.2 mL PCR tube and incubated at 70 °C for 10 min and snapped in the ice at once. Then the master mix, along with the following reagents, is added to the PCR tube.

SL. No.	Component	Amount (μL)
1	$MgCl_2$ (25 mM)	4.0
2	Reverse transcription buffer (10×)	2.0
3	dNTP mixture (10 mM)	2.0

(continued)

SL. No.	Component	Amount (μL)
4	RNAsin (Ribonuclease inhibitor)	0.5
5	AMV Reverse Transcriptase (15 u)	0.5
6	Oligo (dT) 15 primer (0.5 μg)	1.0
7	Nuclease free water	10.0

4.2.1 Application

A study of cDNA is important for studying the coding region and for the functional analysis of the derived protein through in silico tools.

5 Purification of cDNA

Points to Remember
To obtain the purest form of DNA, DNA is reextracted and precipitated with isopropanol.

5.1 Basic Principle

It is the same as the DNA isolation process. cDNA is isolated by organic phase (phenol-chloroform extraction method). DNA is precipitated with absolute alcohol in presence of salt as ammonium acetate.

5.2 Protocol

1. The cDNA reaction mixture is made up to 100 μL through the addition of 80 μL sterile TE buffer (pH 8.0) and extracted with equal volume of phenol: chloroform and centrifuged @ 10,000 rpm for 10 min.
2. The aqueous phase containing cDNA is to be collected in a separate tube. The phenol layer was re-extracted with 50 μL of TE buffer and both the aqueous phases are pooled.
3. cDNA is precipitated by ammonium acetate and ethanol, and centrifuged to pellet out cDNA. The pellet obtained should be air dried and finally dissolved in 20 μL of nuclease free water. Good cDNA is transparent [1–5, 8].

6 Recombinant DNA Technology and Gene Cloning

Points to Remember

1. It is basically a technology when foreign DNA is inserted in a cell.
2. Cloning is the process when multiple copies of gene insert are obtained.
3. Cloning may be of three types: DNA cloning, stem cell cloning, and adult cloning.
4. The major advantages of recombinant DNA technology and cloning are therapeutic application of recombinant products, gene therapy, animal pharming, production improvement through cloning of high-producing animals, and transgenic animal production.
5. Human cloning is banned due to ethical concerns.

6.1 Recombinant DNA Technology

It is the technology when the foreign DNA is inserted in the genetic constitution of the native and a new DNA obtained is called recombinant DNA.

6.1.1 Basic Principle

The desired DNA fragment is inserted in plasmid as vector. The vector enters the host cell (*Escherichia coli*) and multiplies, and hence clones of our inserted DNA fragment are obtained. The basic steps are detailed below:

6.2 Protocol for Gene Cloning

1. Elution of PCR product for cloning:

   The amplified products are run in 0.8% agarose gel with a long combed well. The products can be seen under a transilluminator and the gel containing the DNA fragment of interest is to be cut, followed by isolation of DNA from the gel by phenol-chloroform extraction method as described earlier. The gel has to be homogenized in 0.5 mL of 1× TE buffer, mixed with 0.5 mL of phenol-chloroform, and centrifuged @10,000 rpm for 15 min. The supernatant should be transferred to a fresh tube and 0.5 mL of chloroform should be added to it, mixed properly, centrifuged @10,000 rpm for 10 min, and then the DNA should be precipitated through the addition of double volume of the chilled absolute ethanol after adding 1/20th volume of 3 M sodium acetate (60 μL), then stored at −20 °C overnight for complete precipitation of DNA. The extracted DNA should be washed with 70% ethanol followed by dissolving in 6 μL of distilled water. The concentration should be checked using 1 μL eluted product in 1% agarose gel electrophoresis.

2. "A" tailing of blunt-ended PCR fragments.

   This protocol describes the cloning process in vectors containing the "T" overhang. It is employed for the ligation of the products amplified by Taq DNA polymerase. It is to be noted that Taq DNA polymerase will add adenine (A) residue in 3′ end in the process of DNA amplification. Following purification, the 3′ end would be tailed with adenine (A) residue to ensure ligation.

   The purified PCR product should be added in a 0.2 mL PCR tube and reaction buffer (10×) along with 1.5 mM $MgCl_2$, dATP to obtain a final concentration of 0.2 mM and 0.5 units Taq polymerase has to be added. The reaction volume should be adjusted up to 10 μL with autoclaved triple-distilled water. The reaction mixture should be incubated at 70 °C for 15–30 min and cooled to 4 °C for 5 min.

3. Cloning of cDNA.

   CD14 cDNA amplified from buffalo is cloned using the principle of TA ligation. Briefly, the amplified product is ligated with the vector and it is transformed into a suitable *E. coli* strain.

4. Ligation reaction.

   The ligation reaction is set up in a 0.5 mL PCR tube wherein the following components are to be added.

2× Rapid Ligation buffer	2.5 μL
pGEM®-T Easy Cloning Vector (50 ng/μL)	0.5 μL
A-tailed PCR product (template PCR)	1.5 μL
T4 DNA Ligase (3 units/μL)	0.5 μL

   The ligation mix is to be mixed properly and incubated at 19 °C for 3 h. Then it should be kept at −20 °C until used for the process of transformation.

5. Transformation.

   (a) Preparation of competent cells.

      The frozen culture of *E. coli* strain of DH5α is revived by culturing at 37 °C overnight in an LB broth. Next day, 200 μL of freshly grown cells should be inoculated in the LB medium. The cells are to be incubated at 37 °C under constant shaking (@150 rpm) until its absorbance (OD_{600nm}) is reached between 0.25 and 0.35. The cells are to be centrifuged @3,000 rpm for 10 min. Then the pellet is to be re-suspended in 1/10th volume (2 mL) of TSS buffer and incubated on ice for 1 h. Finally, 200 μL of cell suspension should be aliquoted in 1.5 mL microcentrifuge tubes and preserved at −70 °C till further use.

(b) Transformation of ligation mix into competent cells.

The ligation mix (1 μL) is to be mixed to freshly thawed competent cells (200 μL) with the help of a pre-chilled pipette tip with gentle mixing. Following 1 h of incubation on ice, heat shock is applied to the cells at 42 °C for exactly 90 s and chilled immediately on ice.

Finally, SOC medium should be mixed to the tube and the cells are to be incubated at 37 °C for 1 h. The cells should be centrifuged for 5 min and the supernatant should be discarded. The pellet is to be reconstituted in the SOC medium. Then the cells should be plated on LB agar plates with ampicillin. For blue-white screening of recombinants, the plates are also to be smeared with IPTG and X-gal at least 20 min before the plating of cells. The plates were incubated for 12–16 h at 37 °C and then transferred to 4 °C for 1 h. The plates are to be screened for the presence of blue/white colonies. The recombinant clones should be detected by white color on indicator plates. Successful cloning of an insert as detected by white colonies is due to the interruption of the coding sequence of β-galactosidase (lac-z gene) in the vector and produce white colonies.

(c) Screening of transformants for insert.

The positive clones as detected through blue-white screening need to be rechecked through colony PCR and later confirmed by plasmid-PCR and RE digestion of the plasmid DNA.

(d) Preparation of master plate.

In order to ensure the in situ maintenance of the clones until the analysis was over, a master plate should be prepared with LB agar along with ampicillin and identified properly for identification of the colonies. The single white colony should be picked up and touched onto the master plate, and has to be dipped into the test tube containing 10 mL LB broth with ampicillin. The master plate was incubated at 37 °C in the orbital shaker for overnight. The PCR tube was subjected to colony PCR.

(e) Colony PCR.

Colony PCR should be conducted with the same protocol as employed for the amplification of the CD14 gene. PCR should be performed in 25 μL reaction volume. The required amount (15 μL) of master mix should be added to the PCR tube. A positive control with buffalo cDNA and a negative control without template (genomic DNA or colony) should always be kept. The amplified PCR product (5 μL) should be run in 1% agarose gel in 1× TBE along with amplified product from genomic

DNA as marker. The clones that show the amplification is only propagated and analyzed further.

(f) Isolation of plasmid DNA from PCR-positive clones.

Only those colonies that are found to be positive by colony PCR are to be retained, while the rest should be discarded. A glycerol (15%) stock of the positive clones should be prepared and stored at −70 °C.

(g) Restriction enzyme analysis of plasmid DNA.

The recombinants are to be confirmed using RE digestion of the plasmid DNA with the suitable restriction enzyme to release the insert as well as to ascertain the orientation [1–5, 7–13].

EcoRI possesses the cutting sites on either side of the multiple cloning site (MCS) in the vector. Hence, the digestion of the positive plasmid using *Eco*RI should release the insert from the vector, provided the cloned insert should not have any internal cutting site. Accordingly, plasmid DNA was processed for restriction enzyme digestion using *Eco*RI. Digestion reaction was set up in 1.5 mL microcentrifuge tube as follows:

Plasmid DNA	2.0 μL
10× enzyme buffer	2.0 μL
Restriction Enzyme *Eco*RI (10 U/μL)	0.5 μL
Autoclaved MilliQ Water	15.5 μL

References

1. Pal A, Chatterjee PN (2009) Molecular cloning and characterization of CD14 gene in goat. J Small Ruminant Res 82:84–87
2. Pal A, Sharma A, Bhattacharya TK, Chatterjee PN, Chakravarty AK (2011) Molecular characterization and SNP detection of CD14 gene of crossbred cattle. Mol Biol Int 2011:507346, 13 pages. https://doi.org/10.4061/2011/507346
3. Sambrook J, Russell DW (2001) Molecular cloning: a laboratory manual, vol 1, 3rd edn. Cold Spring Harbor Laboratory Press, New York
4. Pal A, Chatterjee PN, Sharma A (2014) Sequence characterization and polymorphism detection in bubaline CD14 gene. Buffalo Bull 32(2):138–156
5. Pal A, Chakravarty AK, Bhattacharya TK, Joshi BK, Sharma A (2004) Detection of polymorphism of growth hormone gene for the analysis of relationship between allele type and growth traits in Karan Fries cattle. Asian Aust J Anim Sci 17(9):1334–1337
6. Pal A, Chakravarty AK, Bhattacharya TK, Sharma A (2005) Polymorphism of growth hormone gene and its association with expected milk production traits in dairy bulls. J Appl Anim Res 27(9):29–33
7. Pal A, Chakravarty AK, Chatterjee PN (2013) Polymorphism of growth hormone gene and its association with seminal and sexual behavioral traits in crossbred cattle. Theriogenology 81:474–480
8. Pal A, Pal A Jr, Mallick AI, Biswas P, Chatterjee PN (2019) Molecular characterization of Bu-1 and TLR2 gene in Haringhata Black chicken. Genomics 112(1):472–483. https://doi.org/10.1016/j.ygeno.2019.03.010
9. Banerjee S, Pal A, Mandl SC, Chatterjee PN, Chatterjee JK (2021) RIG-I has a role in

immunity against Haemonchus contortus, a gastrointestinal parasite in Ovis aries—a novel report. Front Immunol 11:534705. https://doi.org/10.3389/fimmu.2020.534705

10. Pradhan M, Pal A, Samanta AK, Banerjee S, Samanta R (2018) Mutations in cytochrome B gene effects female reproduction of Ghungroo pig. Theriogenology 119:121–130
11. Pal A, Pal A, Chakravarty AK (2018) Mutations in growth hormone gene affect stability of protein structure leading to reduced growth, reproduction and milk production in cross bred cattle—an insight. Domest Anim Endocrinol 71:106405. https://doi.org/10.1016/j.domaniend.2019.106405
12. Pal A, Pal A, Baviskar P (2020) Molecular characterization of RIGI, TLR7 and TLR3 as immune response gene of indigenous ducks in response to Avian influenza. https://doi.org/10.1101/2020.09.28.316687
13. Rawat K, Pal A, Banerjee S, Pal A, Mandal SC, Batabyal S (2021) Ovine CD14- an immune response gene has a role against gastrointesinal nematode Haemonchus contortus—a novel report.
14. Wobus AM, Boheler KR (2005) Embryonic stem cells: prospects for developmental biology and cell therapy. Physiol Rev 85(2):635–678. https://doi.org/10.1152/physrev.00054.2003

Check for updates

Chapter 3

Mitochondrial Genetics

Abstract

In earlier chapters, discussions were mostly confined to nuclear genome. This chapter discusses mitochondria, its peculiarities, properties, and differences from nuclear genome. The protocol is described for isolation of mitochondria and then the process for mitochondrial DNA isolation. The mitochondrial DNA possesses 37 genes, with two ribosomal RNA genes, 13 protein-coding genes, 22 transfer RNA genes, and one non-coding control region (D-loop region). Apart from the 37 mitochondrial genes, about 200 nuclear genes control mitochondrial function. Numtogenesis is described as the process of the transfer of mtDNA into the nuclear genome, or, less specifically, the transfer of mitochondria or mitochondrial components into the nucleus. The protocols were discussed for assessment of mitochondrial dysfunction. Analysis of mitochondrial genetics is helpful for two practical purposes—to assess health and evolution. "Mystondria" (mysterious diseases of mitochondria) refers to diseases due to defect in mitochondrial genome or the mito-nuclear relationship. This chapter also discusses the concept and methodology for mitochondrial replacement therapy, with two different approaches—pronuclear transfer and spindle transfer. This leads to the concept of "Three Parent baby." This chapter equally discusses the methodology for "Whole mitochondrial genome sequencing"—a next-generation sequencing technology. Some case studies were discussed for the role of numtogenesis in cancer and aging.

Keywords Mitochondrial genetics, Numtogenesis, Mystondria, Mitochondrial replacement therapy, Three parent baby, Whole mitochondrial genome sequencing

Points to Remember

1. Mitochondrial DNA is the extrachromosomal DNA present outside the nucleus.
2. Mitochondrial DNA possesses 37 genes with two rRNA genes, 13 number of protein-coding genes, 22 tRNA genes, and one control region (D-loop region), which reflects non-coding gene.
3. One of the interesting facts is that apart from the 37 genes of mitochondria, nuclear genes as PolG also control the regulation of mitochondrial function.
4. Numtogenesis is described as the process of the incorporation of mtDNA in the nucleus to be integrated in nuclear genome.

Aruna Pal, *Protocols in Advanced Genomics and Allied Techniques*, Springer Protocols Handbooks,
https://doi.org/10.1007/978-1-0716-1818-9_3,

1 Basic Principle of Mitogenetics

Mitochondria are among the most important cellular organelle involved in energy metabolism of the cell by the process of oxidative phosphorylation. Mitochondrial genes are 37 in number, devoid of introns in the mitochondria, which aids in the process of energy metabolism. Mitochondrial genes act in conjunction with nuclear genes in the TCA cycle. Mitochondrial DNA is situated in the cytoplasm and is much smaller in size than the nuclear DNA. For the process of extraction of mitochondrial DNA, lysis of cells is essential, followed by destruction of DNA by vortexing vigorously and using suitable chemicals such as DTT (dithioerythritol). Mitochondrial DNA is similar to bacterial DNA. Mitochondrial DNA is transmitted matrilineal in most organisms, although paternal inheritances of mitochondrial DNA can also be seen in some cases. Mammals inherit their mitochondrial DNA from the oocyte population in the mother present during fertilization. The natural selection of mitochondria only occurs in females, highlighting the fact of "Mother's Curse," which is an evolutionary effect that deleterious mitochondrial genome mutations are inherited from the mother in males, whereas the mutations may be beneficial or neutral or less deleterious to females. Note: Mitochondrial DNA can also be extracted from hair shaft, whereas genomic DNA cannot be extracted, since the excess keratin level breaks the cells, leading to the compression and destruction of the nucleus. On the contrary, since mitochondrial DNA is smaller, it is not destroyed. The protocol is similar but DTT is employed.

2 The Basic Differences Between Mitochondrial and Nuclear DNA

Sl no.	Nuclear DNA	Mitochondrial DNA
1.	It is present in the nucleus of the cell	It is present in the mitochondria of the cell
2.	The genomic structure is $2n$, where $n = 23$	Multiple copies of the single mitochondrial DNA are present in a cell
3	It possesses a double helical structure	Mitochondrial DNA is a circular DNA, resembling prokaryotic DNA
4	A nuclear envelop is present surrounding the entire nucleus containing the nuclear genome	The nuclear envelop is absent

(continued)

Sl no.	Nuclear DNA	Mitochondrial DNA
5	Chromatin contains packed DNA in the nucleus	No role of chromatin in the case of mitochondrial DNA
6	The genomic size of nuclear DNA is 3×10^9bp	Mitochondrial DNA is 16,569 bp on average. It differs by species
7	It possesses approximately 30,000 genes	Mitochondrial gene possesses 37 genes
8.	Non-coding region as intron is present in genomic DNA	Introns are absent
9.	Codon of genomic DNA differs from mitochondrial DNA. Termination codon: UGA. Both AGA and AGG codes for Arginine, whereas AUA codes for isoleucine	UGA codes for Tryptophan, AGA codes for termination codon, AGG codes for termination codon, and AUA codes for methionine
10.	It follows Mendelian inheritance	It follows non-Mendelian inheritance
11.	Histone protein is present along with the genome	Histone protein is absent in case of mitochondrial DNA
12.	DNA repair mechanism is well developed in genomic DNA	DNA repair is poor developed
13.	The mutation rate is low	Mutation is higher (10 times larger)

3 Some Interesting Facts About Mitochondria

There are a lot of similarities between mitochondria and prokaryotes. Both divide by simple fission, acts through splitting into two. However, the DNA replication strategies differ little from that of bacteria—both form displacement or D loop structures. Mitochondria have their own self-replicating DNA. The partitioning of circular DNA of mitochondria takes place in the same way as in bacteria. The mitochondrial proteins are synthesized or controlled by instructions from the cell nucleus.

1. Mitochondria is a double-walled, DNA-containing, semi-autonomous, oxidative energy-transducing organelle.
2. These are found only in Eukaryotic aerobic cells (*except mature RBCs, totally absent in prokaryotes and anerobic eukaryotic cells*).
3. Mitochondrial DNA has been considered semiautonomous since:

(a) Mitochondria contains its own DNA.
(b) Mitochondria possesses cytochromes ETS for ATP synthesis.
(c) Mitochondrial replication is similar to prokaryotes. de novo synthesis of mitochondria is not valid.
(d) DNA-forming proteins are synthesized from the nuclear genome, later employed in preparation of their own membranes.
(e) Mitochondrial enzymes are synthesized from the nuclear DNA.

4. Around 200 nuclear genes are responsible for maintaining mitochondrial function. As, for example, proteins of the outer membrane of the mitochondria are transcribed by the cell nucleus and then synthesized in cytosol and in RER (rough endoplasmic reticulum) and ultimately transported to the mitochondria. Thus, we infer nuclear control of the reproduction and not controlled by process of mitochondrial genes. This implies that the process of mitochondria is controlled by nuclear genes, and it responds to cell division and cellular physiological stresses. Thus, at least 200 nuclear genes are responsible for mitochondrial genome replication, transcription, maintenance, and regulation.
5. The mitochondrial translation process is controlled by both nuclear and mitochondrial genes. The rRNA and tRNA required for the translation of mitochondrial protein-making genes are produced by mitochondria. However, the proteins required for mitochondrial translation complex are encoded by nuclear genes.
6. Mitoribosomes are present in mitochondria. With the help of mitoribosomes, mitochondria synthesize their own structural proteins, although the majority of the mitochondrial proteins are synthesized as per the instructions from cell nuclear genome.
7. However, mitochondria can synthesize a part of the enzymes necessary for their proper function—for example, *succinate dehydrogenase*.
8. They exhibit hypertrophy, *that is*, internal growth.

4 Protocol for Mitochondrial DNA Isolation

As discussed, mitochondrial DNA are smaller, and therefore, the basic protocol is similar to that of genomic DNA isolation, but with some modifications.

1. MtDNA properties are common or similar to certain aspects of bacterial DNA, in terms of its supercoiling and its size. Hence, a common miniprep kit can serve the purpose for extraction and enrichment of mtDNA from eukaryotic cells.
2. The manufacturer's instruction was followed and DNA was eluted into 100 μL of elution buffer. The purification of the mtDNA-enriched fraction was carried out with Agencourt AMPure XP system (Beckman Coulter, Brea, CA).
3. The magnetic beads were used in a 0.4× proportion by volume. Then separation was carried out with the help of a magnetic stand. This is followed by washing with freshly prepared 70% ethanol two times. Following air-drying of the beads, the mtDNA was dissolved in 25 μL of 0.1× TE buffer [1–4].

Figure 1 reflects the schematic representation of mitochondria and its process of isolation.

Figure 2 reflects a diagrammatic representation of mitochondrial genes and Fig. 2a mitochondrial replacement therapy Fig. 2b.

Figure 3 reflects the human mitochondrial DNA map. A schematic representation of nuclear and mitochondrial DNA is being described in Fig. 4.

Different commercial kits are available for extraction of mitochondrial DNA isolation. They may be listed as follows:

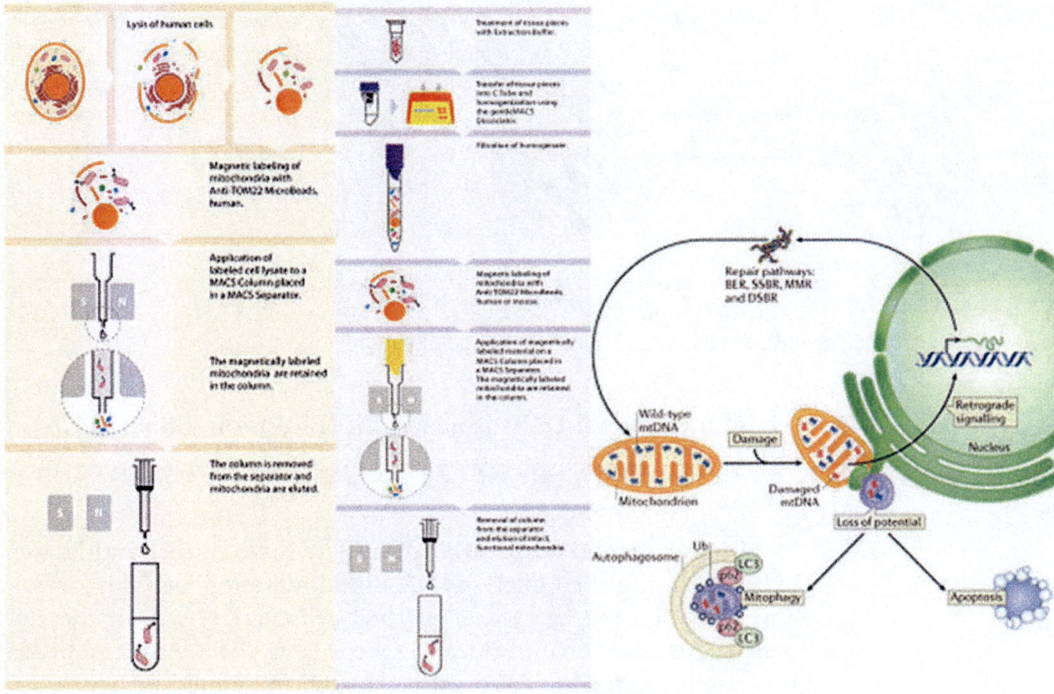

Fig. 1 Schematic representation of mitochondrial DNA and its isolation

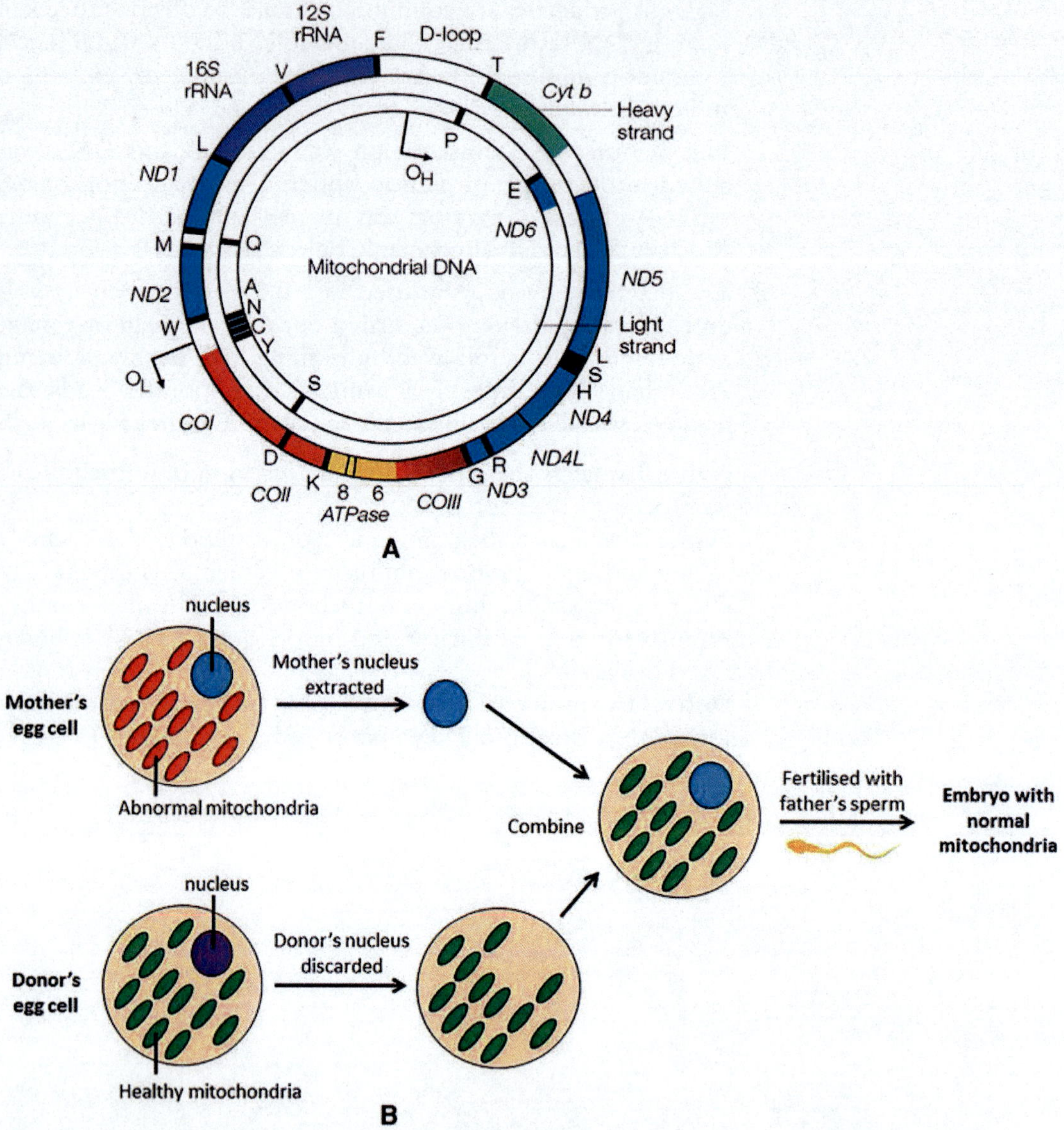

Fig. 2 Mitochondrial genes (left), mitochondrial replacement therapy (right)

1. Mitochondrial DNA Isolation Kit (BioVision, Milpitas, CA).
2. Mitochondria Isolation Kit (MACS) (Miltenyi Biotec Inc., Auburn, CA).

In the second case, DNA from isolated mitochondria was extracted using the Quick-gDNA Blood miniprep kit. This kit was employed for the extraction of total cellular DNA from the cell pellets. An additional purification step using 0.4× AMPure beads was employed for DNA isolation through these methods.

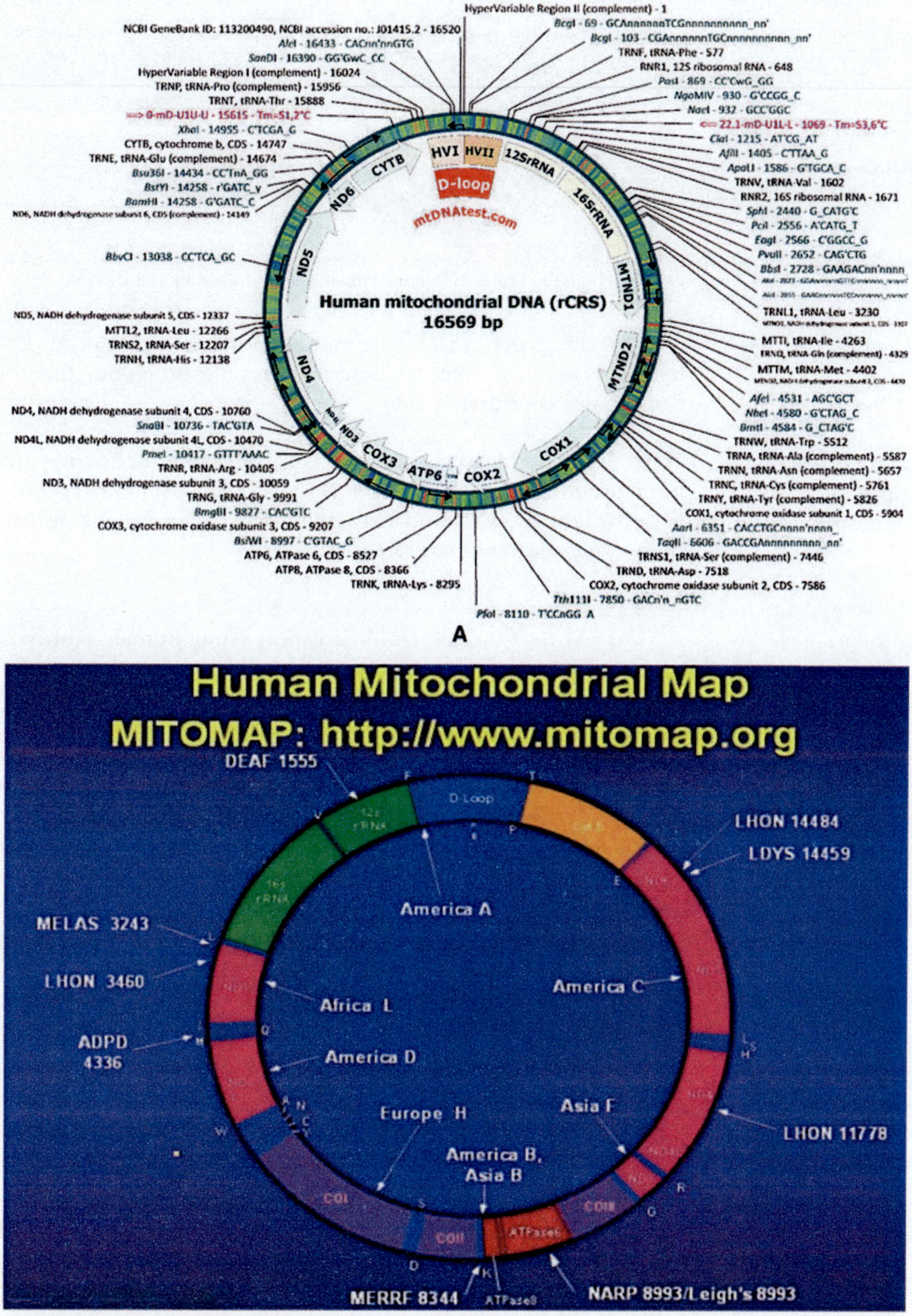

Fig. 3 (**a**) Human mitochondrial DNA map depicting the sites of mutations reported. (**b**) Human mitochondrial DNA map depicting the sites of mutation causing the diseases/disorders due to mutation in different races of human being

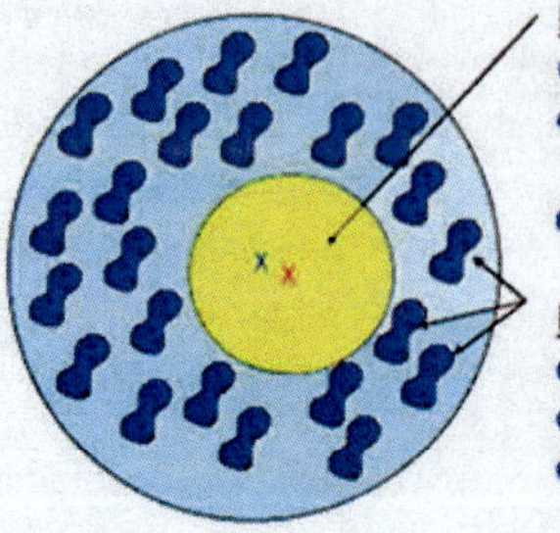

Fig. 4 Schematic representation of nuclear and mitochondrial DNA

It is always better to isolate mitochondria first, then proceed for its DNA isolation. Here we describe certain protocol for the isolation of mitochondria.

4.1 Protocol for Isolation of Mitochondria

Here, we present a protocol for isolation of mitochondria from human embryonic kidney cells (HEK-T293). A similar protocol with a slight modification may be employed for cells from other body organs for any species concerned.

4.1.1 Models for Investigation

1. Cell culture.
   (a) Mitochondrial DNA isolation from human embryonic kidney cells (HEK-T293) which were cultured in DMEM is being discussed in this section. 10% heat-inactivated fetal bovine serum, 1% penicillin-streptomycin, and 1% L-glutamine were added prior to the cell culture procedure. The cell culture was conducted at 37 °C in a reducing environment of 5% CO_2.
   (b) Cells are grown in T75 flasks until they reach 90% confluence.
   (c) Trypsin (0.25% Trypsin-EDTA) is added for cell splitting for a duration of 3–5 min at 37 °C in a reducing environment of 5% CO_2. The next step is inactivation of trypsin by the addition of equal volume of cell media. It is followed by the process of centrifugation of cells for a period of 5 min at 700 × *g*. The entire process is maintained at 4 °C.
   (d) The media is aspirated and resuspended the cell pellet in culture media.
   (e) Cells are seeded in a suitable flask or plate. The cell concentration needs to be 7×10^4 cells, which need to be seeded into a T75 flask 48 h followed by cytokine treatment. This would give rise to ~70% density during treatment.
   (f) Serum starve cells for 24 h. The aim is to have the cells synchronized at the same cell cycle. This step is followed

by the aspiration of the media and replacing it with the same volume of serum-free media (DMEM, sterile).

(g) Solutions are prepared from the lyophilized powder for following cytokine, as interferon gamma (IFN-γ), tumor necrosis factor alpha (TNF-α), and interleukin 1Beta (IL1-β) with ultra-pure water at the following concentrations: 10 ng/μL, 5 pg/μL, and 10 ng/μL respectively.

(h) Serum-starved cells were treated with solubilized cytokines added directly to the cell culture media of suitable flasks. Cell culture media was treated with individual cytokines or a mixture of all three cytokines (referred to as "*3"), or sterile water as a control (referred to as "0").

(i) Such cells treated with solubilized cytokines were incubated for 24, 48, or 72 h prior to assessment and harvesting.

2. Cell viability assessment.

(a) Solution of 3-(4,5-dimethylthiazol-2-yl)-2,5-diphenyltetrazolium bromide (MTT) @5 mg/mL is to be prepared in 1× PBS.

(b) For 12-well plates, 100 μL of MTT solution is added to each well and gently swirled to ensure even distribution.

(c) The plates were incubated for 15 min–1 h at 37 °C in 5% CO_2, and checked under a microscope to detect purple coloration. In case purple coloration is absent, it is to be swirled again and cells are being incubated in 5 min intervals until purple is observed. The amount of time varies according to each researcher.

(d) With the help a repeat pipettor, added 700 μL of MTT solvent (4 mM HCl and 0.1% NP40 in isopropanol) to each well and the plate(s) are rocked at room temperature for 5–10 min, or till all the purple color vanishes. In case the purple color is visible, add an additional 200 μL of solvent and continue to swirl.

(e) For the purpose of analysis, 100 μL is taken from each well and placed into a 96-well plate. It is read on a plate reader using 570 nm with a reference wavelength of 630 nm. Moreover, 595 nm is equally applicable for readings with proper accuracy.

3. Animal model.

Animals were screened and genotyping was performed with standard primers against mutant SOD1.

4.1.2 Isolation of Mitochondria

Note: It is necessary to work quickly and place the working station on ice throughout the procedure.

1. The step involves formation of a mitochondrial isolation buffer (MIB), a spinal cord isolation buffer (SC MIB), and an experimental buffer (EB).
   (a) Prepared the following stock solutions for MIB and EB.
   (b) 1 M Tris/MOPS is prepared by dissolving 12.1 g of Tris base in 70 mL of H_2O. Added dry MOPS to adjust pH up to 7.4. Adjusted the volume up to 100 mL final. Filter, sterilize, and store at 4 °C.
   (c) Prepared 1 M Kphos by mixing 80.2 mL of K_2HPO_4 and 19.8 mL of KH_2PO_4. Stored the above at room temperature.
   (d) Prepared 0.2 M EGTA/Tris by adding 3.8 g of EGTA to 10 mL of H_2O. Added 1 M Tris/MOPS until dissolved, ~30–40 mL. Adjusted up to 50 mL, sterilized the filter, and stored at room temperature. Note that the pH will be ~6.7.
   (e) Prepared 1 M Glutamate by making 10 mL of 1 M solution of glutamic acid. Filtered, sterilized, and stored at 4 °C.
   (f) Prepared 1 M Malate by making 10 mL of 1 M solution of malic acid. Added Tris/MOPS to 50 mM. Filtered, sterilized, and stored at 4 °C.
   (g) MIB is prepared at a concentration of 200 mM sucrose, 10 mM Tris/MOPS, pH 7.4, and 1 mM EGTA/Tris. Filtered, sterilized, and stored at 4 °C.
   (h) Prepared SC MIB at a concentration of 250 mM sucrose, 20 mM HEPES-KOH pH 7.5, 10 mM KCl, 1.5 mM $MgCl_2$, 1 mM EDTA, 1 mM EGTA, 1 mM DTT, and protease inhibitor cocktail.
   (i) EB is prepared at a concentration of 125 mM KCl, 10 mM Tris/MOPS, pH 7.4, 5 mM glutamate, 2.5 mM malate, 1 mM K phosphate, pH 7.4, and 10 mM EGTA/Tris. The freshly prepared solution is filtered, sterilized, and stored at 4 °C.
2. Equipment preparation prior to harvesting the cells.
   (a) Rinsed a small glass vessel and homogenization pestle thrice with sterile water and placed on ice.
   (b) Gathered necessary items such as a standard drill, cell scrapers, 1.5 mL, 15 mL and 50 mL tubes, and solutions.

4.1.3 Mitochondria Isolation

1. ***From cultured cells.***
   (a) For each sample type, used two T175 flasks of cells.
   (b) It is to be checked that cells are ~90% confluent and used in control experiments, or plate and treated as described in earlier steps.
   (c) On the bench top, the media is aspirated and washed the adherent cells twice with 15 mL of 1× PBS each time.
   (d) The buffer is aspirated and scraped the flasks to remove the adherent cells from the bottom of the flask.
   (e) Added 15 mL of 1× PBS to each flask, swirled, and transferred to individual 15 mL conical tubes and placed on ice.
   (f) After scraping is completed, centrifuged the tubes for 5 min at 700 × *g*, 4 °C using a table top centrifuge using a swinging bucket rotor.
   (g) Aspirated the supernatants and resuspended each pellet in 1 mL of MIB.
   (h) Both suspensions are combined and transferred to a small glass vessel for homogenization.
   (i) Added MIB to the vessel until buffer reaches the first line. Homogenized the cells with the pestle attached to a drill at medium speed for three passes.

       It is to be noted that the vessel is kept on ice during the entire step. Care should be taken not to remove the pestle above the liquid and maintain a steady speed with continuous passes.
   (j) Transferred the homogenized solution to a 50 mL conical tube.
   (k) The solution is aspirated into a 3 mL syringe using an 18 gauge 1½ in. needle and is restored back into the conical tube on ice with a 27 gauge ½ in. needle. Proper care should be taken to expel the solution against the inside wall of the tube in order to use that force for cell membrane disruption.
   (l) Repeated the syringe steps for a total of five times.
   (m) Transferred the solution into a 15 mL conical tube and centrifuged for 5 min at 600 × *g*, 4 °C.
   (n) The supernatant is removed carefully and distributed among three 1.5 mL tubes.

       Note: Mitochondria are present in the supernatant following this first, low-speed spin, whereas the cell membranes and unbroken cells are pelleted at the bottom.
   (o) Centrifuge tubes are adjusted in a fixed angle rotor at 10,000 × *g*, 4 °C for 5 min.

(p) Aspirate supernatant and combine the pellets in 100 μL MIB and immediately placed on ice.

Note: Mitochondria have been isolated in the pellet after this high-speed spin. Mitochondria will retain their membrane potential for ~2–3 ion ice following isolation. They are most stable as a concentrated stock solution in MIB.

2. ***Mitochondria Isolation from Mouse Tissue.***

(a) Anesthetize the mouse maintaining bioethical guidelines. The vaporizer is set at 5.0% to induce anesthesia. Confirmation of anesthesia is approached with the absence of reflex with foot pinch and blink reflex of eye. Kept the mouse under anesthesia condition for the entire surgical procedure with constant support of the vaporizer between 1.5 and 2.0%.

(b) Excise the liver and spinal cord from each animal and placed separately in ice cold 1× PBS. The liver tissue is transferred to a weigh dish on ice and finely chopped with a fresh razor blade for 1 min.

(c) Added tissue and suitable buffer (MIB for liver or SC MIB for the spinal cord) to the vessel until the buffer reaches the first line. The tissue is homogenized with the pestle by hand for five passes.

It is to ensure that the vessel is maintained on ice during this step.

(d) Transferred the homogenates to clean 15 mL tubes.

(e) Then the tubes are centrifuged in a fixed angle rotor at 750 × *g*, 4 °C for 10 min.

(f) Saved the supernatants into clean tubes and placed on ice.

Note: Mitochondria are present in the supernatant following the first, low-speed spin, whereas the cell membranes and unbroken cells are pelleted.

(g) Re-suspended the spinal cord pellets in 500 μL of SC MIB.

(h) Homogenization is essential for proper mixing of spinal cord pellets in SC MIB buffer solution. The process of homogenization needs to be repeated for three times

(i) The new homogenate is transferred to a fresh tube.

(j) It is centrifuged in a fixed angle rotor at 750 × *g*, 4 °C for 10 min.

(k) These new supernatants containing more mitochondria are pooled, with the first supernatant from each sample.

(l) The tubes were centrifuged in a fixed angle rotor at 10,000 × *g*, 4 °C for 5 min.

(m) Supernatants are aspirated and the liver pellets are resuspended in 500 μL of MIB and the spinal cord pellets in 50 μL SC MIB and quickly placed on ice.

Note: This time, mitochondria are present in the pellet following the high-speed spin. The membrane potential of such isolated mitochondria will be retained for ~2–3 ion ice after isolation. It will be most stable as a concentrated stock solution in MIB.

(n) For the estimation of the concentration of mitochondria in solution, a protein concentration assay is performed. A commercial protein assay kit or similar method is usually followed. Usual concentrations of mitochondria received: 2 T175 flasks yields ~2 mg/mL, 1 mouse liver ~3–5 mg/mL, and 1 mouse spinal cord ~1–3 mg/mL [2].

4.1.4 Cytochrome c Assay Using an R&D Rat/Mouse Cytochrome c Quantikine ELISA Kit

1. The initial steps include dilution of mitochondria to 0.5 mg/mL working concentration with EB and diluted mitochondria is distributed in 1.5 mL tubes for reactions (typically 30–50 μL of mitochondria per tube).
2. Added EB or DMSO (@ 1% of the total volume) to the diluted mitochondria and incubated the reactions for 7–10 min. These reactions are stable for about 30 min at room temperature if a longer time frame is needed.
3. Mitochondria are pelleted by centrifugation in a fixed-angle rotor at 10,000 × *g*, 4 °C for 5 min, and carefully separated the supernatant and placed each in a separate 1.5 mL tubes.

   Note: Tubes can be frozen at −20 °C for further analysis or used immediately.
4. The release of cytochrome *c* is assessed through comparison of the concentration of cytochrome *c* in the pellet and supernatant.

   Steps:

   (a) Prepared the samples by resuspending the pellets in the original volume of the reaction with 0.5% Tx-100 in 1× PBS.

   (b) For each sample, prepared 2 wells on the ELISA plate. The first one is for the now-solubilized pellet and the other for the supernatant from the reaction by the addition of 100 μL of cytochrome *c* conjugate (use directly from the bottle) along with 100 μL of 0.5% Tx-100 in 1× PBS, followed by all of the pellet or supernatant sample.

   (c) Gently vortexed the plate at a low setting (level 2–4) for 20 s to mix, covered and incubated for 1 h, 37 °C.

   (d) In the next step, washed the plate four times with wash buffer diluted as per the manufacturer's instructions.

Removed any excess solution by tapping the wells on paper towel.

(e) Next, added 150 μL of 1:1 A+B developing solution to each well and incubated the plate for 20–30 min in the dark at room temperature.

(f) Finally, added 50 μL of stop solution to each well and absorbance readings at 540 nm is estimated with a microplate reader.

The amount of cytochrome *c* released is being compared with the amount of cytochrome *c* in the pellet vs. the supernatant for each reaction.

4.1.5 Assessment of Mitochondrial (Dys)function

1. Initially prior to setting up of reactions, mitochondria are diluted to 0.5 mg/mL working concentration with EB, followed by placing in 1.5 mL tubes for reactions (typically 30–50 μL of mitochondria are used per tube).
2. Mitochondrial treatments.

(a) To the diluted mitochondria, appropriate treatments are added at the rate of 1/10th the total volume of diluted mitochondria. The treatments include EB control, 1 μM FCCP mixed with 50 nM Valinomycin, 10 μM rotenone, 5 μM oligomycin, 2 mM KCN, or 200 μM ADP, final concentrations. Incubated the reactions for 7 min.

Note: Proper care should be taken while handling these substances to avoid accidental ingestion or absorption through the skin, which may otherwise be harmful.

(b) Diluted TMRE, reconstituted in sterile water at a working concentration of 100 μM and stored at −20 °C, to 2 μM. Then added a volume equal to the mitochondrial reaction volume.

(c) Incubated the reactions on the bench top for 7 min.

(d) Pelleted mitochondria through centrifugation in a fixed-angle rotor at 10,000 × *g*, 4 °C for 5 min.

(e) Half of the supernatant volume of each sample is loaded onto a 396-well plate and read fluorescence is estimated.

Note: Optimization for the excitation and emission wavelengths of TMRE need to be conducted as per the manufacturer's instructions with the volume and plate that will be used for the assay. The strongest signal was obtained with excitation/emission wavelengths 485 nm/ 535 nm. A standard 384-well black plate with 25 μL of 1 μM TMRE (final reaction concentration) was used for the purpose.

(f) The fold difference is calculated in fluorescence between control (EB) and each treatment. It is obtained by dividing the relative fluorescence value (RFU) obtained from the plate reader for the experimental sample by the RFU from the control sample.

4.2 Application of Mitochondrial DNA

1. There exist certain similarities between the mitochondrial DNA in mammalian cells with the plasmid of bacteria having its independent existence.
2. The polypeptide coding mitochondrial gene is Cytochrome B. Mutations in the Cytochrome B gene affect the basic biological process adversely, leading to cardiovascular disorders, exercise intolerance, renal disorder, and so on.
3. Mitochondrial genes have proved to be very effective for studying evolutionary significance. The most common mitochondrial genes studied include Cytochrome B, D loop, Cox-I (DNA bar coding), for evolutionary analysis.
4. High throughput whole mitochondrial genome sequencing (advanced next generation sequencing studies) is greatly helpful for evolutionary and biological studies.
5. The current studies depicted the role of mitochondria in host immunity.
6. Mitochondrion has an active role in cancer.

4.3 Mitochondria and Diseases

The tendency for mitochondrial mutations has a direct effect with environmental pollution. Heteroplasmy is a common phenomenon in mitochondrial mutation, with levels close to or above threshold (>60%) possessing a very high risk of transmission. Heteroplasmy is the condition where cells possess both normal and mutant mitochondria. Heteroplasmic mitochondrial DNA mutations affect normal and tumor cells, aging, brain, and neurodegenerative disorder. It was detected from next-generation sequencing of complete human mitochondrial DNA genomes.

Mitochondrial diseases can affect an organ in the body irrespective of its age. "Mystondria" (mysterious diseases of mitochondria) is the term used for referring to newer mitochondrial diseases, not yet mentioned in the medical textbooks or in the medical literature. These diseases may arise due to unpredictable epigenetic, genetic, and genotypic and phenotypic mito-nuclear relationships. Mitochondrial diseases often occur due to defects in mitochondrial genome. They cause debility, often fatal and characteristically complex in nature. The role of mitochondria in immunity has been depicted by a number of researchers. They are inherited through the mother. Since, there is no scope for cure or treatment, the only remedy is through mitochondrial replacement therapy (MRT) [5–9]. We shall discuss MRT technology in detail subsequently.

4.4 Mitochondria and Evolution

Mitochondrial DNA, having a separate evolutionary origin from nuclear DNA, also evolves faster than the nuclear genetic markers.

5 Studies on Cytochrome B in Pig and Sheep in Relation to Reproduction and Diseases: A Case Study

In our laboratory, we have studied the Cytochrome B gene of pig and sheep, and used them as a model for studying reproduction and diseases pertaining to humans. We have characterized the Cytochrome B gene of pig and sheep [1, 3, 9]. Let us share the methodology for characterization, including bioinformatics study, and finally genetic association with economically important traits.

5.1 Animals

The present study was conducted at our university (West Bengal University of Animal & Fishery Sciences) farm, funded by the Biocare Project of the Department of Biotechnology, Ministry of Science and Technology, Govt. of India. The study comprises 521 adult Garole sheep (1–2 years of age) maintained in the same agroclimatic region and reared under similar managemental conditions. They graze for 9–10 h daily and are offered green fodder with appropriate vitamin and mineral supplementation. They are housed in flocks.

Blood samples (5 mL) were collected from the jugular vein through aseptic means. Tissue samples were collected following post-mortem examination after natural death. Proper ethical clearance and biosafety approval have been obtained from University BioEthics and Biosafety Committee maintaining the relevant guidelines and regulations.

5.2 Cytochrome B Genotyping

Mitochondrial DNA was isolated from the blood and the tissues of interest of the sheep (Hair and Tissue extraction Kit, Promega). The Cytochrome B gene was amplified with the following primers at annealing at 61 °C:

Forward: 5′CATTGATCTCCCAGCTCCA3′.

Reverse: 5′GATGTAGGGGTGTTCAACTGG3′.

The reaction conditions were optimized in 50 μL volume reaction comprising of 1× PCR buffer, 1.5 mM $MgCl_2$, 100 μM dNTP mix, 1.25 U Taq DNA polymerase, 100 nM of each primer and 50 ng of DNA. The amplification conditions comprise initial denaturation 95 °C for 5 min (one cycle) followed by denaturation at 95 °C for 30 s, annealing at 61 °C for 30 s, extension at 72 °C for 45 s for 35 cycles with a final extension at 72 °C for 5 min. DNA sequencing was conducted for the purified PCR product with deduction of amino acids.

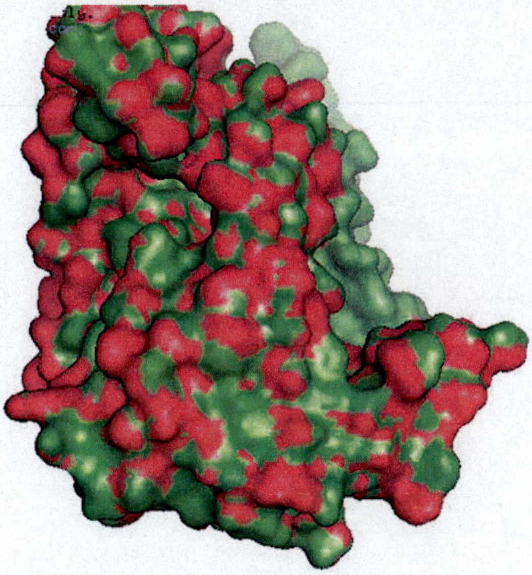

Fig. 5 3 D structure of Cytochrome B of sheep

5.3 SNP Detection of Cytochrome B Gene in Garole Sheep

Certain SNPs were detected in the ovine Cytochrome B gene through a set of bioinformatic analysis and these are screened for thermodynamical stability and deleteriousness [1]. The software employed were the MegAlign Programme of DNASTAR package, I-Mutant3.0 (http://folding.biofold.org/cgi-bin/i-mutant3.0, for depiction of DDG value in Kcal/Mol), PROVEAN (Protein Variation Effect Analyzer, http://provean.jcvi.org/index.php). The SVM-based server was employed for the prediction of the protein stability change by single mutations.

5.3.1 Three-Dimensional Structure Prediction and Model Quality Assessment

A three-dimensional structure was predicted and visualized through certain bioinformatics software such as PSI-BLAST (http://blast.ncbi.nlm.nih.gov/Blast), PHYRE2 server, PSI-Blast, and constructs based on a hidden Markov model (HMM). Visualization of 3D structures was carried out with the PyMOL (http://www.pymol.org/) tool. Other bioinformatic tools employed were Swiss PDB Viewer, SAVES (Structural Analysis and Verification Server, http://nihserver.mbi.ucla.edu/SAVES/)), ProSA (Protein Structure Analysis) web server (https://prosa.services.came.sbg.ac.at/prosa, for refinement and validation of protein structure), NetSurfP server (http://www.cbs.dtu.dk/services/NetSurfP/), TM align software, Ramachandran plot, and KEGG analysis in oxidative phosphorylation. Figure 5 shows the three-dimensional structure of the ovine Cytochrome B gene, whereas Fig. 6 is the schematic representation of the involvement of mitochondrial proteins in oxidative phosphorylation.

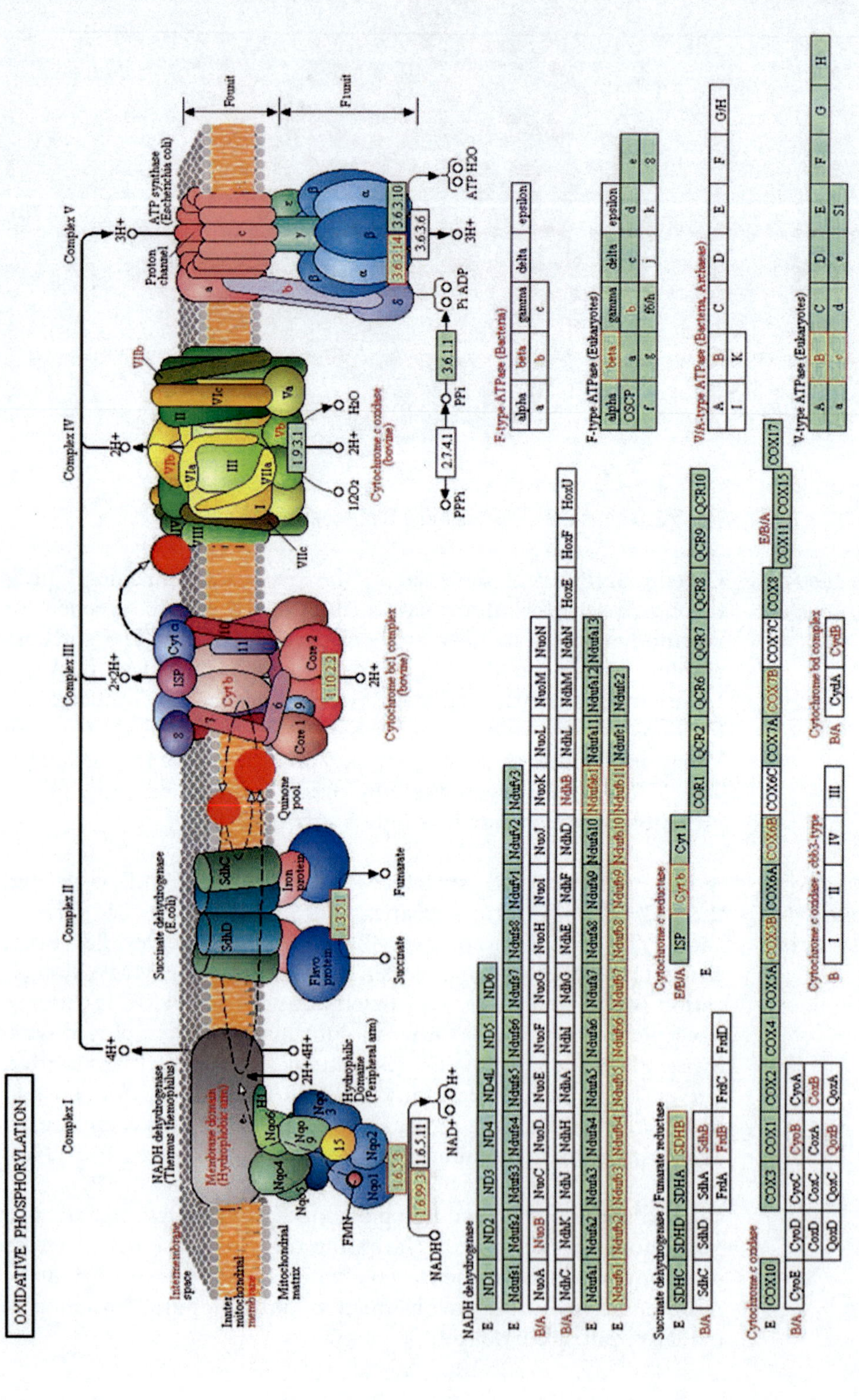

Fig. 6 Oxidative phosphorylation and involvement of mitochondrial proteins

5.3.2 Protein-Protein Interaction Network Depiction

The pathway for other proteins interacting with Cytochrome B (CytB) is important in order to understand the function of CytB. We conducted network analysis with submitting FASTA sequences to STRING 9.1. The functional partners were depicted.

5.3.3 Prediction of Post-translational Modification Sites in Mutant Cytochrome B

The post-translational modification of proteins is essential for the functioning and protein dynamics study for the protein under consideration. We have employed computational approaches to detect modification sites in the mutant protein of Cytochrome B. The bioinformatics tools employed were NetGlycate 1.0 server (http://www.cbs.dtu.dk/services/NetGlycate/, for the prediction of the glycation site), Netphos2.0 server (http://www.cbs.dtu.dk/services/NetPhos/for phosphorylation site), UbPred server (http://www.ubpred.org/ for prediction of ubiquitylation sites).

Other analysis for post-translational modifications includes Das TM filter, DisEMBL, ExPASY Peptide cutter, ExPASY_get pl_MW, GENO 3D, HCA, HHpred-homology detection and structure prediction, LipoP, Net Acet 1.0 server, Net C GlyC 1, Net NES 1, Net Phos 3, Net Turn P-prediction of Beta-turns, NPS @-GOR4 secondary structure prediction results, Phobius prediction, Interpro (Cytochrome B protein family relationship), HMMTOP, SAPS, TMHMM, and UbPrediction.

5.3.4 Health Examination of the Animals

The general health parameters were assessed as body temperature, pulse rate, heart rate, and respiration rate, and sheep were grouped as debilitated or healthy. Other tests include hematological and blood biochemical tests for the assessment of liver, heart, kidney, and the immune system. ECG was performed for confirmation of cardiomyopathical cases. Other parameters assessed were exercise intolerance and extreme debility. The functional assessment of different body organs was conducted through biochemical assessment. The physiology of heart functioning was checked by lipid profile estimation, heart rate, pulse rate, and final confirmation of disorder through ECG (electrocardiogram). The lipid profile includes the measurement of serum level of total cholesterol, triglyceride (TG), HDL cholesterol, LDL cholesterol, VLDL cholesterol, total cholesterol: HDL and HDL: LDL. Electrolyte imbalance may lead to certain disorders in the proper functioning of muscles. Muscle functioning was assessed from serum calcium, magnesium, phosphorus, sodium (Na), potassium (K), and chloride (Cl).

Liver function tests include estimation of serum level of total protein (TP), albumin (ALB), globulin (GLB), albumin:globulin (A:G) ratio, total bilirubin, direct bilirubin (conjugated), indirect bilirubin (unconjugated), SGPT (ALT), SGPT (AST), alkaline phosphatase (ALP).

Kidney function tests include estimation of serum level of glucose (fasting/random), urea, blood urea nitrogen (BUN), creatinine, and uric acid.

Hematological parameters were detected through complete hemogram test (hemoglobin percentage), packed cell volume PCV (hematocrit), total erythrocyte count (TEC), total leucocyte count (TLC). Differential leucocyte count (DLC) was estimated as neutrophil, eosinophil, basophil, lymphocyte, monocyte percentage estimation. Platelet count, MCV, MCH, MCHC, and ESR (first hour) are the other parameters estimated.

Phylogenetic Relationship

Phylogenetic relationship of sheep Cytochrome B was assessed with other livestock species and humans (Mega 6).

Statistical Analysis

The identified SNPs were associated with the phenotypic traits in terms of hematological, biochemical parameters, general health parameters as heart rate, pulse rate, body temperature, respiration rate, and occurrence of cardiac myopathy cases. ANOVA was used for finding the effect of the genotype on the above traits related to disease resistance traits. The model used for analysis was:

$$Y_{eijkl} = \mu + G_i + A_j + F_k + e_{ijkl}$$

where Y_{eijkl} = *l*th observation of the target trait, μ = Overall mean, G_i = Fixed effect of *i*th genotype, A_j = Fixed effect of *A*th animals, F_k = Fixed effect of *k*th farm, e_{ijkl} = Random error.

The analysis includes BLAST search using MegAlign Programme of Lasergene Software, DNASTAR, Edit sequence Programme of Lasergene Software, DNASTAR (DNASTAR Version 4.0, Inc., USA).

Differential mRNA expression levels of mtDNA were assessed with real-time PCR using genomic GAPDH as a reference. The primers used were:

GAPDH (amplicon size 84 bp): 5′-GGCTCCCTAGGCCCCTCCTG-3′ and 5′-TCCCAACTCGGCCCCCAACA-3′.

mtDNA (amplicon size 119 bp): 5′-CCCAGCTACTACCATCATTCAAGT-3′ and 5′-GATGGTTTGGGAGATTGGTTGATGT-3′.

The mitochondria have an active role in almost every vital organ of the body. Defects in mitochondria affect most organs such as heart, kidney, liver, brain, and digestive system. Heteroplasmy is observed for mitochondrial DNA, where an increased percentage of defective mitochondria causes diseases. Since the cytoplasm is mostly contributed by the ovum, defective mitochondria in the mother contribute a lot to the offspring.

Using pig as a model, it has been observed that defects or a deleterious mutation in the Cytochrome B gene of pig is responsible for causing defects in reproduction, such as litter size and age at first furrowing. Three parent baby (***mitochondrial replacement therapy***) implies an offspring from a normal male and the nuclear genome from a female and a cytoplasm from a healthy female with normal mitochondria. Accordingly, an offspring is born with normal mitochondria.

5.4 Mitochondrial Replacement Therapy

Mitochondrial replacement therapy is a special form of in vitro fertilization process in which the future baby's mitochondria comes from another individual. This process is carried out when the mother carries defective or dysfunctional mitochondria. Several vital mitochondrial diseases are caused due to dysfunctional or mutated mitochondria in the offspring since it is maternally inherited. Mitochondrial dysfunctions can lead to several multiorgan defects and failures. Several facts have been highlighted, where we have focused on the effect of reproduction in animals due to this dysfunctional mitochondrial DNA. Scanty informations are available on the role of mitochondrial genes on animal reproduction are available. Similarly we could detect for the first time that mutation in mitochondrial gene (genetic defect in mitochondria) has adverse effect on reproduction of farm animals [13]. It has been conducted successfully by some workers in primates. Based on our practical experience while working with animals in the farm, we have seen that 30% of the total population of the animals suffer from debility, which further leads to reproductive failure and gradually multiorgan failure, ultimately mortality in adverse cases [5, 10, 11].

5.5 Three Parent Baby Concept

The mitochondrial replacement therapy was first carried out in 2016 when a baby was born via this technique to prevent the passing of mitochondrial diseases from the mother to the child. He was referred to as the first "three parent baby" since the nuclear DNA was inherited from the parents and the mitochondrial DNA was inherited from an unrelated woman. Earlier the babies born from this particular parent did not survive since the mother carried a defective mitochondria and since the mitochondria is inherited from the mother's cytoplasm, the dysfunctional mitochondria were carried by the earlier offsprings, which led to their death. The current baby born from the three parent concept became free from all the inherited genetic disorders. The technique was carried out via in vitro fertilization.

5.5.1 Pronuclear Transfer of Mitochondrial Replacement Therapy

The first attempt made by the scientists to keep away faulty mitochondria from being passed to the offspring from the mother is the pronuclear transfer. The first "mitochondrial replacement" technique developed to prevent mitochondrial diseases is known as pronuclear transfer. It was conducted in mouse embryos in 1983 for the first time. Pronuclei are nuclei from the egg and sperm that are in the fertilized egg, called a zygote, but have not yet fused into a single nucleus (Fig. 7).

In this technique, the egg from the mother with the abnormal mitochondria and the egg from the donor with the normal

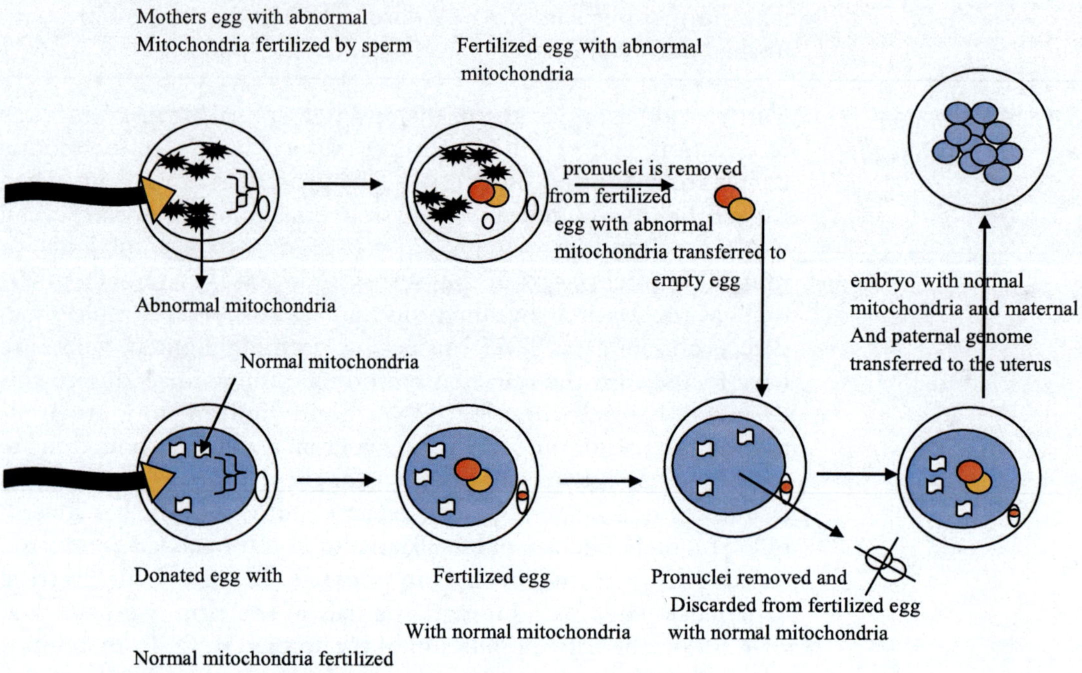

Fig. 7 Pronuclear transfer of mitochondrial replacement therapy

mitochondria were fertilized at the same time. The fertilized pronuclei from the donor's egg was sucked out and discarded, and then the pronuclei from the mother's fertilized egg was transferred to the empty donor's egg with the normal mitochondria and finally the embryo with normal mitochondria, and the maternal and paternal genomes were transferred to the uterus [11].

Limitations of the Technique

Firstly, this technique is considered unethical since it is destroying two embryos. The second llimitation is that the abnormal mitochondria can also gloom into the nuclei during the transfer process, leading to the transfer of unacceptably higher number of mitochondria from the mother's egg to the donor's egg including the disease carrying ones. Then, scientists observed that refinements in the technique produced embryos in which less than 2% of the mitochondria were carried from the mother's egg into the donor egg. It was suggested that even 1% carryover may be dangerous since the mutant mitochondria may replicate, eventually taking over the cell and crippling its energy production.

Rather than considering this technique as a curative treatment for babies who are to be born with defective mitochondria, it is considered a type of infertility treatment since these babies can have health issues related to dysfunctional mitochondria later on.

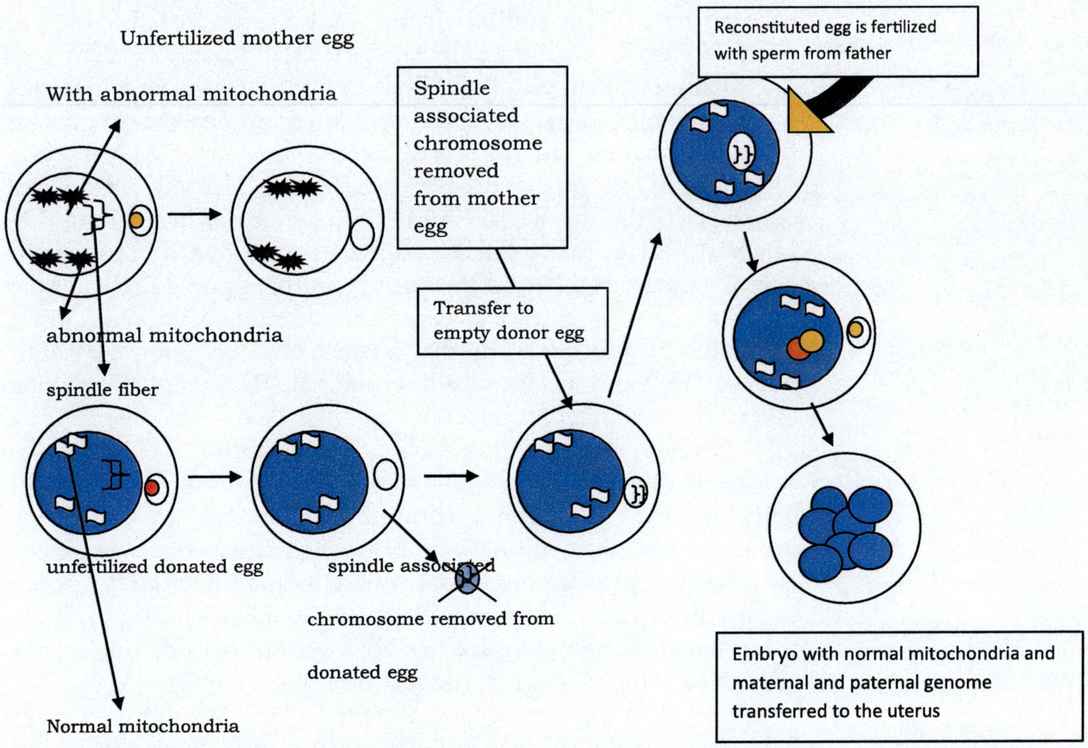

Fig. 8 Spindle transfer for mitochondrial replacement therapy

5.5.2 Spindle Transfer

DNA resides in the body's 46 chromosomes. Cell division leads to the splitting of 46 chromosomes to two sets of 23 each. The chromosomes attach themselves to several protein fiber called spindles. In this technique, two unfertilized eggs, one from the mother and the other from the donor, are taken. The nuclear membrane surrounding the nucleus is broken. By that time, the spindle does not separate itself into the chromosomes. The spindle and its attached chromosomes are removed from the donor's egg and discarded. The same thing is done with the mother's egg, except that the spindle attached to the chromosomes is kept. This spindle attached to the chromosome is injected into the donor's empty egg and the father's sperm is taken and the egg is fertilized. Then finally the embryo is with normal paternal and maternal genome and the normal mitochondria is transferred to the mother's uterus (Fig. 8).

The first three parent baby boy was born in 2016 by this method, that is, the spindle transfer method. The pioneer of spindle transfer is Mitalipov in Portland. In 2009, he showed that healthy baby monkeys can born by this technique [11].

Limitations of the Technique

The spindle fiber transfer technique also has some downside. During the transfer of the spindle attached with the chromosomes, the chromosomes may fall off, which may result in the transfer of fewer chromosomes in the embryo or too many if some are left in the egg of the donor. In both the cases, this may lead to abnormality.

Zhang performed spindle fiber in five embryos, of which one was successful and by which the baby boy was born last year. It was found out that he had only 1% of his mother's mtDNA. At the age of 3 months, he appeared to be healthy. But it is still a matter of concern whether his health will be affected or not in the long run. His health may also be affected due to the mismatch between the parent's nuclear DNA and the donor's mitochondrial DNA. Other than the risk of even trace levels of mitochondria ballooning, it was suggested by another study that mismatches between the parents' nuclear DNA and the donor mitochondrial DNA could affect aging in a study conducted in mice.

However, several controversies exist regarding the techniques as well as the term. According to Cohen, the term "three parent baby" is wrong since the offspring does not carry any trait of the donor because mitochondria do not contribute in transferring any basic trait to the offspring from the parents and therefore the donor can hardly be called as a parent. Andrew R. La Barbera also agrees to this point since mitochondria are not a genetic material and a baby conceived using this technique just has two parents.

A panel of experts also commented that it is ethical to make baby boys through this technique but not girls. Since mitochondrial DNA is passed through the mother and not the father, the baby boys born through this technique would never pass the donor's mitochondrial DNA [11].

6 The Relevance of Three Parent Offsprings in Animal Science

It was observed from the present study that mitochondria are among the most important cell organelle and have various important functions in supplying energy. Any defect in mitochondrial DNA leads to diseases and affects reproduction. A way in the correction of the problem in animal science is twofold—the first is marker-assisted selection and the second is mitochondrial replacement therapy.

6.1 Marker-Assisted Selection

First is detection of the animals carrying defective mitochondria, culling them, and selecting only the animals carrying healthy mitochondria. One important issue is the heteroplasmy of defective mitochondria. When the percentage of defective mitochondria (with mutant DNA) is more than 60%, it is considered defective or deleterious. During the selection process, individuals with more than 60% defective mitochondria will be identified and culled. When the selection will be based on all 37 mitochondrial genes, the process may be regarded as genomic selection. This method is applicable for livestock reared in farms. Thus, a healthy stock of livestock will be created.

6.2 Mitochondrial Replacement Therapy or Three Parent Baby Concept

It is considered a costly technique although if it is massively applied, it can be a cost-effective one. In the field of animal science, high-producing elite animals with mitochondrial defect need not be culled, but can easily be treated using this technique. Donor mitochondrial DNA may be collected from a female with healthy mitochondria. Thus, the unique traits of high production of the animal can be conserved easily. Disease-resistant animals can also be produced via this technique.

Endangered animals in the wild life can also be produced by this method. There is no scope to cull an endangered animal. But when defective mitochondrial mutation is present in endangered animals, it can be treated with mitochondrial replacement therapy. Mitochondrial defects cause extreme debility, exercise intolerance, and ultimately death due to mitochondrial replacement therapy.

To date, there are lacunae in the study of defects in mitochondrial genes in animals and their effects on debility and diseases. In our lab, we are working on mutations in mitochondrial gene causing debility, exercise intolerance, and having deleterious effects on the vital functions of body organs, such as liver, kidney, and heart in sheep and pig maintained in our farm. Initially we started with the Cytochrome B gene, but are now working on all 37 mitochondrial genes (whole mitochondrial genome sequencing) through the next-generation sequencing approach. Still there is a lot of scope for working out on mitochondrial defects as it causes substantial mortality and morbidity in animals. Care should be taken to include the parameter on marker-assisted selection.

Very limited research and studies have been undertaken in the field of mitochondrial genes and its effects on reproduction and diseases in animal science. In our lab, we have initiated it. Although most of the cases remain undetected in a farm, to our practical experience, we could observe that about 30% of the animals suffer from mitochondrial defect, symptomatically revealed as extreme debility with death after multi-organ failure. Three parent baby is one of the most recent concepts and is highly effective in correcting mitochondrial defects leading to reproductive disorders and other diseases without losing our indigenous germplasm. Lack of curative treatments and prenatal diagnosis had led to this alarming situation which can be prevented through the mitochondrial replacement therapy.

7 Whole Mitochondrial Genome Sequencing

Currently, with the advent of high throughput technologies, as next-generation sequencing (NGS), it is possible to sequence all 37 genes. Thus, it will be possible to detect any mutation in 37 genes causing disease. This technique is equally important for studying evolutionary significance.

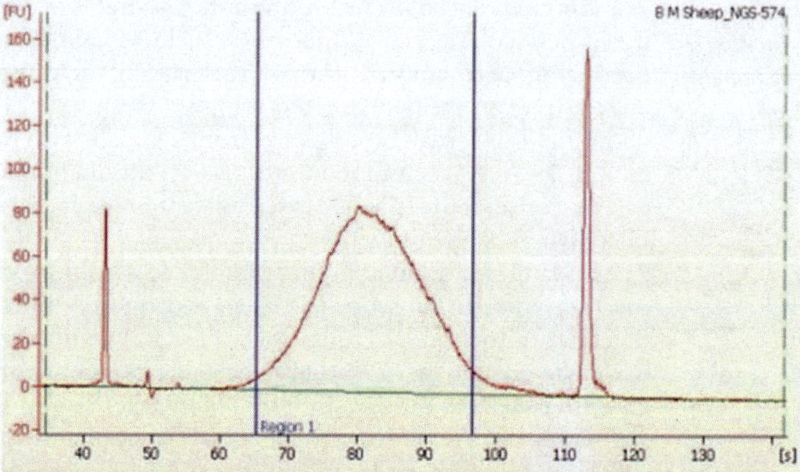

From [bp]	To [bp]	Corr. Area	% of Total	Average Size [bp]	Size distribution in CV [%]	Conc. [pg/µl]	Molarity [pmol/l]
249	1,173	1,832.4	95	494	30.1	1,379.80	4,715.4

Fig. 9 Whole mitochondrial genome sequencing of indigenous sheep (Gene Bank Accession number: MN011573, MN025257)

In our laboratory, we have studied whole mitochondrial genome sequencing for the sheep breeds of our state (Fig. 9). We had obtained a very interesting observation that we could detect a new breed of sheep "Birbhum," which is genetically distinct from other breeds.

For the first time, we have conducted the whole mitochondrial genome sequencing for indigenous ducks and geese. Similarly, we have also characterized the whole mitochondrial genome of Ghungroo pig for the first time.

The protocol is being described in brief here; the details will be discussed while discussing NGS:

- Isolation of mtDNA.
- Qualitative and quantitative analysis of g-DNA: PCR amplified with COX-2(mt specific), GAPDH and Beta actin primers to validate.
- Preparation of library: The paired-end sequencing library (NEB-Next Ultra DNA Library Preparation Kit).
- Quantity and quality check (QC) of library on Bioanalyzer: Bioanalyzer 2100 (Agilent Technologies) using High-Sensitivity (HS) DNA chip.
- Cluster generation and sequencing: Adapters are designed to allow selective cleavage.
- The mean size of the libraries is in the range of 454–715 bp. The libraries were sequenced on illumina platform (2 × 150 bp chemistry) to generate about ~2 Gb data per sample.

Sample ID	# of PE reads	Total data in GB
Garole	15,106,337	4.53
Bonpala	22,311,646	6.69
Birbhum	16,815,507	5.04
Chotanagpuri	21,890,370	6.57

8 Numtogenesis

It may be defined as the process when mtDNA is transferred to the nuclear genome. It may also be discussed as the transfer of mitochondria or mitochondrial components to the nucleus. There are 37 genes in mitochondrial genome controlling important functions of mitochondria. Apart from these genes, there are other nuclear genes controlling mitochondrial functions. It has a role in the process of evolution, aging, and human pathologies, for example, the case of cancer. The mitochondrial DNA fragments which are being transferred to nuclear genome are known as nuclear mtDNA or NUMTs.

8.1 Role of Numtogenesis for the Development of Cancer: A Case Study

As discussed, the process of numtogenesis has implications for evolution, aging, and certain pathological conditions such as cancer. Mitochondrial DNA (mtDNA) fragments transferred to the nuclear genome are called nuclear mtDNA or NUMTs. The process is referred to as numtogenesis; the term is defined as the transfer of mtDNA into the nuclear genome. The involvement of NUMTs in human biology and pathology is considered more important. Not much information pertaining to NUMTs and numtogenesis is available; the role of numtogenesis human cancers is reported. There are definite effects of NUMTs in the initiation, progression, and survival of cancer patients. There exists a suppressor mechanism for numtogenesis and mechanisms involving evolution and conservation for numtogenesis in cancer [12].

8.2 Mitochondrial DNA Escape Through BAK/BAX Macropores: Implications in Numtogenesis Underlying Aging and Cancer—A Case Study

Mitochondria possess remarkable properties with an active role in immunity. Mitochondria do possess definite roles in viral infections. At the onset of viral infections, host cells give rise to antiviral defense mechanisms such as shutting down of translation, editing of foreign RNA and degradation, and formation of interferon. However, viruses have developed the strategies to bypass the host defense system for viral propagation. For the majority of RNA viruses, the cytosolic pathogen recognition receptors (i.e., RIG-I and MDA5) recognize viral RNAs. In the next step, they give rise to conformational change and oligomerization and cause transduction of the signal to the downstream signaling partner MAVS. MAVS is an antiviral adaptor protein tethered to the OMM and

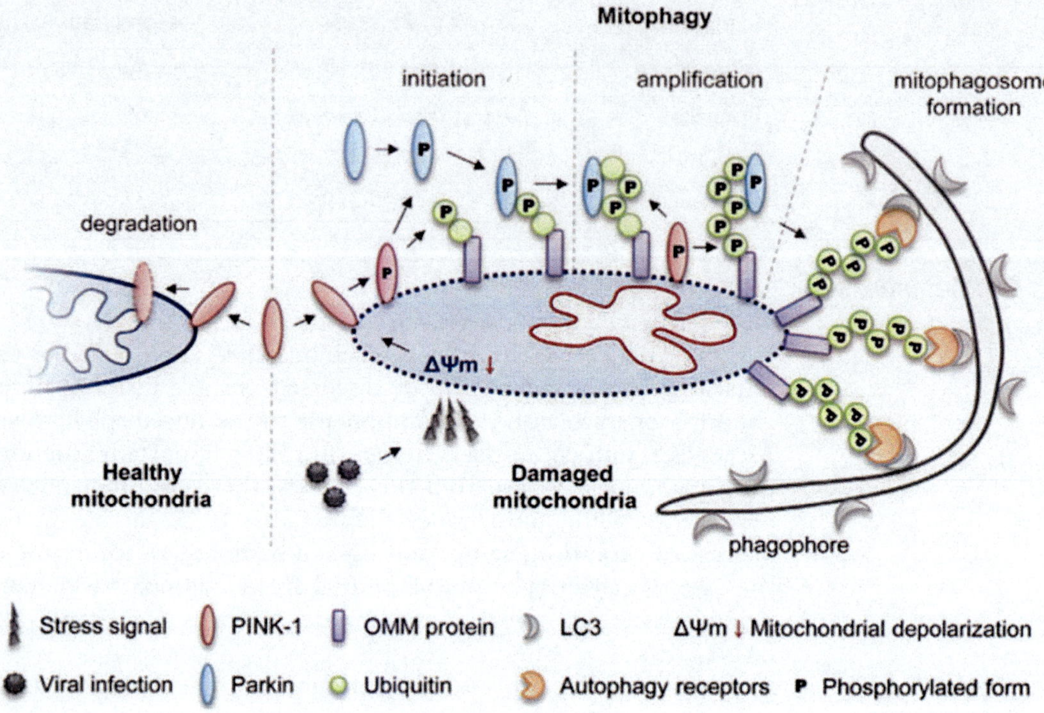

Fig. 10 Process of mitophagy through PINK1 and Parkin pathway

mitochondria-associated membranes (MAM). Activated MAVS as a next step coordinates the assembly of a multimeric signaling complex called MAVS signalosome through the recruitment of other host proteins (e.g., TRAFs, TBK1, and IRFs). The MAVS signalosome gives rise to a highly cooperative context-dependent signal resulting in the biogenesis of interferons (IFNs).

Certain viruses can suppress the host antiviral response by the process of cleaving MAVS. Some viruses [e.g., hepatitis C virus (HCV)] cleave the MAVS protein, leading to suppression of the host antiviral response. This is one way exploited by the viruses to target the mitochondria, thereby affecting host defense strategies [13]. Autophagy has been employed to affect viral propagation at multiple steps of viral life cycle. As autophagy rapidly clears the damaged cellular organelles, it generally blocks the induction of cell.

In case of a healthy mitochondria, PINK1 (a mitochondrial serine/threonine protein kinase) is transported to IMM. Subsequently, they are degraded by mitochondrial proteases and proteasome (Fig. 10). In the process of viral infection, virus-induced stress triggers mitochondrial depolarization, resulting in reduced import of PINK1 to IMM. It results in the stabilization and activation of PINK1. Activated PINK1 leads to selective tagging of damaged mitochondria by phosphorylation of ubiquitin and Parkin.

Activated Parkin leads to the formation of ubiquitin chains, which are, later, phosphorylated by PINK1, amplifying the signal. The damaged mitochondria associated with the unique signature of phospho-ubiquitin act through PINK1 and Parkin activities. They are then recognized by the mitophagy receptors, interacting with LC3. This formation causes the autophagy machinery toward the damaged mitochondria for engulfing with the expanding phagophore. This causes the formation of mitophagosome death. Thus, autophagy has a central role in determining the cell fate during viral infection.

On the other hand, viruses can employ autophagy for their replication in order to inhibit cell death as a result of virus-induced cellular stress. For example, Dengue and Zika viruses undergo autophagy to improve their replication, leading to the induction of autophagy by pharmacological agents (e.g., rapamycin). It ultimately enhances viral dissemination. In the case of the Chi-kungunya virus, autophagy controls virus-induced cell death and in vivo mortality. It has been reported that mitochondrial dynamics, mitophagy, and interaction with MAM can regulate MAVS signalosome formation. Further, there is a report suggesting that MAVS can regulate mitochondrial homeostasis via autophagy, suggesting the intricate interplay or feedback loop to control mitochondria-mediated innate immunity. Thus, it is important to understand the role of mitochondrial dynamics and mitophagy in the two closely connected aspects, acting as a determinant of cell fate and as a determinant of innate immune signaling during viral infection. These phenomena are tightly regulated by many viruses to promote viral persistence. During hepatitis B virus (HBV) or HCV infections, these viruses modulate mitochondrial dynamics to promote mitochondrial fission and mitophagy, to keeping virus-induced mitochondrial injury in check. HBV/HCV induced mitochondrial dynamics also lead to the attenuation of IFN signaling in which Parkin-MAVS interactions affect MAVS downstream signaling and the final IFN production, thus crippling innate immunity. Both HBV and HCV cause chronic infection and the persistence of the virus in the infected hepatocytes is a major reason underlying the chronic hepatic inflammation leading to liver diseases. In general, it has been analyzed that bulk autophagy blocks apoptosis. However, recent studies showed that selective mitophagy can affect both cytoprotective and pro-apoptotic conditions. This might be due to the fact that mitochondria regulates both pro- and anti-apoptotic proteins such as Bcl-2 family, Bax, and Bak. Thus, mitochondrial dynamics, such as mitophagy, can act as a major factor of cell fate during viral infection. Due to the functional significance of mitophagy in viral infections, viruses may modulate mitophagy in distinct fashions independent of autophagy.

For example, dengue virus-induced autophagy inhibits apoptosis to enhance virus replication. On the other hand, dengue restricts mitochondrial fission, causing mitophagy failure. This is related to antiviral immune evasion during dengue infection. Scientists have reported that viruses induce mitophagy to inhibit apoptosis. Transmissible gastroenteritis virus (TGEV), a porcine enteropathogenic coronavirus, induces complete mitophagy to promote cell survival and infection. Zhu et al. show that TGEV-induced mitophagy attenuates cell apoptosis by eliminating virus-induced ROS and enhances TGEV infection. TGEV stimulates DJ-1 protein deglycase to induce mitophagy. DJ-1 is also referred to as Parkinson disease protein 7. Knockdown of DJ-1 inhibits mitophagy, resulting in enhanced apoptosis after TGEV infection. Virus-induced mitophagy is a major factor of cell viability; it is considered a critical factor in virotherapy with oncolytic viruses.

Newcastle disease virus (NDV) and measles virus are known as the promising cancer-killing oncolytic viruses. NDV and measles virus induce autophagy simultaneously with mitophagy in the non-small cell lung cancer cells (NSCLCs). These exploit mitophagy favoring viral replication by blocking cytochrome c release-triggered apoptosis. Inhibition of autophagy and mitophagy with the help of pharmacological inhibitors such as 3-methyladenine (3-MA) leads to better oncolysis in NDV-infected NSCLCs. It has been observed that the delayed administration of 3-MA leads to more effective oncolysis, because autophagy is needed for efficient NDV replication. Thus, in case the 3-MA treatment is late, it leads to the inhibition of the early stages of the autophagy process and may result in efficient NDV propagation, resulting in robust oncolysis in NSCLCs. On the contrary, the measles virus does not cause cell death via apoptosis rather via necrosis or other mechanisms. Oncolysis induced by the measles virus was evaded in autophagy-impaired NSCLCs. Thus, it is clear that cell viability is differentially affected through the induction of autophagy in cells infected with virus. Hence, there is an urgent need to know the role of autophagy for individual viruses and accordingly design strategies for maximizing the oncolytic effect. This understanding will aid the development of effective strategies for therapy in response to viral pathogen apart from strategies which target autophagy or mitophagy.

We explain the effect of mitochondrial dynamics involved in antiviral immunity—case study reports.

Viruses change the physiology and function of the cell to induce a cellular environment appropriate for viral proliferation. Viral immune evasion is a necessary factor for persistence and proliferation of virus. Mitochondria can effect many cellular events including metabolism, bioenergetics, cell death, and innate immune signaling. Recently, it has been observed that viruses have a role in the process of regulation of function and dynamics of mitochondria to aid viral proliferation [7].

9 Most Recent and Interesting Finding

One of the most recent and interesting findings reveals the absence of mitochondria. A cnidarian parasite of salmon (Myxozoa: *Henneguya*) was observed to lack a mitochondrial genome.

Till now our discussion was confined to the fact that mitochondria are among the most essential organelle conferring energy to the cell. But this recent finding establishes a different fact.

It has been observed that the parasite *Henneguya salminicola* infects salmon as part of its life cycle and uses the host's energy (Fig. 11). Since there is no necessity for this organism for energy, it is postulated that perhaps due to lack of function, it (mitochondria) has been removed through evolution [14].

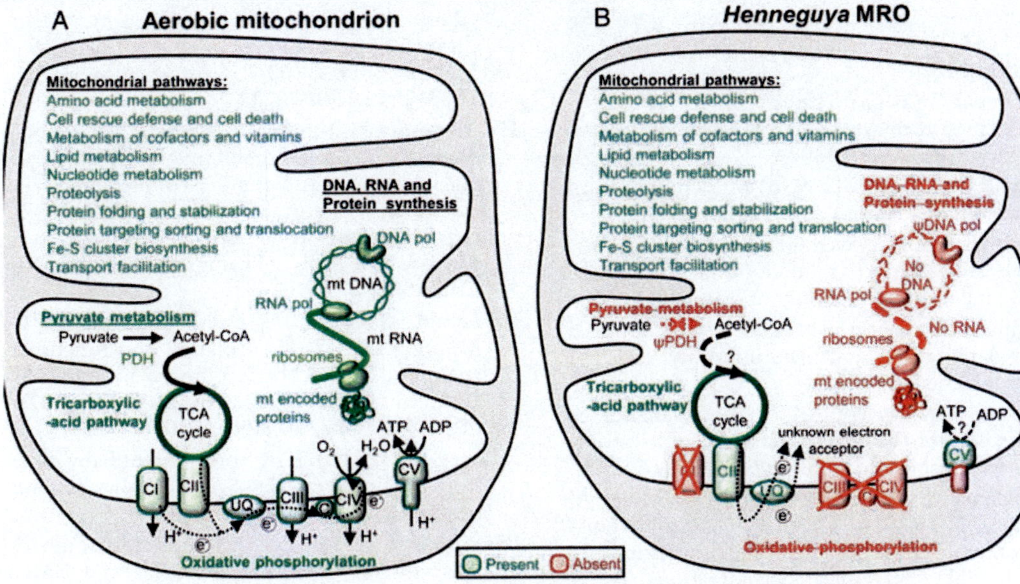

C **Mitochondrial / MRO pathways present in selected species**

		DNA	Cristae	CI	CII	CIII	CIV	CV	TCA cycle	PDH	PFO	PNO	PFL	AOX	H_2 synthesis	ASCT	SCS	ACS	Propionate	Glycine
Aerobic mitochondria	*Hydra magnipapillata* (Metazoa)	+	+	+	+	+	+	+	+	+	-	-	c	-	-	+	+	-	+	+
	Kudoa iwatai (Metazoa)	+	+	+	+	+	+	+	+	+	-	-	-	-	-	-	+	-	+	p
	Thelohanellus kitauei (Metazoa)	+	+	+	+	+	+	+	+	+	-	-	-	-	-	-	+	-	p	p
	Myxobolus squamalis (Metazoa)	?	+	+	+	+	?	+	+	+	-	-	-	-	-	-	+	-	-	p
	Myxobolus cerebralis (Metazoa)	?	+	+	+	+	?	+	+	+	-	-	-	-	-	-	+	-	p	p
Hydrogen-producing mitochondria	*Bastocystis* sp. (Stramenopiles)	+	+	+	+	-	-	-	p	+	+	+	-	+	+	+	+	c	+	+
	Acanthamoeba castellani (Amoebozoa)	+	+	+	+	+	+	+	+	+	+	-	-	+	+	+	+	-	+	+
Hydrogenosomes	*Trichomonas vaginalis* (Metamonada)	-	-	p	-	-	-	-	-	-	+	-	-	-	+	+	+	-	-	p
	Piromyces sp. (Nucletmyceta)	-	-	+	+	+	+	+	+	-	-	+	+	-	+	+	+	-	-	p
Mitosomes	*Giardia intestinalis* (Metamonada)	-	-	-	-	-	-	-	-	-	c	-	-	-	c	-	-	c	-	-
	Nematocida parisii (Microsporidia)	-	-	-	-	-	-	-	-	+	-	-	-	-	-	-	-	-	-	-
	Cryptosporidium muris (Alveolata)	-	+	-	+	-	-	+	+	-	-	c	-	+	-	-	+	c	-	p
	***Henneguya salminicola* (Metazoa)**	-	+	-	+	-	-	p	+	-	-	-	-	-	-	-	+	-	-	p

Legend: + = present; - = absent; p = partial; c = in cytosol; ? = unknown; Myxozoa

Fig. 11 Diagrammatic representation of lack of mitochondria in *Henneguya salminicola*. (**a**) Aerobic mitochondrion. (**b**) *Henneguya* MRO. (**c**) Mitochondrial/MRO pathways present in selected species

References

1. Pal A, Pal A, Banerjee S, Batobyal S, Chatterjee PN (2019) Mutation in cytochrome B gene causes debility and adverse effects on health of sheep. Mitochondrion 46:393–404. https://doi.org/10.1016/j.mito.2018.10.003
2. Peloquin JJ, Bird DM, Platzer EG (1993) Rapid miniprep isolation of mitochondrial DNA from metacestodes, and free-living and parasitic nematodes. J Parasitol 79(6):964–967
3. Pradhan M, Pal A, Samanta AK, Banerjee S, Samanta R (2018) Mutations in cytochrome B gene effects female reproduction of Ghungroo pig. Theriogenology 119:121–130
4. Quispe-Tintaya W, White RR, Popov VN, Vijg J, Maslov AY (2013) Fast mitochondrial DNA isolation from mammalian cells for next-generation sequencing. Biotechniques 55 (3):133–136. https://doi.org/10.2144/000114077
5. Chappel S (2013) The role of mitochondria from mature oocyte to viable blastocyst. Obstet Gynecol Int 2013:183024
6. Frank SA, Hurst LD (1996) Mitochondria and male disease. Nature 383:224
7. Kim SJ, Ahn DG, Syed GH, Siddiqui A (2018) The essential role of mitochondrial dynamics in antiviral immunity. Mitochondrion 41:21–27
8. Kumar DP, Sangeetha N (2009) Mitochondrial DNA mutations and male infertility. Indian J Hum Genet 15(3):93–97
9. Pal A, Chakravarty AK (2019) Genetics & breeding for the disease resistance of livestock. Elsevier Publication. ISBN: 9780128164068
10. Fogleman S, Santana C, Bishop C, Miller A, Capco DG (2016) CRISPR/Cas9 and mitochondrial gene replacement therapy: promising techniques and ethical considerations. Am J Stem Cells 5(2):39–52. eCollection 2016
11. Pal A, Banerjee S (2018) Mitochondrial replacement therapy—a new remedy for defects in reproduction. Indian J Anim Sci 88 (6):637–644
12. Singh KK, Roy Choudhury A, Tiwari HK (2017) Numtogenesis as a mechanism for development of cancer. Semin Cancer Biol 47:101–109. https://doi.org/10.1016/j.semcancer.2017.05.003
13. Horner SM, Liu HM, Park HS, Briley J, Gale M Jr (2011) Mitochondrial-associated endoplasmic reticulum membranes (MAM) form innate immune synapses and are targeted by hepatitis C virus. Proc Natl Acad Sci U S A 108:14590–14595
14. Yahalomi D, Atkinson SD, Neuhof M, Chang ES, Philippe H, Cartwright P, Bartholomew JL, Huchon D (2020) A cnidarian parasite of salmon (Myxozoa: *Henneguya*) lacks a mitochondrial genome. Proc Natl Acad Sci U S A 117(10):5358–5363. https://doi.org/10.1073/pnas.1909907117
15. Alexandre H, Junghwa A, Anne R, Thanh NT, Arnaud C (2013) Combining multiple autosomal introns for studying shallow phylogeny and taxonomy of Laurasiatherian mammals: application to the tribe Bovini(Cetartiodactyla, Bovidae). Mol Phylogenet Evol 66 (3):766–775
16. Lampl T, Crum JA, Davis TA, Milligan C, Moore VDG (2015) Isolation and functional analysis of mitochondria from cultured cells and mouse tissue. J Vis Exp 2015(97):52076. https://doi.org/10.3791/52076
17. Saey TH (2016). First 'three-parent baby' born from nuclear transfer. Science News Online. https://www.sciencenews.org
18. Schwartz M, Vissing J (2002) Paternal inheritance of mitochondrial DNA. N Engl J Med 347(8):576–580

Chapter 4

Cloning

Abstract

In this chapter on cloning, the three basic types of cloning have been described as embryo cloning, adult cloning, and therapeutic cloning. The detailed lab protocol for adult cloning has been described with a suitable practical example. Adult clones of different species of animals such as horse, deer, cattle, cat, camel, goat, sheep, mule, dog, monkey, and buffalo have been described. Issues regarding the ban on human cloning are discussed, where therapeutic cloning was observed to be very much promising. Another cloning technique, the Honolulu technique of cloning, is described.

Keywords Somatic cell nuclear transfer, Donor, Recipient, Electroporation, Clone, Enucleation, Nuclear transfer

Points to Remember

1. The cloning process is basically of three types: embryo cloning, adult cloning, and therapeutic cloning.
2. Dolly was the first report of adult cloning through somatic cell nuclear transfer.
3. Adult clones have been developed for different species such as horse, deer, cattle, cat, camel, goat, sheep, mule, dog, monkey, and buffalo.
4. Human cloning is banned due to ethical concerns.
5. Therapeutic cloning is very promising for transplant generation.
6. The detailed lab protocol for somatic cell nuclear transfer is described.

1 Basic Principle of Cloning

Cloning is the process of preparing genetically identical copies. Naturally, cloning is observed in case of identical twins. The term

Aruna Pal, *Protocols in Advanced Genomics and Allied Techniques*, Springer Protocols Handbooks,
https://doi.org/10.1007/978-1-0716-1818-9_4,

clone, coined by Herbert J. Webber, is derived from the Ancient Greek word κλών *klōn*, "twig". It refers to the process of creation of a new plant from a twig.

Through the process of asexual reproduction, clones are formed in nature.

2 Cloning Procedures

Different types of cloning procedures are described as follows:

2.1 Embryo Cloning

This technique is similar to that of natural production of monozygotic twins and triplets. In the process of fertilization, the egg is fertilized by the sperm, leading to the production of a fertilized embryo. One or more cells are removed from this fertilized embryo, followed by implantation in the uterus of different mothers, leading to the production of one or more duplicate embryos. The process results in the formation of twins and triplets with identical DNA. Figure 1 diagrammatically presents the process of embryo cloning in a flow chart. This process is very effective on various species of animals. It is to be noted the process is effective only when the cell separation is conducted before cell differentiation, the process when cells begin to specialize for future development of different organs.

2.2 Adult Cloning (Reproductive Cloning)

This is an effective technique for the production of a duplicate of an existing animal. Several successful attempts have been made to clone different livestock species. In this process, an ovum is enucleated and inserted with another nucleus from a cell removed from an adult animal, which is to be cloned. The next process comprises the development of the fertilized ovum (also called as pre-embryo), which is to be implanted in the uterus of the surrogate mother, leading to the development of a new animal with genetic material identical to that of the animal from which the nucleus has been taken. This process is also known as nucleus transfer with the involvement of both the donor cell and the egg cell. Figure 2 presents the work flow of adult cloning.

During the process initially, the cell is passed to the dormant stage or the gap zero stage. When the cell's metabolism shuts down but there is no cell death, the nucleus of the donor cell will be inserted into the egg cell through the process of electroporation or fusion. Electric waves are used for simulating the egg cell to develop into an embryo followed by implantation. Adult cloning in humans is banned in many countries because of ethical concerns. Moreover, the process may have the potentiality of producing several genetic defects in human embryo. Hence human cloning has been considered as an immoral procedure.

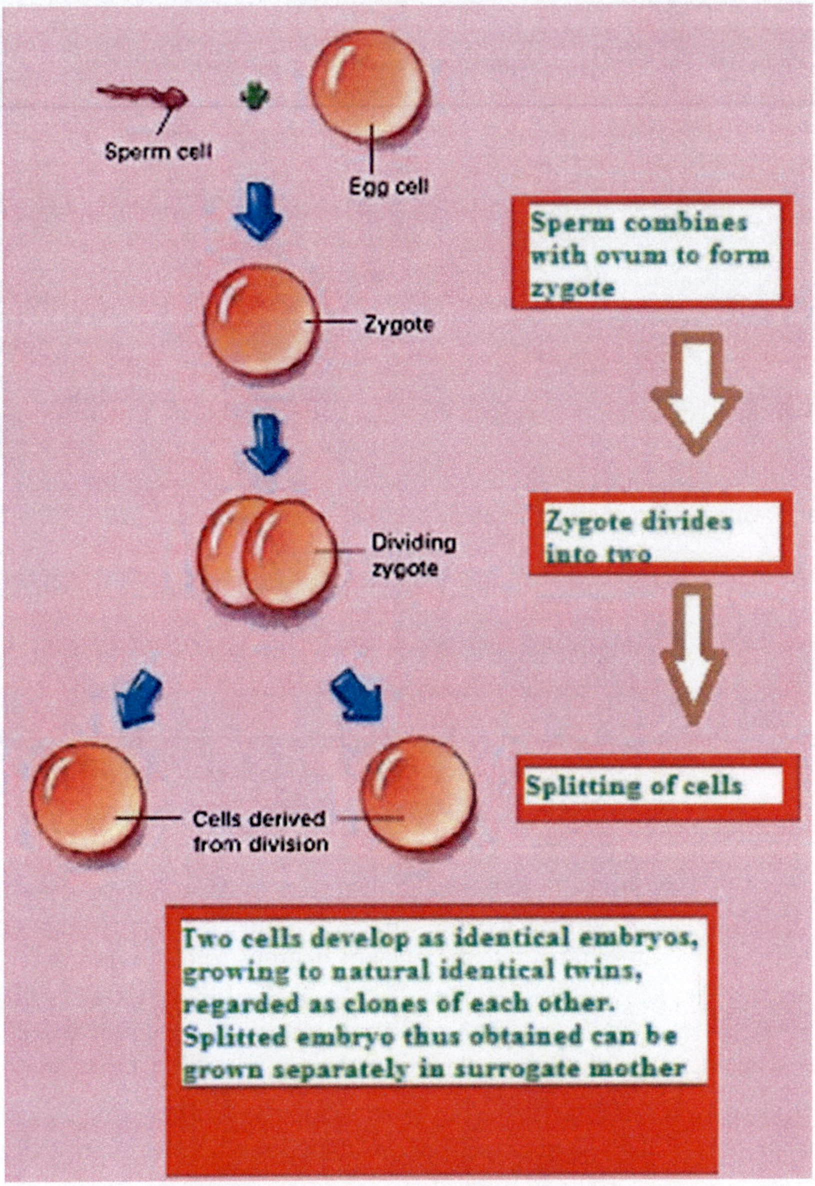

Fig. 1 Embryo cloning (Embryo splitting)

2.3 Adult Cell Cloning

An unfertilized egg is enucleated and replaced with a diploid cell ($2n$). The source of such a diploid nucleus may be either an embryo or an adult cell. Whenever the enucleated ovum cell is replaced with the nucleus derived from an adult cell, it is known as adult cloning. The ovum cell upon enucleation, that is, when the nucleus is removed, passes to G_0 stage, where the cell does not die—it is in a dormant stage and cell metabolism is at its minimum.

The first cloned mammal was "Dolly." It was a major breakthrough in the scientific world. It was evolved through the somatic

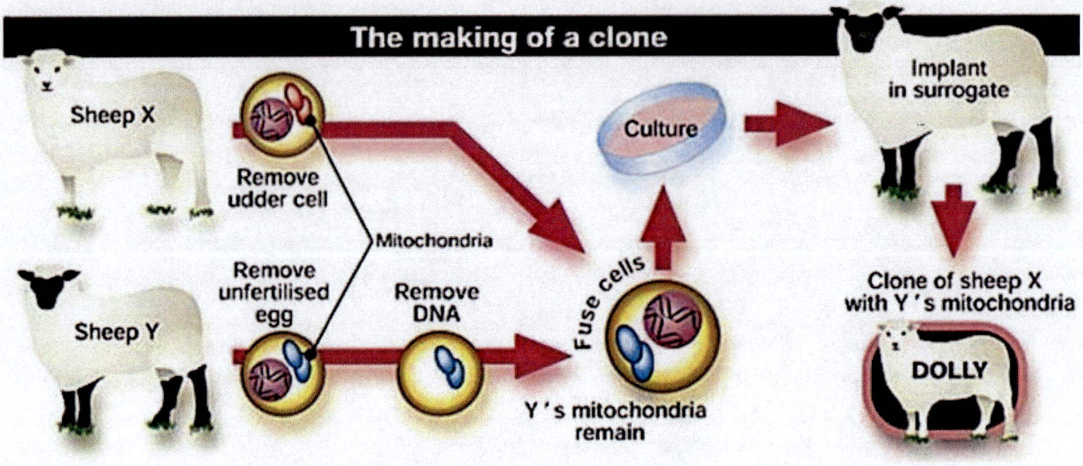

Fig. 2 Schematic representation of adult cloning (Birth of Dolly)

cell nuclear transfer technique or adult cloning technique in 1996. But unfortunately, Dolly died in 2003.

The basic steps involved in the procedure of cloning are described as follows:

Step 1: In this process, as discussed earlier, the ovum (egg cell) was removed from an ovary and it is enucleated.

Step 2: The nucleus from the udder cell (from whom the duplicate is to be prepared) is inserted in an enucleated cell.

Step 3: The fused cell divides and develops naturally. It contains cytoplasm from an adult female sheep (which donated the ovum), but the nucleus belongs to the donor sheep.

Step 4: The embryo is implanted in the uterus of the foster mother and the embryo develops into a sheep (Dolly), genetically identical to the donor sheep [1] (Table 1).

2.3.1 Merits and Demerits of Adult Cell Cloning

Merits

1. The best and healthy animal can be produced with the desired traits of interest.
2. It is helpful in biopharming. Cloned animals with suitable genetic modifications have the ability to produce economically useful proteins in their milk or other products such as wool. The process is usually referred to as biopharming.
3. Healthy animals would be produced; hence, hereditary diseases can be prevented.
4. Through adult cloning, endangered animals can be saved from extinction. Even the dead animals can be revived.
5. The technique is helpful for couples with infertility, but the complications of a ban on human cloning have posed certain restrictions.

Table 1
Advances in animal cloning for different species

Sl no.	Species	Name of cloned animal	Year
1	Sheep	Dolly	1996
2	Cattle, calf	Gene	1997
3	Cow	Daisy	1999
4.	Rhesus monkey	Tetra	2000
5.	Cat	CC (Copycat)	2002
6.	Horse	Prometa	2003
7.	Deer	Dewey	2003
8.	Mule	Idaho Gem	2003
9.	Dog	Snuppy	2005
10.	Buffalo	Samrupa (died after lung infection)	2009
11.	Buffalo	Garima	2009
12.	Camel	Injaz	2009

6. Testing of newer drugs on cloned animal or identical animals is possible.
7. Adult cell cloning is helpful in organ transplantation and gene therapy.

Demerits

1. There exist certain religious constraints that the process is not natural.
2. The technique requires great expense and technical skill.
3. There is a loss of genetic variability among the individuals developed through cloning, hence further genetic improvement is not possible.
4. It is a difficult technique. It is an equally wasteful technique as most of the attempts for cloning have largely been unsuccessful.
5. It causes a reduction in genetic variability, so any change in environmental conditions can lead to difficulties in survivability.

3 Somatic Cell Cloning Lab Protocol in Rabbit

3.1 Stage 1: Superovulation and Oocyte Collection

1. For the superovulation of mature female rabbits, two injections of 0.3 mg and four 0.4 mg s/c injections at 12 h interval is essential.

2. After 12 h of the first injection, @ 100 IU hCG, I/V was administered.
3. After an interval of 13.5 h of the hCG injection, the rabbits were laparotomized.
4. Oocytes were collected through flushing of the oviduct with Dulbecco's modified PBS (D-PBS) and BSA@ 3 mg/mL. 10 mg/mL of hyaluronidase in DPBS solution was used for the removal of the cumulus cells and pipetted with a small-bore Pasteur pipette. Oocytes thus obtained were processed through parthenogenetic activation or nuclear transfer experiments.
5. Zygotes were used as a control. It involves a similar sequence of processes as described earlier. It includes superovulation, administration of FSH injection followed by artificial insemination (AI) and hCG injection. In vivo, zygotes were collected after an interval of 16 h of AI.

3.2 Stage 2: Enucleation of Oocytes

1. Metaphase chromosomes were visualized in oocytes with the aid of an inverted microscope and removed or enucleated through micromanipulation.
2. Oocytes were placed in a small drop of PBS and 15% FBS, and 5mg/mL cytochalasin B and visualized under oil immersion.
3. Staining of the sides with oocytes aid in better visualization. Hoechst staining for proper visualization, when cytoplasm is colored and has granules, is beneficial.
4. During the process of enucleation, the metaphase plate was removed along with 10% of the adjacent cytoplasm, along with the first polar body.
5. The enucleation process was confirmed by a short UV light exposure (2–3 s) under an epifluorescent microscope.

3.3 Stage 3: Donor Cell Preparation

1. Ear-skin biopsy was employed for an adult male rabbit for the harvesting of fibroblast cells. The biopsy area was cleared by shaving and cleaned with 70% alcohol. A small section of skin tissue was cut, washed with PBS and antibiotic-antimycotic solution. The tissue was cut into small pieces (1 mm cubes) and set for fibroblast cell culture in DMEM (Dulbecco's modified Eagle's medium) and 10% FBS.
2. After an interval of 14 days, cells were washed twice with Ca^{2+} and Mg^{2+} free PBS and cells were detached from the basement by trypsin. Serial processes were employed for centrifugation and the cells were resuspended through 15 passages in DMEM and 10% FBS.
3. Thus, mononuclear nuclear donor cells were trypsinized and ready for use. Cells with a smooth membrane surface were selected for nuclear transfer.

3.4 Stage 4: Nuclear Transfer, Embryo Culture, and Transfer

1. Nuclear transfer: It is a process of nuclear transfer to the enucleated oocyte (present at the G_0 phase). During the process, not only the nucleus but the intact donor cells were introduced in the perivitelline space of enucleated oocytes and electrofusion was practiced. It is the process when the oocyte-fibroblast complex was placed in an electro cell manipulator and manually oriented in a 3.5 mm gap chamber. In the electro cell manipulator in 0.3 M mannitol solution with 0.1 mM calcium chloride and 0.1 mM magnesium chloride. It was then exposed to AC pulse (0.1 KV/cm) for 2–3 s followed immediately by a 60 ms DC pulse (2.4 KV/cm).
2. Embryo culture: The media frequently used for embryo culture in the laboratory were TCM 100 and KSOM, both of which can support greater rates of in vitro blastocyst development. KSOM culture is excellent for better in vivo survival of the cultured embryos. Embryos were cultured for 4.5 days in 5% CO_2 in air at 39 °C in a humified atmosphere with 100 mL drops of TCM199 with 15% FBS under oil. During embryo transfer evaluation, the embryos were cultured in KSOM with 0.1% BSA for 2 days, then replaced with KSOM and 1% BSA for the remaining culture again in a humified atmosphere of 5% O_2, 5% CO_2, 90% N_2.

   Blastocysts were visualized and counted following Hoechst 33342 epifluorescence staining.
3. Embryo transfer: Embryos were made ready preferably at the zygote stage or the two-cell stage after their overnight culture in the KSOM medium.

   Synchronization of recipients with oocyte donors was done through 1.2 mg GnRH analog.

3.5 Stage 5: Characterization of Donor Cells

1. Cell-specific marker staining:

   The fibroblast phenotype was confirmed in the cultured cells used for nuclear transfer through an immunocytochemical technique. Monoclonal antibodies were directed against cytoskeletal filament vimentin (fibroblast-specific cell marker) or cytokeratin (epithelial cell-specific cell marker) to confirm fibroblast in cell culture. Charged slides were used. The process, in brief, may be cited as washing with PBS, followed by fixation in methanol at 48 °C for 20 min, followed by washing thrice more in PBS and blocked with 3% BSA in PBS for 15 min. After the removal of the block, monoclonal antibodies of 1:40 dilution of antivimentin or 1:40 dilution of anti-cytokeratin were added followed by incubation for 1 h. The cells were washed thrice with PBS, followed by incubation for 1 h with 1:300 dilution of fluorescent isothiocyanate-labeled, anti-mouse immunoglobulin G. The cells were again washed thrice with PBS and visualized under a fluorescence microscope with 50% glycerol in PBS.

2. Cell cycle determination by flow cytometry:

   Flow cytometry to determine the cell-cycle stage profile was conducted as discussed earlier. Briefly, cells were trypsinized and resuspended in DMEM with 10% FBS @ of approximately 5×10^5 cells/tube. Pelleted cells were resuspended in 1 mL of 48 °C "saline GM" (6.1 mM glucose, 137 mM NaCl, 5.4 mM KCl, 1.5 mM $Na_2HPO_4 \cdot 7H_2O$, 0.9 mM KH_2PO_4, and 0.5 mM EDTA). Cells were fixed by slowly adding 48 °C ethanol while gentle vortexing and incubated overnight at 48 °C. The cells were pelleted after washing thrice with PBS and 0.5 mM EDTA. For the purpose of visualization under flow cytochemistry, the cells were pelleted and stained with 30 mg/mL of proprium iodide with 1 h incubation at room temperature. Cell population existing at different stages of the cell cycle were discriminated through single parameter histogram of DNA [2]. The cell was differentiated for the G_0/G_1 (2C DNA), S (between 2C and 4C), and G2 M (4C) phases of the cell cycle.

4 Therapeutic Cloning (or Known as Biomedical Cloning)

It is a process of cloning where the earlier phases are similar to that of adult cloning. In this technique, the sequential processes are described as follows:

1. Stem cells are collected from pre-embryo.
2. The DNA or nucleus of the sick person is inserted in it.
3. Desired tissue or whole organ is produced for the transplantation of the sick person who has supplied the DNA or nucleus.
4. There is the death of pre-embryo in this process.

Figure 3 presents the flow diagram of therapeutic cloning pertaining to different body organs.

Advantages

1. A healthy copy of a sick person tissue or organ can be readily developed.
2. There would be unlimited supply since there is no need to rely on another person's tissue.
3. There is no risk of immuno-rejection as in the case of organ transplantation from other people.
4. There is no necessity of immunosuppressive drugs as in case of organ transplantation from other people.

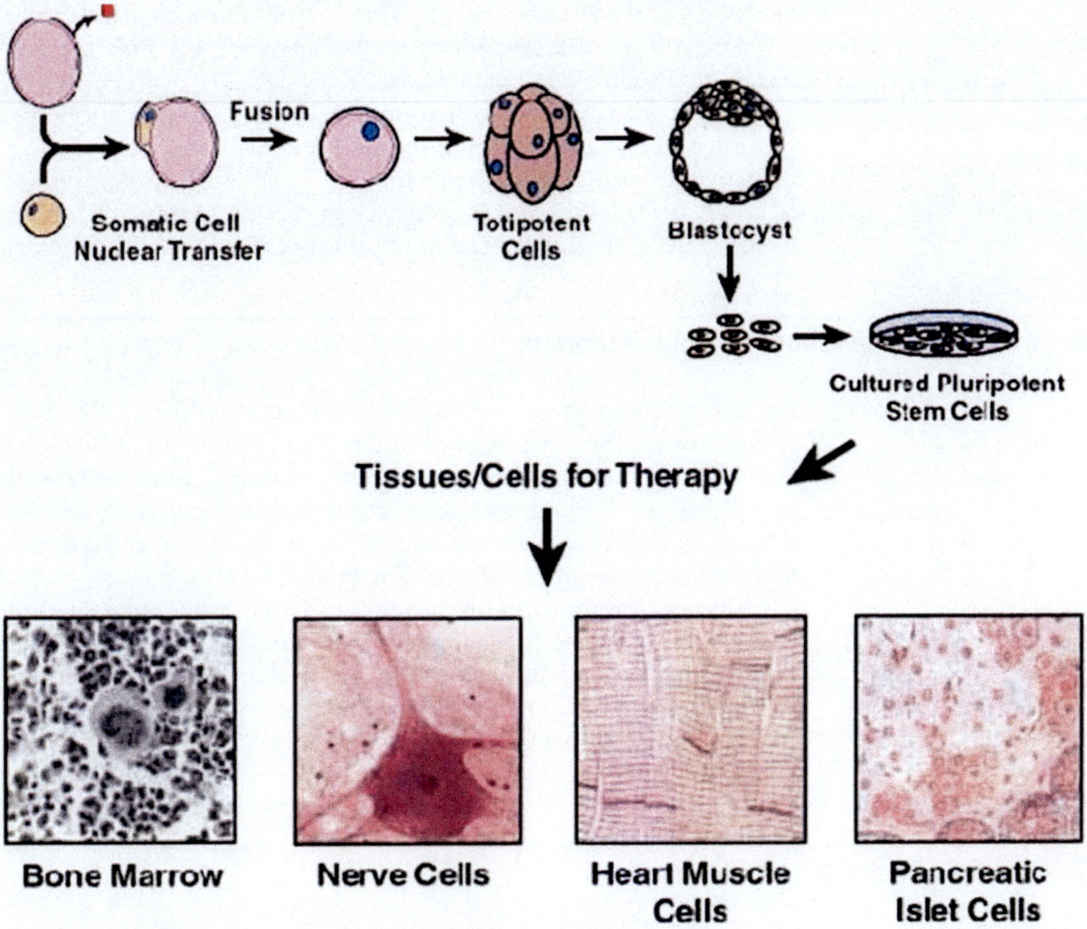

Fig. 3 Therapeutic cloning pertaining to different body organ

Therapeutic cloning is similar to reproductive cloning, but there exist certain differences. In case of therapeutic cloning, the blastocyst is implanted into a surrogate mother, and the stem cells are taken out of the blastocyst and used to develop whatever type of cells are needed.

5 Ethical Considerations in Human Cloning

There is an ethical ban on adult cloning in humans. However, in therapeutic cloning, when embryonic stem cells are removed from the blastocyst, the embryo is destroyed. There are certain ethical considerations: "Is this the same as abortion?" "Does the blastocyst deserve the right to survive?"

Administrative processes were initiated to impose certain restrictions or ban on the human cloning process. In 2001, in the USA, a ban on federal funding for embryonic stem cell research was imposed. Again after 8 years, in March 2009, an Executive Order reversing the government's ban on funding stem-cell research was signed by President Barack Obama.

Similar initiatives were also undertaken in Europe. The Convention on Human Rights and Biomedicine was conducted with an objective to prohibit human cloning in 1998, considering the ethical reasons, as discussed earlier. Prohibition of Human Cloning has been developed as per the Application of Biology and Medicine.

6 Ethical Considerations in Animal Cloning

An ethical committee involved with animal bioethics (CPCSEA) committee of the purpose of control and supervision of experiments on animals has been constituted. Ethical issues are raised against animal cloning such as whether or not humans should be allowed to interfere in the production of new life. Currently, scientific research on human cloning is restricted by regulations.

Animal cloning is highly beneficial for the production of elite animals or high-producing animals. It is equally beneficial to clone wild and endangered animals which are on the verge of extinction. With this technique, the extinct woolly mammoth can be revived. The dodo bird can equally be revived if we have preserved the nucleus.

7 Honolulu Technique of Cloning

This technique employs adult cells and nuclei for cloning. The cell's nucleus is extracted and a donor nucleus is injected into an egg, leading to a new cell followed by chemical culturing to start cell growth.

Thus, the Honolulu technique differs from nuclear transfer in one major way. In nuclear transfer technique used cumulus cells, which were injected directly into an enucleated egg cell, contrary to the Honolulu technique.

In this technique, unfertilized eggs were used as a recipient of the donor nucleus. The nuclei are removed from the recipient cell and the nucleus from the donor is inserted. The main contrasting point is that no culturing was done on the cell. After 1 h, the cell accepts the new nucleus. Again after 5 h, the egg cell is placed in a chemical culture media; cell growth starts as in natural fertilization. Soon the egg cell develops into an embryo, which can be transplanted to the surrogated mother to follow the gestation and give birth (https://www.omicsonline.org/cloning-techniques-peer-reviewed-open-access-journals.php).

Figure 4 presents a diagrammatic view of the different stages of cloning, which include blue-white screening, recombinant DNA

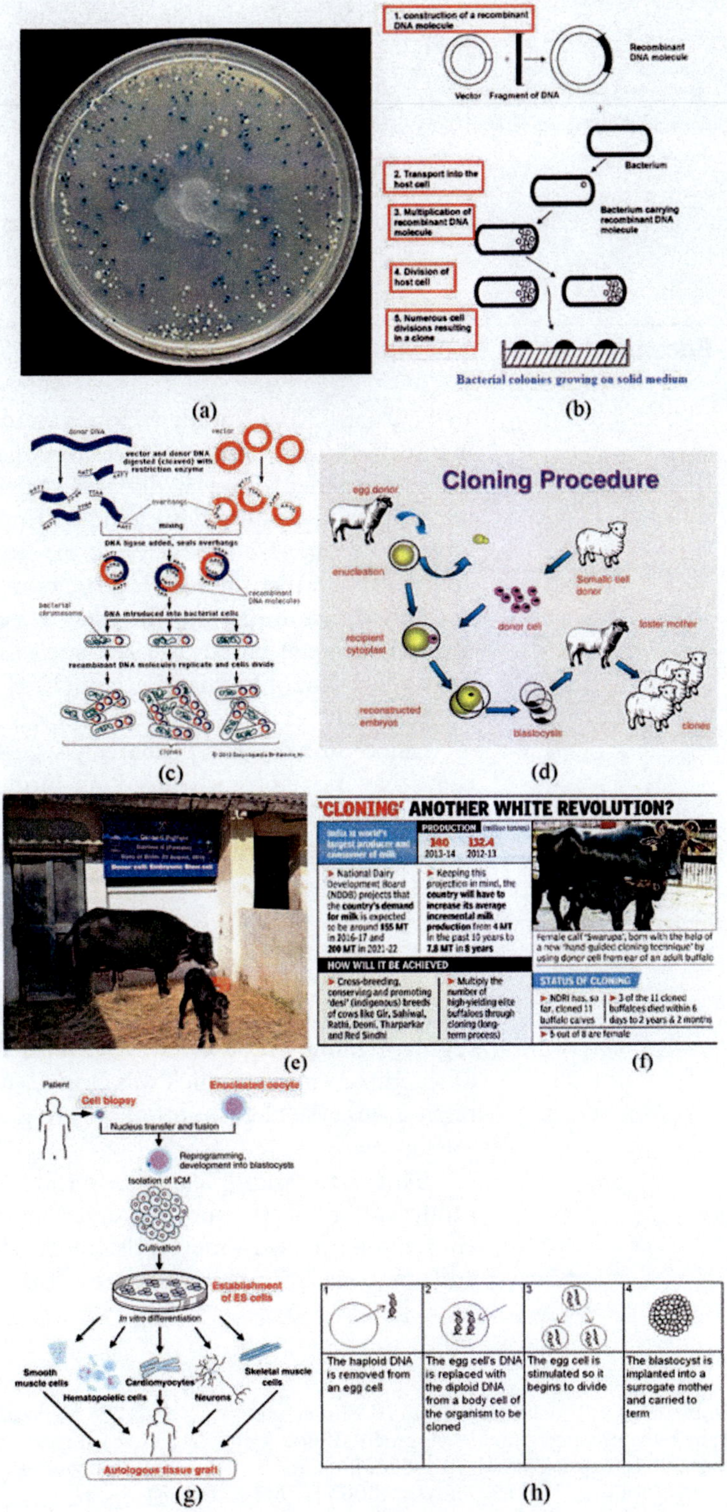

Fig. 4 Cloning. (**a**) Photographs from gene cloning, blue-white screening [3–6]. (**b**, **c**) Diagrammatic representation of recombinant DNA technology [3–6]. (**d**) Representation of adult cloning procedure. (**e**, **f**) Photograph of the cloned buffalo developed at NDRI, Karnal. (**g**, **h**) Flow chart of embryonic cloning

technology, adult cloning procedure, photograph of the cloned buffalo, and a flow chart for embryonic cloning.

Limitations for Cloning

1. It produces identical copies of the parent, and thus reduces biodiversity.
2. There are certain ethical regulations for cloning. Cloning is not permissible in humans.

8 Recent Advances in Cloning

In February 2019, monkey clones were developed with gene editing technologies simulating various diseases, particularly psychological and neurological diseases. Scientists have amulgated the techniques as cloning with gene editing for the production of the world's first gene-edited, cloned macaque monkeys (Macaca fascicularis) at the Institute of Neuroscience in Shanghai, China. These combined techniques may be used to generate a variety of monkey models for gene-based diseases, such as brain or neurological diseases (like schizophrenia, Alzheimer's disease, Parkinson's disease), immune-related diseases, metabolic disorders, and cancer.

Since monkey (or primates) is genetically closer to humans, they may be efficiently used as model organisms for studying human diseases. Due to certain ethical concerns as already discussed earlier, human cloning has already been banned.

9 A Case Study

The study involved cutting out a gene, which was involved in regulating the sleep/wake cycle. This results in the development of an edited embryo which was cloned (copied) to produce five live-born monkeys. The five monkeys are genetically identical, with the missing gene.

This causes multiple effects in the edited monkeys, for example, reduction in sleep time, increased movement during the night, altered blood hormone levels, increased anxiety and depression, and some schizophrenia-like behaviors.

References

1. Wobus AM, Boheler KR (2005) Embryonic stem cells: prospects for developmental biology and cell therapy. Physiol Rev 85(2):635–678. https://doi.org/10.1152/physrev.00054.2003
2. Dinnyes A, Dai Y, Barber M, Liu L, Xu J, Zhou P, Yang X (2001) Development of cloned embryos from adult rabbit fibroblasts: effect of activation treatment and donor cell preparation. Biol Reprod 64:257–263
3. Pal A, Chatterjee PN (2009) Molecular cloning and characterization of CD14 gene in goat. J Small Ruminant Res 82:84–87

4. Pal A, Sharma A, Bhattacharya TK, Chatterjee PN, Chakravarty AK (2011) Molecular characterization and SNP detection of CD14 gene of crossbred cattle. Mol Biol Int 2011:507346, 13 pages. https://doi.org/10.4061/2011/507346
5. Pal A, Pal A Jr, Mallick AI, Biswas P, Chatterjee PN (2019) Molecular characterization of Bu-1 and TLR2 gene in Haringhata Black chicken. Genomics 112(1):472–483. https://doi.org/10.1016/j.ygeno.2019.03.010
6. Sambrook J, Russell DW (2001) Molecular cloning: a laboratory manual, vol 1, 3rd edn. Cold Spring Harbor Laboratory Press, New York

Check for updates

Chapter 5

Immunohistochemistry

Abstract

Immunohistochemistry (IHC) is the technique of assessing the expressed protein in vivo through antigen-antibody reaction. In this technique, a more expressed protein is darkly stained than a less expressed one. Immunohistochemistry (IHC) is a technique that involves histological, immunological, and biochemical techniques. IHC with both direct (involving primary antibody) and indirect (involving both primary and secondary antibodies) methods was discussed. A detailed protocol for IHC has been discussed with practical examples of IHC as conducted in our laboratory of growth hormone in pituitary gland, and CD4 and CD8 expressions in uterus. The basic steps for IHC include fixing and embedding the tissue; cutting and mounting the section; deparaffinizing and rehydrating the section; antigen retrieval; immunohistochemical staining; counterstaining (if desired); dehydrating and stabilizing with mounting medium; and viewing the staining under the microscope.

Keywords Immunohistochemistry, Primary antibody, Secondary antibody, Antigen retrieval, Staining and counterstaining, Mounting medium

Points to Remember

1. It is the technique for assessing the expressed protein in vivo through antigen-antibody reaction.
2. A more expressed protein is more darkly stained than a less expressed one.

1 Principle

Immunohistochemistry is a technique when the tissue is cross-sectioned for histological examination and fixed on slides. Suitable proteins are labeled with antibody tagged with fluorescent dye.

Through the genomic techniques, we detect the genes responsible for a particular function. *In silico* studies (bio-informatics) aid in the prediction of the derived amino acid, 3D structure, protein domains leading to critical functional analysis. Mutations may be

Aruna Pal, *Protocols in Advanced Genomics and Allied Techniques*, Springer Protocols Handbooks, https://doi.org/10.1007/978-1-0716-1818-9_5,

detected in terms of SNP (single nucleotide polymorphism), CNV (copy number variation), Indel (insertion and deletion).

Quantitative mRNA expression can be detected, which need to be validated and confirmed through techniques involved in the detection of protein expression as immunohistochemistry and Western blotting.

Common methods of protein detection include:

- ELISA.
- Gel electrophoresis.
- Western blot.
- Immunoprecipitation.
- Spectrophotometry.
- Enzyme assays.
- X-ray crystallography.
- NMR.
- Immunohistochemistry.

The advantages of immunohistochemistry include:

1. Antibodies bind to the antigen in a specific manner.
2. The technique gives a *spatial location.*
3. It can be used to locate particular cells and proteins.
4. It can be used to identify cellular events—for example, apoptosis.

Immunohistochemistry (IHC) is a technique that is a combination of histological, immunological, and biochemical techniques. It is employed for the identification of specific tissue components/proteins by means of a specific antigen/antibody reaction tagged with a visible label. IHC helps to visualize the distribution and localization of specific cellular components or proteins within a cell or tissue in situ.

2 History of Immunohistochemistry

The principle of IHC has existed since the 1930s. It started in 1941 when Coons identified pneumococci using a direct fluorescent method. It is an indirect method and involves the addition of horseradish peroxidase (HRP). The peroxidase anti-peroxidase technique of IHC evolved in 1979. The use of Avidin & Biotin complex started in the early 1980s.

The cellular antigens that we can target include those present in the cytoplasm, nucleus, cell membrane, lipids, and proteins.

Let us discuss four ***basic approaches of immunohistochemistry*** as follows:

1. **Direct method**: It involves binding with the primary antibody only (Fig. 1).
2. **Indirect method**: This technique involves both primary and secondary antibodies (Fig. 2). It is the most commonly employed technique. We shall discuss the detailed protocol of this technique in this chapter.
3. **Immunohistochemistry with enzyme-linked indirect method**: Figure 3 diagrammatically presents the technique of immunohistochemistry with enzyme-linked indirect method.

Fig. 1 Schematic representation of immunohistochemistry with primary antibody only

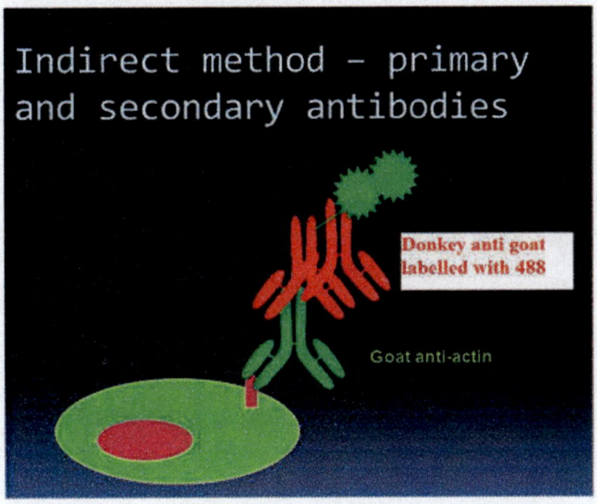

Fig. 2 Indirect method of Immunohistochemistry with both primary and secondary antibodies

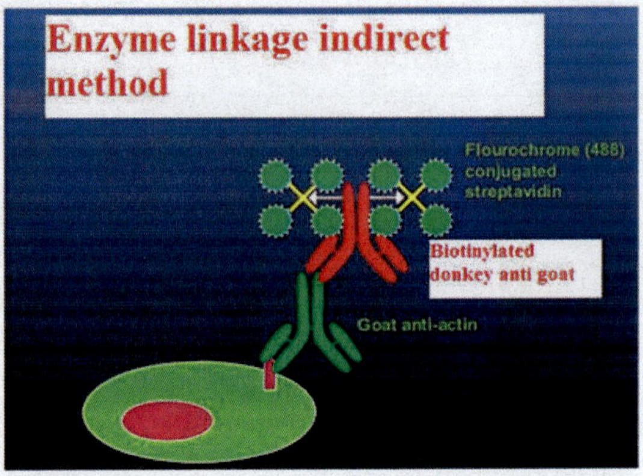

Fig. 3 Immunohistochemistry with enzyme-linked indirect method

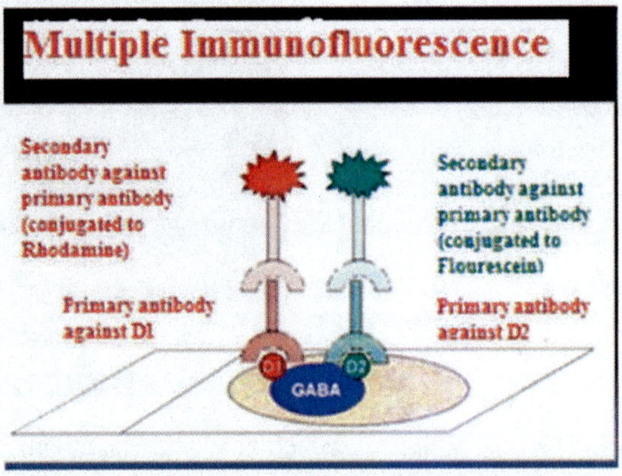

Fig. 4 Immunohistochemistry with multiple immunofluorescence

4. **Immunohistochemistry with multiple immunofluorescence**: The efficacy of immunohistochemistry may be improved with multiple immunofluorescence (Fig. 4).

3 Outline of the Protocol of IHC

1. Tissue samples should be collected and fixed with 10% neutral-buffered formalin and embedded in paraffin.
2. For antigen retrieval, thin sections of tissues (~3 μm) should be deparaffinized with xylene and rehydrated by washing with ethanol (95%) and MiliQ water.
3. Non-specific sites should be blocked by overnight incubation of sections with a blocking reagent (1% BSA, 10% FBS-PBS) at 4 °C.

4. Tissue sections are then to be stained with anti-protein, that is, antibody raised against that particular protein under consideration (1:100 dilution), for 1 h at room temperature.
5. Slides are to be washed three times with Tris-buffered saline (pH 7.6) supplemented with 0.05% Tween-20.
6. Thereafter, the slides should be treated with horseradish peroxidase-labeled rabbit anti-goat antibody for 1 h and washed as described above.
7. The slides were then to be mounted with DAB (3, 3-′-diaminobenzidine) substrate chromogen solution for 5 min.
8. Counter-staining should be performed with Mayer's hematoxylin. The slides should be rinsed with water, dried, and visualized.

4 Detailed Protocol of Immunohistochemistry

4.1 IHC-Paraffin Protocol (IHC-P)

Immunohistochemistry (or IHC) is a technique for the detection of the presence and location of proteins in tissue sections. Although a less sensitive technique quantitatively compared to other immunoassays such as Western blotting or ELISA, it allows for the visualization of processes in the context of intact tissue. This is particularly beneficial for assessing the progression and treatment of diseases such as cancer. Immunohistochemical staining is processed with antibodies recognizing the target protein. As antibodies are highly specific, they will bind to the particular protein of interest in the tissue section. This is the basis for IHC. The antibody-antigen interaction is later on visualized with the help of either chromogenic detection, where an enzyme conjugated to the antibody cleaves a substrate for the production of a colored precipitate at the location of the protein, or fluorescent detection, when a fluorophore is conjugated to the antibody and visualized using fluorescence microscopy.

IHC-P involves a series of processes including the staining of tissues after being fixed in neutral buffered formalin then embedding in paraffin then, sectioned.

IHC-P Protocol

- Fixation and embedding the tissue in paraffin.
- Cutting and mounting the section.
- Deparaffinizing and rehydrating the section.
- Antigen retrieval.
- Immunohistochemical staining.
- Counterstaining (if necessary).

- Dehydrating and stabilizing with mounting medium.
- Viewing the staining under the microscope.

1. ***Optimizing a new antibody for IHC-P.***
   (a) Antigen retrieval.
   (b) Primary antibody concentration.
   (c) Detection.
2. ***Fixation.***
3. ***Deparaffinization.***
4. ***Antigen retrieval.***
   (a) Buffer solutions for heat-induced epitope retrieval.
   (b) Heat-induced epitope retrieval methods.
   (c) Enzymatic antigen retrieval.
5. ***Immunohistochemical staining.***
   (a) General guidelines.
   (b) Protocol.
   (c) Controls.
   (d) Signal amplification.
6. ***Resources.***
7. ***Additional buffer compositions:***

4.1.1 Optimizing a New Antibody for IHC-P

When using a new antibody in IHC-P, the antibody needs to be checked to find the optimal staining conditions. The method of antigen retrieval and optimal dilution of the antibody need to be standardized.

Antigen Retrieval

Heat-induced: Sodium citrate 10 mM, pH 6.0; The reagents useful for antigen retrieval are: Tris/EDTA pH 9.0; Enzymatic: trypsin, pepsin, or other protease

Primary Antibody Concentration

It is recommended to use 0.5 μg/mL and 5 μg/mL antibody overnight at 4 °C. In case the antibody is unpurified, we recommend to start with the initially low starting dilutions and also test a 20-fold higher dilution.

Detection

It is recommended to use horseradish peroxidase (HRP) for visible light microscopy. The recommended chromogen for horseradish peroxidase is peroxide or DAB. Different fluorochrome-conjugated antibodies are present for fluorescent microscopy; the choice will be dependent on the needs of the experiment.

4.1.2 Fixation

It is an important step for fixation. 10% neutral buffered formalin (NBF) is mostly used. Other fixatives are paraformaldehyde (PFA) or Bouin solution (formalin/picric acid).

The fixation time depends on the size of the tissue block and the type of tissue. The ideal time for fixation is between 18 and 24 h, which seems ideal for most applications. Both under-fixation and over-fixation are detrimental. Under-fixation can lead to staining of the edge, with strong signal on the edges of the section and no signal in the middle. Over-fixation may cause masking of the epitope. Antigen retrieval is important for overcoming this masking, but if the tissue has been fixed for a long period (i.e., over a weekend), signal may be absent even after antigen retrieval.

Followed by fixation, the tissue block is embedded in paraffin, and cut in a microtome to the desired thickness. Approximately 5 μm is ideal for IHC. The next step is affixing onto the slide. Tissue sections are recommended to be mounted on positively charged or APES (amino-propyl-tri-ethoxy-silane)-coated slides, popularly known as charged slides. After mounting, the slides should be dried to remove water which may have been trapped under the section. This is conducted by keeping the slide at room temperature overnight. In case there exists any problem with the section adhering to the slide, the slide may be incubated at 60 °C for a few hours.

4.1.3 Deparaffinization

It is one of the most important steps in immunohistochemistry. All the paraffin must be removed and rehydrated; otherwise, it will hamper the staining process.

4.2 Materials and Reagents

Xylene, 100% ethanol, 95% ethanol.

4.3 Method

To perform the washing step, place the slides in the slides rack.

1. Xylene: 2 × 3 min.
2. Xylene 1:1 with 100% ethanol: 3 min.
3. 100% ethanol: 2 × 3 min.
4. 95% ethanol: 3 min.
5. 70% ethanol: 3 min.
6. 50% ethanol: 3 min.
7. Running cold tap water to rinse.

Until the slides are ready for antigen retrieval, they need to be kept under tap water. The slides should not be allowed to dry from this stage onward. Antigen retrieval during drying out phase will cause non-specific antibody binding.

Formalin-fixed tissue is required for immunohistochemical test. This is essential for the formation of methylene bridges during fixation, which cross-link proteins and therefore mask antigenic sites. Heat-mediated (also known as heat-induced epitope retrieval, or HIER) and enzymatic are the two methods of antigen retrieval.

Both antigen retrieval methods help break the methylene bridges and expose the antigenic sites, which allows the antibodies to bind. Some antigens prefer enzymatic to heat-mediated antigen retrieval and vice versa. Excess concentration of reagents can affect the morphology of the section enzymatic retrieval can sometimes damage on the other way. Tris/EDTA pH 9.0 buffer is suitable for most antigens. Sodium citrate pH 6.0 is also widely used.

Heat-induced epitope retrieval is commonly performed with a pressure cooker, a microwave, or a vegetable steamer. For overnight incubation, some labs use water bath to 60 °C and incubate the slides in retrieval solution. If the antigen retrieval method is not stated on the antibody datasheet, the best method for each antigen must be found experimentally. As per Abcam, there are several methods to find the retrieval that give good staining results.

5 Buffer Solutions for Heat-Induced Epitope Retrieval

Three important and popular buffers have been listed. In the absence of advice from other researchers for a particular antibody, the choice of retrieval buffer is best accomplished by experiment.

- Sodium Citrate Buffer (10 mM Sodium Citrate, 0.05% Tween 20, pH 6.0).

Tri-sodium citrate (dihydrate) 2.94 g Distilled water 1000 mL Mix to dissolve. Adjust pH to 6.0 with 1N HCl. Mix well after adding 0.5 mL of Tween 20. The above mixture is stored at room temperature for 3 months or at 4 °C for longer storage.

1 mM EDTA, adjusted to pH 8.0.

EDTA 0.37 g. Distilled water 1000 mL. Store at room temperature for 3 months.

Tris-EDTA Buffer (10 mM Tris Base, 1mM EDTA Solution, 0.05% Tween 20, pH 9.0).

Tris 1.21 g EDTA 0.37 g Distilled water 1000 mL (100 mL to make 10×, 50 mL to make 20×) Mixed to dissolve. pH is kept normally at 9.0. Added 0.5 mL of Tween 20 and mixed well properly. Stored at room temperature for 3 months or at 4 °C for longer storage.

6 Heat-Induced Epitope Retrieval Methods

(a) Pressure cooker slides should be placed in a metal rack.

Materials and reagents.

Domestic stainless steel pressure cooker.

Hot plate.

To hold approximately 400–500 mL, a vessel with a slide rack is required.

Antigen retrieval buffer (i.e., Tris/EDTA pH 9.0, sodium citrate pH 6.0).

6.1 Method

1. To the pressure cooker add appropriate antigen retrieval buffer. Place the pressure cooker on the hotplate and turn it on full power. Do not secure the lid of the pressure cooker at this point; place the lid on it.
2. To deparaffinize and rehydrate the sections as above, we need to wait till it starts boiling. Once the boiling starts, transfer the slides from the tap water to the pressure cooker. Secure the pressure cooker lid.
3. As soon as the cooker has reached full pressure, keep it up to 3 min.
4. When 3 min have elapsed, turn off the hotplate and place the pressure cooker in an empty sink.
5. The pressure release valve is activated and cold water is passed over the cooker. Once it is de-pressurized, open the lid and run cold water into the cooker for 10 min.
6. Continue with the immunohistochemical staining protocol.

6.2 Notes and Tips

1. Antigen-retrieval time is 3 min, whereas less than 3 min may give weak staining. More than 3 min may cause over-retrieval, resulting in non-specific background staining. Simultaneously, it also increases the chances of sections dissociating from the slides. Hence, a controlled experiment is recommended, when slides of the same tissue section are retrieved for 1, 2, 3, 4, and 5 min before being immunohistochemically stained. It is performed to evaluate the optimum antigen retrieval time for the particular antibody being used.
2. The slides are allowed to cool, so that they may be handled. The antigenic sites are allowed to re-form after being exposed to such high temperature.
   (a) Microwave: It is used usually for uneven antigen retrieval, where we will get hot and cold spots and the antigen retrieval will become longer due to the absence of pressure. A scientific microwave is much more appropriate. On-board pressurized vessels are an added advantage that aid in keeping the temperature at a constant 98 °C to avoid section dissociation. In this method, the buffer may boil, and a significant amount of the retrieval buffer may evaporate. If needed, add more buffer. Do not allow the slides to dry out.

The slides need to be placed in a plastic rack and vessel after this step. Sometimes some glass racks and vessels crack due to heat treatment.

6.3 Materials and Reagents

Domestic (850 W) or scientific microwave.

Coplin jar or Microwaveable vessel with a slide rack is required to hold.

Antigen retrieval buffer (e.g., Tris/EDTA pH 9.0 and sodium citrate pH 6.0).

6.4 Method

1. The sections are deparaffinized and rehydrated as above.
2. Add the appropriate antigen retrieval buffer to the microwaveable vessel.
3. Remove the slides from the tap water and place them in the microwaveable vessel. Place the vessel inside the microwave. Wait until the solution boils if working on a domestic microwave and also set it to full power. Boil for 20 min from this point. Set the temperature at 98 °C if working on a scientific microwave, for 20 min.
4. Remove the vessel and run cold tap water into it for 10 min after 20 min. Use care with the hot solution.
5. Continue with the immunohistochemical staining protocol.

6.5 Notes

1. Use a sufficient volume of antigen retrieval solution in order to cover the slides by at least a few centimeters if using a non-sealed vessel to allow for evaporation during the boil. Be sure to watch for evaporation and for boiling over during the procedure, and do not allow the slides to dry out. Twenty minutes is required for antigen retrieval; otherwise, it will give weak staining. If the slides are kept for more than 20 min, these may become over-retrieved, causing non-specific background staining. Simultaneously, it also increases the chances of sections dissociating from the slides. Hence, optimum antigen retrieval time is one of the prime criteria for immunohistochemical staining to evaluate the particular antibody used. A controlled experiment is suggested, where slides of the same tissue section are retrieved for 5, 10, 15, 20, 25, and 30 min.
2. This allows the slides to cool enough for handling, and allows the antigenic site to re-form in high temperature.
   (a) Vegetable steamer: Some labs use a vegetable steamer or rice cooker for heat-mediated antigen retrieval treatment. It is similar to microwave, as it maintains the temperature of the buffer at 100 °C without vigorous boiling of the microwave method. This method can be done in water bath at 100 °C in place of the steamer.

The slides need to be placed in a plastic or metal rack and vessel for this procedure. Standard glass histology staining racks and vessels may crack upon heating.

6.6 Materials and Reagents

Vegetable steamer—A vessel with a slide rack to hold approximately 400–500 mL (or 250 mL if using Tissue—Tek containers) will be required.

- Antigen retrieval buffer (e.g., Tris/EDTA pH 9.0 and sodium citrate pH 6.0).

6.7 Method

1. Deparaffinize and rehydrate the sections as above.
2. Set up the vegetable steamer as per the manufacturer's instructions and preheat.
3. Pre-heat the appropriate antigen retrieval buffer to boiling in a flask. A microwave may be effectively used for this purpose.
4. Put the container which will hold the slides rack into the vegetable steamer.
5. Add the hot buffer to the container with care. It is more convenient to add the buffer to the container prior to putting the container in the steamer.
6. Close the lid of the steamer. The container with lid should be used for buffer. The rack of slides may vary as per the temperature of the antigen retrieval (AR) solution but it will return to 95–100 °C within several minutes.
7. Keep the container in the steamer for 20 min from this point.
8. After 20 min have passed, remove the vessel and wash thoroughly with cold tap water into it for 10 min. Use care with the hot solution.
9. Continue with the immunohistochemical staining protocol.

6.8 Enzymatic Antigen Retrieval

The choice of enzyme will depend on the datasheet for the antibody. If not, trypsin has been shown to be useful for a wide range of antigens which require retrieval post-formalin/PFA fixation.

There exist at least two methods for the application of the enzyme solution to the tissue: First is the direct method through pipetting the solution onto the tissue on the slide. Second is to place a rack of tissue slides into a container of enzyme solution. Since each slide needs to be handled individually, the first method uses less reagent, and the incubation time is to be cautiously monitored for each slide, so that all the slides get the same treatment. As a result, it is easier to work with large batches of slides (e.g., >5) by immersing them in a container of enzyme solution. Instead, an automated staining system may also be used; in that case, an appropriate enzymatic retrieval protocol needs to be standardized.

7 Enzymatic Retrieval, Pipetting Method

7.1 Materials and Reagents

37 °C incubator.

Humidified chamber (either the incubator itself or a container with a wet paper towel) is needed.

Two slide rack containers of TBS with slide rack are also needed.

Enzymatic antigen retrieval solution (For trypsin).

The methodology employs trypsin. There exist commercially available trypsin preparations optimized for IHC; otherwise, it can also be prepared as follows:

Trypsin Stock Solution (0.5% in Distilled Water)
Trypsin 50 mg.

Distilled water 10 mL

Mix to dissolve. Store at −20 °C.

Calcium Chloride Stock Solution (1%)
Calcium chloride 0.1 g.

Distilled water 10 mL.

Mix well and store at 4 °C.

Trypsin Working Solution (0.05%)
Trypsin stock solution (0.5%) 1 mL.

Calcium chloride stock solution 1% 1 mL.

Distilled Water 8 mL.

Adjust pH to 7.8 with 1N NaOH. Store at 4 °C for 1 month or −20 °C for long-term storage.

7.2 Method 1

1. Prepare the trypsin and pre-heat to 37 °C. Carefully blot excess water and pipet the enzyme solution (generally 50–100 μL will suffice) onto the section. If necessary, the solution is spread around the section with the pipette tip. Care should be taken to prevent damage to the tissue.
2. Place the slides in a humidified container and then into the 37 °C incubator. Do not keep slides directly on the incubator shelves. The container holding the slides is preferred to be pre-heated in the incubator.

3. After 10–20 min, remove the slides from the incubator and transfer to a rack in a container of tap water. However, the exact time needs to be standardized. Rinse thoroughly by running tap water for 3 min.
4. Continue with immunohistochemical staining protocol.

8 Enzymatic Retrieval, Immersion Method

8.1 Materials and Reagents

37 °C water bath.

Slide racks and slide rack containers.

Enzymatic antigen retrieval solution.

8.2 Method

1. Set water bath to the required temperature as per the enzyme being used. Add ultrapure water to the two containers which hold slide racks. Place the containers into the water bath to warm.
2. Deparaffinize and rehydrate sections as above. In order to warm the slides, place them in water container.
3. Prepare the enzymatic antigen retrieval buffer by using the warm water in the other container, and then to allow the solution to re-heat, return the container to the water bath.
4. Pre-heat the solution for 10–20 min, then transfer the warmed slides into the enzyme with an intermittent gentle agitation. The next step is to remove the slides and keep them in running tap water for 3 min in order to rinse off the excess enzyme.
5. Continue with immunohistochemical staining protocol.

Some important points need to be considered:

1. For instructions, the manufacture literature needs to be considered carefully.
2. Sufficient volume of water or buffer is to be used to properly cover the slides.
3. Maintenance of proper temperature is a prime criterion. In case the temperature of the solution is lower, enzyme activity will be reduced and it leads to under-retrieval of the antigenic site.
4. To avoid impairing the activity of the enzyme, prepare the enzymatic antigen retrieval solution as quickly as possible. Before introducing the slides, allow this solution to return to temperature.
5. Ten to twenty minutes is only suggested as a starting-point incubation time. The antigens which were retrieved for less than 10 min, leading to weak staining.

Excess time, as more than 20 min cause over retrieval, causing non-specific background staining. This leads to an increase in the chances of sections dissociating from the slides. It may even cause damage to the morphology of the tissue. Evaluation of optimum antigen retrieval time for the particular antibody is important. Control experiment is necessary for standardization, where slides of the same tissue section are incubated in the enzyme solution for 10, 15, 20, 25, and 30 min.

9 Immunohistochemical Staining

9.1 General Guidelines

The following protocol is for the laboratory which does not have an automated stainer or other capillary gap system that allows rapid application and rinsing of reagents. Reagents may be applied manually through pipetting. Otherwise, the sequence of the protocol can be adapted to automated and semi-automated systems depending on availability.

A humidified chamber is required to avoid drying of the tissue. Drying at any stage will lead to high background staining. To avoid drying a shallow, plastic box with a sealed lid and wet tissue paper in the bottom in an adequate chamber can be used. Dilutions of the primary of the tissue, and secondary antibody are required. Adjust dilutions appropriately from the results obtained. Maintain all incubation time as per protocol.

Horseradish peroxidase (HRP) or alkaline phosphatase (AP) are the enzymes most commonly employed for IHC. However, there exist a number of chromogens that are used.

9.2 Protocol

Antigen retrieval may be processed prior to commencement with the following protocols.

Day 1

1. We recommend waiting until the primary antibody incubation period if you are using HRP conjugate for detection, blocking of endogenous peroxidase.
2. By TBS plus 0.025% Triton X-100 with gentle agitation wash the slides.
3. Block in 10% normal serum with 1% BSA in TBS for 2 h at room temperature.
4. Drain slides for a few seconds (do not rinse) and wipe around the sections with tissue paper for better result.
5. Apply primary antibody diluted in TBS with 1% BSA.
6. Incubate overnight at 4 °C.

Day 2

1. By TBS 0.025% Triton rinse 2 × 5 min with gentle agitation.
2. When an HRP conjugate is employed for detection, the slides are incubated in 0.3% H_2O_2 in TBS for 15 min.
3. For enzymatic detection (HRP or AP secondary conjugates): Applied enzyme-conjugated secondary antibody to the slide. The dilution is maintained to the concentration as per the instruction of the manufacturer in TBS with 1% BSA, and incubate for 1 h at room temperature.

   For fluorescent detection:

Apply fluorophore-conjugated secondary antibody to the slide. Concentration is maintained with suitable dilution as per the instruction of the manufacturer in TBS with 1% BSA, and incubate for 1 h at room temperature.

To avoid photobleaching, this step should be done in a dark room or place.

1. Rinse 3 × 5 min TBS. The next step is covering with coverslip which contain mounting medium if using fluorescent detection. Continue with the following steps while visualizing the protein with a chromogen.
2. Develop with chromogen at room temperature for 10 min.
3. Rinse with running tap water for 5 min.
4. Counterstain (if required).
5. Dehydrate, clear, and mount.

9.3 Controls

For non-specific interaction and Fc receptor binding estimation the contribution of staining protocols are important. It employs an antibody directed to an irrelevant antigen (for example, BrdU) with the same isotype as the antibody of interest can be analyzed. The antibody directed to the irrelevant antigen is known as the isotype control. Use normal serum from an unimmunized animal of the same species as the primary antibody for whole-serum antibodies.

A negative antibody control is recommended if an isotype control is not available. Simply replace the primary antibody with antibody diluent.

To ensure that the antibody is performing as expected, a positive tissue control is strongly recommended. It may also be necessary to include a negative tissue control depending on the experiment: a tissue in which the protein of interest is not expected to be found.

9.4 Notes and Tips

1. To reduce surface tension, the use of 0.025% Triton X-100 in the TBS is helpful. Cover the whole tissue with the reagent. It is helpful for dissolving Fc receptors, hence reducing non-specific

binding. TBS is recommended to give a cleaner background than PBS.

2. The endogenous immunoglobulins in the tissue may cross-react with the secondary antibody. In order to minimize this phenomenon, it is advocated for pre-treating the tissue with the same normal serum from the species in which the secondary antibody was raised. This eliminates Fc receptor binding of both the primary and secondary antibody by the use of normal serum also before the application of the primary. BSA is used to reduce non-specific binding, caused by hydrophobic interactions.

3. As per protocol, the primary antibody is diluted. At a concentration between 0.5 and 10 μg/mL, most antibodies are used in IHC-P.

   For tissue staining, make sure the primary antibody is raised in a species. For example, if we have mouse tissue and our primary antibody was raised in a mouse, an anti-mouse IgG secondary antibody would effectively bind to all the endogenous IgG in the mouse tissue leading to high resolution.

4. Antibodies of lower titer or affinity can be used by simply allowing more time for the antibodies to bind; preferably, overnight incubation is required. Also, once the tissue has reached the saturation point, no more binding can take place regardless of the antibody's titer or affinity for its target. Overnight incubation ensures that this occurs.

5. Endogenous peroxidase activity can be suppressed by H_2O_2 and reduces background staining. To detect the presence of endogenous peroxidases, a tissue slide is incubated after being re-hydrated in a solution of DAB. Under the microscope, if areas of the section appear brown, a blocking step should help reduce this staining. To reduce antibody:antigen binding Some epitopes are modified by peroxide by Incubating with peroxide after the primary incubation which avoids this problem.

   Peroxides can be diluted in TBS or water. Methanol is preferred as a diluent for peroxides, which have the ability to reduce damage to the tissue as in aqueous solutions. However, some constraints have been identified as reduction in the binding ability of some antibody:antigen pairs, as in case of cell surface proteins, after methanol/peroxide incubation. Peroxidases need to be avoided in case AP or fluorescence is used in detection.

6. With the appropriate chromogen solution, we can develop the colored product of the enzyme. The determining factors are the enzyme label of interest, the colored end product preferred, and the use of aqueous or organic mounting media. Some commonly used substrates are listed below:

Enzyme substrate color advantages and disadvantages.

3,3′-Diaminobenzidine (DAB) gives rise to brown intense color; Horseradish peroxidase (HRP) 3-Amino-9-ethyl carbazol (AEC) gives ed intense color and contrasts well with blue. Double staining endogenous peroxidase activity in the tissue can lead to false positive staining. AEC is alcohol soluble and incompatible with organic mounting media. 5-bromo-4-chloro-3-indoyl phosphate; nitroblue tetrazolium (BCIP/NBT) gives blue intense color. Alkaline phosphatase (AP): Endogenous alkaline phosphatase activity in the tissue can lead to false positive staining. Glucose oxidase nitroblue tetrazolium (NBT) gives rise to no blue, no endogenous enzyme activity, and low staining intensity (high concentration of primary and secondary antibodies required for effectiveness).

7. Common counterstains employed are hematoxylin (blue), nuclear fast red, or methyl green. DAPI (blue) or propidium iodide/PI (red) can be used as counterstain in fluorescent detection.
8. DAB is a suspected carcinogen, so utmost care needs to be taken while handling. Wear the appropriate protective clothing. Deactivation is usually followed with chloros in a sealed container overnight. Noxious fumes are evolved when chloros is added, aiding in disposal following laboratory guidelines.

   Add 0.24 mg/mL Levamisole if using AP(Sigma L9756) to the chromogen solution. Levamisole causes less endogenous phosphatase activity, which leads to reduction in background staining.
9. In case AEC, Fast Red, INT or any other aqueous chromogen are also alcohol soluble. It is always recommended to use a suitable aqueous mounting media to avoid dehydration.
10. The next step is dehydration and clearance of DAB, New Fuchsin, Vega Red, NBT, TNBT, or any other organic chromogen developed by following through the rehydration protocol. Mount sections in an appropriate organic mounting media. Mounting with organic mounting media sections would have a better refractive index than those mounted in aqueous mounting media. Organic mounting media will aid in better clarity and sharpness of the the image seen down the microscope.

9.5 Signal Amplification

To achieve better signal strength, strategies have been developed for the addition of more enzymes or fluorophores to the target of interest.

(a) Avidin-biotin complex (ABC).

Su-Ming Hsu and colleagues (J) Histochem Cytochem developed this technique. It employs the high affinity of avidin, (a protein found in chicken egg white), which binds against biotin (an enzyme co-factor in carboxylation reactions). Avidin has four irreversible binding sites for biotin and binding is essentially.

In brief, in the protein of interest the primary antibody is bound. To the primary antibody a biotinylated secondary antibody is then bound. A complex of avidin and biotinylated enzyme is formed in a separate reaction by mixing the two in a ratio that leaves some of the binding sites on avidin which is unoccupied. This complex is further incubated with the tissue section. The unoccupied biotin-binding sites of the complex will bind to the biotinylated secondary antibody. This results in better enzyme accumulation to the target than enzyme-conjugated secondary or primary antibody.

The commercially available components of the avidin-biotin complex are available in kits. This provides two reagents and for optimal ratio instructions for combining them. The complex can be employed with any biotinylated antibodies. The presence of endogenous biotin in tissues such as kidney, liver, brain, prostate, colon, intestines, and testes, would bind with the avidin-biotin complex leading to background staining.

(b) Labeled streptavidin biotin (LSAB).

The of the interaction of streptavidin (similar to avidin in binding affinity) and biotin is similar to the ABC method. Streptavidin conjugate to an enzyme or fluorophore, followed by the binding of the primary antibody to biotinylated anti-Ig followed by the binding of secondary antibody. Abcam offers a streptavidin–HRP conjugate, ab7403.

Streptavidin leads to less non-specific background staining than avidin as it is non-glycosylated (unlike avidin). It does not show any interaction with lectins or other carbohydrate-binding proteins.

(c) ***HRP polymer.***

Both avidin-biotin methods (ABC and LSAB) are losing favor. Newer method consist of a secondary antibody polymer-enzyme-antibody products (for example, anti-mouse and/or rabbit IgG) attached to a polymer-enzyme complex. If we compared with the AB methods, one step is eliminated and the issue of endogenous biotin is avoided. Abcam offers ab2891, a goat anti-rabbit/mouse IgG HRP polymer.

(d) ***Tyramide signal enhancing (TSE).***

One of the most effective amplification procedures as the patented and licensed method is TSE (also known as TSA or

CSA, depending on the manufacturer of the commercially available kits). To improve the obtained results with poorly performing antibodies, the detection of relatively sparse antigens is essential.

After first applying a primary antibody and secondary-HRP conjugate to the area around the protein of interest, this method relies on a peroxidase-catalyzed reaction to covalently attach the tyramide portion of tyramine-protein conjugates. Due to covalent bond between tyramides proteins are not washed out even if the slides are treated to remove the antibodies. An antibody-enzyme or -fluorophore conjugate is employed against the protein portion of the tyramine-protein conjugate to obtain a signal. The protein as biotin and a streptavidin-enzyme conjugate is applied in one commercially available version of the method, in place of an antibody-conjugate. The limitations of the procedure are the cost of the kits and the time requirement to perform the multiple steps.

(e) ***Resources.***

We can collect information from IHC world (www.ihcworld.com) on trouble-shooting the immunohistochemical procedure, antigen retrieval, and positive control.

Histonet (www.histosearch.com/histonet.html) is another useful server from scientists around the world.

Nordic immunohistochemical quality control (www.nordiqc.org) gives information regarding positive controls and antigen retrieval steps for many target proteins.

(f) ***Buffer compositions.***

Fixation:

Formalin solution (10%, buffered neutral):

Formaldehyde (37–40%) 100 mL.

Distilled water 900 mL.

NaH_2PO_4 4.0 g.

Na_2HPO_4 (anhydrous) 6.5 g.

Mix to dissolve.

Paraformaldehyde (4%, buffered neutral):

8% Paraformaldehyde:

Paraformaldehyde 40 g.

Distilled water 500 mL.

0.2M Phosphate Buffer (PB), pH 7.4:

Na_2HPO_4 10.9 g.

NaH_2PO_4 3.2 g.

Distilled water 500 mL.

The solution pH should be adjusted to 7.4. Make up a separate 0.2 M solution of either the monobasic or dibasic sodium phosphate (depending on how you need to adjust the pH) to adjust the pH.

Heat 8% PFA solution at 60 C through stirring (do not allow the solution to go above 60 °C). As the solution has reached 60 °C and the PFA is dissolved, added 500 mL of 0.2 M phosphate buffer to bring the solution to 4% PFA in 0.1 M phosphate. Add 1N NaOH dropwise until the solution becomes clear (try 1–2 drops per 500 mL). Add a few more drops if the solution is not clear; simultaneously, add two pellets of solid NaOH in 1–2 L of solution. Cool the solution and filter.

The PFA needs to be prepared fresh on the same day of use. Storage overnight at 4 °C is possible. But it will not fix as well the second day PFA solution can be frozen at −20 °C. However, for best results, tissue should either be fixed invariably with fresh PFA or freshly thawed PFA.

Bouin Solution (especially for preserving soft and delicate structures such as brain tissues): Picric acid (saturated) 75 mL.

Formaldehyde (37–40%) 25 mL.

Glacial acetic acid 5 mL, mix well.

Immunocytochemistry is a similar technique, where instead of tissue section, cells from cell culture are employed. Rest of the protocols are same [1, 2].

10 Application for Immunohistochemistry

1. In the detection of diseases. It is employed to detect the tissue distribution of an antigen of interest in case of healthy and diseased individuals.
2. IHC is widely employed for the diagnosis of cancers; specific tumor antigens are expressed de novo or up-regulated in certain cancers. IHC plays an important role in diagnostic and research laboratories.
3. Recently we are using the technique of IHC for the detection of the immunological marker for maintaining pregnancy in our laboratory. We could detect a better expression profile of CD4 and CD8 gene in pregnant female sheep compared to that of non-pregnant. Both CD4 and CD8 are regarded as surface markers, being expressed in endometrium.
4. IHC can equally be used for assessing the growth parameters. For example, growth hormone expression can be assessed in the pituitary gland [1–5].

The following table lists out the IHC markers for cancer or carcinoma:

Primary site	IHC
Lung carcinoma	TTF-1, Napsin A
Breast carcinoma	GATA3, ER, GCDFP-15
Urothelial carcinoma	GATA3, uroplakin II, S100P, CK5/6, p6
Squamous cell carcinoma	P40, CK5/6, p63, SOX2
Renal cell carcinoma, clear cell type	PAX-8/PAX-2, RCCma, pVHL, KIM-1
Papillary renal cell carcinoma	P504S, RCCma, pVHL, PAX 8, KIM-1
Translocation renal cell carcinoma	TFE3
Hepatocellular carcinoma	Arginase-1, glypican-3
Adrenal cortical neoplasm	SF-1, Mart-1, HMB-45, MiTF, SOX10
Melanoma	S100, Mart-1, HMB-45, MiTF, SOX10
Merkel cell carcinoma	CK20 (perinuclear dot staining), MCPy
Mesothelial origin	Calretinin, WT1,D2-40,CK5/6, mesot
Neuroendocrine origin	Chromogranin, synaptophysin, CD56
Upper gastrointestinal tract	CDH17, CDX2, CK20

Let us discuss certain cases of immunohistochemistry of different proteins principally expressed in various tissues.

10.1 Immunohistochemistry for Growth Hormone in Cattle Pituitary: A Case Study

Paraffin-embedded sections of human pituitary tissue were studied for the detection of the human growth hormone expression profile. Antigen affinity-purified polyclonal antibody was employed @ 10 μg/mL overnight at 4 °C. The tissue was stained with the Anti-Goat HRP-DAB Cell & Tissue Staining Kit followed by counterstaining with hematoxylin (blue) (Fig. 5).

10.2 Immunohistochemistry for CD4 in Sheep Uterus

Paraffin tissue blocks were prepared by standard manual alcohol-acetone protocol. The 5–6 μm thick paraffin sections obtained from different segments of the uterus (middlehorn of uterus and tip of the horn) of non-pregnant ewes were taken on Millennia 2.0 adhesion slides (Cat. No. 71863-01, Abcam). The tissue sections were de-paraffinized and hydrated in distilled water. The tissue slides were covered with trypsin enzymatic antigen retrieval solution (Cat. No. ab970, Abcam) and were kept in the incubator in a humid environment at 37 °C for 5–10 min. The sections were then incubated for 60 min in a peroxidase-blocking solution (Lot. No. 00065614, Dako) at room temperature to block non-specific antibody-binding activity. After subsequent washing with phosphate buffer saline (PBS), the sections were incubated at 37 °C for 2 h in a humid environment with CD4 antibody (Mouse anti-Sheep CD4 antibody Cat no. MCA2213GA, Abcam) in 1:200

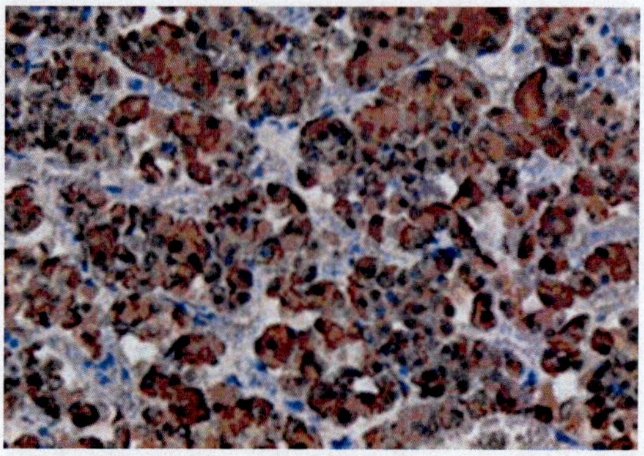

Fig. 5 Immunohistochemistry for growth hormone in cattle pituitary [4]

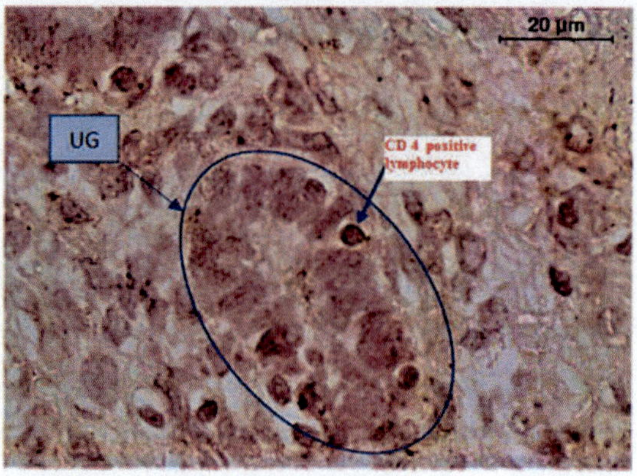

Fig. 6 Photomicrograph showing CD4 positive lymphocyte (arrow) localize at the glandular epithelium of the inter-caruncular area of tip of the horn of non-gravid uterus. Uterine gland (UG). IHC 100× [6]

dilution. Immunoreactivity was detected after 1-h incubation at 37 °C with secondary antibody (rabbit anti-mouse IgG H&L) in dilution 1:200. Slides were then rinsed three times in PBS for 5 min each, followed by treatment with freshly prepared DAB solution for 3 min (DAB substrate, Cat. No. 34001, Thermo Fisher Scientific). The sections were counter-stained with Mayer's hematoxylin, hydrated in ethanol, cleared in xylene, and then mounted in DPX.

The microscopic view of the slides is depicted in Fig. 6.

10.3 Immunohistochemistry for CD8 in Sheep Uterus

In this process, the entire steps as discussed in earlier sections were repeated when mouse monoclonal anti CD8 antibody (Cat. No. MA1-80900, Thermo Fisher Scientific) was used instead of CD4 antibody.

A microscopic view of IHC of sheep has been depicted in Fig. 7.

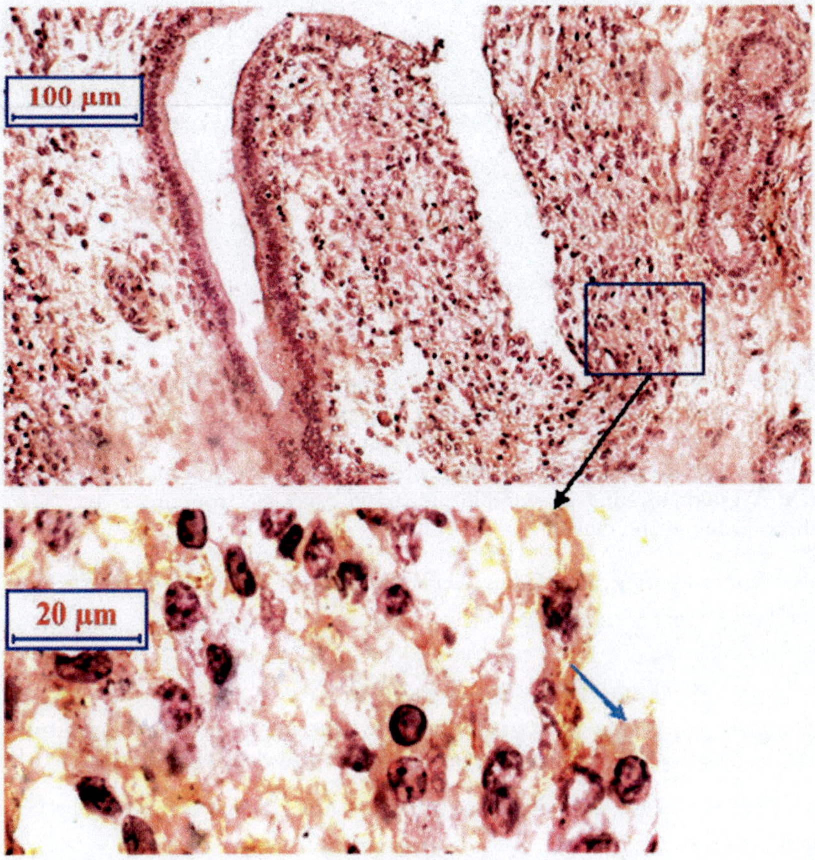

Fig. 7 IHC of CD8 expression profile in uterus of pregnant sheep [6]

References

1. Wang H, Pevsner J (1999) Detection of endogenous biotin in various tissues: novel functions in the hippocampus and implications for its use in avidin-biotin technology. Cell Tissue Res 296 (3):511–516
2. Giorno R (1984) A comparison of two immunoperoxidase staining methods based on the avidin-biotin interaction. Diagn Immunol 2 (3):161–166
3. Duraiyan J, Govindarajan R, Kaliyappan K, Palanisamy M (2012) Applications of immunohistochemistry. J Pharm Bioallied Sci 4(Suppl 2): S307–S309. https://doi.org/10.4103/0975-7406.100281
4. Pal A, Pal A, Chakravarty AK (2020) Mutations in growth hormone gene affect stability of protein structure leading to reduced growth, reproduction and milk production in crossbred cattle—an insight. Domest Anim Endocrinol 71:106405. https://doi.org/10.1016/j.domaniend.2019.106405
5. Kandukuri SR, Lin F, Gui L, Gong Y, Fan F, Chen L, Cai G, Liu H (2017) Application of immunohistochemistry in undifferentiated neoplasms: a practical Approach. Arch Pathol Lab Med 141(8):1014–1032. https://doi.org/10.5858/arpa.2016-0518-RA
6. Islam M, Pal A, Banerjee S, Das P (2020) Better expression profile of CD8 and CD4 gene in uterus of pregnant ewe in comparison to non-pregnant—a novel report. https://doi.org/10.1101/2020.10.07.329730
7. Pal A, Chakravarty AK (2019) Genetics & breeding for the disease resistance of livestock. Elsevier Publication. ISBN: 9780128164068

Check for updates

Chapter 6

Western Blotting

Abstract

Western blotting is a widely used analytic technique employed in molecular biology to detect unknown proteins using a hybridization-based methodology. The basic factors involved in Western blotting are separation of protein mixtures according to size using gel electrophoresis; effective transfer of separated proteins to a solid support; and detection of a target protein by appropriately matched antibodies. The basic steps involved in Western blotting are sample preparation, gel electrophoresis, blotting (or transfer), blocking, antibody probing, and detection.

Keywords Western blotting, Antibodies, Gel electrophoresis, Blotting (or transfer), Blocking, Antibody probing and detection

Points to Remember

1. Western blotting is a molecular biology technique to detect unknown proteins using hybridization-based procedures.

1 Introduction

Blotting is a technique involving certain procedures for resolving of macromolecules (DNA, RNA, or protein), transfer to a solid support followed by detection with a specific probe. As discussed in earlier chapters about a number of techniques for protein identification, Western blotting is one of the important techniques for protein detection. It is one of the most widely used techniques adopted for protein expression and identification analysis.

2 Principle

Western blotting is one of the most employed technique for protein identification. Often it is also known as the protein immunoblot. It is a widely used analytical technique applied to molecular biology to

Aruna Pal, *Protocols in Advanced Genomics and Allied Techniques*, Springer Protocols Handbooks,
https://doi.org/10.1007/978-1-0716-1818-9_6,

detect specific proteins in a sample of tissue homogenate or extract. The basic principle is the detection of protein expression from different samples/tissue under consideration.

- Detection of viral proteins.
- Characterization of monoclonal and polyclonal antibody preparations.
- Determination of the specificity of the immune response to viral antigens.

The basic principle of Western blotting procedure relies upon three basic factors:

1. The separation of protein mixtures according to size using gel electrophoresis.
2. The effective transfer of separated proteins to a solid support.
3. The specific detection of a target protein by perfectly matched antibodies.

   After detection, the target protein will be detected as a band on a blotting membrane, X-ray film, or an imaging system. Figure 1 represents the flow diagram for the process of Western blotting.

2.1 Basic Steps for Western Blotting

1. Sample preparation.
2. Gel electrophoresis.
3. Blotting (or transfer).
4. Blocking.
5. Antibody probing.
6. Detection.

3 Protocol

1. About 100 µg of total proteins should be extracted from cell lysate of representative experimental groups and should be resolved by 10% SDS-PAGE gel at a constant voltage of 40 V.
2. Proteins on the gel are to be transferred to Immobilon-P polyvinylidene difluoride (PVDF) membranes (Millipore, Bedford, MA).
3. Blocking should be done with 5% (w/v) non-fat dried milk (prepared in 20 mM Tris, 150 mM NaCl, pH 7.6; 0.1% (v/v) Tween-20) overnight at 4 °C. Primary antibodies are used at the following dilutions: 1:250 anti-Protein1, 1:250 Protein2, and 1:5000 anti-β-Actin.

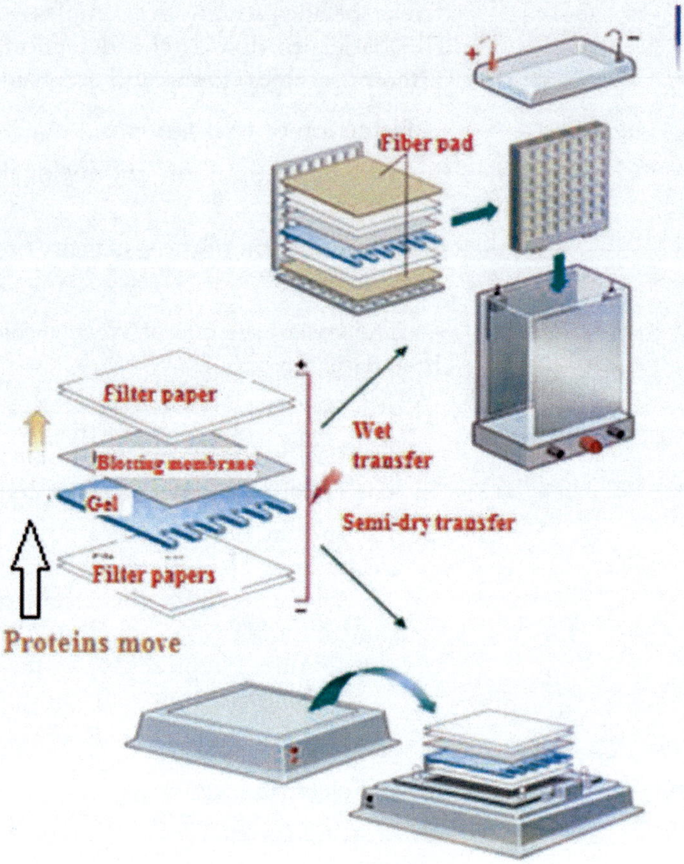

Fig. 1 Schematic representation of Western blotting

4. Membranes are to be washed thrice and incubated with secondary antibodies (dilution 1:50,000 for anti-goat IgG and 1:70,000 for anti-mouse IgG).
5. The target protein is to be detected by chemiluminescence (Immobilon Western chemiluminescent HRP substrate; (Millipore, Bedford, MA) captured on X-ray films).

3.1 Sample Preparation

- Western blotting is applicable for all sources of protein, ranging from single cells to whole tissues, biological fluids, and proteins secreted in vitro.
- In majority of the cases, the cells are harvested, washed, and lysed for releasing the target protein.
- For best results, all these steps should be carried out in cold condition, preferably on ice.
- This will reduce proteolysis, dephosphorylation, and denaturation, after the disruption of the cells.

- The selection of extraction method relies basically on the sample. Other factors to be considered are whether the analysis is targeting all the proteins in a cell or only a particular subcellular fraction.
- The endogenous proteases will be discharged upon cell disruption, leading to the degradation of the target molecule.

The sample needs to be protected during the process of cell disruption and purification using certain protease inhibitors to prevent uncontrolled protein losses.

3.2 Gel Electrophoresis: Gel Preparation

Reagent	8% (running gel)	5% (stacking gel)
Acrylamide/Bisacrylamide (40%)[a]	4.0 mL	2.5 mL
1 M Tris–HCl	7.5 mL	7.5 mL
Distilled water	8.2 mL	9.7 mL
10% SDS	200 μL	200 μL
10% ammonium persulfate	100 μL	100 μL
TEMED (added last)	10 μL	10 μL

[a] 19:1 w:w ratio of acrylamide to *N*,*N*′-methylene bis-acrylamide

- Ingredients are mixed properly as indicated, without any air bubbles.
- The separating gel is poured into glass plate assembly.
- The gel is overlaid with water to make it a flat surface and to exclude air.
- It is left for 20 min to ensure proper polymerization.
- The stacking gel is prepared and poured properly on the running gel. The comb is inserted and left for 20 min.

3.3 Gel Electrophoresis: Sample Buffer

Boiling of the sample of protein in a sample buffer (at 95 °C for 5 min) is conducted, containing:

1. The β-mercaptoethanol reducing the disulfide bonds.
2. SDS for the disruption of the secondary and tertiary protein structure.
3. Glycerol to allow the samples to sink into wells.
4. Bromophenol blue dye for the visualization of the samples.

The end result has two important features:

5. All proteins possess only a primary structure; and
6. All proteins contain a large negative charge. This implies that they will all move toward the positive pole upon being placed in an electric field.
7. Their rate of migration toward the positive pole through the gel is dependent on their linear size.

3.3.1 Loading Samples and Running the Gel

The samples are loaded into separate wells. A protein marker is also loaded. It is run at 200 V for 30–40 min. The running buffer (pH 8.3) contains tris base and glycine.

4 SDS

4.1 Blotting

- After the process of gel electrophoresis, the separated protein mixtures are transferred to solid support for further analysis.
- The transfer can be conducted in wet or semi-dry conditions.
- Semi-dry transfer is preferable and is normally faster.
- Wet transfer is advisable in case of large proteins, >100 kD.
- The membrane is placed next to the gel for every kind of transfer.
- The membrane and the gel are sandwiched between absorbent materials. The sandwich is clamped between solid supports in such a manner so as to maintain proximity between the gel and the membrane.
- The entire system is connected to a power supply and blotting is allowed for 1 h at 80 mA.
- The buffer for wet transfer contains:
  - Tris.
  - Methanol (20%).
  - Glycine.
  - 0.1% SDS.

4.2 Blotting Membranes

Two types of solid supports are available where the separated proteins are transferred. Both of them bind proteins with high affinity:

- *Nitrocellulose membrane.*
  - It possesses excellent protein-binding and -retention capabilities.
  - It is brittle, and as a result it is usually less effective, in cases when it becomes necessary to reuse the blots.

- *Polyvinylidene fluoride (PVDF) membrane.*
  - PVDF demonstrates superior mechanical strength, which makes it suitable for stripping/reprobing.

4.3 Visualization of Proteins in Membranes: Ponceau Red Stain

Ponceau Red, a reversible stain with poor sensitivity, is one of the earlier used dyes. Figure 2 shows the photograph of staining of the slide with Ponceau Red stain. Another advantage of using Ponceau S is that it is compatible with both nitrocellulose and PVDF membranes. This is a quick and easy way to visualize proteins transferred to membranes. Ponceau S can be easily washed away with water. The process is regarded as a "gentle" treatment, which possesses no interference with subsequent immunological detection steps.

4.4 Blocking

Blocking is one of the most important processes in the immunodetection phase of Western blotting. It aids in the prevention of non-specific binding of the antibody to the blotting membrane. The most commonly used blocking solutions contain 3–5% BSA or non-fat dried milk in a solution of PBS (phosphate buffered saline) or TBS (tris buffered saline).

Usually, a minute quantity of Tween 20 detergent is mixed to the blocking and washing solutions to reduce background staining, and the buffer is known as PBST or TBST.

4.5 Antibody Probing

- Followed by the process of separation and transfer of the protein samples onto a membrane, the protein of interest is identified followed by localization with a specific antibody.
- The blot will be incubated in a dilute solution of antibody, usually for a few hours at room temperature then overnight at 4 °C.

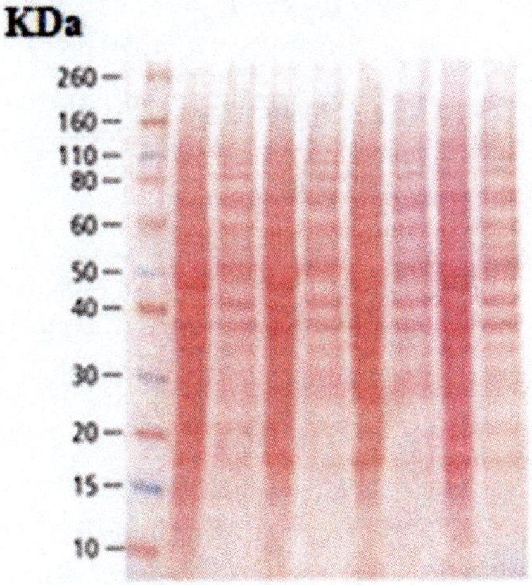

Fig. 2 Ponceau red staining of Western blot

- The dilution of the antibody is carried out in wash buffer (PBST or TBST) or in the blocking solution. We need to decide depending on the antibody employed.
- Depending on the level of purity and specific binding properties, different levels of dilution for antibody preparation are required. The appropriate dilution recommendations will be followed as per the manufacturer's datasheet.

Usually, Western blotting protocols use a non-labeled primary antibody directed against the target protein. The membrane is washed several times in TBST. In order to remove the residual primary antibody, it is to be agitated, usually 5 min or more per wash.

A species-specific, labeled secondary antibody directed against the constant region of the primary antibody is then used. The secondary antibody employs as a mechanism for the amplification of the emitted signals, depending on the principle of binding of secondary antibodies to primary antibodies. It also serves as a carrier of the label. Dilution of the secondary antibody is carried out according to the manufacturer's instructions and incubated for 1 h at room temperature.

4.6 Detection with Substrate

- The most widely used antibody label employed in Western blots is HRP. HRP is a small stable enzyme but possesses high specificity and rapid turnover.
- The detection of signal is achieved in the final step of the immunodetection procedure, after HRP is exposed to as a substrate solution.
- Substrate solutions employed for Western blotting are chemical reagents; the enzymes act on these and yield a signal. Signal can be easily detected and measured.
- The HRP label can be detected through colorimetric or chemiluminescent substrates.
- Colorimetric substrates for HRP {e.g., Tetramethylbenzidine (TMB)} produce purple/black bands directly on the surface of the blot.

These substrates are very easy to use and take from minutes to a few hours to produce visible bands. Figure 3 depicts how the signal is detected. Figure 4 detects Western blotting with the help of secondary antibodies and an actual photograph of a Western blot.

- Detection limits for colorimetric substrates are present in the low nanogram range.
- More routinely, HRP is employed with enhanced chemiluminescence (ECL) detection.
- For ECL detection, the substrate is luminal, that is oxidized by HHRP, in the presence of H_2O_2 to produce light.

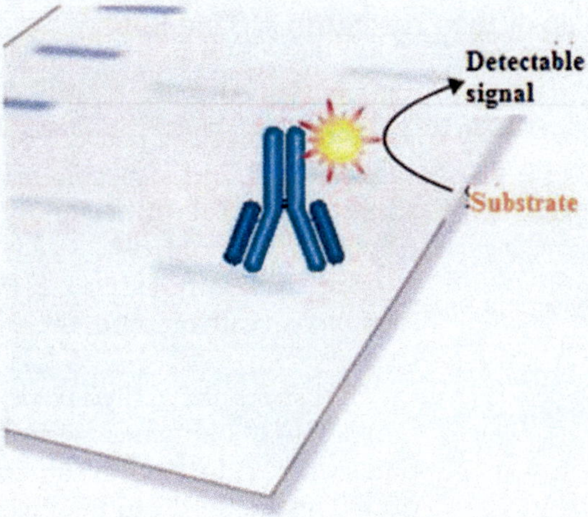

Fig. 3 Signal detection in Western blotting with the help of substrate

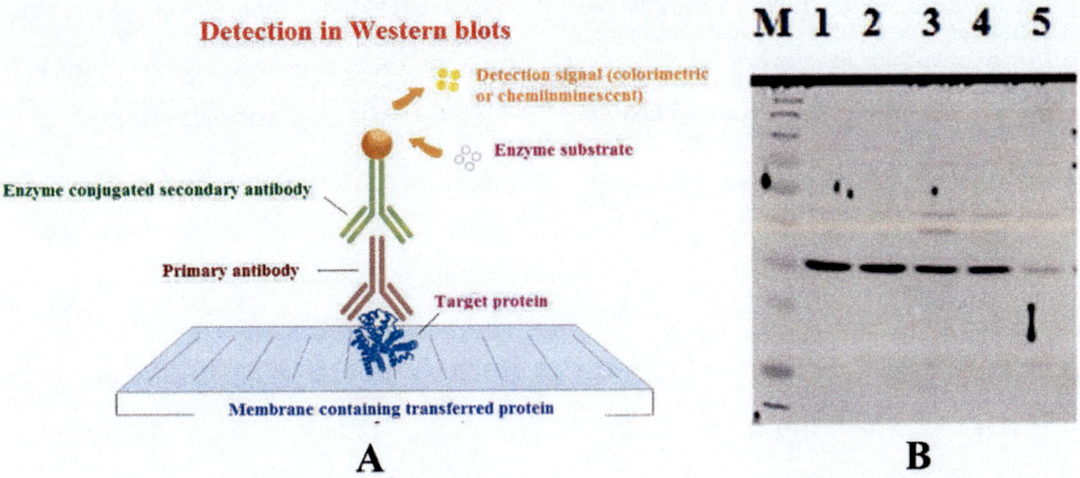

Fig. 4 (**a**) Schematic representation of Western blotting with the help of secondary antibody. (**b**) The actual photograph of Western blot with protein marker

Detection of the emitted light is conceived though the exposure of the Western blot to an X-Ray film, or with a CCD camera for light capture.

- The emitted light gives rise to a band on the film. Band is observed on the screen of the imaging system if used, indicative of the HRP-labeled antibody binding with the target protein.
- ECL detection of HRP is extraordinarily sensitive, allowing for the visualization of pictogram to femtogram amounts of target protein [1, 2].

5 Principles of the Procedure: A Case Study

The nitrocellulose strips are employed with separated, bound antigenic proteins. The source may be either partially purified in activated HIV-1 and a specific HIV-2 synthetic peptide on the same strips. Individual nitrocellulose strips are used with diluted serum or plasma and controls.

Specific antibodies to HIV-1 and HIV-2, if present in the specimens, will bind to the HIV-1 and HIV-2 proteins on the strips.

The strips are washed to remove unbound materials. Goat antihuman IgG conjugated with alkaline phosphates is used as a secondary Ab and then the substrate BCIP/NBT is added for detection. This method can be effectively employed to detect marginal amounts of HIV-specific antibodies in serum or plasma.

References

1. Brahma B, Kumar S, De BC, Mishra P, Patra MC, Gaur D et al (2015) Comparative genomic analysis of buffalo (*Bubalus bubalis*) NOD1 and NOD2 receptors and their functional role in *in-vitro* cellular immune response. PLoS One 10(3):e0119178. https://doi.org/10.1371/journal.pone.0119178
2. Pal A, Chakravarty AK (2019) Genetics & breeding for the disease resistance of livestock. Elsevier Publication. ISBN: 9780128164068

Check for updates

Chapter 7

Introduction to Bioinformatics

Abstract

Bioinformatics is an interdisciplinary field using ***dry lab***, that is, methods and software tools for understanding biological data. Bioinformatics includes experts from computer science, statistics, even mathematics, and of course biologists for a better understanding of biological data. This chapter provides a glimpse into the application of bioinformatics in the field of proteomics, genomics, structural analysis, system biology, evolutionary biology, population genetics, transcriptomics, biophysics, and imaging with the list of software involved. This chapter also discusses the software for proteomics with suitable examples, illustrations, and actual photographs and images as conducted in our laboratory.

Keywords Bioinformatics, Dry lab, In silico study, Homology modeling

1 Basic Principle

The term bioinformatics is derived from the words "Biology" and "information technology." A detailed study for any biological problem is not always possible for any wet lab with limited infrastructure. It is an interdisciplinary field that uses ***dry lab***, that is, methods and software tools for understanding biological data. The field of bioinformatics includes experts from computer science, statistics, even mathematics, and of course biologists for a better understanding of biological data.

Apart from genomic and proteomic analysis, bioinformatics is very much essential for molecular evolutionary studies, including phylogenetic analysis. For a better understanding of genomics, high-throughput sequencing such as next-generation sequencing generates a huge volume of data, popularly referred to as "***Big Data.***" Bioinformatics analysis is extremely helpful in deriving a valid conclusion from Big data analysis.

In our "Animal Genomics Laboratory," we deal with genomic analysis with molecular biology tools. For further validation, we deal with the analysis of the derived protein through bioinformatics tools. In silico studies for protein include translation and derivation of primary, secondary, tertiary and 3D structures of proteins. It is

Aruna Pal, *Protocols in Advanced Genomics and Allied Techniques*, Springer Protocols Handbooks,
https://doi.org/10.1007/978-1-0716-1818-9_7,

known that the 3D structure of protein with particular protein folding and domains are functionally active. The posttranslational modifications of the protein are also equally important for the function of the protein. For example, when we work with immune response genes, say a receptor which binds with the pathogen, in silico studies as molecular docking are extremely helpful [1, 2].

Through bioinformatics, we get a primary idea, which later on can be validated through ***wet lab*** as experimentation, as for example, protein analysis can be conducted in proteomics lab. In silico studies include prediction of protein structures through computer-generated software, with certain probabilities. In wet labs, we clone the particular gene and insert an expression vector, where the protein is expressed. One of the major limitations is that unless the expression is carried out in a mammalian expression vector, the proper post-translational modification does not take place. Generally, a baculo virus vector in insect cell line is used. The expressed protein will be isolated and 3D structure is predicted through X-ray crystallography, NMR, and so on. Protein detection is also later on confirmed through a variety of techniques such as through wet lab techniques such as ELISA, Western blotting, immunohistochemistry, and so on. The first step involves the detection of mutations at the genomic level (as detected through single nucleotide polymorphism, SNP or SNV, copy number variation or CNV). The next step is to determine the amino acid changes in the protein, which arise due to non-synonymous mutations. Amino changes will effect the protein folding and 3D structural conformation, which is actually the biologically active protein. The entire process of wet lab as described is very much cumbersome, costly, and time consuming. Hence, in silico studies may be regarded as an alternative and easy way for proteomics studies.

2 Sections of Bioinformatics

1. Computational molecular biology.
2. Computational genomics.
   2.1 Structural genomics.
   2.2 Functional genomics.
      2.2.1. Gene expression.
      2.2.2. Proteomics.
3. Comparative genomics.
4. Pan genomics is a concept introduced in 2005 by Tettelin and Medini which eventually took root in bioinformatics. Pan genome is referred to as the complete gene sequence or repertoire for a particular taxonomic group. The pan genomics concept is applicable for closely related strains of a species. It

can also be equally applied to a larger context like genus and phylum. It is divided into two parts. The core genome refers to the set of genes common to all the genomes under study (these are often housekeeping genes vital for survival). The dispensable/flexible genome refers to the set of genes not present in all but in one or some genomes under study.

A bioinformatics tool BPGA is appropriate to characterize the pan genome.

List of the publicly available software for the major categories for bioinformatics:

3 Proteomics

It includes protein sequences and identification, proteomics experiment, function analysis, sequence sites, features and motifs, protein modifications, protein structure, protein interactions, similarity search/alignment.

Let us discuss some of the most important bioinformatics tools. We are providing a few actual diagrams or photographs or glimpses for various output panes of different bioinformatics software pertaining to proteomics study. Figure 1 determines the 3D model of different variants of the Cytochrome gene from Swiss Homology Modelling Software. The output for the detection of various post-translational modification sites are depicted as NetNes software (prediction of the site for leucine-rich nuclear export signal for different variants e.g. Cytochrome B gene), NetPhos (prediction of phosphorylation sites of for different variants of Cytochrome B) in Figs. 2 and 3, respectively. The secondary structure of CytB Variant A and Variant B is being studied using GOR software as revealed in Fig. 4a, b, respectively. Prediction of arginine and lysine propeptide cleavage sites in eukaryotic proteins Variant A from

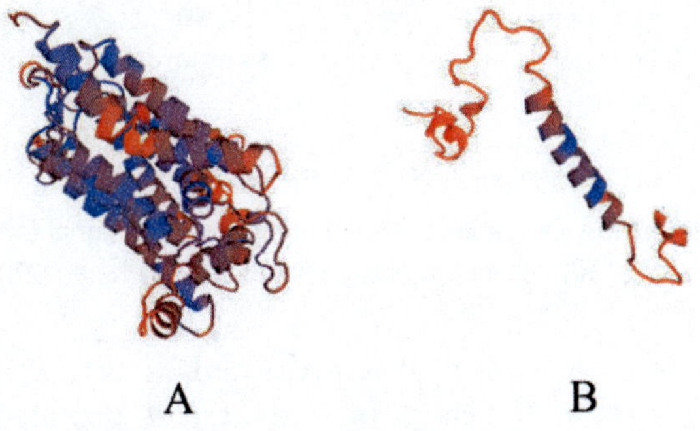

Fig. 1 The figure represents the 3D model of different variants of Cytochrome gene from Swiss Homology Modelling Software. (**a**) Represents variant 1, (**b**) represents variant 2

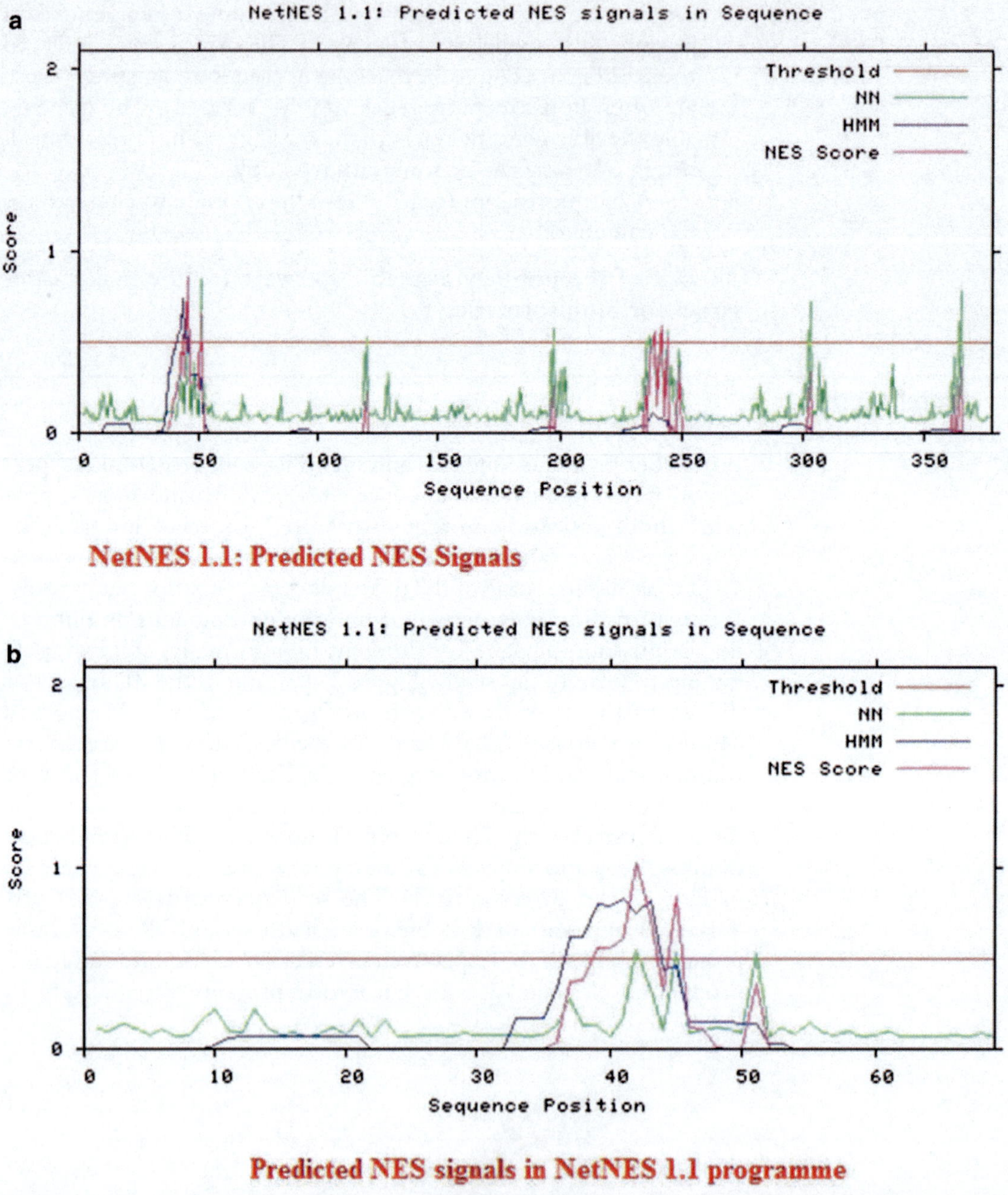

Fig. 2 Prediction of the site for leucine-rich nuclear export signal for different variants (Cytochrome B gene) by NetNes software. (**a**) Represents analysis for variant 1, whereas (**b**) represents analysis for variant 2

PropP is presented in Fig. 5. Determination of transmembrane helices of Variant A is depicted in Fig. 6. A 3D structural view through molecular visualization tools such as the PyMOL software is represented in Figs. 7 and 8 for **Cytochrome B and** GH protein, respectively. Spheres indicate disulfide bonds. Sticks represent the

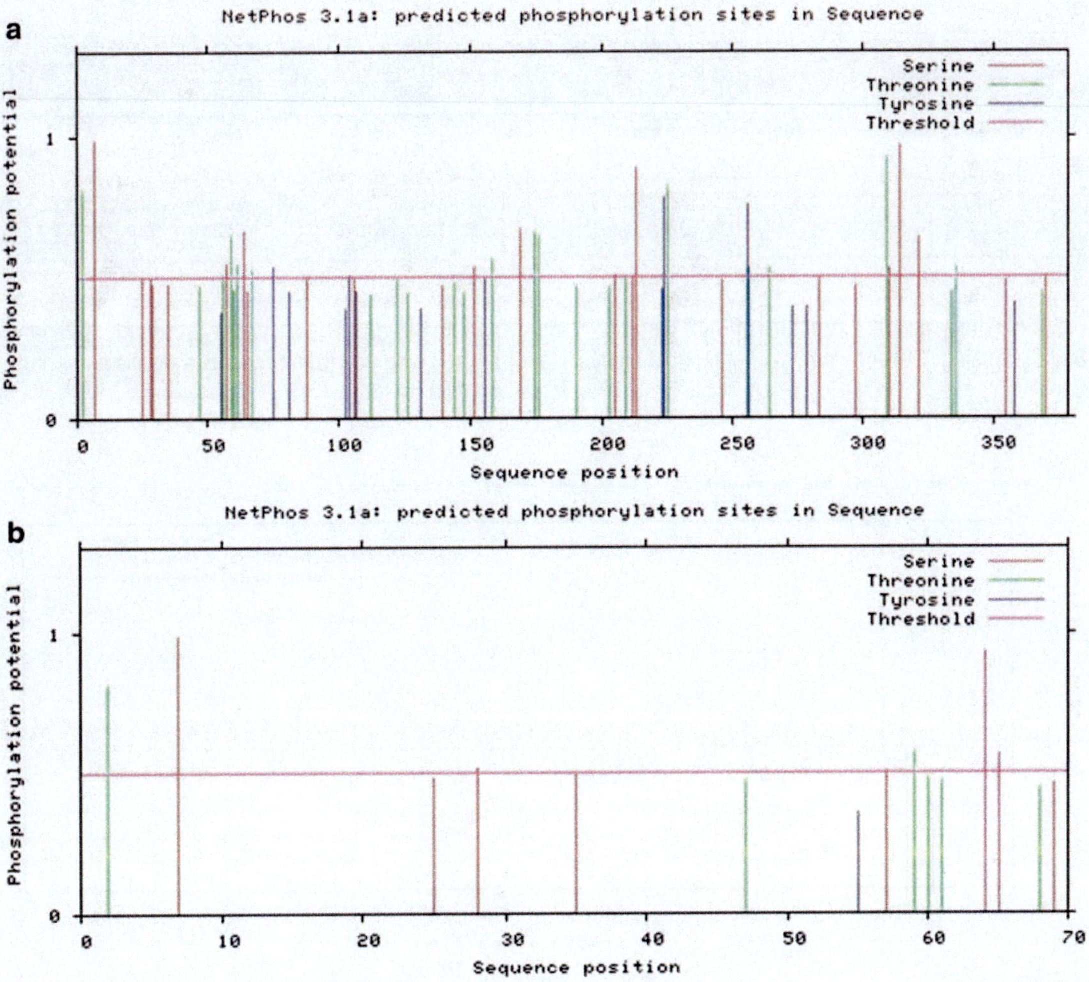

Fig. 3 Prediction of phosphorylation sites of for different variants of Cytochrome B by NetPhos. (**a**) Represents analysis for variant 1, whereas (**b**) represents analysis for variant 2

main chain and the side chain is represented by lines. Structural alignment for two different protein ligands is often necessary for their comparison or other purposes. Figure 8 presents the alignment of two structures (different variants of Cytochrome B) through TM-Align software, and the basic principle involves RMSD (Root mean square deviation). It is important to detect the mutations and its functional relevance to proteins. Two commonly important software are I-mutant and Provean. I-mutant works on the principle of thermodynamics. Figure 9a–d gives a pictorial representation of the principles of thermodynamics, while Fig. 10 presents the output pane for I-mutant based on thermodynamic principles. Figure 11: Output pane for Provean in original. Figure 12 presents molecular interaction growth hormone molecule with other associated molecules. Figure 13 reflects the

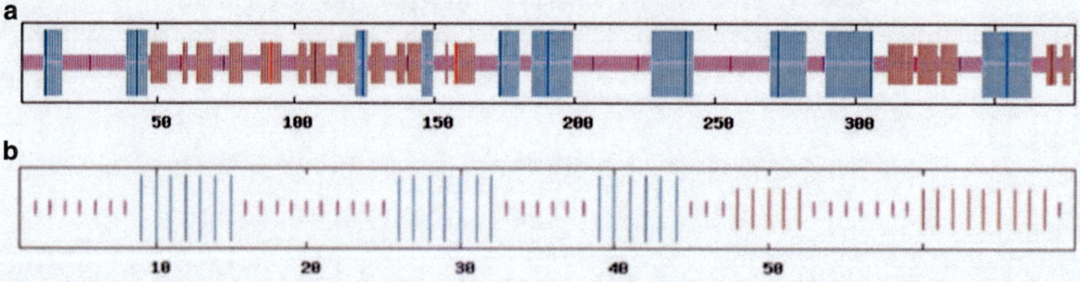

Fig. 4 (**a**) The secondary structure of CytB Variant A Using GOR software. (**b**) The secondary structure of Variant B of CytB using GOR software. (**a**) Represents analysis for variant 1, whereas (**b**) represents analysis for variant 2

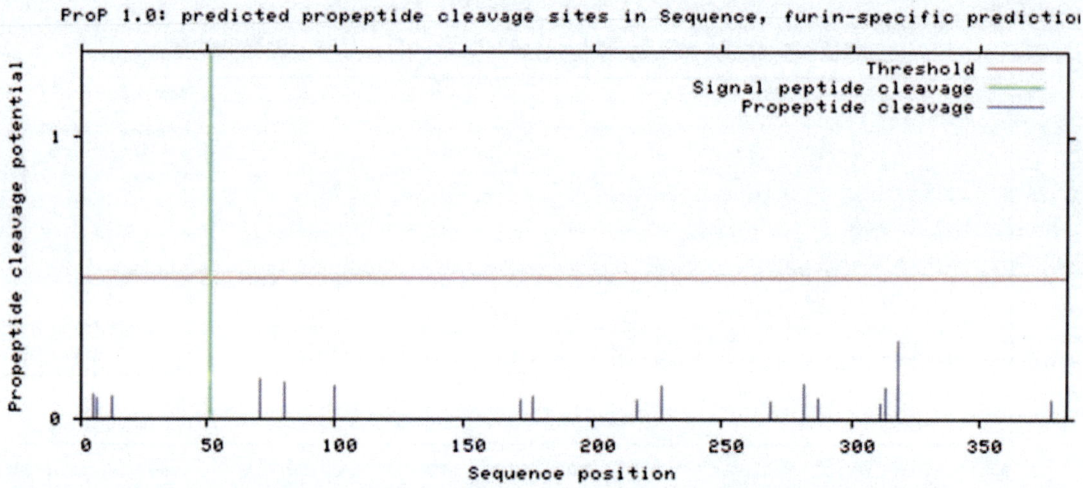

Fig. 5 Prediction of arginine and lysine propeptide cleavage sites in eukaryotic proteins Variant A from PropP. (**a**) Represents analysis for variant 1,whereas (**b**) represents analysis for variant 2

Ramachandran plot depicting the structural analysis and stability of the protein.

It is important to detect the PDB structure of the protein for providing as input for most of the bioinformatics tools. There are many tools available; in this chapter, we have discussed PHYRE. Figure 14 depicts the prediction of the secondary structure by Phyre software [1–13].

UniProtKB • functional information on proteins.

UniProtKB/Swiss-Prot • protein sequence database.

STRING • protein-protein interactions.

SWISS-MODEL Repository • protein structure homology models.

PROSITE • protein domains and families.

ViralZone • portal to viral UniProtKB entries.

neXtProt • human proteins.

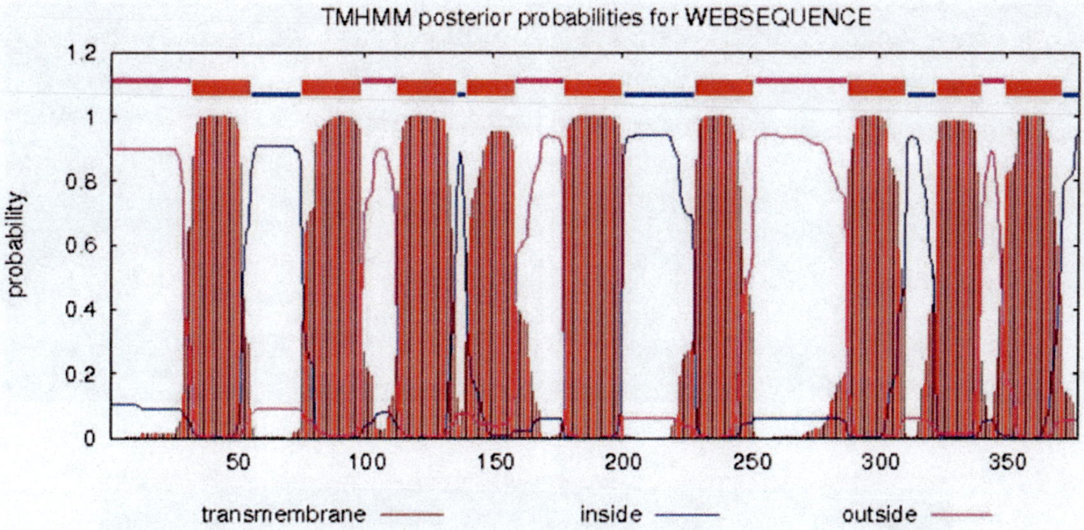

Fig. 6 Determination of transmembrane helices of Variant A

SWISS-MODEL Workspace • structure homology-modeling.
PyMOL.
Vital-IT • life science informatics initiative.
SwissDock • protein ligand docking server.
PatchDock.
FireDock.
MyHits • protein domains database and tools.
PhyreRisk • Map genetic variants to protein structures.
HAMAP • UniProtKB family classification and annotation.
2ZIP • Prediction of leucine zipper domains.

- 3of5 • find user-defined patterns in protein sequences.
- AACompIdent • protein identification by aa composition.
- AACompSim • amino acid composition comparison.
- Agadir • Prediction of the helical content of peptides.
- ALF • simulation of genome evolution.
- Alignment tools • Four tools for multiple alignments.
- APSSP • Advanced Protein Secondary Structure Prediction.
- Ascalaph • Molecular modeling software.
- big-PI • predict GPI modification sites.
- Biochemical Pathways • Biochemical Pathways.
- BLAST • sequence similarity search.
- BLAST (UniProt) • BLAST search on the UniProt web site.
- BLAST—NCBI • Biological sequence similarity search.
- BLAST—PBIL • BLAST search on protein sequence databases.

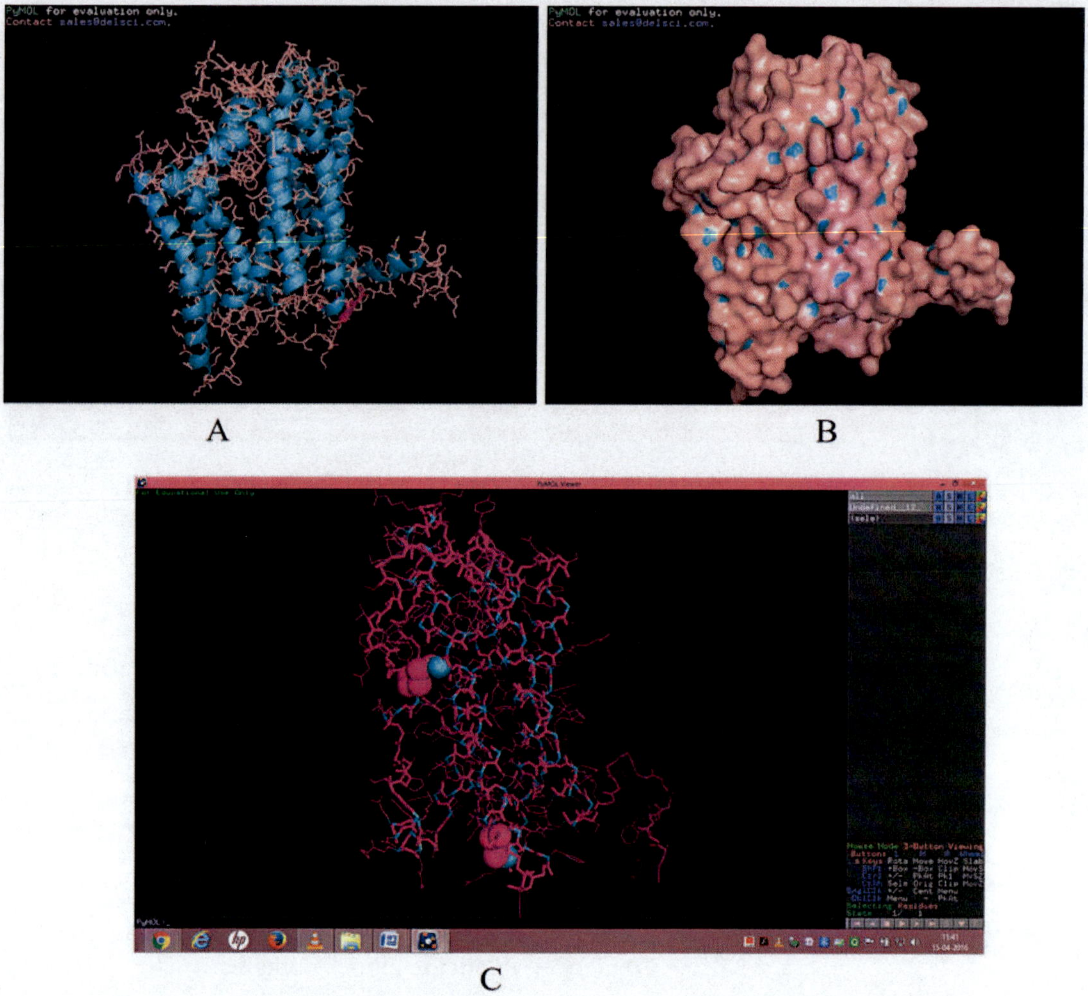

Fig. 7 PyMOL view of Cytochrome B protein. (**a**) Cartoon. (**b**) Surface view. (**c**) 3D structure of GH gene. Spheres indicate disulfide bonds. Sticks represents main chain and side chain represented by lines

- Blast2Fasta • Blast to Fasta conversion.
- boxshade • MSA pretty printer.
- CFSSP • Protein secondary structure prediction.
- ChloroP • chloroplast transit peptides & cleavage sites.
- Click2Drug • Directory of computational drug design tools.
- ClustalO (UniProt) • Align two or more protein sequences.
- Clustal Omega (EBI) • Multiple sequence alignment program.
- ClustalW • Multiple sequence alignment.
- ClustalW—PBIL • Multiple sequence alignment program.
- Coiled-Coils prediction • Prediction of coiled coils regions.
- COILS • Prediction of Coiled Coil Regions in Proteins.

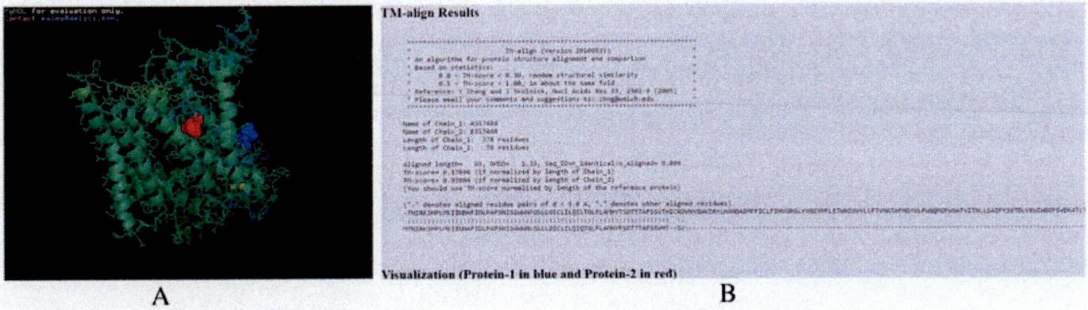

Fig. 8 (**A**) Overlapping of two structures (different variants of Cytochrome B) (left), (**B**) TM-Align software. RMSD calculated (right)

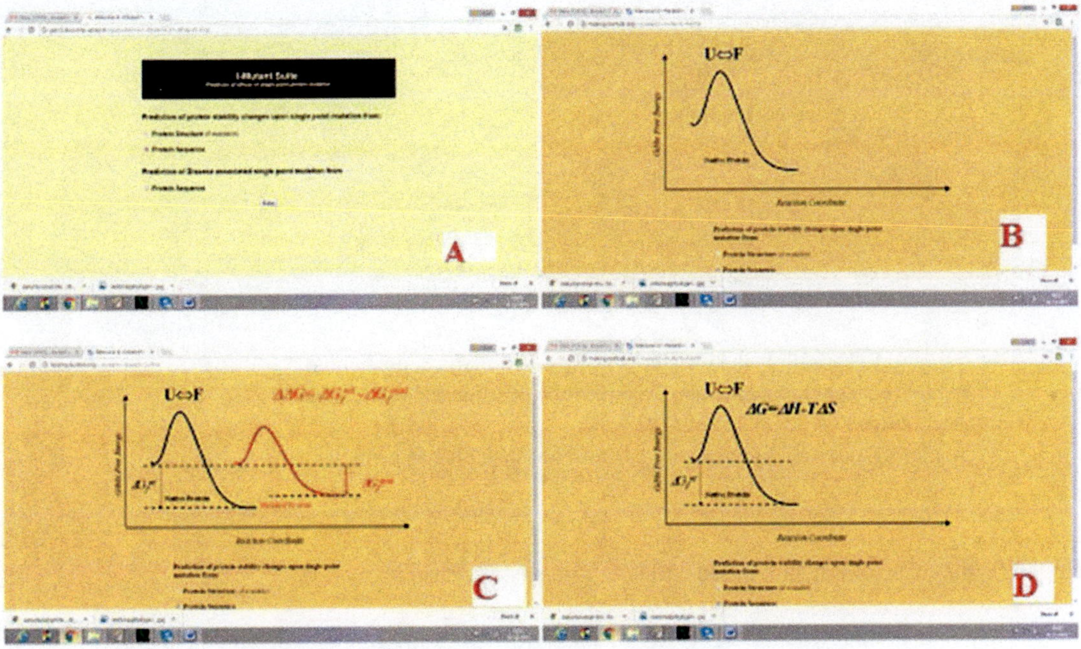

Fig. 9 (**a–d**) Pictorial representation of the principles of thermodynamics

- ColorSeq • Color Protein Sequence.
- Compute pI/MW • theoretical pI and Mw computation.
- CPHmodels • Protein homology modeling.
- CRUNCH • Automated pipe-line for ChIP-Seq data analysis.
- CSS-Palm • Prediction of palmitoylation sites in proteins.
- DAS-TMfilter • Prediction of transmembrane regions.
- Decrease redundancy • Sequence redundancy reduction.
- DIALIGN • Local multiple sequence alignment.
- DictyOGlyc • GlcNAc O-glycosylation sites in *D. discoideum*.

```
***************************************************************************
**                                        **
**                    I-Mutant v2.0                    **
**       Predictor of Protein Stability Changes upon Mutations        **
**                                        **
***************************************************************************


   SEQ File: fileseq.txt

   Position  WT NEW   DDG  pH  T
       46   L   Q  -1.25  7.0  25


   WT:  Aminoacid in Wild-Type Protein
   NEW: New Aminoacid after Mutation
   DDG: DG(NewProtein)-DG(WildType) in Kcal/mol
        DDG<0: Decrease Stability
        DDG>0: Increase Stability
   T:   Temperature in Celsius degrees
   pH:  -log[H+]


***************************************************************************
*                                        *
* Capriotti E, Fariselli P and Casadio R (2005). I-Mutant2.0: predicting  *
* stability changes upon mutation from the protein sequence or structure. *
* Nucl. Acids Res. 33: W306-W310.                                    *
* http://gpcr.biocomp.unibo.it/cgi/predictors/I-Mutant2.0/I-Mutant2.0.cgi *
*                                        *
***************************************************************************
```

Fig. 10 Output pane for I mutant based on thermodynamic principles

- DisEMBL • Prediction of disordered protein regions.
- DLP-SVM • Domain linker predictor.
- Dotlet • sequence similarity plots.
- ELM • Eukaryotic Linear Motifs.
- EMBnet services • bioinformatics tools, databases and courses.
- EMBOSS translation tools • sequence translation tools.
- epestfind • Identification of PEST motifs.
- EzMOL • A wizard for protein display and image production.
- FASTA/SSEARCH/GGSEARCH/GLSEARCH • Sequence similarity searching of protein db.
- FindMod • protein post-translational modification prediction.
- FindPept • peptide identification from unspecific cleavage.

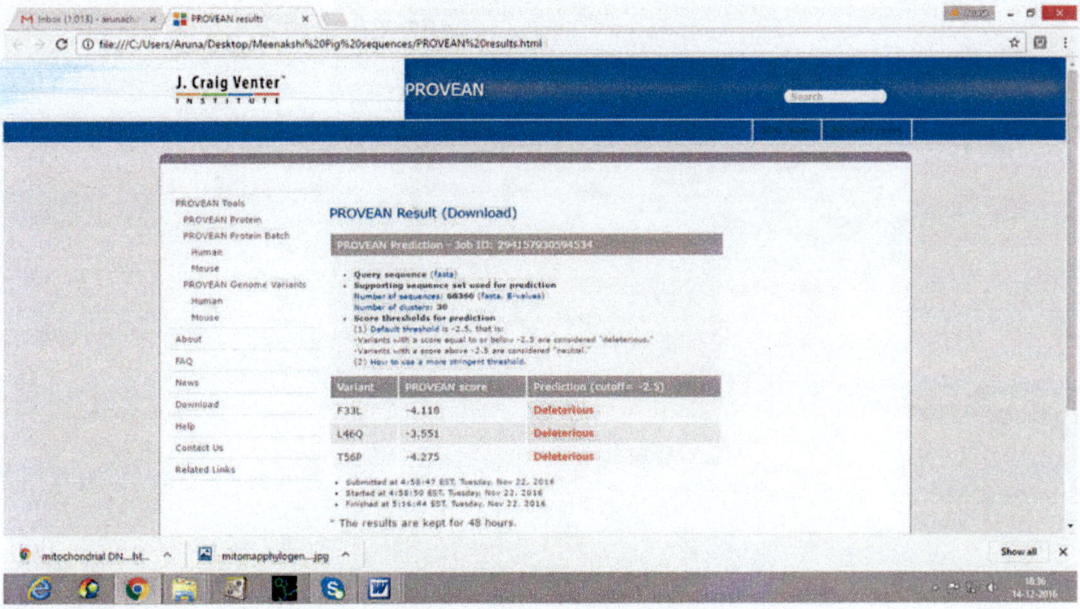

Fig. 11 Output pane for Provean in original

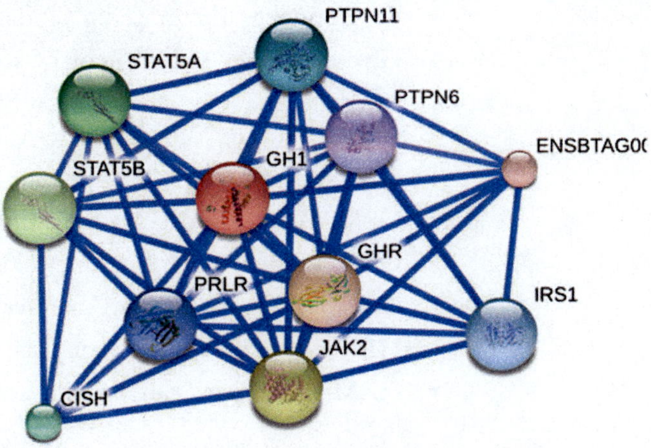

Fig. 12 Molecular interaction growth hormone molecule with other associated molecules

- FingerPRINTScan • scan sequences against PRINTS.
- FUGUE • Sequence-structure homology recognition.
- GENIO/logo • RNA/DNA & Amino Acid Sequence Logos.
- Geno3D • Protein molecular modeling.
- GlobPlot • Protein disorder/globularity/domain predictor.
- GlycoDomain Viewer • visual browser for glycoproteomic data.
- GlycoMod • oligosaccharide structure prediction.

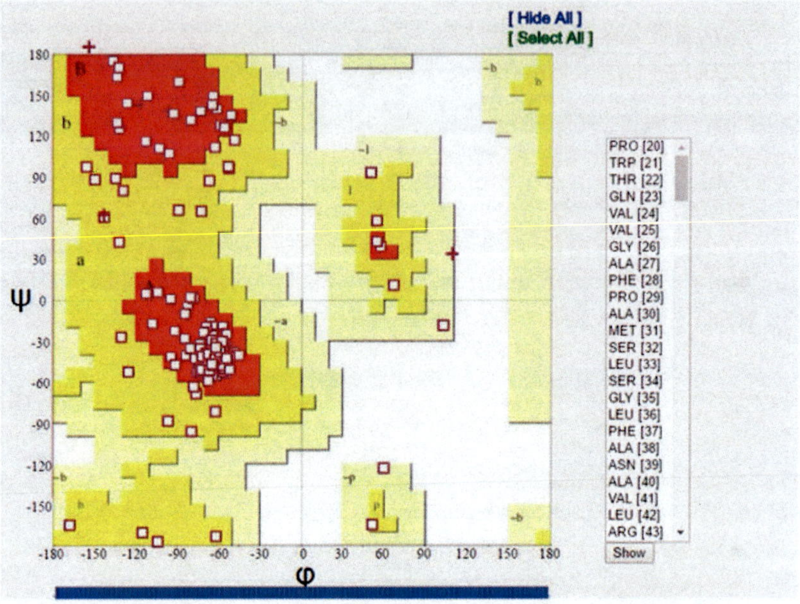

Fig. 13 Ramachandran plot depicting the structural analysis and stability of the protein

- GlyConnect • Integrated glycodata platform.
- GlycoSiteAlign • alignment of sequences around glycosylation sites.
- GOR • Protein secondary structure prediction.
- GPI-SOM • identify GPI-anchor signals.
- GPMAW lite • protein physical and chemical parameters.
- GPS • Prediction of kinase-specific phosphorylation site.
- Graphical Codon Usage Analyser • Codon bias.
- HAMAP • UniProtKB family classification and annotation.
- HAMAP-Scan • Scan protein sequences against the HAMAP families.
- HCA (hydrophobic cluster analysis) • hydrophobic cluster analysis.
- HeliQuest • alpha-helical properties screening.
- HHpred • Homology detection & structure prediction.
- HMMTOP • Prediction of transmembranes helices and topology.
- HTMSRAP • Helical TransMembrane Prediction.
- ImageMaster/Melanie • software for 2-D PAGE analysis.
- InsPecT • MS/MS tool to identify modified peptides.
- InterProScan • Family and domain database search.

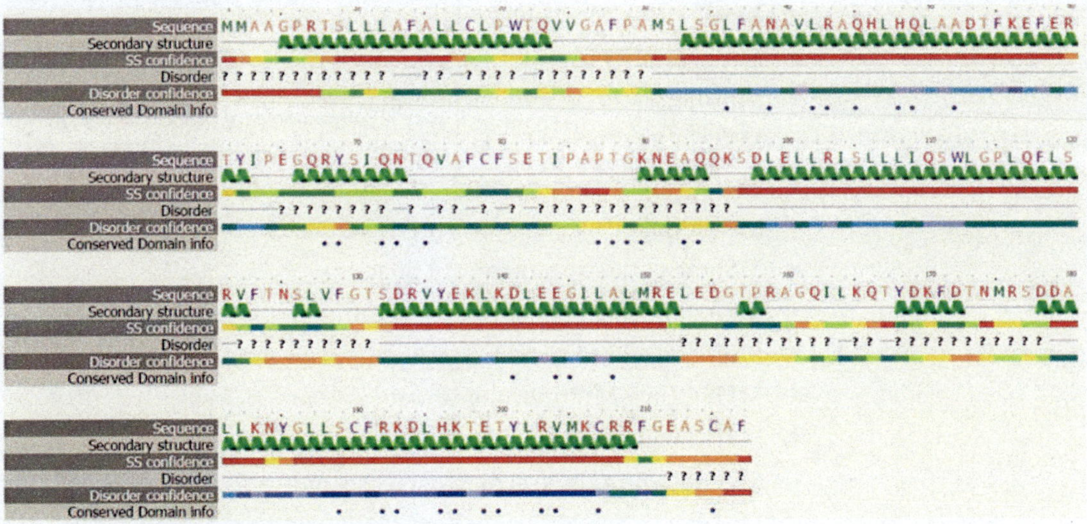

Fig. 14 Prediction of Secondary structure by Phyre software

- iPtgxDBs • integrated proteogenomics search databases.
- ISMARA • genome-wide expression data modeling.
- IsotopIdent • theoretical isotopic distribution.
- Jmol • 3D molecule visualization tool.
- Jpred • Secondary Structure Prediction Server.
- Kalign—EBI • Fast and accurate multiple sequence alignment.
- Kalign—SBC • Fast and accurate multiple sequence alignment.
- LALIGN • Pairwise alignment.
- LipoP • predict lipoproteins and signal peptides.
- MAFFT—CBRC • Multiple sequence alignment.
- MAFFT—EBI • Multiple sequence alignment.
- Make2D-DB II • package to build web-based proteomics database.
- MakeMultimer.py • Reconstruction of multimeric molecules in crystals.
- MALDIPepQuant • quantify MALDI peptides.
- MARCOIL • coiled-coils prediction.
- MarvinSpace • Visualization, modeling and interactive rendering.
- Mascot • protein identification from mass spectrometry data.
- MaxAlign • Gap removal from alignments.
- Medline Ranker • Ranking the biomedical literature.
- MeDor • Metaserver for protein disorder prediction.

- MetaNetX • Metabolic Network Repository & Analysis.
- Missense3D • impact of a missense variant on protein structure.
- MitoFates • Predict mitochondrial presequences + cleavage sites.
- MITOPROT • predict mitochondrial targeting sequences.
- MolTalk • Environment for structural bioinformatics.
- MovieMaker • Generates movies of protein dynamics.
- MSight • mass spectrometry imager.
- Multalin • Multiple sequence alignment.
- Multicoil • Prediction of two- and three-stranded coiled coils.
- MUSCLE • Multiple alignment server.
- MyDomains • Image Creator.
- MyHits • protein domains database and tools.
- Myristoylator • N-terminal myristoylation prediction.
- MzJava • Java library for processing mass spectral data.
- MzVar • Compilation of variant protein databases.
- NetAcet • predict N-terminal acetylation sites.
- NetCGlyc • C-mannosylation sites in mammalian proteins.
- NetCorona • Coronavirus 3C-like proteinase cleavage sites.
- NetGlycate • glycation of epsilon amino groups of lysines.
- NetNES • Prediction of leucine-rich nuclear export signals.
- NetNGlyc • N-glycosylation sites in human proteins.
- NetOGlyc • mammalian mucin type GalNAc O-glycosylation sites.
- NetPhos • predict phosphorylation in eukaryotes.
- NetPhosK • Prediction of kinase-specific phosphorylation site.
- NetPhosYeast • predict protein phosphorylation sites in yeast.
- NetPicoRNA • Prediction of picornaviral protease cleavage sites.
- NetSurfP • Protein Surface Accessibility Prediction.
- NetTurnP • Prediction of Beta-turn regions in proteins.
- NMT • Predict N-terminal N-myristoylation of proteins.
- NQ-Flipper • Correction of unfavorable rotamers of Asn and Gln.
- One to Three • 1- to 3-letter amino acid converter.
- OpenStructure • molecular modeling and visualization.
- Paircoil2 • Prediction of coiled coils regions.
- PATS • predict apicoplast targeted sequences.
- PattInProt • scan sequences against patterns, mismatch allowed.
- PDBePISA • Protein Interfaces, Surfaces and Assemblies.

- PepDraw • Peptide primary structure drawing.
- PepFrag • Identify proteins from a tandem mass spectrum.
- PepSweetener • interactive glycopeptide map for user-entered mass.
- PeptideCutter • protein cleavage sites prediction.
- PeptideMass • peptides from protein cleavage.
- PEPVAC • Prediction of MHCI-restricted epitopes.
- Pfam HMM Search • Pfam database search.
- pftools • Generalized profiles build and search tools.
- Phobius • predict transmembrane topology and signal peptides.
- Phyre2 • 3D structure prediction with HMM.
- pIcarver • theoretical distributions of peptide pI.
- PLOGO • Sequence logos.
- Poodle • Prediction of disordered protein regions.
- PoPMuSiC • Thermodynamic stability prediction upon mutations.
- PoPS • Prediction of Protease Specificity.
- PORTER • Protein Secondary Structure Prediction.
- PPSearch • search a sequence against PROSITE patterns.
- PRATT • find patterns in unaligned protein sequences.
- PRATT (EBI) • find conserved patterns in unaligned sequences.
- PredictProtein • Prediction of physico-chemical protein properties.
- Predotar • predict putative N-terminal targeting sequences.
- PrePS • predict protein prenylation.
- pROC • visualize, smooth and compare ROC curves.
- PROCHECK • Stereochemical quality of a protein structure.
- ProDom • Scan a sequence against ProDom domain families.
- PROF • Secondary Structure Prediction System.
- ProFound • Search protein databases with peptide mass maps.
- ProP • Prediction of Arg & Lys propeptide cleavage sites.
- PROPSEARCH • Functional and/or structural homolog search.
- ProSA-web • Program of error recognition in 3D structures.
- ProtBud • Comparison of asymmetric units and biological unit.
- Protein Colourer • Tool for coloring amino acid sequences.
- Protein Disorder Predictors • Protein Disorder Predictors.
- ProteinProspector • Mass spectrometry database search tools.

- Protein Sequence Logos • Protein sequence logo method.
- ProtParam • protein physical and chemical parameters.
- ProtScale • protein profile computation and representation.
- PSIPRED • Various protein structure prediction methods.
- PSORT • Protein subcellular location prediction.
- PTS1 • peroxisomal targeting signal 1 containing proteins.
- PVS—Protein Variability Server • Protein sequence variability in MSA.
- PyMOL • Molecular graphics visualization.
- QMEAN • estimate quality of protein models.
- QuickMod • identification of ms/ms data.
- Radar • De novo repeat detection in protein sequences.
- RandSeq • random protein sequence generator.
- Rankpep • Prediction of MHC type I and II peptide binding.
- RasMol • Molecular graphics visualization.
- RAxML • ML inference of large phylogenetic trees.
- REP • Protein search for repeats.
- REPRO • De novo repeat detection in protein sequences.
- Reverse Transcription and Translation Tool • Transcription, translation, reverse transcription.
- Reverse Translate • Reverse translation.
- SAM-T08 • HMM-based Protein Structure Prediction.
- SAMBA • Systolic Accelerator for Molecular Biological Appl.
- SAPS • Statistical analysis of protein sequences.
- ScanProsite • scan protein sequence(s) against PROSITE.
- Scratch Protein Predictor • protein structure & structural features prediction.
- SecretomeP • Prediction of non-classical protein secretion.
- Seq2Logo • Protein sequence logo method.
- SeqLogos • Sequence logo from amino acid alignment.
- Sequence Similarity Maps (SSM) • data mining of viral isolates.
- Sequence Variability Server • Protein sequence variability in MSA.
- Sequerome • BLAST similarity search and sequence profiling.
- SHOPS • Genomic operon context analysis.
- SignalP • Prediction of signal peptide cleavage sites.
- SIM • binary sequence alignment.
- SMART • genetically mobile domains and domain architecture.

- SmileMS • LC-MSMS software for small molecule identification.
- SOPMA • Secondary structure prediction.
- SOSUI • Classification and secondary structure prediction.
- SPARQL-playground • SPARQL tutorial and test service.
- Strap • Structural alignments of proteins.
- Sulfinator • tyrosine sulfation site prediction.
- SulfoSite • Prediction of tyrosine sulfations sites.
- SUMOplot • Prediction of sumoylation sites in proteins.
- SUMOsp • Prediction of sumoylation sites.
- SUPERFAMILY Sequence Search • Assign SCOP domains to sequence using SUPERFAMILY.
- Superfamily Sequence search • Search Sequences for SCOP domains
- SUSHI • Pipeline framework.
- Swiss-PdbViewer • analyze protein 3D structures.
- SwissADME • Physicochemical descriptors, pharmacokinetics.
- Swiss Mass Abacus • Calculator of peptide and glycopeptide masses.
- SwissParam • topology, parameters for small organic molecules.
- SwissSimilarity • Ligand-based virtual screening.
- SwissTargetPrediction • Target prediction for bioactive small molecules.
- SwissTree • Gold standard reference gene trees.
- T-Coffee • sequence and structure multiple alignments.
- T-Coffee—EBI • Multiple sequence alignment program.
- T-Coffee—WUR • Multiple sequence alignment program.
- T-REKS • De novo repeat detection in protein sequences.
- TagIdent • protein identification with PI, Mw and tag.
- TargetP • subcellular location prediction.
- TatP • Twin-arginine signal peptides prediction.
- TEIRESIAS • discover patterns in nucleic or amino acid sequence.
- TermiNator • Prediction of N-terminal modification.
- Terminus • N-Acetylation and initial methionine predictor.
- Three- / one-letter code • 3- to 1-letter amino acid converter and inversely.
- Three to One • 3- to 1-letter amino acid converter.

- TLS Motion Determination • Translation/Libration/Screw Motion Determination.
- TMHMM • Prediction of transmembrane helices in proteins.
- TMPred • membrane-spanning region prediction.
- TopMatch • Protein structure comparison.
- TopPred • Topology prediction of membrane proteins.
- TopSearch • Gateway to protein structures.
- Translate • nucleotide sequence translation.
- TRUST • De novo repeat detection in protein sequences.
- VMD—Visual Molecular Dynamics • Molecular visualization program.
- WebLogo • Sequence logos.
- WeStBESel • Webserver for Sequential B-cell Epitope Selection.
- WHAT IF • Molecular modeling package.
- Wise2 • Genomic sequence to protein sequence comparison.
- WU BLAST • Sequence similarity search in protein databases.
- xComb • compute all possible crosslinks between proteins.
- XplorMed • Relationship search in MEDLINE abstracts.
- xQuest • search cross-linked peptides from complex samples.
- XSTREAM • De novo tandem repeat detection.
- YASARA • Molecular graphics, modeling, simulations.
- YinOYang • O-beta-GlcNAc attachment sites in eukaryotes.

(Source: https://www.expasy.org/proteomics).

Other important tools are I-mutant and Provean for detection of thermodynamical stability and the deleterious nature of detected mutations, respectively [5, 8–11].

4 Prediction of Secondary Structure

4.1 PyMOL View (3 D Structural Analysis) of Protein

Three sites for mutation have been indicated by spheres in three different colors (PyMOL view).

4.1.1 I-mutant

It involves the principle of thermodynamics. The basic principle is that for any stable protein, the entropy should be minimum.

4.2 Provean software

It analyzes if the SNP identified is deleterious or not.

4.3 String

It analyzes the functional association of the protein of interest with other related proteins. Example of growth hormone gene.

(a) Identification of SNPs—analysis of deleterious mutation site with I-mutant, Provean software.

(b) Proteomics study of secondary, tertiary, and quaternary structures, post-translation modification site prediction, 3D structure prediction, and functional prediction. Alignment of the protein structure, Ramachandran plot prediction (PyMOL software), prediction of the interaction of the protein (String software) [5, 7–9, 12, 13].

5 Application of Bioinformatics with Suitable Examples

5.1 Functional Assessment How Mutations in the Gene Effects Its Functional Properties: A Case with Cytochrome B Gene in Sheep

Cytochrome B gene analysis in sheep has depicted three variants—two non-synonymous substitutions and one deletion for a variant. Analyses have been conducted for the detection of the nature of SNP if it is deleterious or not by using suitable software such as SIFT and Provean.

SNPs are identified for the Cytochrome B gene and analysis is done through the MegAlign Programme of DNASTAR package. The SNPs present in the protein coding region may be prone to deleterious effects on gene function. Thus, the identified SNPs were evaluated and screened using two different computational tools/algorithms for improving accuracy of predicted results. I-Mutant3.0 (http://folding.biofold.org/cgi-bin/i-mutant3.0). An SVM-based server was used for the prediction of the protein stability change due to single mutations. The FASTA sequence of Cytochrome B protein was used as input querry. It predicts a change in the respective residue with position useful for the depiction of DDG value (Kcal/Mol). The DDG >0 indicates that the protein is with high stability and vice versa. The PROVEAN (Protein Variation Effect Analyzer) is an algorithm that predicts the impact of single mutations via a substitution on the protein function (http://provean.jcvi.org/index.php). In PROVEAN software, score prediction is based on default threshold −2.5. The score <−2.5 designates the mutation is deleterious, while >−2.5 deliberated as a neutral.

5.2 Three-Dimensional Structure Prediction and Model Quality Assessment

The templates which possess highest sequence identity with our target template are identified using PSI-BLAST (http://blast.ncbi.nlm.nih.gov/Blast). The homology modeling is utilize to build Cytochrome B native 3D structure based on homologous template structures using PHYRE2 server. The 3D structures are visualized by PyMOL (http://www.pymol.org/), which is an open source molecular visualization tool. Subsequently, the mutant model is generated using the PyMOL tool. The SwissPDBViewer is employed for controlling energy minimization. The structural evaluation, along with a stereo-chemical quality assessment of the predicted model, is carried out by using SAVES (Structural Analysis and Verification Server), which is an integrated server (http://nihserver.mbi.ucla.edu/SAVES/). The ProSA (Protein Structure Analysis) web server (https://prosa.services.came.sbg.ac.at/prosa) is used for the refinement and validation of the protein structure. The ProSA is used for checking model structural quality with potential errors. This program depicts a plot of its residue energies and Z-scores determining overall quality of model. The solvent accessibility surface area of the Cytochrome B is generated by using the NetSurfP server (http://www.cbs.dtu.dk/services/NetSurfP/). It calculates relative surface accessibility, Z-fit score, probability for Alpha-Helix, probability for beta-strand and coil score, and so on.

5.3 H-Bonding Pattern Analysis and Protein-Protein Interaction Network Depiction

We have also calculated the H-bonding pattern of native and mutant residues of Cytochrome B protein using the PyMOL tool. In order to understand the network of the Cytochrome B protein, we have conducted analysis with submitting FASTA sequences to STRING 9.1.

5.4 Prediction of Post-translational Modification Sites in Mutant Cytochrome B

The post-translational modification of proteins is a vital event for the functioning and protein dynamics. We have used computational approaches for investigating modification sites in the mutant protein of Cytochrome B.

The protein phosphorylation is analyzed using tool, specifically, Ser, Thr, and Tyr residues which are required for catalyzing its role. The NetGlycate 1.0 server (http://www.cbs.dtu.dk/services/NetGlycate/) is used for the prediction of glycation site. The serine (Ser), threonine (Thr), and tyrosine (Tyr) residues with a score of >0.5 are depicted as to be glycated. The phosphorylation sites in mutant proteins are predicted by the Netphos2.0 server (http://www.cbs.dtu.dk/services/NetPhos/). The UbPred server (http://www.ubpred.org/) is used for the prediction of ubiqutylation sites, where lysine (Lys) residues with scores of > 0.62 are considered ubiqutylated.

Note: So far as the protocol is concerned, since bioinformatics is based on computer-generated software, the basic input sequence in case of proteomics is the amino acid sequence [5, 7–9, 11].

6 Genomics

It includes sequence alignment, similarity search, characterization/annotation.

Database

- EPD • collection of eukaryotic promoters.
- Bgee • gene expression patterns comparison.
- SwissRegulon • annotations of regulatory sites.
- OMA • orthology inference among complete genomes.
- arrayMap • It deals with the curated array data repository for cancer genomics.
- CEGA • Conserved Elements from Genomic Alignments.
- ElMMo • miRNA target predictions.
- ImmunoDB • dealing with insect immune-related genes and gene families.
- iPtgxDBs •dealing with integrated proteogenomics search databases.
- miROrtho • catalogue of animal microRNA genes.
- OpenFlu • Influenza genetic and epidemiological data.
- Progenetix • genomic copy number aberrations in cancer.
- SNP2TFBS • It deals with mapping of SNPs to Transcription Factor Binding Sites.
- SwissTree • Gold standard reference gene trees.

Tools

- EPD • collection of eukaryotic promoters.
- Bgee • gene expression patterns comparison.
- OMA • orthology inference among complete genomes.
- ALF • simulation of genome evolution.
- Alignment tools •It includes four tools for multiple alignments
- arrayMap • It deals with curated array data repository for cancer genomics.
- Association Viewer • It displays SNPs in a genetic context.
- BayeScan •It identifies natural selection.
- BLAST—NCBI • dealing with biological sequence similarity search.
- BLAST—PBIL • dealing with BLAST search on protein sequence databases.
- Blast2Fasta • dealing with Blast to Fasta conversion.
- boxshade • deals with MSA pretty printer.
- BUSCO • refers to Benchmarking Universal Single-Copy Orthologs.
- CEGA • Conserved Elements from Genomic Alignments.
- ChIP-Seq • ChIP-Seq data analysis tools.

- Clustal Omega (EBI) • Multiple sequence alignment program.
- ClustalW • refers to multiple sequence alignment.
- ClustalW—PBIL • deals with multiple sequence alignment program.
- COILS • It deals with the Prediction of Coiled Coil Regions in Proteins.
- Decrease redundancy • Sequence redundancy reduction.
- DIALIGN • Local multiple sequence alignment.
- ElMMo • miRNA target predictions.
- EMBOSS translation tools • sequence translation tools.
- ESTscan • coding region detection.
- FASTA/SSEARCH/GGSEARCH/GLSEARCH • deals with sequence similarity searching of protein db.
- FastEpistasis •It deals with the test for epistasis effects.
- fastsimcoal • It refers to coalescent simulation of genomic data.
- FetchGWI/tagger •refers to short sequence mapping.
- GENIO/logo •dealing with RNA/DNA & Amino Acid Sequence Logos.
- GMM • copy number variation detection.
- Graphical Codon Usage Analyser • Codon bias.
- iPtgxDBs • integrated proteogenomics search databases.
- IScan •It deals with the insertion sequence identification.
- ISMARA • It refers to the modeling of genome-wide expression data.
- IsotopIdent •deals with the theoretical isotopic distribution.
- Kalign—EBI • It refers to the fast and accurate multiple sequence alignment.
- Kalign—SBC • It refers to the fast and accurate multiple sequence alignment.
- LALIGN • refers to Pairwise alignment.
- MADAP • refers to clustering for genome annotation data.
- MAFFT—CBRC • deals with Multiple sequence alignment.
- MAFFT—EBI • deals with Multiple sequence alignment.
- MAMOT • HMM models.
- MaxAlign • Gap removal from alignments.
- miRmap • Search and predict miRNA targets with miRmap.
- mOTUs •refers to the tool for profiling microbial community composition.
- Multalin • Multiple sequence alignment.

- MUSCLE • It is also a Multiple alignment server.
- MzVar • refers to the compilation of variant protein databases.
- Newick Utilities • high-throughput phylogenetic tree processing.
- PACMAN • PACMAN: Pacific Biosciences methylation analyzer.
- Phylogibbs • regulatory sites discovery.
- Progenetix • genomic copy number aberrations in cancer.
- QuasR •It quantifies and annotate Short Reads in R.
- Reverse Transcription and Translation Tool • dealing with Transcription, translation, reverse transcription.
- Reverse Translate • refers to the process of Reverse translation.
- SAMBA • Systolic Accelerator for Molecular Biological Appl.
- Sequence Similarity Maps (SSM) • refers to the data mining of viral isolates.
- Sequerome • refers to the BLAST similarity search and sequence profiling.
- SHOPS • Genomic operon context analysis.
- ShoRAH • tools for the analysis of NGS data.
- SIBsim4 • spliced sequence alignment.
- SNP2TFBS •refers to mapping SNPs to Transcription Factor Binding Sites.
- SSA • nucleic acid sequence motifs.
- SwissTree • Gold standard reference gene trees.
- T-Coffee • refers to sequence and structure multiple alignments.
- T-Coffee—EBI •deals with Multiple sequence alignment program.
- T-Coffee—WUR • deals with Multiple sequence alignment program.
- TagScan • genome-wide sequence tag scanner.
- Translate • refers to nucleotide sequence translation.
- TriFLe • TRFLP species identification and simulation tromer • transcriptome analyzer.
- V-Pipe • refers to the Pipeline analyzing NGS data of short viral genomes.
- WebLogo • Sequence logos.
- Wise2 • Genomic sequence to protein sequence comparison.
- WU BLAST • Sequence similarity search in protein databases.
- ZFN-Site • zinc finger nuclease (off-)target sites search.

DNASTAR is a user friendly software involved in nucleotide analysis.

7 Structure Analysis

7.1 Database

SWISS-MODEL Repository

- Protein structure homology models.
- PhyreRisk • Map genetic variants to protein structures.
- SwissBioIsostere • bioIsosteres for small molecules.
- SwissSidechain • non-natural amino-acid side chains.

Tools

- SWISS-MODEL Workspace • structure homology-modeling.
- SwissDock • protein ligand docking server.
- CAMEO • Continuous Automated Model EvaluatiOn.
- Click2Drug • It is a directory of computational drug design tools.
- COILS • It deals with the prediction of Coiled Coil Regions in Proteins.
- MARCOIL • coiled-coils prediction.
- Missense3D • It deals with the impact of a missense variant on protein structure.
- OpenStructure • molecular modeling and visualization.
- QMEAN • estimate quality of protein models.
- rBAN • non-ribosomal peptides annotation tool.
- Swiss-PdbViewer •It analyzes protein 3D structures.
- SwissParam • topology, parameters for small organic molecules.
- SwissTargetPrediction • Target prediction for bioactive small molecules.

8 Systems Biology

Databases

- SwissRegulon • annotations of regulatory sites.
- arrayMap • deals with curated array data repository for cancer genomics.
- Cellosaurus • Knowledge resource on cell lines.
- iPtgxDBs • integrated proteogenomics search databases.

- MetaNetX • Metabolic Network Repository & Analysis.
- Progenetix • genomic copy number aberrations in cancer.
- Rhea • expert curated resource of biochemical reactions.
- SwissLipids • knowledge resource for lipid biology.
- TCS • interaction specificity in two-component systems.

Tools

- arrayMap • deals with curated array data repository for cancer genomics.
- Biochemical Pathways • Biochemical Pathways.
- CRUNCH • It is an automated pipe-line for ChIP-Seq data analysis.
- efmtool • elementary flux modes of metabolic networks.
- Genome History • It duplicate genes from complete genomes.
- Genonets • dealing with genotype network analysis.
- iPtgxDBs • integrated proteogenomics search databases.
- ISMARA • genome-wide expression data modeling.
- MetaNetX • Metabolic Network Repository & Analysis.
- Phylogibbs • regulatory sites discovery.
- Progenetix • genomic copy number aberrations in cancer.
- STR similarity search tool • The tool deals with Short Tandem Repeat (STR) similarity search tool.
- The Systems Biology Research Tool • software dealing with API for systems biologists.

9 Evolutionary Biology

Databases

- Bgee • gene expression patterns comparison.
- OrthoDB • Hierarchical catalog of eukaryotic orthologs.
- OMA • orthology inference among complete genomes.
- CEGA • Conserved Elements from Genomic Alignments.
- EBI Metagenomics Portal • Analysis and archive of metagenomic data.
- ImmunoDB • insect immune-related genes and gene families.
- miROrtho • catalogue of animal microRNA genes.
- Selectome • positive selection.
- SwissTree • Gold standard reference gene trees.

Tools

- Bgee • gene expression patterns comparison.
- OMA • orthology inference among complete genomes.
- ALF • simulation of genome evolution.
- BayeScan • identify natural selection.
- BIONJ • Neighbor Joining (NJ) phylogenetic analysis.
- BLASTO • BLAST on orthologous groups.
- BUSCO • Benchmarking Universal Single-Copy Orthologs.
- CEGA • Conserved Elements from Genomic Alignments.
- Coev-Web • The Coev-Web is a web interface to Coev program.
- CT-CBN • estimate conjunctive Bayesian networks.
- DendroUPGMA • Dendrogram creation using the UPGMA algorithm.
- Evolutionary Trace Server (TraceSuite II) • Maps evolutionary traces to structures.
- fastsimcoal • coalescent simulation of genomic data.
- Genonets • Genotype network analysis.
- MLtree • maximum likelihood optimization.
- MLTreeMap • phylogenetics and functionalities of metagenomes.
- Newick Utilities • high-throughput phylogenetic tree processing.
- PHYLIP • Package of programs for phyologenetic analysis.
- Phylo.io • Phylogenetic trees visualization and comparison [7, 13].

Example (Fig. 15):

Phylogeny.fr • Simple phylogenetic analysis.

Phylogeny programs • Links to phylogeny programs.

RAxML • ML inference of large phylogenetic trees.

REALPHY • Automatic inference of phylogenetic trees.

SuperTree • assemble phylogenetic trees.

SwissTree • Gold standard reference gene trees.

The PhylOgenetic Web Repeater (POWER) •deals with Phylogenetic analysis.

MEGA tool for phylogenetic analysis.

Network analysis for phylogenetic tree generation.

A detail of this analysis has been discussed in the next chapter.

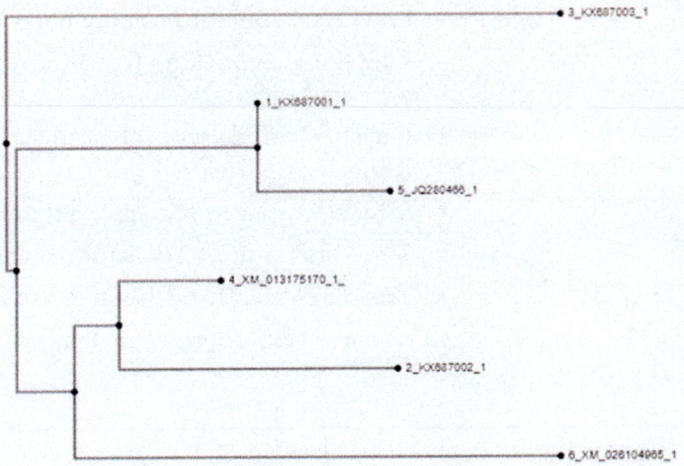

Fig. 15 Phylogenetic tree of HB chicken with other poultry spp (example for Phylo.io) [7]

10 Population Genetics

Tools

- Arlequin • software for population genetics.
- BayeScan • identify natural selection.
- CT-CBN • estimate conjunctive Bayesian networks.
- fastsimcoal • coalescent simulation of genomic data.
- mOTUs • Tool for profiling microbial community composition.
- TriFLe • TRFLP species identification and simulation.
- V-Pipe • It deals with Pipeline analyzing of NGS data of short viral genomes.

11 Transcriptomics

11.1 Databases

- Bgee • gene expression patterns comparison.
- SwissRegulon • annotations of regulatory sites.
- ElMMo • miRNA target predictions.
- Bgee • gene expression patterns comparison.
- ESTscan • coding region detection.
- ExpressionView • explore biclusters in gene expression data.
- ISA • gene expression module discovery.
- ISMARA • genome-wide expression data modeling.

- mOTUs • Tool for profiling microbial community composition.
- Ping pong algorithm • It analyses coherent patterns across paired datasets.
- QuasR •It deals with Quantification and Annotation of Short Reads in R.
- SIBsim4 • spliced sequence alignment.
- The Miner Suite • Tools for data analysis.
- Translate • nucleotide sequence translation.
- tromer • transcriptome analyzer.

12 Biophysics

We will list out certain databases and tools related to Biophysical applications.

Databases involved are as follows:

SWISS-MODEL Repository.

Protein structure homology models.

Swiss Side chain.

Non-natural amino-acid side chains.

Tools

SWISS-MODEL Workspace.

Structure homology-modeling.

OpenStructure.

Molecular modeling and visualization.

QMEAN.

Estimate quality of protein models.

SwissParam.

Topology, parameters for small organic molecules.

13 Imaging

It is an important molecular visualization tool helpful for understanding the molecular mechanisms involved in biological systems.

Some of the commonly used software are listed below:

Ascalaph—It is a molecular modeling software.

ImageMaster/Melanie—It is a software for 2-D PAGE analysis.

Jmol—It refers to the three-dimensional (3D) molecule visualization tool.

MarvinSpace—It is involved with visualization, modeling and interactive rendering.

MovieMaker —It generates movies of protein dynamics.

MSight—It is a software dealing in images related to mass spectrometry.

PyMOL—It refers to molecular graphics visualization tool (refer to Figs. 7 and 8).

RasMol—It is also a molecular graphics visualization tool.

VMD—It is a software based on visual Molecular Dynamics based on the principle of imaging. It is a molecular visualization program.

Key Tools

(a) Gene Bank (e.g., NCBI).

(b) Designing of primers.

(c) BLAST analysis or Homology search.

(d) Phylogenetic analysis- Biodiversity & evolutionary study Mega 6, Network.

References

1. Rawat K, Pal A, Banerjee S, Pal A, Mandal SC, Batabyal S (2021) Ovine CD14- an immune response gene has a role against gastrointesinal nematode Haemonchus contortus—a novel report. https://doi.org/10.3389/fimmu.2021.664877
2. Banerjee S, Pal A, Mandl SC, Chatterjee PN, Chatterjee JK (2021) RIG-I has a role in immunity against Haemonchus contortus, a gastrointestinal parasite in Ovis aries—a novel report. Front Immunol 11:534705. https://doi.org/10.3389/fimmu.2020.534705
3. Pal A, Chatterjee PN (2009) Molecular cloning and characterization of CD14 gene in goat. J Small Ruminant Res 82:84–87
4. Pal A, Sharma A, Bhattacharya TK, Chatterjee PN, Chakravarty AK (2011) Molecular characterization and SNP detection of CD14 gene of crossbred cattle. Mol Biol Int 2011:507346, 13 pages. https://doi.org/10.4061/2011/507346
5. Pal A, Pal A, Banerjee S, Batobyal S, Chatterjee PN (2018) Mutation in cytochrome B gene causes debility and adverse effects on health of sheep. Mitochondrion 46:393–404. https://doi.org/10.1016/j.mito.2018.10.003
6. Pal A, Chatterjee PN, Sharma A (2013) Molecular evolution and structural analysis of Caprine CD14 deduced from cDNA clones. Indian J Anim Sci 83(10):1062–1067
7. Pal A, Pal A Jr, Mallick AI, Biswas P, Chatterjee PN (2019) Molecular characterization of Bu-1 and TLR2 gene in Haringhata Black chicken. Genomics 112(1):472–483. https://doi.org/10.1016/j.ygeno.2019.03.010
8. Pradhan M, Pal A, Samanta AK, Banerjee S, Samanta R (2018) Mutations in cytochrome B gene effects female reproduction of Ghungroo pig. Theriogenology 119:121–130
9. Pal A, Pal A, Chakravarty AK (2018) Mutations in growth hormone gene affect stability of protein structure leading to reduced growth, reproduction and milk production in cross bred cattle—an insight. Domestic Anim Endocrinol 71:106405. https://doi.org/10.1016/j.domaniend.2019.106405
10. Pal A, Pal A, Sharma A, Bhattacharya TK (2020) Mutations in CD14 gene may have significant effect on mastitis resistance/susceptibility of buffalo. BMC Genet. https://doi.org/10.21203/rs.3.rs-15720/v1

11. Pal A, Chakravarty AK (2019) Genetics & breeding for the disease resistance of livestock. Elsevier Publication. ISBN: 9780128164068
12. Pal A, Pal A, Baviskar P (2020) Molecular characterization of RIGI, TLR7 and TLR3 as immune response gene of indigenous ducks in response to Avian influenza. https://doi.org/10.1101/2020.09.28.316687
13. Pal A, Pal A (2020) Wide variabilities identified among spike proteins of SARS Cov2 globally-dominant variant identified. https://doi.org/10.1101/2020.09.26.314385

Check for updates

Chapter 8

Molecular Evolutionary Study: Phylogenetic Tree

Abstract

This chapter discusses mostly the molecular evolutionary studies with biodiversity analysis through phylogenetic tree construction. The commonly used software for phylogenetic tree construction are as follows: Phylip (general package, protdist, NJ, parsimony, maximum likelihood, etc.); PAUP (parsimony); PAML (maximum likelihood); TreePuzzle (quartet based); PhyML (maximum likelihood); MyBayes; MEGA (biologist-centric); DNASTAR; and Network analysis. Structurally, a phylogenetic tree may be classified in two types—rooted and unrooted. Functionally, the tree may be classified as based on distance or characters. Common techniques of a phylogenetic tree involve MAFFT, MEGA (Molecular Evolutionary Genetics Analysis), PHYLIP, reduced median network theory, and neighbor joining tree. The details of these techniques have been illustrated with suitable examples and diagrams. Some common indicators for effective population size have been discussed.

Keywords Molecular evolution, Phylogenetic tree, MEGA, PHYLIP, MAFFT, Network analysis

1 Basic Principle

The term phylogenetic tree is intimately related to the biodiversity analysis and to evolutionary studies.

In this chapter, we shall describe some of the tools for studying molecular evolution. It comes to our mind why evolutionary studies are important and what their practical implications are. Evolutionary studies are important for studying speciation and domestication. Evolutionary studies are equally important for the identification of the genetic closeness of humans to other species, so that rare and deadly diseases may be studied with animal models, which have genetic closeness with humans.

Speciation is the process of evolution of new organisms. The process is driven by mutations and selection bias. Mutation is the process of change in the DNA sequence due to single base changes, deletion/insertion of DNA segments, and so on. Speciation events lead to the creation of different species. Speciation is the process that is triggered by physical separation into groups where different genetic variants become dominant. Thus, any two species share a

Aruna Pal, *Protocols in Advanced Genomics and Allied Techniques*, Springer Protocols Handbooks,
https://doi.org/10.1007/978-1-0716-1818-9_8,

(possibly distant) common ancestor. Thus, the process of speciation through evolution can be studied through phylogenetic tree construction. It can be studied well through the molecular clock hypothesis.

Another important aspect in animal or livestock science is breed documentation. With the advent of domestication, newer breeds appear. Breeds may be defined as a group of individuals within a same species that are morphologically or phenotypically similar, reared in the same geographic region, and that can interbreed among themselves.

Molecular evolutionary studies are very helpful in the detection of newer breeds and the evolutionary patterns of breeds globally.

A phylogenetic tree is represented by a graph depicting the approximate distances between a set of objects (species, genes, proteins, families) in a hierarchical fashion.

Molecular evolutionary studies are usually conducted conventionally with variable markers as RAPD and microsatellites. Currently, we are progressing with molecular studies with functional genes, mitochondrial genes, and with complete mitochondrial genome sequence, an NGS (next generation sequencing) approach.

2 Types of Phylogenetic Tree

1. **Rooted tree**: A rooted tree satisfies the "molecular clock" hypothesis. All leaves are at the same distance from the root (Fig. 1).
2. **Unrooted tree**: All leaves are at different distances from the root. The edge lengths reflect evolutionary distances (Fig. 2).

Traditionally, morphological features were used for the construction of the phylogenetic tree, but currently a molecular approach has been observed to be more appropriate. Molecular features include both nucleotide and protein sequences. The nucleotide sequence includes mostly non-coding nucleotide sequences as microsatellite and RAPD markers among the nuclear genes. Mitochondrial DNA forms one of the main bases of such evolutionary analysis. Nuclear genes as functional genes are also widely applicable for such evolutionary studies, where high genetic variability is observed.

In order to determine the best tree, an optimality criterion may be decided. Optimality criteria may be parsimony, maximum likelihood, minimum evolution, Bayesian, and so on.

Two Types of Methods for Phylogenetic Tree Reconstruction

1. Distance based (input: distance matrix; UPGMA and NJ).

   UPGMA: Unweighted Pair Group Method with Arithmetic Mean.

   NJ: Neighbor joining method.

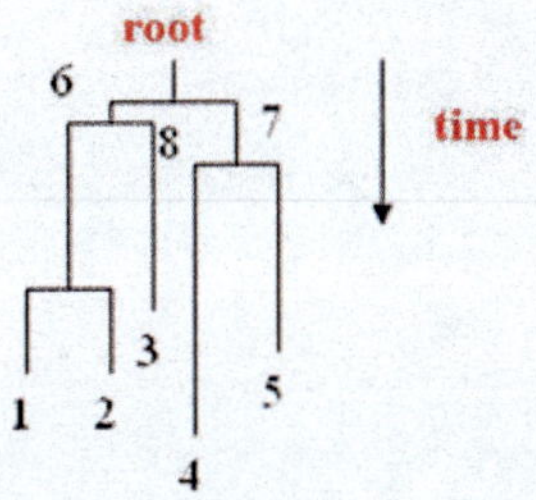

Fig. 1 Rooted tree for phylogenetic analysis

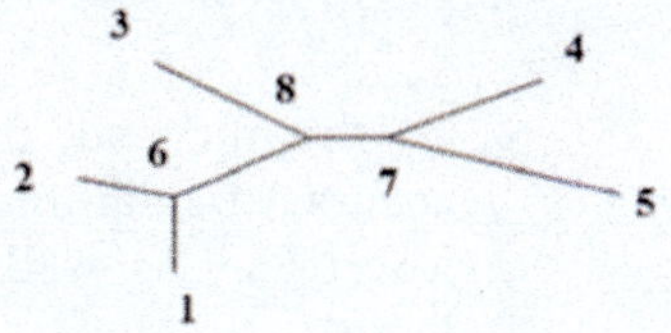

Fig. 2 Unrooted phylogenetic tree. 1–5 are called leaves, or leaf nodes. 6–8 are inferred nodes corresponding to ancestral species or molecules. Branches are also known as edges

2. Character based (input: multiple alignment).

The commonly used software for phylogenetic tree construction are as follows:

- Phylip (general package, protdist, NJ, parsimony, maximum likelihood, etc.).
- PAUP (parsimony).
- PAML (maximum likelihood).
- TreePuzzle (quartet based).
- PhyML (maximum likelihood).
- MyBayes.
- MEGA (biologist-centric).
- DNASTAR.
- Network analysis.
- MAFFT [1–6].

Phylogenetic analysis with different software case studies:

2.1 MAFFT

It is a multiple sequence alignment. It requires less CPU time than other methods. MAFT includes two novel techniques:

1. By this method, homologous regions are easily recognized by the help of FFT (Fast Fourier Transform) in where an amino acid sequence is transformed into a sequence composed of volume and polarity of each amino acid residue.

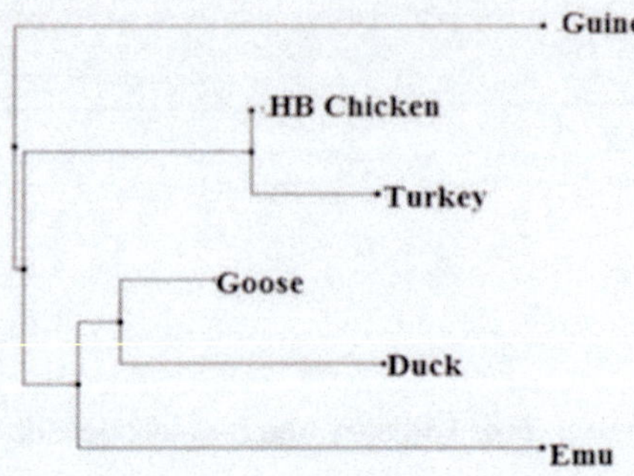

Fig. 3 Phylogenetic tree of HB chicken with other poultry spp [7]

2. By this method, the CPU time is minimized, increasing the accuracy of alignments for sequences which have exertions and insertions, as well as distantly related sequences of similar length.

The progressive method (FFT-NS-2) and the iterative refinements method (FFT-NS-i) are the two heuristics which are used here. Their CPU time is minimized compared with CLUSTALW with comparable accuracy. When the sequence number exceeds 60, the FFT-NS-i is more than 100 times faster than T-COFFEE, without compromising accuracy.

In our lab, we have conducted phylogenetic studies with chicken with respect to other poultry breeds, as depicted in Fig. 3, an example of neighbor joining tree.

Phylogenomics is the combination of the fields of *evolution* and *genomics*. This term is used to involve genome data, and evolutionary reconstruction is a technique in the field of phylogenetics and genomics. It gives us information about large genomes after comparing entire genomes and also after analyzing other small genes. Four major areas fall under phylogenomics:

1. Prediction of gene function.
2. Establishment and clarification of evolutionary relationships.
3. Gene family evolution.
4. Prediction and retracing *lateral gene transfer*.

2.2 MEGA (Molecular Evolutionary Genetics Analysis)

This is advanced software involved with sequence alignments, molecular evolutionary and phylogenetic tree analysis by interfering other phylogenetic histories.

In version 6.0, MEGA allows the inference of time trees, implementing the RelTime method for estimating divergence times for all branching points in a phylogeny.

It provides us with a graphical user interface (GUI) to specify the step-by-step phylogeny and calibration constraints. This software makes us aware of the algorithms of the evolutionary trees with some advanced memory management, which can double the size of the sequence data sets. These two software (command line version of MEGA6 and GUI) are easily accessible from www.megasoftware.net free of charge.

2.3 Protocol

The relationship between taxa or sequences and their ancestors are estimated by this phylogenetic tree. Most of the molecular data (DNA or proteins sequences) are made from this phylogenetic tree. This is used to estimate the relationship between species represented by given sequences, but nowadays the use of this software is expanded to include the relationships among the sequences themselves with the host species by interfering with the functions of genes that have not been experimentally studied and also to explore the microbial outbreaks among others.

The following are the steps to prepare a phylogenetic tree:

1. To identify the target sequence.
2. Alignment of those target sequence.
3. From the aligned sequence estimate a phylogenetic tree.
4. The phylogenetic tree was presented with related information.

A suitable web browser can be used to identify and download the homologous sequences from a national database such as GenBank. One of the several alignment programs may be employed to align the sequences to estimate the tree with all the information. This program has its own file format, which will force us to interconvert our files, which is intimidating sometimes.

MEGA5 does its work in a single environment by interconverting file formats. It is easily available. Steps are simple. MEGA5 is user friendly, particularly for new users.

2.3.1 Step 1: Acquiring the Sequences

Care should be taken that the first step is the most intellectually demanding one, but usually least attention is paid to it. All the steps are equally essential. The rest of the steps are easy, essentially mechanical, and will lead to a meaningful tree.

Usually, the investigator is interested in a particular gene or protein which has been the subject of investigation and aim to determine the relationship of that gene or protein to its homologs. The word "homologs" is key here. The main assumption of a phylogenetic tree is to analyze the relationship between homologous and common ancestors. Alignment programs will align sequences, homologous or not. The programs are all set in the software. In case the sequences do not descend from a common ancestor, the tree will be without meaning and be misleading. BLAST (Basic Local Alignment Search Tool) search is the way to identify the homologous of the target sequence.

Step 1.1

Upon starting MEGA5, the main MEGA5 window is opened. From the **Align** menu, we need to choose **Do Blast Search.** It has its own browser for nucleotide BLAST from NCBI (National Center for Biotechnology Information). In this software, blastn (by default, it is selected), blastx, tblastn, and tblastx are seen in the top 5 (five) tabs. For protein, blastp is used.

It is to be noted that NCBI frequently alters the appearance of the BLAST page, so it may differ in some details from that described here. In the software, one text box is there where we have to put the accession number of the target sequence. After that we need to paste the target sequence in the appropriate box. If the target sequence is in the database, then only the accession number is required. The middle section of the page allows us to choose the databases which will be searched. For default nucleotide collection (nr/nt), we need to choose from among a large number of alternatives, for example, NCBI genomes.

There is a limit for the optional text box to identify particular organisms. For example, if our desired sequence is of human and we want to exclude humans from the search, then we should not pick a lot of human variants in case we are really interested in homologs in other species. To know more about the target sequence of the organism, it becomes necessary to click on the + sign box. One excluded option is there for instance and environmental samples.

Step 1.2: Which BLAST Algorithm to Use?

The particular BLAST algorithm can be choosen which best explain our hypothesis. Blastn is somehow similar for nucleotides, megablast also can be chosen for highly similar sequences, and discontiguous megablast for more dissimilar sequences. Blastn is the default option. If we are looking for closely similar sequences, then megablast can be selected.

The sequences that will be on our tree are very much determined by the choice we make at this point. This is followed by clicking on the BLAST button to start the procedure and care should be taken not to click on "Show results in a new window" box. A results window will be visible, with a graphic illustrating domain that have been identified, as with a statement similar to "this page will be automatically updated in 5 s." Then, the final results window will appear. The top panel describes all about the data panel. The graphical representation is also there for proper illustration of the alignment for the top 100 "hits" (sequences identified by the search). The result will be shown below with each accession number which is a clickable link, a description, a max Score which is a clickable link, a total score, a query coverage, and the E value, and a Max ident.

We need to decide the particular sequence to be added to the alignment and to be included in the tree. In one species, there are many sequences which can represent a species or genus. Only a sequence of interest will be at query coverage. If not covering the complete sequence, we need to ignore that sequence and move on. If we are interested in a sequence that is 100% identical to our query? If we are only interested in more distantly related homologs. In such cases, tree is possible but with no algorithms. Let us discuss with two practical extreme cases. If we decide that we are interested in a hit sequence, click the "**Max score**" link to go to the series of alignments. We need to check if our query was a DNA sequence or a protein sequence.

Step 1.3: DNA Sequences

We can start the analysis with input as the accession number and the query to the hit begin with a link sequence file. In case the link is to a genome sequence, or even to a large file including sequences of several genes, we will not want to include the entire sequence in our alignment. There are two ways to deal with the issue. (1) Notice the alignment itself and note the range of nucleotides in the subject. For notification, we have to click the link to bring up the sequence file with the subject sequence itself (stand = plus/plus) or with its complement (stand = plus/minus).

At the top right, click the triangle in the gray. **Change region shown** box, followed by entering the first and last nucleotides of the range. The next step is to click the **Update View** button. There exists gray customized region just underneath of which there is box for sequence input. We need to mention if the strand is plus or minus. If the strand is minus tick on reverse complement box and update view will be visible below. For the final step, click to alignment bottom (a red cross sign) at the top of the window. (2) If our query is a coding sequence or is some other notable feature, we may see Features in this part of the subject sequence: just below the sequence description with a link to the feature. To bring up the target sequence file, click on the link. Check to verify if the sequence shown is the reverse complement of the query. The next step is to tick the **Show reverse complement** box located in the **Customize view** region, update the view, then click the **Add to Alignment** button near the top of the window.

Step 1.3.1. When we click the **Add to Alignment** button, MEGA5's **Alignment Explorer** window opens and the sequence is added to that window. Back arrow is used for returning the list of homology sequences and other target sequences after adding a sequence to the alignment explorer.

Step 1.4: Protein Sequences

Accession number link is the main difference from nucleotide search or various protein sequence files. Only their underlying coding sequences will differ although they all have same amino acid sequences. Again you have to click on the Add to Alignment button to bring up the protein sequences.

1. When we try to return to the list of hits, we may get a page that says "**How Embarrassing!** Error: −400 Cache Miss. The "Add to alignment" button is next to the circular arrow. Then the BLAST page will displaced. And result will be displaced in the top of the right section.

The top link in the list is our most recent search. Just click that link to get back to your results.

When we have added all the sequences that we want to, just close the MEGA5 browser window. After all we need to save the data by selecting save session in the Alignment Editor window. The file will have the extension .mas.

Step 1.5: Alternatives to MEGA5 for Identifying and Acquiring Sequences

Step 1.5.1. With the help of http://blast.ncbi.nlm.nih.gov/Blast.cgi, you can access NCBI BLAST. To get the identical page, you have to click on the Basic BLAST section or protein blast link, which is described earlier. This is the same as when using MEGA5's browser except that a convenient button cannot be clicked to add the sequences to the Alignment Editor.

Step 1.5.2. Open a new file in a text editor. We can use MEGA5's built-in text editor by choosing **Edit a Text File** from the File menu. Editor is used for reverse complementing or converting to other formats like fasta in addition to other functions for editing molecule sequences.

Otherwise, we need to use Notepad for Windows or TextWrangler for Mac (http://www.barebones.com/products/textwrangler/). The file is saved with a meaningful name with the extension .fasta, for example, myfile.fasta. Microsoft Word, Word Pad, TextEdit (Mac), or another word processor need not be used.

Step 1.5.3. When we have identified the sequence that we want to add and clicked the link to take us to the page for that sequence file, adjust the Region Shown and Customize View if necessary. Display settings will be notified. The default setting is GenBank (full). Change that to **Fasta (text)**, select everything, copy it then paste into the text editor file. Don't leave any blank line between the sequences when we are adding sequences.

2.3.2 Step 2: Aligning the Sequences

In MEGA5's main window, choose **Open a File/Session** from the **File menu,** if the Alignment Explorer window is not already open. Choose the MEGA5 alignment file (.mas) or the sequence file (.fasta) that we had already saved in **step 1**. In the resulting dialog, choose **Align**. By color residues, the Alignment explorer will be displayed for each sequence.

Typically, the name is very long. The name will be displayed for each sequence. The editing of names can be processed at this stage. To change something, you have to double click each name. In case of DNA sequence, DNA Sequences and Translated Protein Sequences are the two tabs. The DNA sequences tab is chosen by default. The translated Protein Sequences tab is checked to visualize the corresponding protein sequence.

Step 2.1

The current step is to align the sequences. ClustalW [8] and MUSCLE [9] are the two alignment methods. MUSCLE is preferable.

Clustal alignment is symbolized by the **W** button, and MUSCLE by an arm with clenched fist to "show a muscle" in the top of the window. Choose **Clustal** or **Muscle** from the **Alignment** menu. If your sequence is DNA, you will see two choices: **Align DNA** and **Align Codons**. In case we deal with a DNA coding sequence, it is *very important* to choose **Align Codons**. That will ensure that the sequences are aligned by codons, in a more realistic approach than direct alignment of the DNA sequences.

Step 2.2

For alignment methods, open settings in the Windows. For MUSCLE, I recommend that you accept the default settings. Preferably, for DNA, sequences, ClustalW is used, for proteins Multiple Alignment Gap and the Multiple Alignment Gap Extension are more preferred.

Step 2.3

To start the alignment process, click O.K.

It depends on the number of sequences involved and the method chosen; the alignment may take anywhere from a few seconds to a few hours. When the alignment is complete, **save** the session. After that, save your original file by the name Myfile_unaligned_mas and then MYfile_unaligned.mas will come as output.

Step 2.4

MEGA5 cannot use the .mas file directly to estimate a phylogenetic tree, so we must also choose **Export Alignment** from the **Data** menu and export the file in MEGA5 format when it will get a .meg extension. It will be needed to input a title for the data. The title can be left blank, but addition will be helpful. In case of an alignment sequence, it is to be mentioned whether the data is coding sequent or not.

Step 2.5: An Alternative to Aligning with MEGA5

After a complete alignment of the sequence, check the gaps; otherwise, it will represent previous insertions or deletions. Only a phylogenetic tree can generate the relationships among sequences. MEGA5 gives us best quality data. http://guidance.tau.ac.il/ provides us with knowledge about Guidance.

Guidance is needed for the unaligned sequences as provided in a file in FASTA format. After our file is downloaded, it will be saved in FASTA format. With the help of MEGA5, download the sequences into the Alignment Explorer; we can find the unaligned sequences in FASTA format by choosing Export Alignment from the Data menu, and then choosing the **FASTA** format. If we forgot to keep the unaligned sequences, we can select all the sequences

(Control-A), then choose **Delete Gaps** from the **Edit** menu before we export the sequences in FASTA format.

2.3.3 Step 3: Estimate the Tree

Neighbor Joining, UPGMA Maximum Parsimony, Bayesian Inference, and Maximum Likelihood [ML] are the methods used to generate phylogenetic tree.

Step 3.1

In MEGA5's main window, choose the option **Open a File/Session** from the **File** menu and open the .meg file that you saved in **step 2**.

Step 3.2

For correction, ML uses various models for multiple changes at the same evolutionary site. MEGA5 provides a feature that chooses the best model. In the model menu, the **Find Best DNA/Protein Models (ML)** option will be chosen. A preferences dialog will appear, but we are safe enough accepting the default setting. Click the **Compute** button to start the run. Models will take some time to provide the details. A completed list will appear in the order of preference. The preferred model is noted, followed by the estimation of the tree using that model. WAG + G + I is the best example.

Step 3.3

Select construct/Test Maximum likelihood Tree, the details are below. ML analysis preferences. The yellow areas are parameters that you can modify.

Three parameters:

1. Model/Method.
2. Rates among Sites.
3. Gap/Missing Data Treatment.

From the drop down menu we need to click on the yellow area in the right end with the perfect option as per the choice.

For **Model/Method**, the WAG model would be selected.

For **Rates among Sites**, the Gamma distributed with Invariant Sites (G + I) option would be chosen. Best matched models function will be displayed by models/methods selection.

Gaps/Missing Data Treatment determines how gaps are handled. MEGA5 ignores all gap columns sequence in complete deletion cases. Due to gaps in the sequence, we can lose information. **Partial Deletion** option exists, where the which sites with missing data are removed only as per the need arises. After the preferences are set, the compute button is clicked to compute the tree.

Step 3.4

Don't forget to save the above generated tree. Save the tree from the File menu. The file will have the extension .mts. We can use the generated tree to explore the tree (**step 4** for reference) by choosing Export Current Tree (Newick).

Step 3.5: Estimating the Reliability of the Tree

The tree which we estimated is almost certainly not the right representation of the historical relationships among the taxa and their ancestors. However, it is an estimate of those relationships. It has become essential to assess the reliability of the estimate. The "bootstrap method" is the commonest way of estimating reliability. For the phylogeny test, set test of phylogeny through "Bootstrap Method," and then set the number of Bootstrap Replicates to an integer between 100 and 2000. As the number is greater, it will take longer to perform the test. Click the compute button. A window with a progress bar reflects how the analysis is progressing. A tree will be reflected with the numbers on every node after analysis. Those numbers, bootstrap percentages, determine the reliability of the cluster descending from that node. The reliability is assessed through the number descending from the node. Generally, we do not take seriously nodes with <70% reliability. The bootstrap test is an indicator for the reliability of each node, not the reliability of the tree. It is advantageous since it tells us which parts of the tree we need to trust and which parts may be ignored.

Step 3.6: Alternatives to MEGA5 for Estimating the Tree

The other alternative program that estimates ML trees is PhyML (http://www.atgc-montpellier.fr/phyml/binaries.php) (Guindon et al. 2010), which can also be used over the web http://www.atgc-montpellier.fr/phyml/. SeaView (http://pbil.univ-lyon1.fr/software/seaview.html) (Gouy et al. 2010) is another important program which can aid in the alignment of sequences and estimation of trees by several methods, and draws trees.

2.3.4 Step 4: Present the Tree

A drawing of a phylogenetic tree imparts a lot of information, both explicit and implicit. Some of that implicit information may be confusing, so it is up to the investigator to assure that the information conveyed, both explicit and implicit, is correct.

A phylogenetic tree consists of external nodes (the tips) representing the actual sequences, internal nodes representing hypothetical ancestors, and branches connecting nodes to each other. The lengths of the branches depict the amount of change which is estimated to have occurred between a pair of nodes. The tree in Fig. 2 is in the "rectangular phylogram" format in which internal nodes are reflected by vertical lines.

In Fig. 2, the node on the extreme left represents the root, which implies the common ancestor from which all sequences have descended. That implicit information may be incorrect and misleading. The ML method, like the Neighbor Joining, Parsimony, and Bayesian Inference methods, is unable to determine the root of a tree. All those methods are based on estimated **unrooted** trees.

Step 4.1 The **Radiation** or **Unrooted** format is a better way to draw an unrooted tree as it does not permit the viewer to assume an unknown root. In order to display the tree in the Radiation format, in the Tree Explorer window, initially we need to choose **Tree/ Branch Style** from the **View** menu, the next step is the selection of **Radiation** from the submenu. Because the Radiation format is uncommon to many readers, the default Rectangular Phylogram format is often published, although it misleadingly implies a rooted tree. A rooted tree indicates the direction to the evolutionary process. The order of descent from the root points toward the tips. We presume that directionality can easily lead to incorrect assumptions about the evolutionary history of those sequences. To avoid the unjustified implication of directionality, it becomes important to specify in the figure legend or in the text that the tree is unrooted.

Step 4.2 Usually, we do want to present a rooted tree in order to draw conclusions that depend on the order of descent. For that we need additional information about the sequences; it refers to the information that is external to the sequences themselves, regarded as an **outgroup**. An outgroup is a sequence which is more distantly related to the remaining (ingroup) sequences than they are to each other. For example, it is known that *Pseudomonas aeruginosa* belongs to the order *Pseudomonadales.* However the remaining organisms belong to the order *Oceanospirillales*, both belong to the class Gammaproteobacteria. Thus, *P. aeruginosa* is a legitimate outgroup for the remaining sequences.

Step 4.2.1. We can root the tree on *P. aeruginosa* with the help of the rooting tool that is found in the sidebar of the Tree Explorer window. In the Rectangular Phylogram view or the Radiation view, when the rooting tool is selected, it branches resulting in *P. aeruginosa* to root the tree on that sequence.

The rooted tree now perfectly implies the direction of evolution of those sequences. When the tree will be published, it would be important to clearly mention that the tree was rooted on *P. aeruginosa.*

Step 4.3 MEGA5 gives rise to a variety of tools for manipulating the appearance of the tree. Rectangular Phylogram and Radiation formats have already been documented. Although those formats seem to be very different, but in fact they are drawings of exactly the same tree. In both the cases, branches are drawn in such a way that the lengths of the lines are proportional to the branch lengths. Those formats depict that there has been much more change between *Hahella chejuensis* KCTC 2396 and *Oceanospirillum* sp. MED292*a* than there has been between *Halomonas boliviensis* LC1 and *Halomonas* sp. HAL1.

Step 4.4

Another important format or representation is the cladogram, or **Topology Only** format. We need to choose **Topology only** from the **View** menu to see the tree drawn, so that the lengths of the branch lines are unrelated to branch lengths. These is sufficiently important information. In some trees, there are some nodes which are separated by very short branches; however, others are separated by very long branches. In case the branches are too short, it may not be possible to see the branching order or topology. The Topology Only format makes it possible to visualize the branching order of the entire tree.

Step 4.4.1. Branch length provides us with valuable information. We need to label the branches with their branch lengths. For that, we need to choose **Show/Hide** from the **View** menu and select **Branch Lengths** from the submenu.

Step 4.5: Publishing the Tree

Although we can print the tree for our own purposes, but in order to publish it, we must save it in a graphics file format that is acceptable to the journal. The portable document format (PDF) is universally acceptable. Choose **Save as PDF File** from the **Image menu**.

We may want to modify the drawing in ways that MEGA5 does not provide. These may include boldfacing some sequence names in order to draw attention to them, and adding an arrow. These manipulations are done through a graphics drawing program. Most of the drawing programs will allow files in PDF format. But if they fail, MEGA5 also allows us to save the image in PNG and Enhanced Meta File formats.

Step 4.6: Alternatives to Drawing Trees Within MEGA5

The tree drawing program FigTree (http://tree.bio.ed.ac.uk/software/figtree/) is a full-featured program which offers many capabilities of the MEGA5 system and many other capabilities.

Usually, MEGA5 is available in Windows and Mac operating systems. However, the Java executable is flexible to run on any operating system such as Linux. In order to import a tree into FigTree, export it as a Newick file as described in **step 3**.

2.4 Using MEGA5 on Macintosh Computers

The **Save** and **Open** dialogs are Windows-like and this may be unfamiliar to Mac users. A document detailing the navigation in MEGA5 for Mac can be downloaded from http://bellinghamresearchinstitute.com/NavigatingMEGA5/index.html.

There are several "virtual machine" programs such as Parallels (http://www.parallels.com/products/desktop/) and VMFusion (http://www.vmware.com/products/fusion/overview.html) which will allow a Macintosh to run the Windows operating system. It is necessary to buy a copy of Windows and install it in the virtual machine. After this, MEGA5 for Windows is installed on that virtual machine. MEGA5 or an advanced version such as MEGA6 is user friendly as it is based on Windows. The user can then get access to the Windows programs, as well as Macintosh programs.

2.5 MEGA: A Case Study

The phylogeny tree was constructed with the DNA amplification of the T7 promoter universal primer in Awassi sheep (Fig. 4).

2.6 PHYLIP

PHYLIP (the *PHYLogeny Inference Package*) is another important package of program.

It is used for studying phylogenies (evolutionary trees). It is freely available over the Internet and written to work on varieties of computer systems as possible. Figures 5 and 6 depict the phylogenetic tree to reveal evolutionary relationships of taxa among the studied sheep population with the neighbor joining method and UPGMA tree, respectively.

Already-compiled executables are available for Windows (95/98/NT/2000/me/xp/Vista), Mac OS X, and Linux systems. Earlier executables are also available for Mac OS 8 or 9 systems. Complete documentation is present on the documentation files which come with the package.

Methods which are present in the package are parsimony, distance matrix, and likelihood methods, involving bootstrapping and consensus trees. Data types which can be handled are molecular sequences, gene frequencies, restriction sites and fragments, distance matrices, and discrete characters.

The programs are regulated through a menu, which requires the options they want to set, and permits them to start the computation. The data are read into the program from a text file, which the user can prepare with the help of any Word processor or text editor. But it is important to note that this text file was *not* in the special format of that Word processor. Care should be taken to ensure that the "flat ASCII" or "Text Only" format is employed. Certain sequence analysis software such as the ClustalW alignment program can write data files in the PHYLIP format. Most of the programs look for the data in a file termed "infile." If the file is not found, it is necessary for the user to type in the file name of the data file.

Output is represented onto special files with names like "outfile" and "outtree." Trees written onto "outtree" are represented in the Newick format.

PHYLIP is another most widely distributed phylogeny package. It may be regarded as the sixth most frequently cited phylogeny package, after MrBayes, PAUP*, RAxML, Phyml, and MEGA. PHYLIP may be regarded as the oldest widely distributed package.

2.7 Reduced Median Network Theory: A Case

Given the complexity of lineage A in Indian sheep, as observed by phylogenetic analysis (Fig.. 7), we found the relationships among various haplotypes of 279 animals of lineage A and their distribution in various breeds. Reduced median network analysis revealed

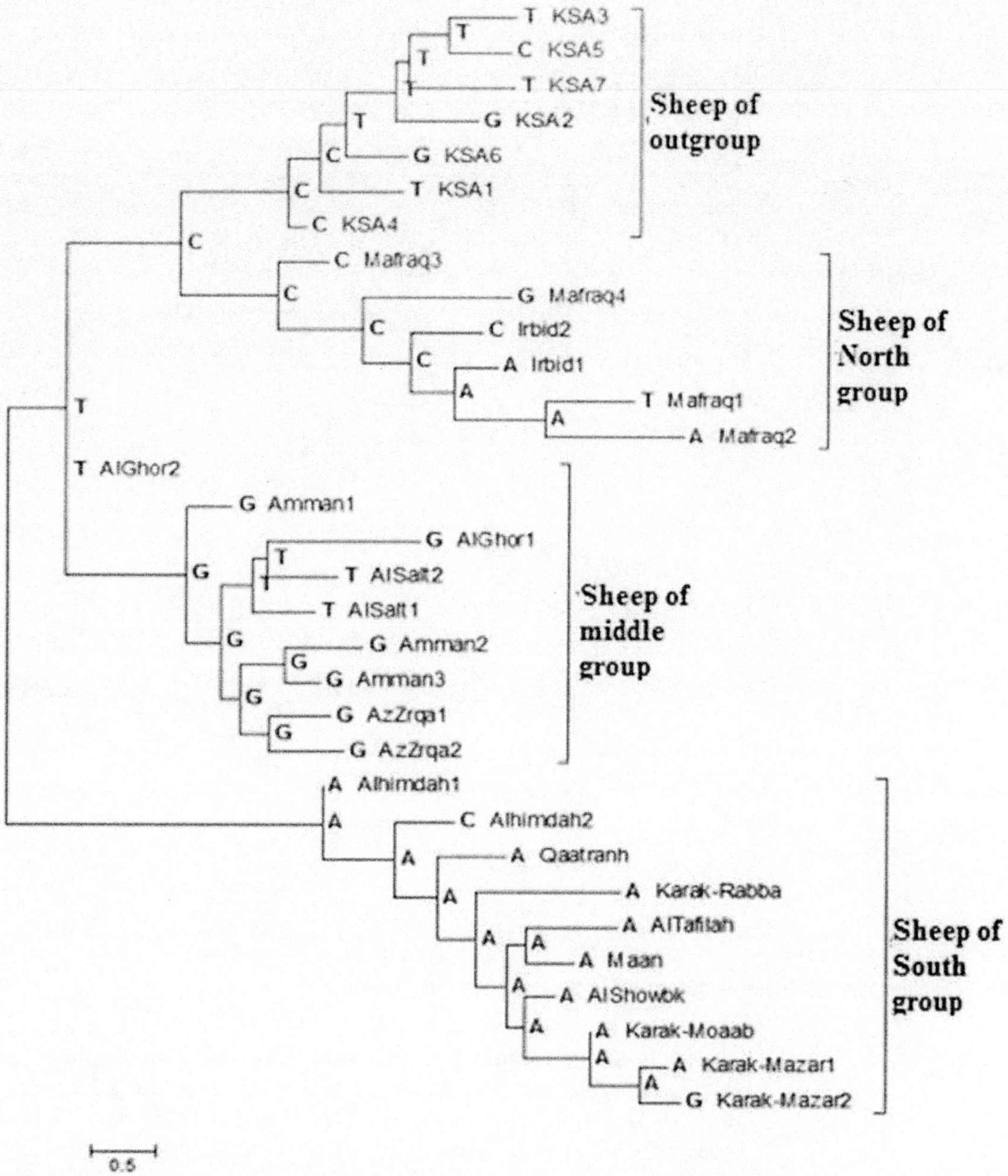

Fig. 4 Molecular phylogenetic analysis by maximum likelihood method with probable ancestors of nucleotides (A, T, G, C) among the sheep population studied

multiple expanding haplotypes and several singletons radiating from these expanding haplotypes, which indicates the expansion of lineage A type sheep populations in India. There was a main expanding haplotype as well as many additional haplotypes indicating expansion both a few mutation steps away from the main

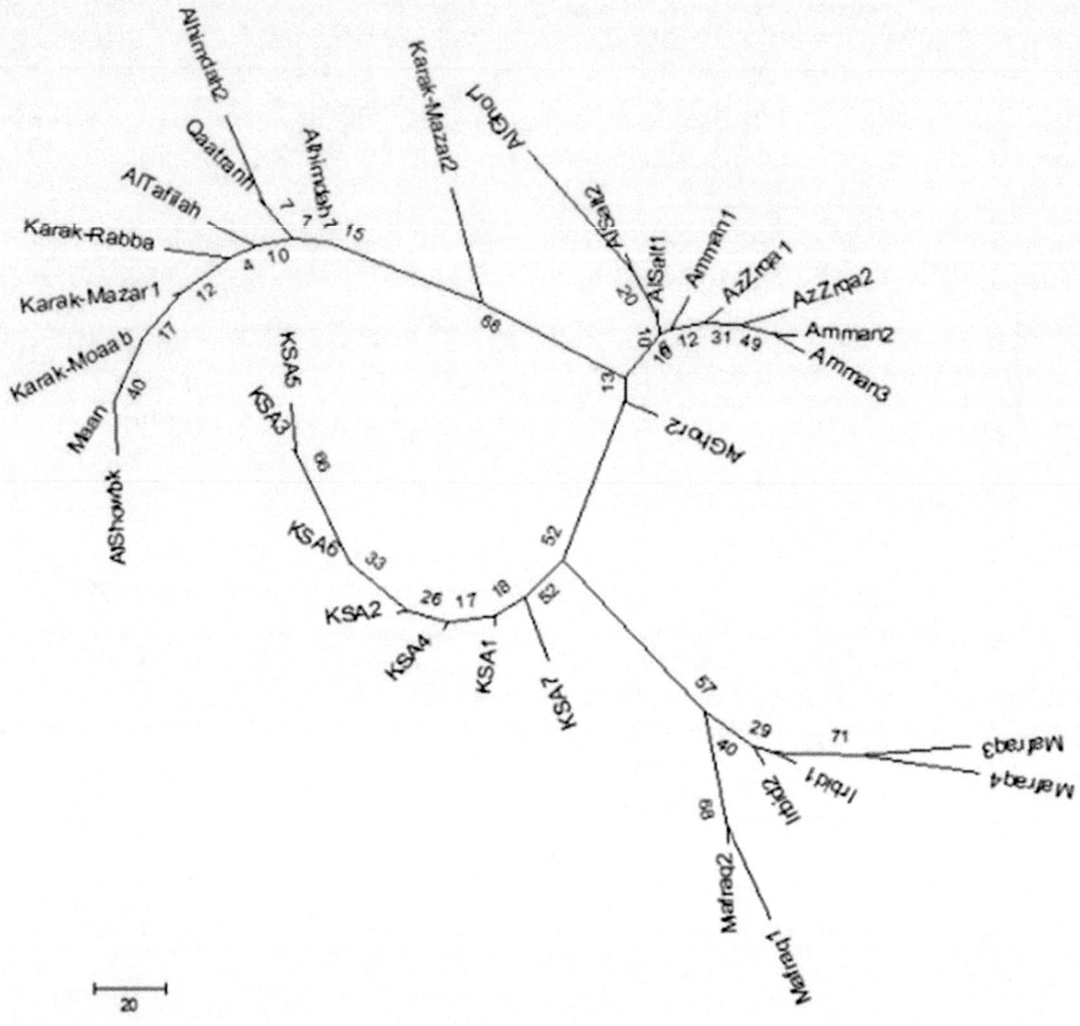

Fig. 5 Evolutionary relationships of taxa using the neighbor joining method

haplotype and 10 mutation steps away from the main haplotype. The main haplotype was observed in seven out of the 12 breeds studied. Different contributions of the breeds to different haplotypes were observed. One of the major findings was that Deccani and Ganjam breeds had largely a cluster of haplotypes which were isolated from the main expanding haplotypes (shaded area; Fig. 7a). When all Indian lineage A type sequences were clustered in two categories, that is, isolated haplotypes (shaded area; Fig. 7a) and the remaining sequences. AMOVA analysis reflected that 66.48 % of the total variation was explained by this classification, which ultimately confirms the divergence of the isolated haplotypes from the rest of the haplotypes of Indian sheep. In order to analyze the relationship of lineage A type haplotypes of Indian sheep with that of other regions of the world, reduced median network

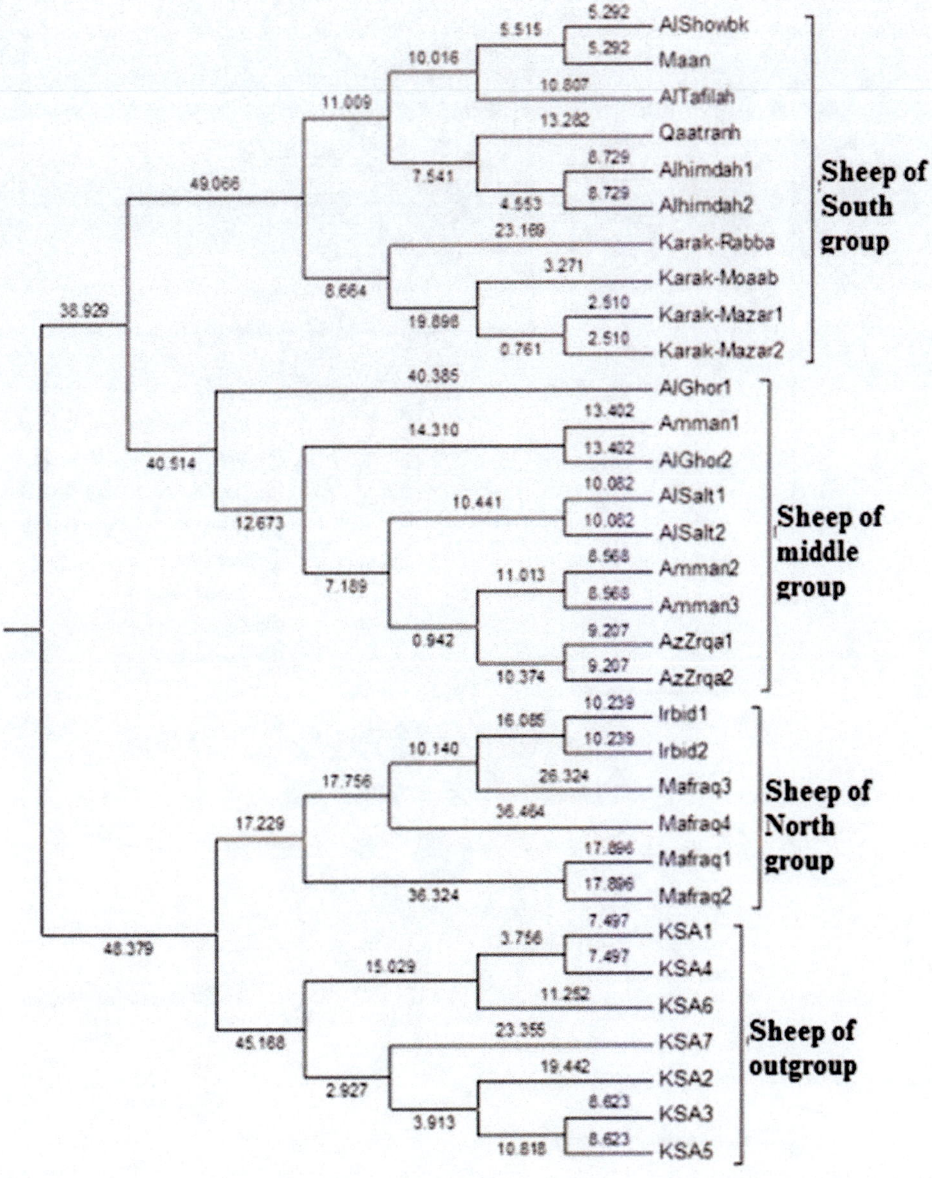

Fig. 6 Evolutionary relationship among the taxa using UPGMA tree among the sheep population studied [1]

analyses of 433 sequences including 279 Indian sequences were performed (Fig. 7b). Three major radiating nodes at a single mutation step denoting 19, 28, and 36 samples each, respectively, were observed in the network. In the first two smaller radiating nodes, samples from all geographic regions were clustered, while the largest node contains only Indian samples. It is interesting to note that, out of a total of 172 haplotypes, 118 were observed to be India specific. Four Indian haplotypes were observed to be shared with those of other regions of the world. It shows extreme diversity and isolation of Indian lineage A type haplotypes. The nucleotide

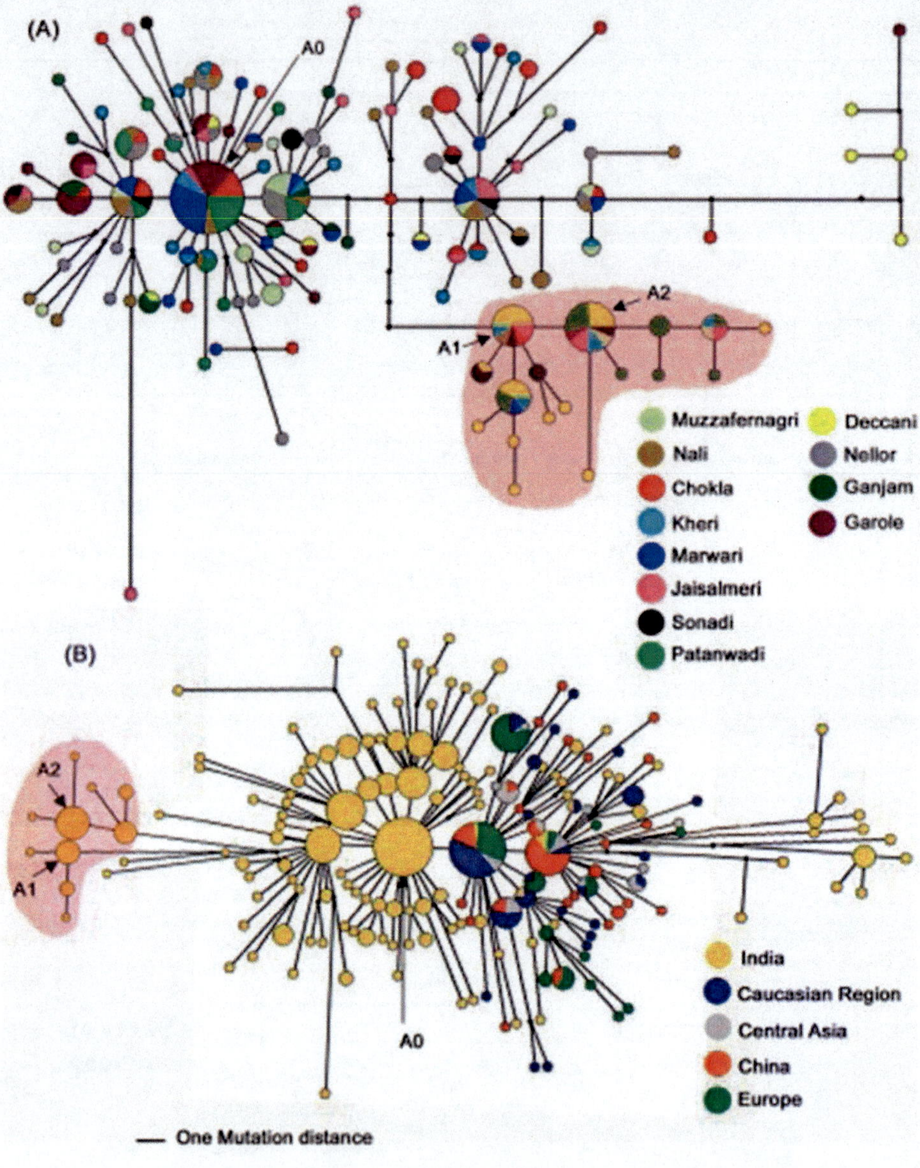

Fig. 7 Phylogenetic analysis among sheep population globally using reduced median network analysis. (**a**) Refers to phylogenetic tree with respect to different sheep breeds; (**b**) refers to phylogenetic tree with respect to geographical distribution of sheep [6]

diversity of lineage A in Indian sheep was observed to be 0.014 ± 0.007 compared to this value ranging from 0.009 ± 0.005 to 0.01 ± 0.005 in other regions.

2.8 Neighbor Joining Tree

It is a method of generation of phylogenetic tree with the neighbor joining method as depicted in Fig. 8.

Apart from the phylogenetic tree, population statistics is also important for understanding the evolution. The parameters studied under population statistics include haplotype, nucleotide diversity

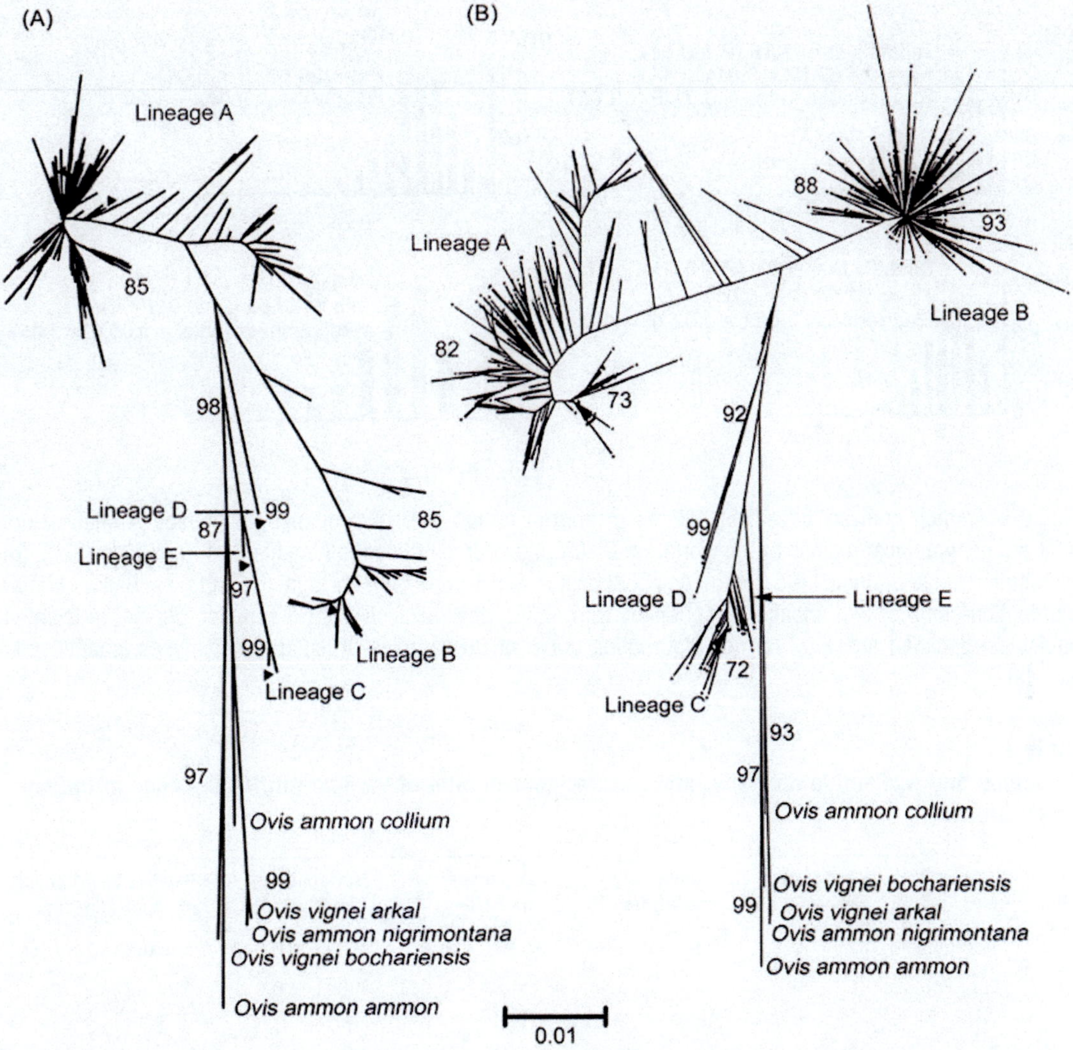

Fig. 8 Neighbor-joining tree of domestic sheep constructed with 432 bp of control region mtDNA. (**a**) It represents the neighbor-joining tree of Indian sheep (330) based on mtDNA sequences and the representative samples of five lineages (▲), such as A, B, C, D, and E. Indian sheep reveal three lineages, namely, A, B, and C. (**b**) It represents the neighbor-joining tree of mtDNA sequences of the Indian (330), Chinese (129), Central Asian, Caucasian and European (406), Portuguese (161), and West Balkan (60) sheep with the representative samples of five lineages (▲), such as A, B, C, D, and E. The sequences of wild Ovis species have been regarded as outgroups. MEGA5 version 5.0.1.102 was used for the construction of the trees using Tamura-Nei model with 10,000 bootstrap. Numbers above a given branch give the bootstrap support for the branch as a percentage out of 10,000 re-samplings [4]

and its standard error, Fu's Fs statistics, mismatch distribution, AMOVA (analysis of molecular variance), and the population pairwise differences (FST) values. Arlequin version 3.5.1.2 software was observed to be efficient in the calculation of these parameters. Figure 9 reflects the diagrammatic illustration of such parameters (Table 1).

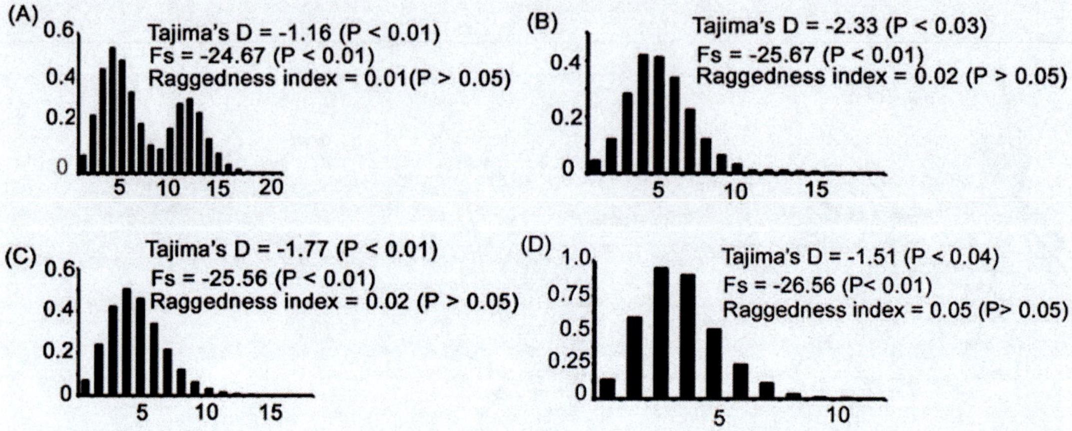

Fig. 9 Mismatch analysis based on 432 bp of control region mtDNA sequences if lineage A sheep from different geographical regions of the world. (**a**) Distribution curve of lineage A sequences from India (279). (**b**) Distribution curve of China (46), Central Asia, Caucasian and Europe (98), Portugal (6), and West Balkan (4). (**c**) Distribution curve of 223 lineage A sequences from India, after exclusive of 56 samples present in isolated haplotypes (shaded are; Fig. 4a). (**d**) Distribution curve of 56 samples of isolated haplotypes (shaded are; Fig. 4a)

Table 1
Haplotype and nucleotide diversity, and relative contribution of various mtDNA lineages in Indian sheep breeds

Sr No.	Breed	Number of samples	Number of haplotypes	Haplotype diversity	Nucleotide diversity	Relative contribution of various lineages
1	Muzzafarnagri	22	18	0.980 ± 0.021	0.022 ± 0.012	A = 0.83; B = 0.16
2	Nali	31	26	0.989 ± 0.011	0.021 ± 0.011	A = 0.87; B = 0.10; C = 0.03
3	Chokla	38	30	0.980 ± 0.013	0.022 ± 0.012	A = 0.85; B = 0.15
4	Kheri	29	26	0.993 ± 0.011	0.031 ± 0.016	A = 0.66; B = 0.34
5	Marwari	31	21	0.918 ± 0.043	0.018 ± 0.010	A = 0.88; B = 0.12
6	Jaisalmeri	25	23	0.993 ± 0.013	0.030 ± 0.016	A = 0.80; B = 0.12; C = 0.08
7	Sonadi	29	21	0.974 ± 0.015	0.036 ± 0.018	A = 0.52; B = 0.48
8	Patanwadi	20	14	0.916 ± 0.055	0.015 ± 0.008	A = 0.90; B = 0.10
9	Deccani	28	23	0.979 ± 0.018	0.015 ± 0.008	A = 1.0
10	Nellore	28	25	0.992 ± 0.012	0.015 ± 0.008	A = 0.93; B = 0.07
11	Ganjam	23	19	0.980 ± 0.020	0.016 ± 0.009	A = 1.0
12	Garole	26	14	0.920 ± 0.034	0.020 ± 0.011	A = 0.88; B = 0.12
		330	193	0.987 ± 0.003	0.025 ± 0.013	A = 0.84; B = 0.15; C = 0.009

A closely related topic is wild life conservation. In this current purview, since there is not sufficient scope for discussing this topic separately, it is being discussed in brief here. Now the question comes to our mind that what is the need for conservation?

There are certain terminologies pertaining to wild life conservation such as extinct, critically endangered, endangered, vulnerable, and at risk.

3 Effective Population Size

It is not sufficient to maintain a sizable population for wild animal or bird. The effective population size is calculated based on the population of both male and female. It is denoted by N_e.

N_e = 4 × Number of female × Number of male/(Number of males + Number of females).

$N_e = 4N_fN_m/(N_m + N_f)$.

3.1 How to Revive an Extinct Animal

It comes to our mind if we can revive any animal, which is already extinct with the recent advances in biotechnology.

If any ancient sample is available, (for example, any part of the fossil) from any cell, nucleus may be collected which may be inoculated in the living cell (ovum) of genetically related species. The cell may be multiplied to form a zygote and implanted accordingly. Thus, the individual may be revived. This is a hypothesis and perceived by the author. However, no references are available to date.

Gene editing is a similar technique for individual revival.

References

1. Al-Atiyat RM, Aljumaah RS (2014) Genetic distances and phylogenetic trees of different Awassi sheep populations based on DNA sequencing. Genet Mol Res 13(3):6557–6568. https://doi.org/10.4238/2014
2. Excoffier L, Lischer HE (2010) Arlequin suite ver 3.5: a new series of programs to perform population genetics analyses under Linux and Windows. Mol Ecol Resour 10:564–567. https://doi.org/10.1111/j.1755-0998.2010.02847.x. PubMed: 21565059
3. Fu YX (1997) Statistical tests of neutrality of mutations against population growth, hitchhiking and background selection. Genetics 147:915–925. PubMed: 9335623
4. Liu J, Ding X, Zeng Y, Guo X, Sun X, Yuan C (2018) Phylogenetic evolution and phylogeography of Tibetan sheep based on mtDNA D-loop sequences. Mitochondrial DNA-New insight. https://doi.org/10.5772/intechopen.76583
5. Schneider S, Excoffier L (1999) Estimation of past demographic parameters from the distribution of pairwise differences when the mutation rates vary among sites: application to human mitochondrial DNA. Genetics 152:1079–1089. PubMed: 10388826
6. Singh S, Kumar S Jr, Kolte AP, Kumar S (2013) Extensive variation and sub structuring in lineage A mtDNA in Indian sheep: genetic evidence

for domestication of sheep in India. PLoS One 8 (11):e77858

7. Pal A, Pal A Jr, Mallick AI, Biswas P, Chatterjee PN (2019) Molecular characterization of Bu- 1 and TLR2 gene in Haringhata Black chicken. Genomics 112(1):472–483. https://doi.org/10.1016/j.ygeno.2019.03.010
8. Thompson JD, Higgins DG, Gibson TJ (1994) CLUSTAL W: improving the sensitivity of progressive multiple sequence alignment through sequence weighting, position-specific gap penalties and weight matrix choice. Nucleic Acids Res 22(22):4673–4680. https://doi.org/10.1093/nar/22.22.4673
9. Edgar RC (2004) MUSCLE: multiple sequence alignment with high accuracy and high throughput. Nucleic Acids Res 32(5):1792–1797. https://doi.org/10.1093/nar/gkh340

Chapter 9

Real-Time or Quantitative PCR

Abstract

Real-time PCR is employed for the estimation of differential mRNA expression profile of genes. It is the quantitative expression of a series of genes from cDNA. The basic principle behind the technique is the progressive recording of PCR-based DNA amplification by continuous recording change in florescence. Real-time PCR primers are designed by aligning gene sequences of several mammals. QPCR technique may involve either Sybergreen chemistry or Taqman probing. The former is simpler to analyze, while later gives more accurate results. QPCR study is conducted from cDNA derived from different sources under study. Equal amount of RNA (quantified by Qubit fluorometer, Invitrogen), wherever applicable, are used for cDNA preparation (Superscript III cDNA synthesis kit; Invitrogen). All qRT-PCR reactions are conducted on ABI 7500 fast system. Each reaction consisted of 2 μL cDNA template, 5 μL of 2× SYBR Green PCR Master Mix, 0.25 μL each of forward and reverse primers (10 pmol/μL), and nuclease-free water for a final volume of 10 μL. Each sample is run in duplicate or triplicate. Analysis of real-time PCR (qRT-PCR) is performed by delta-delta-Ct (ΔΔCt) method . Expression level of mRNA is detected by $2^{-\Delta\Delta Ct}$

Keywords Real-time PCR, Quantitative PCR, Florescence, Differential mRNA expression profile, Delta-delta-Ct (ΔΔCt) method, Sybergreen chemistry, Taqman probe, Melting curve, Amplification plot

1 Principles

It is used for measuring the differential mRNA expression profile of genes. The method is involved with the quantitative expression of a series of genes from cDNA. The most basic principle involved with the technique is the progressive recording of PCR-based DNA amplification through a process of continuous recording change in florescence.

2 Protocol

1. Real-time PCR primers are designed through the process of alignment of gene sequences of several mammals including cow, pig, mouse, buffalo, and human with suitable software.

Aruna Pal, *Protocols in Advanced Genomics and Allied Techniques*, Springer Protocols Handbooks,
https://doi.org/10.1007/978-1-0716-1818-9_9,

2. QPCR is studied from cDNA obtained from different organs collected from slaughter house during the process of slaughter.
3. Care should be taken to use equal amount of RNA (quantified by nanodrop) as cDNA preparation.
4. All qRT-PCR reactions are conducted on Q-PCR from various globally reputed companies as from Applied Biosystem, Roche, Biorad, etc.
5. Each reaction consisted of 2 μL of cDNA template, 5 μL of 2× SYBR Green PCR Master Mix, 0.25 μL each of forward and reverse primers (10 pmol/μL) and nuclease-free water for a final volume of 10 μL.

Each sample is run in triplicate. Analysis of real-time PCR (qRT-PCR) is performed by delta-delta-*Ct* ($\Delta\Delta Ct$) method.
Expression level = $2^{-\Delta\Delta Ct}$.
The process for estimation of $\Delta\Delta Ct$.
Probe (e.g., Taqman probe) is commonly applicable.

3 Common Parameters of Primer Design

3.1 Primer Length

The recommended length of primers is 18–24 bp in length. Longer primers will require more time for hybridization and extension, leading to production of less amplicon.

3.2 Primer Melting Temperature (T_m)

This is the temperature when 50% of the primer and its complements are being hybridized. Optimization for qPCR requires primers of minimal length having melting temperatures (T_m) between 59 and 68 °C, with an optimal T_m of 63–64 °C. Also, the T_m of the primer pair needs to be within 1 °CC of each other. The primers must also have a T_m higher than the T_m of any template secondary structures designed with suitable software.

3.3 Annealing Temperature

Optimal real-time PCR annealing temperatures for any real-time PCR are 59 or 60 °C.

3.4 Product Size

An ideal amplicon would be between 80 and 150 bp. In case multiple genes are employed, (while comparing the relative expression of several genes), the size of all amplicons should be close in length. SYBR® Green detection will produce a more intense fluorescence in larger products than smaller; multiple products length should be closer.

3.5 Mg^{2+} Concentration

The default is adjusted to zero on most primer design software. SYBR® Green buffer mixes contain 3–6 mM of $MgCl_2$.

3.6 Repeats

A repeat is a nucleotide sequence (a dinucleotide) that is repeated (e.g., TCTCTCTCTC). These must be avoided as they are one of the major reasons behind mispriming. When the repeat is unavoidable, the maximum number should be restricted to 4 dinucleotides.

3.7 Runs

Runs are repeated nucleotides (e.g., TAAAAAGC has a 5 bp run of Adenine). Runs need to be avoided as they may lead to mispriming. The maximum run allowed is 3–4 bp.

3.8 3′ Stability

It refers to the maximum ΔG of the five bases from the 3′ end of the primers (ΔG is the Gibbs free energy, the energy needed to break the bonds present at the 3′ end). A greater 3′ stability will increase the efficiency of the primer.

3.9 GC Clamp

This implies to the maximum ΔG of the five bases from the 5′ end of the primers. Often known as a GC clamp, the 5′ stability implies the degree of stability at the 5′ end is due to the amount of Gs or Cs present at the 5′ end of the primer. It is ideal to have 1–2 GC clamps, which allows the primer to bind strongly to the template strand, leading it to be more specific. It is advisable to avoid more than 2 GC clamps.

3.10 Graphs: Melt Curve ABI

Note: The two ways to represent RT-PCR data are:

1. Absolute expression.
2. Relative expression.

Absolute quantification: The expression represents the exact copy number after transformation of data through standard curve and data represented as copy number per cell or per μg RNA.

Relative quantification: The real-time PCR data are presented relative to another gene designated as internal control.

Figure 1 reflects the melting curve for muc2 gene with respect to 18S rRNA (regarded as housekeeping gene).

Figure 2 reflects the melting peak for two different genes.

Real-time PCR with Taqman probe is also useful for SNP detection.

Steps for Real-Time PCR

1. Ready for plate-reaction set up. Plate preparation manually.
2. Switch on the machine and the laptop.
3. Click the icon 7500 installed on the desktop.
4. The screen will display. Click on advanced set up.
5. Experiment properties will appear.
   Name the experiment. Xxxx.
   Select the option: Standard curve.

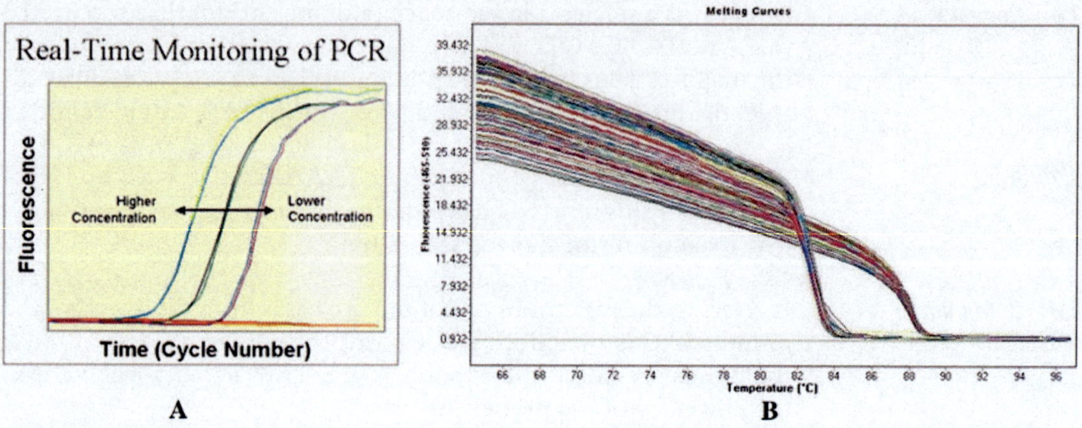

Fig. 1 Melting curves of *18S rRNA* and *muc2* in all 20 tissues. The melting curves of both gene are shown by their names. (**a**) Represents the schematic figure; (**b**) represents the actual image of graph obtained from real-time PCR

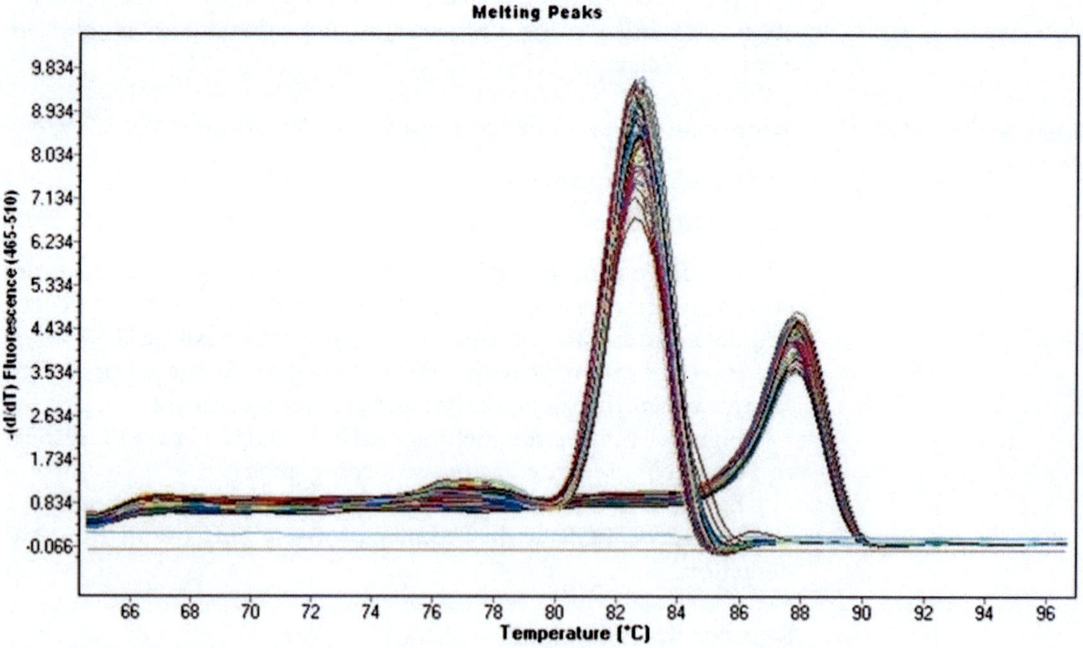

Fig. 2 Melting peak for two different genes

The options will appear as Genotyping, Standard curve. . . . .

Select the dye: Taqman, Sybergreen, ...

6. Plate set up:

   Choose the target. Target name is to be given.

   Assign the target name corresponding to each well.

7. Go to run method: Assign the reaction conditions.

   Three stages will appear: Holding stage, Cycling stage, Melt curve stage.

   Holding Stage:

   Holding stage: 95 °C × 10 min.

   Cycling stage: 95 °C × 30 s.

   60 °C × 30 s.

   72 °C × 30 s.

   Extension: 72 °C × 30 s.

   Melt Curve Stage:

   95 °C × 15 s.

   60 °C × 1 min.

   95 × 30 s.

   60 × 15 s.

   Two options appear under: Continuous and Step and Hold.

   Select the option "continuous."

8. Click the "Start run" option. The machine will start.
9. Now three graphs will appear on the screen:
   (a) Amplification Plot: The graph will appear as sigmoid curve.
   (b) Temperature Plot:
   (c) Melting curve.
10. Follow the default option.

Until now, we were describing the working protocol. In the subsequent section of the chapter, we shall describe the principles and detail trouble shooting protocols.

4 Basics of Real-Time PCR

4.1 Introduction

The polymerase chain reaction (PCR) has been regarded as one of the most powerful technologies in molecular biology. PCR has already been discussed in Chapter 2, Subheading 1.

The basics for PCR are that using PCR, specific sequences within a DNA or cDNA template can be copied, or "amplified," many thousand- to a million-fold using sequence-specific oligonucleotides, heat-stable DNA polymerase, and thermal cycling.

There are certain differences between traditional PCR and real-time PCR. In case of traditional (endpoint) PCR, we realize the detection and quantification of the amplified sequence through a series of analysis involving gel electrophoresis and image analysis through gel documentation. Quantification can equally be achieved

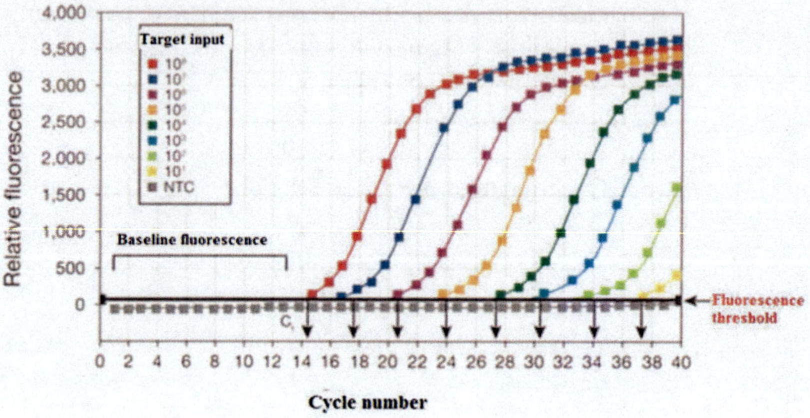

Fig. 3 Relative fluorescence versus cycle number

through nanodrop technique (spectrophotometry). It is to be noted that the entire process of detection and quantification can only be achieved after completion of PCR process. On the contrary, in case of real-time PCR, the amplified product can be measured at each cycle through the process of fluorescent detection with the help of Sybergreen or Taqman dye. Monitoring of reactions are achieved during the exponential–amplification phase of the reaction, accordingly the users can determine the initial quantity of the target with great precision. At present, endpoint PCR is employed to amplify specific DNA for sequencing, cloning, and use in other molecular biology techniques. As discussed, in real-time (quantitative) PCR, the amount of DNA is measured after each cycle via fluorescent dyes, and the intensity of fluorescent signal increases in direct proportion to the number of PCR product molecules (amplicons) generated.

Fluorescent reporters employed in real-time PCR involve double-stranded DNA (dsDNA)–binding dyes or dye molecules tagged to PCR primers or probes hybridizing with PCR products at the process of PCR amplification. The variation in the intensity in fluorescence over the course of the reaction is assessed by an instrument that combines thermal cycling with fluorescent dye scanning ability. Graph represents the plotting of fluorescence against the cycle number. The real-time PCR instrument leads to an amplification plot that is indicative of the accumulation of product over the duration of the entire PCR reaction. Figure 3 depicts the graph of relative fluorescence versus cycle number as detected in QPCR.

4.2 The Advantages of Real-Time PCR

1. It has the capability for monitoring the progress of the PCR reaction as it occurs over time.
2. It has the ability to precisely estimate the amount of amplicon at each cycle, which aids in exact quantification of the amount of starting material in samples.

3. Real-time PCR possesses an increased dynamic range of detection.
4. The greatest advantage of using this technique is that in this case, amplification and detection of amplified product occur in a single tube, unlike traditional PCR. It eliminates post-PCR manipulations. Currently, real-time PCR may be regarded as the leading tool for the identification and quantification of DNA or RNA. Detection can be achieved with accuracy within a twofold range, with a dynamic range of input material covering 6–8 orders of magnitude.

Amplification plots are developed where the fluorescent signal from each sample is plotted against cycle number; hence, amplification plots depict the accumulation of product over the duration of the real-time PCR experiment. The samples employed to prepare the plots in this figure are a dilution series of the target DNA sequence.

4.3 Overview of Real-Time PCR

Real-time PCR is an advancement of the standard PCR technique and is useful for quantification of DNA or RNA in a sample. With the help of sequence-specific primers, the number of copies of a particular DNA or RNA sequence is assessed. Quantification of the amount of amplified product is attained through each stage of PCR cycle. When a particular sequence (DNA or RNA) is abundant in the sample, amplification is visualized in earlier cycles. However, when the sequence is scarce, amplification is visualized in later cycles. Quantification of amplified product is conceived employing fluorescent probes or fluorescent DNA-binding dyes. Real-time PCR instruments are employed measuring fluorescence during the thermal cycling needed for the PCR reaction [1, 2].

4.4 Real-Time PCR Steps

There are three major steps running for 35–40 cycles.

1. Denaturation: It is a process when the double-stranded DNA is broken to single stranded due to breakage of hydrogen bonds, known as melting of DNA. The temperature used for denaturation is 95 °C. The denaturation time need to be increased if template GC content is high.
2. Annealing: Annealing is the process when complementary sequences hybridize. An appropriate temperature is employed, which is calculated based on the melting temperature (T_m) of the primers (typically 5 °C below the T_m of the primer).

   Melting temperature or T_m is calculated as

$$T_m = 4\,(G + C) + 2(A + T)$$

   where G, C, A, and T correspond to number of bases guanine, cytosine, adenine, and thymine bases.

3. Extension: The optimum temperature for DNA polymerase to act is at 70–72 °C. Hence, primer extension happens at rates of up to 100 bases per second. In case, an amplicon in real-time PCR is small, extension step may be easily combined with the annealing step, using 60 °C as the temperature.

4.5 Real-Time PCR May Be Grouped as of Two Types

1. *Two-Step qRT-PCR*.

   As the name indicates, first step involves two stages. The first step involves first strand cDNA preparation with reverse transcriptase and primed by oligo DT or random primers from total RNA or poly(A) RNA. The temperature employed for cDNA synthesis depends on the RT enzyme selected. After reverse transcription process, about 10% of the cDNA needs to be transferred to a separate tube for the real-time PCR reaction.

2. **One-Step qRT-PCR**.

   As the name implies, one-step qRT-PCR involves both the first-stranded cDNA synthesis reaction and real-time PCR reaction in the same tube, simplifying reaction setup. This process leads to the reduction of the possibility of contamination. Final reaction will be governed by gene-specific primers (GSP).

4.6 Overview of Real-Time PCR Components

It is very important to discuss about the components for real-time PCR.

1. DNA polymerase:

   Selection of thermostable DNA polymerase is critical for PCR performance. One of the main issues related to PCR specificity is that Taq DNA polymerase has residual activity at low temperatures, which implies that Primers can anneal nonspecifically to DNA during reaction setup. As a result, the polymerase synthesizes nonspecific product. This problem of production of nonspecific products resulting from mis-priming can be reduced through a “hot-start” enzyme. A hot-start enzyme for PCR reaction is effective to ensure that DNA polymerase (Taq) always is not effective for reaction setup and the initial DNA denaturation step.

2. Reverse transcriptase:

   Reverse transcriptase is chosen in such a way so that it has the ability to generate high yields of full-length cDNA that too at high temperature. High-temperature performance is also essential for denaturation of RNA with secondary structure.

3. dNTP.
4. Magnesium.

   It is advisable to purchase both the dNTPs and the thermostable DNA polymerase from the common vendor. However, it is very common to see a loss in sensitivity of one full threshold cycle (Ct) in experiments that employ these reagents from separate vendors or manufacturers..

In real-time PCR, magnesium chloride or magnesium sulfate is generally utilized at a final concentration of 3 mM. This concentration is effective for most targets; however, the optimal magnesium concentration lies between 3 and 6 mM.

5 Good Experimental Technique

It is very much essential to follow scientific experimental technique. It is advisable to use standard dedicated instruments. Solutions should be perfectly prepared for each stage of the cycle from sample reparation to post PCR analysis.

It is recommended to use of aerosol-barrier tips and screw cap tubes, which can help decrease cross-contamination problems. The ideal practice is the preparation of a master mix that contains all the reaction components except sample. This will reduce the number of pipetting steps and, consequently, reduces the chances of cross-well contamination and other pipetting errors.

Template: Using 10–1000 copies of template nucleic acid for each real-time PCR reaction is absolutely important. Approximately 100 pg to 1 μg of genomic DNA or cDNA generated from 1 pg to 100 ng of total RNA is generally employed. Excess template may also bring higher contaminant levels that can significantly decrease PCR efficiency. Depending on the specificity of the PCR primers for cDNA rather than genomic DNA, it may be important to treat RNA templates to reduce the possibility that they contain genomic DNA contamination. One alternative way is to treat the template with DNase I. Pure, intact RNA is necessary for full-length, high-quality cDNA synthesis and may be important for accurate mRNA quantification. RNA should be lack of any RNase contamination, and aseptic conditions should be maintained. Total RNA typically works well in qRT-PCR; isolation of mRNA is typically not necessary, although it may progress the yield of specific cDNAs.

6 Real-Time PCR Primer Design

Good level of primer design is one of the most important parameters in real-time PCR. That is why many researchers prefer to purchase TaqMan® Assay products—primers and probes for real-time PCR designed using a proven algorithm and trusted by scientists around the world. While designing the primer, keep in mind that the amplicon length should be approximately 50–150 bp, since longer products do not amplify as efficiently. In general, primers should be 18–24 nucleotides in length, perfect for annealing temperatures. Primers should be designed according to standard PCR guidelines including target sequence and be free of internal secondary structure. Primers should keep away from stretches of homopolymer sequences (e.g., poly(dG)) or repeating motifs, as these can hybridize inappropriately. Primer pairs should have exact melting temperatures (within 1 °C) and contain approximately 50% GC content. Primers with high GC content can form stable defective hybrids. Conversely, high AT content depresses the T_m of completely matched hybrids.

For qRT-PCR, design primers anneal to exons on both sides of an intron (or span an exon/exon boundary of the mRNA) to allow separation between amplification of cDNA and potent contaminating genomic DNA through melting curve analysis. To confirm the specificity of your primers, perform a BLAST® search against public databases to be sure that your primers only identify the target of interest. For optimal results a titration is often required. A final concentration of 200 nM for each primer is successful for most reactions.

6.1 Primer Design Software

Primer design software programs, such as OligoPerfect™ designer and Primer Express® software, in addition to sequence analysis software, such as Vector NTI® Software, can automatically evaluate a target sequence and design primers, according to the criteria discussed earlier. At a minimum, using primer design software will ensure that primers are specific for the target sequence and lack of internal secondary structure and avoid complementary hybridization at 3' ends within each primer and with each other. As mentioned previously, good primer design is especially essential when using DNA-binding dyes for amplicon detection.

7 Real-Time PCR Analysis Technology

The baseline of the real-time PCR reaction refers to the signal level during the preliminary cycles of PCR, usually cycles 3–15, in which there is little change in fluorescent signal. The low-level signal of the baseline can be equated to the environment or the "noise" of the reaction (Fig. 4). The baseline in real-time PCR is determined by user analysis or automated analysis of the amplification plot. The baseline should be set carefully to allow exact determination of the threshold cycle (Ct) defined below. The baseline determination needs to take into account enough cycles to remove the background found in the early cycles of amplification. It should not include the cycles in which the amplification signal begins to rise above background. When comparing different real-time PCR reactions or experiments, the baseline should be defined in the equal way for each (Fig. 4).

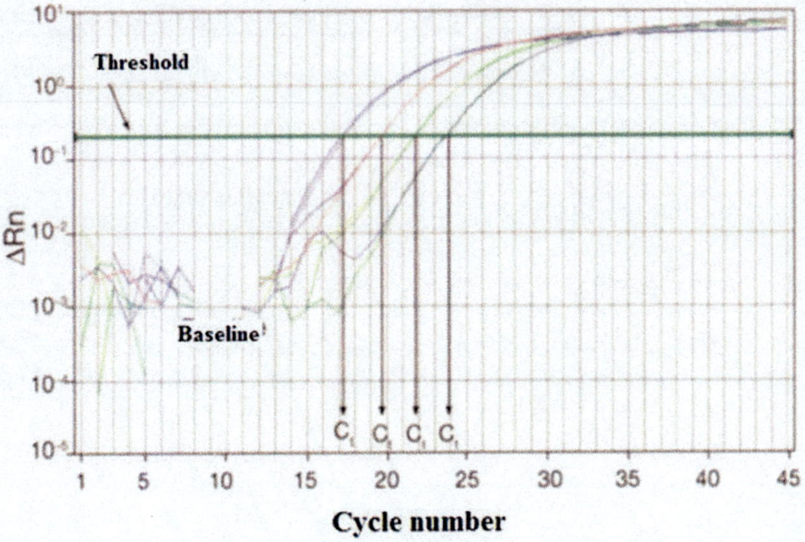

Fig. 4 The baseline and threshold of a real-time PCR reaction

7.1 Threshold

The threshold of the real-time PCR reaction is the level of signal that reflects a statistically significant enhance over the calculated baseline signal (Fig. 4). It is set to differentiate relevant amplification signal from the background. By default, real-time PCR instrument software automatically sets the threshold at 10 times the standard deviation of the fluorescence value of the baseline. Figure 5 represents the amplification plot for a tenfold dilution series in quantitative PCR. However, the location of the threshold can be adjusted at any point in the exponential phase of PCR.

7.1.1 Ct (Threshold Cycle)

The *Ct* is used to calculate the initial DNA copy number because the Ct value is related to the starting amount of target. As for example, while comparing real-time PCR results from samples possessing different amounts of target, a sample with twice the starting amount will produce a *Ct* one cycle earlier than a sample with twice the number of copies of the target in relation to a second sample. As a result, the PCR is functional at 100% efficiency (i.e., the amount of product doubles during each cycle) in both reactions.

With a tenfold dilution series, the *Ct* values are ~3.3 cycles apart thus the standard curve. A dilution series of known template concentrations can be employed to have a standard curve for the determination of the initial starting amount of the target template in experimental samples or for the assessment of the reaction efficiency (Fig. 6). The log of each known concentration in the dilution series (*x*-axis) is plotted against the *Ct* value for that concentration to understand it clearly.

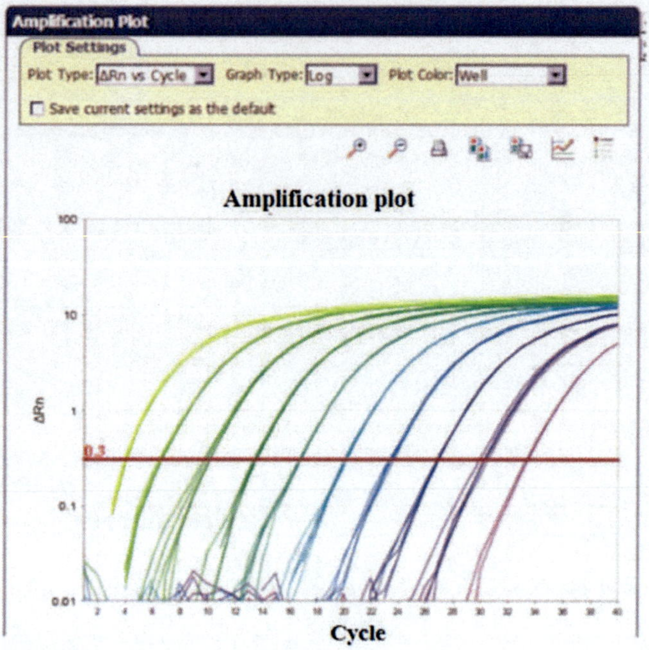

Fig. 5 Amplification plot for a tenfold dilution series

From this standard curve, information can be obtained about the performance of the reaction. Similarly, information are equally obtained about different reaction parameters (such as slope, *y*-intercept, and correlation coefficient). The concentrations selected for the standard curve would include the expected concentration range of the target in the experimental samples.

Correlation coefficient (R^2): The correlation coefficient is defined as a measure of how well the data fit the standard curve. The R^2 value shows the linearity of the standard curve. Ideally, $R^2 = 1$, although 0.999 is generally the maximum value.

***Y*-intercept**: The *y*-intercept refers to the theoretical limit of recognition of the reaction or the Ct value expected whenever the lowest copy number of target molecules denoted on the *x*-axis leads to statistically significant amplification. Though PCR is theoretically able to identify a single copy of a target, a copy number of 2–10 is generally specified as the lowest target level that can be reliably quantified in real-time PCR applications. It possesses the limitations to the effectiveness of the *y*-intercept value as a direct measure of sensitivity. However, the *y*-intercept value can be valuable for the comparison of different amplification systems and targets.

Exponential Phase: It is essential for the quantification of our real-time PCR reaction in the early part of the exponential phase being different to in the later cycles or whenever the reaction reaches the plateau. At the beginning of the exponential phase, all

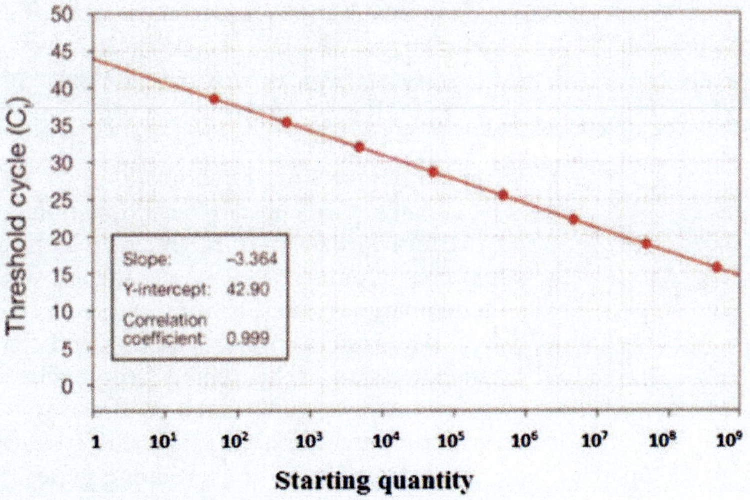

Fig. 6 Example of a standard curve of real-time PCR data

reagents are in sufficiently high quantity. The DNA polymerase remains highly efficient, and the amplification product is present in a small amount, which will not compete with the primers' annealing capabilities.

Slope: The slope of the log-linear phase of the amplification reaction is a measure of reaction efficiency accurately. To obtain accurate and reproducible data, reactions would have an efficiency as close to 100% as possible, equivalent to a slope of −3.32.

Efficiency: We can determine the PCR efficiency of 100% corresponds to a slope of −3.32, as described by the following equation: Efficiency = 10(−1/slope) − 1 generally, the efficiency (E) of a PCR reaction will be 100%, which means the template is doubled following each thermal cycle for the exponential amplification. The actual efficiency can give important information about the reaction. The length, secondary structure, and GC content of these experimental factors content of the amplicon can influence efficiency. Few conditions influencing the efficiency are the dynamics of the reaction, the utilization of nonoptimal reagent concentrations, and enzyme quality lead to efficiencies below 90%. The presence of PCR inhibitors in one or more of the reagents can lead to the efficiencies of greater than 110% in major cases.

Dynamic range: This is the range above which there exists raise in starting material concentration leading to a corresponding increase in amplification product. In ideal conditions, the dynamic range for real-time PCR would be 7–8 orders of size for plasmid DNA and at least a 3–4 log range for cDNA or genomic DNA.

Absolute quantification: Absolute quantification explains a real-time PCR experiment where samples of known quantity are serially diluted followed by amplification to create a standard curve.

Unknown samples are later on quantified by comparison with this curve.

Relative quantification: Relative quantification explains a real-time PCR experiment where the expression of a gene of interest in one sample (treatment) is compared to expression of the same gene in a different sample (control).

A standard curve reflects threshold cycle (*Ct*) on the *y*-axis and the initial quantity of RNA or DNA target on the *x*-axis. Slope, *y*-intercept, and correlation coefficient values are utilized to help the information related to the performance of the reaction.

The results are measured as fold change (increase or decrease) in expression of the treated sample related to the untreated sample. A normalizer gene (such as β-actin) is utilized as a control for experimental variability for this type of quantification, which is also referred to as housekeeping gene.

7.2 Melting Curve (Dissociation Curve)

A melting curve reflects the change in fluorescence seen when double-stranded DNA (dsDNA) with incorporated dye molecules dissociates ("melts") into single-stranded DNA (ssDNA) when the temperature of the reaction is raised.

As an example, when double-stranded DNA bound with SYBR® Green I dye is heated, a sudden fall in fluorescence is detected when the melting point (T_m) is reached because of dissociation of the DNA strands and subsequent release of the dye.

The fluorescence is plotted against temperature, followed by the plotting of the $-\Delta F/\Delta T$ (change in fluorescence/change in temperature) against temperature to achieve a clear view of the melting dynamics. Post-amplification melting-curve analysis is a simple, uncomplicated process to check real-time PCR reactions for primer-dimer artifacts to ensure reaction specificity. Since the melting temperature of nucleic acids is affected by length, GC content, and the presence of base mismatches, among other factors, different PCR products can regularly be differentiated through their melting characteristics. The characterization of reaction products (e.g., primer-dimers vs. amplicons) via melting curve analysis decreases the necessity for time-consuming gel electrophoresis. The typical real-time PCR data set shown in Fig. 7 describes many terms. Figure 5 visualizes a typical real-time PCR amplification plot. During the earlier cycles, PCR reaction leads to small change in the fluorescent signal. As the reaction continues, the level of fluorescence begins to enhance with each cycle. The reaction threshold is arranged over the baseline in the exponential portion of the plot. This threshold is utilized to assign the threshold cycle, or *Ct* value, of each amplification reaction. *Ct* values for a series of reactions possessing a known quantity of target can be utilized to generate a standard curve. Quantification is conducted by comparing *Ct* values for unknown samples against this standard curve or, in the case of relative quantification, against each other, with the

A

B

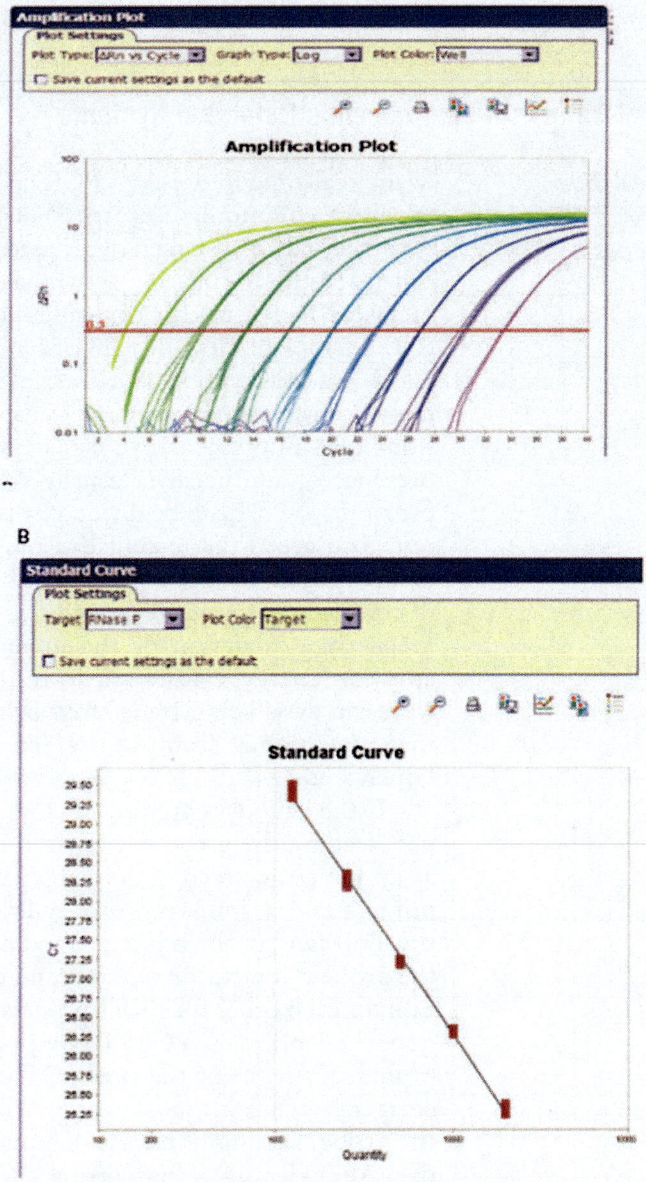

Fig. 7 Amplification of RNase P DNA starting from 1.25×10^3 to 2×10^4 copies. Real-time PCR of twofold serial dilutions of human RNase P DNA was performed using a FAM™ dye–labeled TaqMan® Assay with TaqMan® Universal Master Mix II, under standard thermal cycling conditions on a ViiA™ 7 Real-Time PCR System. (**a**) Amplification plot. (**b**) Standard curve visualizing copy number of template versus threshold cycle (*Ct*)

standard curve providing as an efficiency check. *Ct* values are inversely linked to the quantity of starting template: the higher the amount of starting template in a reaction, the lower the *Ct* value for that reaction. Figure 7 reflects the standard curve obtained from the *Ct* values in the amplification plot. The standard curve gives significant information regarding the amplification efficiency, replicate consistency, and theoretical detection limit of the reaction.

8 Real-Time PCR Fluorescence Detection Systems

8.1 Real-Time Fluorescent PCR Chemistries

Many real-time fluorescent PCR chemistries are present; however, the most commonly used are 5′ nuclease assays such as TaqMan® Assays and SYBR® Green dye–based assays. The 5′ nuclease assay is called for the 5′ nuclease activity associated with Taq DNA polymerase. The 5′ nuclease domain is able to breakdown DNA bound to the template, downstream of DNA synthesis.

A second key element in the 5′ nuclease assay is known as fluorescence resonance energy transfer (FRET). In FRET, the emissions of a fluorescent dye can be strongly decreased due to the presence of another dye, usually known as the quencher, in close proximity. FRET can be divided by two fluorescent dyes: green and red. The green fluorescent dye has a higher energy of release in comparison to the red because green light has a shorter wavelength in comparison to red. If the red dye is in close association with the green dye, excitation of the green dye will produce the green emission energy, transferred to the red dye. If described in other way, energy is being transferred from a higher to a lower level so that the signal from the green dye will be suppressed or "quenched."

If the two dyes are not in close proximity, FRET cannot happen, leading the green dye to produce its full signal. A 5′ nuclease assay for target recognition or quantification contains two PCR primers and a TaqMan® probe. Prior to the beginning of PCR, the TaqMan® probe is integral and has a degree of flexibility. While the probe is intact, the reporter and quencher have a natural attraction for each other for FRET process. The reporter signal is actually quenched prior to PCR. DNA polymerase enzyme extends the primer upstream of the probe. When the probe is bound to the correct target sequence, the polymerase's 5′ nuclease activity breaks the probe, leading to release of a fragment containing the reporter dye. After cleavage, the reporter and quencher dyes are no longer involved to each other; the released reporter molecule will no longer be quenched.

8.2 5′ Nuclease Assay Specificity

The basic principle for assay includes the signal from the target and excludes signal from nontarget as obtained from the results. Specificity is perhaps the most important aspect of any assay. The greatest risk to assay specificity for 5′ nuclease assays is homologs. Homologs are genes parallel in sequence to that of the target but they are not. Homologs are really common within species and across related species.

A mismatch between the target and homolog located at the 3′ most bases of the primer has maximal impact on the specificity of the primer. A match further missing from the 3′ end will have less impact on specificity. On the contrary, mismatches across most of the length of a TaqMan® MGB probe, which is shorter than a TaqMan® TAMRA™ probe, can have a strong force on specificity—TaqMan®. MGB probes are observed to be the stronger tools for specificity compared to that of primers.

As for example, a 1- or 2-base random mismatch in the primer binding site will let the DNA polymerase to act and allow extension with high efficiency. At that point, the DNA polymerase will continue its synthesis to produce a copy of the homolog. On the contrary, mismatches on the 5′ end of the TaqMan® probe binding site cannot be stabilized by the DNA polymerase because of the presence of the quencher block on the 3′ end. Mismatches in a TaqMan® MGB probe binding site will decrease the tightness of the probe bound, in such a way that instead of cleavage, the intact probe is displaced. The integral probe returns to its quenched configuration in such a manner that when data are collected at the end of the PCR cycle, signal is produced from the target but not from the homolog, although the homolog is being amplified. PCR may also cause amplification of nonspecific products produced by primers binding to apparently random locations in the sample DNA, popularly termed as "primer-dimers." Since nonspecific products are not related to the target, they do not have TaqMan® probe binding sites and thus are not visible in the real-time PCR data.

TaqMan® probe types TaqMan® probes may be categorized into two types: MGB and non-MGB. The first TaqMan® probes may be regarded as non-MGB. For this reason, TAMRA™ dye is used as the quencher. Early in the development of real-time PCR, extensive testing exposed that TaqMan® probes are essential for annealing temperature significantly higher than that of PCR primers to let the cleavage to take place. As a result, TaqMan® probes were longer than primers.

A 1-base mismatch in such long probes had a reasonably mild result on probe binding, permitting cleavage to take place. For many applications linking high genetic complexity, such as eukaryotic gene expression and single-nucleotide polymorphisms (SNPs), a higher degree of specificity was important. TaqMan® MGB probes were a later improvement of the TaqMan® probe technology. TaqMan® MGB probes acquire a minor-groove binding (MGB) molecule on the 3′ end. When the probe binds to the target, a short minor groove is produced in the DNA, leading the MGB molecule to bind and increase the melting temperature, thereby strengthening the probe binding. Subsequently, TaqMan® MGB probes can be quite shorter than PCR primers. Owing to the MGB moiety, these probes can be shorter than TaqMan® probes and still attain a high melting temperature. This provides TaqMan® MGB probes to bind to the target with better efficiency than primers at higher temperatures. With the smaller probe size, a 1-base mismatch has a much larger impact on TaqMan® MGB probe binding. And because of this higher level of specificity, TaqMan® MGB probes are advised for most techniques involving high genetic complexity.

TaqMan® probe. The TaqMan® probe has a gene-specific sequence and is estimated to bind the target between the two PCR primers. At the 5′ end of the TaqMan®, a fluorescent dye is attached that will report the amplification of the target, known as the "reporter." A "quencher" is attached at the 3′ end of the probe, which quenches fluorescence from the reporter in intact probes. In addition, the 3′ end of the probe is also blocked by the quencher such that thermostable DNA polymerase cannot act and proceed the extension process.

R is the reporter dye, and Q is the quencher molecule; TaqMan® probe signal production. Both an MGB or non-MGB probe follows the same pattern for signal production.

In the early PCR cycles, only the low, quenched reporter signal is identified, termed as "baseline". This early signal is automatically subtracted to zero in the real-time PCR software. As soon as the system detect that the sample contains a target, then enough of the cleaved probe will accumulate leading to amplification signal to emerge from the baseline. The point, at which amplification signal is detected, is inversely related to the initial target quantity.

SYBR ® Green dye SYBR® Green I dye is a fluorescent DNA-binding dye that binds to the minor groove of any double-stranded DNA. DNA-bound SYBR® Green dye produces a much intensive fluorescent signal due to excitation compared to unbound dye. A SYBR® Green dye–based assay generally comprises two PCR primers. Under ideal conditions, a SYBR® Green assay leads to an amplification pattern same as that of a TaqMan® probe–based assay. In the early PCR cycles, a horizontal baseline is visualized. In case the target was present in the sample, adequately accumulated PCR product will be produced at some point, thereby amplification signal becomes visible.

8.3 SYBR® Green Assay Specificity

Assay specificity testing is essential for all assays but especially for those most susceptible to specificity problems. SYBR® Green assays do not help from the specificity of a TaqMan® probe, leading them more prone to specificity problems. SYBR® Green dye will attach to any amplified product, target or non-target, leading to summing up of signals and production of a single amplification plot. SYBR® Green amplification plot shape cannot be utilized to consider specificity. Plots generally appear similar, irrespective of the amplification consisting of target, nontarget, or a mixture. The fact that a SYBR® Green assay shaped an amplification would not be automatically taken to assume that the majority of any of the signal is derived from target. Since amplification of nontarget can vary from sample to sample, at least one type of specificity measurement would be performed for every SYBR® Green reaction. Generally, this ongoing assessment is the dissociation analysis.

8.4 SYBR® Green Dye Dissociation

SYBR® Green dissociation is the gradual melting of the PCR products following the process of PCR when employing SYBR® Green–based identification. Dissociation is important for specificity assessment since it does not involve additional cost to the experiment and can be conducted in the PCR reaction vessel. However, dissociation needs more time to the thermal protocol, needs additional analysis time, and is somewhat subjective with limited resolution. The concept of SYBR® Green dissociation reflects that in case the target is one defined genetic sequence, it would have one specific melting temperature (T_m) useful to aid in the recognition of the target in samples.

Some nontarget products will have T_m values significantly dissimilar from that of the target, leading to the identification of those nontarget amplifications. The dissociation procedure is associated after the final PCR cycle.

Following the melting process, the real-time PCR software will plot the data as the negative first derivative that converts the melt profile into a peak. Accurate recognition of the target peak relies on amplification of pure target. Many samples, e.g., cellular RNA and genomic DNA show high genetic complexity, creating opportunities for nontarget amplification that may suppress the amplification of the target or, in other cases, change the shape of the melt peak.

By starting with pure target, the researcher will be able to combine a peak T_m and shape with a specific target after amplification. The presumptive target peak should be narrow, symmetrical, and lack of other anomalies, such as shoulders, humps, or splits. These anomalies are strong signs that multiple products of similar T_m values were generated, casting strong doubts related to the specificity of those reactions. Wells with dissociation anomalies would be absent from further analysis. SYBR® Green dissociation is low resolution and may not distinguish between target and nontarget with similar T_m values (e.g., homologs). Therefore, one, narrow symmetric peak would not be predicted to be the target, nor one product, without extra supporting information. Dissociation data would be analyzed for each well where amplification was observed. If the sample possesses a peak that does not match up to the pure target peak, it can be concluded that target was not detected in that reaction. In case the sample contains a peak which matches the T_m and shape of the pure target peak, target may have amplified in that reaction. Dissociation data cannot be taken as definitive when studied separately. However, the confidence improves in combination with other information, such as data from target-negative samples, sequencing, or gels.

Different models of real-time PCR machine are present in the market. Each model must have an excitation source, which excites the fluorescent dyes, and a detector to sense the fluorescent emissions. The basic function of each model should have a thermal cycler. The thermal block can be fixed (StepOnePlus™ system) or user interchangeable (ViiA™ 7 system, the QuantStudio® 6 and 7 Flex systems, and QuantStudio® 12K Flex system). Blocks are existing for the acceptance of a variety of PCR reaction vessels: 48-well plates, 96-well plates, 384-well plates, 384-microwell cards, 3072–through-hole plates, etc. All real-time PCR instruments are launched with software for data collection and analysis.

8.4.1 Dye Differentiation

Most real-time PCR reactions consist of multiple dyes, including one or more reporter dyes, a quencher dye, and a passive reference dye. Multiple dyes in the similar well can be estimated independently, either through optimized combinations of excitation and emission filters or through a process referred to as multicomponenting. Multicomponenting is a mathematical process for the measurement of dye intensity for each dye in the reaction. Multicomponenting offers the benefits of easy correction for dye tag errors, refreshing optical performance to factory standard devoid of hardware control and gives a source of troubleshooting information.

9 Melting Curve Analysis

9.1 Melting Curve Investigation and Detection Systems

Melting curve analysis can only be conducted with real-time PCR finding technologies in which the fluorophore remains connected with the amplicon. Amplifications that used SYBR® Green I or SYBR® GreenER™ dye can be subjected to melting curve analysis. Dual-labeled probe recognition systems like TaqMan® probes are not well-matched since they give rise an irreversible change in signal through cleavage and release of the fluorophore into solution during the PCR. Moreover, the increased specificity of this process causes this less of a concern. The level of fluorescence of both SYBR® Green I and SYBR® GreenER™ dyes significantly enhances upon binding to dsDNA. By monitoring the dsDNA as it melts, a decrease in fluorescence will be observed as soon as the DNA becomes single-stranded and the dye dissociates from the DNA.

9.2 Significance of Melting Curve Analysis

The specificity of a real-time PCR assay is identified by the primers and reaction conditions used. However, there is constantly the scope that even well-designed primers can form primer-dimers or amplify a nonspecific product. There is also scope while conducting qRT-PCR that the RNA sample possesses genomic DNA, which may also be multiplied. The specificity of the real-time PCR reaction can be established with melting curve analysis. When melting curve analysis is not possible, extra care need to be undertaken to establish the fact that differences observed in Ct values between reactions are valid and are not due to the presence of nonspecific products.

9.2.1 Melting Curve Analysis and Primer-Dimers

Primer-dimers occur when two PCR primers (either same-sense primers or sense and antisense primers) bind to each other than the target. Melting curve analysis can detect the presence of primer-dimers since they represent a lower melting temperature than the amplicon. The presence of primer-dimers is not worthwhile in samples containing template, since it decreases PCR effectiveness and thereby obscures analysis. The arrangement of primer-dimers generally evident in no-template controls (NTCs), where there is a great quantity of primer and no template. The existence of primer-dimers in the NTC would act as an alert to the user that they may also be available in reactions including template. If there are primer-dimers in the NTC, the primers need to be redesigned. Melting curve study of NTCs can differentiate between primer-dimers and spurious amplification because of contaminated nucleic acids in the reagent components.

9.3 How to Perform Melting Curve Analysis

To conduct melting curve analysis, the real-time PCR instrument can be programmed to comprise a melting profile instantly following the thermal cycling protocol. After amplification is complete, the instrument will reheat the amplified products for yielding a complete melting curve data. Most real-time PCR instrument platforms now include this feature into their analysis packages.

9.4 Passive Reference Dyes

Passive reference dyes are regularly used in real-time PCR to regularize the fluorescent signal of reporter dyes and correct for fluctuations in fluorescence that are not PCR-based. Standardization is essential for the correction for fluctuations from well to well resulted by changes in reaction concentration or volume and corrections for variations in instrument scanning. Most real-time PCR instruments utilize ROX™ dyes as the passive reference dye because ROX™ dye has no effect on the real-time PCR reaction and has a fluorescent signal that can be renowned from that of many reporters or quencher dyes used.

A passive reference dye, e.g., ROX™ dye is employed to standardize the fluorescent reporter signal in real-time PCR on compatible instruments like Applied Biosystems® instruments. The use of a passive reference dye is an efficient tool for the normalization of fluorescent reporter signal, without changing the instrument's default analysis parameters. TaqMan® real-time PCR master mixes possess a passive reference dye that serves as an internal control to:

- Standardize for non-PCR-related fluctuations in fluorescence (e.g., caused by variation in pipetting).
- Regularize for fluctuations in fluorescence resulting from machine "noise."
- Compensate for variations in instrument excitation and finding.
- Supply a stable baseline for multiplex real-time PCR and qRT-PCR.

Fluorescein reference dye Bio-Rad iCycler® instruments need the collection of "well factors" prior to each run to compensate for any instrument or pipetting nonuniformity. Well factors for experiments using SYBR® Green I or SYBR® GreenER™ dyes are planned using an additional fluorophore, fluorescein. Well factors are composed employing either a separate plate containing fluorescein dye in each well (external well factors) or the experimental plate with fluorescein spiked into the real-time PCR master combine (dynamic well factors). We must choose the method when we start each run using the iCycler® instrument. The iCycler® iQ™5 and MyiQ™ systems permit us to save the data from an external well factor reading as a separate file that can then be referenced for future readings.

10 Contamination Prevention

As with usual PCR, real-time PCR reactions can be affected by nucleic acid contamination, resulting in fake positive results.

Some of the possible sources of contamination are:

- Cross-contamination between the samples under study.
- Contamination may occur from laboratory equipment.
- Carryover contamination of amplification products and primers from previous PCRs. This is regarded as the major source of fake positive PCR results.

10.1 Uracil N-Glycosylase (UNG)

Uracil N-glycosylase (UNG) is employed to decrease or avoid DNA carryover contamination between PCR reactions through prevention of the amplification of DNA from previous reactions. The use of UNG in PCR reactions decrease fake positives, in turn rising the efficiency of the real-time PCR reaction and the reliability of data.

10.2 How UNG Carryover Hindrance Works

UNG for carryover prevention starts with the substitution of dUTP for dTTP in real-time PCR master mixes. Later on, real-time PCR reaction mixes are then acted with UNG that result in degradation of any contaminating uracil-containing PCR products, quitting the natural (thymine-containing) target DNA template unaffected. With standard UNG, a short incubation at 50°C is conducted before the PCR thermal cycling with an aim that the enzyme can break the uracil residues in any contaminating DNA. The deletion of the uracil bases results in fragmentation of the DNA, thereby not allowing its use as a template in PCR. The UNG is later inactivated in the incline up to 95 °C in PCR. A heat-labile form of the enzyme is also available, that is inactivated at 50 °C, in order to utilize it in one-step qRT-PCR reaction mixes.

11 Multiplex Real-Time PCR

11.1 Introduction to Multiplexing

PCR multiplexing is the amplification and exact identification of two or more genetic sequences in the same reaction. For successful result, PCR multiplexing should be able to produce enough amplified product for the identification of all of the intended sequences. Real-time PCR multiplexing may be utilized to make quantitative or qualitative results. In case of quantitative PCR multiplexing, all of the planned sequences must produce sufficient geometric-phase signal. In case of qualitative results, if amplified products are adequate an endpoint detection method such as gel electrophoresis may be employed. The suffix "plex" is utilized in multiple terms. Singleplex is an assay planned for the amplification of a single genetic sequence. Duplex is a combination of two assays considered for the amplification of two genetic sequences. The most common kind of multiplex is a duplex, wherein the assay for the target gene is conducted in the same well as that for the control or standardize gene. It is equally possible to have higher-order multiplexes. Some business-related real-time PCR kits are designed and validated as a multiplex. For example, the MicroSEQ® E. coli O157:H7 Kit multiplexes the *Escherichia coli* target assay with an inside positive control assay. For research applications, the scientists generally have to choice to select which assays to multiplex and conducts for multiplex validation. While considering the creation of a multiplex assay, it is significant to weigh the profits of multiplexing versus the degree of effort needed for validation.

11.2 Multiplexing Benefits

Three benefits of multiplexing are as follows:

1. Raised throughput (more samples potentially assayed per plate),
2. Reduced sample usage, and
3. Condensed reagent usage.

   They are dependent on the number of targets in the experiment.

   For example, if a quantitative experiment consists of simply one target assay, running the target assay as a duplex with the standardizer assay, such as an endogenous control assay, will enlarge throughput, decrease sample required, and reduce reagent usage by half. If a quantitative experiment consists of two target assays, it may be possible to join two target assays and the standardizer assay in a triplex reaction. In that case, the throughput raise, sample reduction, and reagent reduction will be even greater. If the target assay is multiplexed with the standardizer assay, another benefit of multiplexing is minimizing pipette precision errors.

Target and standardizer data from the same well are derived from a single sample addition, so any pipette accuracy error should affect both the target and normalizer results equally. In order to gain this exact benefit, target data must be normalized by the normalizer data from the identical well before calculating technical replicate precision. Comparison of the multiplex data studied in a singleplex manner (without well-based normalization) to an investigation done in a multiplex manner reveals that the multiplex precision benefit will be substantial, depending on the singleplex error. For example, for samples with minimal singleplex precision error, the multiplex precision advantage will be minimal as well. The precision advantage of multiplexing is especially valuable for quantitative experiments requiring a higher degree of precision. For example, in copy number variation experiments, discriminating 1 copy from 2 copies of the gene is a twofold difference, which needs good precision. However, discriminating 2 copies from 3 copies is only a 1.5-fold difference, which needs even better precision. Multiplexing is one recommended method to help attain the necessary degree of precision for this type of experiment.

11.3 Instrumentation for Multiplexing

Multiplex assays typically include multiple dyes in the same well. The real-time PCR instrument should be proficient of measuring those different dye signals in the same well with accuracy. These capacities should remain specific for each dye, although one dye signal is significantly higher than another. Proper instrument calibration is essential for accurate measurement of each dye contribution within a multiplex assay.

11.4 Chemistry Recommendations for Multiplexing

The best fluorescent chemistry for real-time PCR multiplexing is those that can assign diverse dyes to identify each genetic sequence in the multiplex. The huge majority of multiplexing is conducted with multi-dye, high- specificity chemistries, as TaqMan® probe-based assays. For multiplex assays involving RNA, two-step RT-PCR is normally suggested over one-step RT-PCR. One-step RT-PCR needs the same primer concentration for reverse transcription and PCR, decreasing flexibility in primer concentrations optimal for multiplexing. In two-step RT-PCR, the PCR primer concentration can be optimized for multiplexing, in such a way to have no poor effect on reverse transcription.

11.5 Dye Choices for Multiplexing

Assuming a multi-dye real-time PCR fluorescence chemistry is being employed, each genetic sequence being identified in the multiplex will need a different reporter dye. The reporter dyes chosen should be adequately excited and perfectly detected when together in the same well by the real-time PCR instrument. The instrument manufacturer would be capable to recommend the dyes.

Note that Applied Biosystems® real-time PCR master mixes compose a red passive reference dye whereas blue-only excitation instruments are able to excite this ROX™ dye-based reference adequately to act as a passive reference dye. Blue excitation is usually not enough for red dyes to act as a reporter. Reporter dyes do not have to be assigned on the basis of the type of target gene or gene product. However, following an outline in assigning dyes is able to simplify the creation of a multiplex assay. For example, FAM™ dye is the most frequent reporter dye used in TaqMan® probes. We follow the pattern of assigning FAM™ dye as the reporter for the target assay and assigning VIC® dye as the reporter for the standardizer assay. With the help of this pattern, multiple duplex assays can be created by pairing a different FAM™ dye-labeled target assay with the same VIC® dye for the standardizer assay. In a triplex assay, a third dye, such as ABY® or JUN® dye, can be combined with FAM™ dye and VIC® dyes.

11.6 Multiplex PCR Saturation

Multiplex PCR saturation is an unwanted phenomenon that can occur in a multiplex assay, wherever the amplification of the more abundant gene saturates the thermostable DNA polymerase, leading to the suppression of the amplification of the less abundant gene. The remedy for saturation is a decrease of the PCR primer concentration for the more abundant target, termed "primer limitation." The primer-limited concentration would be enough to allow geometric amplification, but of sufficiently less in such a way that the primer is shattered before the PCR product accumulates to a level that starves amplification of the less abundant target. While preparing a multiplex assay, the researcher should recognize which gene or genes in the multiplex have the possible to cause saturation that is based on the absolute DNA or cDNA abundance of every gene or gene product in the PCR reaction. In relation to this, the three most common duplex scenarios are written below.

11.6.1 Duplex Scenario 1

In this most general scenario, the more abundant gene or gene product is the same in all samples. Only the assay for the more abundant target needs primer limitation. As for example, the normalizer exist 18S ribosomal RNA that is 20% of eukaryotic total RNA. The 18S rRNA cDNA would be more plentiful than any mRNA cDNA in every sample. Therefore, only the 18S assay would need primer limitation.

11.6.2 Duplex Scenario 2

In this scenario, the two genes possess about equal quantity in all samples. Usually, a Ct difference between the two genes of 3 or higher, predicting the same threshold, would not qualify for equal abundance. Genomic DNA applications, like copy number variations, are most likely classified into this category. Primer limitation is not essential for scenario 2, since the two genes are progressing through the geometric phase at approximately the same time.

11.6.3 Duplex Scenario 3

In this scenario, either the gene or gene product has the possibility to be significantly more abundant than the other. In this case, previously discussed assays should be primer limited.

11.7 Multiplex Primer Interactions

Another possible threat to multiplex assay performance is unexpected primer interactions among the primers from different assays. The risk of primer interaction increases with the number of assays in the reaction, since the number of select primer pairs increases with the more number of assays in the multiplex.

In singleplex each assay may perform well, but in a multiplex reaction the primer interactions can make competitive products, suppressing amplification. The probability of primer interactions increases when the assays being multiplexed have homology. In case, primer interaction happens on the basis of the observation of drastically different Ct values in the singleplex versus multiplex reaction, the solution is to employ a different assay in the multiplex reaction.

11.8 Internal Controls and Reference Genes

Real-time PCR has become a procedure of choice for gene expression analysis. To attain accurate and reproducible expression profiling of selected genes using real-time PCR, it is important to use reliable internal control gene products for the standardization of expression levels between experiments—typically expression products from housekeeping genes are employed. The target chosen to be the internal standard (or endogenous control) would be expressed at approximately the same level as the experimental gene product. By employing an endogenous control as an active reference, quantification of an mRNA target can be standardized for differences in the amount of total RNA mixed to each reaction. Regardless of the gene that is chosen to act as the endogenous control, the gene should be tested under all of one's experimental conditions, for the confirmation of consistent expression of the control gene under all conditions.

11.9 Relative Gene Expression Analysis Using Housekeeping Genes

Relative gene expression comparisons are best assessed while the expression level of the chosen housekeeping gene is constant. Ideally, the expression level of the chosen housekeeping gene would be validated for each target cell or tissue type for the confirmation that it remains similar at all points of the experiment. For example, GAPDH expression has been exposed to be upregulated in proliferating cells, and 18S ribosomal RNA (rRNA) may not always express the overall cellular mRNA population.

12 Real-Time PCR Instrument Calibration

Timely, exact calibration is critical for the proper performance of any real-time PCR instrument. It conserves data integrity and consistency over time. Real-time PCR instruments would be calibrated as part of a regular maintenance routine and before utilizing newer dyes for the first time, following the manufacturer's instructions.

12.1 Excitation/Emission Difference Corrections

The optical elements in real-time PCR instruments can be separated into two main categories: the excitation source, e.g., halogen lamps or LEDs, and the emission detector, such as a CCD camera or photodiode. Although manufacturers may attain excellent uniformity for excitation strength and emission sensitivity across the wells of the block, variations will still exist. This difference can increase with age and usage of the instrument. Uncorrected excitation/emission differences across the plate may lead to shifts in Ct values. In case a passive reference dye is there in the reaction, these differences will influence the reporter and passive reference signals to the same intensity. This causes the standardization of the reporter to the passive reference, which corrects the difference.

12.2 Universal Optical Fluctuations

In usual plastic PCR plates and tubes, the liquid reagents are at the bottom of the well, air space is above the liquid, along with a plastic seal is above the well. With this configuration, a number of temperature-related phenomena happen. During cycling, temperatures arrive at 95 °C. This water vapor or steam will condense on the cooler walls of the tube and form water droplets that return to the reagents at the bottom. This entire process, called "refluxing," is nonstop during PCR. Second, at high temperature, air dissolved within the liquid reagents will become less soluble, making small air bubbles.

The pressure of the steam will apply force on the plastic seal, leading it to change shape slightly during PCR. All of these temperature-related phenomena are in the excitation and emission light path and can responsible for fluctuations in fluorescent signal. The degree of these fluctuations can depend on factors such as how much air was dissolved in the reagents and how well the plate was sealed. Simply, universal fluctuations do not make obvious distortions in the reporter signal but they do influence the precision of replicates. Standardization of reporter to passive reference signals rectifies the fluctuations occurred due to the presence of passive reference dye in the same light path as the reporter.

12.3 Precision Improvement

The correction effect of passive reference normalization will progress the precision of real-time PCR data. The degree of development will vary depend on a number of factors, such as how the reagents and plate were prepared.

12.4 Atypical Optical Fluctuations

Atypical optical fluctuations are thermal-related anomalies that are uncommon across all reactions in the run and they produce a distortion in the reporter signal. One example of an atypical optical fluctuation is a significant configuration that alters in the plate seal, which may be termed "optical warping." Optical warping takes place when a well is inadequately sealed, and then, during PCR, the heat and pressure of the heated lid causes the seal to seat properly. A second example is the presence of large bubbles that may burst during PCR. This causes distortions in the amplification plot, which may cause baseline problems affecting *Ct* values. This may be corrected to normalization to a passive reference dye, which provides excellent correction for optical wrapping. Thus, the resulting corrected amplification plot may come into view completely anomaly-free. Standardization does not fully correct for a large bubble bursting, but it can help minimize the data distortion caused.

The crucial points discussed in this chapter consist of the following points:

1. Introduction.
2. Real-time PCR assay types.
3. Amplicon and primer design considerations.
4. Nucleic acid purification and quantitation.
5. Reverse transcription considerations.
6. Controls.
7. Normalization methods.
8. Using a standard curve to know the efficiency, sensitivity, and reproducibility.

13 Application for Real-Time PCR

1. Differential mRNA expression analysis [1–6].
2. Genotyping with Taqman probe.

14 Evaluation of TaqMan qPCR System Integrating Two Identically Labeled Hydrolysis Probes in Single Assay: A Case Study

Evolution of viral pathogens is an important aspect in diagnostic virology using TaqMan qPCR/RT-qPCR. Certain concerns are connected to false negativity because of failure in probe binding. For the purpose of compensation for such deficiency is to incorporate a second identically labeled probe in the assay. Moreover, the exact process of change influencing the reaction parameters has not been demonstrated in detail.

In this study, a TaqMan protocol is employed utilizing two identically labeled hydrolysis probes—simple, LNA (locked-nucleic-acid) and MGB (minor-groove-binder) modified probes or their combinations in a single assay. The results based on a synthetic amplicon assume that the second probe does not compromise the TaqMan qPCR/RT-qPCR parameters. This frequently and reproducibly comparable to the corresponding single-probe assays. It does not depend on the relative probe orientation, if reverse or tandem, and probe modifications or combinations of them. Whereas, the second probe contribute to the overall fluorescence signal in an additive manner. We expect that the current study would act as a theoretical basis for the development or improvement of TaqMan qPCR/RT-qPCR assays for the identification of highly variable nucleic acid templates [7].

References

1. Pal A, Pal A, Mallick AI, Biswas P, Chatterjee PN (2020) Molecular characterization of Bu-1 and TLR2 gene in Haringhata Black chicken. Genomics 112(1):472–483. https://doi.org/10.1016/j.ygeno.2019.03.010
2. Pal A, Pal A, Chakravarty AK (2020) Mutations in growth hormone gene affect stability of protein structure leading to reduced growth, reproduction and milk production in cross bred cattle—an insight. Domestic Anim Endocrinol 71(6):106405. https://doi.org/10.1016/j.domaniend.2019.106405
3. Banerjee S, Pal A, Mandl SC, Chatterjee PN, Chatterjee JK (2021) RIG-I has a role in immunity against Haemonchus contortus, a gastrointestinal parasite in Ovis aries—a novel report. Front Immunol 11:534705. https://doi.org/10.3389/fimmu.2020.534705
4. Rawat K, Pal A, Banerjee S, Pal A, Mandal SC, Batabyal S (2021) Ovine CD14- an immune response gene has a role against gastrointesinal nematode Haemonchus contortus—a novel report. https://doi.org/10.3389/fimmu.2021.664877
5. Pal A, Pal A, Baviskar P (2020) Molecular characterization of RIGI, TLR7 and TLR3 as immune response gene of indigenous ducks in response to Avian influenza. https://doi.org/10.1101/2020.09.28.316687
6. Islam M, Pal A, Banerjee S, Das P (2020) Better expression profile of CD8 and CD4 gene in uterus of pregnant ewe in comparison to non-pregnant—a novel report. https://doi.org/10.1101/2020.10.07.329730
7. Nagy A, Vitásková E, Černíková L, Křivda V, Jiřincová H, Sedlák K, Horníčková J, Havlíčková M (2017) Evaluation of TaqMan qPCR system integrating two identically labelled hydrolysis probes in single assay. Sci Rep 7:41392

Chapter 10

Tissue or Cell Cultures

Abstract

Tissue culture is the process of in vitro culture of mammalian cells in a suitable medium. The detailed protocol for mammalian cell culture has been described for fibroblast subculture or lymphocyte culture. Although the protocol for cell culture differs with different cells, the basic mechanism involves detachment of cells (preferably through trypsinization as in fibroblast culture) and culture in a suitable culture medium, incubated at 37 °C in a CO_2 (with 5% CO_2) incubator. Some salient features to be considered for cell culture are that neuronal cells cannot grow, fibroblast continues to divide and grow in culture up to certain generations, and stopping further growth as they reach the wall of a container, the mechanism is known as contact inhibition. In cell culture, the cancer cells exhibit uncontrolled growth and more rounded shape; they lose contact inhibition and are piled on each other; and features are exploited by cancer specialists, i.e., the serologist to identify cancer cells from the normal cells. This chapter discusses different types of cell culture (primary/secondary) and cell lines (finite or continuous). This chapter also discusses the cryopreservation of animal cell culture, the prospect of cell culture, and different culture media for common cell lines available.

Keywords Tissue culture, Primary cell culture, Secondary cell culture, Fibroblast subculture, Lymphocyte culture, CO_2 incubator, Cell line

1 Principle

Tissue culture is the process of in vitro culture of mammalian cells in a suitable medium. Tissue culture is the process of the growth of tissues or cells separated from the organism. Tissue or cell culture employs the use of a liquid, semi-solid, or solid growth media, such as broth or agar. Tissue culture means the culture of animal cells and tissues. The term "tissue culture" was given by American pathologist Montrose Thomas Burrows, M.D.

2 History of Tissue Culture

1. In 1885, Wilhelm Roux took a section of the medullary plate of an embryonic chicken and placed it in a warm saline solution

Aruna Pal, *Protocols in Advanced Genomics and Allied Techniques*, Springer Protocols Handbooks,
https://doi.org/10.1007/978-1-0716-1818-9_10,

for several days. This established the basic principle of tissue culture.

2. In 1903, Jolly studied in vitro cell survival and cell division with salamander leukocytes. In the early experimentations, tissues were studied in fragments, and this gave rise to the name "tissue culture." The term tissue culture is applicable where the cells are maintained or cultured in vitro for more than 24 h.
3. In 1907, the zoologist Ross Granville Harrison showed the growth of frog embryonic cells developing to nerve cells in a medium of clotted lymph. The cell growth that can be visualized in clotted lymph fluid in a depression slide is often designated as the foundation of animal cell culture as a science. In the technique of "the hanging drop" mechanism, the tissue that was isolated was suspended on the underside of a coverslip. This coverslip was then sealed over a depression in a microscope slide. Later on, this technique was used with plasma clots that proved to be of better efficiency for the growth of cells from warm-blooded animals. The matrix made of insoluble protein formed the basis of the necessary anchor for cell growth. The necessary nutrients were supplied by the enclosed fluid.
4. In 1913, E. Steinhardt, C. Israeli, and R. A. Lambert developed the vaccinia virus in fragments of corneal tissue in a guinea pig.
5. In 1996, regenerative tissue was used for the first time with the aim to replace a small length of the urethra. This ultimately has led to the concept of the technique of isolating samples of tissue and growing it in in vitro conditions (i.e., outside the body) without scaffold.

In the present context, some commonly used protocols for mammalian cell culture have been described. Figure 1a represents the actual photograph of the tissue culture flask with media, and Fig. 1b depicts the microscopic view of cells cultured.

3 Fibroblast Subculturing Protocol

As a protocol of cell culture, fibroblasts should be grown in 90% RPMI 1640 with 10% FBS. Alternative media include alpha-MEM and Dulbecco's modified MEM.

1. Appropriate volumes of fibroblasts growth medium should be prepared: EDTA in HBSS, trypsin in EDTA-HBSS, and stop medium; EDTA in HBSS (0.53 mM) is needed, 0.04% trypsin in EDTA-HBSS, and growth medium with 10% FBS (stop medium). Using a 20–30 mL (T75) flask, the appropriate volumes prepared should be 20, 8–10, 4–5, and 5–7 mL,

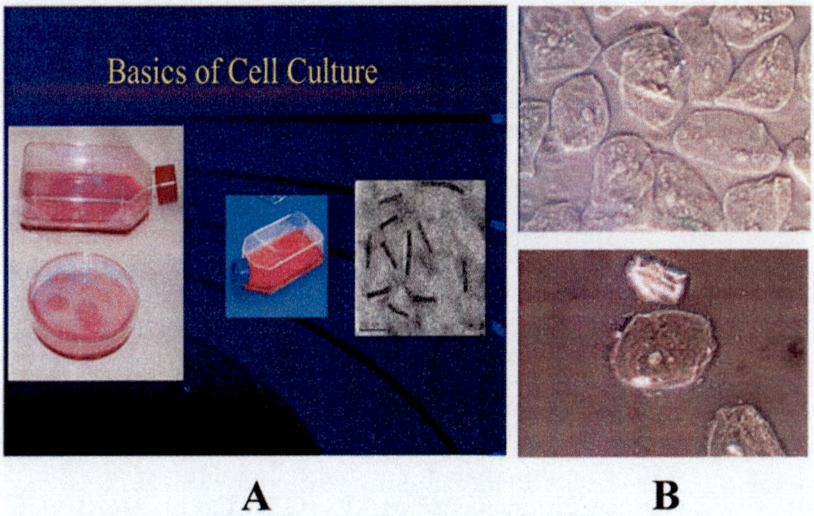

Fig. 1 Photograph of tissue culture plate and flask with culture media (**a**), and visualization of cells under phase-contrast microscope (**b**)

respectively. If different flask sizes are to be used, then the volumes should be adjusted accordingly.

2. The growth medium should be dispensed into the flask(s).
3. The medium should be removed by aspiration and EDTA should be added carefully so as not to dislodge the cell sheet. Cells are to be left in EDTA for up to 10 min, but EDTA must be removed if cells start to loosen (observe through an inverted microscope).
4. EDTA solution should be removed and trypsin is added.
5. Incubation should be at 37 °C with 5% CO_2 for 4–7 min. Incubation should not be done for more than 9 min. Cells need to be dislodged at this point and round up. Detachment is encouraged by lightly tapping the side of the flask.
6. Trypsin is inactivated by washing the sides of the flask with a stop medium.
7. Flasks are reseeded to a volume of 15–30 mL with $1.0–1.4 \times 10^4$ cells/cm^2 and incubated at 37 °C with 5% CO_2 with loose caps to ventilate.
8. Confluent cultures should be split 1:5 to 1:10. After 2–3 days, the monolayer will be confluent; splitting should be done 2–3 times a week. Initially, seeding out is done at about $1.0–2.0 \times 10^6$ cells/25 cm^2. If splitting is less frequent, then replacement of medium every 3–4 days is to be done.

4 Protocol for Lymphocyte Culture

1. Approximately 0.5 mL of heparinized whole blood should be inoculated into a glass or plastic tube with 10 mL of PB-MAX medium (GIBCO Cat. No. 12557).
2. The culture should be incubated at 37 °C in a 5% CO_2 atmosphere for 72 h.
3. 0.5 μg/mL of KaryoMAX™ Colcemid Solution (GIBCO Cat. No. 15212 or 15210) should be added to each culture tube.
4. The culture should be incubated for an additional 15–30 min.
5. The culture is to be transferred to a centrifuge tube and spin at 500 × *g* for 5 min.
6. The supernatant and the cells are resuspended in 5–10 mL of hypotonic 0.075 M KCl (GIBCO Cat. No. 10575-090).
7. Incubation is to be done at 37 °C for 10–12 min.
8. Spinning should be done at 500 × *g* for 5 min.
9. The supernatant should be removed, the cellular sediment should be agitated and added drop-by-drop in 5–10 mL of fresh, ice-cold fixative composed of 1 part acetic acid to 3 parts methanol. Then, it should be left at 4 °C for 10 min.
10. Repeat **steps 7** and **8**.
11. Spin at 500 × *g* for 5 min.
12. The cell pellet is resuspended in a small volume of 0.5–1 mL of fresh fixative. The next step is to drop on to a clean slide and air-dried.
13. In the next step, staining is conducted with Orcein or Giemsa. Giemsa banding has proved to be mostly used technique and the common method. The staining process involves prior treatment of the slides with Trypsin-EDTA 10× [2].

5 Techniques of Animal Cell Culture

Both fibroblast cell culture and lymphocyte culture have been described. In the current section, other details of animal cell culture have been described.

On the basis of morphology (shape and appearance) or on their functional characteristics, there are three types of cell:

1. Epithelial-like cell: The cells are attached to a substrate and appear flattened and polygonal in shape.
2. Lymphoblast-like cell: The cells do not attach remain in suspension with a spherical shape.

3. Fibroblast-like cell: The cells are attached to a substrate and appear elongated and bipolar.

6 Some Characteristics of Animal Cell Growth in Culture

The animal cells are able to grow only up to limited generations, although the best nutrition may be provided in the media. Growth is also dependent on the sources of tissue isolated. The features of different kinds of cell cultures are briefly discussed.

1. Neuronal cells are present in the nervous system. The neuronal cells are not able to divide and grow in cell culture.
2. Fibroblasts are cells derived from connective tissue or skin. Fibroblasts continue to divide and grow in culture up to certain generations, afterward they die. This implies all normal animal cells to be mortal.
3. In cell culture, the animal cells continue to divide and grow, ultimately fill the surface of the container. Later on, they stop further growth as they reach the wall of the container. This mechanism is known as contact inhibition.
4. In cell culture, the cancer cells seem to differ from the normal cells. As a result of uncontrolled growth and more rounded shape, they lose contact inhibition and are piled on each other. These features are exploited by cancer specialists, i.e., the cerologists to identify cancer cells from the normal cells.

7 Primary Cell Cultures

For the establishment of cells in culture, single-cell suspension is essential. This is attained through the dissociation of organs (e.g., kidney or liver) or tissues through the process of mechanical or enzymatic methods. The next step is the transfer of the cells into a special glass or plastic containers with a culture medium. The cells start growing; the maintenance of growth of such cells is called primary cell culture.

The cells are to be enumerated together by a proteinaceous material. Therefore, crude preparation of proteolytic enzymes (trypsin and collagenase) is commonly employed to break the cementing matter and separate cells of a given tissue. The characteristics of the animal cells govern the characteristics of the cells in culture.

The growing cells may be classified into two different kinds:

1. Anchorage-dependent (adherent cells).
2. Anchorage-independent cells (suspension culture).

Commonly, the adherent cells can be obtained from such organs that are fixed at a place (e.g., kidney, liver). The cells too are mobile but remain fixed in connective tissues.

On the contrary, the suspension cells grow continuously in a liquid medium and there is no need to adhere or attach to the surface of the container. It depends entirely upon the source of cells. All suspension cultures are based on culturing the blood cells, as all blood cells are vascular in nature and get dissolved in plasma.

8 Secondary Cell Cultures and Cell Lines

As discussed earlier that as soon as the nutrients of the medium are utilized, the primary cell culture no longer remains viable as the cells die. Hence, subculturing is of utmost essential in a fresh medium. Hence, the primary culture should be removed; adherent cells are to be dissociated enzymatically or by repeated pipetting. In the next step, individual cells thus obtained are diluted with a freshly prepared medium, suitable for the cell type, and transferred to a fresh culture flask. These subcultures are known as secondary cell cultures or cell lines. Subculturing is to be done on a fresh medium at certain intervals. The objective of subculturing is to provide optimum nutrients and space for the growing cell lines. It is very much essential to repeat subculturing in frequent intervals in order to restore the nutrients; otherwise, the cells will die.

During the process of subculturing in in vitro conditions, certain cell types as fibroblasts are selected, while normal diploid cells die. Ultimately, only fibroblast cells grow in culture.

In certain cases, some cells of secondary cell cultures in vitro may be transformed spontaneously or chemically. Such transformed cells are observed to be immortal and ultimately give rise to continuous cell lines, i.e., cancerous cell lines.

These cells proliferate indefinitely and the doubling time is about 10–25 h. Those cultures contain a mixture of cell types when a particular cell type may dominate over the others [1, 2].

9 Type of Cell Lines

There are different types of cell lines produced from different tissues or organs. These can be broadly classified into two types: finite cell lines and continuous cell lines.

1. Finite Cell Lines:

   Finite cell lines are referred to as the cell lines that grow through a limited number of cell generations and have a limited life. The growth of the cells is slow and they form a monolayer. The cells replicate and double in 24–96 h. The properties of cell lines are anchorage dependence, contact inhibition, and density limitation.

2. Continuous Cell Lines:

   In the earlier section, we discussed finite cell lines. On the contrary, continuous cell lines are obtained from either transformed cell lines in vitro or cancerous cells. They have rapid cell division. Their generation time is 12–14 h. They possess no or reduced density limitation. The transformed cells in vitro have the following differences:

   (a) It possesses improved growth and proliferation due to the rapid growth rate.
   (b) There is the existence of altered ploidy, i.e., aneuploidy or heteroploidy due to altered chromosome number.
   (c) They possess different cell shapes and organizations of microfilaments.
   (d) They have the ability to translocate.
   (e) Their energy metabolism is different from that of finite cell lines.
   (f) Continuous cell line possesses no contact inhibition and no anchorage dependence.
   (g) They possess different growth factor requirements and responses to regulate molecules.

10 Environmental Factors Required for Culturing the Animal Cells

Nutritional requirements of the animal cells are specific. The cells are not suitable to survive independently.

Therefore, mechanical, physical, and chemical (nutritional and hormonal) requirements of the cells need to be fulfilled with proper temperature for growth, mammalian proteins, and nucleic acids in a medium. Consequently, cell cultures including human are 37 °C.

11 Cryopreservation of Animal Cells

Cryopreservation is done by using liquid nitrogen at a very low temperature (−470 °C to −197 °C). Cryoprotective agents as glycerol or dimethyl sulfoxide (DMSO) are used with a high concentration of serum in order to prevent the development of ice crystals within the cell and thereby preventing cell damage. It is to be noted that DMSO is soluble in lipids, which enter inside the cell by diffusion through the bilipid layer of the cell membrane. DMSO also leads to protein permeability.

As a result, serum of high concentration is responsible for maintaining the proper concentration of proteins inside the cell.

Simultaneously, it maintains the structural integrity of the animal cells.

In the case of thawing, a suitable temperature needs to be maintained in a water bath. It is advisable for quick thawing by immersing in a water bath, which reduces the ice crystals and cell damage. Such stored cells can directly be transferred on to a fresh culture medium and incubated properly.

12 Equipment Required for Animal Cell Culture

The equipment necessary for animal cell culture are given below:

1. Laminar Air Flow (LAF):

   It is a very important instrument involved in animal cell culture. Sterility and asepsis are the primary criteria for any cell culture work, which is maintained in laminar airflow. In case the stringency for asepsis is not maintained, the contamination of bacteria or fungi is very common. It is usual that the contaminated bacterial/fungal cells have better growth than the cultured animal cell. Hence in case of contamination, the growth of animal cell fails.

   At the initial stage, LAF hood is sterilized with 70% ethanol. The entire cell culture work is conducted with LAF hood in position.

   The LAF hood has basically two functions:

   (a) It maintains a sterile environment for cell manipulation (i.e., protects tissue culture from an operator).

   (b) It equally protects the operator from the potential infection risk from the culture.

2. CO_2 Incubators:

   It is another prime requirement for any cell culture work. The CO_2 incubators maintain the suitable environmental conditions for the growing animal cells. A silicon gasket is applied on the inner door leading the chamber of the incubator airtight. This leads to the separation of the internal environment. The cells are observed on the bottom of the tissue culture vessel (e.g., Petri plates). Incubation facilities need to be provided for the temperature of 25–30 °C for insect culture and 37 °C for mammalian cells, CO_2 (2–5%), 95% air at the level of 99% relative humidity. To avoid cell death, incubators are set at cut-off at 38.5 °C.

3. Inverted Microscope:

   The visualization of the cells in cell culture is of utmost importance. This is aided through an inverted microscope. The culture medium remains above the growing cells in cell culture plates. In case such plates are put over the stage of an ordinary microscope, the growing cells at the bottom may not be visible. Therefore, the inverted microscope with 10× to 100× magnification is used for such a purpose.

An inverted microscope is designed in such a way when the optical system is at the bottom and the light source is at the top. Whenever a plate is observed in the inverted microscope, the cells of the culture growing at the bottom of the plate can easily be visualized. Whenever we wish to count cells with a counting chamber, a standard microscope with a movable slide holder is needed.

4. Refrigerator:

   Liquid media need to be kept at 4 °C, enzymes (e.g., trypsin) and media components (glutamine and serum at −20 °C).

5. Tissue Culture Ware:

   These are the culture plastic ware treated with polystyrene.

   The major limitation encountered for this work was the maintenance of the cultures free from contamination. Another major limitation was low-level contamination in an animal cell culture that leads to bacterial overgrowth [1].

13 Uses of Cell Culture

1. Cell culture may serve as the model system to study basic cell biology, interactions between disease-causing agents and cells, to study the effects of drugs, processes, and triggering of aging and nutritional studies.
2. It is equally helpful in toxicity study [2].
3. Aids in Cancer research.
4. Helpful for virological studies.
5. Prime criteria for Recombinant DNA technology.
6. Aids in studies related to gene therapy.

   Due to human and Animal Ethics limitations, there exist certain restrictions in conducting studies on living individuals as humans or animals.

   Tissue culture is the method of in vitro cultivation of organs, tissues, and cells at defined temperatures using an incubator and supplemented with a medium containing appropriate nutrients and growth factors.

   Different tissues grown in culture include fibroblast, skeletal tissue, and cardiac tissue. Choice of media is dependent on the type of cell to be cultured. The most common media are GMEM, EMEM, DMEM, etc. Media need to be supplemented with antibiotics viz penicillin, epithelial tissue (liver, breast, skin, and kidney), and various types of tumor cells (Table 1).

Table 1
Common cell lines available are presented

Human cell line		
1.	MCF-7	Breast cancer
2.	HL 60	Leukemia
3.	HEK-293	Human embryonic kidney
4.	HeLa	Henrietta lacks
Primate cell line		
1.	Vero	African green monkey epithelial cells
2.	Cos-7	African green monkey kidney cells
Hamster		
1	CHO	

References

1. ACT Laboratory Procedure Manual (1980) Section 2, pp 70–77 and 2nd edn, 1991, Chapter 2, pp 24–30
2. Pal A, Sharma A, Bhattacharya TK, Chatterjee PN, Chakravarty AK (2011) Molecular characterization and SNP detection of CD 14 gene in crossbred cattle. Mol Biol Int 2011:507346, 13 pages. https://doi.org/10.4061/2011/5

Check for updates

Chapter 11

DNA Microarray

Abstract

Microarray is a *hybridization-based technique* where a greater number of genes can be studied at a time. Each microchamber will be allotted to particular gene and they will be tagged by fluorescent dye, which helps in reading. This chapter is involved with brief overview of the protocol which involves extraction of mRNA, conversion of mRNA into colored cDNA (fluorescently labeled), mixing of cDNA together, hybridization of cDNA with array, each cDNA sequence hybridizes specifically with the corresponding gene sequence in the array, washing off the unhybridized cDNA, reading the array with laser, and lastly analyzing images. Important points to be considered for analysis include noise suppression, spot localization, and detection. It includes the extraction of the background intensity, the spot position, and the spot boundary and size, data quantification and quality assessment. The entire protocol for DNA microarray has been grouped in the following processes: protocol for preparing poly-L-Lysine slides for microarrays, protocol for amplifying products to print on array, protocol for cleanup and 384-well arraying of PCR products, protocol for postprocessing microarrays, protocol for total RNA isolation, protocol for reverse transcription and amino allyl coupling, and protocol for array hybridization and array washing.

Keywords DNA microarray, Microchamber, Hybridization-based technique, Poly-L-Lysine slides, Background intensity, Data quantification

1 Basic Principles

Each phenotypic trait is governed by many genes, referred to as polygenic in inheritance. In order to study expression pattern of a greater number of genes at a time, real-time PCR or quantitative PCR is very effective. But it has certain limitations.

Microarray is a ***hybridization-based technique*** where a greater number of genes can be studied at a time. Each microchamber will be allotted to particular gene, and they will be tagged by fluorescent dye that aids in reading. Figure 1 depicts the diagrammatic representation of the outline of DNA microarray.

The limitations in studying the number of genes in quantitative PCR can be easily overcome by hybridization-based DNA microarray.

Aruna Pal, *Protocols in Advanced Genomics and Allied Techniques*, Springer Protocols Handbooks, https://doi.org/10.1007/978-1-0716-1818-9_11, © Springer Science+Business Media, LLC, part of Springer Nature 2022

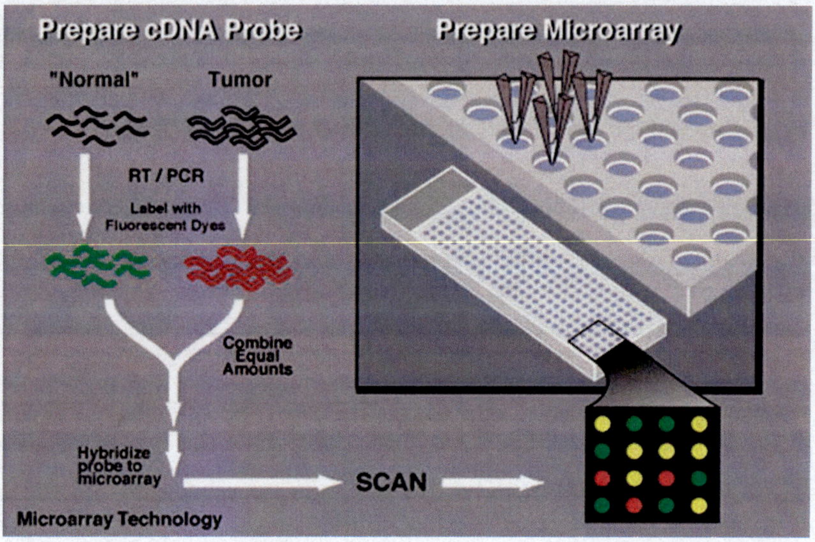

Fig. 1 Schematic representation of the outline of DNA microarray

2 Brief Overview of the Protocol

1. Extraction of mRNA.
2. Conversion of mRNA into colored cDNA (fluorescently labeled).
3. Mixing of cDNA together.
4. Hybridization of cDNA with array.
5. Each cDNA sequence hybridizes specifically with the respective gene sequence in the array.
6. Washing off the unhybridized cDNA.
7. Reading the array with laser.
8. Analyzing images.

It is a very important step to read and analyze an array. Let us discuss the peculiarities of the sample. Laser scans array and produces images. One laser is assigned for each color, e.g., one for green and one for red. Figure 2 depicts the actual photograph of DNA microarray with different colors depicting intensity of expression. Details of color coding have been illustrated in Fig. 3.

For the purpose of image analysis, some important points are taken into consideration, they include:

2.1 Suppression of Noise

1. Important step is to localize and detect spot. It is necessary to extract the background intensity and position the spot including its boundary and size.
2. Quantification of data and assessment of quality.

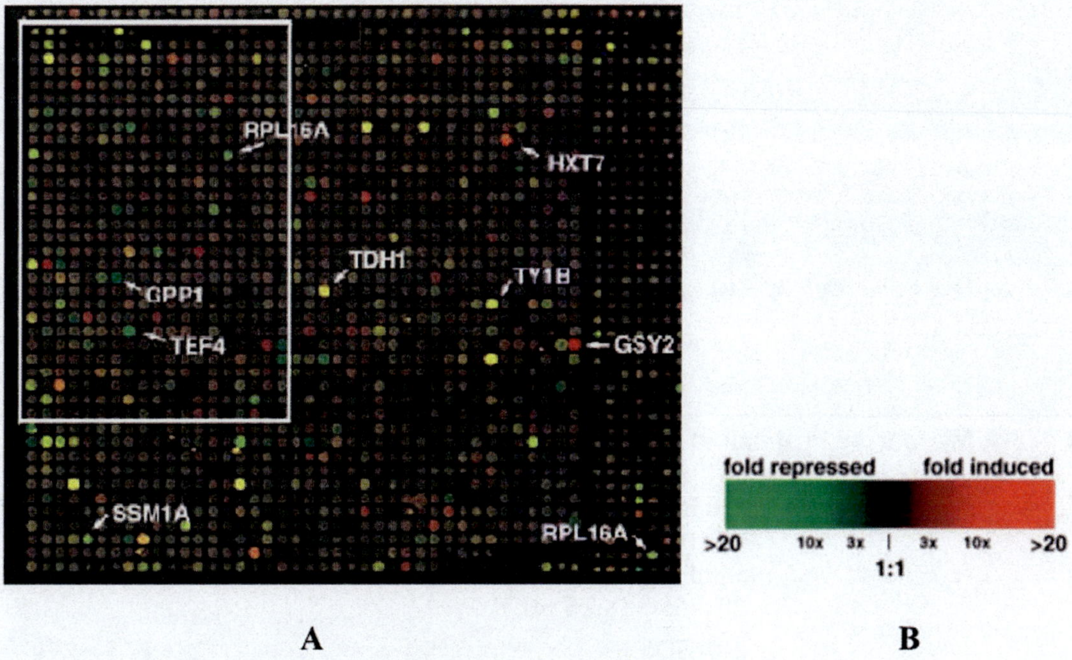

Fig. 2 Actual photograph of DNA microarray (**a**) and color coding (**b**)

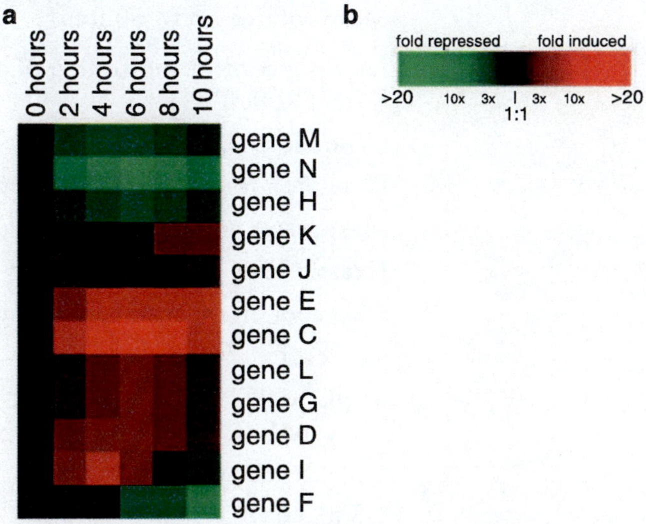

Fig. 3 Color coding of DNA microarray (**a**) A representative example of colour coding of DNA microarray. (**b**) Significance of the representative colours

There exist certain limitations for the data analysis, which are discussed as follows:

1. Difficulty in reading for the table.
2. Data are represented with a color scale.

3. The coding scheme includes:
   (a) Green = repressed (less mRNA) gene in experiment.
   (b) Red = induced (more mRNA) gene in experiment.
   (c) Black = no change (1:1 ratio).
   Otherwise, we can also code as:
   (d) Green = control condition (e.g., aerobic)
   (e) Red = experimental condition (e.g., anaerobic)
4. We only use ratio.

   Let us discuss the protocol in detail.

3 DNA Microarray Protocol in Brief

1. Pre-hybridization solution is to be prepared in a Coplin Jar and incubated at 65 °C during the labeling incubation period to equilibrate.
   20× SSC (8.75 mL).

   20% SDS.

   0.25 mL BSA (100 mg/mL).

   5.0 mL H2O.

   Make volume up to 50.0 mL.
2. Labeling control and testing of genomic DNA:
   CONTROL TEST.

   Genomic DNA ~2 μg.

   ~2 μg Random Hexamers (3 μg/μL) 1 μL.

   Make water volume up to 41.5 μL.

   Heat at 95 °C for 5 min.

   Snap cool on ice and briefly centrifuge.

   10× buffer (5 μL).

   5 μL dNTP's (5 mM each dATP, dGTP and dTTP, 2 mM dCTP).

   1 μL Cy-labeled dCTP.

   1.5 μL (Cy3).

   1.5 μL (Cy5) Klenow fragment (10 U/μL).

   Incubate at 37 °C for 90 min.
3. Incubated the microarray slide(s) in the pre-hybridization solution for a period of 20 min at particular temperature of 65 °C. The process is to be conducted just before the end of the labeling reactions incubation time at 37 °C.

4. The control and test reactions are combined and purified with PCR purification kit. PCR purification is based on two-step washing process: as ethanol washing, 250 μL volumes of Buffer PE and final elution of the labeled cDNA from the MinElute column with 14 μL H_2O. The eluted volume obtained will be 13 μL finally.
5. The rinsing of pre-hybridized microarray slides was followed in H_2O for 1 min, followed by rinsing in isopropanol for 1 min. The slides are dried through the process of spinning at 1500 rpm for 5 min to dry slides.
6. Prepared the hybridization solution as per the following protocol:
   Sample (13 μL).

   H_2O (26 μL).

   20× SSC (12 μL).

   2% SDS (9 μL).

   Heated the above solution at 95 °C for 2 min. Allowed to cool slowly at room temperature and centrifuge for 30 s.

   2× 20 μL H_2O is added to the corners of the hybridization chamber. A slide is placed into the chamber. A LifterSlip™ glass coverslip (22 mm × 25 mm) is placed over the array section on the slide with the help tweezers. The hybridization solution is pipetted on to the slide at the top of the coverslip. Sealed the chamber and incubated in a water bath at 65 °C overnight.
7. Prepare wash solutions as follows:
   Wash A (1× SSC 0.5% SDS).

   Wash B (0.06× SSC) 20× SSC 20 mL.

   2.4 mL 20% SDS.

   H_2O to 400 mL.

   Incubate Wash A solution at 65 °C overnight. Dispensed 400 mL volumes into three glass slide washing dishes. The slides are removed from the hybridization chambers and coverslip(s) are removed with care and rinsed in Wash A. Slides are placed in a slide rack with continuous rinsing with agitation for 5 min. The next step is the transfer of slide(s) to a clean slide rack and rinsed with agitation in Wash B for 2 min. The process of washing in Wash B is repeated for 2 min again. Spinned at 1500 rpm for 5 min to dry slide(s).
8. Scanned the slide(s) using Affymetrix 418 scanner followed by analysis of data.

4 Detail Protocol of DNA Microarray

Let us explain the individual and basic parts of DNA microarray experiment.

1. The RNA:

   It may be regarded as the primary component for DNA microarray experiment. An experiment is performed in which RNA (total or messenger) is collected. Two or more samples of RNA are collected in order to compare gene expression levels.

2. The Array:

   It refers to the glass slide or membrane on which spots representing genes, ORFs, ESTs, etc., are placed.

3. The Probe:

   Fluorescently labeled cDNA is obtained from the harvested RNA samples though the process of reverse transcription. The two RNA populations that need to be compared are labeled with different fluorescent markers.

4. The Hyb • The probe is placed for hybridization to the array.
5. The Scan • The array is washed and scanned two times with lasers specific for the fluorescent markers employed.
6. Labeled cDNA will fluoresce and "light up" the spot with which it has hybridized.

4.1 Analysis

Fluorescent intensities are extracted and read from both of the images, which are scanned. The ratio of these intensities is employed for the determination of the relative expression levels between the two populations of cells.

The entire protocol has been grouped in the following processes:

5 Protocol For Preparing Poly-L-Lysine Slides for Microarrays

First of all, let us discuss some salient and basic points pertaining to experiments on DNA microarray.

Basics: The following protocol is employed for coating 180 slides with poly-L-lysine.

1. *Slide racks are bent slightly inward in the middle in order to hold slides in a more secure manner.*
2. *Powdered gloves should not be used at any time during this protocol.*
3. *The slides need to be covered or submerged in solution at all times in order to avoid dust.*
4. *After coating, slides are usually stable for 4 months.*

5.1 Detail Protocol

1. Prepare wash solution for 180 slides:
   200 g of NaOH pellets.
   1200 mL 95% EtOH.
   800 mL of H_2O.
2. Wash solution is mixed until NaOH pellets are completely dissolved.
3. Rinsed the slide dishes completely with water. Added one rack of Gold Seal Microslides per slide dish. This is followed by Pouring of wash solution over slides and covering it.
4. Slides are shaken gently for 2 h in wash solution. Although the slides come "pre-washed," there are still significant amounts of leftover oils and debris. This basic washing is essential to completely clean the slides before application of the poly-L-lysine.
5. Slides are to be rinsed with several liters of Milli-Q H_2O. Fill a large container with H_2O so that we can thoroughly dump the rinsed over slides. The slides should sit in the water during the preparation of the poly-L-lysine solution.
6. Poly-L-lysine solution is prepared (care is to be taken to use plastic-ware when preparing poly-L-lysine[1]): 180 slides:
   1500 mL H_2O.
   205 mL 1× PBS (filtered).
   352 mL poly-L-lysine.
   Mix ingredients on stir plate in plastic beaker in order listed above.
7. Before adding poly-L-lysine, dump H_2O from slides. Immediately after adding poly-L-lysine, pour solution over slides.
8. The slides are shaken gently in poly-L-lysine for 30 min.
9. The slides are rinsed again with Milli-Q H_2O in same fashion as described above.
10. The slides are spinned in tabletop at 600 rpm until dry (~5 min).
11. To ensure slides are completely dry, racks are placed in 50 °C vacuum oven for at least 5 min. If a vacuum oven is not available, the slides are placed in a standard oven for a longer amount of time.
12. Slides are stored in a clean plastic slide box. Slides need to be stored for a minimum of 14 days before spotting DNA. It is not

[1] Poly-L-lysine solution may be reused in case we need to prepare a large volume of slides. In order to complete the steps, lysine solution is to be filtered after step 9 and stored at 4 °C in a plastic container. Whenever it is ready to use, the solution is distributed to slide dishes and added an additional 10–15 mL of new poly-L-lysine to each dish. This process needs to be repeated a maximum of 6 times.

advisable for using slides that are more than 4 months old since it may cause degradation of poly-L-lysine.

6 Protocol for Amplifying Products to Print on Array

There are many processes existing for the preparation of products for spotting on arrays. Amplification options are largely regulated by many factors. These include sequence availability and availability of commercial primers. For sequenced genomes, specific primers are designed for amplification from a genomic template. In other words, amplification can occur with common primers on sequenced clones or on synthesized templates containing common 5′ and 3′ regions. As a general rule, it is recommended to invest in primer sets not in template synthesis, since primers have greater shelf-life than templates. It has been seen that long template stocks can hydrolyze if kept for long during storage and usage, which lead to unsuccessful subsequent amplifications.

For genomes that are not sequenced, random genomic libraries can be constructed and common primers can be employed in colony PCR. Following hybridizations, the next step is sequencing for clones as necessary. If we are using genomic DNA as a template for the PCR, it is recommended to first run a 25 μL reaction, then use 0.5 μL of this PCR product for the template in our 100 μL reaction. This reduces genomic DNA contamination in our final product. All PCR reactions are conducted in a 96-well format with 100 μL reaction volume. Individual template and primer input need to be optimized for each system. PCR reactions are conducted in 96-well thermocyclers. Reactions are assayed via 96-well agarose gel (4 μL run per sample). The concentration of the final PCR products is not critical. In case we run a 100 μL reaction and a band can be observed on the gel, then it is presumed to have sufficient product to array.

7 Protocol for Cleanup and 384-Well Arraying of PCR Products

Notes

1. Millipore microscreen 96-well filter plates can be successfully used for the purification of PCR products. The cost is more than that of precipitation, but time and effort are saved. The Biorobotics TAS array printer at the AGAC can be effectively used with Nunc 384 well flat-bottom plates.
2. In case the full plate of samples is lacking, they should be loaded on to the plate in 4 × 4 blocks starting with wells A–

D 1–4 and moving down the plate (next E–H 1–4, then I–L 1–4, etc.)

For ethanol precipitation:

1. 96-well reactions are to be run on 1% agarose gel followed by documentation of the results. In the next step, 10 μL sodium acetate, pH 5.5 and 110 μL isopropanol at room temperature were added.
2. Transferred the reactions to a Costar 96-well U-bottom Polypropylene Plate. In order to save on pipette tips, added IPA to Costar plates initially followed by the transfer of PCR + NaOAC to IPA. The next step includes centrifugation of 18 plates simultaneously in an RC3B centrifuge using 6-plate carrier adapters. Care should be taken to tap together the plates correctly.
3. The plates are spinned at 4500 rpm for at least 2 h.
4. A 12-channel Wheaton Aspiration Adapter is generally used to carefully aspirate solution. An 8- or 12-channel pipette is equally applicable.
5. Added 100 μL of 70% EtOH followed by spinning of the plates for another hour at 4500 rpm.
6. It is to be aspirated again and air dried in 96-well speed-vac.
7. PCR products need to be resuspended in 20 μL of H_2O for at least 18 h.
8. Transferred products to 384-well printing plates.
9. Dried plates down in speed-vac, resuspended products in an appropriate volume of 3× SSC. In order to print with the Biorobotics spotter, we need a final volume of at least 20 μL in the Nunc plates.
10. Plates are resuspended for at least overnight prior to printing.

8 Protocol for Post-processing Microarrays

Some of the salient points are to be kept in mind.

Before beginning, water is boiled on heat plate or in microwave, sufficient to cover slide rack.

- The heat block is turned on (highest setting) for the purpose of snap drying.
- Care is to be taken to use gloves without any powder, applicable for the entire protocol.
- No solution should be splashed on array during hydration and heat fixing.

8.1 Hydration/Heat Fixing

1. Take about 20–30 slides for processing.
2. It is necessary to designate array area after processing. So, two lines above and below center need to be sketched on the back of slide. Postprocessing causes the array to disappear.
3. 100 mL 1× SSC is poured into hydration tray followed by warming on slide warmer at medium setting (37–40 °C—a heat block is also equally applicable to warm the SSC).
4. The slide array is placed side down and observed the spots till proper hydration is attained. Hydration times will differ depending on printing. The time may be standardized based on the requirements. If not hydrated properly, it may cause spots to be too small for proper quantitation. Similarly, overhydration may lead to spots to run together. The use of hydration is to "fill in" the spotted DNA/SSC that dries soon in a ring after spotting. Good hydration will lead the DNA to spread out evenly over the entire area of the spot.
5. After proper hydration is attained, the slide is snap dried immediately by setting the slide array side up on the heat block (set at ~140 °C) for a few seconds till the array is properly dried.
6. Placed the slides in rack.

8.2 Surface Blocking

(**Steps 1–5** need to be performed in a fume hood).

It is essential that succinic anhydride is to be stored in vacuum desiccator till ready to use.

1. Measured 335 mL 1-methyl-2-pyrrolidinone into designated clean, dry slide dish with a stir bar. It is to be kept in mind that 1-methyl-2-pyrrolidinone should be colorless, if yellow coloration develops; it needs to be discarded immediately.
2. Dissolved 5.5 g succinic anhydride completely.
3. Soon succinic anhydride dissolves, added 15 mL of 1 M NaBorate (pH 8.0) and submerged the slides in solution. Plunged the slides rapidly in blocking solution and shaken evenly under level of solution for at least 1 min (NaBorate is made up of boric acid and pH is adjusted with NaOH. The succinic anhydride reacts with the poly-L-lysine coating, thereby preventing nonspecific hybridization on the slide).
4. Soaked slides in solution on shaker for 15 min.
5. Fifteen minutes prior to incubation, reduced heat on boiling water to make the temp as ~95 °C. Drained the excess of blocking solution off slides and transferred the slide rack to almost boiling water, plunged for some time followed by incubation for 90 s. It leads to the denaturation of the spotted DNA and makes it accessible for hybridization.

6. Transferred rack to dish of 95% EtOH and plunged 5×, followed by centrifugation.
7. Arrays may be utilized immediately or can be stored for future use.

9 Protocol for Total RNA Isolation

This protocol has already been described earlier.

10 Protocol for Reverse Transcription and Amino-Allyl Coupling

The protocol for reverse transcription has already been discussed in earlier chapter (Chapter 2).

In Chapter 2, we had discussed the protocol with eukaryotic RNA primarily. In case of poly-A RNA removed the random hexamer (pdN6) from the priming and input for at least 2 μg of RNA. Optimizations are essential. For bacterial total RNA, it is desirable to use 10 μg RNA primed with 30 μg of hexamers.

1. Incubated RNA and oligo dT at 70 °C for 10 min (use of PCR tubes and thermocycler for these steps are desirable).
2. Chilled on ice for 10 min.
   (a) cDNA synthesis process is initiated.
   (b) Hydrolysis.
3. Added and Mixed:
   10 μL 1 N NaOH.
   10 μL0.5 M EDTA.
4. Incubated the above solution for 15 min at 65 °C.
5. Neutralized with addition of 25 μL 1 M Tris–HCl pH 7.4, mixing well.
6. Samples can be stored at 4 °C overnight at this point.
   (c) Cleaning up.
7. In order to continue with the amino-allyl dye coupling procedure, it is essential to remove Tris from the reaction for the prevention of the monofunctional NHS-ester Cye-dyes from coupling to free amine groups in solution.
8. Fill up one Microcon 30 concentrator with 450 μL water.
9. Added neutralized reaction.
10. Spinned at 12 K for 8 min.
11. The flo-thru is dumped.
12. Repeated the process 2× with refilling the original filter.

13. Eluted (turned on microcon upside down in clean tube and spin at 4000 rpm for 4 min). At this point, a volume over 150 μL needs to be concentrated again with a second microcon filter.
14. Dried the eluate in speed vac.
15. Samples can be stored at −20 °C indefinitely.
    (d) Coupling.
16. NOTE: Cy-dyes are sensitive to light. Overhead fluorescent lighting is to be avoided as much as possible. Monofunctional Cy-dyes should be stored at 4 °C.
17. Resuspended cDNA pellet in 9 μL NaBicarbonate Buffer (pH 9.0). Kept for 10–15 min at room temperature to ensure resuspension.
18. Transferred the entire 9 μL volume into tube containing the dried Cy-dye aliquot. Cy3 is used for one sample and Cy5 is for the other.
19. Mixed by pipetting.
20. Incubated for 1 h at room temperature in the dark.
    Making dye aliquots:
21. If using fresh tube of Cy3 or Cy5, resuspended the entire tube in 32 μL DMSO.
22. The content was aliquoted in 4 μL × 8 tubes and dried immediately in speed vac. These aliquots may be stored indefinitely at 4 °C.
23. Note: Decreasing the number of aliquots/dye tube may increase the signal strength.
    (e) Quenching and cleanup.
    An important step is to be kept in mind before combining Cy3 and Cy5 samples for hybridizations. The reactions should be quenched to prevent cross-coupling.
23. In order to accomplish the process, added 4.5 μL 4 M hydroxylamine, followed by incubation for 15 min at room in dark.
    In order to remove unincorporated/quenched Cy dyes, PCR purification kits are employed.
24. Mixed Cy3 and Cy5 reactions.
25. Added 70 μL water.
26. Then added 500 μL Buffer PB.
27. Applied to Qia-quick column followed by spinning at 13,000 rpm for 30–60 s.
28. The next step is the process of aspirate off flo-thru.
29. Added 750 μL Buffer PE and spinned at 30–60 s.
30. The aspirate dumped off flo-thru and repeated the step.

31. Again aspirated flo-thru followed by spinning for 1 min at high speed in order to dry column.
32. Transferred to fresh centrifuge tube.
33. Added 30 μL Buffer EB to center of filter and let sit 1 min at RT.
34. Spinned at 13,000 rpm for 1 min.
35. Repeated elution step again.
    (f) Hybridization Preparation.
36. Dried down Qia-quick eluate in speed vac. The volume is adjusted to 15 μL with water.
37. Added 3 μL 20× SSC 1.5 μL polyA (10 mg/mL).

An alternative blocker may be used such as salmon sperm DNA if we are using total bacterial RNA.

Optional: Filter probe in Millipore 0.45 μm membrane: Pre-wet with 10 μL water and spin through at 8K in microcentrifuge. Remove flo-thru and deposit samples on inside tube as drop, not on filter. Spin at 10K and remove sample to 0.5 mL tube.

At this point reactions can be stored for up to 48 h at 4 °C before hybridization. Care should be taken not to store at −20 °C.

11 Protocol for Array Hybridization

1. Prepare probe as described at end of labeling protocol.
   Cy3+Cy5 samples: 15.0 μL.

   20× SSC: 3.0 μL.

   Blocker (10 mg/mL): 1.5 μL.
2. Added 0.45 μL 10%SDS.
3. Boiled probe for 2 min at 100 °C, followed by cooling for 5–10 min at room temp. Boiling aids the sample to denature so that it becomes accessible for hybridization.
4. The slides are labeled with etcher and set in hybridization chamber. About 10 μL drop of 3× SSC is put at one end of the hybridization chamber and another at the other end of the chamber on top of the slide. A constant humidity in the chamber should be maintained during hybridizations. In case the 3× SSC is not used, the array will dry out.
5. Cleaned a LifterSlip or coverslip with EtOH and Kimwipes. Placed slip on array with the help of forceps.

   LifterSlip: Should be applied with dull white strips on the long axis of the chip touching the glass, which creates a platform, allowing even distribution of the hybridization solution across the array. Slowly pipetted the probe under one cover of the slip till the total array surface is covered.

Coverslip: Put the probe on to one edge of the array in a drop (avoiding bubbles). The coverslip needs to be placed on the array starting at the edge next to the probe when we slowly lower the other end on to the slide. It is to be practiced with water first in order to avoid trapping bubbles under the coverslip. If bubbles do get trapped under the coverslip, most will move out once the slide is placed in the water bath.

6. Tightly screwed down hybridization chamber lid and carefully placed chamber in a 63 °C water bath. Precautions should be taken to keep array completely flat during transfer and hybridizations.
7. Allowed hybridization to run for at least 5 h but not more than 16 h. Other systems may benefit by different temperatures and times.

12 Protocol for Array Washing

- It is advisable that all wash solutions be filtered before using.
- Prepared wash solutions in glass slide dishes, care needs to be taken for each dish to have its own rack.

Wash solution I	Wash solution II
340 mL Millipure water	350 mL Millipure water
10 mL 20× SSC	1 mL 20× SSC
1 mL 10% SDS	

1. Carefully removed array from water bath, making sure to keep chamber level. Array is dried with paper towels and attempted to "wick" any water away from chamber seams.
2. Unscrewed the chamber and removed the array. Some water can enter the chamber and pool under slide at this time. In that case, it is helpful to use a pair of forceps to pry array away from chamber.
3. Keep array level while submerging in Wash I. Once submerged tilted the array and dumped off coverslip very carefully. It may be necessary to lightly swish array under solution to dislodge the slip.
4. Once slip is off and lying on bottom of slide dish, put array in rack and removes any additional hybridizations from water

bath. The slides are not allowed to dry out at any point during the washes. When all the slides are in Wash I, plunge racked up and down 10–20 times (at least 1 min).

5. Individually transferred slides to slide dish that contains Wash II; it is not recommended to transfer entire slide rack as this will cause too much SDS carryover. Plunged 10–20 times again.
6. Dried the array in room temperature tabletop centrifuge at 600 rpm for 5 min.
7. Tried to scan array within hours of washing as the Cy dyes are instable and degrade differentially.

Currently, with the rapid evolvement of high-throughput sequencing technique, next-generation sequencing techniques, DNA microarray is losing its importance. In the subsequent chapters, we shall describe NGS techniques in subsequent chapters.

13 Uses of DNA Microarray

DNA microarray is useful for studying gene expression profiling (i.e., upregulation and down regulation) for a biological event. Gene mutations can also be studied efficiently with this technique. Currently, DNA microarrays are useful in clinical diagnostic tests for important diseases. It is equally applicable for pharmacogenomics studies [1–3]. The technique may be used to screen the best drugs that may be applicable. Since the entire matter of action of drug is genetically controlled, DNA microarray is very much applicable for choice of drug.

Large-scale population studies are very much essential for determination of particular mutations for breast cancer or other diseases [1]. DNA microarray technique is equally helpful for studies on patient with tumor (Fig. 4).

DNA microarray: Also known as DNA Chip.

Few salient points:

1. It allows simultaneous measurement of the level of transcription for every gene in a genome (gene expression).
2. Microarray detects mRNA or rather the more stable cDNA.

14 The Birth of the Modern DNA Array

In the late 1990s and 2000s, the progress of DNA array technology was at its peak, as both new methods of production and fluorescent detection were employed. It also increases the knowledge of the DNA sequences of multiple genomes, provided the raw information required to ensure that arrays could be constructed that

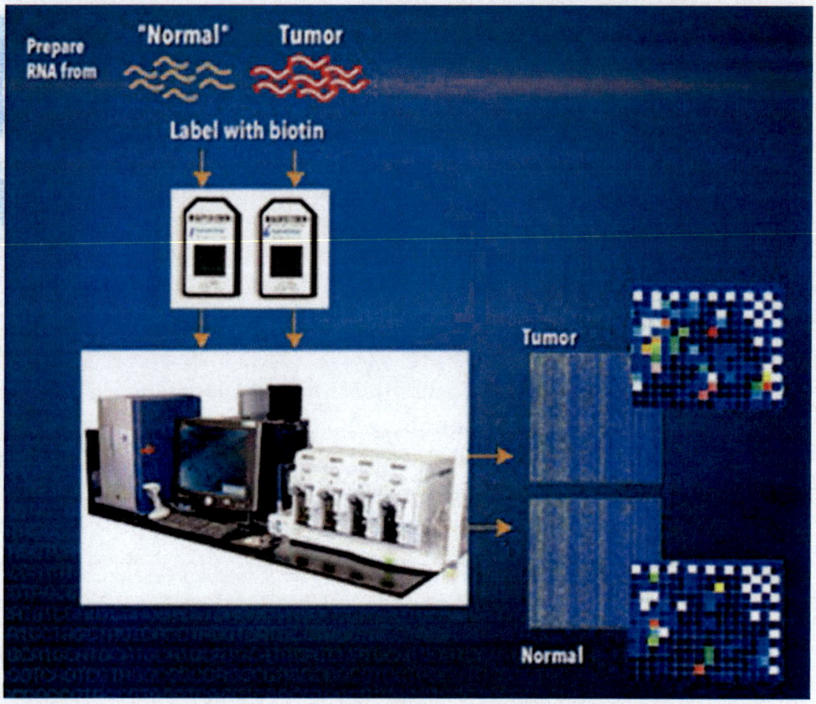

Fig. 4 Schematic representation of different pattern for tumor patient and normal patient

represent the genes in a genome, all the sequence in a genome or a large fraction of the sequence variation in a genome. During this period, there was a gradual transition from spotting relatively long DNAs on arrays for the production of arrays using 25–60 bp oligos. The transition to oligo arrays was possible through the process of increasing amounts of publicly available DNA sequence information. The use of oligos (as opposed to longer sequences) also leads to an increase in specificity for the intended binding target as oligos could be designed in such a way to target regions of genes or the genomes that were most dissimilar from other genes or regions. Three basic types of arrays came into play during this time frame:

(a) Spotted arrays on glass,

(b) In situ synthesized arrays and

(c) Self-assembled arrays (Fig. 5).

14.1 Applications of Microarrays

14.1.1 Gene Expression Analysis

The primary application of DNA microarrays is that it is involved with the measurement of gene expression levels (Fig. 6). During the application, RNA is extracted from the cells of interest and followed by labeling, either directly or converted to a labeled cDNA or converted to a T7 RNA promoter tailed cDNA, which is later converted to cRNA by the Eberwine amplification process (Van Gelder et al. 1990).

A wide range of methods have been developed for labeling of the cDNA or cRNA including:

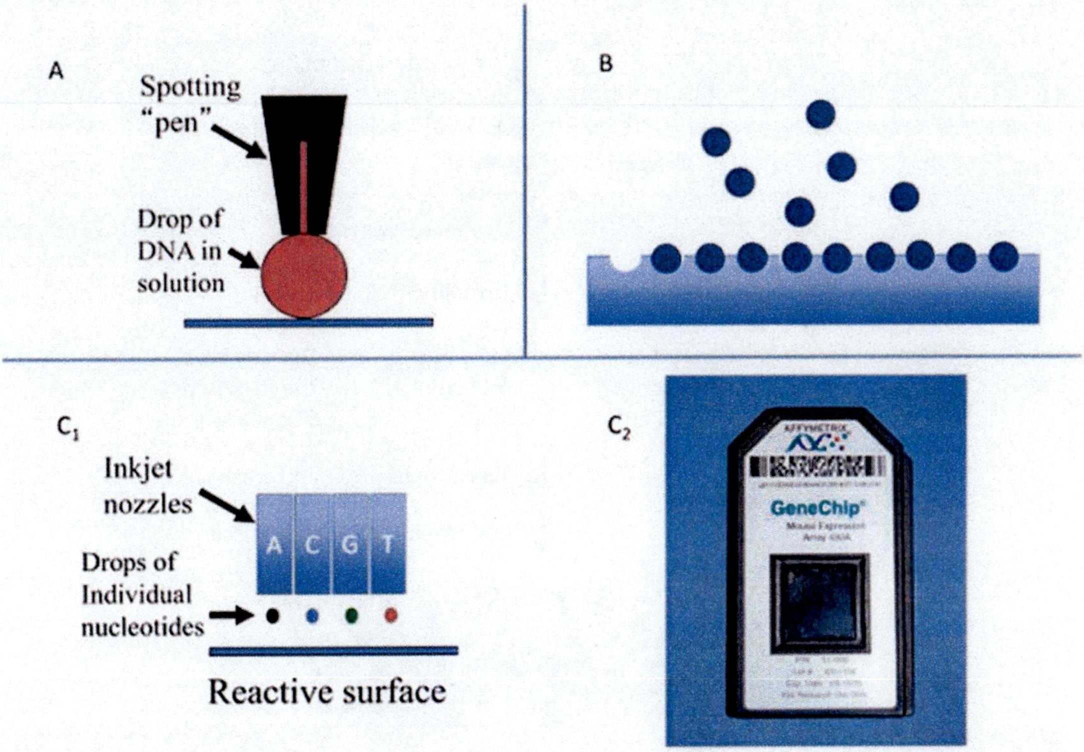

Fig. 5 Three basic types of microarrays: (**a**) Spotted arrays on glass, (**b**) self-assembled *arrays* and (**c**) in situ synthesized arrays. (**a**) With spotted arrays, a "pen" (or multiple pens) is dipped into solutions composed of the DNA of interest and physically deposited on a 1" × 3" glass microscope slide. The glass slide surface is coated with something to help retain the DNA such as polylysine {DeRisi 1997 #28191}, a silane {Call 2001 #28277}, or a chemically reactive surface {Rogers 1999 #28278} (to which chemically reactive oligos or PCR products would be added). (**b**) Self-assembled arrays can be constructed through the application of a collection of beads possessing a diverse set of oligos to a surface with pits the size of the beads. After the array is constructed, a series of hybridizations determine which oligo is in what position on each unique array [4–7]. (**c_1** and **c_2**) In situ synthesized arrays can be constructed by inkjet oligo synthesis methods (**c_1**) or by photolithographic methods such as used by Affymetrix (**c_2**)

1. the incorporation of fluorescently labeled nucleotides during the synthesis;
2. the incorporation of biotin-labeled nucleotide that is subsequently stained fluorescently labeled streptavidin;
3. the incorporation of a modified reactive nucleotide to which a fluorescent tag is added later; and
4. a variety of signal amplification methods.

The two most frequently employed methods are the incorporation of fluorescently labeled nucleotides in the cRNA or cDNA synthesis step or the incorporation of a biotin-labeled nucleotide in the cRNA synthesis step (as is done by Affymetrix).

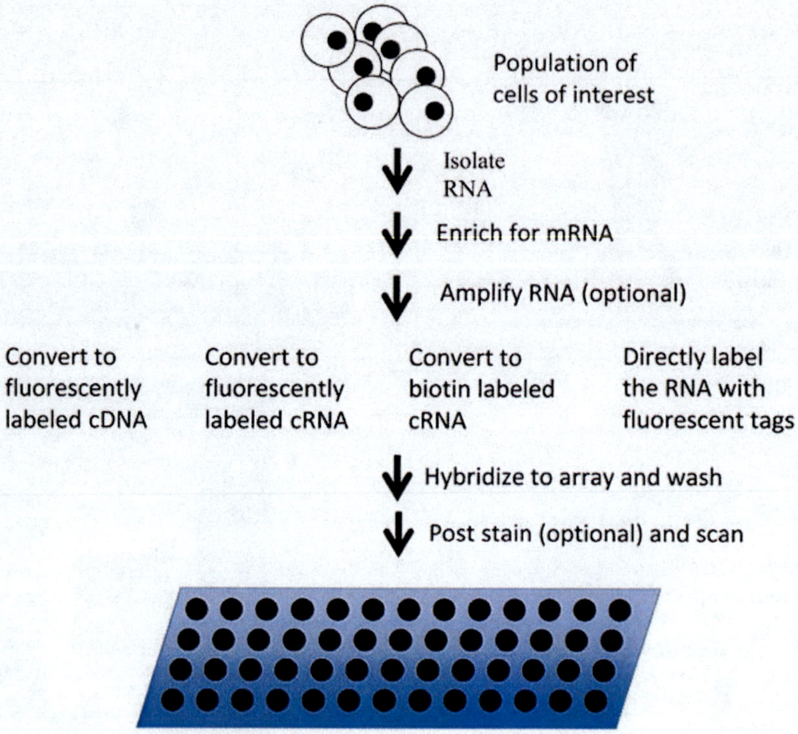

Fig. 6 Schematic representation of DNA microarray

14.2 Gene Expression Analysis via Microarrays

RNA is isolated from the sample of interest and enriched for messenger RNA. In eukaryotes, poly-A tailed mRNAs are enriched with affinity purification with oligo dT beads or columns. However, in prokaryotes, unselected RNA is depleted for ribosomal sequences with bead or columns coated with sequences that are complementary to 16s. As enriched RNA is available, it is optionally amplified and labeled and the resulting labeled sample is hybridized to a microarray. The array is then washed to remove unbound sample. In case, the sample was labeled with biotin, the array is post-stained with fluorescently labeled streptavidin and washed again. The array is then scanned to determine fluorescence signal at each spot on the array. The stepwise protocol has been discussed Fig. 6.

The next step is the labeling of cRNA or cDNA followed hybridization to the microarray, the array is washed and the detection of signal is measured through fluorescence at each spot. In case of biotin labeled samples, the array is stained post-hybridization with fluorescently labeled streptavidin. Laser-induced fluorescence is normally measured with a scanning confocal microscope. The intensity of the signal(s) on each spot is considered as a measure of the expression level of the corresponding gene.

14.2.1 Transcription Factor Binding Analysis

Microarrays have also been used in combination with chromatin immunoprecipitation to determine the binding sites of transcription factors. Transcription factors (TFs) are cross-linked to DNA with formaldehyde and the DNA is fragmented. The TF(s) of interest (with the DNA to which they were bound still attached) are affinity purified either with an antibody to the TF or by tagging the transcription factor with peptide that is amenable to affinity chromatography (e.g., a FLAG-, HIS-, myc, or HA-tag). Following purification, the DNA is released from the transcription factor, amplified, labeled, and hybridized to the array. This technique is generally referred to as ***"ChIP-chip" for Chromatin Immuno-Precipitation on a "chip" or microarray***.

The binding site for TF is generally a distance away from the genes that they regulate, and the design of the array and size distribution of the fragment length are correlated. This means the array should contain probes that will investigate the region of DNA bound to the transcription factor. In case of bacteria or yeast, the intergenic regions are small and the same arrays used for gene expression work can be applied to ChIP-chip. In case of mammalian genomes, the intergenic regions are large and the TF binds many kbp away from the gene of interest sometimes. Hence, for mammalian genomes, oligo arrays with oligos spaced evenly across the entire genome are typically used for ChIP-chip experiments.

14.3 Genotyping

Microarrays is one of the most common methodology to detect single-nucleotide polymorphism (SNP). Several alternative approaches have been utilized to detect SNPs, but the most commonly used are allele discrimination by hybridization as used by Affymetrix. Other common methods are allele specific extension and ligation to a "bar-code" oligo that is hybridized to a universal array. The universal assay may be Illumina "Golden Gate Assay" or approaches where the arrayed DNA is extended across the SNP in a single-nucleotide extension reaction. The example of the later is the Arrayed Primer Extension assay or the Infinium Assay of Illumina. Figure 7 attempts to explain the detection approaches in more detail. Allelic discrimination by hybridization possesses the limitations in background due to nonspecific hybridization in complex genomes. For the reduction of this background, Affymetrix developed a PCR-based approach to reduce genomic complexity (Fig. 7).

To explain briefly, SNPs for their assay are selected to be between restriction sites that are <1 kb apart. Genomic DNA is fragmented with a restriction enzyme followed by end repairing and ligation of the adapters for PCR to the fragments. PCR is performed under conditions that selectively amplify products of <1 kb in size. This methodology decreases genomic complexity by approximately 50-fold resulting in a corresponding increase in signal to noise on the array.

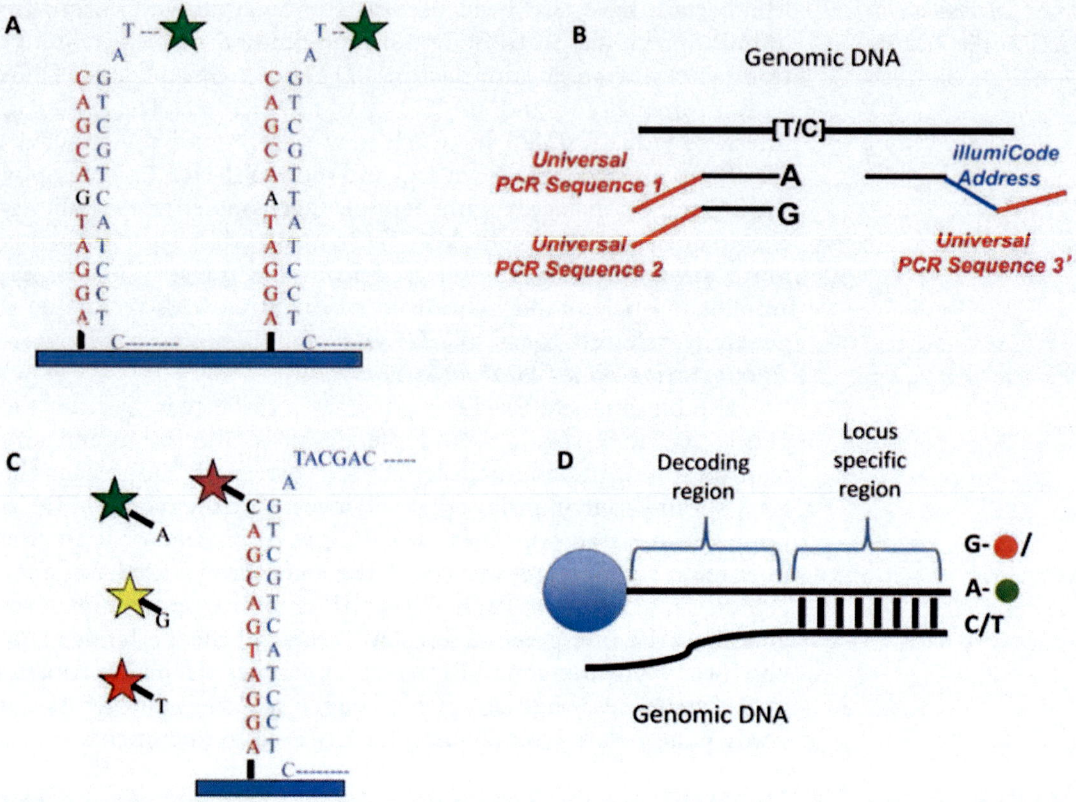

Fig. 7 PCR-based approach to reduce genomic complexity as developed by Affymetrix. In this section, we will describe, different SNP detection strategies for arrays. (**a**) Allele discrimination by hybridization: During this process, oligos that are complimentary to each allele are placed on the array and labeled genomic DNA is hybridized to the array. The variant position is located in the center of the oligo (typically, 25 bp on Affymetrix arrays). It is to be noted that this position has the greatest effect on hybridization. Typically, multiple array positions are employed for each allele to improve signal to noise. (**b**) Illumina's "Golden Gate Assay": In this case, two allele specific oligos are each tailed with a different universal primer (1 and 2) and hybridized in solution to genomic DNA. A third oligo that is complementary to the same locus is tailed with a "barcode" sequence and a third universal primer (3). Polymerase is employed to extend the allele specific primers across the genomic sequence and the extended products are ligated to the third oligo. PCR is conducted using primers complimentary to universal sequences 1, 2, and 3. The PCR primers complimentary to the universal sequences 1 and 2 are labeled with a unique fluorophore. The barcode sequence on the third oligo leads to the PCR product to be uniquely identified on an array consisting of oligos complimentary to the barcode sequence. The utilization of multiple barcodes (one for each locus of interest) permits the assay to be multiplexed to sample many loci. (**c**) Arrayed primer extension (APEX: In this assay, the array consists of DNA oriented with the 5′ end attached to the array and the 3′ end stopping one nucleotide short of the SNP. Genomic DNA is fragmented followed by hybridization to the array. The oligo on the array is extended in single-nucleotide dye terminator sequencing reaction. (**d**) Illumina's Infinium assay: This assay resembles to the APEX assay except that the oligo to be extended is on a bead. The single nucleotide that is added is labeled with a nucleotide specific hapten as opposed to a fluorophore. The haptens are in turn detected by staining with fluorescently labeled proteins binding to each hapten

Both the Affymetrix and the Illumina methods for SNP genotyping were proved to be highly successful and are highly used. Today SNP arrays are able to detect >1 M different human SNPs that are available from both vendors. Call rates, i.e., the fraction of SNPs on the array that can be reliable called, and reproducibility of SNP calls exceed 99.5%. Additionally, the same arrays or variations can also be used to detect copy number variants.

14.4 Data Standards and Data Exchange

Other than DNA sequencing, microarrays were most probably the earliest technology that allowed biologists to vast amounts of complex digital data. With the advent of the technology, it became essential to reproduce a given microarray experiment, a detailing of the array, the sample, the protocols, and the available data analysis methods. Additionally, it also became essential to have easy access to the raw and processed data. These would in turn allow others to perform analyses and meta-analyses (on combinations of data) for the original data .To meet these issues of reproducible science and data exchange, members of the Microarray Gene Expression Data Society (now the Function Genomics Data Society—www.FGED.org) performed and formulated the MIAME (Minimum Information About a Microarray Experiment) standards for the description of microarray experiments and for the exchange of microarray data. These works influenced the creation of public databases for microarray data and subsequent standards efforts in other areas.

15 Limitations of DNA Microarrays

In spite of advantages, there exist certain limitations.

1. First, arrays give an *indirect* measure of relative concentration. This means the signal measured at a certain position on a microarray is assumed to be proportional to the concentration of a single species in solution that can hybridize to that location. However, due to the kinetics of hybridization, the signal level at a particular location on the array is not linearly proportional to concentration of the species hybridizing to the array. At higher concentrations, the array will become saturated and at low concentrations, equilibrium favors no binding. Hence, the signal is linear only over a limited range of concentrations in solution.
2. Second, especially for complex mammalian genomes, it is quite difficult (if not impossible) to design arrays in which multiple related DNA/RNA sequences will not bind to the same probe on the array. A sequence on an array that was designed to detect "gene A" may also detect "genes B, C, and D" if those genes have significant sequence homology to gene A. This can

particularly problematic for gene families and for genes with multiple splice variants. It is considered that it is advisable to design arrays particularly to detect splice variants by making array probes either to each exon in the genome or to exon junctions. Moreover, it is tough to design arrays that will uniquely detect every exon or gene in genomes with multiple related genes.

Finally, a DNA array is able to detect sequences only, which the array was designed to detect. As a result, if the solution being hybridized to the array contains RNA or DNA species for which there is no complimentary sequence on the array, detection is not possible for those species. For gene expression analysis, it implies that genes that have not yet been annotated in a genome will not be represented on the array. Additionally, noncoding RNAs that are not yet recognized as expressed are typically not represented on an array. For highly variable genomes such as those from bacteria, arrays are specifically designed using information from the genome of a reference strain. Such arrays may miss a large fraction of the genes present in a given isolate of the same species. For example, in the bacterial species *Aggregatibacter actinomycetemcomitans*, the gene content differs by as much as 20% between any two isolates [8]. Hence, an array designed using gene annotation from a "reference isolate" will not contain many of the genes found in other isolates.

16 The Future of DNA Arrays

It is essential to have an unbiased methodology for direct measurement for all DNA or RNA species present in a particular sample. The advent of next-generation sequencing technologies along with the rapid decrease in the cost of sequencing (http://www.genome.gov/sequencingcosts/) has now enabled sequencing cost competitive with microarrays for all assays with the possible exception of genotyping. With similar cost, sequencing has many advantages relative to microarrays. Sequencing is a direct mode of measurement where nucleic acids are present in solution. One need to only count the number of a given type of sequences present to detect its abundance. Counting sequences is linear with concentration and the signal-to-noise ratio can be obtained by sequencing is only limited by the number of reads used for each sample.

Sequencing is a relatively unbiased approach to measure which nucleic acids are present in solution. The major limitation for sequencing is that sample preparation and different enzymes can bias sequencing counts. However, it differs from DNA arrays. Sequencing is no way related to prior knowledge of nucleic acids present. Sequencing can independently detect closely related gene sequences, novel splice forms, or RNA editing that may be missed due to cross-hybridization on DNA microarrays. Due to these advantages and the decreasing cost of sequencing, DNA arrays are being rapidly replaced by sequencing for almost every assay that earlier performed on microarrays [9]. As the cost of sequencing is currently dropping by a factor of 2 every 5 months, it's likely that DNA arrays will be fully replaced by sequencing methods within the next 5–10 years.

References

1. Rew DA (2001) DNA microarray technology in cancer research. Eur J Surg Oncol 27 (5):504–508
2. El-Ashker M, Hotzel H, Gwida M, El-Beskawy M, Silaghi C, Tomaso H (2015) Molecular biological identification of Babesia, Theileria, and Anaplasma species in cattle in Egypt using PCR assays, gene sequence analysis and a novel DNA microarray. Vet Parasitol 207 (3-4):329–334
3. Kuchipudi SV, Dunham SP, Chang K-C (2015) DNA microarray global gene expression analysis of influenza virus-infected chicken and duck cells. Genom Data 4:60–64
4. Gunderson KL, Kruglyak S, Graige MS, Garcia F, Kermani BG, Zhao C, Che D, Dickinson T, Wickham E, Bierle J, Doucet D, Milewski M, Yang R, Siegmund C, Haas J, Zhou L, Oliphant A, Fan JB, Barnard S, Chee MS (2004) Decoding randomly ordered DNA arrays. Genome Res 14(5):870–877. https://doi.org/10.1101/gr.2255804
5. Ferguson JA, Steemers FJ, Walt DR (2000) High-density fiber-optic DNA random microsphere array. Anal Chem 72:5618–5624
6. Walt DR (2000) Techview: molecular biology. Bead-based fiber-optic arrays. Science 287:451–452
7. Steemers FJ, Ferguson JA, Walt DR (2000) Screening unlabeled DNA targets with randomly ordered fiber-optic gene arrays. Nat Biotechnol 18:91–94
8. Kittichotirat W, Bumgarner RE, Asikainen S, Chen C. Identification of the pangenome and its components in 14 distinct Aggregatibacter actinomycetemcomitans strains by comparative genomic analysis. PLoS ONE. 2011;6:e22420
9. Wold B, Myers RM. Sequence census methods for functional genomics. Nature methods. 2008;5:19–21

Chapter 12

Epigenetics and DNA Methylation

Abstract

Epigenetics is a wide term representing molecular processes leading to modification of gene expression without bringing any DNA base sequence alterations. Major epigenetic mechanisms involved in gene expression modifications are DNA methylation, histone modification, gene imprinting, epigenetic writers, and transcription regulators. In this chapter, protocols for detection of all the epigenetic mechanisms have been discussed. The principle involved in DNA methylation includes conversion of methylated cytosine into thymine by sodium bisulfite treatment; each CpG site is targeted by two different probes. A histone modification is the process of a covalent post-translational modification (PTM) to histone protein including methylation, phosphorylation, acetylation, ubiquitylation, and sumoylation, which is the cause of alteration chromatin structure or recruiting histone modifiers. Genomic imprinting is an epigenetic phenomenon that leads to genes to be expressed in a parent-of-origin-specific manner. Both genomic region compaction and gene expression are regulated by epigenetic modifications, deposited by specific enzymes (known as "writers") and subsequently recognized by effector proteins ("readers") and removal of these marks by erasers, which all together regulate gene transcription, and defects may cause cancer initiation and progression. Epigenetic regulation through various nutritive substance leads to the development of a separate study referred to as ***nutrigenomics***, which are discussed in Chapter 12.

Keywords Epigenetics, DNA methylation, Histone modification, Gene imprinting, Epigenetic writers, Transcription regulators

1 Principles

Mere presence of gene does not guarantee its expression. It is not necessary that gene will express or even the gene expression varies from tissue to tissue and stage of development or biological process. Let us discuss with an example. Although "insulin gene" is present in each and every cell of the body, it is expressed only in pancreas. This is because of the epigenetic regulation. It has also been observed that the same gene is expressed differentially in different body organ.

Epigenetic processes are the prime factors to determine when and where specific genes are expressed. Epigenetics is a wide term representing molecular processes leading to modification of gene expression without bringing any DNA base sequence alterations.

Aruna Pal, *Protocols in Advanced Genomics and Allied Techniques*, Springer Protocols Handbooks,
https://doi.org/10.1007/978-1-0716-1818-9_12,

Another classical example for epigenetics is in lactation physiology of dairy cattle. Two-time milking could cause less methylation of lactation genes than one-time milking scenario. Let us consider a case of lactation pathology, say during mastitis, when DNA remethylation around a STAT5-binding enhancer site in the αS1-casein promoter could shut down αS1caesin synthesis abruptly.

Major epigenetic mechanisms involved in gene expression modifications are:

1. DNA methylation.
2. Histone modification.
3. Gene imprinting.
4. Epigenetic writers.
5. Transcription regulators.

DNA methylation as an epigenetic factor has a major role in the regulation of gene expression. Reports are available that defects in DNA methylation process are a causal factor for human diseases, e.g., human cancer cells [1].

2 Protocol for Studying Epigenetics

The Infinium Human Methylation 450 Bead-Chip® (Illumina, CA, USA; hereinafter termed the 450k array), which involves >480,000 CpG site probes, was released by Illumina (CA, USA) in 2011. Infinium Methylation EPIC kit is also available for analyzing robust methylation profiling. It is a microarray-based sequencing technology involved with extensive coverage of CpG islands, genes, and enhancers. It is a useful tool for studying epigenome-wide association studies.

2.1 Principles

This platform is dependent on the conversion of methylated cytosine into thymine by sodium bisulfite treatment. Each CpG site is targeted by two different probes (Infinium I design):

1. One probe designed to match the methylated version and
2. The other designed to match the unmethylated version of the corresponding CpG site. Additionally, Infinium II probes are also employed in the 450k array, when a single probe is designed for some target CpG loci. Differentially labeled nucleotides are employed to determine the corresponding methylation status.

During the process of DNA methylation in mammalian cells, a methyl group is added to the cytosine nucleotides leading to form a methylcytosine. These modifications generally occur on CpG dinucleotides; however, abundant DNA methylation in non-CpG contexts has recently been discussed at a genome-wide level in stem cells. At the biochemical level, the methylated cytosine can be differentiated, isolated, or marked by methylation-sensitive restriction digestion, methylcytosine immune precipitation, or bisulfite treatment. During detection, the methylated cytosine sequences can be measured by array hybridization or direct sequencing.

The distribution of DNA methylation is not even throughout the genome. Certain regions of DNA in which the frequency of the CpG sequence is more than in others are known as CpG islands (CGI). CGIs are usually located at the 5′-regions of housekeeping genes or other most expressed genes.

CpG sites within the CGIs of promoters are mostly unmethylated when the downstream genes are expressed. Thus, it differs from the CpG sites in the coding regions of a gene. This report led to the conclusion that methylation of CpG sites in the promoter region of a gene leads to the inhibition of gene expression.

For the occurrence of gene transcription, the gene promoter would be readily accessible to transcription factors and other regulatory units (e.g., enhancers). DNA methylation can directly inhibit transcription factor binding, leading to changes in chromatin structure, which restrict access of transcription factors to the gene promoter. As for example, methylated CpGs attract methyl-CpG-binding domain proteins that thereby involve "repressor complexes," resulting in histone modification.

3 Summary of Epigenetic Gene Regulation

DNA methylation and its pattern in adult cells determine the fate of cells, structure of chromatin, and activation of gene. For any biological process, DNA methylation is removed at fertilization and again established during embryogenesis. Imprinted genes retain their parental pattern of methylation, leading to parental patterns of expression. Patterns of histone modification parallel DNA methylation. Gene regions that are methylated are genetically inactive, highly condensed, and with special histone modifications. Gene regions that are active have little DNA methylation and distinct histone modifications (acetyl groups and H3K4methyl). Inactivation of X chromosome females is directly related to extensive CG island methylation on one chromosome leading to condensation, inactivation, and Barr body formation. In case of aging and cancer, gene alteration and CG island methylation is very common. CG islands methylation are observed only for X chromosome inactivation and tumor suppressors in cancer. Usefulness of

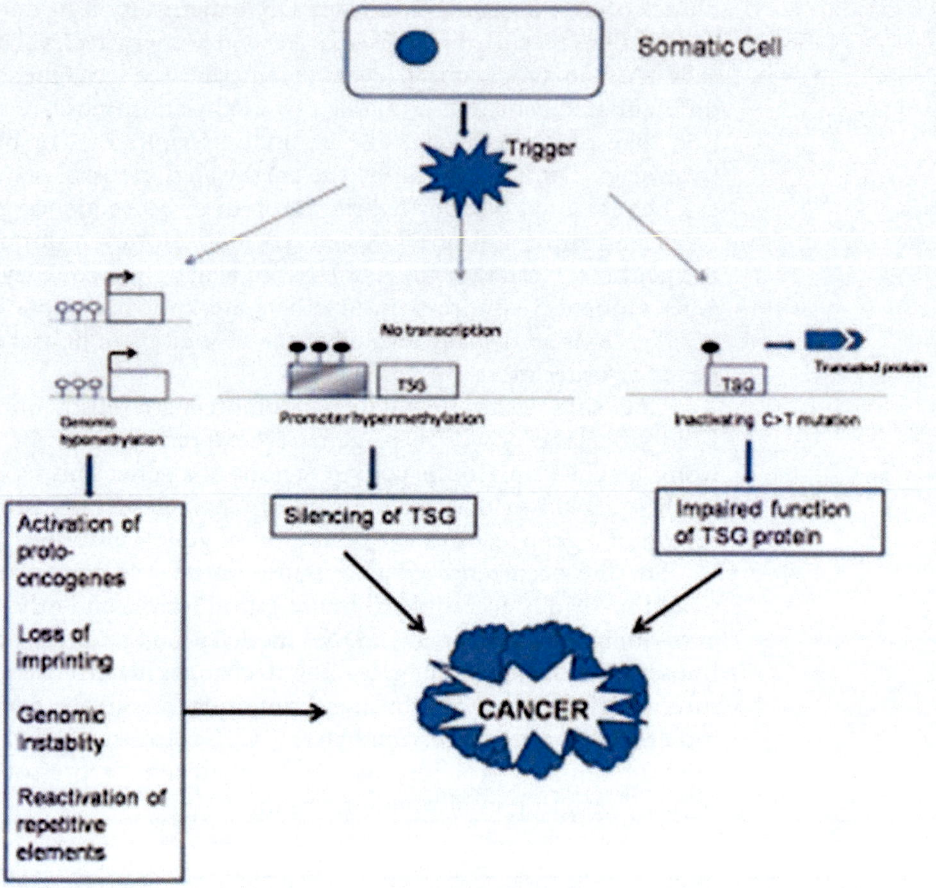

Fig. 1 Schematic representation of epigenetic regulation of cancer

epigenetics has manifold effects. Role of epigenetics in cancer has been detected. Figure 1 describes the role of epigenetic regulation of cancer as a flow diagram.

Epigenetics may be defined as the science for the studying heritable changes of DNA, not involving changes in DNA sequence, that regulate gene expression. DNA methylation and histone modification are the most common modes of epigenetic regulation. Diagrammatic illustrations for DNA methylation and histone modification have been depicted in Fig. 2a, b, respectively. Figure 3a–d explains different modes of cytosine methylation. Figure 4 clearly depicts that DNA methylation and histone modifications help to compartmentalize the genome into domains of different transcriptional potentials. The process of DNA methylation is also helpful in the sense that DNA methylation differentiates totipotent embryonic stem cells from unipotent adult stem cells (Fig. 5).

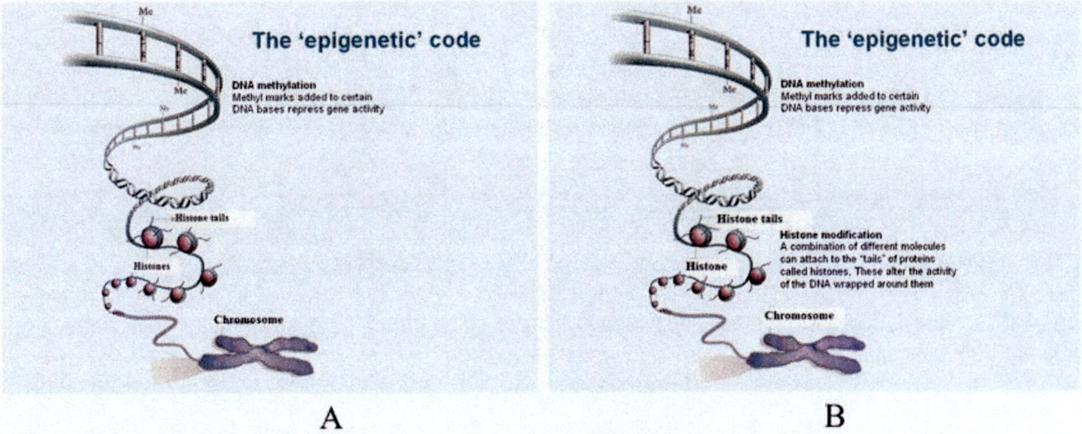

Fig. 2 DNA Methylation and the Epigenetic Code (**a**). Histone Modification (**b**)

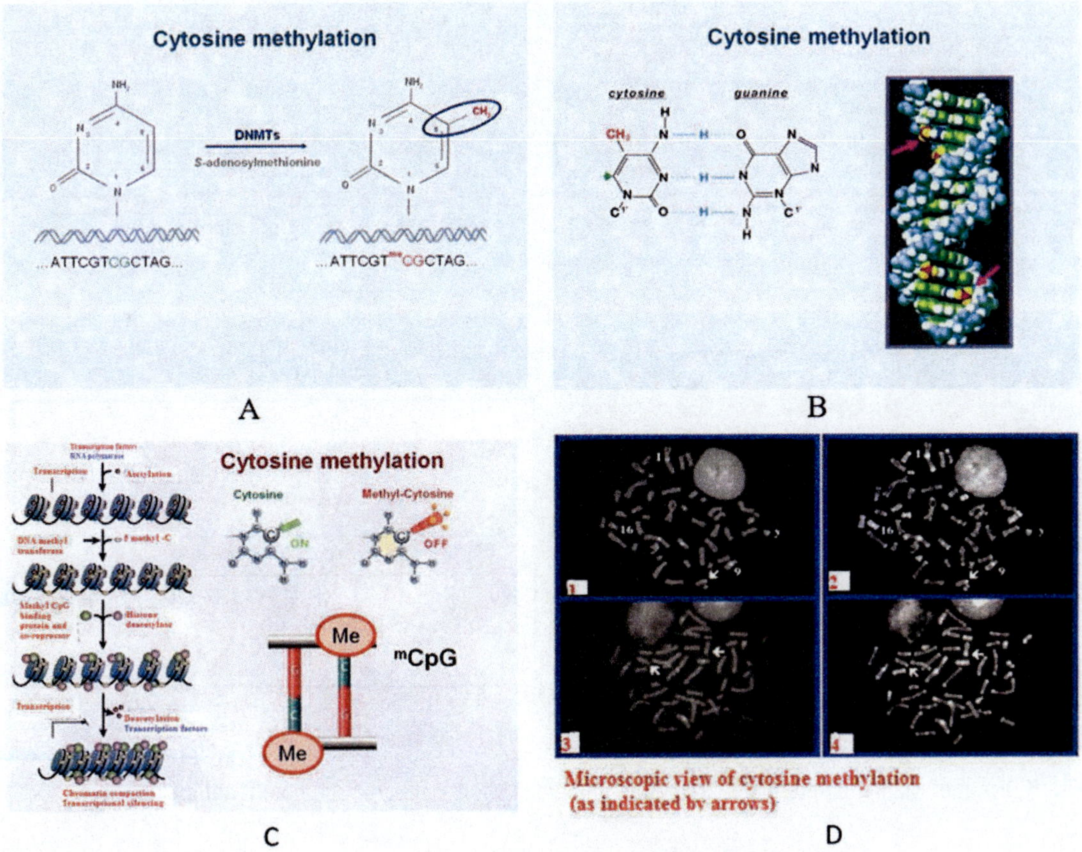

Fig. 3 Cytosine methylation maintains inactive-condensed chromatin state

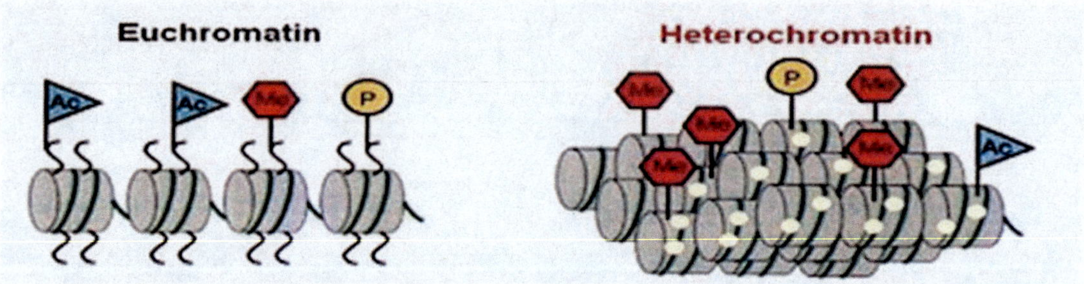

Fig. 4 DNA methylation and histone modifications help to compartmentalize the genome into domains of different transcriptional potentials

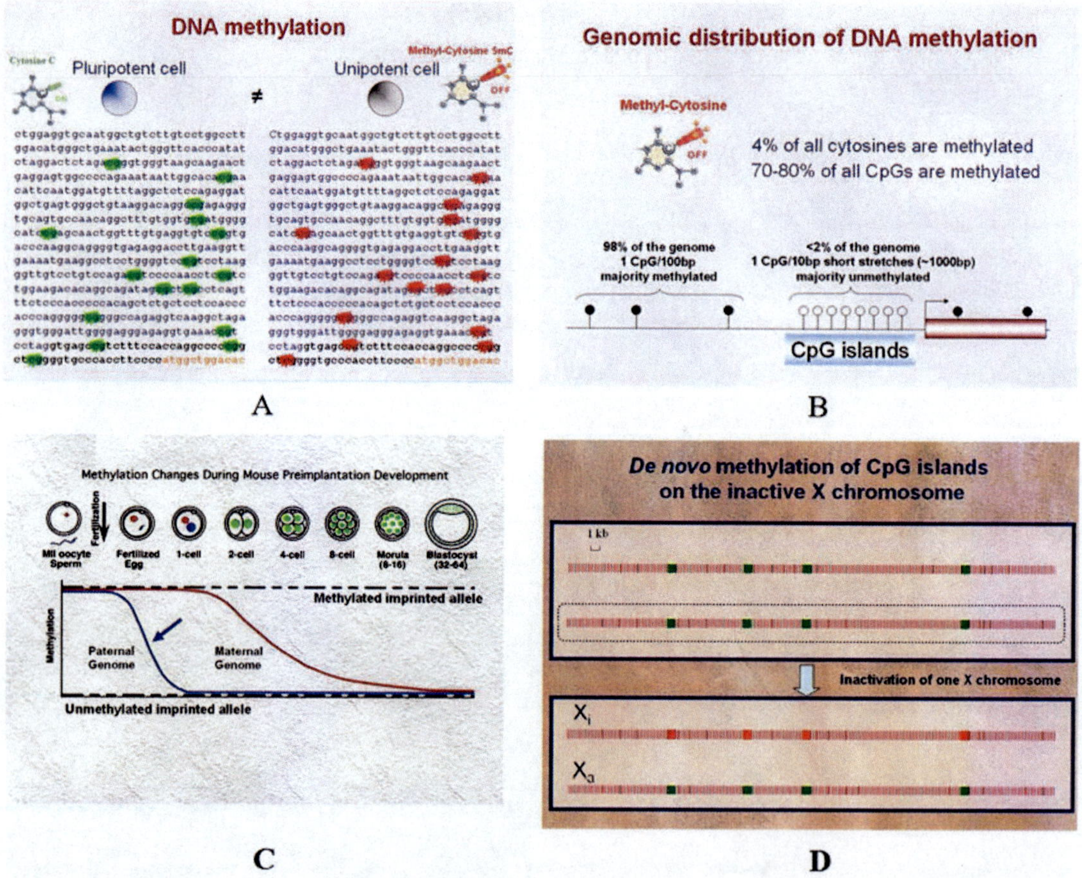

Fig. 5 DNA methylation differentiates totipotent embryonic stem cells from unipotent adult stem cells

4 DNA Methylation

4.1 5-Methyl Cytosine Is Found in Heterochromatic Regions

The distribution of cytosine methylation in mammals. Heterogeneity visible at cytogenetic scale associated with heterochromatic regions (right).

Differences between euchromatin and heterochromatin:

Euchromatin	Heterochromatin
It exhibits high histone acetylation	It exhibits low histone acetylation
It has low DNA methylation	It has low dense DNA methylation
H3-K4 methylation	H3K9 methylation
Stains lightly	Stains darkly

Gene Imprinting: Genomic imprinting is an equally important methodology for epigenetic regulation. Figure 6 depicts the visual representation of genome imprinting.

Nutrigenomics: Diet has an influence on gene expression. It may also be considered as an epigenetic factor. Figure 7 gives an overview.

5 DNA Methylation

5.1 Protocols

5.1.1 Bisulfite Modification (Conversion) of DNA

The basic principle of studying DNA methylation using sodium disulfite method involves conversion of unmethylated cytosine to uracils. The subsequent process involves detection of methylated cytosines with methylation specific PCR (MSP) technique. Bisulfite genomic sequencing method is also employed after PCR amplification with or without cloning. This protocol also explains the process for handling nanogram quantities of DNA.

5.2 Materials and Reagents

The following reagents were freshly prepared immediately before use:

- 10 mM hydroquinone (Sigma; #H 9003).
- 40.5% sodium bisulfite (Sigma; #S 8890, 8.1 g/20 mL).

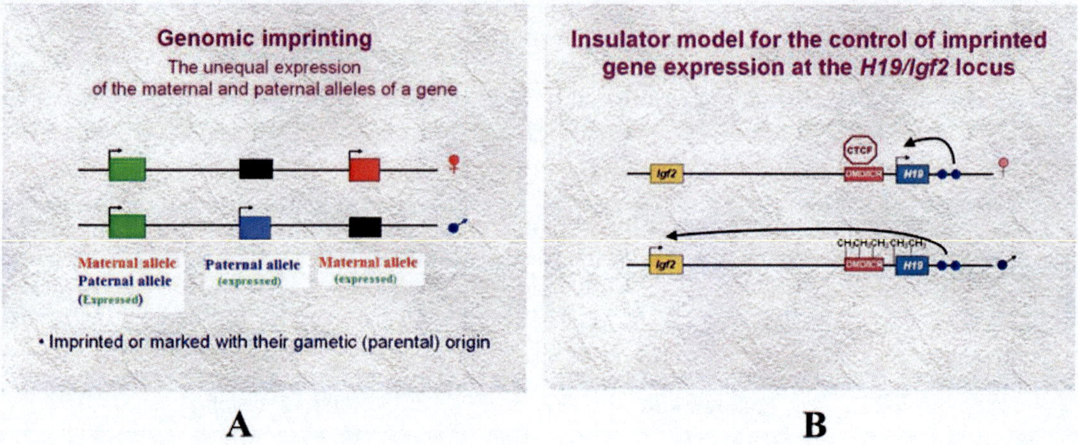

Fig. 6 Insulator model for the imprinting of H19 & Igf2 loci

Fig. 7 Example of nutrigenomics. (Source: Waterland and Jirtle, Mol Cell Boil (2003). Also Wolff and Cooney, Faseb J (1998))

About 8.1 g of sodium bisulfite was dissolved in 17 mL cold water, and pH was adjusted to 5.0 with 10 N NaOH. The volume was adjusted to 20 mL with the addition of water.

The above solutions were dissolved properly through gentle inversion of minimum amount. Care was taken so that the solution was kept in cool and dark place.

5.3 Procedure

1. Restriction digestion is to be employed with 2–5 μg of high molecular weight genomic DNA overnight. This process does not digest DNA within the region of interest.
2. DNA extraction is employed with phenol:chloroform:isoamyl alcohol (PCI; 25:24:1). The supernatant is precipitated with 1/10 volume 5 M ammonium acetate and twice volumes of ethanol for 15 min. The above mixture is centrifuged (14,000 rev/min) at room temperature. The supernatant is removed and the pellet is washed twice with 70% ethanol. The pellet is dried and resuspended in 100 μL TE (10 mM Tris-HCl, pH 7.5, 1 mM EDTA).
3. The denaturation of the DNA is followed by adding freshly prepared NaOH (3 M) to a final concentration of 0.3 M. Incubate the denatured DNA at 42 °C for 30 min.
4. To a siliconized microcentrifuge tube, add the following mixture:
   - 1020 μL 40.5% sodium bisulfite.
   - 60 μL of 10 mM hydroquinone.
   - 110 μL of DNA (+ NaOH).
   - 10 μL of water.
5. The above mixture is mixed gently and overlaid with mineral oil. The tube is covered with aluminum foil to shield from the light.
6. The covered tube is incubated at 55 °C for 16–18 h.
7. Purify DNA using the Geneclean II kit (Intermountain Scientific Corporation) or any other methods of your choice.
8. After purification, resuspend DNA and add TE to a final volume of 100 μL.
9. Denature the sample with freshly prepared NaOH and incubated at 37 °C for 15 min.
10. Neutralize by adding ammonium acetate (pH 7.0) to 3 M.
11. Precipitate the DNA with three volumes of ethanol and centrifuge for 10 min (14,000 rev/min) at room temperature. Wash twice with 70% ethanol and dry under a vacuum. Resuspend in 50 μL TE and store at −20 °C wrapped in foil. Care should be taken to use the treated DNA within 1 month as degradation is common in the cleaned and frozen sample.

6 Bisulfite Modification for Nanogram Quantities of DNA

In cases when smaller quantities of DNA are available, such as DNA from microdissected tissues, certain modifications are necessary.

The basic modifications are included in the method of extraction with the addition of mussel glycogen to precipitations.

The steps are the same as above and only changes are noted.

1. In **step 4**, the bisulfite reaction is scaled down to accommodate the smaller volume of DNA:
   - 255 μL of sodium bisulfate.
   - 15 μL of hydroquinone.
   - 27.5 μL of DNA (+ NaOH).
   - 2.5 μL of water.
2. In **step 8**, TE is added to a final volume of 25 μL.
3. In **step 11**, the precipitation is conducted in the presence of 40 μg of mussel glycogen at −80 °C for 30 min, followed by centrifugation (14,000 rpm) at room temperature for 30 min. The DNA is suspended again in 25 μL of TE.

6.1 Notes

1. It is important to note that the DNA needs to be completely denatured prior to and in the presence of the bisulfite solution, otherwise the modification will not be complete.
2. To ensure complete denaturation, no more than 5 μg (or less) of starting material (DNA) is to be utilized.
3. The initial alkaline denaturation would be at 42 °C for 30 min.
4. The process of DNA digestion with restriction enzymes can be avoided.

7 Bisulfite Genomic DNA Sequencing

This methodology leads to precise analysis of methylation in a definite region through the conversion of all nonmethylated cytosines into thymines, whereas methylated cytosines remain unaltered. Since very small amount of genomic DNA is required, it may be regarded to be useful for the analysis of clinical samples, in case sample amount is limited.

7.1 Bisulfite Sequencing of Small DNA/Cell Samples (Epigenome Network of Excellence)

In molecular biology experimentations, often it becomes difficult to work with small quantities of samples. Here we describe the protocol, when small quantity of DNA is sufficient. The protocol is based on a strategy using agarose embedded DNA. Physical trapping aids in avoiding DNA loss through the process of various incubation steps, at the same time maintaining a good bisulfite conversion rate. Two alternative processes to conduct bisulfite treatment of agarose embedded small DNA aliquots or cells are discussed with critical points in the bisulfite reaction and primer design.

7.2 Bisulfite Sequencing of Very small Samples (Epigenome Network of Excellence)

This protocol is standardized for bisulfite sequencing of small samples when the starting material was a small number of cells. The smallest amount of material leading to several unique clones was recovered was 25 oocytes. This corresponds to 100 DNA molecules in the initial material. This protocol is also useful for analyzing up to 200 ng of purified genomic DNA in one sample.

As discussed, the basic principle involves the fact that bisulfite modification converts unmethylated cytosine to uracil, while methylated cytosine does not react. Denaturation of DNA is followed by bisulfite modification. In the next step, double-stranded DNA is obtained by primer extension followed by PCR amplification of the fragment of interest.

7.3 DNA Methylation

In case of vertebrates, DNA methylation at the 5C position of cytosine residues forms 5-methylcytidine, which occurs almost exclusively within CpG dinucleotides. NonCpG methylation does occur (primarily CpNpG and CpHpH methylation, where H = A, T,C). Other forms of cytosine include 5-hydroxymethylcytosine, 5-formylcytosine, and 5-carboxylcytosine, as intermediates in a pathway for DNA demethylation. It has been observed that in mammalian genomes, approximately 70–80% of CpG dinucleotides are methylated. Stretches of CpG-rich sequences with low levels of DNA methylation referred to as CpG islands are also exist. DNA methylation is an important mechanism associated with epigenetic gene repression.

Targets of de novo DNA methylation during differentiation have been identified as promoters of stem cell- and germline-specific genes. DNA methylation process also involves methyl-CpG-binding proteins. These proteins in turn involve other proteins causing silencing modifications to neighboring histones. The entire coordination process between DNA methylation and silencing histone marks causes compaction of chromatin and gene repression.

7.3.1 A. CpG Islands

CpG islands (CGIs) consist of only 0.7% of the human genome but contain 7% of the CpG dinucleotides. CpG islands are usually more enriched at gene promoters. Around 60% of all mammalian gene promoters are CpG-rich. CpG islands are unmethylated and present in open regions of DNA with low nucleosome occupancy. CpG islands promote relaxed chromatin structure favoring active transcription referred to as euchromatin. It improves accessibility of RNA polymerase II and other components of the basal transcription machinery to the transcription start site. Most CGI promoters have heterogeneous transcription start sites and devoid of TATA boxes. As a result, transcription factors with CpG in their recognition sites (as SP1) aid in recruitment of TATA-binding protein to promoters without TATA boxes.

Transcription from CGI promoters aid in nonproductive, bidirectional cycles of initiation, and premature termination. The regulatory signals required for the progression of this nonproductive state to productive is not understood properly till date. Similarly, directional synthesis of full-length transcripts is not properly characterized till date. The mechanisms associated with CpG islands free of methylation appear to involve binding of transcription factors and other transcriptional machinery or the act of transcription itself. CpG islands may be hypermethylated for silencing specific genes during cellular differentiation, genomic imprinting, and X chromosome inactivation.

7.3.2 DNA Methylases

DNA methylation is catalyzed by DNA methyltransferases (DNMT). DNMT3A and DNMT3B are associated with de novo methylation and are targeted to particular genomic regions through specific histone modifications. During DNA replication, the protein Np95 identifies hemimethylated DNA and directs DNMT1 to the replication fork to adjust patterns of DNA methylation.

7.4 Techniques to Assess DNA Methylation

7.4.1 Methylation-Sensitive Restriction Enzymes

The methylation status of a DNA sequence can be assessed with a variety of techniques like as the use of restriction enzymes (REs). This recognizes short DNA sequences and accordingly cleave double-stranded DNA at specific sites within or adjacent to these sequences. Some REs are sensitive to methylation and cleavage of DNA will not occur in case a cytosine in their recognition sites is methylated. However, there exists other REs, which are insensitive to methylation. The methylation-sensitive RE HpaII was employed in early epigenetic studies for the estimation of 55–70% of all HpaII sites (5′-CCGG-3′) that are methylated in the mammalian genome. It is also employed to detect CpG-rich, hypomethylated DNA regions, referred to as HpaII tiny fragments (HTFs). A list of REs that are sensitive to CpG and CpNpGp methylation can be found.

DNA fragments generated through a methylation-sensitive isoschizomer will differ in size from fragments formed by a methylation-insensitive isoschizomer. The extent of cytosine methylation can be estimated by calculating the ratio of the different DNA fragments. Figure 8 depicts different epigenetic mechanisms as methylation, phosphorylation, acetylation, and ubiquitination diagrammatically.

7.5 Bisulfite Sequencing

Bisulfite sequencing leads to techniques that assess DNA methylation through bisulfite conversion, converting unmethylated cytosine residues to uracil residues. Methylated cytosine residues are unmodified. The target DNA is purified and initially followed denaturation with alkali or heat and treatment with sodium bisulfate. Then, the DNA is again cleaned up, treated with alkali, finally

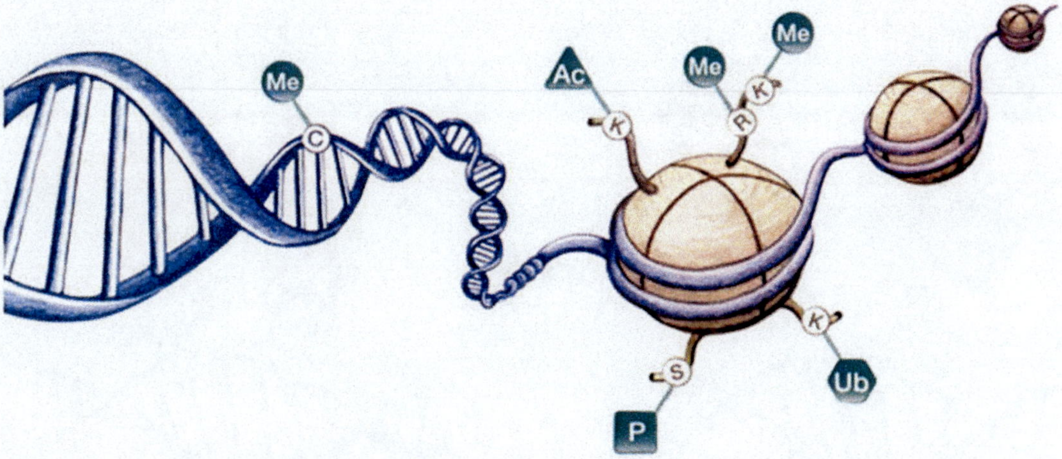

Fig. 8 Epigenetic mechanisms involved in regulation of gene expression. Cytosine residues within DNA can be methylated, and lysine and arginine residues of histone proteins can be modified. *Me* methylation, *Ac* acetylation, *P* phosphorylation, *Ub* ubiquitination

cleaned up again to remove salts and other components that can inhibit downstream applications.

Wide varieties of commercially available kits are available, which are commonly used for this purpose. Followed by bisulfite conversion and DNA cleanup, the DNA is amplified by whole genome PCR. The next step is the analysis of the amplified products, employing a technique distinguishing products derived from unmethylated DNA(containing thymine residues) from products derived from methylated DNA (containing cytosine residues).

These technologies include:

1. Pyrosequencing.
2. Methylation-specific PCR.
3. Methylation-sensitive single-strand conformation analysis (MS-SSCA).
4. High-resolution melting analysis.
5. Methyl cytosine immunoprecipitation (mCIP).
6. Bisulfite methylation profiling (BiMP).
7. MALDI-TOF mass spectrometry.

For high-throughput analysis, bisulfite-treated DNA can be assessed utilizing microarrays with two sets of oligonucleotide probes. One of the oligonucleotide probes is complementary to cytosine-containing DNA, and the other is complementary to thymine-containing DNA. Typical bisulfite conversion protocols include long incubation times under harsh conditions, as a result highly fragmented DNA is obtained. However, advanced kits are

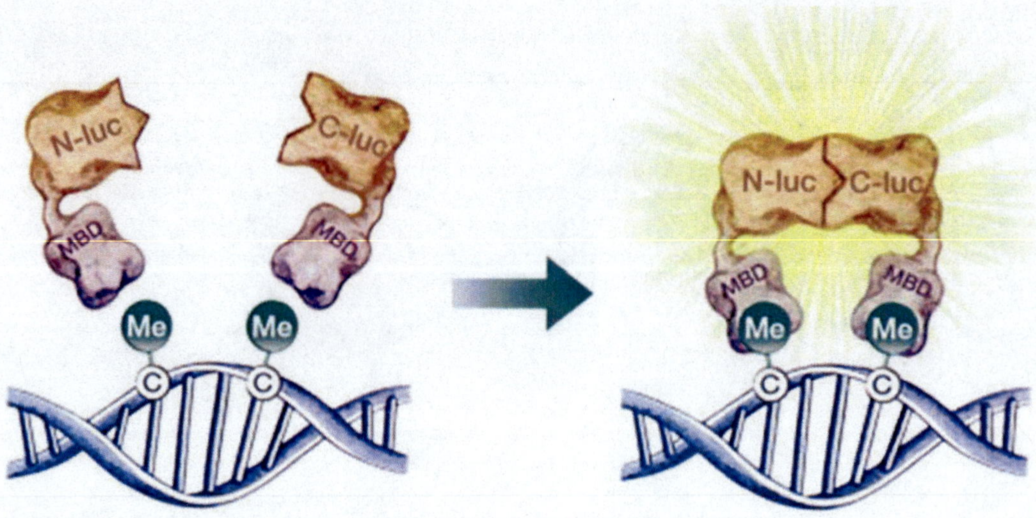

Fig. 9 A schematic diagram showing the use of a split-luciferase biosensor to assess DNA methylation at the genome level. To assess DNA methylation at a specific DNA sequence, one of the methyl-CpG-binding domains (MBD) is replaced with a sequence-specific DNA-binding domain. N-luc is the N-terminal portion of firefly luciferase; C-luc is the C-terminal portion of firefly luciferase

violable resulting in efficient DNA conversion and recovery with reduced template fragmentation and that too time effective.

7.5.1 Luciferase-Based Sensors of DNA Methylation

The firefly luciferase reporter protein (Fluc) is also equally efficient to assess DNA methylation at the genome level or at specific DNA sequences (Fig. 9). Scientists have created split-luciferase biosensors comprise two fusion proteins:

1. DNA-binding domain fused to the N-terminal portion of Fluc.
2. DNA-binding domain fused to the Fluc C-terminus.

To estimate the levels of global DNA methylation, both of the fusion proteins are constructed with the DNA-binding domain of a methyl-CpG-binding domain protein (such as MBD2). It has a 100-fold preference for methylated CpG sites over unmethylated CpG sites. The fusion proteins are expressed in a cell-free expression system, followed by incubation with the target DNA to allow DNA binding. In case multiple methylated CpG sites exist, the N-terminal and C-terminal portions of Fluc will interact. The level of restored Fluc activity is estimated using a firefly luciferase assay, such as the Steady-Glo® or Dual-Glo® Luciferase Assay System. Luminescence levels indicate DNA methylation levels throughout the genome. To estimate site-specific DNA methylation levels, the N-terminus of Fluc is coupled to the MBD DNA-binding domain, whereas the C-terminus is coupled to a sequence-specific DNA-binding domain.

8 Histone Modification

A histone modification is the process of a covalent post-translational modification (PTM) to histone protein including methylation, phosphorylation, acetylation, ubiquitylation, and sumoylation. The PTMs made to histones can affect gene expression by the means of alteration chromatin structure or recruiting histone modifiers.

8.1 Histone Modification and Histone Variants

Epigenetic gene regulation is also carried out by the changes in histones that make up the nucleosome and histone modification. Canonical nucleosomes are octamers consisting of H2A, H2B, H3, and H4 proteins. There are several histone variants varying by a small number of amino acids or which involve large insertions.

Generally, these histone variants are found at specific locations within the chromatin or are used for demarcation of the boundaries between heterochromatin and euchromatin regions. Most of the histone-mediated regulation originates from histone modification, usually modification of the exposed amino termini of histones protruding from the nucleosome core. The majority of histone modifications include acetylation, methylation, phosphorylation, ubiquitination, and sumoylation, with various combinations of modifications within a single nucleosome. Out of these, histone acetylation and methylation were presumed to be better. Trimethylation of histone H3, specifically the lysine at position 4 (H3K4me3), is a mark related to transcriptionally active chromatin, whereas H3K27me3 causes compact chromatin, suppressing gene expression.

The term "histone code" is used to explain how different combinations of histone modifications affect transcription levels. Identification of proteins that can read, write, or erase these marks is essential to explore the complexities of epigenetic regulation. Chromatin immunoprecipitation (ChIP) is a powerful assay to detect proteins that bind to chromatin and map protein binding throughout the genome with the techniques such as microarray analysis or high-throughput sequencing. In ChIP analysis, protein: protein and protein:DNA complexes are crosslinked and immunoprecipitated with an antibody developed against the protein of interest and purified. The DNA sequence of interest is later on amplified from the immunoprecipitated material with PCR. Otherwise, the immunoprecipitated DNA may be sequenced (ChIP-seq) or analyzed with the help of microarrays (ChIP-chip) with an aim to identify target sequences. One limitation of the traditional ChIP method is the presence of specific antibodies that recognize crosslinked epitopes. To overcome the need for suitable antibodies, HaloCHIP™ System is suitable. This system employs HaloTag® protein, which is a mutated hydrolase.

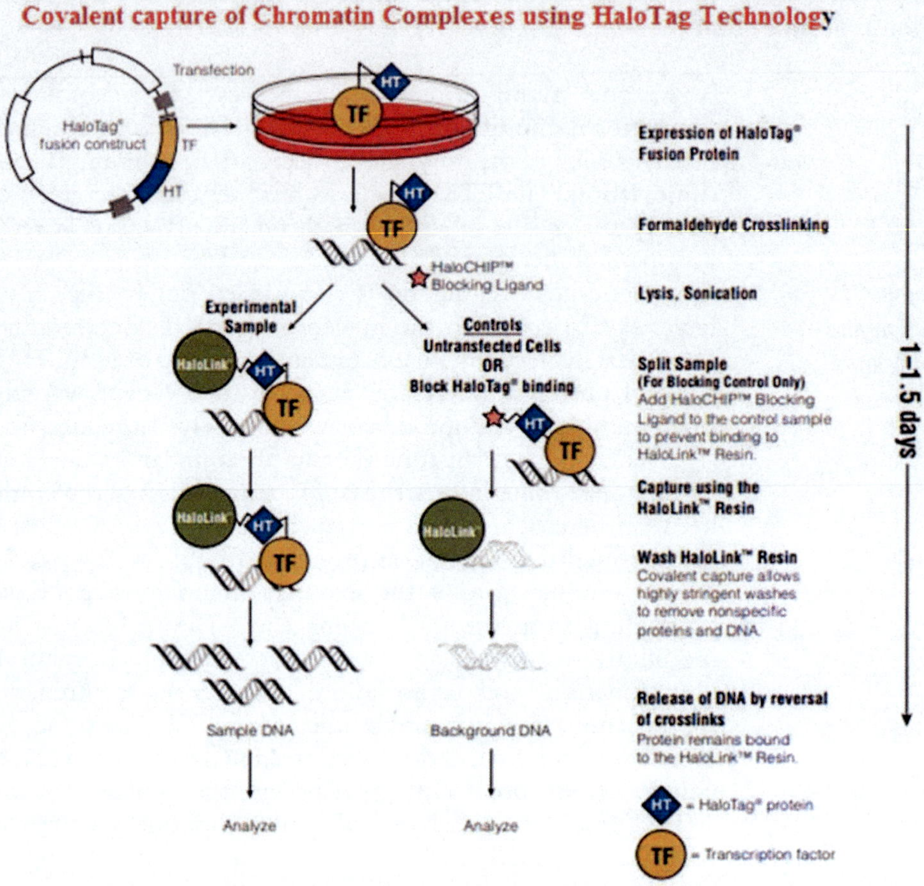

Fig. 10 Schematic diagram of the HaloCHIP™ System

It catalyzes a covalent attachment to a variety of ligands, e.g., a resin-based ligand for immobilization. This tag is able to get fused to any protein. In case of ChIP, the DNA-binding protein of interest is fused to the HaloTag® protein by a process of cloning the protein-coding region into a HaloTag® vector. The recombinant construct is transfected into cells for stable or transient expression, followed by the treatment of the cells with formaldehyde to have covalent protein:DNA and protein:protein crosslinks, lysed, and sonicated to shear the DNA into smaller fragments. The crosslinked complexes are captured from the lysate through covalent binding of the HaloTag® moiety to the HaloLink™ Resin (Fig. 10).

Covalent binding leads to more extensive and stringent washing than that is possible with noncovalent interactions, leading to reduced background and increased signal-to-noise ratio. Later on heating of the purified complexes reverses the crosslinks, thereby releasing captured DNA fragments that can be purified and analyzed using PCR, sequencing, or microarray analysis.

8.1.1 Histone Acetylation and Deacetylation

Acetylation of a lysine residue can neutralize a positive charge of histone protein through the process of reduction of the electrostatic interaction with negatively charged DNA. This reduction in affinity causes increased accessibility of the DNA to protein complexes, ultimately leading to increased gene expression. Lysine acetylation promotes and maintains euchromatin structure through nucleosome-remodeling complexes, such as Swi2/Snf2, via their bromodomains. However, the factors controlling gene expression are complex, and histone acetylation also can lead to reduced gene expression through indirect mechanisms. The process of lysine acetylation of the N-terminal tails of core histones is regulated mostly by two enzyme families: histone acetyl transferases (HATs) and histone deacetylases (HDACs). HATs utilize acetyl CoA as a coenzyme for the process of transfer an acetyl group to the epsilon amino group of the lysine side chain. These enzymes are classified into three families: GNAT, p300/CBP, and MYST. HDACs reverse histone acetylation and promote gene silencing. HDACs are usually components of large protein complexes and are present at the site of DNA methylation by methyl DNA-binding proteins.

HDACs fall into four categories:

1. Class I, which includes HDAC 1, 2, 3, and 8;
2. Class, II, which includes HDAC 4, 5, 6, 7, 9, and 10;
3. Class III, which includes the NAD+-dependent sirtuins (SIRTs); and
4. Class IV, which includes HDAC11.

Misregulation of HATs and HDACs is involved with development and progression of cancer and other diseases such as neurodegenerative disorders and cardiovascular diseases. There is ample scope for making these enzymes attractive therapeutic drug targets. Many HDAC inhibitors cause cell cycle arrest at the G1/S phase, and studies revealed that tumor cells generally are more sensitive to HDAC inhibitors than normal cells [2]. HDAC inhibitors have gained the capability to restore the ability of animals to recall memory that had been lost in Alzheimer's and Parkinson's disease models, probably by changing chromatin structure in neurons. Screening of potential HDAC inhibitors can be aided through certain commercially available kits (HDAC-Glo™ I/II Assays and Screening Systems and SIRT-Glo™ Assays and Screening Systems). The HDAC-Glo™ I/II and SIRT-Glo™ Assays are single-reagent-addition, homogeneous, and luminescent assays. These estimate relative activities of HDAC class I and II enzymes and sirtuins, respectively. The HDAC-Glo™ I/II Assays utilize an acetylated, live-cell-permeant, luminogenic peptide substrate that is deacetylated by HDAC activities from cells, extracts, or purified enzyme sources (Fig. 11). The SIRT-Glo™ Assay utilizes a similar substrate for the detection of SIRT activities from purified enzyme sources

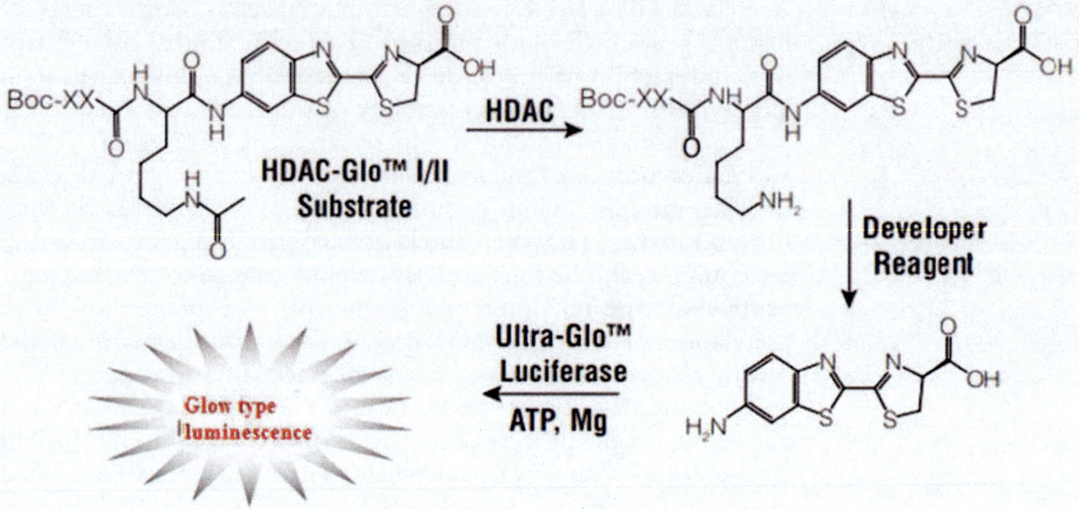

Fig. 11 The single-reagent-addition HDAC-Glo™ I/II Assay

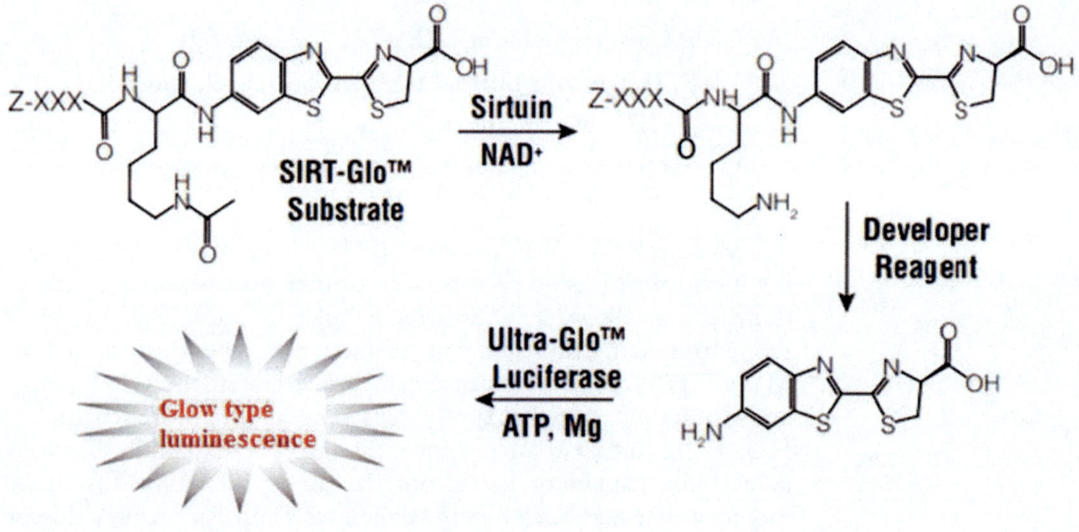

Fig. 12 The single-reagent-addition SIRT-Glo™ Assay

(Fig. 12). Deacetylation of the peptide substrate is noted with enzymatic systems. Here, a protease present in the Developer Reagent breaks the deacetylated peptide from aminoluciferin. The step is followed by quantification of aminoluciferin in a luciferase-based reaction. The HDAC-mediated luminescent signal increases in proportional to enzyme activity. It remains persistent, which allows batch processing of multiwell plates in high-throughput screening.

HDAC activity causes deacetylation of the luminogenic HDAC-Glo™ I/II Substrate, enabling the peptide sensitive to specific proteolytic cleavage. The process is mediated by the HDAC-Glo™ I/II Developer Reagent that generates aminoluciferin. Free aminoluciferin can be measured using Ultra-Glo™ firefly luciferase leading stable, persistent light emission. Boc represents an amino-terminal blocking group that protects the substrate from nonspecific cleavage. XXLysine is an HDAC I/II-optimized consensus sequence obtained from histone 4.

SIRT activity deacetylates the luminogenic SIRT-Glo™ Substrate, which enables the peptide sensitive to a particular proteolytic cleavage process. This is conducted through the SIRT-Glo™ Developer Reagent and ultimately liberating aminoluciferin (Fig. 12). This is followed by free aminoluciferin, which can be measured using Ultra-Glo™ firefly luciferase for the production of stable, persistent light emission. Z represents an amino-terminal blocking group, which protects the substrate from nonspecific cleavage. XXXLysine is a SIRT-optimized amino acid sequence, which is based on a consensus sequence derived.

8.2 Protocol to Determine HDAC Inhibitor Potency Using the HDAC-Glo™ I/II Assays and Cell Extracts or Purified Enzyme

8.2.1 Materials Required

HDAC-Glo™ I/II Assay (Cat.# G6420) or HDAC-Glo™ I/II Screening System (Cat.# G6430). Both of these systems contain the known HDAC inhibitor Trichostatin A. The other items needed are multichannel pipette or liquid-dispensing robot, orbital shaker, reagent reservoirs, nonacetylated HDAC-Glo™ I/II Control Substrate (optional), multiwell, white-walled, opaque-, or clear-bottom tissue culture plates compatible with luminometer.

1. Initial dilutions of the putative HDAC inhibitor and known HDAC inhibitor Trichostatin A are prepared. Only HDAC-Glo™ I/II Buffer is to be added to the no-inhibitor and no-HDAC control wells.
2. Diluted the HDAC enzyme source using HDAC-Glo™ I/II Buffer to the desired concentration. In case, we use the HeLa Nuclear Extract supplied with the HDAC-Glo™ Screening System, the extract is to be diluted to 1:3000.
3. Dispensed the HDAC enzyme source to each well of inhibitor dilutions prepared in step 1 and no-inhibitor controls. Added HDAC-Glo™ I/II Buffer to the no-HDAC controls (dispensed 50 μL for 96-well plates, 10 μL for 384-well plates or 2.5 μL for 1536-well plates).
4. Mixed the plate at room temperature for 30–60 s with an orbital shaker at 500–700 rpm to ensure homogeneity.
5. Incubated enzyme/inhibitor mixes at room temperature for at least 30 min but not more than approximately 2 h.

6. Prepared the HDAC-Glo™ I/II Reagent as described in the HDAC-Glo™ I/II Assay and Screening System Technical Manual #TM335.
7. Added an equal volume of HDAC-Glo™ I/II Reagent to each assay well (100 μL for 96-well, 20 μL for 384-well, or 5 μL for 1536-well plates).
8. Mixed the plate at room temperature for 30–60 s with an orbital shaker at 500–700 rpm to assure homogeneity.
9. Estimated luminescence at signal steady-state (15–45 min after adding the HDAC-Glo™ I/II Reagent).

8.2.2 Histone Methylation

Histone methylation occurs at lysine residues, which can be mono- or dimethylated. Histone methylation process is catalyzed through protein lysine methyltransferases (PKMTs) and protein arginine methyltransferases (PRMTs). However, it can be reversed through protein demethylases. Till date, researchers have identified more than 30 demethylating enzymes, more than 50 protein lysine methyltransferases, and more than 10 protein arginine methyltransferases that suggest that protein methylation is a dynamic and complex process. The effect of histone methylation differs on transcriptional activity and the number of methyl groups, leading to position of the amino acid being modified. In general, the H3K9me1 mark is activating; however, H3K9me2 and H3K9me3 are repressive. H3K4me3 and H3K36me3 are linked with active chromatin; whereas H3K9me3, H3K27me3, H3K36me2, and H4K20me1 are usually found in transcriptionally repressed heterochromatin.

The downstream effects of histone methylation are largely determined by proteins that bind to the modified histones. For example, H3K9me3 acts as a binding site for heterochromatin protein 1 (HP1), which later recruit histone methyltransferases, histone deacetylases, and other proteins affecting chromatin structure. H3K4me3 recruits proteins that promote euchromatin, whereas H3K9me1, H3K9me2, and H3K27me3 interact with proteins that promote heterochromatin. Two classes of proteins are the polycomb group (PcG) proteins and the trithorax (trxG) group proteins, which are the antagonists of earlier one. These were first identified as regulators of hox gene expression in Drosophila. It has been observed recently that related proteins exist in mammals and plants. PcG proteins repress transcription; whereas trxG proteins activate transcription. Some PcG and trxG proteins possess histone methyltransferase activity having the ability to modify histones directly, while others bind to and interpret histone modifications.

In embryonic stem (ES) cells, CpG islands that are regulated by PcG proteins, commonly are "bivalent," which implies that they not only retain the permissive H3K36me2-depleted and

H3K4me3-enriched environment but also exhibit H3K27me3. Genes with bivalent promoters usually are actively silenced in ES cells. They lose the repressive H3K27me3 mark at the same time retain the activating H3K4me3 mark later during differentiation.

8.2.3 Histone Phosphorylation

Histones are phosphorylated on serine, threonine, and tyrosine residues. Serine and threonine phosphorylation may aid in DNA repair or DNA condensation, segregation, and decondensation during mitosis. While others are involved in epigenetic regulation of transcription, including H3T3ph, H3T6ph, H3T11ph, H2. AS1ph, H3S10ph, and H4S41ph. H3S10ph is one of the best characterized molecules for histone modifications. H3S10ph has active role in DNA-restructuring responsibilities during mitosis and is important for chromatin decondensation associated with transcriptional activation of target genes. H3S10ph aids in recruitment of chromatin-modifying enzymes and chromatin-remodeling complexes. It also prevents binding of HP1 to neighboring H3K9me3 marks at the onset of mitosis. H3S10ph, H3T3ph, and H3T11ph can block binding of DNMT3a to H3, leading to reduction of methylation of nearby chromatin.

Several kinases are responsible for phosphorylation of H3S10, including IkB kinase α (IKKα), proviral integration site for Moloney murine leukemia virus 1 (PIM1), and ribosomal S6 kinase 2 (RSK2). Addition of the H3S10ph mark to H3K9me3 is being catalyzed by Aurora B kinase. This also modulates chromosome structure during mitosis and mediates chromosome alignment and attachment to microtubules of the mitotic spindle.

Histones possess many highly conserved tyrosine residues, most of which can be phosphorylated. Phosphorylation of H3Y99 is crucial for polyubiquitination and subsequent proteolysis of excess histones. The process causes an increase in cell's sensitivity to DNA-damaging agents, causing genomic instability and inducing apoptosis. Another tyrosine residue, H3Y41, is essential for chromatin structure and oncogenesis. In human hematopoietic cell lines, phosphorylation of H3Y41 by Janus kinase 2 (JAK2) destabilizes binding of HP1α to histone H3, causing a more open chromatin structure around certain gene promoters such as leukemia oncogene LMO2 that can trigger oncogenesis in hematopoietic cells. Overexpression or aberrant activation of JAK2 activity causes a higher levels of H3Y41, loss of HP1α binding, and higher expression of LMO2.

A number of commercially available kinase enzyme systems are available for monitoring the activity or identification of inhibitors of different kinases involved in histone phosphorylation, including IKKα, PIM1, RSK2, and several cyclin-dependent kinases (CDKs), e.g., CDK1, CDK2, and CDK5. These luminescent assays convert can convert ADP produced by these kinases to ATP, which is then converted to light by Ultra-Glo™ Luciferase. The luminescent signal is positively correlated with ADP amount and kinase activity. Given below is an example of protocol.

8.3 Protocol to Determine CDK1 Inhibitor Potency Using the ADP-Glo™ Kinase Assay + CDK1/CyclinA2 Kinase Enzyme System

8.3.1 Materials Required

- ADP-Glo™ Kinase Assay + CDK1/CyclinA2 Kinase Enzyme System, which includes:

  ADP-Glo™ Reagent.
  Kinase Detection Buffer.
  Kinase Detection Substrate.
  Ultra Pure ATP, ADP.
  Purified CDK1/CyclinA2.
  Histone H1 substrate and
  5× Kinase Buffer A [40 mM Tris (pH 7.5), 20 mM $MgCl_2$, 0.1 mg/mL BSA].
- Solid white, 384-well plate.
- Multichannel pipette or automated pipetting station.

  Kinase enzyme is to be prepared in 1× Kinase Buffer A at 2.5 times the desired final concentration. An amount that will convert 5–10% of the ATP to ADP is recommend.
- CDK1 inhibitor of interest.
- 5% DMSO prepared in 1× Kinase Buffer A, as a no-inhibitor control.
- Water.
- Luminometer capable of reading multiwell plates.
- Plate shaker.

8.3.2 Reagent Preparation

1. Kinase detection Buffer is thawed at room temperature. In case a precipitate is present, the buffer is incubated at 37 °C with constant swirling for 15 min. However, the precipitate is removed from the Kinase Detection Buffer through carefully pipetting the supernatant from the bottle.
2. The Kinase Detection Buffer and Kinase Detection Substrate are equilibrated to room temperature. Transferred the entire volume of Kinase Detection Buffer to the amber bottle containing Kinase Detection Substrate to form the Kinase Detection Reagent. Mixed by gently vortexing, swirling, or inverting.
3. The desired inhibitor concentration range is prepared, including a series of 5× CDK1 inhibitor solutions in 1× Kinase Buffer A.
4. Prepared the ATP+ADP standards.
5. Prepared the ATP/Substrate Mix:

Prepared 200 μL of 2.5× ATP/Substrate Mix in a 1.5 mL tube using 1× Kinase Buffer A. Use the example below as a guideline.

Component Volume 5× Kinase Buffer A 40 μL.
100 μM ATP (10×).
50 μL of water.
60 μL of Histone H1 (1 mg/mL).
50 μL of total volume.
200 μL of Kinase.

6. Added the following reaction components to the wells of a low-volume 384-well plate:
   1 μL of CDK1 inhibitor or 5% DMSO.
   2 μL of kinase enzyme.
   2 μL of 2.5× substrate/ATP mix.
7. Incubated at room temperature for 60 min.
8. Added 5 μL of ADP-Glo™ Reagent.
9. Incubated at room temperature for 40 min.
10. Added 10 μL of Kinase Detection Reagent.
11. Incubated at room temperature for 30 min.
12. Record luminescence (with an integration time of 0.5–1 s).

8.4 Histone Ubiquitination

Conjugation of ubiquitin with a 76-amino acid protein with the lysine residues of histone proteins leads to an effect on transcription activity and nucleosome stability resulting into gene accessibility. The fate of histone ubiquitination is dependent on the histone substrate and degree of ubiquitination. Mono-ubiquitination of histone H2A (H2Aub1) is usually regarded as a repressive mark. However, H2B mono-ubiquitination has role in both transcriptional activation and silencing. There are reports of cross-talk between histone ubiquitination and other forms of histone modification. For example, ubiquitinated H2B has been documented as a docking site for the COMPASS protein complex, including the histone methyltransferase responsible for H3K4 methylation. Also, H2Aub (but not H2A), specifically represses di- and tri-methylation of H3K4, and ubiquitin-specific protease 21 (USP21) relieve this repression.

The process of ubiquitination of histones can be reversed by cleaving the peptide bond between ubiquitin and the ubiquitinated protein. Several deubiquitinases (DUBs) have been identified to deubiquitylate histones 2A, 2A.Z, and 2B, which also includes USP3, USP10, USP21, USP22, and Bap1. Histone deubiquination has been linked with both transcription activation and repression.

To measure the activity of deconjugating enzymes such as deubiquitinating (DUB), deSUMOylating (SENP), and deneddylating (NEDP) proteases, the homogeneous, bioluminescent

DUB-Glo™ Protease Assay (SUB/SENP/NEDP) kit is available with companies. This assay utilizes a luminogenic substrate that contains the C-terminal pentapeptide of ubiquitin: Z-RLRGG-aminoluciferin. By the addition of the DUB-Glo™ Reagent to the test sample, the substrate is cleaved to form the luciferase substrate aminoluciferin. This results in a glow-type luminescent signal that is proportional to the amount of DUB, SENP, and NEDP1 activity present.

8.5 Sumoylation as a Mechanism of Epigenetic Regulation

Another post-translational modification that plays an important role in epigenetic regulation is sumoylation. The process involves the addition of the small ubiquitin-related modifier SUMO. This modification leads to stabilization of proteins, alteration of subcellular localization, affecting enzyme activity, and mediating interactions with other proteins. Many transcription factors and cofactors can be sumoylated that are generally indicative of transcription repression. In Drosophila, the sumoylated form of Sp3 harbors the polycomb protein Sfmbt and HP1α, β, and γ to repress transcription.

Many histone-modifying enzymes, nucleosome-remodeling complexes and their associated enzyme cofactors possess one or more SUMO interaction motifs (SIMs). This motif aids these proteins for interaction with sumoylated transcription factors and cofactors, thus directing these enzymes to specific promoters. Two such groups of histone-modifying enzymes recruited by SUMO are histone deacetylases (decrease histone acetylation at the target promoter) and histone demethylases (lysine-specific demethylase 1), catalyzing the removal of methyl groups from H3K4.

8.5.1 Interpreting DNA Methylation and the Histone Code

Many important regulatory proteins consist of domains binding to modified residues, such as plant homeobox domain (PHD) fingers, bromodomains, chromatin organization modifier (chromo) domains, WD40 repeat, and Tudor domains. Transcription-friendly H3K4me3 acts as a binding site for effector proteins that contain a PHD finger, e.g., nucleosome remodeling factor (NURF) and the ING4-containing histone acetyltransferase complex. The H3K36me2 modification that interferes with transcription initiation acts as a binding site for the chromodomain of the RPD3S histone deacetylase complex.

Histone modifications may also occur in cooperation. The specific combination of histone modifications at a particular site usually determines which protein complexes and accessory proteins are involved in the process of activation or repression of transcription directly, catalyzing additional histone modifications or involve other histone-modifying proteins. This sort of cooperation can be explained by the fact that these proteins may possess one or more modified-histone-binding domains. Example includes, the TFIID protein complex containing both PHD finger and bromodomain and thus binding more strongly to H3K4me3 marks near acetylated H3K9 and H3K14 residues. The absence of any domain through gene mutation or rearrangement can lead to serious lapses in gene regulation and diseases such as cancer. Very recently, scientists have characterized a translocation regarding histone demethylase KDM5A that leads to fusion of the H3K4me3-binding PHD finger of KDM5A to the transcriptional activator NUP98, which is a general leukemia translocation partner, in an acute myeloid leukemia patient. In a similar way, mixed lineage leukemia (MLL) family members that act as histone methyltransferases are involved in translocations in MLL.

8.5.2 Noncoding RNAs

Expression of noncoding RNAs, such as microRNAs, small RNAs, and large RNAs, play a role in epigenetic gene regulation. Noncoding RNAs can affect both cytosine methylation and histone modification for silencing DNA repeats in the genome. A class of 29-nucleotide RNAs that was first discovered through their interaction with the spermatogenesis-specific PIWI protein (piwi-interacting RNAs; piRNAs) mapping to repetitive DNA sequences and is essential for silencing short interspersed elements (SINEs), long interspersed elements (LINEs), and long terminal repeat (LTR) retrotransposons. Certain noncoding RNAs also are involved in X chromosome inactivation. The 17 kb X-inactive specific transcript (XIST) RNA binds and coats the inactive X chromosome and forming complexes that modify chromatin structure to suppress transcription. XIST levels are controlled by another large noncoding RNA, TSIX that is transcribed from the strand opposite of XIST. Noncoding RNAs are involved with genomic imprinting via a similar mechanism.

9 Epigenetic Inheritance

Maintenance and inheritance of epigenetic marks during cell division are crucial for maintaining a committed cell lineage and cellular phenotype in progeny cells and set a memory of transcriptional status. The transmission of epigenetic information by multiple cell divisions involves many of the mechanisms described in this chapter: DNA methylation, histone modification, histone variants, and expression of noncoding RNAs. These similar mechanisms govern the inheritance of epimutations that can cause changes in chromatin structure and transcription levels of genes important to diseases like cancer and imprinting disorders.

10 Epigenetics and Disease

Aberrant regulation of epigenetic mechanisms can lead to genomic imprinting disorders, such as Angelman syndrome and Prader-Willi syndrome, and may contribute to the heritability of many forms of cancer, asthma, Alzheimer's disease, and autoimmune diseases such as systemic lupus erythematosus, rheumatoid arthritis, and multiple sclerosis. Epimutations have the ability to interfere with epigenetic regulation at many levels, involving DNA methylation, histone modification, and noncoding RNAs. Some of the epimutations are inherited, while others accumulate because of environmental factors or with age. For example, although monozygotic twins are epigenetically identical at birth, their patterns of DNA methylation and histone acetylation may differ dramatically with their progression of age.

11 Gene Imprinting

Genomic imprinting is a process of epigenetics that causes the **genes** to be expressed in a parent-of-origin-specific manner. Forms of genomic imprinting have been depicted in fungi, plants, and animals. As of 2014, about 150 imprinted genes are known in the mouse and about half of that in humans.

Genomic imprinting is a form of epigenetic inheritance in which the regulation of a gene or chromosomal region is dependent on the sex of the transmitting parent. During the process of gametogenesis, imprinted regions of DNA are differentially marked in accordance to the sex of the parent, which result in parent-specific expression. Mice are the primary research model employed for studying genomic imprinting. Later on, imprinted regions have been studied in a broad variety of organisms, such as mammals, plants, and insects. Each of these organisms have multiple, interrelated, epigenetic mechanisms to maintain parent-specific expression. Imprinted genes and imprint control regions are normally species specific as well as locus-specific. The similar process of epigenetic mechanisms is usually employed to achieve imprinted expression.

Epigenetic regulation of the genome is an important phenomenon of development. Epigenetic regulation of gene expression may cause heritable changes in gene expression, even when in the absence of alterations in DNA sequence. It is due to the molecular processes that assist transcription, block transcription, or degrade existing transcripts. Hence, genomic imprinting has been regarded as an epigenetic process marking DNA according to sex, resulting in the differential expression of a gene based on its parent of origin. It is necessary for an imprint is to establish meiotically stable male and female imprints during the process of gametogenesis. It simultaneously maintains the imprinted state by the process of DNA replication in the somatic cells of the embryo. Omitting of the preceding generation's imprint occurs in the germ line, which is followed by imprint re-establishment, as per the sex of the organism. There is requirement of epigenetic marks to be interpreted by the genome in each step of the imprinting process. Accordingly, epigenetic marks need to be acted upon resulting in parent specific gene expression.

Genomic imprinting has been documented in eutherian mammals and marsupials. Mice form the basic and the primary research model organism for the study of genomic imprinting. Approximately 100 imprinted genes have been documented in mice with a probability of many more to be present. Thus, it has been observed that imprinting includes chromosomal domains that direct imprinted epigenetic regulation, although endogenous transcriptional units need to be identified as imprinting targets. Many imprinted genes in mice are found to be developmentally important for the developmental process. They are associated with the formation of the placenta or the function of brain. It has equally been observed that noncoding transcriptional units like noncoding RNA are being imprinted and normally form imprinted domains with imprinted genes that are of developmental importance. Imprinted genes found in mice are usually employed as candidates for investigating imprinted genes in other mammals. The status conservation of imprinted gene varies.

Some imprinted genes are conserved in mammals, while not others. For example, only a portion of the imprinted genes identified in mice are also observed to be imprinted in humans, with placental-specific imprinted genes standing out in this discordance. This implies that imprinting cannot be prior predicted in nonmodel species through monitoring homologous genes. It is important to study how imprinting is lost in orthologous genes. It is equally important to detect the epigenetic changes found within these regions, which can lead to a better understanding of how imprinted domains might be regulated.

11.1 Higher-Order Chromatin Structures

Transcriptional inactivation of an imprinted allele is generally maintained through the formation of heterochromatin, which a compacted chromatin structure with the ability to spread in *cis* and generally impose transcriptional silencing. Heterochromatic regions become stable throughout the development and they are propagated by cell division in late replication in S-phase of the cell cycle. Heterochromatic protein 1 (HP1) is a highly conserved nonhistone chromatin protein that is able to recruit other heterochromatic proteins and accessory factors like histone methyltransferases. It aids in reinforcing the structure of heterochromatin and thereby initiating the spreading in *cis*. Polycomb group proteins also form a silencing pathway largely parallel to heterochromatic silencing that targets homeotic genes. Polycomb group silencing also comprise of histone deacetylases and histone methyltransferases. Only modest overlap has been visualized between Polycomb group and heterochromatic silencing.

11.2 Noncoding RNA, Antisense RNA, and RNA Interference

RNA interference (RNAi) is a highly conserved post-transcriptional silencing mechanism. In this process, double-stranded RNA (dsRNA) is channelized to form guides leading to the degradation of complementary RNA transcripts through an RNA silencing complex (RISC). The production of noncoding RNA has been explained at multiple imprinted regions in both mammals and plants. In many organisms, components of the RNAi silencing pathway are observed to be associated with the recruitment DNA methyltransferases and other factors facilitating higher-order chromatin structure. Imprinted domains in diverse organisms become characterized, leading to significant role in the regulation of genomic imprinting for noncoding RNA and RNAi.

11.3 Imprinting in Mammals

In mammals, most of the common imprinted genes are grouped into clusters, sharing common ICRs, which target the parent-specific regulation of multiple genes within the cluster. Many mammalian ICRs consist of differentially methylated regions (DMRs) that gain parent-specific DNA methylation marks. This may happen in either stages at the germline stage for imprint establishment or in somatic cells for imprint maintenance. During a comparative survey in human and mouse genomes depicted, more tandem repeats were observed in methylated regions of imprinted genes than methylated regions of nonimprinted genes. The presence of these repeats gives an insight on the additional structural elements in imprinted regions, leading to chromatin alterations or recruitment of additional epigenetic mechanisms. The "presence of noncoding RNA" is a common feature of mammalian imprinting. In mice, extensive transcription of noncoding RNA has been observed at multiple imprinted loci. Most of these transcripts extend beyond the previously established boundaries of imprinted regions.

11.4 DNA Methylation and Igf2-H19 Imprinting in Mammals

The mouse *insulin-like growth factor 2* (*Igf2*) and *H19* genes were the first imprinted genes characterized. Later on, imprinting pattern was observed for the human *Igf2*iand *H19* genes. The imprinted status of *Igf2* become a standard assay for the determination of the presence of genomic imprinting in other vertebrates such as fish, birds, marsupials, sheep, and cattle. The reciprocal imprinting of the Igf2 and H19 genes is mechanistically coupled. H19 is maternally expressed, whereas Igf2 gene was observed to be paternally expressed. Two ICRs are observed for *Igf2* and both are found to be paternally methylated. DMR1, which is upstream of *Igf2* promoter 1, is a silencer that is inactivated by methylation. DMR2, the other ICR, is located in exon 6 of *Igf2*, which acts as an enhancer activated by methylation. *H19* possesses one ICR, located upstream of the *H19* gene, which is also paternally methylated. Regulation of the *Igf2* and *H19* imprinted domains are dependent on paternal-specific DNA methylation within the DMRs in order to maintain monoallelic expression. Deletions of the *H19* DMR and *Igf2* DMR1 or alterations to Dnmts control the biallelic expression of both *H19* and *Igf2*. In order to establish *Igf2/H19* DMR methylation, passage through the germline is essential. This would be done by the Dnmt3a methyltransferase, which in turn is assisted by the Dnmt cofactor, Dnmt3L. After the establishment, paternal-specific methylation is then detected and maintained in somatic cells by Dnmt1. Dnmt1 is not able to re-establish parent-specific DNA methylation patterns in case prior methylation marks are lost.

11.5 Chromatin Domains and the CTCF Insulator

The evolutionarily conserved CCCTC-binding factor (CTCF) is also associated with *Igf2* and *H19* imprinting. Within the *H19* ICR, there exists a CCCTC-binding site that is only functional on the maternal allele that is unmethylated. When CTCF binds the maternally unmethylated *H19* ICR, it acts as an insulator and blocks the access of the *Igf2* promoter to enhancers. Paternal methylation of the *H19* ICR restricts CTCF binding, causing enhancers access to the *Igf2* promoter on the paternal chromosome. Silencing of the *Igf2* maternal allele is also done by CTCF, which insulates maternal DMR1 and DMR2 from methylation when bound to the maternal *H19* ICR. A loss of CTCF function leads to de novo methylation of the maternal *H19*ICR, which effectively removes imprinted expression of *H19* and *Igf2*. Recent phylogenetic and mutational analysis has observed that the CTCF-binding sites, and not DNA methylation of ICRs, are the better reliable predictor of the imprinted expression of *Igf2*. CTCF-binding sites are conserved in humans, mice, and marsupials, which all have imprinted *Igf2* and *H19*. CTCF-binding sites lack monotremes that do not imprint *Igf2* or *H19*. It has been detected that *Igf2* DMR2 is biallelically methylated in both marsupials and monotremes. It is only biallelically expressed in monotremes, depicting that methylation alone does not lead to imprinted expression.

CTCF binds at several sites within mammalian genomes, when it is identified both as a transcriptional regulator and a chromatin insulator, and are able to obstruct the spread of heterochromatin. It also mediates long-range chromosomal interactions. CTCF-directed intrachromosomal loops are believed to contribute to parent-specific expression of *Igf2* and *H19*. Self-association between CTCF proteins bound to ICRs can generate a chromosomal loop that isolates *H19* to maintain maternal expression. Simultaneously, it reinforces *Igf2* silencing by the creation of a repressive domain. Disruption of CTCF binding to the maternal *H19* ICR leads to the de novo DNA methylation of maternal *Igf2*

DMR1 and DMR2. It suggests that intrachromosomal looping causes regulation of the entire maternal *Igf2/H19* imprinted region. Isolation of imprinted alleles by CTCF has been documented at various other mammalian imprinted domains. Here, parent-specific binding of CTCF is crucial for maintaining active expression from an imprinted allele. Furthermore, it remains to be confirmed if the initiation of higher-order chromatin structures via CTCF-mediated intrachromosomal looping is a general feature of other imprinted domains.

11.5.1 Antisense Transcripts and Mammalian Imprinting

Noncoding RNA transcripts are usually associated with the imprinted regions in mammals. The examples of such noncoding RNA transcripts include the *H19* and *Airn* RNAs. Deletion of the DMR2 *Airn* promoter, or the truncation of the *Airn* transcript, results in paternal activation and biallelic expression of *Igf2r* and the neighboring gene clusters. Additionally, the *Airn* transcript is capable of maintaining paternal silencing in this gene cluster even if the paternal *Igf2r* promoter is experimentally activated or if DNA methylation of DMR2 is lost. Part of the silencing function of *Airn* may be responsible for recruiting additional silencing complexes to the imprinted region. In the mouse placenta *Airn* can recruit the histone H3K9 methyltransferase G9a, which contributes to the imprinted silencing of the gene *Slc22a3* within the *Igf2r* imprinted cluster. Another important aspect of regulation by noncoding RNAs is through the process of transcription and the interference caused by transcription. It has been predicted that transcription of *Airn* with the aid of neighboring genes in *cis* causes their silencing. In addition, the *Airn* transcript is overlapped with its own promoter and active transcription of *Airn* is needed for the prevention of de novo methylation of this promoter on the paternal allele. Currently, the transcriptional importance of noncoding RNAs has been observed for the *Kcnq1* imprinted domain. In stem cells, targeted depletion of the *Kcnq1ot1* noncoding RNA did not relieve silencing of the paternally silenced genes, suggesting transcription through these genes during the production of *Kcnq1ot1* leads to their silencing more so than the presence of the *Kcnq1ot1* transcript.

MicroRNAs (miRNAs) are endogenous 21–25 nt RNA transcripts that target complementary sequences designated for silencing. Two miRNA genes, *miR-127* and *miR-136*, are found to be part of an imprinted domain. These are in turn helpful for the imprinted expression of the *Rtl1* of mice (retrotransposon-like gene) and the orthologous *PEG11* gene in sheep and humans. Imprinted expression is related to an unmethylated maternal ICR, causing the miRNA genes only being maternally expressed which drives maternal-specific silencing of *Rtl1*. In case of sheep, *PEG11* leads to the production of a functional protein as well as an antisense *PEG11* transcript. Imprinted silencing is governed by maternally produced antisense miRNAs that act as guides for RISC-mediated destruction of maternal *PEG11* transcript. However, complex modulations of maternal miRNA generation reflect that maternal gene expression levels are balanced for dosage and complete silencing is not present. It is not completely clear if RNAi processing of *PEG11* transcripts by RNAi machinery leads to additional chromatic remodelers for the regulation of expression from the maternal allele.

Genomic imprinting has been related to dosage compensation in some mammals. The silencing is usually directed towards the paternal X chromosome. In female mice, the selective silencing of the paternal X chromosome occurs in extraembryonic tissues, through the production of the noncoding RNA *Xist*. Transcription of *Xist* spreads from an initial transcription site to involve most of the paternal X chromosome, leading to the involvement of additional epigenetic silencing factors, like histone methyltransferases and heterochromatic proteins. Preferential silencing of the paternal X chromosome exists if *Xist* noncoding RNA is lost, and silencing is destabilized. This may be related to the observation that the RNAi component Dicer is required for the spread of *Xist* and recruitment of the H3K27 trimethylation silencing in somatic cell X inactivation. In extraembryonic mouse tissues, imprinted silencing of the paternal X chromosome originates from the imprinted silencing of specific target genes or regions. There exists every possibility that imprinted silencing of the paternal X chromosome in extraembryonic mouse tissues has the origin from the imprinted silencing of specific target genes or regions. These specific target genes or regions then act as nucleation sites for RNAi-directed spreading of silencing through the whole chromosome.

Genomic imprinting is an epigenetic regulatory mechanism which operates through expression of certain genes from maternal or paternal in a parent-of-origin-specific manner. Imprinted genes have been detected in diverse biological systems which are implicated in some human diseases and in embryonic and seed developmental programs in plants.

12 Epigenome Writers

12.1 Writers, Readers, and Erasers of Epigenetic Marks

Changes in gene expression and activity were reported to have role in the formation of cancerous tumors. Earlier, alterations (deletions, point mutations, translocations, etc.) in the DNA genome of the cell were thought to be the basis for tumor formation. Recent studies indicate that epigenetic modifications, involving the regulated addition of chemical groups to DNA and histones, have profound effect on gene expression leading to cancer. Both genomic region compaction and gene expression are controlled by epigenetic modifications. These are deposited by specific enzymes (known as "***writers***"), and then recognized by effector proteins ("***readers***"). Most of the epigenetic marks are reversible, and various enzymes ("***erasers***") remove these marks. These three classes of proteins regulates gene transcription, and defects in this system contribute to cancer initiation and progression. The key concepts regarding these three types of proteins, with a particular focus on methyl and acetyl marks will be discussed.

Addition of chemical tags on the DNA and modification of histone proteins leads to a distinct feature on chromatin architecture The major factors underlying these changes have been identified as epigenetic modifiers of the chromatin. The series of enzymes catalyzing these modifications depicts the diversity of epigenetic space and the intricacy in regulating gene expression. These epigenetic players are classified as *writers:* that introduce various chemical modifications on DNA and histones, *readers:* the specialized domain containing proteins identifying and interpreting those modifications and erasers, which may be grouped as the dedicated group of enzymes effective in removal of these chemical tags.

Past research has established that these epigenetic tools are associated with numerous disease conditions especially cancer. In addition to the involvement of epigenetics in cancer, these enzymes and protein domains provide new targets for cancer drug develop ment Different types of epigenetic enzymes and protein domains have been discussed with an emphasis on methylation and acetyla tion. It is important to study the recent developments in small molecule inhibitors as potential anti-cancer drugs targeting the epigenetic space.

13 Transcription Regulator

The diagrammatic representation (Fig. 13) clearly depicts transcription regulators, which equally have important role in epigenetics.

Diet or compounds found in food effecting gene expression programs through epigenetic mechanisms, leads to an opening a new avenue in exploring therapies based on these mechanisms. This branch is often referred to as Nutrigenomics. It is discussed in detail in later chapters.

One possibility is through directly affecting catalytic activities of the enzymes responsible for "writing" or "erasing" the epigenetic modifications. Wang et al. identified two phytochemicals—dihydrocaffeic acid (DHCA) and malvidin-3′-O-glucoside (Mal-gluc), metabolic intermediates derived from Concord grape juice, grape seed extract, and trans-resveratrol—that attenuate depression-like behaviors in mice. Using a mouse model, Wang et al. reported that treating mice with DHCA and Mal-gluc, which were added to drinking water, increased their resilience to stress and reduced depression-like behaviors. Specifically, the

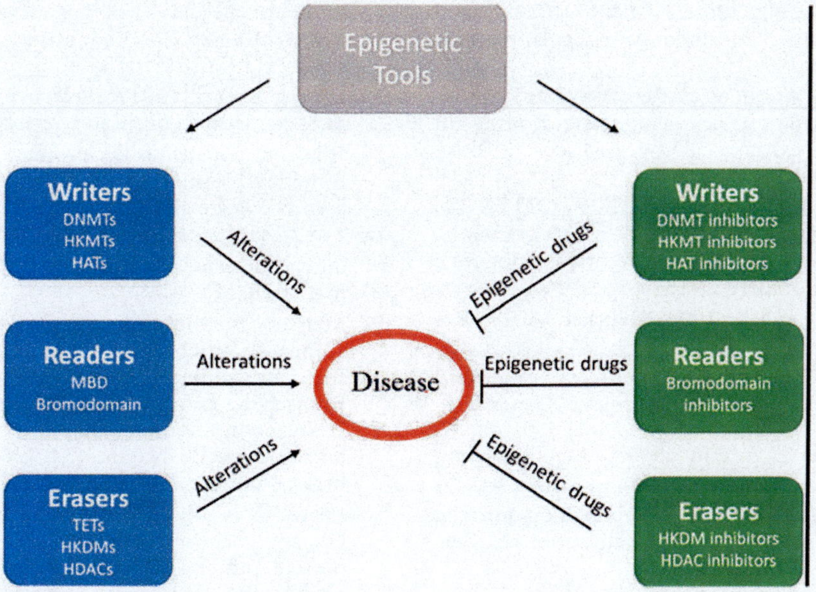

Fig. 13 Transcription regulators

authors observed that DHCA reduced expression of "methyl-DNA writer"—the DNA methyltransferase 1 (DNMT1). DNMT1 methylates intronic sequences of interleukin 6 (*IL-6*) genes, thereby reduce level of this pro-inflammatory cytokine that has previously been implicated in the development of depressive disorders. Mal-gluc, however, decreased the expression of an "acetyllysine eraser," histone deacetylase 2 (HDAC2), which cause a significant increase in histone H3 acetylation on promotor of the RAS-related botulinum toxin substrate 1 (*Rac1)* gene, whose expression level was observed to be downregulated under chronic stress. When combined, the treatment of DHCA and Mal-gluc decreased IL-6 level to baseline and simultaneously improve *Rac1* expression, that contribute to the resilience against the development of depression-like phenotypes in mice. On the whole, the study by Wang et al. provides definite evidence of how grape-derived phytochemicals can affect inflammation and brain synaptic plasticity, promoting resilience against stress in mice. This increases the possibility that these compounds, combined with currently available antidepressants, may act as therapeutic agents in the treatment of patients suffering from depression and anxiety.

References

1. Badran AH et al (2011) Evaluating the global CpG methylation status of native DNA utilizing a bipartite split-luciferase sensor. Anal Chem 83:7151–7157
2. Johnstone RW, Ruefli AA, Lowe SW (2002) Apoptosis: a link between cancer genetics and chemotherapy. Cancer Rev 108(2):153–164

3. Blackledge NP, Klose RJ (2011) CpG island chromatin: a platform for gene regulation. Epigenetics 6:147–152
4. Carey N, Marques J, Reik W (2011) DNA demethylases: a new epigenetic frontier in drug discovery. Drug Discov Today 16:683–690
5. Chen QW, Zhu XY, Li YY, Meng ZQ (2014) Epigenetic regulation and cancer (review). Oncol Rep 31(2):523–532. https://doi.org/10.3892/or.2013.2913
6. Clark SJ et al (1994) Nucleic Acids Res 22:1827
7. Deaton AM, Bird A (2011) CpG islands and the regulation of transcription. Genes Dev 25:1010–1022
8. Draker R, Sarcinella E, Cheung P (2011) USP10 deubiquitylates the histone variant H2A.Z and both are required for androgen receptor-mediated gene activation. Nucleic Acids Res 39:3529–3542
9. Gacias M, Casaccia P (2014) Epigenetic mechanisms in multiple sclerosis. Rev Esp Escler Mult 6(29):25–35
10. Goldman SL, Hassan C, Khunte M, Soldatenko A, Jong Y, Afshinnekoo E, Mason CE (2019) Epigenetic modifications in acute myeloid leukemia: prognosis, treatment, and heterogeneity. Front Genet 10:133. https://doi.org/10.3389/fgene.2019.00133
11. Gutiérrez L et al (2012) The role of the histone H2A ubiquitinase Sce in Polycomb repression. Development 139:117–127
12. Hammaker D, Firestein GS (2018) Epigenetics of inflammatory arthritis. Curr Opin Rheumatol 30(2):188–196. https://doi.org/10.1097/BOR.0000000000000471
13. Hedrich CM (2017) Epigenetics in SLE. Curr Rheumatol Rep 19(9):58. https://doi.org/10.1007/s11926-017-0685-1
14. Islam AB et al (2011) Selective targeting of histone methylation. Cell Cycle 10:413–424
15. Ito S et al (2011) Tet proteins can convert 5-methylcytosine to 5-formylcytosine and 5-carboxylcytosine. Science 333:1300–1303
16. Li LC, Dahiya R (2002) MethPrimer: designing primers for methylation PCRs. Bioinformatics 18(11):1427
17. Meissner A, Mikkelsen TS, Gu H, Wernig M, Hanna J, Sivachenko A, Zhang X, Bernstein BE, Nusbaum C, Jaffe DB, Gnirke A, Jaenisch R, Lander ES (2008) Genome-scale DNA methylation maps of pluripotent and differentiated cells. Nature 454(7205):766–770. https://doi.org/10.1038/nature07107
18. Porter JR et al (2008) A general and rapid cell-free approach for the interrogation of protein-protein, protein-DNA, and protein-RNA interactions and their antagonists utilizing split-protein reporters. J Am Chem Soc 130:6488–6497
19. Pradhan S et al (1999) Recombinant human DNA (cytosine-5) methyltransferase. I. Expression, purification, and comparison of de novo and maintenance methylation. J Biol Chem 274:33002–33010
20. Shi DQ, Ali I, Tang J, Yang WC (2017) New insights into 5hmC DNA modification: generation, distribution and function. Front Genet 8:100. https://doi.org/10.3389/fgene.2017.00100
21. Sun WJ et al (2012) Histone acetyltransferases and deacetylases: molecular and clinical implications to gastrointestinal carcinogenesis. Acta Biochim Biophys Sin (Shanghai) 44:80–91

Check for updates

Chapter 13

RNA Sequencing (RNA-seq)

Abstract

RNA sequencing (or popularly known as RNA-seq) or whole transcriptome shortgun sequencing (WTSS) is a useful next-generation sequencing (NGS) technique applicable for studying alternative gene spliced transcripts, post-transcriptional modifications, gene fusion, variations in gene expression, and mutations/ single nucleotide polymorphisms (SNPs). Other than mRNA transcript, RNA-seq is equally applicable for total RNA, small RNA, such as miRNA, tRNA, and ribosomal profiling, identification of exon/intron boundaries and previously annotated 5′ and 3′ gene boundaries. RNA-seq has been documented as a better technique than DNA microarray. The protocols involved for RNA-seq has been discussed. The basic step for RNA-seq includes RNA isolation, cDNA synthesis, attachment of adapter, cDNA library preparation followed by sequencing. The depth up to which the library is sequenced varies based on the techniques which the output data will be used for. The sequencing usually follows either single-read or paired-end sequencing methods; the depth varies as per the technique involved in output data usage. This is followed by data analysis which has been discussed very briefly. RNA-seq methodology has been discussed for different sequencing platform as illumina, Ion torrent, SOLID, etc.

RNA-seq methodology has also been discussed for non-coding RNA like miRNAs, lincRNAs, and craRNA. Single cell RNA sequencing has also been discussed very briefly in this chapter.

Keywords RNA-seq, Whole transcriptome short gun sequencing (WTSS), Next-generation sequencing

1 Basic Principle

RNA sequencing or popularly known as RNA-seq is a next-generation sequencing technique. It aims to detect the presence and quality of RNA in a biological specimen at a given point of time. It is also referred to whole transcriptome shortgun sequencing (WTSS).

RNA-seq is utilized to study the continuously changing cellular transcriptome. RNA-seq is a useful NGS technique applicable for studying alternative gene spliced transcripts, post-transcriptional modifications, gene fusion, variations in gene expression, and mutations/SNPs. Other than mRNA transcript, RNA-seq is equally applicable for total RNA, small RNA, such as miRNA,

Aruna Pal, *Protocols in Advanced Genomics and Allied Techniques*, Springer Protocols Handbooks, https://doi.org/10.1007/978-1-0716-1818-9_13,

tRNA, and ribosomal profiling. The technique is equally applicable to identify exon/intron boundaries and previously annotated 5′ and 3′ gene boundaries.

2 Introduction to RNA Sequencing

RNA sequencing (RNA-seq) is one of the most advanced technology revolutionizing the study of the transcriptome. RNA-seq may be regarded as an efficient, at the same time a sensitive as well as accurate technology for the measurement of gene expression throughout the transcriptome. It has the ability to detect novel changes at nucleotide level arising due to disease conditions, in response to therapeutics, variations in environmental conditions and other related studies.

RNA-seq is unique in the sense that it can detect both previously known and novel or unique features in a single assay. It can detect transcript isoforms, gene fusions, single nucleotide variants, without prior knowledge.

Advantages of RNA-seq technology for studying transcriptome over conventional gene expression arrays are listed below:

1. RNA-seq technique has a broader dynamic range, which allows a better sensitive and accurate measurement of gene expression.
2. There is no need of prior knowledge of sequences. It is capable of capturing both known and novel features.
3. This technique is equally applicable to any species, although reference sequencing is missing.
4. RNA-seq is comparatively better economic technique in comparison to other techniques.

2.1 Basic Mechanism of Action of RNA-seq

At the advent of RNA-seq technique, initially Sanger sequencing technology was utilized, which was observed to be of low-throughput, costly, and of low accuracy. Now with the evolvement of NGS technology, it is possible to utilize RNA-seq with great potential.

Let us discuss the basic steps of RNA-seq technology.

1. The initial step includes the conversion of RNA into cDNA fragments, referred to as cDNA library. This leads the RNA to be sequenced into an NGS workflow.
2. In the second step, adapters are attached to each end of the fragments. These adapters possess functional elements which allow for sequencing, for example, the amplification element and the primary sequencing site.

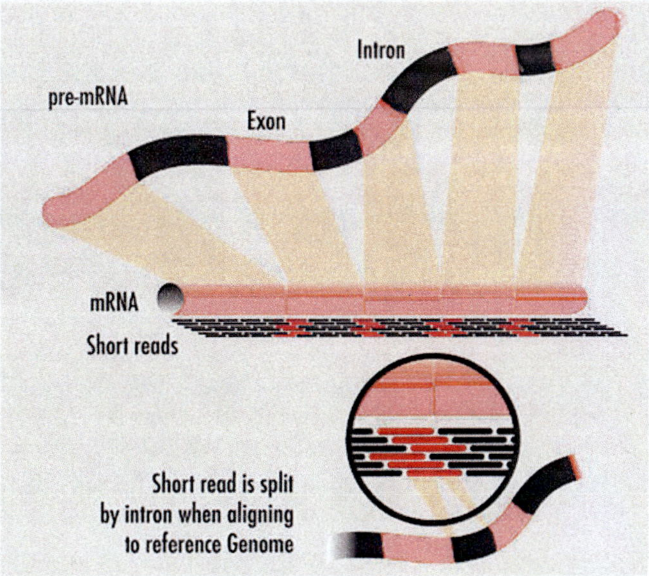

Fig. 1 Schematic representation for RNA-seq technology. The technique utilizes short reads of mRNA free of intronic non-coding DNA. The reads are then be aligned back to the reference genome

3. The cDNA library is then processed for analysis by NGS technique, leading to the production of short sequences corresponding to either one or both ends of the fragment.
4. The depth up to which the library is sequenced depends on the techniques which the output data will be utilized. The sequencing usually follows either single-read or paired-end sequencing method. Single-read sequencing is a comparatively cheaper and faster technique (for reference, about 1% of the cost of Sanger sequencing). Single-read sequencing technique sequences the cDNA from just one end.
5. Paired-end sequencing technique leads to the sequencing from both ends, and are therefore more expensive and time-consuming.

Figures 1 and 2 describe the layout and methods of the technique of RNA-seq. Another important point that is to be discussed includes whether to use strand-specific or non-strand-specific protocol. Strand-specific protocol refers to analyze the DNA stand for which the information about DNA strand is available. Since extra information is available for strand-specific protocol, it seems to be a favorable option.

The analysis process involves management of millions of reads obtained by the end of work flow. Then alignment process involves alignment to a genome of reference followed by genomic assembly for producing RNA sequence map that spans the transcriptome.

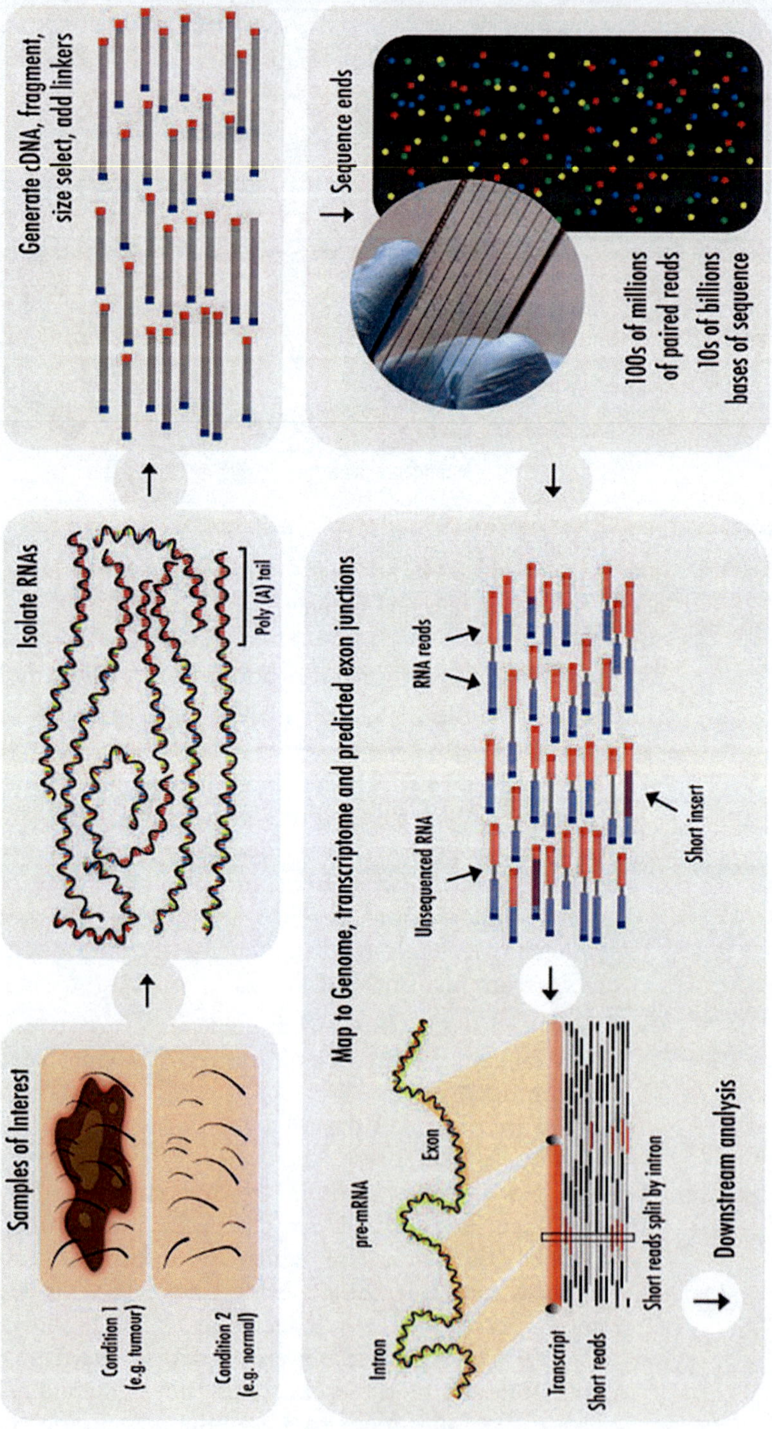

Fig. 2 Schematic representation of RNA-seq

RNA-seq has been considered as a better technique. We have already discussed how RNA-seq as transcriptome is a better technique compared to that of conventional technique.

Now we shall discuss how RNA-seq is superior to microarray hybridization in following ways:

1. **Since RNA-seq technique is not limited to genomic sequences**, it can identify transcripts from organisms with previously undetermined genomic sequences. However, microarray is a hybridization-based approach, requiring species-specific probes, RNA-seq. Hence RNA-seq was proved to be superior for the identification of novel transcripts, SNPs, or other alterations.
2. **Low background signal**—the cDNA sequences utilized in RNA-seq can be mapped to targeted regions on the genome, which enables it easier to remove experimental noise. Cross-hybridization or sub-standard hybridization, having effect on microarray experiments, cannot affect RNA-seq experiments.
3. **More quantifiable**—Microarray data represents the values relative to other signals detected on the array, **whereas RNA-seq data is quantifiable**. RNA-seq also discards the issues microarrays have in detection of very high or very low expression levels.

2.2 An RNA-seq Protocol

2.2.1 Experiment Planning

Certain prior considerations are necessary before initializing RNA-seq experiment, which includes:

1. Method of RNA purification.
2. Reads of RNA needed.
3. Platform to use RNA-seq—whether illumina or Agilent technology.
4. Reference genome to use.
5. Method for assessing the quality of your RNA.
6. Necessity for enriching the target RNA.
7. Barcode for RNA.
8. Sufficient biological and technical replicates.
9. Need for single-read or paired-end sequencing.

2.3 cDNA Library Preparation

The next step is the preparation of cDNA library. The process involves addition of the platform-specific "adapter sequences" and amplification of the DNA. However, the exact procedure will vary with the platform used at this stage. The amplification of the DNA is followed in two steps. The first step involves first strand synthesis by reverse transcriptase followed by a DNA polymerase-mediated second strand synthesis.

2.4 cDNA Sequencing

As the library is prepared, and adapters are added, sequencing is to be performed in suitable platform to sequence cDNA library. Transcript data thus produced, can be mapped to reference genome. The presence of splice variants and modifications will complicate the alignment process. The choice of reference genome may also vary depending upon the situation. Software packages such as STAR and quality control tools like Picard or Qualimap are useful.

2.5 RNA-seq Data Analysis

Data analysis is one of the most crucial steps to be followed. General tools available for quantification of expression level are Tools like Sailfish, RSEM, and BitSeq. Specialized tools are also available, quantifying alternatively spliced genes, such as MISO, are available. It becomes very much essential to select the appropriate tool with reading reviews and roundups. RNA-seq is a superior option to microarrays and is in demand.

Let us discuss the Protocol for RNA-seq:

1. **RNA 'Poly (A)' library**:

   Creation of a sequence library differs as per the platform in high throughput sequencing. Each has many kits designed to build various types of libraries. This helps in the adaptation of the resulting sequences to the particular requirements of their instruments.

   Frequently, in mRNA analysis the 3′ polyadenylated (poly (A)) tail is targeted so that coding RNA is isolated from non-coding RNA. This can be achieved simply with poly (T) oligos covalently attached to a given substrate. Currently, many studies use magnetic beads for this step.

2. **Small RNA/non-coding RNA sequencing**:

   In order to sequence RNA other than mRNA, the library preparation is modified. The cellular RNA is selected according to the desired size range. For small RNA targets, such as miRNA, the RNA is isolated through size selection. This can be achieved with a size exclusion gel, through size selection magnetic beads, or with a commercially developed kit. After isolation, linkers are added to the 3′ and 5′ end followed by purification. The next step is cDNA generation through reverse transcription.

   RNA-seq mapping of short reads in exon-exon junctions.

3. **Direct RNA sequencing**:

   It has been observed that certain biases and artifacts may arise during the process of conversion of RNA into cDNA with reverse transcriptase, interfering with proper characterization and quantification of transcripts. The principle for single-molecule direct RNA sequencing (DRSTM) technology (under development by Helicos) involves sequencing of RNA molecules directly in a massively parallel manner. The greatest

Table 1
Advantages of RNA-Seq compared with other transcriptomics methods

Technology	Tiling microarray	cDNA or EST sequencing	RNA-Seq
Technology specifications			
Principle	Hybridization	Sanger sequencing	High-throughput sequencing
Resolution	From several to 100 bp	Single base	Single base
Throughput	High	Low	High
Reliance on genomic sequence	Yes	No	In some cases
Background noise	High	Low	Low
Application			
Simultaneously map transcribed regions and gene expression	Yes	Limited for gene expression	Yes
Dynamic range to quantify gene expression level	Up to a few-hundredfold	Not practical	>8,000-fold
Ability to distinguish different isoforms	Limited	Yes	Yes
Ability to distinguish allelic expression	Limited	Yes	Yes
Practical issues			
Required amount of RNA	High	High	Low
Cost for mapping transcriptomes of large genomes	High	High	Relatively low

advantage of this technique is that it does not involve the conversion of RNA to cDNA or other biasing sample manipulations such as ligation and amplification (Table 1).

4. **Transcriptome assembly**:

   Different assembly methods are applied for producing a transcriptome from raw sequence reads: de novo and genome-guided.

   The initial approach does not depend on the presence of a reference genome in order to reconstruct the nucleotide sequence. As the reads are of small size, de novo assembly may be difficult, though some software does exist (Velvet (algorithm)), Oases, and Trinity. The deep coverage also leads the computing power to track all the possible alignments prohibitive. This deficiency can be improved with the help of long sequences derived from the same sample utilizing other techniques such as Sanger sequencing.

Thus, we can summarize the differences for RNA-seq versus microarray:

1. RNA-seq can be employed for characterization of novel transcripts and splicing variants as well as to profile the expression levels of known transcripts (but hybridization-based techniques are limited to detect transcripts corresponding to known genomic sequences).
2. RNA-seq possesses higher resolution than whole genome tiling array analysis.
   - In principle, mRNA can lead to single-base resolution, where the resolution of tiling array depends on the density of probes.

3. RNA-seq possess the application of the same experimental protocol to various purposes, but specialized arrays are required for designing these cases.
   - Detection of single nucleotide polymorphisms (require SNP array otherwise).
   - Mapping of exon junctions (require junction array otherwise).
   - Detection of gene fusions (require gene fusion array otherwise).
   - Next-generation sequencing (NGS) technologies overrule microarrays as the tool of choice for genome analysis. Figures 3 and 4 give an idea how next-generation sequencing or high throughput sequencing evolves. Figure 5 gives a flow chart of the mechanism of action of RNA-seq.

Figure 6 gives an overview of analysis workflow for microarray and RNA-seq transcriptional profiling.

2.6 Featured Workflow for RNA-seq Analysis

1. Library Preparation.
2. Sequencing.
3. Data Analysis.

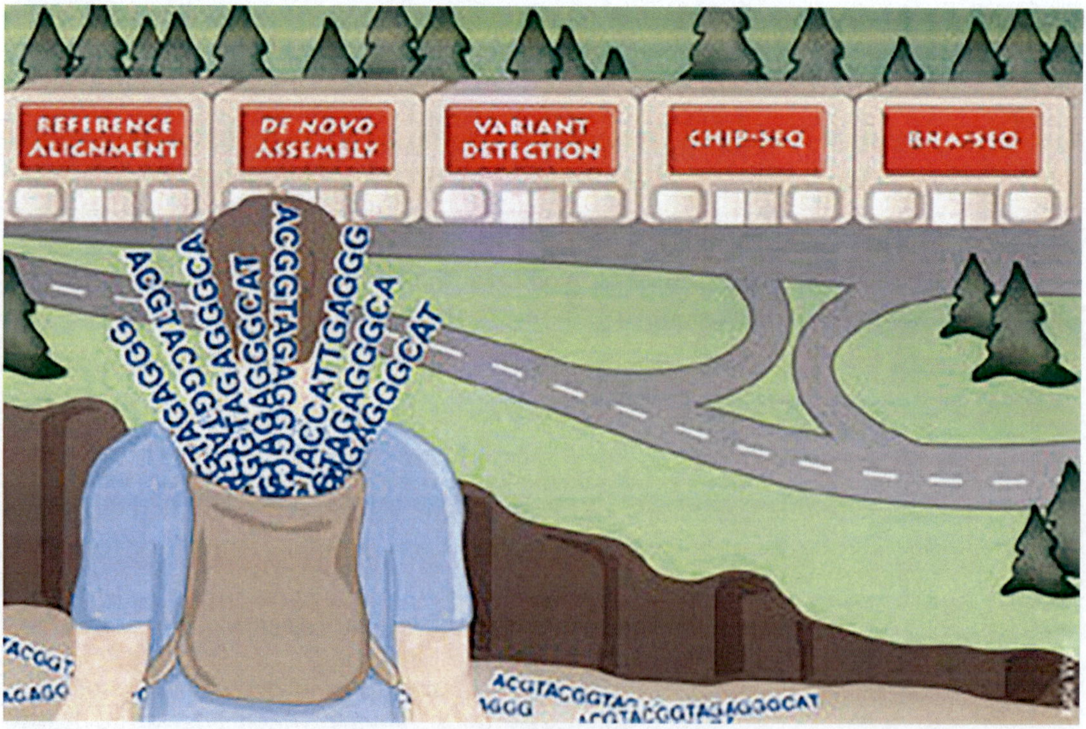

Fig. 3 A glimpse of evolution of high throughput sequencing

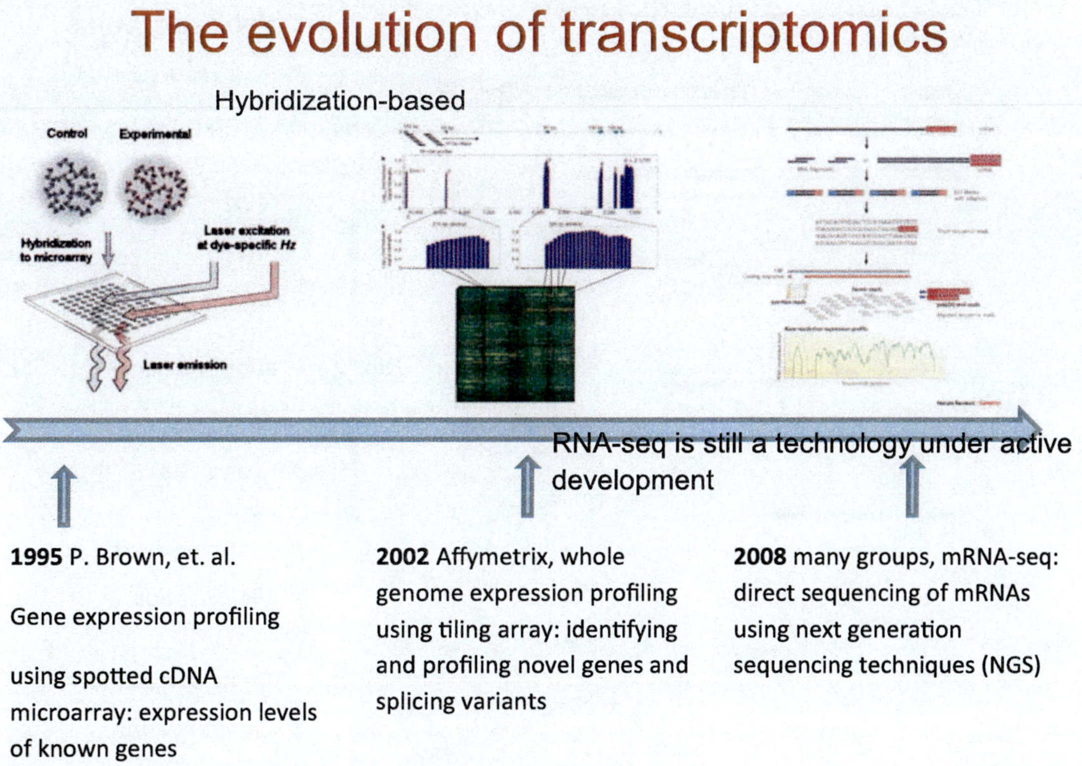

Fig. 4 The schematic representation of evolution of RNA-seq

2.6.1 Library Preparation

The technique is equally applicable for Agrigenomics, cancer research, oncology, complex disease genomics, drug discovery and development, genetic health, immune-oncology research, immunology, microbial genomics, neuroscience, reproductive health.

The systems that are applicable are **HiSeq 1500, HiSeq 2000, HiSeq** 2500, HiSeq 3000, HiSeq 4000, HiSeq X Five, HiSeq X Ten, iScan, iSeq 100, MiniSeq, MiSeq, MiSeqDx, MiSeqDx (Research Mode), NextSeq 500, NextSeq 550, NextSeq 550Dx, NextSeq 550Dx (Research Mode), NovaSeq 6000, HiScanSQ, HiSeq 1000, and Hi-scan.

2.6.2 Sequencing

Benchtop sequencers are iSeq 100, MiniSeq, MiSeq, NextSeq. NextSeq is the most effective technique applicable for a wide range of sequencing application and analysis.

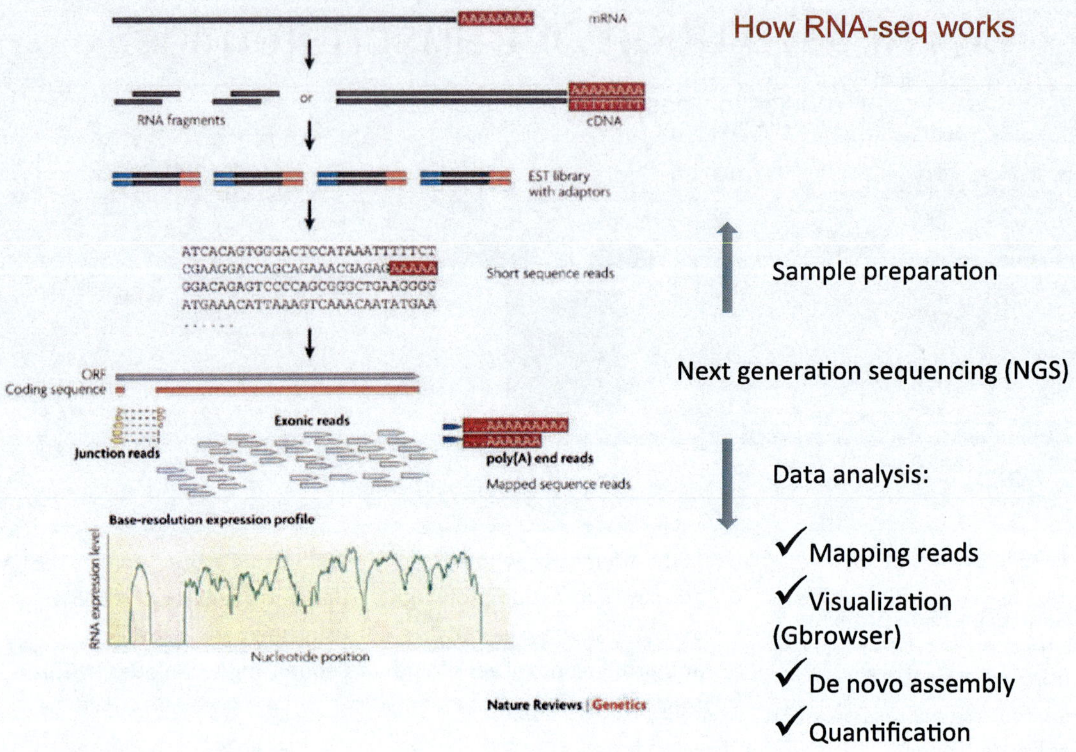

Fig. 5 Mechanism of action of RNA-seq: a revolutionary tool for transcriptome

	iSeq 100	MiniSeq	MiSeq	NextSeq
Run time	9–17.5 h	4–24 h	4–55 h	12–30 h
Maximum output	1.2 Gb	7.5 Gb	15 Gb	120 Gb
Maximum reads per run	4 million	25 million	25 million †	400 million
Maximum read length	2 × 150 bp	2 × 150 bp	2 × 300 bp	2 × 150 bp

Production Scale Sequencers

NextSeq series, HiSeq400 system, HiSeq X series, NovaSeq6000 system.

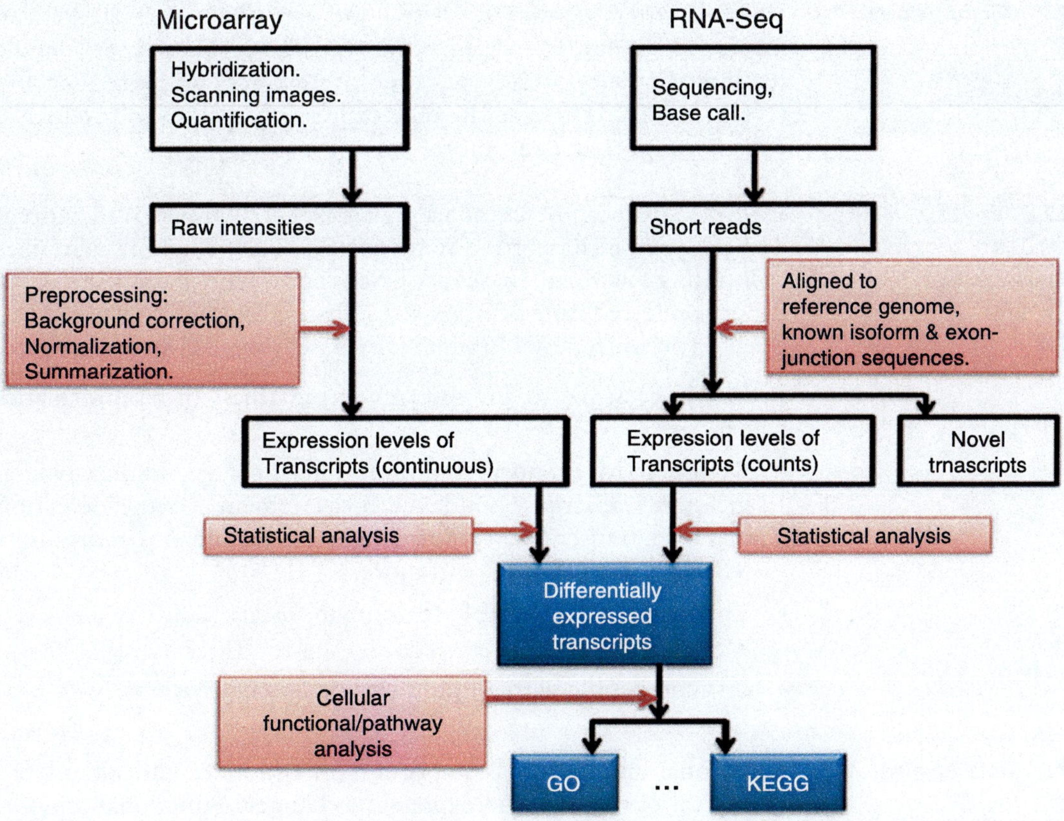

Fig. 6 Overview of analysis workflow for microarray and RNA-seq transcriptional profiling

NextSeq series	HiSeq400 system	HiSeq X series		NovaSeq6000 system
Run time	12–30 h	<1–3.5 days	<3 days	~13–38 h (dual SP flow cells) ~13–25 h (dual S1 flow cells) ~16–36 h (dual S2 flow cells) ~44 h (dual S4 flow cells)
Maximum output	120 Gb	1500 Gb	1800 Gb	6000 Gb
Maximum reads per run	400 million	5 billion	6 billion	20 billion
Maximum read length	2 × 150 bp	2 × 150 bp	2 × 150 bp	2 × 250**

2.7 Data Analysis

2.7.1 Intuitive Analysis of RNA-seq Data

Illumina offers certain RNA-seq software tools packaged in an intuitive user interface designed for biologists. These user-friendly technologies support a wide range of next-generation sequencing (NGS) studies. These aid in gene expression analysis to total RNA expression profiling and more.

2.7.2 Benefits of RNA-seq Data Analysis with BaseSpace Apps

RNA-seq data can be immediately and securely transferred, stored, and analyzed in BaseSpace Sequence Hub, the Illumina genomics computing platform. BaseSpace Sequence Hub comprises of an expert-preferred suite of RNA-seq software tools which were developed or optimized by Illumina. Our BaseSpace RNA-seq Apps are:

- Easily accessible to any researcher, regardless of bioinformatics experience.
- Designed to support common transcriptome studies, which includes transcript abundance measurement for the detection of novel transcripts, coding single nucleotide polymorphisms (cSNPs), gene fusions, and more.
- The technique is suitable for human, mouse, and rat RNA-seq analysis.
- It is compatible with all Illumina sequencing systems.

2.8 Data Analysis

2.8.1 Interpretation of RNA-seq Data

Functional annotation is important in order to understand the biological effects of gene expression changes. Functional annotation can be effectively carried out in BaseSpace Correlation Engine, and is followed after data analysis.

BaseSpace Correlation Engine consists of data sets from thousands of public studies which inform biological interpretation. It aids to connect differential gene expression data from RNA-seq experiments with disease associations, or observe correlated genes and microRNA targets.

TopHat Alignment

It aligns RNA reads and aids in detection of gene fusions employing the industry-standard method, integrated with the Cufflinks Assembly & DE App.

Illumina, Inc.

App Highlights

The BaseSpace apps, TopHat Alignment and Cufflinks Assembly and Differential Expression are created for analysis of RNA sequencing data. The main components of the TopHat and Cufflinks apps are retrieved from the Tuxedo suite.

The TopHat app runs Bowtie, TopHat, the Isaac Variant Caller, and Cufflinks. The app develops aligned reads, variant calls, and FPKM abundance estimates of reference genes and transcripts.

The TopHat Alignment workflow performs the following functions

- Read mapping using the TopHat 2 aligner.
- FPKM estimation of reference genes and transcripts using Cufflinks 2.
- Variant calling (SNVs and small indels) with the Isaac Variant caller.
- Optional fusion calling with TopHat-Fusion.

Available reference genomes include

- *Homo sapiens* UCSC hg19 (RefSeq & Gencode gene annotations).
- *Mus musculus* UCSC mm10 (RefSeq gene annotations).
- *Rattus norvegicus* UCSC rn5 (RefSeq gene annotations).

Current Limitations

- Minimum and maximum of 100,000 and 400,000,000 reads per sample, respectively.
- Maximum of 2,000,000,000 read across all samples in a single analysis.
- Minimum read length 35 bp and maximum read length 125 bp.
- Paired-end reads are required for fusion detection

Reference Documents

http://genomebiology.com/2013/14/4/R36/abstract Title of Paper: TopHat2: accurate alignment of transcriptomes in the presence of insertions, deletions and gene fusions.

http://genomebiology.com/2011/12/8/R72/abstract Title of Paper: TopHat-Fusion: an algorithm for discovery of novel fusion transcripts.

http://res.illumina.com/documents/products/whitepapers/whitepaperiasscworkflow.pdf Title of Paper: Fast, accurate and easy alignment and variant calling with Isaac Genome Alignment Software and Isaac Variant Caller.

App Highlights

The BaseSpace apps like TopHat Alignment and Cufflinks Assembly and Differential Expression are developed for analysis of RNA sequencing data. The main components of the TopHat and Cufflinks apps are derived from the Tuxedo suite. The TopHat app runs Bowtie, TopHat, the Isaac Variant Caller, and Cufflinks. The app produces aligned reads, variant calls, and FPKM abundance estimates of reference genes and transcripts.

3 RNA-seq in Details

3.1 Applications of RNA-seq

RNA-seq aids in the investigation and discovery of the transcriptome, the total cellular content of RNAs including mRNA, rRNA, and tRNA. Understanding the transcriptome is helpful to connect the information on our genome with its functional protein expression. RNA-seq identifies the genes, genes which are turned on in a cell, the level of expression and switch on or switch off mechanism. This allows scientists to understand the biology of a cell and analyze changes which may indicate disease. Some of the most popular techniques employing RNA-seq are transcriptional profiling, SNP identification, RNA editing, and differential gene expression analysis.

This provides researchers about some vital information about the function of genes. As for example, the transcriptome covers all the tissues for which a gene of unknown function is expressed, which might indicate what its role is. It also gathers information regarding alternative splicing events, producing different transcripts from one single gene sequence. It is an advantage, since this information cannot be collected through routine DNA sequencing. This methodology can also identify post-transcriptional modifications occurring during mRNA processing such as polyadenylation and 5′ capping. Figure 7 depicts the sequence logo with observed and expected nucleotide distribution surrounding the 5′ fragmentation site.

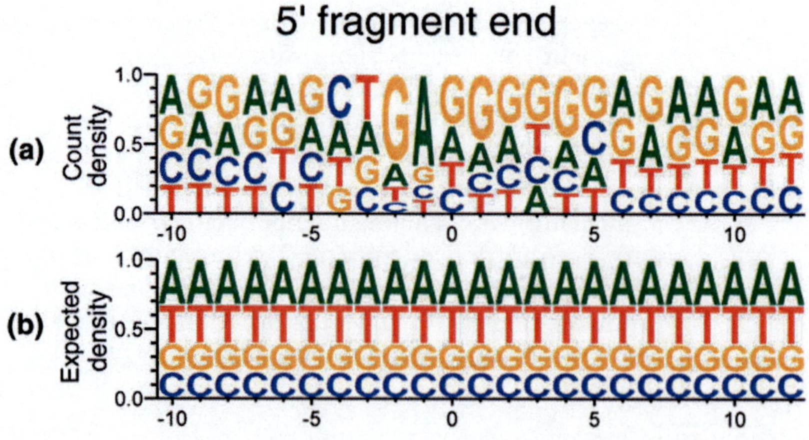

Fig. 7 Sequence logo showing observed and expected nucleotide distribution surrounding the 5′ fragmentation site. Similar biases are present at the 3′ end

4 Protocol for RNA-seq Experiment

The next-generation sequencing platforms most frequently employed for RNA-seq are the Illumina HiSeq (4000, 2500), Ion Torrent, and SOLiD systems. The library preparation and nucleotide detection protocols for each platform differ from each other. Let us discuss the main steps common to these techniques:

4.1 Preparation of Total RNA

It depends on class of RNA to be sequenced (i.e., mRNA, lincRNA, and microRNA), and accordingly enrichment is done. Good quality total RNA is the prime criteria; however, alternative protocols for degraded RNA also exist.

4.2 Library Preparation

Library preparation can be carried out through commercial kits such as Illumina TruSeq Stranded mRNA Library Prep Kit or its variants. Library preparation consists of:

4.2.1 RNA Fragmentation

Unlike short RNAs, mRNAs are usually fragmented to smaller pieces of RNA to allow sequencing.

4.2.2 Reverse Transcription

First and second strand cDNA is subjected to reverse transcription from fragmented RNA with random hexamers or oligo (dT) primers.

4.2.3 Adapter Ligation

The 5′ and/or 3′ ends of cDNA are being repaired. Adapters are ligated, possessing sequences to allow hybridization to a flow cell.

4.2.4 Library Cleanup and Amplification

Libraries are prepared with correctly ligated cDNA fragments and subjected to amplification by PCR for the addition of any remaining sequencing primer sequences.

4.2.5 Library Quantification, Quality Control and Sequencing

Library concentration is analyzed through qRT-PCR and/or Bioanalyzer to make it is ready for sequencing.

4.3 Data Analysis

Downstream data analysis comprises of quality control, for example, trimming of sequencing adapters followed by removal of reads with poor quality scores leading to mapping reads, analysis of differential expression, identification of novel transcripts and pathway analysis.

4.4 Experimental Design

Like other technique, a well-designed RNA-seq experiment comprises of proper replication, randomization, and blocking. Internet NGS user forums, such as Seqanswers (http://www.seqanswers.com/), has been developed to detect statistically significant differential gene expression from an experiment without replicates. Although it is possible to calculate DGE without replicates from a technical point of view, extreme cautions must be undertaken for the interpretation of these experiments.

All experiments including novel transcript discovery, both technical and biological replicates must be carefully considered from the outset. During the initial stages of RNA-seq, technical replicates (libraries prepared from the same RNA sample) were generally used. It has been observed that biological variation far outweighs technical variation, while coverage of at least 5 reads/nucleotide is obtained. Thus, technical replicates are most useful for the comparison of the performance of two or more competing sequencing technologies.

In case our objective is to investigate differences between treatments biological replication is important to generalize the results to a larger population. The required number of replicates will differ based on the amount of biological variability associated with the samples of interest and need to be empirically determined. As a result, the number of replicates used for any prior microarray analysis is generally a good starting point.

Most sequencing platforms support multiplexing of libraries through the introduction of barcodes during library preparation. This leads to simultaneous sequencing of multiple libraries in a single sequencing run, hence enabling more efficient use of a sequencer machine. Multiplexing also helps a balanced block design to reduce potential confounding factors such as PCR amplification and flow cell effects. Let us consider a case when three placebo-treated samples and an equal number of drug-treated samples are sequenced on an Illumina HiSeq 2000 or 4000 instrument. The Illumina HiSeq utilizes a flow cell with eight lanes, one of which is usually reserved for a PhiX sequencing control employed for quality control purposes. Other than using a single lane per sample, all six libraries need to be barcoded during library preparation allowing all libraries to be simultaneously sequenced over six lanes. Reads are demultiplexed after sequencing based upon barcode sequence and analyzed accordingly. Such designs offer insurance against one poor sequencing lane compromising the study. However, if an unblocked design was utilized and one lane was to have strong artifacts, an entire sample may be lost, and we need to compromise the study. As an example, Garrett-Bakelman FE et al. used the NEBNext Multiplex Oligos for Illumina (96 Index Primers) barcode plate (E6609S) and standard barcodes (#10-15, 18-23, 27-28 from Multiplex Oligos for Illumina Index Primers Sets 1,2,3) for both ribodepleted and polyA RNA-seq library preparation.

The major advantage of removing confounding factors, a multiplexed, balanced block design also permits for re-sequencing on an additional lane(s) later on in order to increase the number reads as needed. There is no need to introduce sample-specific biases from flow cell to flow cell variation. Determining the "correct" number of sequencing reads per sample is a challenging problem which is subject to vigorous debate. When profiling the chicken transcriptome, Wang et al. showed that 30 million reads are enough to obtain a reliable measurement of all genes in the genome, whereas 10 million lead to detection of 80% of genes. In order to support a lower number of reads as optimal for transcriptional profiling, Tarazona et al. report that increase of sequencing depth results in more false positives due to increased noise. ENCODE consortium, postulated that between 100 and 200 million reads per sample or greater are required, particularly when novel transcripts or splicing events are of particular interest. The issue of sequencing depth is a rapidly evolving factor without a clear consensus.

The fact of inter-species variation in transcriptome size causes further complications. Sequencing a bacterial transcriptome will need far less number of reads than that needed for a vertebrate transcriptome. Estimation of transcriptome size is problematic as genome size and transcriptome size are not properly correlated, and estimating transcriptome size is especially difficult for non-model organisms without a sequenced genome. As for example, the genome of the laboratory mouse is 2.6 Gb and encodes ~25,000 protein-coding genes, but a similar number of genes are encoded in the 1 Gb chicken genome. Thus, the efforts for sequencing the transcriptome of non-model organisms will particularly benefit from a small pilot study to empirically detect adequate sequencing depth. The condition when alternative splicing is of greater interest, getting paired-end sequence data can be more valuable than increasing number of reads. This is because of the increased probability of a splice junction falling within or between the sequenced ends. An additional advantage of paired-end RNA-seq is the chance to detect chimeric transcripts due to gene fusion events. Paired-end sequence reads cause greater certainty where an individual read can be mapped to multiple loci on the genome, particularly in repetitive regions.

An online tool known as Scotty permits users to design proper sequencing depth and the number of biological replicates, simultaneously satisfying user-defined inputs such as maximum cost and required statistical power. While calculating the required sequencing depth, we should consider the potential for loss of reads to undesired RNA species, chiefly ribosomal RNA (rRNA), as well as reads which are unable to be mapped, since both factors can reduce the number of useable reads for downstream analysis by about 60–80%. A number of hybridization-based rRNA depletion approaches have been undertaken to enrich for less abundant species of RNA. Enrichment strategies either deplete rRNA or permit positive selection for mRNA.

For eukaryote transcriptome analysis using SOLiD or Ion Torrent platforms, polyA+ selection can be analyzed prior to library construction with either magnetic bead-conjugated oligo (dT) oligonucleotides (Dynabeads; Life Technologies) or immobilized oligo(dT) capture probes (mRNA Catcher PLUS; Life Technologies).

On the contrary, Illumina TruSeq libraries utilize two rounds of magnetic bead-conjugated oligo(dT) capture for polyA+ selection. This is associated with the final polyA+ elution step also leading to fragment and prime RNA for downstream cDNA synthesis. In the process of polyA+ enrichment for polyadenylated mRNAs, non-polyadenylated RNA species, involving microRNAs, lincRNAs, and other macro ncRNAs are depleted and not depicted in the resulting libraries. On the contrary, rRNA depletion strategies have been shown to preserve these RNA species. Two of the most commonly employed techniques utilized for rRNA depletion are RiboMinus (Life Technologies) and Ribo-Zero (Epicentre). Both these techniques use a pool of rRNA capture probes followed by spin column or magnetic bead-based collection of the non-rRNA fraction. rRNA capture probes for multicellular eukaryotes (human/mouse compatibility) and for microorganisms (yeast/bacteria) are commercially available. Ribo-Minus has an added advantage as it aids in removing mitochondrial rRNA as well as cytoplasmic rRNA. Finally, rRNA depletion needs to be followed, when sequencing the degraded RNA isolates from formalin-fixed paraffin-embedded samples. PolyA+ selection methods assume the availability of high-quality total RNA to enable isolation of full-length transcripts. Garrett-Bakelman FE et al. the NEBNext Ultra II Directional RNA Library Prep Kit for Illumina (E7760/E7765) and NEBNext rRNA Depletion Kit (NEB #E6310) as well as hemoglobin depletion reagents (New England Biolabs) to prepare ribodepleted RNA-seq library and the NEBNext Ultra II Directional RNA Library Prep Kit for Illumina (E7760/E7765) and NEBNext Poly(A) mRNA Magnetic Isolation Module (NEB #E7490) to prepare polyA RNA-seq library from the NASA twin astronaut blood sample.

Just as rRNA contamination will significantly reduce the number of reads mapping to mRNA, so will highly expressed mRNAs, such as housekeeping genes, reduce the number of reads mapping to weakly expressed genes. For example, 75% of reads from a human mammary epithelial cell line library are responsible for mapping the most abundant 7% of the transcriptome. Clearly, when the number of reads is at a premium, it would be most useful to have them map to regions of interest. In order to remove transcripts responsible for a small number of housekeeping genes, the Ribo-Zero concept was extended by designing capture oligos to remove globin mRNAs. Exome capture microarrays may also be employed to increase sensitivity, although at the cost of reduced quantification accuracy. On the contrary, the Rinn laboratory developed RNA CaptureSeq to specifically sequence weakly expressed regions of the transcriptome. RNA is hybridized to tiling microarrays possessing the probes corresponding to genomic regions of interest. The captured RNAs are eluted and sequenced. CaptureSeq allowed ~380-fold enrichment of reads mapping to targeted regions of the transcriptome in comparison to the conventional RNA-seq without capture. Although conventionally used for exome sequencing or targeted re-sequencing of DNA, microarrays were employed to capture 467 cancer-related genes for targeted RNA sequencing. This approach leads to the identification of mutations and fusion transcripts while largely preserving transcript abundance. An important caveat, however, is that inferring a somatic DNA mutation by cDNA sequencing is problematic without careful validation by Sanger sequencing due to extensive RNA editing. Life Technologies released a newly developed targeted RNA-seq workflow which allows the targeted sequencing of over 6000 RNAs. Illumina is expected to release an equivalent workflow soon. It is based on the sequencing of short PCR-amplified amplicons, not associated with full-length transcripts of targeted RNAs. Hence, directed sequencing will gives rise to comparable information to quantitative RT-PCR.

5 RNA-seq Technologies

The three most widely used NGS platforms for RNA-seq are

1. Illumina.
2. SOLiD.
3. Ion Torrent.

These platforms possess similar sample input requirements and sequences millions of cDNA fragments per run. The procedures for sample preparation and pertinent application-specific advantages and disadvantages are discussed.

5.1 Illumina

Illumina and Ion Torrent both sequences employ a "sequencing by synthesis" (SBS) approach. Incorporation of dNTPs is detected at a time at millions of fixed positions on a flow cell (Fig. 8).

PolyA+ RNA is enriched with oligo (dT) beads. The next step is fragmentation and reverse transcription. The 5′ and 3′ ends of cDNA fragments are the next to prepare permitting efficient ligation of "Y" adapters possessing a unique barcode and primer binding sites. In the next step, ligated cDNAs are PCR-amplified and ready for cluster generation and sequencing.

For Illumina, once TruSeq RNA-seq libraries have been made, they are hybridized to a flow cell. It contains a lawn of covalently bound oligonucleotides that are complementary to the sequencing adapters and which were introduced during library preparation. Once hybridized, the capture oligonucleotide primes DNA polymerase extension activity leading to a covalently bound full-length

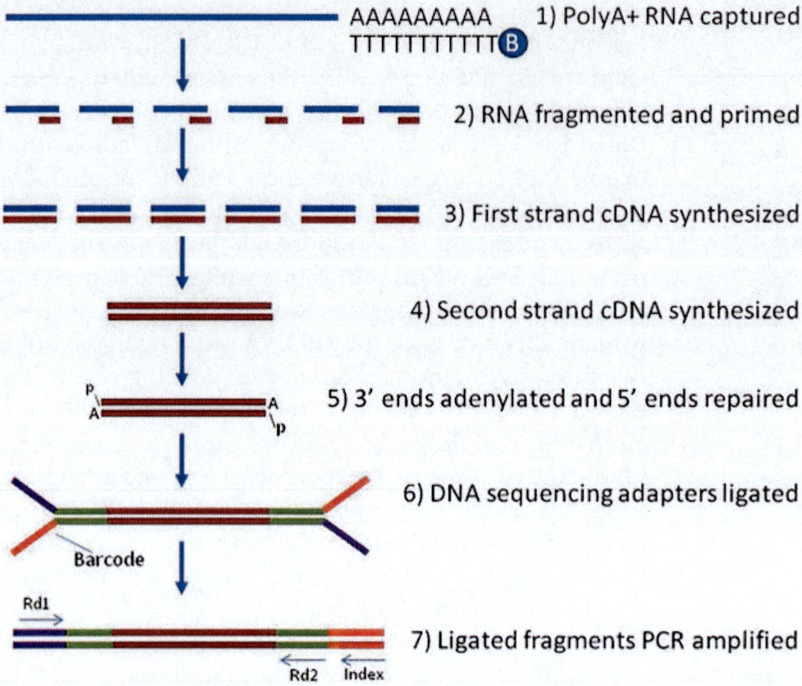

Fig. 8 Illumina RNA library preparation

complimentary copy of the cDNA fragment. This is subjected to several rounds of PCR amplification for the production of discrete clones ~1 μm in diameter which can be optically resolved during sequencing. Obtaining optimal cluster density is important since it will determine the number of reads obtained. Clearly, low density will result in fewer than expected reads, but over-clustering can be just as problematic since dense flow cells are difficult to analyze and to obtain accurate base calling regarding interference and overlap between adjacent clusters. Hence it becomes important to quantify each library accurately using quantitative PCR as quality control.

In the case of Illumina SBS, all four dNTPs are fluorescently labeled and concurrently introduced on to the flow cell (Fig. 8). Figure 9 compares RNA-seq technology through different sequencing platform.

The presence of four dNTPs leads to natural competition for binding between dNTPs, results into minimization of incorporation biases. SBS undergoes multiple cycles of nucleotide incorporation and detection. Only one nucleotide is incorporated per cycle using reversibly terminated dNTPs. Following nucleotide incorporation as detected by fluorescence, the fluorophore is removed leading to regeneration of a 3′ hydroxyl group allowing incorporation of the next dNTP in the subsequent cycle. This reversible terminator chemistry permits sequencing of homopolymeric regions, such as “AAAAAA,” with high confidence.

During base calling, fluorescence intensity values for each nucleotide are converted to nucleotide identity employing a cross-talk matrix controlling spectral overlap. Since spectral overlap is detected during the first four cycles, it implies that approximately equal numbers of each base be present (i.e., to have a balanced library). Hence, it is important to assign barcodes which are well balanced ensuring accurate demultiplexing after sequencing. Accordingly, it is recommended to use PhiX control lane for the estimation of correct spectral overlap, while sequencing unbalanced libraries of AT- or GC-rich genomes. Error rate (which implies incorporation of the incorrect nucleotide) consequently mounts with an increasing number of cycles. Up to 150 cycles are supported and an overall error rate of 0.2%.

In the first iteration of Illumina sequencing technology, sequencing of only one end of each cDNA fragment was conducted. The use of paired-end sequencing is most frequently employed for the sequencing of the transcriptome. The flow cell possesses randomly arrayed capture oligos complementary to the 5′ and 3′ sequencing adapters. During the "bridge amplification" PCR step, cDNA fragments captured by their 5′ adapters are prone to be captured by 3′ capture oligos. This leads to the first sequencing run of up to 150 cycles with the 5′ sequencing primer to be followed by a second sequencing run using the 3′ primer in order to obtain a total of 300 nt of sequence per fragment. Several third-party library preparation kits are also commercially available advantageous for certain experiments. For example, Smart-seq reduces the required input to as little as 100 pg total RNA.

5.2 Ion Torrent

Illumina sequencing and cluster generation is based on solid-phase PCR amplification, whereas emulsion PCR is used for the preparation of Ion Torrent libraries for sequencing. For the first step the library template is prepared from fragmented RNA. Unlike the standard library protocol is strand-specific by default (Fig. 9) in this technique, which is different from Illumina. In the next step, beads with complementary oligonucleotides are mixed with PCR reagents and a dilute solution of cDNA library and oil added leading to an emulsion. Ideally, each microdroplet of emulsion will be consisting of one bead and one cDNA fragment and PCR reagents to allow for clonal amplification. After 16–18 cycles of PCR, the emulsion is broken through organic extraction, beads are purified and loaded on to a disposable semiconductor sequencing chip. The sequencing chip is modeled as a honeycomb, where each bead fits into one of hundreds of millions of tiny wells. These well serve as microreactors during sequencing, while each well possesses their own detector. In contrary to Illumina's fluorescence-based SBS, Ion Torrent detects sequence identity by detecting pH alterations because of hydrogen ion releases due to nucleotide incorporation (Fig 9). In this technique, dNTPs are not differentially labeled

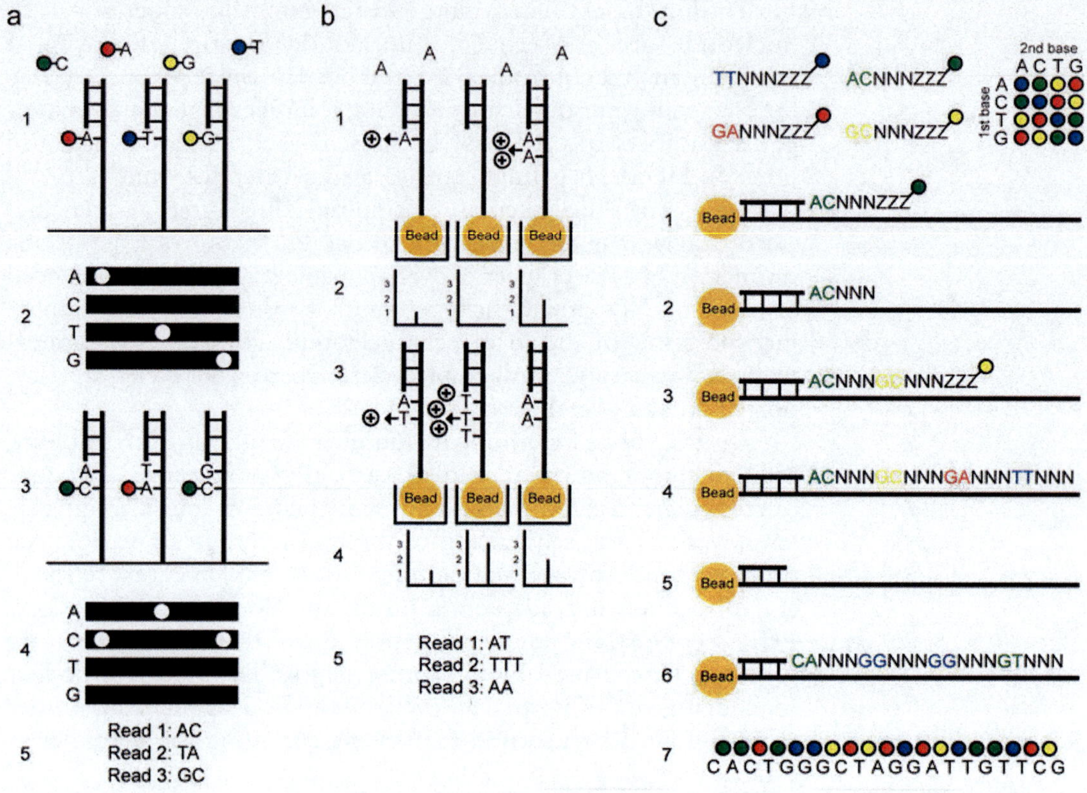

Fig. 9 Sequence detection methods of Illumina, Ion Torrent, and SOLiD. (**a**) Illumina detection is fluorescence-based using reversible terminator dNTPs, leading to one nucleotide incorporation per cycle. cDNA fragments are linked covalently to a flow cell. Fluorescence is detected with the addition of each nucleotide. (**b**) Ion Torrent sequence by synthesis is based on the detection of hydrogen ions (⊕) for base calling. Each pH detector well-possesses one clonally amplified cDNA fragments on a microbead. Nucleotides are added in a sequence. Since nucleotides are not reversibly terminated, incorporation of multiple nucleotides is assessed by an increase in the number of hydrogen ions detected. (**c**) SOLiD sequence detection is special in that fluorescently labeled oligonucleotides are ligated rather than incorporated by a polymerase

by a fluorophore. Hence, they need to be added sequentially in order to facilitate ion release with a particular nucleotide.

Since Ion Torrent sequencing is not dependent upon the optical detection of dNTP incorporation, sequencing reactions are faster, and the number of reads obtainable per sequencing run has been better. There is a basic difference in Illumina and Ion Torrent technologies. Illumina relies on reversible terminator chemistry to restrict dNTP incorporation to once per cycle and sequence through homopolymers. However, Ion Torrent depends on the number of hydrogen ions released with reference to the number of dNTPs incorporated. As a result, "A" can easily be differentiated from "AA" through the detection of a doubling in the number of hydrogen ions released. Care should be taken to differentiate

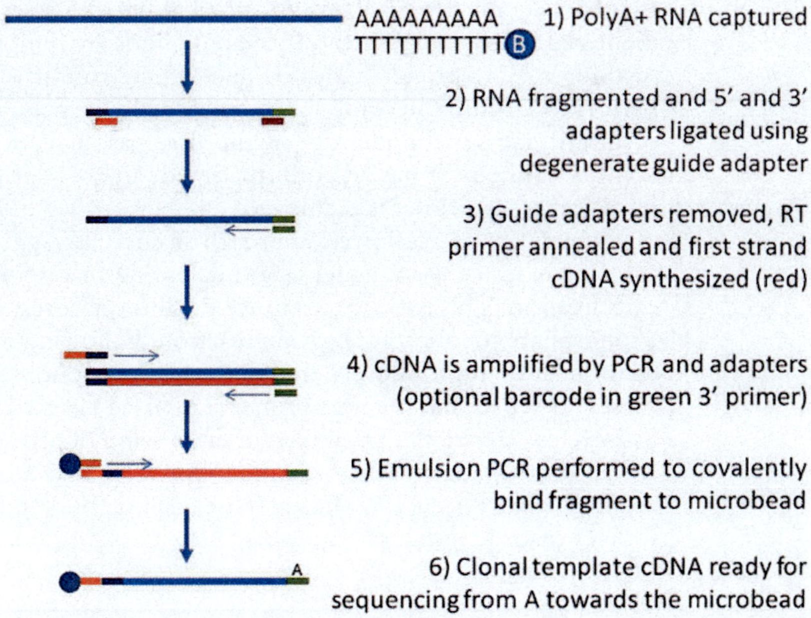

Fig. 10 Ion Torrent and SOLiD libraries are both prepared utilizing similar protocols. Partly degenerate guide adapters hybridize the fragmented target RNA leading to splint ligation of 5′ and 3′ adapter with defined sequences. In the next step, cDNA is synthesized and amplified by PCR for adding additional required sequences. The next step is emulsion PCR on microbeads

between a run of 7 and 8 adenosines is far more challenging, and as a result, the error rate increases (1.7%).

5.3 SOLiD

Ion Torrent and SOLiD RNA libraries preparation possess the similar molecular biology (Fig. 10). The only differentiating factor is the adapter sequences. The basic principle of SOLiD is that it utilizes a "sequencing by ligation" approach to obtain billions of reads per sequencing run, each up to 75 bp in length, which is absent in earlier discussed methodologies (Illumina/Ion Torrent).

The first step involves emulsion PCR and beads consisting of clonally amplified cDNA fragments are attached to the surface of a sequencing flow cell. Several rounds of ligation reactions are involved in the sequencing process.

In the first round, a sequencing primer is annealed. The next step is addition of a mixture of 16 fluorescently labeled 8-mer oligonucleotides. The 16 oligonucleotides present all possible combinations of the first two nucleotides (AC, AG, AT, etc.); however, bases 3–5 are degenerate and not known. The final three 3′ bases are conjugated to one of four fluorescent labels, present with different excitation and emission spectrum.

Therefore, each fluorophore presents four dinucleotides and in each ligation reaction, the identity of only the first two nucleotides is detected. Hence, after one round of ligation, the identity of these two bases is narrowed down but not known. To assess their true identity, the original primer and ligated oligos are removed. Then a second, $n - 1$, primer annealed and a new round of ligation is performed. From the combination of knowledge from two rounds of interrogation, the identity of the first base is confirmed. The identity of the next base is confirmed with an $n - 2$ primer, and so on till an $n - 4$ primer is employed.

Practically the followed approach is the cleavage of the final three 3′ bases of each 8-mer oligonucleotide to be followed after each ligation. This is done to remove the fluorophore and provide a 5′ phosphate for a second ligation reaction. Following 5–7 cycles of ligation and fluorophore detection and fluorophore cleavage, a "reset" is set so that the next primer (i.e., $n - 1$) is used for another 5–7 cycles. Hence the advantage of employing SOLiD sequencing is the interrogation each nucleotide twice. As a result, it leads to reduced errors (<0.1%) during base calling but at the cost of shorter reads length.

6 Non-coding RNA-seq

RNA-seq is equally applicable for studying non-coding RNAs, such as ***microRNAs*** and ***lincRNAs***. It is also employed to detect a new class of non-coding circular transcripts (circRNAs). To ascertain a complete picture of the transcriptome, biologists often combine coding and non-coding RNA-seq data. With regard to experimental design and sequencing chemistry, the requirements for non-coding RNA-seq are almost the same as mRNA-seq.

MicroRNAs (miRNAs) may be regarded as short pieces of RNA directing post-transcriptional gene silencing of their targets with a process of imperfect hybridization to the 3′ UTRs of mRNAs. Mature miRNAs are 19–24 nt in size and are developed by two cleavage events. First, cleavage of a nuclear primary transcript may extend up to several kilobases in length. Second, cleavage of the cytoplasmic intermediate hairpin precursor is approximately 70 nt. Because of rapid processing and turnover, precursor transcripts are sequenced infrequently, the mature form is considered the most. However, identical protocols have been successfully employed for sequencing precursor and mature miRNAs. In contrast to Sanger sequencing, which identified only the most strongly expressed miRNAs, NGS can identify weakly expressed miRNAs as well as reveal heterogeneity in length and sequence.

For mRNA-seq, good quality total RNA is the prime essential criteria for obtaining good miRNA-seq library. It is critical that the RNA isolation procedure preserve the integrity of small RNAs. In fact, commonly employed spin columns did not retain RNAs <200 nt which hindered their discovery. Although Trizol retains small RNAs, it is not advisable when a low number of cells are used

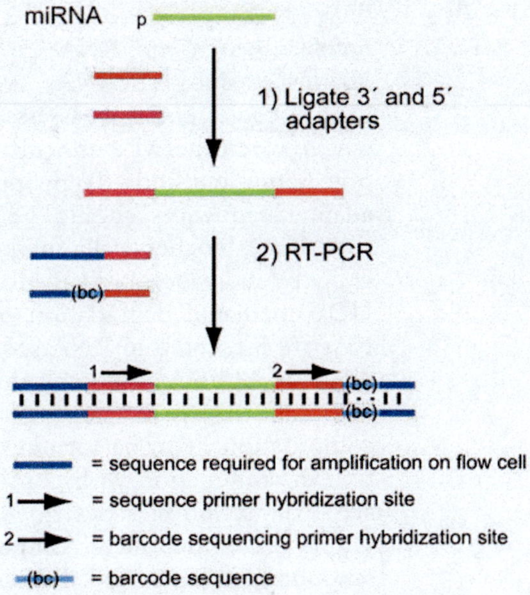

Fig. 11 MicroRNAs are sequenced by ligating RNA adapters to each end of the mature microRNA followed by reverse transcription and PCR (RT-PCR). To allow barcoding, two sequencing reactions are performed with two sequencing primers. The first primer one is to obtain the microRNA sequence and primer two obtains the barcode sequence

as starting material. In that case miRNAs with low GC content are selectively depleted.

In addition to miRNAs, short RNA also involves piwi-interacting RNAs (piRNAs) which is also equally characterized using RNA-seq. The stages involved in small RNA library preparation are similar to conventional RNA-seq, only except is fragmentation (Fig. 11). The latest version of Illumina miRNA library preparation employs a 5′ monophosphate and 3′ hydroxyl groups for ligation of miRNAs in a reaction containing total RNA. For Ion Torrent and SOLiD libraries, it is strongly suggested or employed short RNA enrichment by polyacrylamide gel electrophoresis or magnetic bead purification.

After the process of RNA adapter ligation, miRNAs are reverse transcribed, amplified by PCR and sequenced.

Next step involves analysis. miRNA-focused analysis platforms are freely available. Reads need to be trimmed of barcode and adapter sequences and mapped, either to known miRNAs in miRBase or to a reference genome for novel miRNA discovery. Irrespective to the mapping of mRNA-seq data, there is no use for a splicing-aware aligner, and BWA or Bowtie may be used.

Similar to miRNAs, lincRNAs encompass part of the non-coding assortment of RNAs within eukaryotic cells. However, their function is observed to be more heterogeneous and less well defined compared to miRNAs. The sequencing of lincRNAs involves paired-end RNA-seq and observed to be most useful. However, since lincRNAs are frequently antisense to known genes, it is essential to assess the strandedness of mapped reads, which were otherwise difficult to assess using conventional library preparation methods. To maintain strandedness, either the 5′ or 3′ adapter sequences should be unique. The first/second cDNA strands are biochemically marked during library preparation, generally by the process of substitution of dUTP for dTTP to enable UDG-mediated degradation of dUTP-containing DNA. LincRNAs are presented in RNA-seq libraries earlier subjected to polyA selection, although skipping this step may lead to identification of additional RNAs. Identification of novel lincRNAs is conducted computationally by performing ab initio transcriptome reconstruction in conjunction with consideration for epigenetic markers of active transcription as described earlier. Accordingly, circRNA transcripts were identified computationally by searching for exon scrambling in paired-end RiboMinus RNA-seq data.

Other than the class of non-coding genes, the polyA tails of mRNA transcripts are not translated to protein. After RNA polymerase II-dependent transcription of most mRNAs, a stretch of untemplated adenosine monophosphates is mixed to the transcript following cleavage by cleavage and polyadenylation specificity factor (CPSF) and poly(A) polymerase. A long-standing question was the relationship between polyA tail length, transcript stability, and translatability. However, due to the difficulty to sequence through homopolymeric (<50 nt of consecutive adenosines) regions, by both Sanger and next-generation sequencing, this question has only been indirectly studied. However, a recent innovative approach to use a combination of sequencing technologies and complex statistical analysis by the Kim laboratory has provided much insight. In the method, known as TAIL-seq, the initial process is the depletion of rRNAs and small ncRNAs from total RNA, followed by ligation of a biotinylated 3′ adapter to the remaining mRNAs and long non-coding RNAs. In the next step, the nuclease RNase T1, which is at low concentration specifically cuts after G residues (and not within polyA tails), is incubated with the ligated RNAs, then the process of pull-down with streptavidin beads was followed in order to enrich for 3′ adapter ligated RNA fragments.

After 5′ adapter ligation, reverse transcription and PCR amplification, libraries are sequenced from both ends. The first read gives 51 nt of sequence identification for mapping purposes, whereas the second read, up to 231 nt in length, gives tail length as identified by a stretch of thymine nucleotides (corresponding to the pre-reverse transcription polyA tail). Other than sequencing libraries on the Illumina platform, better matched to sequencing homopolymeric regions, incomplete cleavage of the thymine reversible terminator fluorophores lead to persisting thymine fluorescence signal in subsequent cycles, and make non-poly(T) nucleotides mostly indistinguishable from true poly(T) stretches. However, the transition from poly(T) to non-poly(T) stretches was accomplished by an increase in non-T signal. This is accomplished by a Gaussian mixture hidden Markov model for the detection of the position of this transition poly(A) tail length, that can be measured with high resolution and accuracy and at a genome-wide scale. Finally, this technique shows that tail length has a correlation with mRNA half-life, but not translational efficiency. Furthermore, TAIL-seq has been identified for the first time with widespread uridylation and guanylation of mammalian mRNAs.

7 Single-Cell RNA-seq Indexing/Barcoding

RNA-seq has become a widely used methodology for transcriptomic analysis at single cell level. The major challenge in single-cell (sc)RNA-seq is the barcoding (or indexing) step. A wide number of new technologies have been developed in relation to this. Recently, a new methodology for labeling the transcriptomes of single cells has been developed, termed as single cell combinatorial indexing (sci) applied to scRNA-seq (sci-RNA-seq).

This method includes the following steps:

1. Nuclei are extracted with or without methanol fixation and distributed into 96- or 384-well plate.
2. mRNA is retro-transcribed into the first cDNA strand with poly(T) primers possessing unique molecular identifiers (UMIs) thereby lead to the first level of indexing at single cell level.
3. All the cells are pooled and reallocated through FACS sorting into 96- or 384-well plate (e.g., 10–100 cells per well).
4. The second cDNA strand is synthesized, followed by addition of a tag called transposon 5 (Tn5) by transposase and the cDNA is amplified by PCR. In the process of the cDNA amplification, the primers target the first barcode on one end and the Tn5 adapter on the other end. There is introduction of a second barcode that is specific to each well of the PCR plate.
5. All the amplicons are pooled followed by sequencing. Since each nucleus is characterized by two specific barcodes, mRNA profiling of individual cells can be performed.

An additional method recently introduced is known as sci-CAR, a combinatorial indexing based co-assay that profiles single-cell chromatin accessibility and mRNA (CAR). This methodology combines the Assay for Transposase-Accessible Chromatin using sequencing (ATAC-seq) and sci-RNA-seq that has been explained above. Following the synthesis of the first cDNA strand, with the Tn5 transposase, a first ATAC-seq index is introduced. The rest of the steps are similar to the ones described above.

Finally, both amplicons from ATAC-seq and RNA-seq are pooled and sequenced. Since each ATAC-seq and RNA-seq libraries are tagged with unique barcodes corresponding to specific wells, the mRNA and chromatin accessibility for single cells can be linked and studied.

Latest advances in droplet microfluidic system have developed good methods for barcoding single cells. However, indexing errors may happen during the cell being encapsulated in the droplet microfluidics. Random encapsulation of cells into a droplet can happen causing errors in the cell indexing. As a result, more platforms have been developed to reach a precise and efficient expression profiling of single cells during scRNA-seq. A new droplet-microfluidics system for precise barcoding has been created. The system is based on a spiral system when highly concentrated beads containing the barcoded oligonucleotides are ordered linearly through an inertial effect followed by encapsulation one by one into the droplets avoiding multiple beads into a droplet. As a result, the researchers found a decrease in the barcoding system errors.

For the encapsulation step, Rosenberg et al. at Washington University have developed a separate methodology known as a split-pool ligation transcriptome sequencing (SPLiT-seq) tagging the cellular origin of RNA through a combinatorial barcoding system. The advantage of this system is that it does not need encapsulation steps into droplets, micro-wells or wells but it utilizes the cells themselves as compartments.

In SPLiT-seq single cell-individual transcripts are labeled in a unique fashion via four rounds of combinatorial barcoding:

1. Cells are divided into a 96-well plate, cDNA is developed using well-specific barcoded primers.
2. Cells from all the wells are pooled and re-divided into a new 96-well plate wherein a second barcode is designated to the cDNA.
3. The cells are later on pooled, divided, and a UMI is added through ligation.
4. The last barcode is added to cDNA by PCR during the sequencing preparation.

This combinatorial indexing is advantageous to uniquely label over one million cells by generating 21,233,664 barcode combinations for a 96-well plate.

At the end, an alternate methodology to the droplet-based techniques has been presented by Gierahn. They created a Seq-Well methodology based on a nanowell-based system. Seq-Well puts single cells and barcoded poly(dT) mRNA capture beads in an array of around 86,000 subnanoliter wells through a semi-permeable membrane that allows proficient cell lysis and transcript capture. After cell lysis, mRNA is captured by the beads containing a cell barcode and a UMI. As a next step, the membrane and beads are removed leading to bulk reverse transcription, amplification, and library preparation for sequencing. More recently, Butler A et al. formulated an analytical strategy for integration of single-cell transcriptomic data across different conditions, technologies, and species to identify shared cell populations.

7.1 Gene Quantification and Differential Expression Testing

One of the most frequent applications of RNA-seq, analogous to microarray experiments, is to detect differentially expressed genes between two or more groups. The number of reads mapped to each RNA species is related to its abundance within the cell in a linear fashion. As a result, the number of reads discretely mapping to each gene or isoform may be utilized to infer the level of expression. However, a normalization step must first be conducted to find out the differences between libraries. The main bias which normalization seeks to resolve is the total library size (i.e., the number of aligned reads/sequencing depth) as this will vary sample-to-sample. Further, as mentioned earlier, longer transcripts are more predicted to be sequenced than a short transcript and read coverage is often not steady.

Several normalization procedures have been created in recent years, relative advantages and disadvantages depending upon the situation. The simplest normalization methods scale read counts which map to each locus through the total number of reads. These involve total count, upper quartile, and median normalization methods, wherein the number of gene reads is divided by the total number of mapped reads and multiplied by total, upper quartile, or median number of reads from all sequenced libraries under analysis in the experiment, respectively. Two R/Bioconductor packages, DE-Seq and edgeR, involve similar normalization methods for the calculation of a gene-specific scaling factor, and assumed that the majority of genes are not differentially expressed. The final normalization methods tested by Dillies et al. were quantile normalization, that is typically used for normalization of microarray data, and FPKM normalization. The major disadvantages for the most widely utilized FPKM-based normalization failed to sufficiently normalize variation between samples, had high false-positive rate and did not sufficiently reduce the coefficient of variance of a pool of 30 housekeeping genes. On the contrary, both DE-Seq/edgeR normalization lead to the lowest coefficient of variance of all methods tested and low false-positive rate followed by testing of simulated and real data. Whether the underlying assumption of DE-Seq/edgeR on the basis of the fact that the majority of genes are not differentially expressed holds true in all cases is not clear. Overexpressed and underexpressed genes cannot balance out.

Recent observations of widespread transcriptional amplification in cells overexpressing c-Myc emphasize the need for careful decision making while choosing a normalization method.

Other strategies could be to re-visit normalization after differential expression testing for the removal of differentially expressed genes before estimation of scaling factors or to make use of synthetic RNA spike-in transcripts, e.g., Universal Human reference RNA (UHRR) with ERCC spike-in.

After library normalization, statistical testing for differential expression can be performed. Two widely used counts-based workflows employ the R/Bioconductor statistical computing environment. Other than performing count normalization, EdgeR and DE-Seq both test for differential gene expression with a negative binomial distribution. For the first step, a matrix possessing a number of reads corresponding to each gene of interest for every sample is constructed using a Python script called HTSeq-count. Next, edgeR or DE-Seq utilizes the counts table with biological replicates for the calculation of variation and test for statistically significant differential expression. These tools can be functional at the command line or in the MultiExperiment Viewer (MeV) software, having a convenient graphical user interface. These methods are able to make use of ANOVA-like generalized linear models (GLMs) for the analysis of complex experimental designs. This allows users for controlling any known batch effects produced during library preparation, simultaneously analyzing time course experiments and experiments with greater than two groups. Besides, both methods use the Benjamini-Hochberg procedure for controlling the false discovery rate (FDR) associated with multiple hypothesis testing. For instance, De Cecco M et al. used featureCounts for counting mappings and EdgeR for normalization of data and derive differential expression. Garrett-Bakelman FE et al. used featureCounts and Kallisto.

However, Cufflinks/Cuffdiff, part of the Tuxedo suite, is FPKM-based. The Cufflinks component is involved to assemble aligned reads into transcripts, isoforms, and genes while simultaneously identifying transcriptional start sites (TSSs). Cuffdiff leads to tests for statistically significant differences in expression in the Cufflinks output. Although not able to apply GLMs, Cuffdiff has a number of advantages over DE-Seq/edgeR. It is particularly suitable for those who are uncomfortable using R/Bioconductor and the command line. Cufflinks/Cuffdiff, in association with all of the other tools in the Tuxedo suite, is implemented in Galaxy. A major technical advantage of Cuffdiff is that it can detect and test differential isoform expression. However, the significant disadvantage is that GLMs and complex multi-group comparisons are not supported. Quantification of isoforms may be one of the most challenging, but important, aspects of RNA-seq experiments, since quantifying expression at the gene level may mask alterations at the transcript level when two or more isoforms have opposite expression patterns.

During alignment, reads frequently map to multiple regions leading to ambiguity while deciding the true origin of the read. By default, Cufflinks will uniformly divide multi-mapped reads among all the positions which it maps to. Optionally, multi-read correction can be conducted to assign reads probabilistically. However, ambiguity may arise while considering multiple isoforms transcribed from a single open reading frame (ORF). Considering the example of a full-length transcript and truncated transcript is resulting from alternative polyadenylation. Since assignment of reads mapping to the 3′ exons of the full-length transcript is unambiguous, 5′ exon reads would originate from either the full-length or truncated transcript. Cuffdiff can be employed to resolve this problem by probabilistic deconvolution assigning reads to the correct isoform. Cuffdiff may be regarded as more appropriate than edgeR/DE-Seq while characterizing transcriptomes with frequent alternative splicing, as a constant number of reads per gene could mask differential expression of two or more isoforms.

However, the tools available for isoform quantification is rapidly evolving, and the authors of DE-Seq have recently published their negative binomial method called DEXSeq. The basic principle is that it tests for a significant difference in the number of reads per gene between samples. DEXSeq tests for differential expression of individual exons. During testing against Cuffdiff (version 1.3), a far higher number of false positives had been identified by Cuffdiff. A newer version of Cuffdiff (version 2.0) has been published and released. Till date, a detailed and direct comparison of DEXSeq to Cuffdiff 2.0 is not available (Fig. 12).

7.2 Data Visualization and Higher Level Analysis

Following a list of differentially expressed genes and/or transcripts, visualization, and higher level analysis can process in analogous to microarray experiments with minor modifications. The newer versions of MeV, that was designed for analysis of microarrays, has evolved certain modules particularly designed for RNA-seq data. One such MeV module, based on the GOSeq R/Bioconductor package, tests for enrichment of Gene Ontology (GO) terms linked with a list of significantly over- or underexpressed genes.

Unlike microarray GO analysis, lists of differentially expressed genes identified by RNA-seq are biased towards longer transcripts, as there is greater statistical power calling strongly expressed genes being differentially expressed. Simultaneously, Young et al. revealed that certain GO categories are enriched for short or long transcripts. The GOSeq method regulates length bias and occasionally selection bias. Microarray analysis is not subjected to the same length bias; accordingly, samples studied by microarray were compared with respect to RNA-seq gene lists with and without bias correction. The GOSeq bias correction method lead to more GO

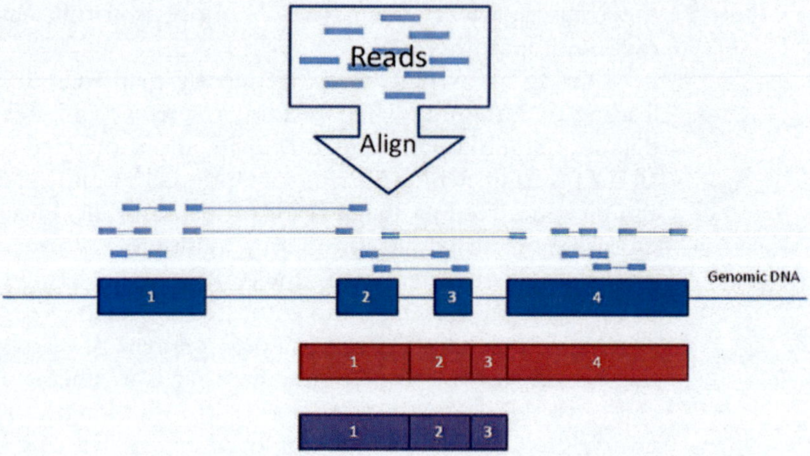

Fig. 12 Paired-end reads are obtained and mapped to a reference genome with a splice-aware aligner, e.g., TopHat or STAR. The example reveals two isoforms as transcribed from a single gene. Although reads mapping to exon 4 are certain to be originating from the long isoform, the assignment of remaining reads is ambiguous and must be handled probabilistically, for example using Cufflinks

categories consistent with the microarray interrogation. The same GOSeq method is present as a stand-alone package in R/Bioconductor, together with numerous packages for hierarchical clustering, preparation of heatmaps, principal component analysis (PCA), and visualization. CummeRbund, the final component of the Tuxedo suite, is an R/Bioconductor package particularly designed to visualize the Cuffdiff output and offers many of the same functionality.

8 Emerging Technologies

Recent advances in NGS productivity violate Moore's Law, the prediction originally applied to computing which states that the number of transistors per microprocessor will double every 2 years (Fig. 13). Per base sequencing costs have reduced significantly associated with increased sequencing output. Output increases are mostly due to the more number of reads, and longer read length and this trend continues apace.

Most exciting emerging technologies have evolved to counteract the known biases and disadvantages of RNA-seq, e.g., eliminating fragmentation, reverse transcription, and PCR amplification biases. This will lead to the reduction of input requirements. Sample input need is currently at the stage when RNA of a single eukaryotic cell can be sequenced with the aid of existing technologies. However, these approaches are based on various degrees of amplification by PCR or in vitro transcription.

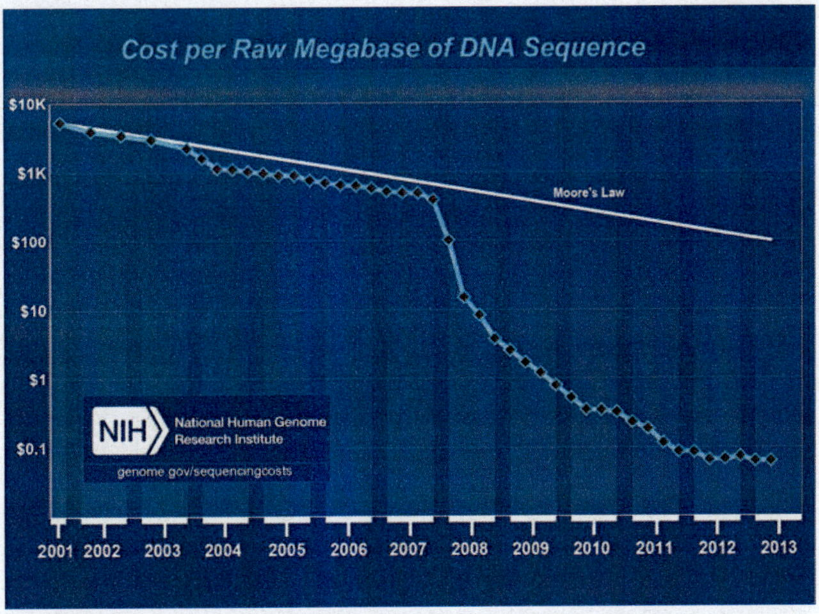

Fig. 13 In recent years, sequencing costs per megabase have decreased faster than Moore's Law. (Image: National Human Genome Research Institute)

A sequencing method which works with the need for the amplification to convert RNA to cDNA will lead to an undistorted picture of the transcriptome. Direct RNA sequencing (DRS) from Helicos BioSciences has opened the way to achieve this feat through capturing polyadenylated RNAs on a flow cell leading to sequencing by synthesis using fluorescently labeled nucleotide analogs, similar to Illumina sequencing. Thus, the DRS system has not found to be commercially viable because of a combination of high error rate, a short read length, and an inability to perform paired-end sequencing. However, Helicos intellectual property rights have been licensed to Illumina and Life Technologies leading to the exciting possibility of Illumina/Ion Torrent-like DRS in the near future.

Sequencing longer transcripts, and eventually full-length transcripts, would decrease the uncertainty while quantifying alternatively spliced genes. Long sequencing reactions, nanopore sequencing can also allow the sequencing of RNA without amplification. In nanopore sequencing either biological (e.g., α-hemolysin) or synthetic (e.g., graphene) nanopores are embedded in a synthetic polymer membrane. An electric current is used across the nanopore. Since each nucleotide passes through the nanopore, a nucleotide-specific disruption in charge is detected. Elimination of the need to use fluorescence permits a far higher density of nanopores, faster sequencing, and as a result, a greater sequence output. Oxford Nanopore Technologies is constructing an α-hemolysin nanopore-based sequencer which has already been demonstrated to detect sequence identity of single-stranded DNA. Moreover, sequencing of only up to 85-mer DNA oligos was reported in their paper. In a 2012 press release, sequencing of the entire 48 kb lambda genome as a single, complete fragment was revealed. Whether sequencing of full-length RNA or cDNA fragments can be received, is yet to be determined. In the meantime, long reads can be received by 454 and Pacific Biosciences sequencing platforms (1 kb and >10 kb, respectively), although fewer reads are generated and are accordingly better suited to genome sequencing.

8.1 Automatic Data Analysis Workflows

As per the earlier discussion, RNA-seq data analysis is conducted in a number of consecutive steps where numerous tools, computational and statistical approaches have been developed with continuous availability. Trimmomatic for read trimming and adapter removal; FastQC for datasets quality control; Bowtie and TopHat to align reads to a reference genome (Fig. 12), SAMtools to sort the BAM files, HTSeq to count transcript abundance; and Kallisto, DEseq2, Cufflinks and edgeR for the identification and quantitation of expressed genes and transcripts, Sleuth, Quantitative Set Analysis for Gene Expression (QuSAGE) for the calculation of gene set enrichment, are only a few examples, as already described. A more comprehensive list, continuously updated can be observed here (https://en.wikipedia.org/wiki/List_of_RNA-Seq_bioinformatics_tools) for reference.

The choice of the right combination of tools for each particular experimental goal exists as a problem. Since 2015, development of user-friendly analysis pipelines became the choice of more and more research groups. Efforts were been directed for the creation of toolboxes for building such as automatic analysis paths (ViennaNGS, NEAT), and guides for the creation and use of computational workflows and RNASeqGUI, or detailed demonstrations of data analysis. Moreover, most of these instruments still need validation on real data sets and confirmation with experimental biological results.

Applications for which tools have been developed or improved recently are being discussed. The major applications of RNA-Seq are:

1. Quantification of gene expression under different experimental conditions.
2. Identification of novel transcripts.
3. Alternatively spliced sites or editing events.
4. The detection of RNA-fusion transcripts.

The other points as detection of non-coding and small RNAs, single-cell RNA-seq, and various others may be added.

9 Pipelines for the Identification of Alternative Splicing and RNA Editing Sites

Many of the RNA-seq data analysis packages involve tools for the identification of alternate splice or RNA editing sites. Certain specialized software suites were developed recently.

Among the applications developed for the identification of alternative splicing events are:

1. SplicingViewer (http://bioinformatics.zj.cn/splicingviewer).
2. JETTA (http://igenomed.stanford.edu/~junhee/JETTA/index.html).

   These available websites are available since 2012.
3. SplicingTypesAnno (http://genome.sdau.edu.cn/research/software/SplicingTypesAnno.html).
4. SplicingExpress (http://www.bioinformatics-brazil.org/splicingexpress).
5. SpliceJumper (https://github.com/Reedwarbler/SpliceJumper).

   These are available since 2015.
6. CLASS2 (http://sourceforge.net/projects/Splicebox) since 2016.

For the detection of editing sites, we can mention two recently published methods—ICE-seq and the genome sequence-independent method of Zhang and Xiao that work among other several new tools—REDItools, RCARE, RASER, RDDpres, and RED.

10 Pipelines for Detection of Fusion Transcripts

Computational schemes for the identification of chimeric RNAs have been developed. The fusion transcripts possess exons from different genes which have the potential of encoding for new proteins with new functions leading to abnormal cell proliferation and cancer. Table 2 lists some of the most used fusion gene detection software which were studied on comparative basis. EricScript was observed to be the most efficient tool, in terms of computational memory and time consumption, high sensitivity (78%) and positive predictive value (100%). The performances of the tools rely on the properties of the RNA-seq datasets (read length, read number, and the quality of the reads), and as a result users would choose the analysis tool which bests fits their datasets. Another comparative analysis of 15 tools implied that no single method performed best, even though SOAPfuse, FusionCatcher, and JAFFA were equally efficient. Multiple tools need to be utilized for one project. Four different tools (Tophat2, FusionMap, SOAPfuse, and

Table 2
Brief description of most recent fusion-detection tools and citations within

Tool	Year	URL	Description
EricScript	2012	http://sourceforge.net/projects/ericscript/	Perl based tool, using R, ada, BWA, SAMtools, Bedtools, seqtk, and BLAT for the identification of chimeric transcripts
FusionCatcher	2014	http://code.google.com/p/fusioncatcher/	Uses Bowtie, Bowtie2, BWA, BLAT, Liftover, STAR, Velvet, Fatotwobit, SAMtools, Seqtk, Numpy, Biopython, Picard, and Parallel for fusion identification
FusionQ	2013	http://www.wakehealth.edu/CTSB/Software/Software.htm	Uses paired-end RNA-seq data to detect gene fusions, construct the structures of chimerical transcripts, and estimate their abundances
JAFFA	2015	https://github.com/Oshlack/JAFFA/wiki	Uses Bpipe, Velvet, Oases, SAMtools, Bowtie2, BLAT, Dedupe, Reformat, and R packages, for the detection of fusions
nFuse	2012	https://code.google.com/p/nfuse/	Advanced version of deFuse. Uses Python and R scripts, and external software like BLAT, Bowtie, Bowtie2 and Gmap
PRADA	2014	http://sourceforge.net/projects/prada/	Starting from raw paired-end RNA-seq data analysis gene expression levels, quality metrics, detects fusion transcripts, and intragenic fusion variants
SOAPfuse	2013	http://soap.genomics.org.cn/soapfuse.html	Developed in Perl, aligns RNA-Seq paired-end reads to annotated genes and a human genome reference
TopHat-Fusion	2013	http://tophat.cbcb.umd.edu/fusion_index.html	Uses two scripts "Tophat" and "Tophat-fusion-post" for gene fusion analysis

Chimerascan) were designed to detect gene fusion events in human cancers from a compilation of a large number of RNA-seq data sets. They developed a comprehensive fusion-gene database: FusionCancer, available at http://donglab.ecnu.edu.cn/databases/FusionCancer/, which most likely facilitates the cancer gene fusion research.

11 Pipelines for the Analysis of Small, Non-coding RNAs

For the last years, a wide range of micro-RNAs and small non-coding RNAs have been observed to have vital roles. So, the interest in discovery of newer molecules and to understand their functions increased together with the necessity for bioinformatics tool needed for their analysis. Analysis pipelines, as well as toolkits like the UEA sRNA workbench or the sRNA toolbox aid in the analysis of expression profiling, differential expression, target gene prediction, identification of novel functional RNA, and many more. Many of which we included a selection in Table 3 have been developed and are freely available.

12 Tools for Single-Cell RNA Profiling Experiments

Single-cell RNA profiling (scRNA-seq) experiments became more popular for the last few years, and a detailed review of the specific design and computational analysis methods has been published. The smartseq2 protocol is popular for full-length coverage of polyadenylated RNAs (Fig. 14). Most of the methodologies permit the detection of major cell types and states, by the identification and quantitation of cell-type specific transcripts. As for example, SINCERA (https://research.cchmc.org/pbge/sincera.html) is a newer analytic pipeline for the process of single-cell RNA-seq data from organs or sorted cells. This leads to distinguish the major cell types (epithelial, endothelial, smooth muscle, and fibroblast-like cells), and for the identification of the cell type-specific gene activities by conducting RNA-seq analysis of single cells isolated from embryonic mouse lung.

A scRNA-seq survey of 53,193 epithelial cells of the mouse small intestine in homeostasis and at infected state identified markers for all small intestinal differentiated cell types and major subtypes.

Most recently a new approach for scRNA-seq analysis is in vogue, wherein single-cells are not clustered based on transcript read counts, rather based on transcript-compatibility counts acting as "distinctive signatures for each cell's gene expression." The methodology reveals higher speed and similar results with better "conventional" approaches. NICHE-seq, which combines photoactivatable fluorescent markers, two-photon laser scanning microscopy, and flow cytometry-based fluorescence-activated cell sorting coupled to massively parallel scRNA-seq, allows spatially defined RNA-seq analysis. The other approach involves use of array-based barcoded solid-phase RNA capture for spatial transcriptomics profiling, though not at the single-cell level.

The major techniques allowing the preparation of a scRNA-seq library, and focusing mainly on single cell isolation methods and library preparation, are explained below.

13 Single-Cell RNA Sequencing: Sample Preparation

The transcriptomic analysis is an important tool for potentially mapping genotypes to phenotypes. Single-cell RNA sequencing (scRNA-seq) was observed to be a powerful technique for addressing this type of analysis. Eberwine et al. were the beginner

Table 3
Analysis pipelines for the identification and analysis of microRNA and small RNA from RNA-seq datasets

Tool	Year	URL	Features
CAP-miRSeq	2014	http://bioinformaticstools.mayo.edu/research/cap-mirseq/	Read pre-processing, alignment, miRNA detection and quantification, data visualization, variant detection in miRNA coding region, and differential expression analysis between experimental conditions
Chimira	2015	http://www.ebi.ac.uk/research/enright/software/chimira	Web-based system for microRNA (miRNA) analysis from small RNA-Seq data. Automatic sequences cleaning, trimming, size selection and mapping directly to miRNA hairpins; identifies epi-transcriptomic modifications (chemical, editing, single nucleotide polymorphisms); visual study of differential expression between two specific samples or sets of samples; the identification of the most highly expressed miRNAs within sample pairs (or sets of samples) and also the projection of the modification profile for specific miRNAs across all samples
FlaiMapper	2015	https://github.com/yhoogstrate/flaimapper	Extracts and annotates the locations of sncRNA-derived RNAs (sncdRNAs)
ISRNA [162]	2014	http://omicslab.genetics.ac.cn/ISRNA/	Genome mapping function, statistics for genomic location, length distribution and nucleotide composition; number of reads for known microRNAs and short non-coding RNAs, coverage of short reads on genes, expression abundance; management and comparison among multiple datasets
MAGI	2014	http://magi.ucsd.edu	Prediction of microRNA target genes, statistics for differential expression; diagnostic plot generation for quality control; interactive visualization and exploration of results
miRseqViewer	2015	https://github.com/insoo078/mirseqviewer or at http://msv.kobic.re.kr	Interactive; visualizes the sequence alignment, secondary structure and normalized read counts in multipanel windows; handles multiple samples of multiple groups; allows the sorting, zooming, highlighting, editing, filtering, saving, and exporting of data
OASIS	2015	http://oasis.dzne.de	Web-based; differential gene expression; biomarker detection; batch submission of jobs; analysis of novel miRNAs, miRNA targets and functional analyses including Gene Ontology and pathway enrichment;

(continued)

Table 3 (continued)

Tool	Year	URL	Features
			downloadable interactive web reports; easy visualization, exploration and analysis of data
piPipes	2015	http://bowhan.github.io/piPipes/	Analyzes piRNA and transposon-derived RNAs from high-throughput sequencing libraries; produce figures and tables suitable for publication
Transcriptator	2015	http://www-labgtp.na.icar.cnr.it/Transcriptator	Web application; helps the users to characterize the de novo assembled reads, for non-referenced organisms, performs the functional enrichment analysis of differentially expressed transcripts/genes for both RNA-seq and micro-array experiments; easy to read tables and interactive charts; modular pipeline allows additions of new plugins in the future

scientists to work on the single cell transcriptome sequencing, and Iscove et al. were the first to amplify the cDNA of a single cell by PCR. These techniques opened up the newer area for scRNA-seq. In 2009, the first next-generation sequencing platform for scRNA-seq was introduced for the first time. Various methodologies have been initiated to conduct single-cell transcriptomic analysis, but for all of the experimental design (i.e., which scientific question to address), sample preparation (i.e., processing the sample), and downstream data analysis (i.e., which computational tool) are the basic important steps.

14 Single-Cell Isolation Methods for scRNA-seq

14.1 Limiting Dilution

Cells are isolated individually with a pipette through serial dilution. It is an easy and fast method but not very efficient since the probability to have only one cell in an aliquot follows a statistical distribution (Poisson's distribution).

14.2 Micromanipulation or Manual Cell Picking

Cells are isolated individually with an inverted microscope possessing a movable stage with micro-pipettes. The separation is conducted through a microscope observation for picking a specific cell. Moreover, this method is time-consuming, and it allows isolation of not many single cells.

It is one of the most common methods for the isolation of highly pure single cells. Cells can be sorted individually directly into tubes

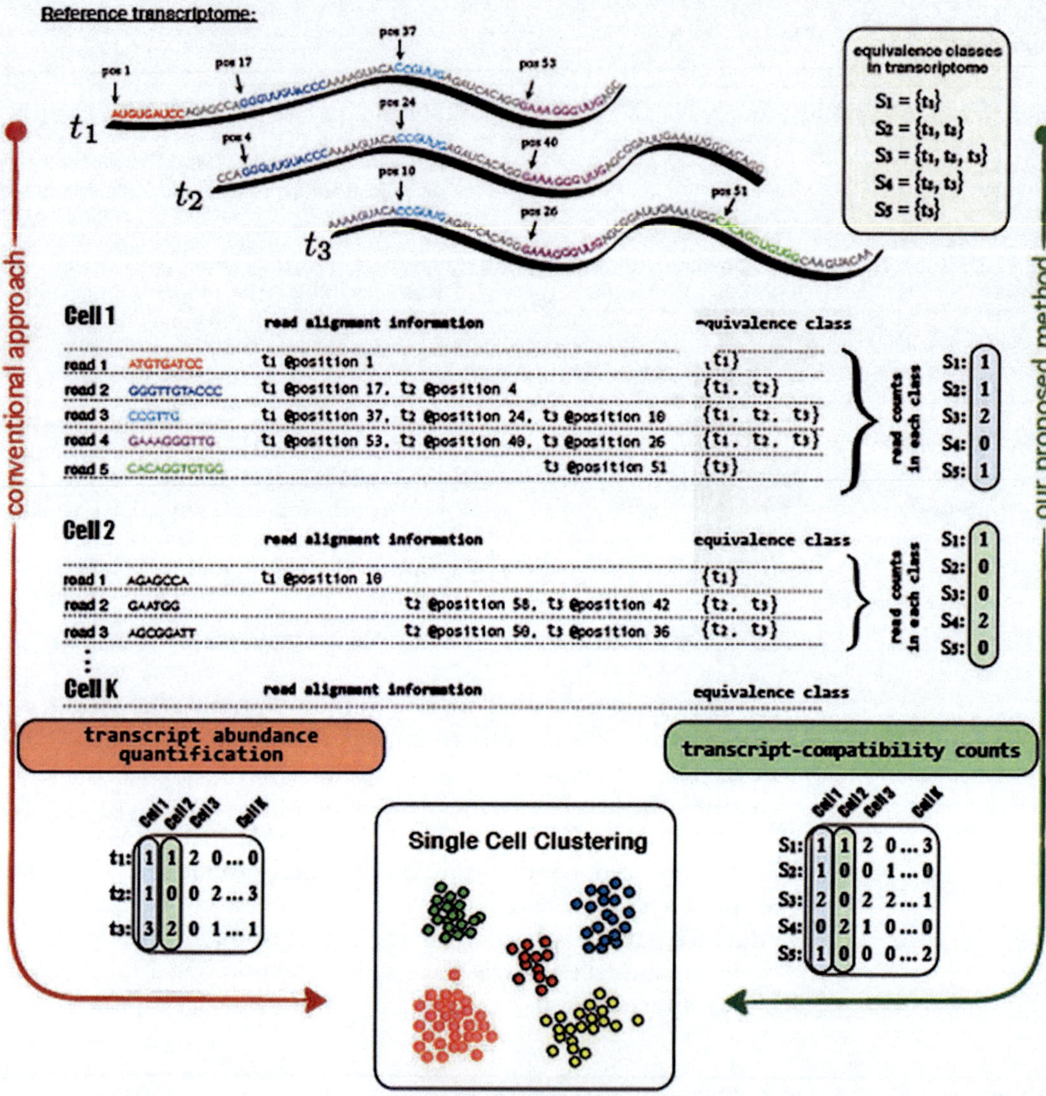

Fig. 14 Overview of the transcript-compatibility counts (TCC) method. "An scRNA-Seq example with K cells (only the reads coming from Cell1 and Cell2 are shown here) and a reference transcriptome consisting of three transcripts, t1, t2 and t3" are used for exemplification. "Conventional approach: Single cells are clustered based on their transcript or gene abundances (here we only focus on transcripts for concreteness). This widely adopted pipeline requires computing a (#transcripts × #cells) expression matrix through first alignment of each cell's reads to the reference. The corresponding alignment information is next to each read. For the purpose of illustration only reads possess the mapped positions (the aligned reads of Cell1 are also annotated directly on the transcripts). Reads 1 and 5 are uniquely mapped to transcripts 1 and 3, whereas reads 2, 3 and 4 are mapped to multiple transcripts (multi-mapped reads). The quantification step must, therefore, take into account a specific read-generating model and handle multi-mapped reads accordingly. The proposed method of employing single cells as clustered are based on their transcript-compatibility counts. Our method assigns the reads of each cell to equivalence classes via the process of pseudo-alignment and simply counts the number of reads that fall in each class to construct a (#eq.classes × #cells) matrix of transcript-compatibility counts. In the next step, the method employs direct use of the transcript-compatibility counts for downstream processing and single cell clustering. The basic idea is that although equivalence classes may not have an explicit biological interpretation, their read counts give a distinct signature of each cell's gene expression.

14.3 Fluorescence--Activated Cell Sorting or FACS

or 96/384-well plates. The basis of FACS sorting is the particular light scattering and fluorescent characteristics of each cell. During the sorting, the stream of cells is broken into individual droplets. Whenever a particle is recognized based on previous sorting criteria, an electrical charge is placed to the stream since the droplet possessing that particle break off the stream. In the next step, the charged droplets fall through two charged plates which separate droplets into containers depending on the droplet charge. The limitation of this method is the large starting materials necessary to conduct the sorting.

14.4 Laser-Assisted Tissue Microdissection

The basis of laser-assisted tissue microdissection system is the capture of single cells or groups of cells onto a thermoplastic membrane from stained tissue histological sections (frozen sections or fixed wax-embedded sections) or cytological preparations.

There are two major systems which have been created for performing laser-assisted tissue microdissection, such as laser capture microdissection and laser beam microdissection. In the laser capture microdissection, the plastic membrane which overlies the tissue section is attached to a particular "cap" ensuring that the plastic membrane is held in direct contact with the tissue section. The plastic is gradually melted by a low-power narrow-beam infrared laser directed at the cells of interest under the control of microscope. Soon the plastic cools and solidifies again. The cells that are embedded into the plastic membrane are removed from the tissue section through lifting the cap along with the plastic membrane off the tissue sections. In the laser beam microdissection, a pulsed very narrow-beam ultraviolet light laser is employed to "draw" around the cell of interest and unwanted tissue is photoablated. In certain systems, the laser can be employed to "catapult" the microdissected cell falling into the collecting tube. However, in certain other system the section is inverted, with an idea that after microdissection the microdissected cell falls into the collecting tube by gravity. The major limitation of laser-assisted tissue microdissection is the requirement for the identification of the cells of interest based on morphological characteristics.

14.5 Microfluidics

The microfluidic technology is becoming very popular as it needs a little amount of sample and that too cost-effective. This method is based on various principles. The most popular are the hydrodynamic cell traps (Fig. 5a) and the microdroplet-in-oil based isolation.

The hydrodynamic cell trap involves passive elements which fit only one cell in the trap. The microdroplet-in-oil based isolation utilizes channel filled with oil to grip aqueous droplets when single

Fig. 14 (continued) Transcript-compatibility counts can be assessed as feature vectors and cells may be identified by their differential expression over these features. In comparison to the conventional approach, current methodology does not resolve multi-mapped reads (no need for an assay-specific read-generating model). It needs only transcript compatibility information for each read (no need for exact read alignment)" cells can be confined and isolated as a single cell. The benefit of this system is that it needs a low volume, leading to the screening of up to thousands of single cells per second. Commercial platforms which take the advantage of the microfluidic technologies are the Fluidigm C1 platform (hydrodynamic cell traps) and the Chromium system from 10× Genomics (microdroplet-in-oil based), currently used for scRNA-seq. As an example, He M et al. processed epidermal cells with 10× Genomics Chromium Single Cell 3′ Library and Gel bead kit v2 followed by the quantification of the scRNA library with Kapa Library Quantification Kit (Kapa Biosystems).

14.6 Library Preparation for scRNA-seq

Following the isolation of single cells, the steps necessary for the preparation of a scRNA-seq library are as follows:

1. Lysis of the cells.
2. Reverse transcription into complementary (c)DNA.
3. Second-strand cDNA synthesis.
4. Amplification of the cDNA.

Normally, the lysis buffer is a hypotonic solution possessing RNase inhibitors. The first strand of cDNA is synthesized with poly (dT) primers which bind the mRNA. The transcriptase employed at this stage is characterized usually by a low RNase H activity and increased thermostability. The reverse transcription primers may have extra nucleotides attached to one end such as adapter sequences or unique molecular identifiers (UMIs) that are random *n*-mer oligonucleotides. This system leads to the assignment of each read to its original cells.

In the next step, the second strand of cDNA is generated and the cDNA amplified by conventional PCR or in vitro transcription. The second method needs an additional reverse transcription step, hence more time-consuming. The amplified and tagged cDNA from different cell types is then pooled followed by sequencing with next-generation sequencing (NGS).

14.7 Analysis of DATA After NGS

Currently, experimental methods for scRNA-seq are more accessible due to the presence of platforms which provide all the reagents (e.g., Fluidigm and 10× Genomics). The analysis of the raw data generated need bioinformatic support. Companies, such as Fluidigm and 10× Genomics, provide free software to undergo the steps like quality control, analysis, and interpretation of scRNA-seq data. However, the limitation of this software is that the algorithmic and parameters used are non-totally transparent.

14.8 Other Applications

Analysis of workflows is to be developed for a diverse array and important research topics. As for example, VIP was developed for the detection and discovery of viruses, and it is available at https://github.com/keylabivdc/VIP. Fully automated pipelines were created for the identification of sex chromosome-linked genes in species without an existing reference genome or determination of the expression levels of SINE retrotransposons, and Allelome. Pro was created for identification of allele-specific genomic features.

15 Summary

Transcriptional profiling with NGS platforms has provided a new window of the inner workings of the cell in development and disease; thus, it is the new avenue for the researchers. The field is advancing fast, with each new generation of sequencer needs less input requirements, simplified library preparation protocols, and better reads of a longer length. This chapter discusses in detail the protocol, advantages, disadvantages, and finally lead to an unbiased view of the transcriptome. It discusses the improvements at the level of each step necessary for the generation and analysis of RNA-seq data. New and improved automatic and user-friendlier workflows are coming up. The latest introduction of RNA-seq at single cell level will lead to newer avenues of research allowing a better analysis of complex biological systems.

Widely used RNA-Seq software packages

Primary category	Tool name	Notes
Splice-aware read alignment	GEM	Filtration-based approach to approximate string matching for alignment
	GSNAP	Based on seed and extend alignment algorithm aware of complex variants
	MapSplice	Based on Burrows-Wheeler Transform (BWT) algorithm
	RUM	Integrates alignment tools Blat and Bowtie to increase accuracy
	STAR	Based on seed searching in an uncompressed suffix arrays followed by seed clustering and stitching procedure; fast but memory-intensive
	TopHat	Uses Bowtie, based on BWT, to align reads; resolves spliced reads using exons by split read mapping
Transcript assembly and quantification	Cufflinks	Assembles transcripts to reference annotations or de novo and quantifies abundance
	FluxCapacitor	Quantifies transcripts using reference annotations
	iReckon	Models novel isoforms and estimates their abundance
Differential expression (DE)	BaySeq	Count-based approach using empirical Bayesian method to estimate posterior likelihoods
	Cuffdiff2	Isoform-based approach based on beta negative binomial distribution
	DESeq	Exon-based approach using the negative binomial model
	DEGSeq	Isoform-based approach using the Poisson model
	EdgeR	Count-based approach using empirical Bayes method based on the

References

1. Corney DC, Basturea GN (2014) RNA seq. Mater Methods 3:203. https://doi.org/10.13070/mm.en.3.203
2. Wang Z et al (2009) RNA-Seq: a revolutionary tool for transcriptomics. Nat Rev Genet 10:57–63

Check for updates

Chapter 14

Next-Generation Sequencing

Abstract

Next-generation sequencing (NGS) technologies refer to the high-throughput sequencing techniques. "*Next-generation*" sequencing technologies lead to rapid generation of data by sequencing massive amounts of DNA in parallel using diverse methodologies with definite sequencing platform, say Illumina or Agilent. The basic methodologies involve, RNA isolation, cDNA preparation followed by cDNA library preparation. NGS experiments basically consist of four phases: Sample collection, Template generation, Sequencing reactions and detection, and Data analysis. NGS experiments can be grouped broadly into two general categories: de novo assembly and resequencing. Evolution of NGS technology has led to advances in genomics far better than DNA microarray. In this book, we have discussed a number of technologies pertaining to NGS. The examples include epigenetics, metagenomics, exome sequencing, genome wide association studies, transcriptomics, and RNA-seq.

Keywords Next-generation sequencing, High throughput sequencing, cDNA library, De novo assembly, Sequencing platform

1 Basic Principle

Next-generation sequencing (NGS) technologies refer to the high throughput sequencing techniques. With the advent of NGS technologies, there was a marked revolution in Genomics studies.

The advantages of NGS over traditional Sanger sequencing was that NGS technologies are fast, inexpensive, and led to accurate DNA sequencing. "***Next-generation***" sequencing technologies result in rapid generation of data through sequencing massive amounts of DNA in parallel using diverse methodologies. NGS technologies involve commercially available sequencing platforms, say for example from Illumina or Agilent.

The collection of a variety of technologies involving high throughput technologies is collectively known as next-generation sequencing technologies. The basic methodologies involve, RNA isolation, cDNA preparation followed by cDNA library preparation. They proceed the technique with suitable adapters.

Aruna Pal, *Protocols in Advanced Genomics and Allied Techniques*, Springer Protocols Handbooks, https://doi.org/10.1007/978-1-0716-1818-9_14,

2 A Brief History or Genesis of Next-Generation Sequencing

Human genome project (HGP) was based primarily on Sanger sequencing methodology, introduced by Sanger in 1977. The sequencing process was time consuming and even costly. The time consumed for initial sequencing of human genome in human genome project was more than a decade with an expenditure of an estimated $70 million. Hence it became a necessity to create a technology platform for robust sequencing enabling to generate a large amount of data quickly and that too at lower cost. Considering the fact, the National Human Genome Research Institute (NHGRI) started a funding program in 2004 with an objective to develop sequencing technology and at the same time reducing the cost of genome sequencing to ~$100,000 in 5 years and, ultimately, $1000 in 10 years. New technologies enable the sequencing of large amounts of DNA in parallel and at comparatively lower costs than conventional methods. The synonyms for "*next-generation sequencing*" were used as "*massive-parallel sequencing*" for these new high-throughput technologies, whereas automated methodologies utilize fluorescently labeled terminators followed by capillary electrophoresis separation, leading to automated laser signal detection for improved nucleotide sequence detection. DNA fragments of 500 bp to 1 kb in length can be easily sequenced through Sanger sequencing. This method is helpful for sequencing small amounts of DNA fragments and is useful for clinical cytogenetics.

As mentioned earlier, "*Next-generation*" and "*massive-parallel*" DNA sequencing are the terms utilized for high-throughput DNA sequencing technologies available to sequence large numbers of different DNA sequences in a single reaction (i.e., in parallel). All next-generation sequencing (NGS) technologies examine the sequential addition of nucleotides to immobilized and spatially arrayed DNA templates. The different NGS technologies differ in certain terms of generation of the templates, methodology to identify their sequences. A basic workflow for NGS sequencing technologies is presented in Fig. 1. Figure 2 shows the photographs for commonly employed next-generation sequencing platform, as Roche/454 FLX: 2004, Illumina Solexa Genome Analyzer: 2006, Applied Biosystems SOLiDTM System: 2007, Helicos HeliscopeTM, Pacific Biosciencies SMRT: launching 2010.

The diagrammatic workflow for overview of next-generation sequencing (NGS) technologies has been depicted in Fig. 3, while the details have been depicted in Fig. 4. Photographic representation of signal detection in NGS studies has been depicted in Fig. 5. Figure 6 depicts the detection of non-coding RNA through next-generation sequencing. The basics for next-generation sequencing machines have been depicted in Fig. 7. Figure 8 explains the basic difference of DNA microarray and NGS.

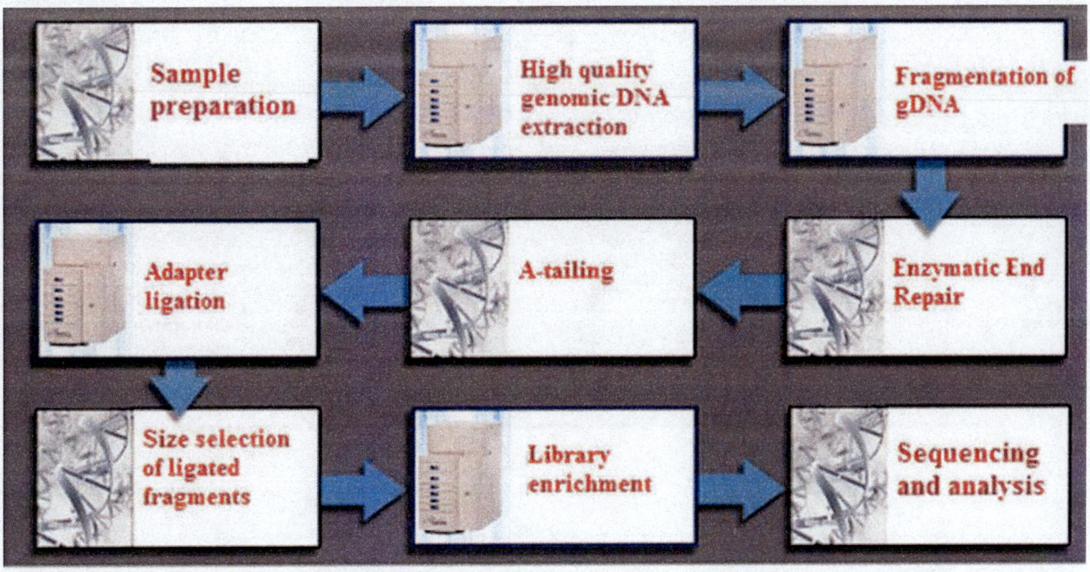

Fig. 1 Basic workflow for next-generation sequencing

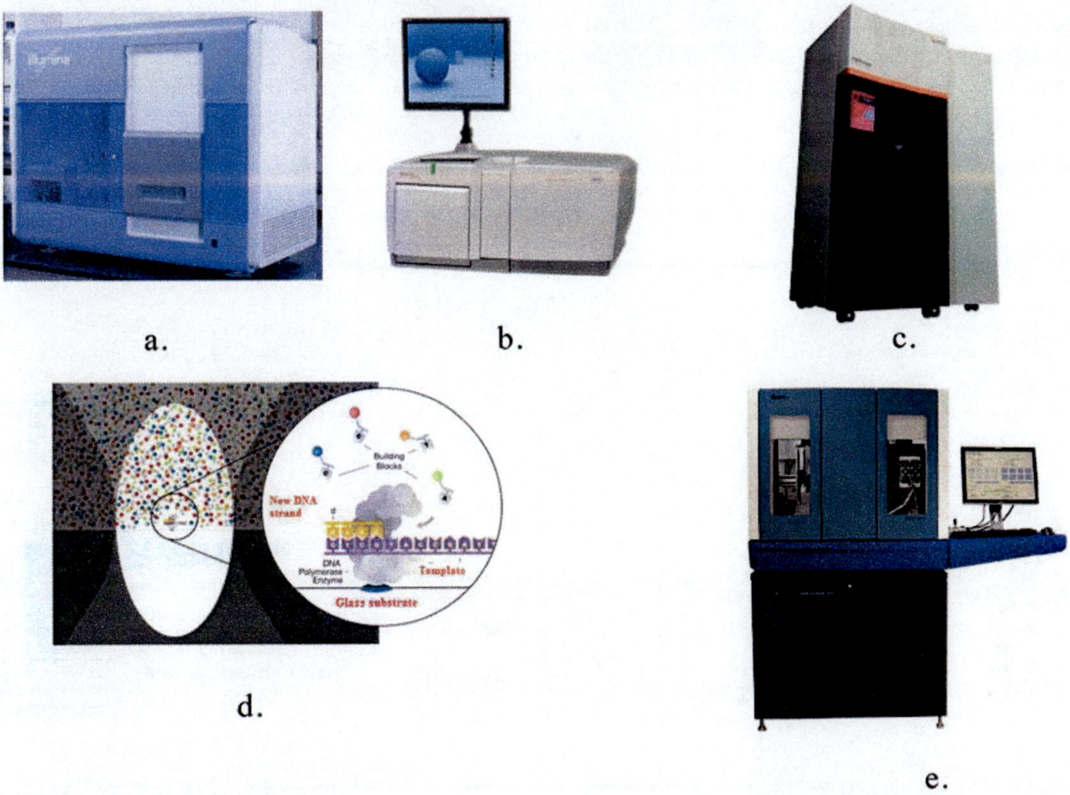

Fig. 2 Photographs of commonly employed next-generation sequencing platform, (**a**) Roche/454 FLX: 2004, (**b**) Illumina Solexa Genome Analyzer: 2006, (**c**) Applied Biosystems SOLiDTM System: 2007, (**d**) Helicos HeliscopeTM, (**e**) Pacific Biosciences SMRT: launching 2010

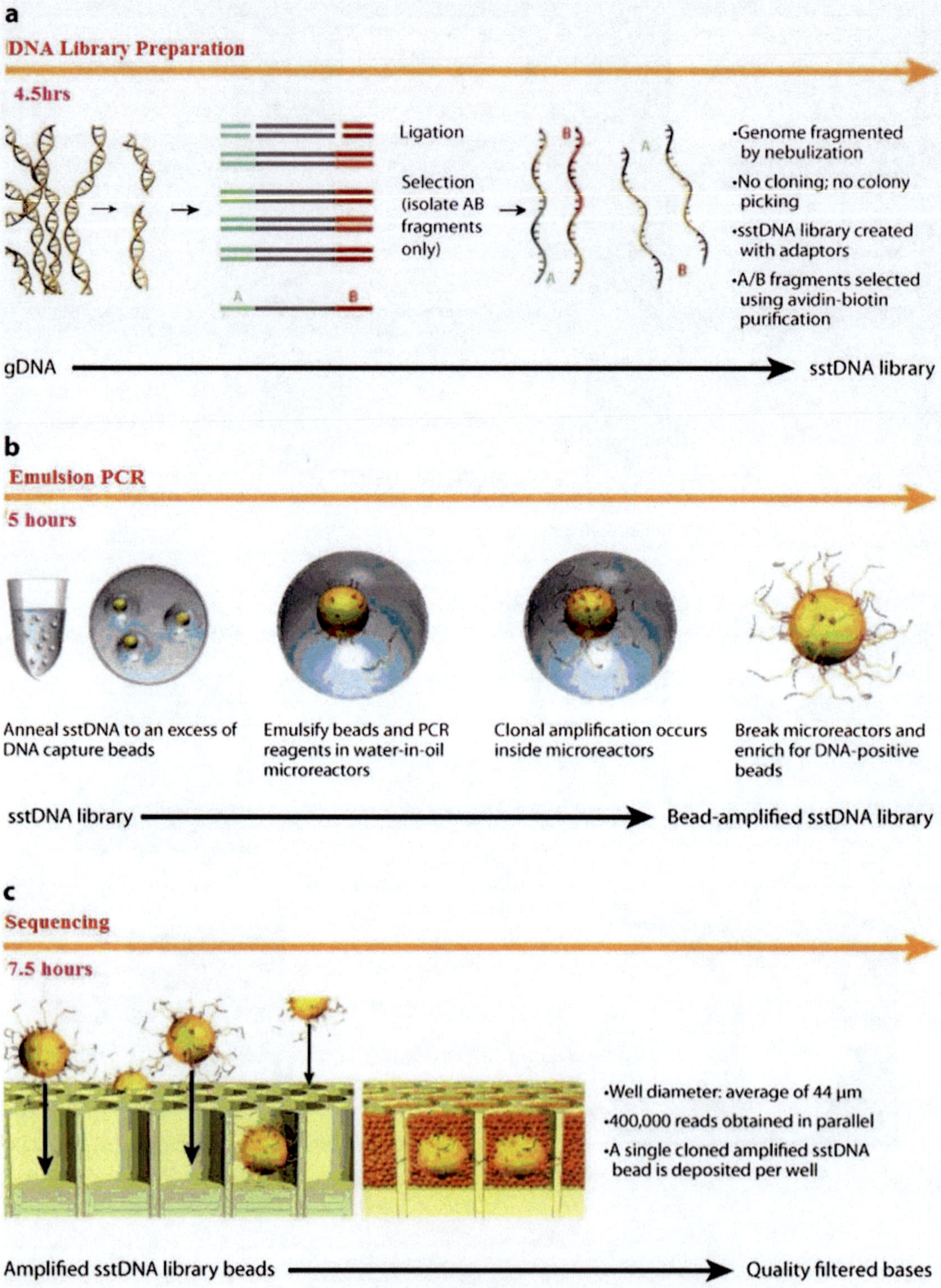

Fig. 3 (**a–c**) Diagrammatic workflow for next-generation sequencing technologies

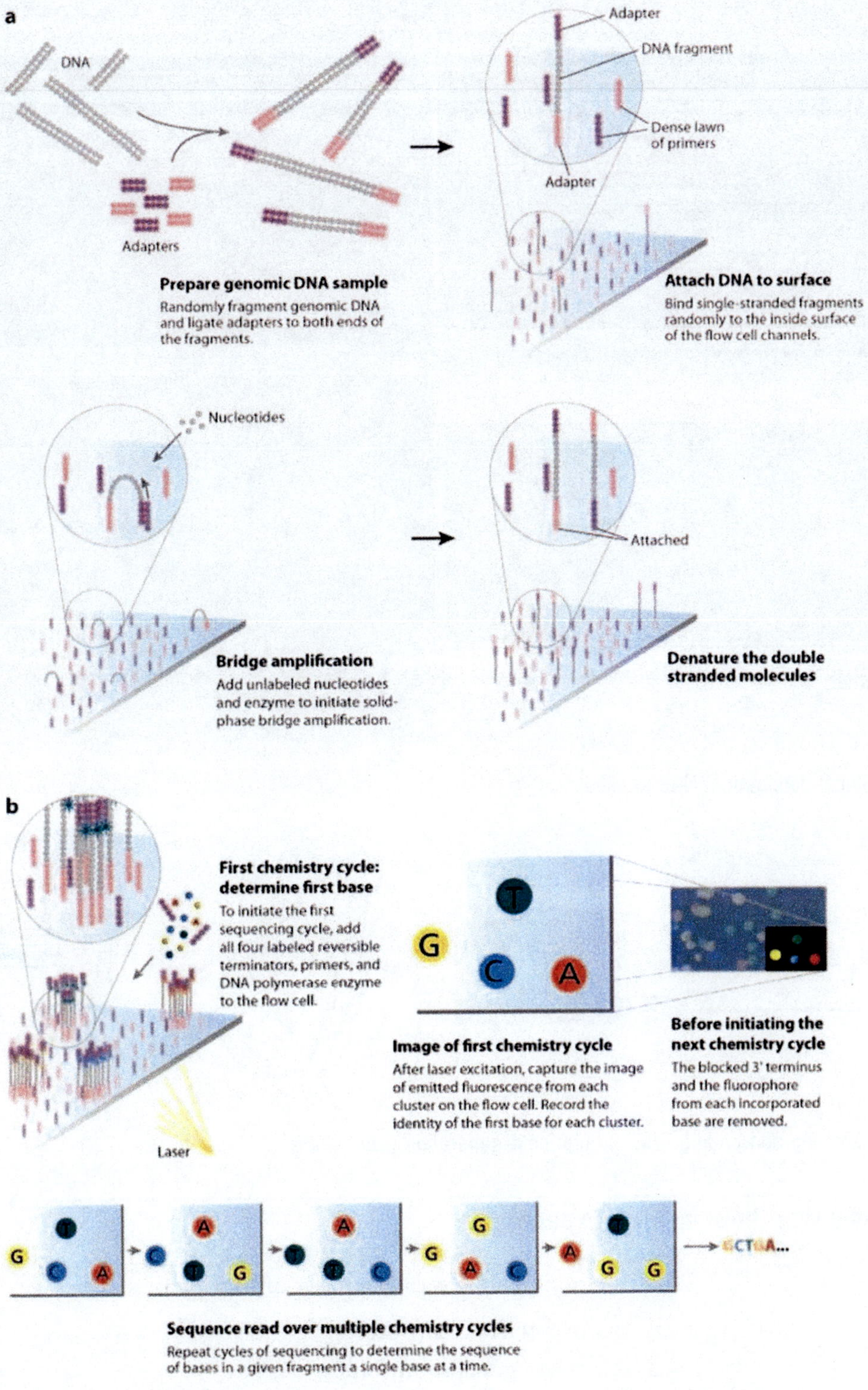

Fig. 4 (**a**, **b**) Next-generation sequencing processing in detail—diagrammatic representation

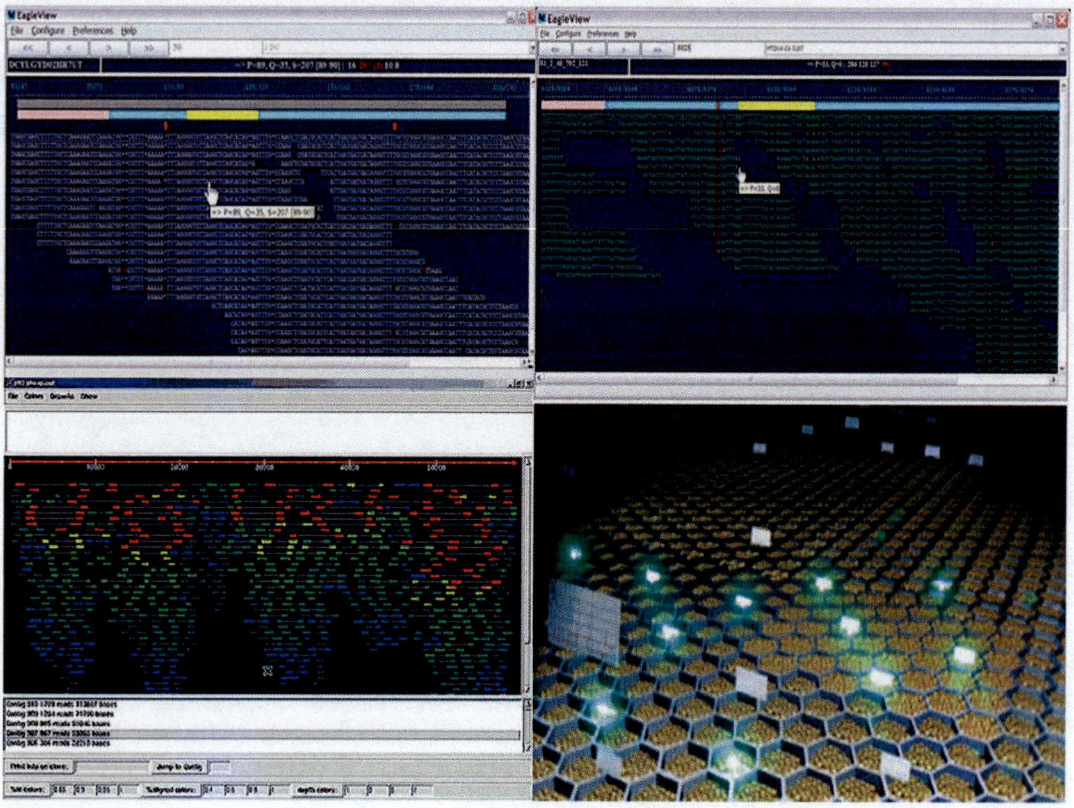

Fig. 5 Signal detection in NGS studies

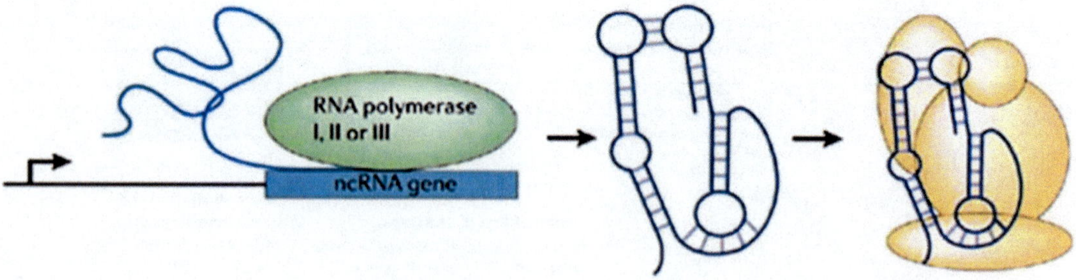

Fig. 6 Detecting non-coding RNA through next-generation sequencing

3 Basic Workflow for NGS Experiments

NGS experiments basically consist of four phases:

1. Sample collection (purple).
2. Template generation (blue).
3. Sequencing reactions and detection (green).
4. Data analysis (orange) [1, 2].

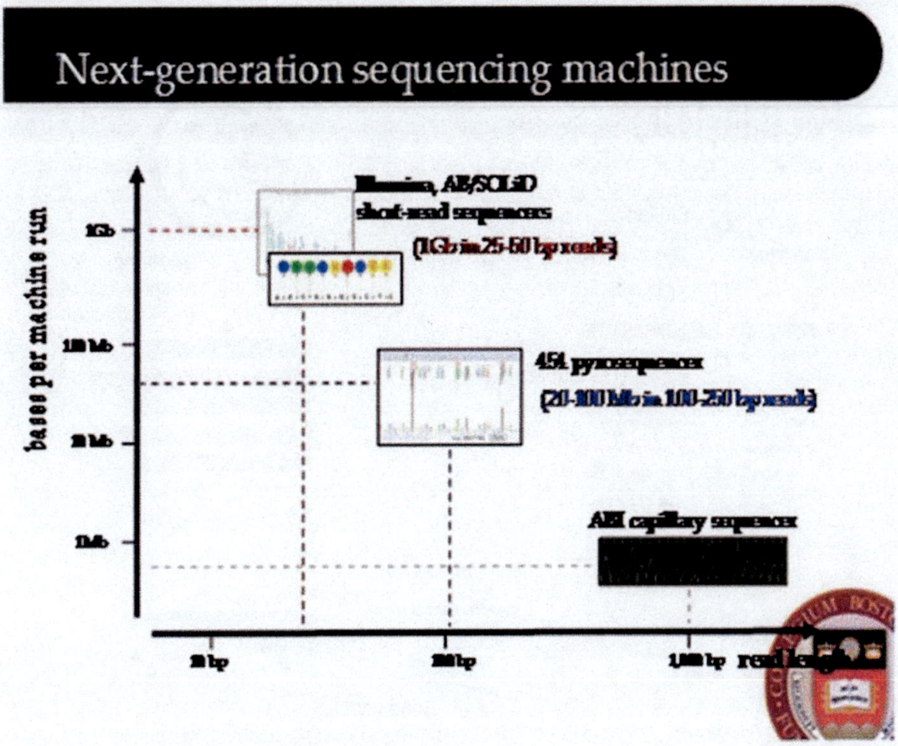

Fig. 7 Basics for next-generation sequencing machines

Experiments lead to broad applications, based on the source and nature of input DNA utilized for sequencing. Source materials involve normal and diseased tissues, from which NGS experiments are able to sequence the whole genome (WGS) or targeted genomic/epigenetic elements.

NGS experiments can be classified in two major general categories:

1. de novo assembly.
2. resequencing.

Assembled genomes are prepared from scratch, without the help of an existing scaffold. However, resequencing experiments align sequence reads with reference to a reference genome. One of the limitations encountered is that in the Human Genome Project (HGP), all human genome sequencing efforts were resequencing, as a result it is not cost-effective at all and not easy to employ and of limited (immediate) value to reassemble a human genome. However, smaller genomes such as those of novel bacteria are routinely assembled de novo.

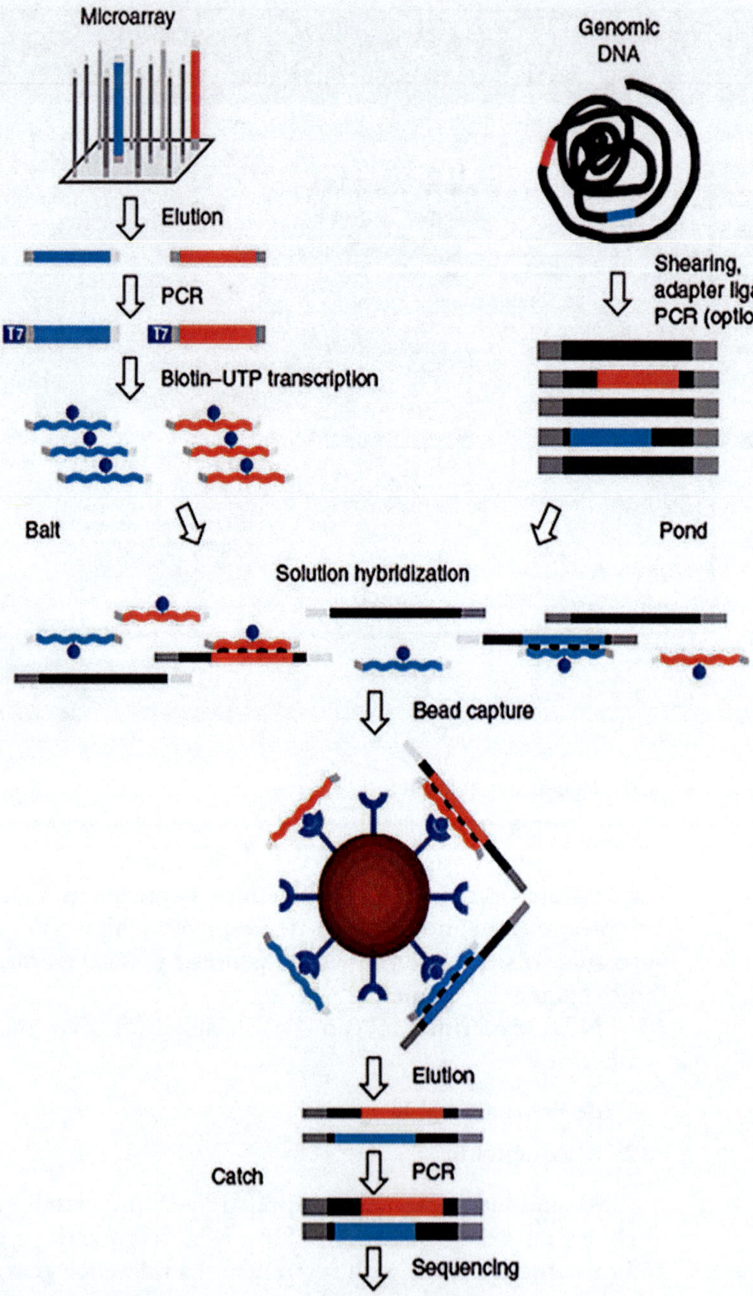

Fig. 8 Elucidating the basic difference of DNA microarray and NGS

4 Template Generation

It is one of the basic and initial steps for a NGS study. As a general rule, the starting material for all NGS experiments is double-stranded DNA; although, the source of this material may vary (e.g., genomic DNA, reverse-transcribed RNA or cDNA, immunoprecipitated DNA). Whatever be the starting material, it needs to be converted into a library of sequencing reaction templates (sequencing library), which need the stages of fragmentation, size selection, and adapter ligation. Fragmentation and size selection steps lead to the breakage of the DNA templates into smaller fragments, which can be readily sequenced. The size of DNA fragments depends on each sequencing platform's specifications.

Adapter ligation aids in platform-specific synthetic DNAs to the ends of library fragments, which act as primers for downstream amplification and/or sequencing reactions. Ideally, the above steps lead to an unbiased sequencing library which accurately represents the sample's DNA population. Depending on the NGS technology employed, a library is either sequenced directly (single-molecule templates) or is amplified then sequenced (clonally amplified templates). Template generation aids to spatially separate and immobilize DNA fragment populations for sequencing, usually by attachment to solid surfaces or beads. This leads to the downstream sequencing reaction to act as millions of microreactions conducted in parallel on each spatially distinct template.

The main difference of NGS with Sanger sequencing is that NGS can process millions of sequence reads in parallel in contrast to 96 at a time in case of Sanger sequencing that too at lower cost (1/6 of the cost).

5 Commonly Used NGS Platforms

- Roche/454 FLX: 2004.
- Illumina Solexa Genome Analyzer: 2006.
- Applied Biosystems SOLiDTM System: 2007.
- Helicos HeliscopeTM: recently available.
- Pacific Biosciencies SMRT: launching 2010.

6 Applications for NGS

1. It can act on ancient DNA samples.
2. DNA mixtures from diverse ecosystems and metagenomics are applicable.
3. Resequencing earlier published reference strains is possible.
4. Detection of all mutations in an organism is possible.
5. Errors in published literature can be crosschecked.
6. Expansion of the number of available genomes.
7. Comparative studies.

8. Deciphering cell's transcripts at sequence level without knowledge of the genome sequence.
9. Genomic/transcriptomic/Epigenetic sequencing for extremely large genomes, crop plants.
10. Detection of cancer specific alleles ignoring traditional cloning.
11. Chip-seq: interactions protein–DNA.
12. Epigenomics.
13. Detecting ncRNA.
14. Genetic human variation: single nucleotide polymorphism (SNP), copy number variation (CNV) (diseases) [1–5].

7 DNA–Protein Interaction

- They play a key part in regulating gene expression.
- Chip: technique for studying DNA–protein interactions.
- Genome-wide ChIP-based studies for DNA–protein interactions.
- Readout of ChIP-derived DNA sequences in NGS platforms.
- Details of transcription factor/histone binding sites in the human genome.
- Increase our knowledge of the gene expression in the context of specific environmental stimuli.

8 Detection of Non-coding RNA

- The presence of ncRNA in genome is difficult to predict through computational methods with high certainty due to the evolutionary diversity.
- Detection of expression level changes which correlate with changes in environmental factors, with disease occurrence and progression, complex disease set or severity.
- Improve the annotation of sequenced genomes (impact of mutations more interpretable).

9 Mutation Discovery

Detection of mutations/SNVs in the genome is one of the most important functions of NGS. Multiplexing the amplification of 10,000 human exons with primers from a programmable microarray and sequencing them with NGS.

References

1. Kumar KR, Cowley MJ, Davis RL (2019) Next-generation sequencing and emerging technologies. Semin Thromb Hemost 45(7):661–673. https://doi.org/10.1055/s-0039-1688446
2. Qin L, Nie Y, Zhang H et al (2020) Identification of new mutations in patients with hereditary spherocytosis by next-generation sequencing. J Hum Genet 65:427. https://doi.org/10.1038/s10038-020-0724-z
3. Pal A et al (2017a) Next-generation sequencing confirms the distinct breed identity of Birbhum sheep - an indigenous breed of sheep identified from the Birbhum district of West Bengal. In: The National Symposium on Biodynamic Animal Farming for the Management of Livestock Diversity Under Changing Climatic Scenario and 14th Annual Convention of SOCDAB, 2017, CVAS, Mannuthy, 8–10 February
4. Pal A et al (2017b) Birbhum - an indigenous breed of sheep identified from the Birbhum district of West Bengal and confirmed through Next generation sequencing. In: The 24th West Bengal State Science and Technology Congress, 2017 conducted by Department of Higher Education, Science and Technology and Biotechnology, Govt. of West Bengal
5. Pal A et al (2017c) Molecular characterization by Next generation sequencing and study of the Genetic basis of antiviral resistance of Indigenous ducks. In: The National Symposium on Biodynamic Animal Farming for the Management of Livestock Diversity Under Changing Climatic Scenario and 14th Annual Convention of SOCDAB, 2017, CVAS, Mannuthy, 8–10 February

Chapter 15

Metagenomics

Abstract

Metagenomics is a branch of science dealing with the study of the microbial population (bacterial, protozoal, or viral) through genomic approach (next-generation sequencing), which is a culture independent method utilizing DNA sequencing techniques to study DNA extracted directly from environmental samples. Various avenues for metagenomics include clinical metagenomics, identification of newer probiotics acting against antibiotic sensitivity, gut microbial biodiversity and evolution. We have discussed in detail about the protocol with suitable illustrations or case studies pertaining to certain cases. They include clinical metagenomics (bone and joint infection), exploration of gut metagenome for increasing the feed conversion efficiency and growth promotion, metabolism, and increased immunity through the use of Probiotics identified from native chicken/duck, metagenomic analysis for detection of viral diversity. The detailed protocol for 16S Metagenomic Sequencing Library Preparation has been discussed.

Keywords Metagenomics, Next-generation sequencing, Gut microbial biodiversity, 16S metagenomic sequencing library

1 Basic Principles

Metagenomics is a stream of genomics dealing with the study of the microbial population (bacterial, protozoal, or viral) through genomic approach (next-generation sequencing). The process does not involve any culture of bacterial, viral, or protozoal population, rather it is based on the DNA sequencing techniques and differential quantification of the DNA extracted directly from environmental samples.

In Greek, *meta* means "transcendent." The process involves metagenomics based on the genomic diversity of most microbes and that too without involving microbial culture. The technique has revolutionized the advanced research in microbiology involving clinical and environmental aspects. "Meta" refers to the need to recognize the necessity for developing computational methods, which maximize the understanding of the genetic composition and activities of communities. The processes are complex, as result enabling only sampling, but complete characterization is obviously difficult.

Aruna Pal, *Protocols in Advanced Genomics and Allied Techniques*, Springer Protocols Handbooks,
https://doi.org/10.1007/978-1-0716-1818-9_15,

In the second concept, *meta* refers to the technology for conceptualization of biology at the aggregate level, surpassing the individual organism focusing on the gene pool in the community and the influence of the gene on activities involving various functions. It is evident that individual organisms are the basis for community. Metagenomics has the ability to flourish and explore the research activities on individuals at genomic level. With the advent of metagenomics in near future, it is expected that top-down approach of metagenomics will replace the bottom-up approach of classical microbiology genomics of individual organism. Understanding of the community and biosphere comprising of the communities will predominate the research activities for the benefit of the mankind.

The impact of metagenomics is far beyond its application. The term describes the following aspects:

1. Genome-level characterization of communities or their members. It is important to note that the technique does not involve cultivation or culture of organism.
2. It involves high-throughput techniques to study at genomic level of communities. It involves techniques for genomics, and other "omics" studies.
3. Metagenomics involves techniques aimed to understand trans-organismal approaches and the biosphere at the genomic stage.

Although in its current early implementation metagenomics focuses on non-eukaryotic microbes, eventually, its concepts and methods will ultimately transform all biology. Future concept includes the application of metagenomics in various field including biotechnology, epidemiology, virology, agriculture, clinical microbiology, forestry, fisheries, and microbial forensics.

2 Metagenomics Is an Important Field of Current Study with Many Practical Outcomes

1. Metagenome has evolved as important aspect of clinical metagenomics. It involves relation of gut microbiota with diseases. It has proved to be effective with chicken as a model, exploring association of harmful gut microbes with diseases.
2. Exploration of gut metagenome for increasing the feed conversion efficiency and growth promotion, metabolism and increased immunity through the use of probiotics identified from native chicken/duck.

3. Gut microbial biodiversity is still greatly unexplored. Understanding the gut microbiota composition and its function associated with diseases through metagenomic analysis-short gen metagenomics.
4. Genome and metagenome analysis may be utilized for revealing the evolutionary concept of the host interacting with the gut microbiota or microbiome in the livestock species including poultry [1].
5. Targeted metagenomics approach studies the functional, ecological, and evolutionary patterns of important genes present in microorganisms derived from various ecosystems.
6. Solution to the problem of *Antibiotic resistance in chicken*, which has emerged as a Global problem through exploration of gut microbes and identification of probiotics as alternatives to antibiotics.

In this chapter we shall describe metagenomics based on different applications. Since bacterial metagenomics is entirely different from viral metagenomics, there exists completely different protocol.

In simple words, we can describe metagenomic approach as assessment or detection of microbial population existing in the environment under study. In the current we shall discuss about the metagenomic studies with practical example with gut microbes existing in chicken or duck for solving various problems associated.

As discussed, clinical metagenomics is dealing with disease association with the microbial population in the gut. Traditionally we describe any disease as deviation from normal health. Infectious diseases are referred to as diseases caused by infectious agents as bacteria, virus, or protozoa. It had been reported that when the population of beneficial microbes decline in the gut, the pathogenic microbes gain and cause the disease.

Note: As a general rule, beneficial microbes, commonly termed as probiotics are present in the lower GI tract (e.g., Caecum). The upper portion of the gut (e.g., duodenum) possesses pathogenic strains.

3 A Case Study with Classical Example in Case of Bone and Joint Infection *Depicting the Need of Clinical Metagenomics* as an Important Support to Traditional Microbial Culture

Bone and joint infection (BJI) is one of the most severe infections requiring antibiotic treatment. But it is difficult to culture some bacteria, designated as hardly culturable bacteria. Metagenomics may serve as a supplementary approach to address the problem. In describing a case study, where 24 BJI samples were studied, both routine microbial culture and metagenomic analysis were conducted.

In case of monomicrobial samples in culture ($n = 8$), the presence of the pathogen was documented by metagenomics in all cases. In case of polymicrobial samples ($n = 16$) revealed 32/55 bacteria (58.2%) and 41/55 [74.5%] at the species level and genus level respectively.

Two hundred and seventy three bacteria were revealed through metagenomics, which were not detected in culture. Out of 273 bacteria, 182 being possible pathogens and 91 as contaminants. Altogether, we observed that clinical metagenomics applicable to BJI samples is an important tool to supplement conventional culture.

3.1 Material and Methods

3.1.1 Samples

In the current study, 179 pre-operative samples were collected from 47 patients (ranging from 1 to 8 samples per patient), out of them, two samples (swabs) were solid specimens. Grouping was conducted for each non-swab sample ($n = 177$) macroscopically as:

First (less than 1 mL, $n = 100$), second (from 1 to 10 mL, $n = 60$), and third (more than 10 mL, $n = 17$). The samples were collected for a period of 5 months and stored at −80 °C and transported in dry ice to the Genomic Research Laboratory. The samples were cultured on (1) one sheep blood agar plate was incubated at 36 ± 1 °C under aerobic conditions for a period of Day 1 and 2 for microbial growth, later discarded; (2) two chocolate blood agar plates were cultured under reducing conditions (kept under a 5% CO_2-enriched aerobic atmosphere) at 36 ± 1 °C. The first plate was observed for 2 consecutive days and discarded, whereas the second plate was conserved opened and observed following 10 days; (3) two Schaedler agar plates were incubated in anaerobic conditions: the first plate was noted on Day 2 and 5 and discarded; the second was observed after 15 days and read; (4) one Schaedler broth supplemented with Vitamin Kl was cultured for 15 days in anaerobic condition with careful observation. Gram staining was performed as opacity is observed followed by sub-culturing of the broth on agar plates (i.e., one chocolate agar and one Schaedler agar in anaerobic condition) and incubated for 3 days. Subculture of the broth was systematically conducted on Day 15 even if opacity is not present. In case of positive culture, each microorganism was detected using matrix-assisted laser desorption ionization-time of flight mass spectrometry and antimicrobial susceptibility. The quantification of bacteria were conducted and noted up to 100 colony-forming units (CFUs) followed by expression as $>10^2$. As real numbers were necessary for the estimation of the proportion of bacteria in culture, we regarded the value of 101 when $>10^2$ CFUs were counted on the plate. Negative sample in culture was not included in this study. A single bacteria or yeast was recovered for 104 out of 179 samples (58.1%), the remaining yielding 2 (24/179, 13.4%), 3 (26/179, 14.5%), 4 (14/179, 7.8%), or 5 (11/179, 6.1%) bacteria and yeasts.

Ultimately, two negative controls were sequenced (Control sample analysis is the same process as samples but without any biological material).

3.2 DNA Manipulations

DNA was extracted from tissue samples and swabs in 1 mL physiological solution with suitable kit and vortexed properly. The objective of this method was to decrease the eukaryotes DNA by a differential lysis of human cells, resulting in the elimination of free DNA. The concentration of bacterial and human DNA was estimated through qPCR experiments with 16S rRNA and beta-actin reference genes, respectively.

The reference curves for bacterial and human DNA quantitation were constructed with known concentrations of *Escherichia coli* DH5α genomic DNA and human genomic DNA from the TaqMan beta-Actin Detection Reagent kit, respectively. About 3 ng of DNA (the sum of bacterial and human DNA determined by qPCR) in a 50 μL volume were subjected to DNA purification, which results in a 10 μL purified solution followed by sequencing experiments. The Nextera XT DNA Sample Preparation Kit was employed as per the instruction of the Illumina (San Diego, CA) except that 16 (instead of 12) PCR enrichment cycles were considered. In case the samples having less than 3 ng total DNA in 50 μL of non-purified extracts, the volume of Nextera tagmentase was decreased from 5 to 3 μL. The libraries were sequenced in Rapid Run mode for 2 × 250 + 8 cycles on an Illumina HiSeq 2500 instrument (with a HiSeq Rapid Flow Cell v2) at Fasteris with HiSeq Rapid SBS Kit v2 and HiSeq Control Software 2.2.58. Demultiplexed fastq files were generated using CASAVA-1.8.2 from on-instrument base-calling by Real-Time Analysis (RTA) software 1.18.64.0. The Trimmomatic package was employed to remove bases which correspond to the standard Illumina adapters.

3.3 Bioinformatic Methods

The quality of the reads was analyzed through FastQC (http://www.bioinformatics.babraham.ac.uk/projects/fastqc/). The reads (fastq format) were assessed with MetaPhlAn2 to attain the taxonomic profile of the microbial community. MetaPhlAn2 employs a database of clade-specific marker genes, and normalizes the abundances of the species with reference to the genome size of the bacteria to which reads are assigned. For the purpose of filtering out the human reads, Kraken25 with default parameters was used and the classified-out option on the miniKraken database, which embeds the genomes of bacteria, archaea, and viruses, except those of eukaryotes. Along with the seqtk subset command (https://github.com/lh3/seqtk), it was possible to recover the pairs of reads which were found as from bacteria, archaea, and viruses, and to assemble them with metaSPAdes with default parameters. The contigs generated were annotated with PROKKA. The genome coverage was analyzed by processing the metagenomic contigs by QUAST against the reference genome of the species downloaded from the RefSeq database (http://www.ncbi.nlm.nih.gov/genome/). The total quality-filtered reads (fastq format) were mapped with the help of Bowtie2 (using the local argument for accepting soft clipping) onto a database made of the ARDs from the ResFinder database (downloaded in July 2016) and ARDs from functional metagenomic studies, finally clustered at a 95% nucleic acid identity level (using CD-HIT). The next step involves assembling mapped reads with metaSPAdes with default parameters. The open reading frames (ORFs) and the amino acid sequences were obtained through Prodigal. The identification of ARDs was performed by blastp (with a 10^{-30} *e*-value) utilizing the aforementioned ARD database, with an 80% amino acid identity threshold over 80% of the reference ARD sequence. To obtain the depth of sequencing of the bacterial species and of the ARDs in samples, the reads were mapped separately with Bowtie2 against the contigs assigned to one particular species and against the ARDs identified in this sample. The depth of sequencing (expressed in ×, i.e., the number of time each nucleotide was sequenced) was calculated using Samtools and BBMap (https://sourceforge.net/projects/bbmap). The single nucleotide variants (SNVs) were identified using Samtools, Freebayes, and Tablet.

The high cost for metagenomic sequencing is a major limitation to its application in the clinical setting. In order to find out a solution to it, multiplexing of samples opens up a solution to decrease the cost but with a lower number of reads per sample. To analyze the possibility of utilizing a lower throughput for our samples, we downsized our samples to 2 M quality-filtered reads and applied the same pipeline as for full-reads samples. The same pipeline was applicable to all samples after downsizing to 2 M paired reads with the seqtk sample command (https://github.com/lh3/seqtk). The figures were constructed using R v3.2.3 and the ggplot2 package, with colors usually chosen from colorbrewer2.org. The raw reads (just adapter-trimmed) are present under the Bioproject PRJNA382079.

3.4 Identification of Multiple Clones Within Species

Besides the identification of species which were not recovered in culture, we checked if more than one clone could be detected within the species. For this purpose, we considered the contigs >1000 bp and assigned them a taxonomy with Kraken. For the main species identified in the sample (species with >0.1% total bacterial reads with Kraken) and which were not considered contaminants (see section above), with at least ten contigs available, ($n = 74$). The reads were mapped against their respective contigs and the distribution of single nucleotide variants (SNVs) were detected with suitable analysis. For all species, we observed some regions containing SNVs, but we presumed if more than one clone would be present within one species. In that case, the number of

SNVs per contigs would be homogeneously distributed on the contigs, resulting a positive correlation between the number of SNVs per contigs and their size [2].

4 Metagenomic Analysis for Detection of Probiotics (Bacterial) for Growth, Feed Conversion Efficiency, Metabolism and Immunity

Exploration of **gut metagenome** for **increasing the feed conversion efficiency and growth promotion, metabolism, and increased immunity** through the use of **probiotics** identified from native chicken/duck.

Use of probiotics in poultry sector as growth promoter and immunomodulator is of immense importance. Probiotic species such as *Lactobacillus*, *Streptococcus*, *Bacillus*, *Bifidobacterium*, *Enterococcus*, *Aspergillus*, *Candida*, and *Saccharomyces* were observed to have beneficial effect on broiler performance. These probiotics modulate the intestinal microflora, cause pathogen inhibition, intestinal histological changes, immunomodulation, effect certain haemato-biochemical parameters, improve the sensory characteristics of dressed broiler meat and promote microbiological meat quality of broilers. Anticipated mechanisms of pathogen inhibition by the probiotic microorganisms include competition for nutrients, production of antimicrobial conditions, antimicrobial compounds (volatile fatty acids and bacteriocin), lowering GIT pH, and competition for binding sites on the intestinal epithelium and stimulation of the immune system. Subsequently, modes of action of probiotics for majority of these proposed probiotics have not been studied completely. Thus, there is an urgent need for the thorough study in this area.

Metagenomics is a technique which is independent of culture. It involves DNA sequencing directly from environmental samples. We describe the protocol with a case study. In the current study, DNA was extracted from directly from the gastrointestinal tract of chicken or guineafowl. The hypervariable region encoding 16S rRNA is sequenced through metagenomics approach to detect the composition of microbial populations in organisms. The nine variable regions (V1–V9) of the 16S rRNA bears significant degrees of sequence diversity and has been reported to effectively distinguish maximally the bacterial communities within the gastrointestinal tract of various organisms. Technological advances such as the next-generation sequencing platform provides an opportunity to evaluate the host specific microbial diversity of GIT with 16S metagenomics approach. The objective of the current study includes exploration of microbial biodiversity within the GI tract as well as exploration of newer microbe which may act as beneficial probiotic [3].

4.1 Protocol

All animal studies were approved by the institutional animal care and use committee's (IACUC) guidelines and were approved by IACUC.

1. The study comprises of chickens and ten Guinea fowls of 20 weeks old ($n = 10$), sampled from poultry flocks raised at the Tennessee State University poultry research farm. The broiler chickens were chosen from the same parental line and the guinea fowl were of the French variety derived also from same parental line. Uniformity needs to be maintained, to avoid environmental deviation for the birds. The birds were raised under standard management techniques. They were fed isocaloric and isonitrogenous (3150 ME Kcal ME/kg diet and 21% CP, respectively) diets for 20 weeks. The mash feed and water should be free from probiotics and were provided at free choice.
2. The birds were sacrificed through cervical dislocation and the gut content was collected with the help of sterile forceps or knife as the case may be. The gut content and epithelial lining of intestinal epithelium were scrapped and collected in a sterile 50 mL polystyrene tube containing 30 mL of sterile 1× PBS solution. The lining of the epithelium was scraped to capture the microbes which were attached to the GIT epithelium. The diluted samples of GIT contents were placed in ice and immediately used for DNA extraction or stored at −80 °C until use.
3. DNA was extracted using the pure link Genomic DNA mini kit (Life Technologies, Waltham, MA, USA) protocol and the concentration of DNA was measured using a NanoDrop spectrophotometer (Thermo Fisher, Waltham, MA, USA) and agarose gel electrophoresis. From the quantification results the high-quality DNA samples ($A_{260}/A_{280} = 1.85–1.90$) of ten chicken and ten Guinea fowl were obtained. Five DNA samples were pooled to make two separate pooled samples each of chickens and Guinea fowls and these pooled samples were used further for DNA library construction and NGS.
4. The 16S DNA library was constructed by following the instructions from 16S metagenomic kit and ion plus fragment library kit (Life Technologies). Primers were provided in the kit to amplify the 16S region of the DNA samples. Two sets of primers, respectively, were designed to amplify V2-4-8 regions and V3-6, 7–9 regions of 16S gene. PCR amplicons of equal volume and concentration from the two primers were pooled and employed in library construction.
5. The quality of the constructed library was analyzed using Agilent 2100 Bio-analyzer (Agilent Technologies, Santa Clara, CA, USA) following the protocol of Agilent DNA7500 kit

reagents (Agilent Technologies) and the library concentration was diluted up to 26–30 pM using nuclease free water.

6. The diluted 16S library was used for template preparation using IonOneTouch-2 system by following instructions from ion PGM Template OT2 400 kit. The DNA template quality was analyzed using the qubit 2.0 fluorometer following instructions from IonOT2 400 kit using Ion sphere quality control kit (Life technologies). DNA quality scores were checked for each of the two birds selected for the NGS.
7. The DNA template was enriched with Ion one Touch ES system (life technologies) as per instructions from IonOT2 400 kit.
8. Following the enrichment of DNA library, samples were prepared for amplification reaction using the protocol from Ion PGM 400 sequencing kit (Life technologies). The amplified sample was loaded onto 316V2 (life technologies) chip for sequencing with the Ion Torrent PGM system as per the instructions from Ion PGM 400 sequencing kit. The sequencing requires about 6–7 h and being monitored through the ion torrent server. The detailed laboratory protocols can be accessed at https://doi.org/10.17504/protocols.10.mdgc23w. Sequencing data were analyzed with the help of the ion reporter software provided by Life Technologies.
9. The Ion reporter software consists of a bundle of bioinformatics tools which is helpful in streamlining and simplifying the analysis of Ion PGM sequencing data. The 16S metagenomic workflow in ion reporter employed Core QIIME pipeline and the GreenGenes and Microseq ID databases for phylogenetic diversity. In QIIME pipeline, first step includes clustering of all sequences from all test samples into Operational Taxonomic Units (OTUs) according to their sequence similarity. The OTUs in QIIME represented some extent of taxonomic relatedness. In the successive stage, QIIME was employed to select representative sequences from each OTU for downstream analysis. These representative sequences were utilized for taxonomic identification of the OTUs and phylogenetic alignment. In the next step, the QIIME used the OTU file created above to extract representative sequences from the fasta file by one of several methods and in Stage 3, provided information on the microbial lineages observed in microbial samples. By default, QIIME employed the Ribosomal Database Project (RDP) classifier to assign taxonomic data to each representative sequence from stage two. Later on, the module classified individual reads from the sequencing data with the help of three reference library options.

(a) The first one being the Basic Local Alignment Search Tool (BLAST) to the curated Green Genes database.

(b) secondly by BLAST alignment to the premium curated MicroSEQ ID database.

(c) thirdly by optimal two step BLAST alignment to both reference libraries.

The alignment at different levels of taxonomy followed the clinical and laboratory standards institute (CLSI) guidelines. It requires the family, genus, and species level to have <97%, >97%, and >99% identity, respectively.

5 Metagenomic Analysis for Detection of Viral Diversity

This is an important approach to assess viral diversity, particularly in animals observed to be resistant to diseases. We could explore from our earlier study that duck and sheep are comparatively better resistant to other important livestock species [3]. Ducks were reported to be resistant to most of the common viral diseases affecting poultry. Ducks are asymptotic to avian influenza. Hence it is essential to assess the viral diversity in healthy duck to establish it.

5.1 Protocol in Brief

1. Sample collection and processing. Cloacal swabs were collected from the flock of 23 ducks and the swabs were placed in the Eppendorf tubes containing 200 μL Hank's Balanced Salt Solution (HBSS). The Eppendorf tubes were vortexed and centrifuged at 6000 × *g* for 10 s. The swabs were repeatedly washed with 200 μL HBSS at 7400 × *g* for 15 s.

2. The total volume of washings was pooled and passed through 0.45 μm filter. The filtrate was ultra centrifuged at 37,000 rpm for 4 h at 4 °C. The supernatant was discarded and the pellet was resuspended in 50 μL HBSS for overnight softening, then resuspended in 200 μL HBSS and stored at −80 °C.

3. The next step includes extraction of viral nucleic acids and next generation sequencing.

   The filtrate was treated with a mixture of nucleases (TURBO DNase, Benzonase, RNase I) at 37 °C in dry bath for 2 h. Viral nucleic acid was extracted from the nuclease treated filtrate using High Pure Viral Nucleic acid Kit (Roche) and QIAamp viral RNA mini kit (Qiagen).

4. NanoDrop spectrophotometer was observed to be useful for quantification of viral nucleic acids.

5. The viral nucleic acids were reverse transcribed with SuperScript III reverse transcriptase (Invitrogen) as per manufacturer's protocol with random hexamer primer. In the next

step, double strand cDNA was synthesized with appropriate kit.

6. The paired-end library was prepared with Illumina TruSeq RNA Library Preparation v2 Kit. Fragmentation of DNA was conducted with Covaris S2 shearing which leads to double-stranded DNA fragments consisting of 3′ or 5′ overhangs.
7. Further, the fragmented DNA was end-repaired followed by A-tail adapter ligation. Selective enrichment of adapter ligated DNA fragments was done through PCR amplification and validation of library was done on Bioanalyzer 2100 using HS DNA Chip. The DNA library was sequenced for paired end 2×150 sequences on Illumina MiSeq platform. The paired end reads were trimmed of their set primer sequences and the sequences were then de novo assembled into contigs using CLC genomics workbench software, with criteria of at least 95% identity over 35-bp to merge two fragments and minimum contig length of 100 bases.
8. The assembled contigs and singleton sequences greater than 100 bp were compared to the NCBI GenBank nonredundant nucleotide and reference viral protein databases (http://www.ncbi.nlm.nih.gov/refseq/) using BLASTn and BLASTx, respectively. On the basis of the best BLASTx result, sequences were classified into their likely taxonomic groups of origin based on the best hit (lowest *E* score) sequence match.

6 Phylogenetic Analysis

All alignments and phylogenetic analyses were based on the translated amino acid sequences. NCBI ORF finder is a powerful tool to predict putative open reading frames (ORFs) in the genome. Identity was determined by BLASTn and BLASTp program. Phylogenetic analyses were conducted with duck viral sequences, based on their best BLASTp hits, and representative members of related viral species or genera. Sequence alignment was conducted with CLUSTALW with the default settings. Aligned sequences were subjected to trimming in order to match the genomic regions of the viral sequences obtained in the study. A phylogenetic tree with 1000 bootstrap resamples of the alignment data sets was generated employing the neighbor-joining method based on the p-distance model in MEGA version 6.06 [1].

7 Metagenomic Work May Be Described in the Following Layout

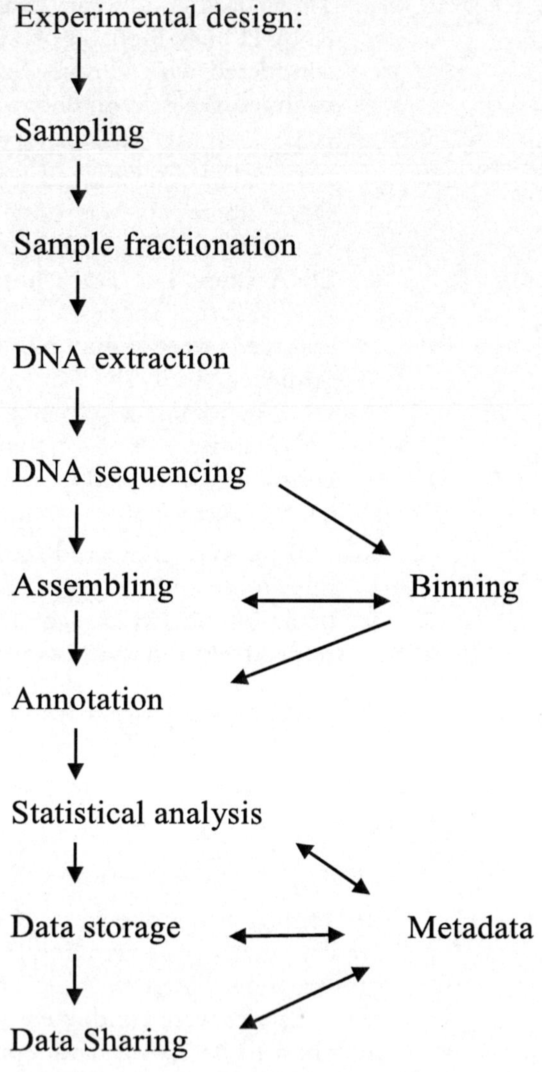

8 Detailed Protocol for 16S Metagenomic Sequencing Library Preparation

Preparing 16S Ribosomal RNA Gene Amplicons for the Illumina MiSeq System:

Metagenomic studies are usually conducted through analysis of the prokaryotic 16S ribosomal RNA gene (16S rRNA), which is about 1500 bp long containing nine variable regions interspersed between conserved regions. Variable regions of 16S rRNA are commonly employed in phylogenetic classifications, like genus or species in diverse microbial populations. It remains always variable as which 16S rRNA region to sequence. It depends upon experimental objectives, design, and sample type. This protocol explains a method for the preparation of samples for sequencing the variable V3 and V4 regions of the 16S rRNA gene. This protocol may also be employed for the sequencing other regions with different region-specific primers. This protocol utilizes a number of procedures including a benchtop sequencing system, on-board primary analysis, and secondary analysis with MiSeq Reporter or BaseSpace. It gives rise to a comprehensive workflow for 16S rRNA amplicon sequencing.

9 Workflow Summary

1. Order amplicon primers—The protocol involves the primer pair sequences for the V3 and V4 region, which develops a single amplicon of approximately ~460 bp. The protocol also includes overhang adapter sequences which must be appended to the primer pair sequences for compatibility with index and sequencing adapters.
2. Prepare library—The protocol describes the steps to amplify the V3 and V4 region and with a limited cycle PCR, added sequencing adapters and dual-index barcodes to the amplicon target. Utilizing the full complement of Nextera XT indices, up to 96 libraries may be pooled together for sequencing.
3. Sequence on MiSeq—Utilizing paired 300-bp reads, and MiSeq v3 reagents, the ends of each read are overlapped to give rise to high-quality, full-length reads of the V3 and V4 region in a single 65-h run. The MiSeq run output is approximately more than 20 million reads and, assumes 96 indexed samples. It is able to produce more than 100,000 reads per sample, accordingly, they are commonly recognized as sufficient for metagenomic surveys.

10 Analyze on MSR or BaseSpace

The Metagenomics workflow is a secondary analysis option, which is based on the MiSeq Reporter (on-system software) or BaseSpace (cloud-based software). The Metagenomics Workflow conducts a taxonomic classification utilizing the Greengenes database depicting genus or species level classification in a graphical format. This protocol is useful to sequence alternative regions of the 16S rRNA gene and other targeted amplicon sequences of interest. When using this protocol for amplicon sequencing other than 16S rRNA, use the Generate FASTQ Workflow (secondary analysis option).

User-defined, forward and reverse primers which are complementary upstream and downstream of the region of interest are designed with overhang adapters, and employed to amplify templates from genomic DNA. A subsequent limited-cycle amplification step is conducted for the addition of multiplexing indices and Illumina sequencing adapters. Libraries are normalized and pooled followed by sequencing on the MiSeq system with v3 reagents.

11 Amplicon Primers

The gene-specific sequences employed in this protocol target the 16S V3 and V4 region as depicted in Fig. 1. They are selected from the Klindworth et al. [1] publication as the most promising bacterial primer pair. Illumina adapter overhang nucleotide sequences are added to the gene-specific sequences. The full-length primer sequences, using standard IUPAC nucleotide nomenclature, to follow the protocol targeting this region are: 16S Amplicon PCR Forward Primer = 5′TCGTCGGCAGCGTCAGATGTGTATA AGAGACAGCCTACGGGNGGCWGCAG 16SAmplicon PCR Reverse Primer = 5′GTCTCGTGGGCTCGGAGATGTGTA TAAGAGACAGGACTACHVGGGTATCTAATCC. This method

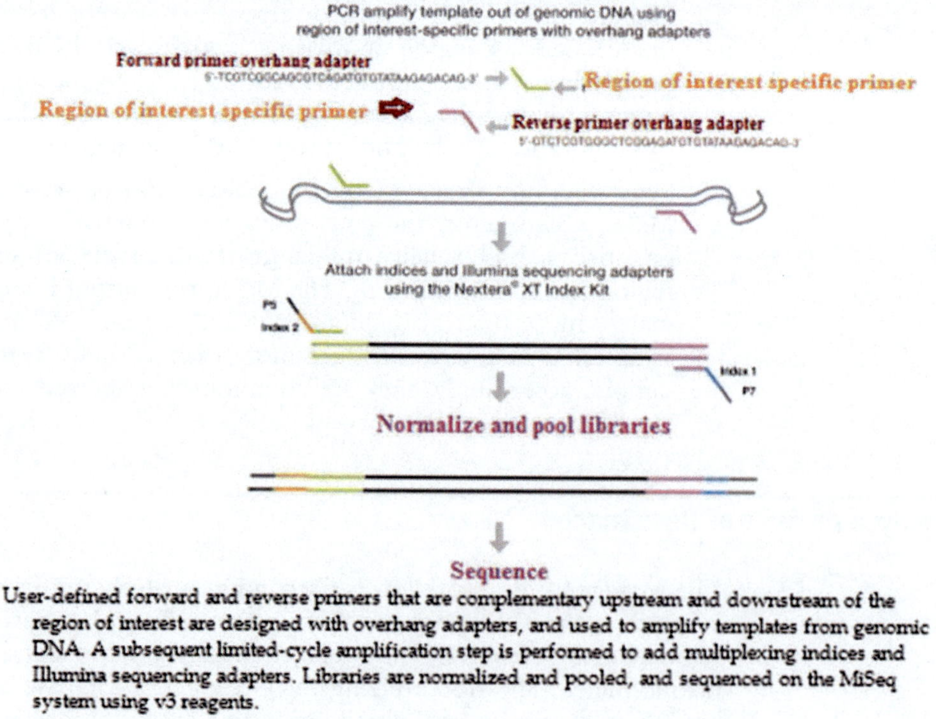

Fig. 1 16S V3 and V4 amplicon workflow

can also be utilized to target other regions on the genome (either for 16S with other sets of primer pairs, or non-16S regions throughout the genome; i.e., any amplicon). The overhang adapter sequence needs to be added to the locus-specific primer for the region to be targeted. The Illumina overhang adapter sequences to be added to locus-specific sequences are: Forward overhang: 5′TCGTCGGCAGCGTCAGATGTGTATAAGAGACAG-[locus-specific sequence] Reverse overhang: 5′GTCTCGTGGGCTCGGA GATGTGTATAAGAGACAG-[locus-specific sequence].

The following facts are recommended for designing other locus-specific primers: It is recommended that the targeting regions resulting in an amplicon which when sequenced with paired-end reads possess at least ~50 bp of overlapping sequence in the middle. As for example, while running 2 × 300 bp paired-end reads, it is recommended to have an insert size of 550 bp or smaller so that the bases sequenced at the end of each read overlap. The locus-specific portion of primer (not including overhang sequence) should have a melting temperature (T_m) of 60–65 °C. Online PCR primer sequence analysis tools need to be employed to check the properties of primer designs. For the T_m calculation only, the gene-specific portion should be used in calculation. For the purpose of hairpin and dimer calculations, the fully assembled primer sequence (including the overhang) should be utilized. It is recommended to use standard desalting purification while ordering oligo primer sets.

The following diagram depicts the workflow using the 16S Library Preparation Protocol (Fig. 2). Safe stopping points are marked.

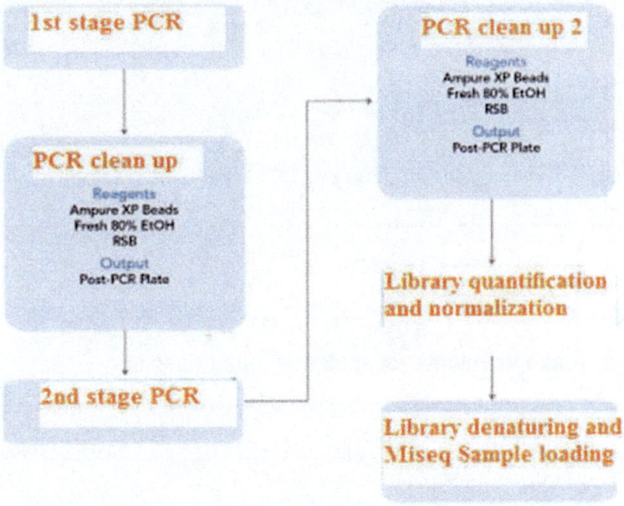

Fig. 2 16S Library preparation workflow

12 Amplicon PCR

This step utilizes PCR to amplify template out of a DNA sample with the region of interest-specific primers with overhang adapters attached (Table 1).

12.1 Procedure

1. Set up the following reaction of DNA, 2× KAPA HiFi Hot-Short ReadyMix, and primers (Table 2).
2. The plate is sealed properly and performed PCR in a thermal cycler using the following program:
   (a) 95 °C for 3 min.
   (b) 25 cycles of:
      - 95 °C for 30 s.
      - 55 °C for 30 s.
      - 72 °C for 30 s.

Table 1
List of consumables for conducting metagenomic studies

Item	Quantity	Storage
Microbial genomic DNA (5 ng/μL in 10 mM Tris pH 5.5)	2.5 μL per sample	−15 to −25 °C
Amplicon PCR reverse primer (1 μM)	5 μL per sample	−15 to −25 °C
Amplicon PCR forward primer (1 μM)	5 μL per sample	−15 to −25 °C
2× KAFA HiFi HotStart ReadyMix	12.5 μL per sample	−15 to −25 °C
Microseal 'A' film		
96-well 0.2 mL PCR plate	1 plate	
[Optional] Bioanalyzer chip (Agilent DNA 1000 kit catalog # 5067-1504)		

Table 2
Set up of reaction volume for metagenomic analysis

	Volume
Microbial DNA (5 ng/μL)	2.5 μL
Amplicon PCR forward primer 1 μM	5 μL
Amplicon PCR reverse primer 1 μM	5 μL
2× KAPA HiFi HotStart ReadyMix	12.5 μL
Total	25 μL

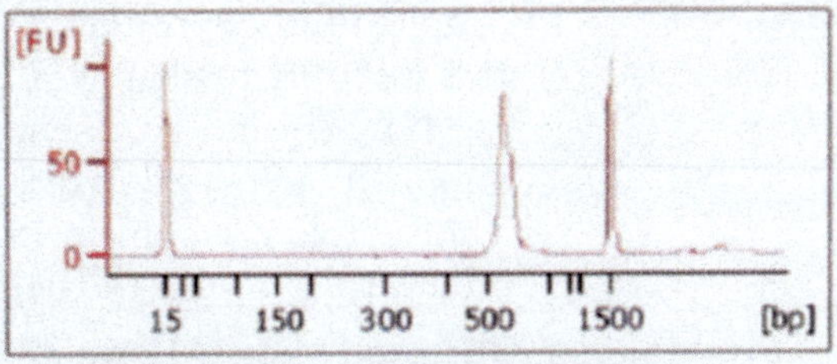

Fig. 3 Example bioanalyzer trace after amplicon PCR step

(c) 72 °C for 5 min.

(d) Hold at 4 °C.

3. It is recommended to run 1 μL of the PCR product on a Bioanalyzer DNA 1000 chip for the verification of the size. With the help of the V3 and V4 primer pairs in the protocol, the expected size on a Bioanalyzer can only trace since the Amplicon PCR step is ~550 bp. An example of graphical representation of Bioanalyzer Trace from Amplicon PCR Step is represented in Fig. 3.

13 PCR Clean-Up

This step utilizes AMPure XP beads for the purification of the 16S V3 and V4 amplicon away from free primers and primer dimer species.

Consumables

Item	Quantity	Storage
10 mM Tris pH 8.5	52.5 μL per sample	–15 to –25°C
AMPure XP beads	20 μL per sample	2 to 8°C
Freshly prepared 80% Ethanol (EtOH)	400 μL per sample	
96-well 0.2 mL PCR plate	1 plate	
[Optional] Microseal 'B' film		
[Optional] 96-well MIDI plate	1 plate	

13.1 Preparation

Bring the AMPure XP beads to room temperature.

13.2 Procedure

1. The Amplicon PCR plate is centrifuged at 1000 × *g* at 20 °C for 1 min then removed the seal carefully.
2. The shaker may be employed for mixing, with the help of a multichannel pipette set to 25 μL, transferred the entire

Amplicon PCR product from the PCR plate to the MIDI plate. It is essential to change the tips between samples.

Care should be taken to transfer the entire content of the sample to a 96-well MIDI plate while using shaker for mixing. The sample can remain in the 96-well PCR plate, in case we intend to mix with the help of pipette.

3. Vortexed the AMPure XP beads for 30 s to assure that the beads are evenly dispersed. An appropriate volume of beads is added to a trough based on the number of samples processing.
4. With the help of a multichannel pipette, added 20 μL of AMPure XP beads to each well of the Amplicon PCR plate. Tips are changed between columns.
5. For proper mixing of the ingredients, it is generally pipette entire volume up and down 10 times while using a 96-well PCR sealed plate and shake at 1800 rpm for 2 min when using a MIDI plate.
6. Incubated at room temperature for 5 min. Shaking is avoided at this stage.
7. The plate is kept on a magnetic stand for 2 min or until the supernatant has been cleared.
8. With the Amplicon PCR plate on the magnetic stand, a multichannel pipette is used to remove and discard the supernatant. Tips are changed between samples.
9. Wash the beads with freshly prepared 80% ethanol as follows:

   Using a multichannel pipette, added 200 μL of freshly prepared 80% ethanol to each sample well. The next step is incubation on a magnetic strand for 30 s. The supernatant is removed carefully and discarded.
10. With the Amplicon PCR plate on the magnetic stand, performed a second ethanol wash. Added 200 μL of freshly prepared 80% ethanol to each sample well with the help of a multichannel pipette. This is followed by the incubation of the plate on the magnetic stand for 30 s. Removal and discarding of the supernatant is the next step. Excess ethanol is removed using a P20 multichannel pipette with fine pipette tips.
11. With the Amplicon PCR plate still on the magnetic stand, the beads are air-dried for 10 min.
12. Removed the Amplicon PCR plate from the magnetic stand. Added 52.5 μL of 10 mM Tris pH 8.5 to each well of the Amplicon PCR plate, preferably with multichannel pipette.
13. Proper mixing is essential with pipette up and down 10 times. It is necessary to change tips after each column (or seal plate and shake at 1800 rpm for 2 min). Take care to resuspend beads fully.

14. It is kept for incubation at room temperature for 2 min.
15. The plate is placed on the magnetic stand for 2 min till the supernatant has cleared.
16. 50 μL of the supernatant from the Amplicon PCR plate is transferred to a new 96-well PCR plate, preferably with multichannel pipette tip. Change tips between samples to avoid cross-contamination.

14 Index PCR

This step attaches dual indices and Illumina sequencing adapters with the Nextera XT Index Kit.

14.1 Consumables

Item	Quantity	Storage
2x KAPA HiFi HotStart ReadyMix	25 μL per sample	–15 to –25°C
Nextera XT Index 1 primers (N7XX) from the Nextera XT Index kit (FC-131-1001 or FC-131-1002)	5 μL per sample	–15 to –25°C
Nextera XT Index 2 primers (S5XX) from the Nextera XT Index kit (FC-131-1001 or FC-131-1002)	5 μL per sample	–15 to –25°C
PCR grade water	10 μL per sample	
TruSeq index plate fixture (FC-130-1005)	1	
96-well 0.2 mL PCR plate	1 plate	
Microseal 'A' film	1	

14.2 Procedure

1. Transferred 5 μL from each well to a new 96-well plate, with multichannel pipette preferably. The remaining 45 μL is not utilized in the protocol and can be stored for other uses.
2. Arranged the Index 1 and 2 primers in a rack (i.e., the TruSeq Index Plate Fixture) with the following arrangements as necessary:
   (a) Arrange Index 2 primer tubes (white caps, clear solution) vertically, aligned with rows A through H.
   (b) Arrange Index 1 primer tubes (orange caps, yellow solution) horizontally, aligned with columns 1 through 12.

   Figure 4 shows the TruSeq Index Plate Fixture as required in NGS studies.
3. Placed the 96-well PCR plate with the 5 μL of resuspended PCR product DNA in the TruSeq Index Plate Fixture.

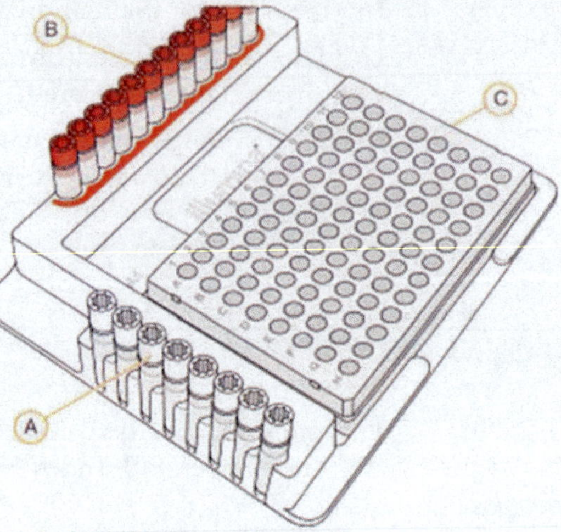

Fig. 4 TruSeq index plate fixture

4. The reaction is set up with the following reaction of DNA, Index 1 and 2 primers, 2× KAPA, HiFi HotStart ReadyMix, and PCR Grade water:

	Volume
DNA	5 μL
Nextera XT Index primer 1 (N7xx)	5 μL
Nextera XT Index primer 2 (S5xx)	5 μL
2x KAPA HiFi HotStart ReadyMix	25 μL
PCR grade water	10 μL
Total	50 μL

5. Gently pipetted up and down 10 times to mix.
6. Covered the plate with Microseal "A."
7. Centrifuged the plate at 1000 × *g* at 20 °C for 1 min.
8. Performed PCR on a thermal cycler using the following program:
   (a) 95 °C for 3 min.
   (b) 8 cycles of:

- 95 °C for 30 s.
- 55 °C for 30 s.
- 72 °C for 30 s.

(c) 72 °C for 5 min.

(d) Hold at 4 °C.

15 PCR Clean Up 2

This step uses AMPure XP beads to clean up the final library before quantification.

15.1 Consumables

Item	Quantity	Storage
10 mM Tris pH 8.5	27.5 μL per sample	−15 to −25 °C
AMPure XP beads	56 μL per sample	2–8 °C
Freshly prepared 80% ethanol (EtOH)	400 μL per sample	
96-well 0.2 mL PCR plate	1 plate	
[Optional] Microseal "B" film		
[Optional] 96-well MIDI plate	1 plate	

15.2 Procedure

1. The Index PCR plate is centrifuged at 280 × *g* at 20 °C for 1 min to collect condensation.
2. Transfer the entire Index PCR product from the PCR plate to the MIDI plate with the help of a multichannel pipette set to 50 μL. Change the tips between samples.

   If we plan to use a shaker for mixing, it recommended to transfer the sample to a 96-well MIDI plate. In case of mixing by pipette, the sample can remain in the 96-well PCR plate.
3. Vortexed the AMPure XP beads for 30 s to ensure the even dispersion. Add an appropriate volume of beads to a trough.
4. With the help of a multichannel pipette, add 56 μL of AMPure XP beads to each well of the Index PCR plate.
5. While using 96 well PCR plate or seal plate, pipetting is recommended for proper mixing of the content up and down 10 times gently. When a MIDI plate is used, shake at 1800 rpm for 2 min.
6. Incubated at room temperature without shaking for 5 min.
7. Placed the plate on a magnetic stand for 2 min or till the supernatant is clear.

8. In case of the Index PCR plate on the magnetic stand, a multichannel pipette is employed to remove and discard the supernatant. Change the tips between samples.
9. With the Index PCR plate on the magnetic stand, washed the beads with freshly prepared 80% ethanol as follows:
   (a) Added 200 μL of freshly prepared 80% ethanol to each sample well, preferably using a multichannel pipette.
   (b) Incubated the plate on the magnetic stand for 30 s.
   (c) Removed and discarded the supernatant.
10. A second phase of ethanol washing is conducted as follows:
    (a) With the help of a multichannel pipette, added 200 μL of freshly prepared 80% ethanol to each sample well.
    (b) Incubated the plate on the magnetic stand for 30 s.
    (c) Carefully removed and discarded the supernatant.
    (d) Excess ethanol is removed with the help of a P20 multichannel pipette with fine pipette tips.
11. With the Index PCR plate still on the magnetic stand, allowed the beads to air-dry for 10 min.
12. In the next step, removed the Index PCR plate from the magnetic stand. Added 27.5 μL of 10 mM Tris pH 8.5 to each well of the Index PCR plate.
13. While employing a 96-well PCR plate, gently pipette mix up and down 10 times till the beads are fully resuspended, changing tips after each column. In case of a MIDI plate, seal plate and shake at 1800 rpm for 2 min.
14. Incubated at room temperature for 2 min.
15. Placed the plate on the magnetic stand for 2 min or till the supernatant has been cleared.
16. Carefully transferred 25 μL of the supernatant from the Index PCR plate to a new 96-well PCR plate. Tips are changed between samples to avoid cross-contamination.

16 Validation of Library

For the verification of the size, run 1 μL of a 1:50 dilution of the final library on a Bioanalyzer DNA 1000 chip. The expected size on a Bioanalyzer trace of the final library is ~630 bp. V3 and V4 primer pairs are employed in this protocol.

17 Library Quantification, Normalization, and Pooling

It is recommended to quantify the prepared libraries with a fluorometric quantification method which utilizes dsDNA binding dyes.

Calculation of DNA concentration in nM is based on the size of DNA amplicons as determined by an Agilent Technologies 2100 Bioanalyzer trace (Fig. 5):

(concentration in ng/μL)/(660 g/mol × average library size) × 10^6 = concentration in nM.

For example: 15 ng/μL/(660 g/mol × 500) × 10^6 = 45 nM.

The concentrated final library is diluted using resuspension Buffer (RSB) or 10 mM Tris pH 8.5 to 4 nM. Aliquoted 5 μL of diluted DNA from each library and mixed the aliquots for pooling libraries with unique indices. Based on coverage needs, up to 96 libraries may be pooled for one MiSeq run.

In case of metagenomics samples, more than 100,000 reads per sample is enough for complete survey of the bacterial composition. This number of reads permits for sample pooling of up to 96 libraries, given the MiSeq output of >20 million reads.

18 Library Denaturing and MiSeq Sample Loading

For the preparation for cluster generation and sequencing before MiSeq sequencing, following steps are followed:

1. Denaturation of pooled libraries with NaOH.
2. Dilution with hybridization buffer.
3. Heat denaturation.
4. Each run should include a minimum of 5% PhiX to serve as an internal control for these low-diversity libraries.

It is recommended to use MiSeq v3 reagent kits for improved run metrics (e.g., Illumina).

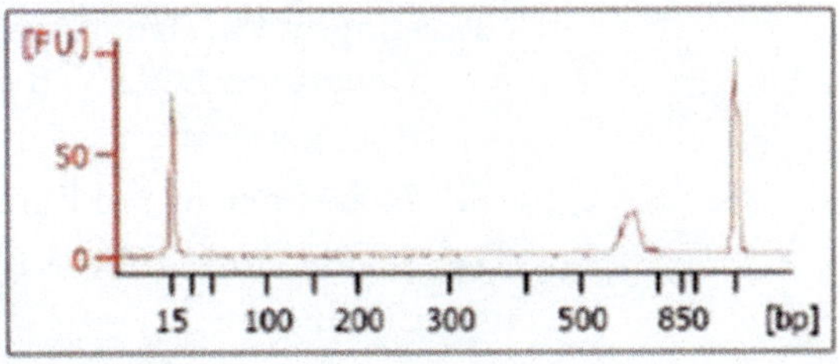

Fig. 5 Example bioanalyzer trace of final library

18.1 Consumables

Item	Quantity	Storage
10 mM Tris pH 8.5 or RSB (resuspension buffer)	6 μL	−15 to −25°C
HT1 (hybridization buffer)	1540 μL	−15 to −25°C
0.2 N NaOH (less than a week old)	10 μL	−15 to −25°C
PhiX Control Kit v3 (FC-110-3001)	4 μL	−15 to −25°C
MiSeq reagent cartridge	1 cartridge	−15 to −25°C
1.7 mL microcentrifuge tubes (screw cap recommended)	3 tubes	
2.5 L ice bucket		

18.2 Preparation

1. A heat block is set suitable for 1.7 mL microcentrifuge tubes to 96 °C
2. Removed a MiSeq reagent cartridge from −15 to −25 °C storage and thawed at room temperature.
3. An ice-water bath is prepared by combining 3 parts ice and 1 part water in an ice bucket.

18.3 Denaturation of DNA

1. The following volumes of pooled final DNA library and freshly diluted 0.2 N NaOH are combined in a microcentrifuge tube:
   (a) 4 nM pooled library (5 μL).
   (b) 0.2 N NaOH (5 μL).
2. Kept the remaining dilution of 0.2 N NaOH for the preparation of a PhiX control for the next 12 h.
3. In order to mix the sample solution the solution is vortexed briefly, and then centrifuged the sample solution at 280 × *g* at 20 °C for 1 min.
4. Incubated for 5 min at room temperature for the denaturation of the DNA into single strands.
5. Added the following volume of pre-chilled HT1 to the tube consisting of the denatured DNA:
   (a) Denatured DNA (10 μL).
   (b) Pre-chilled HT1 (990 μL) Adding the HT1 results in a 20 pM denatured library in 1 mM NaOH.
6. Kept the denatured DNA on ice till it is needed for to proceed to final dilution.

18.4 Dilution of Denatured DNA

1. Dilution of the denatured DNA to the desired concentration is conducted as per the following example:

Final concentration	2 pM	4 pM	6 pM	8 pM	10 pM
20 pM denatured library	60 μL	120 μL	180 μL	240 μL	300 μL
Pre-chilled HT1	540 μL	480 μL	420 μL	360 μL	300 μL

It is recommended to target 800–1000 K/mm^2 raw cluster densities using MiSeq v3 reagents. It is suggested to initiate the first run with the help of a 4 pM loading concentration and adjust subsequent runs appropriately.

2. Inversion is conducted several times to mix and then proceed for pulse centrifugation of the DNA solution.
3. Placed the denatured and diluted DNA on ice.

18.5 Denaturation and Dilution of PhiX Control

The undermentioned instructions need to be followed to denature and dilute the 10 nM PhiX library to the same loading concentration as the Amplicon library. The final library mixture should contain at least 5% PhiX.

1. Combined the following volumes to dilute the PhiX library to 4 nM:
   (a) 10 nM PhiX library (2 μL).
   (b) 10 mM Tris pH 8.5 (3 μL).
2. Combined the following volumes of 4 nM PhiX and 0.2 N NaOH in a microcentrifuge tube:
   (a) 4 nM PhiX library (5 μL).
   (b) 0.2 N NaOH (5 μL).
3. Vortexed briefly to mix the 2 nM PhiX library solution.
4. Incubated for 5 min at room temperature for the denaturation of the PhiX library into single strands.
5. Added the following volumes of pre-chilled HT1 to the tube containing denatured PhiX library resulting in a 20 pM PhiX library, comprising of denatured PhiX library (10 μL) and pre-chilled HT1 (990 μL).
6. Diluted the denatured 20 pM PhiX library to the same loading concentration as the Amplicon library as follows:

Final concentration	2 pM	4 pM	6 pM	8 pM	10 pM
20 pM denatured library	60 μL	120 μL	180 μL	240 μL	300 μL
Pre-chilled HT1	540 μL	480 μL	420 μL	360 μL	300 μL

7. Inverted several times to mix and then pulse centrifuge the DNA solution.
8. Placed the denatured and diluted PhiX on ice.

18.6 Combine Amplicon Library and PhiX Control

PhiX control is recommended of greater than or equal to 5% for low diversity libraries. It is possible with RTA v1.17.28 or later, that is bundled with MCS v2.2. In order to attain optimal performance, update to v3 software (MCS 2.3). While using an older version of the MiSeq software or sequencing these libraries on the GA or HiSeq, Illumina recommends using ≥25% PhiX control spike-in.

1. Combined the following volumes of denatured PhiX control library and denatured amplicon library in a microcentrifuge tube:
   (a) Denatured and diluted PhiX control (30 μL).
   (b) Denatured and diluted amplicon library (570 μL).
2. The combined sample library and PhiX control are set aside on ice till the heat denaturation of the mixture immediately prior to loading it onto the MiSeq v3 reagent cartridge.
3. Utilizing a heat block, incubated the combined library and PhiX control tube at 96 °C for 2 min.
4. Following the incubation, inverted the tube 1–2 times for mixing and immediately transferred to the ice-water bath.
5. Placed the tube in the ice-water bath for 5 min.

It is recommended to conduct the heat denaturation step immediately prior to loading of the library into the MiSeq reagent cartridge in order to ensure efficient template loading on the MiSeq flow cell.

19 MiSeq Reporter Metagenomics Workflow

Following the loading of the samples, the MiSeq system gives the scope for instrument on secondary analysis with the MiSeq Reporter software (MSR), with many options for analyzing MiSeq sequencing data.

By following this 16S metagenomics protocol, the metagenomics workflow classifies organisms with V3 and V4 amplicon utilizing a database of 16S rRNA data. The classification is according to the Greengenes database (http://greengenes.lbl.gov/). The output of this workflow comprises of classification of reads at several taxonomic levels as kingdom, phylum, class, order, family, genus, and species. The analysis output involves:

- Clusters Graph—It represents numbers of raw cluster, clusters passing filter, clusters that did not align, clusters not associated with an index, and duplicates.
- Sample Table—Comprises of the sequencing results for each sample.
- Cluster Pie Chart—It involves a graphical representation of the classification breakdown for each sample.

The methodology explained in this procedure for 16S metagenomics analysis can be employed for any targeted amplicon sequencing, pertinent to research on virology, exploring mutation, or other microbiology-related studies. If it is required to employ the protocol for other amplicon sequencing studies, it is preferable to use the MiSeq Reporter. This is followed by generation of FASTQ workflow leading to FASTQ files for downstream analysis.

Some of the commonly used Acronyms are:

HT1: Hybridization Buffer, IEM: Illumina Experiment Manager, MSR: MiSeq Reporter, PCR: Polymerase Chain Reaction, rRNA: Ribosomal RNA, RSB: Resuspension Buffer.

19.1 Consumables and Equipment

1.7 mL microcentrifuge tubes, 10 μL barrier pipette tips, 10 μL multichannel pipettes, 10 μL single channel pipettes, 20 μL barrier pipette tips, 20 μL multichannel pipettes, 20 μL single channel pipettes, 200 μL barrier pipette tips, 200 μL multichannel pipettes, 200 μL single channel pipettes, 1000 μL barrier pipette tips, 1000 μL multichannel pipettes, 1000 μL single channel pipettes, 96-well 0.2 mL skirtless PCR plates or Twin.Tec 96-well PCR plates, Agencourt AMPure XP 60 mL kit Beckman Coulter, Ethanol 200 proof (absolute) for molecular biology (500 mL), PCR Forward Primer (Standard desalting), PCR Reverse Primer (Standard desalting), KAPA HiFi HotStart ReadyMix (2×) KAPA Biosystems, part # KK2601, Microseal "A" adhesive seals, Microseal "B" adhesive seals, MiSeq Reagent Kit v3 (600 cycle), Nextera XT Index Kit, PhiX Control Kit v3, PCR grade water, Fluorometric quantitation with dsDNA binding dye reagent, RNase/DNase-free 8-well PCR strip tubes and caps, RNase/DNase-free multichannel reagent reservoirs, disposable VWR, Tris–HCl 10 mM, pH 8.5, 96-well storage plates, round well, 0.8 mL ("MIDI" plate, Fisher Scientific, part # AB-0859).

Equipment requirement includes L ice bucket, 96-well thermal cycler (with heated lid), Fluorometric determination for

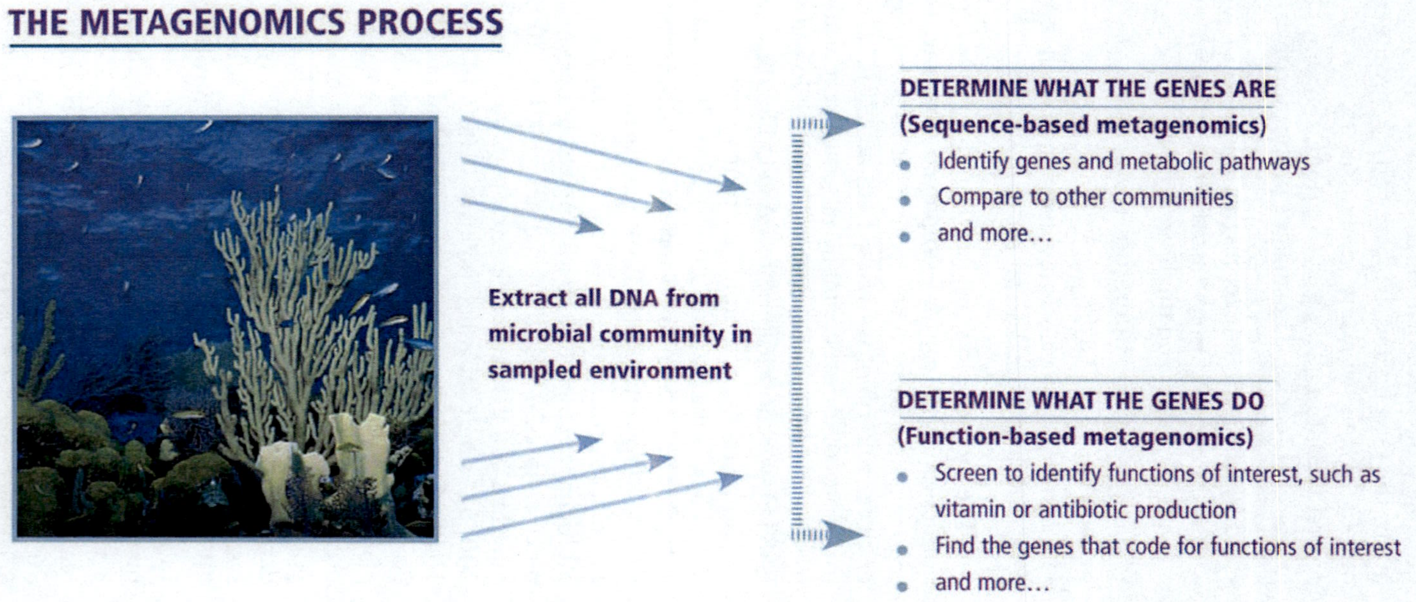

Fig. 6 Metagenomic process as sequence based metagenomics and function based metagenomics

quantification with dsDNA binding dyes, Magnetic stand-96, Microplate centrifuge, TruSeq Index Plate Fixture Kit (reusable) (Illumina), 2100 Bioanalyzer Desktop System Agilent, Agilent DNA 1000 Kit Agilent, High Speed Micro Plate Shaker VWR.

20 Applications of Metagenomics

- It aids in the process of characterization of the biodiversity existing on Earth.
- This technique enables the researcher to interpret sequenced genomes even the partial sequences received through direct sampling of specific environmental niches.
- Examples include samples from different environmental conditions: ocean, acid mine site, coral reefs, human, soil microbiome which may vary according to the health status of the subject. Figure 6 depicts the overview of metagenomic process as sequence-based metagenomics and function-based metagenomics.

References

1. Klindworth A, Pruesse E, Schweer T, Peplles J, Quast C et al (2013) Evaluation of general 16S ribosomal RNA gene PCR primers for classical and next-generation sequencing-based diversity studies. Nucleic Acids Res 41(1):e1
2. Ruppé E, Lazarevic V, Girard M, Mouton W, Ferry T, Laurent F, Schrenzel J (2017) Clinical metagenomics of bone and joint infections: a proof of concept study. Sci Rep 7:7718. https://doi.org/10.1038/s41598-017-07546-5
3. Pal A, Chakravarty AK (2019) Genetics & breeding for the disease resistance of livestock. Elsevier Publication, Amsterdam. ISBN: 9780128164068

Chapter 16

Genome-Wide Association Studies/SNP Chips

Abstract

Genome-wide association study (GWA study or GWAS), or whole genome association study (WGA study or WGAS), has been proved to be an effective technique for identifying common genetic variants in different individuals and its association with a trait. This technique is very much useful for identification of suitable markers for disease traits as well as production traits, leading to efficient marker assisted selection or genomic selection. This chapter lists out some SNP chips commonly employed for human as well as for livestock species. In the later section, a common overview of the analysis of data for GWAS has been discussed. Considering huge data size, it is commonly known as "Big Data." Different case studies with suitable examples from livestock species have been discussed.

Keywords Genome-wide association studies (GWAS), Genomic selection, SNP chips, Big data, Base-Space engine, Sample annotation, QC filtering

1 Basic Principle

Genomic selection (or Marker assisted selection with limited number of genes) is very much helpful to exploit the genetic variation at genome or nucleotide level. Identification of nucleotide variation or mutations through conventional genetic markers leads to limited data. Now with the advent of SNP chips and high throughput sequencing technology, it is possible to detect variation for many genes at a time. Genetic association with suitable statistical packages leads to genomic selection. We can select for many genes at a time.

Genome-wide association study (GWAS) is also referred to as whole genome association study (WGAS), has been proved to be an effective technique for the identification of the genetic variants existing in different individuals and its association with the phenotypic traits. GWASs are associated with single-nucleotide polymorphisms (SNPs) linked to the phenotypic traits.

GWAS is employed for detection of SNPs pertaining to different traits of economic interest.

GWAS has been regarded as a useful methodology to identify the existing genetic variants of a trait. It aids for the detection of the association with common single-nucleotide polymorphisms (SNPs) associated to the common human diseases such as type 2 diabetes,

Aruna Pal, *Protocols in Advanced Genomics and Allied Techniques*, Springer Protocols Handbooks,
https://doi.org/10.1007/978-1-0716-1818-9_16,

heart disease, inflammatory bowel disease, and psychiatric disorders. Figure 1 depicts schematic representation of GWAS in human, while Fig. 2 represents the workflow for it. In this chapter, we shall be dealing with the strategies of case-control studies for GWAS based on case control and an overview of the analysis with suitable statistical methodologies. However, in Chapter 29, we attempted to discuss about big data analysis.

Three important points need to be considered in relation to genome-wide association studies:

1. GWAS involves analysis through Bayesian statistics.
2. GWAS is effectively applicable for genomic selection.
3. This is equally useful for human genomic studies, including disease resistant traits.

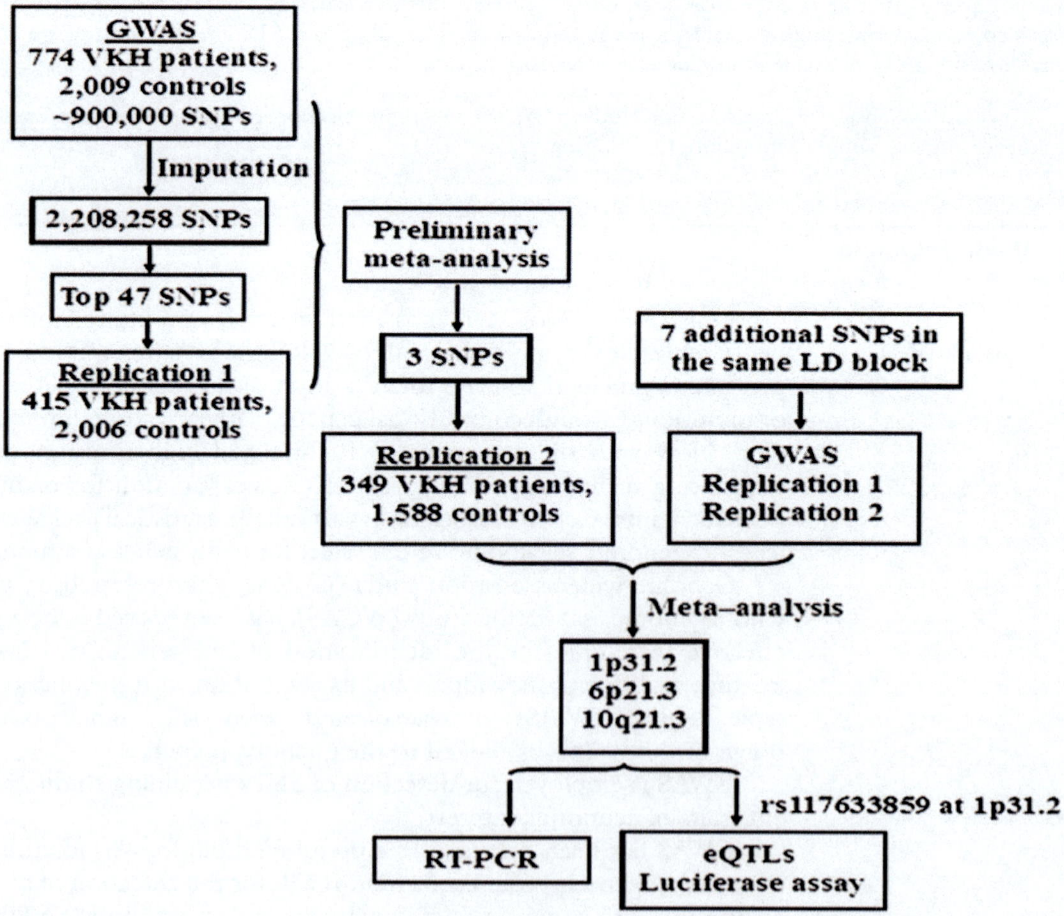

SNP chips are commercially available for human and others livestock species (Table 1):

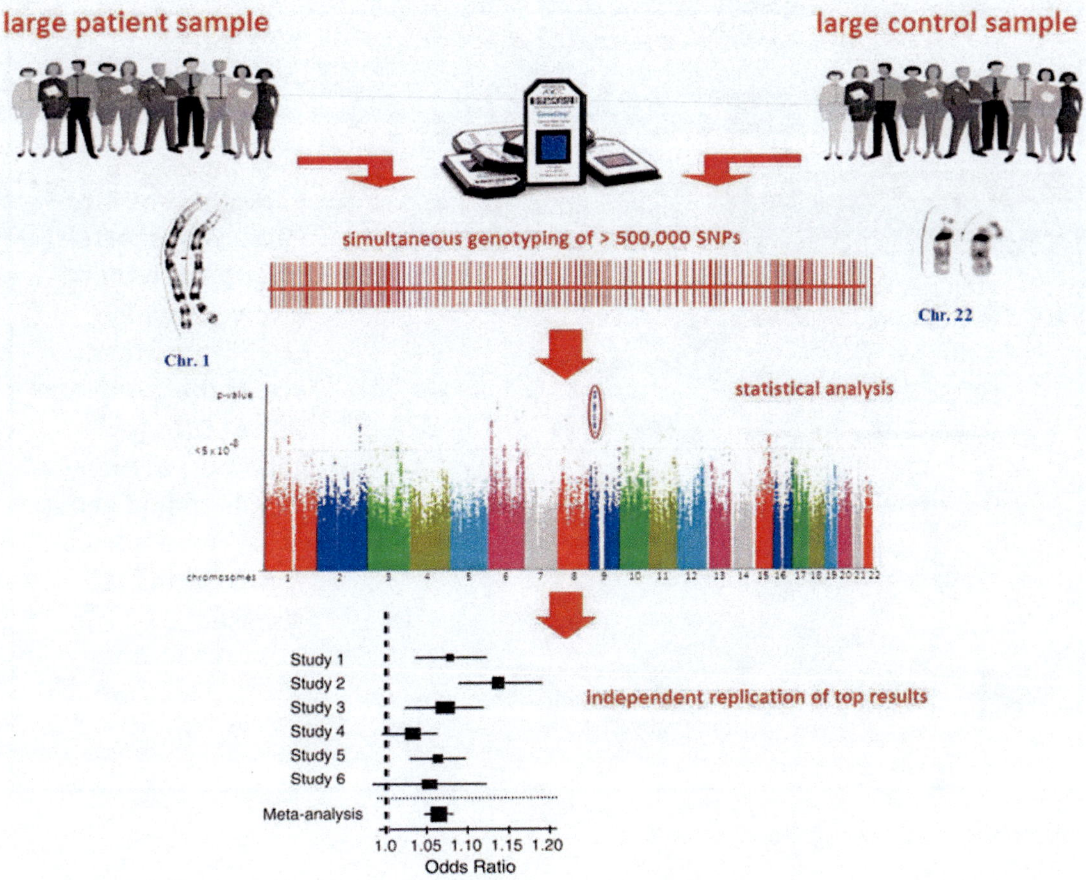

Fig. 1 Diagrammatic overview of GWAS for human

After the SNP identification process is over, we get a huge volume of data, which need to be analyzed. For a definite conclusion, data need to be analyzed. We will discuss Big data analysis in Chapter 29.

2 Data Analysis for GWAS

Data Analysis—GWAS Processing: An overview of data processing using the BaseSpace Correlation Engine for GWAS (ref Illumina):

The Illumina Base Space Correlation Engine consists of over 25,000 biosets procured by mining the vast amounts of publicly available genomic data. An important point is to be noted is that for the analysis of such huge data commonly termed as Big data is conducted through Linux-based systems.

The various sources being described as the Gene Expression Omnibus, Array Express, and other repositories. These data are processed through a systematic screening, curation, and analysis

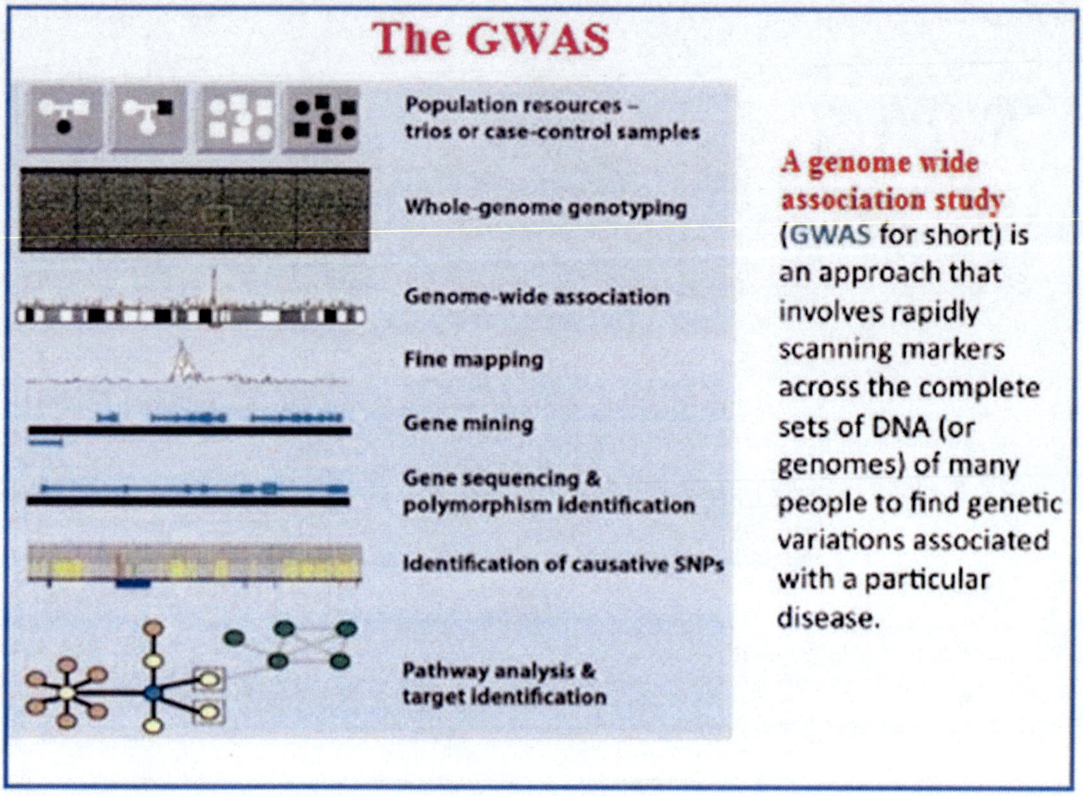

Fig. 2 Diagrammatic overview of workflow GWAS

process (Fig. 3). This technical note explicits the utilization of the BaseSpace Engine for analysis of data derived from genome-wide association studies (GWAS).

Biosets for the BaseSpace Engine are obtained by processing the following types of source data from GWAS:

3 Precomputed Summary Statistics Utilizing Standard Single-Marker Statistical Analysis Methods for GWAS

These initial-phase of biosets were obtained from data analyzed from external sources (e.g., from the database of Genotypes and Phenotypes [dbGaP]). This is conducted to analyze through sequence-centric application. Care needs to be undertaken to meet Illumina quality control criteria and significance thresholds. Afterwards, these studies can again be processed from raw data according to the standard data analysis protocol for GWAS [1–3].

Criteria for inclusion of studies involving precomputed results include:

1. Data collection and methodologies described in peer-reviewed journals.

Table 1
SNP Chips commercially available for Human and other livestock species

Species	Name of SNP chip	Specification of SNP chip	Remarks
Porcine	**Porcine SNP60 DNA Analysis Kit v2**	It is available for different sample sizes, ranging from 24, 48, 288, and 1152 by Illumina.	
		The Analysis Kit v2 consists of 65,000 number of probes, which are spaced evenly. It possesses good SNP density for GWAS. It acts as an efficient determinator of genetic merit. It can equally identify QTL (quantitative trait loci) and genetic studies on comparative basis. It can aid in investigating the breed-wise comparison for different breeds as Duroc, Landrace, Pietran, and Large White Yorkshire.	
		A semi-custom version of this chip is available as the Porcine SNP60+ DNA Analysis Kit v2. It can allow up to 25,000 additional custom probes for targeted studies. The Porcine SNP Chip Consortium allows for selecting the highest quality markers with more than 510,000 single nucleotide polymorphisms (SNPs). It combines novel discovered SNPs with existing databases and study results.	
		This is useful for marker-based selection, with minor allele frequency being assessed through sequencing of sample, counting of allele, quality score, spacing, location, and validation status.	
	GGP Porcine Arrays	The array targets desired breeding traits and genetic makeup. It includes study of disease predisposition, with GeneSeek Genomic Profiler (GGP) porcine microarrays.	
Bovine	**Bovine 777K SNP BeadChip**		
	BovineSNP50 v3 BeadChip	The BovineSNP50 v3 BeadChip microarray kit consists of 53,714 highly informative SNPs uniformly distributed across the entire genome of major cattle breed types. The chip is applicable for genome-wide enabled selection, identification of QTL (quantitative trait loci), evaluation of genetic merit of individuals, and comparative genetic studies.	

(continued)

Table 1
(continued)

Species	Name of SNP chip	Specification of SNP chip	Remarks
		This BeadChip was prepared by Illumina, collaborating with the USDA-ARS, University of Missouri, and the University of Alberta. More than 22,000 SNP probes target novel SNP loci which were discovered by Illumina. It was prepared by sequencing of three pooled populations of economically important beef and dairy cattle.	
		Additional content is obtained from publicly available sources such as the bovine reference genome, Btau, and the Bovine HapMap Consortium data set. Validation of the SNP probes have been conducted in 18 common beef and dairy breeds. This product is targeted for evenly distributed SNPs which are polymorphic across the breeds tested. It provides a median spacing of 37.4 kb.	
		This 24-sample BeadChip depicts a high-density genotyping solution for the characterization of the genome in dairy and beef cattle.	
		The PCR-free, single tube sample preparation is easier to use and reduces potential sample handling errors.	
		An array-based Laboratory Information Management System and robotic automation are appropriate to accurately and efficiently tracking of samples throughout analysis.	
		For high-throughput, cost-effective genetic screening, the BovineSNP50 BeadChip array carries more than 53,000 evenly spaced SNP probes spanning the bovine genome. Customization is essential for greater flexibility of the BovineSNP50+ version of the BeadChip. This can be customized to include up to 600,000 additional markers.	
	BovineLD Genotyping BeadChip	It imparts genomic selection to the entire herd with this expert-designed genotyping array featuring scalable content at an economical price.	

(continued)

Table 1 (continued)

Species	Name of SNP chip	Specification of SNP chip	Remarks
	BovineHD DNA Analysis Kit	This genome-wide bovine genotyping array kit covers over 777,000 SNPs. It is compatible with any breed of beef or dairy cattle.	
	GGP Bovine & Bos Indicus Arrays	This chip accurately predicts key traits such as marbling and rib-eye area with GeneSeek Genomic Profiler (GGP) bovine microarrays.	
Ovine	Ovine 600K SNP BeadChip, or Ovine 50K BeadChip	The OvineSNP50 BeadChip array covers over 54,241 evenly spaced probes targeting SNPs. It provides more than sufficient SNP density for genome-wide association studies. It also aids in other applications such as genome-wide selection, determination of genetic merit, identification of quantitative trait loci, and comparative genetic studies.	
		The capability of OvineSNP50 BeadChip is to provide uniform genome-wide coverage with an estimated one marker per 46 kb on average.	
		The BeadChip gives high call rates, permits for flexible content deployment, and allows the detection, followed by measuring copy number variation.	
		The BeadChip affects experimental variability so that scientists can work up to 12 samples in parallel.	
		The assay's single-tube sample preparation when there is no need for PCR or ligation steps significantly reduces labor and potential sample handling errors.	
Caprine	Caprine 50K SNP BeadChip	The GoatSNP50 chip employs more than 52,000 SNP variants providing uniform genomic coverage. The array was developed by the International Goat Genome Consortium screening whole genome sequencing data in the goat breeds as Alpine, Boer, Creole, Katjang, Saanen, and Savanna. A total of ten breeds were utilized to validate the 52k SNP content.	
Murine (Mouse)	GGP Mouse (GIGA-MUGA) Arrays	GeneSeek Genomic Profiler (GGP) mouse genotyping microarrays use marker-assisted selection and parentage applications.	

(continued)

Table 1 (continued)

Species	Name of SNP chip	Specification of SNP chip	Remarks
Equine	GGP Equine Arrays	Desired traits are selected and verified animal pedigree with the aid of GeneSeek Genomic Profiler (GGP) equine microarray.	
Cannine	CanineHD Whole-Genome Genotyping BeadChip	This array allows genotyping of any domestic dog breed, and allows ample SNP density for within-breed association and CNV studies.	
Shrimp (Aqua culture)	*Infinium ShrimpLD-24 BeadChip*	This BeadChip microarray provides a high-powered genotyping solution to advance aquaculture breeding programs of Pacific white shrimp.	
Human	*Infinium iSelect HD and HTS Custom Genotyping BeadChips*	The chip is able to interrogate virtually any target across any species to create a fully customized genotyping array pertaining to study needs.	
	Infinium XT	It is a comprehensive microarray which allows production-scale genotyping of up to 50,000 single or multi-species custom variants.	
	Infinium QC Array-24 Kit	It is applicable for high-throughput, cost-effective genotyping for quality control, tracking, and stratification applications along with biobanking.	
	Infinium PsychArray-24 Kit	It is a cost-effective, high-density Illumina microarray for mass-scale genetic studies related to psychiatric predisposition and risk.	
	Infinium OncoArray-500K BeadChip	This comprehensive, high-density microarray kit allows research related to cancer predisposition and risk.	
	Infinium OmniZhongHua-8 Kit	It is an advance Chinese population GWAS. This microarray is particularly designed to provide coverage of common, intermediate, and rare variation specific to Chinese population.	
	Infinium OmniExpress-24 Kit	This 24-sample BeadChip array allows exceptional throughput of thousands of samples per week. It enables powerful human genome-wide association studies (GWAS).	
	Infinium Omni5-4 Kit	The Infinium Omni5-4 BeadChip array leads to comprehensive coverage of the genome, possesses powerful tagSNPs selected from the International HapMap and 1000 Genomes Projects which target genetic variation down to 1% minor allele frequency (MAF).	

(continued)

Table 1
(continued)

Species	Name of SNP chip	Specification of SNP chip	Remarks
		With the help of the proven iScan System, this four-sample BeadChip allows high-throughput sample processing, with standardized content for whole-genome genotyping and copy number variation (CNV) studies.	
	Infinium Omni2.5-8 Kit	This 8-sample BeadChip array covers ~2.4 million markers, and gives exceptional genomic coverage rates across diverse populations.	
	Infinium Multi-Ethnic Global-8 Kit	It is a cost-effective array for analyzing complex disease in diverse human populations.	
	Infinium Multi-Ethnic EUR/EAS/SAS-8 Kit	It may be regarded as a cost-effective array for analyzing complex disease in diverse human populations, focused on European, East Asian, and South Asian populations.	
	Infinium Multi-Ethnic AMR/AFR-8 Kit	It is a cost-effective array for studying complex disease in diverse human populations, focused on Hispanic and African American populations.	
	Infinium ImmunoArray-24 v2 BeadChip Kit	The Infinium ImmunoArray-24 v2 BeadChip is the next-generation genotyping array for the detection and analysis of genetic variation in the immune system.	
	Infinium Global Screening Array-24 Kit	The Infinium Global Screening Array-24 v3.0 BeadChip is a next-generation genotyping array used for studying population-scale genetics, variant screening, pharmacogenomics studies, and precision medicine research.	
	Infinium Global Diversity Array-8 v1.0 Kit	The Global Diversity Array-8 (GDA) v1.0 BeadChip comprises of entire coverage of clinical research variants with proper multi-ethnic, genome-wide content.	
	Infinium CytoSNP-850K v1.2 BeadChip	This consortium-built array gives rise to comprehensive coverage of cytogenetically relevant genes for congenital disorders and cancer research.	
	Infinium Core-24 Kit	These 24-sample arrays allow economical large-scale human genotyping studies, with high-throughput processing abilities, and the ability to add up to 300K semi-custom markers.	

(continued)

Table 1
(continued)

Species	Name of SNP chip	Specification of SNP chip	Remarks
	Infinium Asian Screening Array-24 v1.0 BeadChip	A powerful, cost-effective genotyping array for large-scale genetic studies and pharmacogenomics in East Asian population.	
	HumanCytoSNP-12 BeadChip	This 12-sample BeadChip array allows the analysis of genetic and structural variation in the human genome, like duplications, deletions, amplifications, copy-neutral LOH, and mosaicism.	
	Human CytoSNP FFPE-12 BeadChip	This BeadChip array is standardized for cytogenetic studies employing formalin-fixed paraffin-embedded (FFPE) paired tumor-normal samples.	

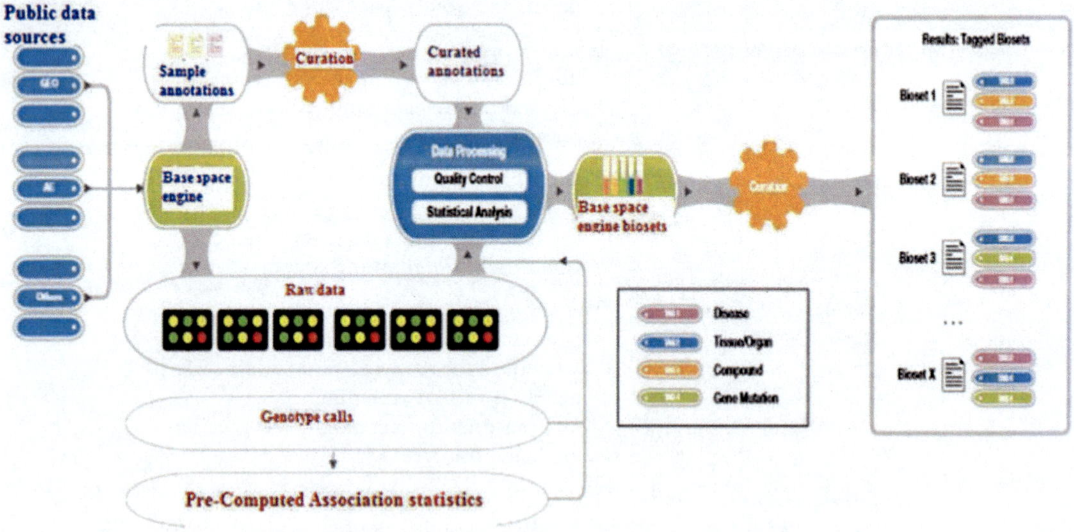

Fig. 3 Outline of data analysis for GWAS

2. Reasonable investigation into population stratification.
3. Quality control (QC) filtering is essential for individual mutations and single-nucleotide variants (SNVs), or summary statistics (minor allele frequencies [MAF], Hardy-Weinberg Equilibrium [HWE] p-value, call rates).
4. Valid association-testing methodology.

5. Literature-derived and curated SNV biosets from GWAS publications.

A list of GWAS studies was derived from the National Institutes of Health (NIH) Catalog of Published Genome-wide Association Studies, excluding the studies where data are already available via dbGaP.

3.1 Raw Data or Genotype-Level Data

It refers to data in the form of genotype calls or data in the form of raw files (e.g., CELs). It is applicable where genotype calls are not made available.

The following sections discuss the analysis procedure employed to create SNV biosets from the various sources of data. Standard statistical tests and analysis procedures are employed throughout the analysis protocol.

3.2 Raw Data Processing

Genotype calls are obtained from raw data utilizing standard methods advised by the platform manufacturer. For example, Chiamo3,4 or BRLMM5 can be employed to obtain genotype calls from the 500k Affymetrix chip. GenCall6 can be employed for raw data from Illumina chips. When raw data exists in the form of genotype calls, they are used directly. A brief overview of diagrammatic representation of data analysis and processing is involved in Fig. 3.

3.3 Data Annotation

3.3.1 Study Annotation Comprises

1. Study design.
2. Phenotype.
3. Sample inclusion criteria (e.g., ancestry, clinical features).
4. Study description.

3.3.2 Sample Annotation

When raw or genotype level data are present, the parser extracts the following information for each sample from study annotation files:

1. Family ID.
2. Sample ID.
3. Paternal ID (for family-based studies).
4. Maternal ID (for family-based studies).
5. Sex.
6. Phenotype (discrete or quantitative).
7. Group/cluster (e.g., geographical region) to assess possible effects of population stratification.

3.3.3 Platform Annotation

When raw or genotype-level data are present, the parser extracts the following platform information:

1. SNV ID (rs# or identifier).
2. Alleles.
3. Chromosome.
4. Position (bp).
5. Genetic distance (Morgans).

3.3.4 Exclusion Lists

When provided by the data source, exclusion lists for individual samples and SNVs are processed. For example, samples can be excluded for various reasons, such as:

1. Discordance in genotyping when multiple platforms are used.
2. Discordance in clinical features (diagnosis, gender).
3. Ancestry.
4. Duplications.
5. Related samples for population-based GWAS.

4 Data Exploration and Generation of Summary Statistics

Genotype calls are analyzed following thorough basic preprocessing. Various summary statistics are generated before QC filtering as follows:

- Setting invalid genotypes to "missing"—e.g., female Y genotype, heterozygous haploid chromosome.
- Minor allele frequency (haploid chromosomes counted only one time).
- Missing genotype rate.
- Missing rates by case or control status.
- HWE failures.
- Mendel errors (family-based data only).

4.1 QC Filtering

When whole-genome association statistics are present (precomputed or computed by the BaseSpace Engine), the below mentioned filtering criteria are applicable to individual samples or markers.

Individual Data Are Removed from Further Analysis in the Following Conditions:

1. When the missing genotype data are more than 10%.
2. In case Mendel errors are better than 5%, based exclusively on family-based data.
3. When the individuals are in an exclusion list.

4. Whenever, gender discrepancy is present between chromosome X data and reported sex.

SNVs Are Discarded from Further Analysis If:

1. MAF in both Cases and Controls or overall MAF is less than 1%.
2. HWE p-value in Controls or overall HWE is less than 1×10^{-6}.
3. The average Call Rate in Cases and Controls or overall Call Rate is less than 95%.
4. Mendel errors exceed 10% based on family-based data.
5. SNVs are in an exclusion list.

The experiment type determines the filtering criteria applicable to data sets derived from GWAS publications.

4.2 Stratification Analysis

In order to obtain confounding effects of population stratification, PLINK7 is employed. It offers the BaseSpace Engine for complete-linkage agglomerative clustering. It is based on pairwise genome-wide identify-by-state (IBS) distance.

4.3 Multi-Dimensional Scaling (Fig. 4)

Figure 4 represents the graphical representation plot of multidimensional scaling.

The BaseSpace Correlation Engine is responsible for standard multidimensional scaling analysis on an $N \times N$ (N = total number of samples) matrix of IBS pairwise distances. Plotting of the various dimensions against each other is beneficial for the identification of any clustering of samples. The process involves plotting the first dimension vs. second dimension, may be regarded as visualization exercise. It is also essential to color-code the individuals as per the cluster information (e.g., ancestry and geographical location).

5 Genomic Control

Genomic inflation factor was calculated based on median chi-squared and analyzed through genomic control method.

5.1 Association Testing

Case-control analysis: Multiple association tests are conducted with respect to all markers in the data set.

5.1.1 Population-Based Association Testing

- Allelic association test.
- Cochran-Armitage trend test.
- Dominant gene action (1 degree of freedom [df]) test.
- Recessive gene action (1 df) test.
- Genotypic (2 df) test.

BaseSpace engine is applicable for determining the maximum test statistics from above-mentioned tests. It estimates the

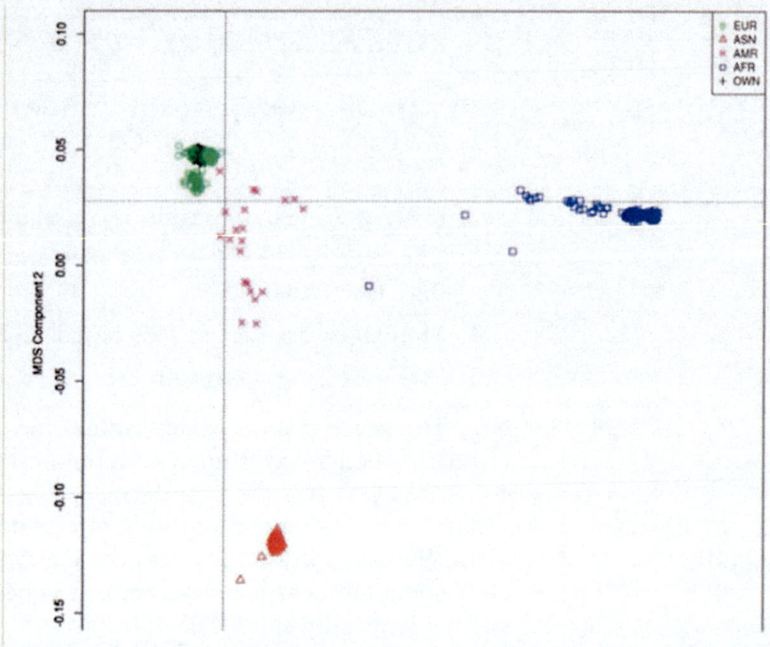

Fig. 4 Multidimensional scaling (MDS) plot of 1KG against the CEU of the HapMap data (which could be seen as your "own" data in this example, as it is being used in the online tutorial at https://github.com/MareesAT/GWA_tutorial/). The black crosses (+ = "OWN") in the upper left part represent the first two MDS components of the individuals in the HapMap sample (the colored symbols represent the 1KG data (○ = European; □ = African; X = Ad Mixed American; Δ = Asian). The MDS components representing the European samples (○) are located in the upper left, the African samples (□) are located in the upper right, the Ad Mixed. American samples (X) are located near the intersection point of the dashed lines, the Asian components (Δ) are located in the lower left part

maximum test statistic from the tests above for a thorough study into various effects: additive, dominant, and recessive.

However, in case stratification information (e.g., geographical location) is present, the Cochran–Mantel–Haenszel test for $2 \times 2 \times K$ stratified tables is employed. Association analysis statistics are documented using Manhattan plots, histograms visualizing the p-value distribution, and Q-Q plots (Figs. 5, 6, and 7).

5.2 Quantitative Analysis

Quantitative traits are analyzed for association using the Wald test.

5.3 Family-Based Association Testing

Family-based association testing for disease traits is performed with the standard transmission disequilibrium test (TDT) association test.

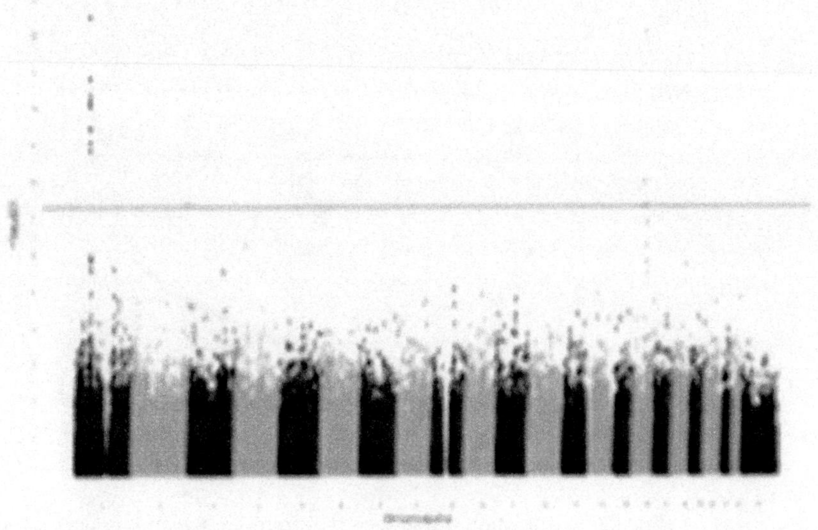

Fig. 5 Association analysis: Manhattan plot. Manhattan plot of $-\log_{10}$ p-values of SNVs from association analyses of case control samples in a GWAS study

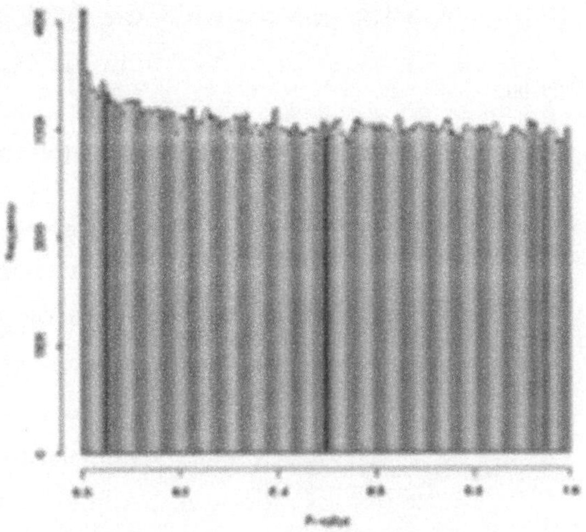

Fig. 6 Association analysis: histogram-histogram showing p-value distribution of SNVs from association analyses of case-control; samples in a GWAS study

5.4 Adjusted Test Statistics

To address the confounding effects of population stratification in family-based association studies, corrections for p-values are conducted using Genomic Control.

To adjust for multiple testing, a false-discovery rate (FDR) analysis is employed. SNVs which do not meet a minimum FDR (default q value $= 0.2$) are disqualified and discarded from the bioset.

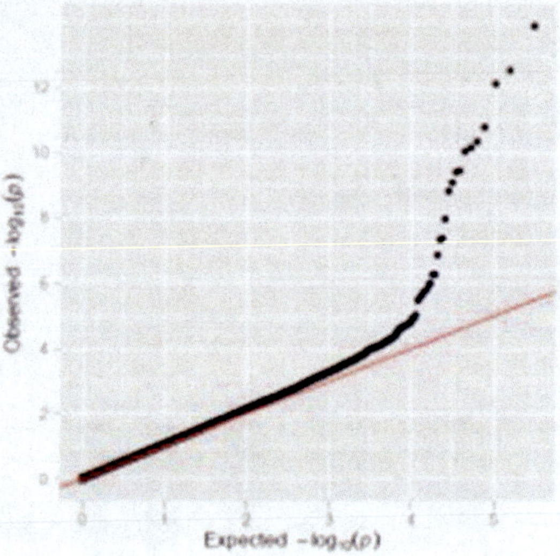

Fig. 7 Association analysis: Q-Q plot-log quantile-quantile (Q-Q) plot of observed vs. expected: $\log_{10}$ *p* values of SNVs from association analyses of case-control samples in a GWAS study

5.5 Constructing a Bioset

5.5.1 From Pre-computed Analyses and Analyses of Genotype-Level or Raw Data

GWAS biosets possess significant SNVs with the following criteria:

- to pass or qualify the QC Filtering criteria as discussed.
- to satisfy a minimum FDR (default *q* value = 0.2).

The following statistics related to each SNV can be included in the bioset:

- Genotype call rates for cases and controls.
- Genotype counts for cases and controls.
- MAF for cases and controls.
- HWE *p*-value for cases and controls, or combined HWE *p*-value.
- Association test statistic.
- Association test *p*-value (used for ranking SNVs within a bioset).
- Odds ratio with confidence intervals (default 95%).
- Risk allele.
- Strand.

6 GWAS Studies in General Consist of Discovery and Replication Phases

1. Discovery Phase: Genome-wide scans of thousands and hundreds of thousands of SNVs are utilized for the identification of SNVs as linked with the phenotype to be studied. SNV biosets obtained from the discovery phase comprises of top-ranked SNVs in the association study, or SNVs with *p*-value of 1×10^{-5} or less.

2. Replication Phase: Individual genotyping technologies are utilized generally for the confirmation of the association, usually in similar population cohorts and rarely in various population cohorts. SNV biosets obtained from the replication phase comprises of top-ranked SNVs in the association study. SNVs with *p*-value less than a cutoff assessed by considering the platform size, when provided, or the number of SNVs as reported:
   (a) If platform size is unavailable or number of SNVs reported < 500, correction factor is set as equal to the number of SNVs reported.
   (b) If platform size is available, set correction factor is equal to the platform size.
   (c) *p*-value cutoff = max $((1 \times 10^{-5}, 0.05)/\text{correction factor} \times 100)$, with the constraint that *p*-value cutoff is ≤ 0.05.
3. **Combined Analysis**: Many articles describe combined analysis of discovery and replication cohorts. A separate bioset for the combined analysis is prepared.

   Top-ranked SNVs satisfy a p-value cutoff as follows:
   (a) Set correction factor as equal to number of SNVs reported.
   (b) *p*-value cutoff = max $((1 \times 10^{-5}, 0.05)/\text{correction factor} \times 100)$, with the constraint that *p*-value cutoff is ≤ 0.05.

An FDR-like analysis is also done where the expected (by chance) and observed SNVs at various *p*-value thresholds (10^{-5}, 10^{-4}, 10^{-3}, 10^{-2}, 0.05) are calculated and observed. The *p*-value cutoff is revised to the highest threshold which satisfies an expected/observed ratio ≤ 0.5, when exceeding, the *p*-value cutoff computed using **steps 1–3**.

BaseSpace Correlation Engine Upload: An upload package into the BaseSpace Engine comprises of:

- A study description document, comprising of informative data about the study design, authors, data collection, platform, and phenotypes under consideration.
- Biosets consisting of significant SNVs with associated statistics. The bioset summary consists of information about the composition, ancestry, number of samples in each cohort, platform employed, and any test conducted to qualify participants (e.g., affected vs. unaffected). It consists of analysis relevant to the generation of association *p*-values, population stratification, imputation, and meta-analysis. In case the researchers have not applied imputation, the NextBio platform calculates estimates of association statistics of untyped markers with reference to the reference population and platform used for the study.
- Bioset tags.

6.1 Imputation

When an SNV bioset is uploaded, the BaseSpace Engine is able to estimate ranks or *p*-values for highly correlated SNVs which do not exist on the platform. As a result, the results of the data set being imported can be extended efficiently to other searchable SNVs in the system. For SNV sets where imputation has not already been performed while processing, this is done as follows:

SNVs lying outside the study platform that are in high linkage disequilibrium (LD) with SNVs in a bioset are assigned a rank (or *p*-value). It is based on the ranks (or *p*-values) of the highly correlated SNVs. Population-specific LD information (r^2) is received from HapMap 3.12. The platform employs the pairwise correlation values with reference to the reference population closest to the population under consideration in the study.

- If LD $r^2 \geq 0.8$, the rank of the imputed SNV is regarded as being estimated with high confidence.
- If LD $r^2 \geq 0.5$ but <0.8, the rank of the imputed SNV is tagged as being estimated with low confidence.

SNVs are mapped to genes as far as practicable, and a gene-set enrichment analysis is conducted for the correlation of the study to all the other BaseSpace Engine content.

6.2 Tagging Biosets with Ontology Terms

The ultimate step in the preparation of data for use in the BaseSpace Correlation Engine is the tagging of biosets with standard ontology terms depicting disease, tissue, treatment, experimental design, and sample source. This step is marked as important and gives a semantic backbone to all the analyzed data. This enables the BaseSpace Correlation Engine to perform additional high-level correlations and also leads to easier ways for users to find relevant data sets.

6.3 Ranking of SNVs

SNVs within a bioset are ranked by *p*-value. The absolute rank value is calculated from a transformation function which enables a more accurate comparison of SNV biosets to biosets from other data types.

Let us consider certain case studies for genome-wide association studies pertaining to different economic traits [1–3].

7 Genome-Wide Association Studies for Candidate Genes for Growth Relevant Traits in Pigs: A Case Study

7.1 Animals and Phenotypes

Pigs under this current analysis were from two distinct farms. Four pure breeds, including Duroc, Yorkshire, Landrace, and Pietrain were taken into consideration. ADG (post-weaning average daily gain) and AGE were measured from 70 to 115 kg, and later adjusted to 100 kg. AGE was adjusted to 100 kg with the following formula:

AGE adjusted to 100 kg = Measured age – [(Measured weight – 100 kg)Correction factor]

where correction factors differ for sire and dam, and the formulas are shown below:

Sire: Correction factor = Measured weightMeasured age*1.826Dam: Correction factor=Measured weightMeasured age*1.715

ADG was calculated by following equation:

ADG adjusted to 100 kg = 100 kg AGE adjusted to 100 kg

Out of total, 31,173 phenotypic observations from farm 1 and 24,374 phenotypic observations from farm 2 were documented. Tail and ear tissue samples were collected and preserved with 75% alcohol and were stored in −20 °C freezers.

7.2 Genotyping, Imputation, and Quality Control

Genomic DNA was isolated from frozen collected ear tissue samples with appropriate technique (Tecan Freedom EVO NGS workstation and TIANGEN magnetic animal tissue genomic DNA kit). DNA samples were screened for concentration ≥40 ng/μL, quantity ≥1 μg, followed by genotyping through gel electrophoresis test. A total of 4865 DNA samples (1595 samples from farm 1 and 3270 samples from farm 2) were subjected to genotyping through Illumina PorcineSNP50 Bead Chip, including 50,697 SNPs, and all SNP markers were mapped to Sus scrofa genome build 11.1 [4]. The data in PLINK binary format is available at https://figshare.com/articles/pig-growth-data_zip/7533020.

Missing genotype data was calculated by FImpute version 2.2 [5]. Data quality control was conducted through PLINK and G2P software before and after imputation [6, 7]. The data was filtered through genotype call rate >0.9, Minor Allele Frequency (MAF) <0.01, and individuals with SNP marker call rate <0.9 and accordingly missing phenotypic records were also removed. The remaining data was employed for the subsequent analysis.

7.3 Estimation of Genetic Parameters

Genetic parameters, such as genetic variance, residual variance, and heritability were measured with Average Information—Restricted Maximum Likelihood algorithm (AI-REML) in an R procedure (Gilmour et al. 1995). The parameters were measured by three methods, such as pedigree-based Best Linear Unbiased Prediction (aBLUP), genomic BLUP (gBLUP), and single-step BLUP (ssBLUP) [8–10]. All BLUP models can be written as:

y=Xb + Zu + ey=Xb + Zu + e

where ***y*** is a vector of phenotypic observations; ***b*** and ***u*** represent fixed effects and breeding values, respectively; ***X*** and ***Z*** were design matrices for ***b*** and ***u***, respectively; ***e*** represents the residual error vector with a normal distribution of e~N(0,Iσ2e)e~N(0, Iσe2), where ***I*** was an identity matrix and σ2e σe2 is residual variance. In a BLUP model, u~N(0,Aσ2u)u~N(0,Aσu2), in which σ2uσu2 is additive genetic variance and ***A*** is an additive genetic relationship matrix that derived from pedigree records; In gBLUP model, ***A*** is replaced by ***G*** and u~N(0,Gσ2u)u~N(0,Gσu2), ***G*** is derived from genomic information and can be constructed by VanRaden method:

G = M′M2 Σmi=1pi(1−pi)G = M′M2 Σi=1mpi(1−pi)

where ***M*** is a ***m*** × ***n*** standardized genotype matrix, ***m*** is the marker size and ***n*** is the number of genotyped individuals, p_i is the minor allele frequency of the *i*th genetic marker; In ssBLUP method, u~(0,Hσ2u)u~(0,Hσu2) and the relationship matrix ***H*** is derived from both pedigree records and genomic information simultaneously using the following equation:

H = (| | A11−A12A−122A21+A12A−122GA−122A21GA−122A21A12A−122GG+αA22) | | H = (A11-A12A22-1A21+A12A22-1GA22-1A21A12A22-1GGA22-1A21G+αA22)

Individuals are assigned to different groups based on available information. The group with footer *1* is designated as the individuals which only have pedigree information. The group with footer *2* is designated as the individuals having both pedigree and genomic information. The A_{11} and A_{22} represent the relationship among individuals within group *1* and group *2*, respectively, A_{12} represents the relationship among individuals between group *1* and group *2* and A_{21} is the transpose of A_{12}, $\boldsymbol{\alpha}$ is the ratio for combing ***G*** and A_{22} matrix and set to 0.05 in this study.

7.4 Genome-Wide Association Study

Association tests were conducted by a multi loci model, FarmCPU [11]. FarmCPU model iteratively utilizes fixed effect model and random effect model. The top three identified columns of principal components such as, sex, farm effects, and pseudo QTNs (Quantitative Trait Nucleotides) were employed as covariates in the fixed effect model for association tests and the model can be expressed as:

y=Pbp+ Mtbt+Sjdj+ey=Pbp+ Mtbt+Sjdj+e

where, ***y*** is phenotypic observation vector; ***P*** is a matrix of fixed effects, including top three principal components, sex, and farm effects; $\boldsymbol{M_t}$ is the genotype matrix of ***t*** pseudo QTNs which are employed as fixed effects; $\boldsymbol{b_p}$ and $\boldsymbol{b_t}$ are the relevant design matrices for ***P*** and $\boldsymbol{M_t}$, respectively. $\boldsymbol{S_i}$ is the *i*th marker to be tested and $\boldsymbol{d_j}$ is the corresponding effect; ***e*** is the residual effect vector and e~N(0, Iσ2e)e~N(0,Iσe2). Random effect model is employed for the selection of the most appropriate pseudo QTNs. The model is written as:

y=u+ey

where ***y*** and ***e*** stay the same as in fixed effect model; ***u*** is the genetic effect and u~N(0,Kσ2u)u~N(0,Kσu2), in which ***K*** is the relationship matrix that defined by pseudo QTNs.

The Bonferroni correction threshold for multiple tests was employed for detecting the genome-wide significant SNPs, which defined as α/K ($\alpha = 0.05$ and K is the number of SNPs) [12].

7.5 Variance Explained by Candidate Regions

Candidate region is designated as the genomic region located within 1 Mb upstream and downstream of the genome-wide significant SNPs. The proportion of genetic variance and phenotypic variance depicted by candidate regions are measured in a mixed linear model with multiple random effects. The model can be written as:

y=Xb + Ziui + ey=Xb + Ziui + e

where *y* is a vector of phenotypic observations, *b* and *u* represents fixed effects and breeding values, respectively. *X* and *Z* are the design matrices for *b* and *u* respectively. *e* represents the residual error vector with a normal distribution of *e*~N(0, I2e), where I is an identity matrix and 2*e* is a residual variance

In the current study, k_1 is a relationship matrix which is defined by all SNPs within the candidate regions and k_2 is constructed by the rest of SNPs. The ratio between σ2u1σu12 and the sum of σ2u1σu12 and σ2u2σu22 is defined as the proportion of genetic variance explained by candidate regions. The proportion of phenotypic variance explained by candidate regions can be measured as the ratio between σ2u1σu12 and the phenotypic variance. In addition, the variance explained by randomly selected genomic regions was designated as null distribution and compared with the variance explained by the candidate regions.

7.6 Identification of Candidate Genes

The candidate genes adjacent to the genome-wide significant SNPs were detected by Ensembl database with the gene annotation information of Sus scrofa genome version v11.1 (www.ensembl.org/biomart/). The genomic locations were downloaded from https://www.animalgenome.org/pig/ and genes within the candidate region are considered as candidate genes.

8 Another Case Study: Genome-Wide Association Studies for Candidate Genes for Growth Relevant Traits in Pigs

8.1 Animals and Traits

The study comprises of gilts (n = 820) which were commercial breeding stock. The gilts were derived from two genetic lines including 412 Large White line pigs and 408 pigs from a Large White × Landrace cross. Traits considered under present study include body weight, body composition and structural soundness evaluations. Females have an average body weight of 124 ± 11 kg and age of 190 ± 7 days. Body composition traits consists of ultrasonically measured loin muscle area at the tenth rib, tenth rib backfat, and last rib backfat. Body structural soundness traits (total 17 traits) were measured including body conformation (body length, depth and width, top line, rib shape, and hip structure), front feet and leg structure (legs turned, buck knees, pastern posture, foot size, and uneven toes), rear feet and leg structure (legs turned, weak/upright legs, pastern posture, foot size, and uneven toes) and overall leg action. This is a general observation indicating both structural soundness and freedom of other defects affecting the gait and locomotion [13].

The structural soundness traits were evaluated by two experienced scorers employing nine-point scales separately. In this case, one and nine indicated the extreme phenotypes of the traits. The intermediate score is the most desirable for four of the scoring traits along with top line, turned front legs, turned rear legs, and weak/upright rear legs. We can briefly discuss the main steps as follows:

1. SNP Array Genotyping.

   This includes DNA isolation followed by quantification.

   The DNA was isolated from dry ear tissue with the DNeasy 96 Blood & Tissue Kit (Qiagen, Valencia, CA, USA). DNA quantification was obtained using a NanoDrop 1000 (Thermo Fisher Scientific Inc., Wilmington, DE, USA). Porcine 60K BeadChip (Illumina, San Deigo, CA, USA) was employed.
2. Phenotypic Statistics.

   The statistical analyses for descriptive analyses involve population mean. Normal distribution testing was performed using the UNIVARIATE procedure of the SAS software package (SAS Institute, release 9.2, Cary, NC, USA).
3. Population Stratification Analysis.

   The animals in the study were derived from two genetic lines. They are originated from Large White or Large White × Landrace interbreed crossing. Population stratification was analyzed with an identical-by-state (IBS) distance clustering method in PLINK program.
4. Bayesian Methods for GWA Analyses.

   New Bayesian method for genomic selection (Bayes C) is preferred, developed from the Bayes B and GBLUP approaches.
5. Bootstrap Analysis.
6. Haplotype Block and Association Analysis.
7. Gene Ontology.

9 Study Design: Case Study

We used a three-stage study design. The discovery GWAS stage included 556,134 autosomal SNPs in 542 cases and 1050 controls of Han Chinese origin. The replication I stage contained 33 promising SNPs in 665 cases and 864 controls; the replication II stage included 17 (out of the 33) SNPs in 638 cases and 1863 controls. We also performed fine mapping around the newly identified susceptibility locus by genotyping tag SNPs and performing imputation. In fine mapping, we used 600 cases and 1100 controls (all Han Chinese), of which 496 cases and 530 controls were case-control samples at the discovery GWAS stage and 104 cases and 570 controls were newly collected. Collecting study samples and genotyping in GWAS.

All the cases were included through cooperation of 40 centers in China. All cases were diagnosed with primary Sjögren's syndrome by at least two rheumatologists as per the American-European Consensus Group (AECG) criteria for primary Sjögren's syndrome 1 and devoid of other autoimmune disease. All controls were included from the Peking Union Medical College Hospital as per the following rules:

1. no significant history of disease.
2. no family history of rheumatologic diseases.
3. normal biochemical and immunological profile.
4. negative serology of anti-Ro (SSA) and anti-La (SSB).

All the ethical and biosafety formalities have been completed in terms of informed consent given by each individual at recruitment. All case and control individuals included in our study, anti-Ro (SSA)/anti-La (SSB) antibodies were detected by ELISA (EUROIMMUN AG, Germany). Genomic DNA was extracted from peripheral blood samples using the whole blood DNA extraction kit (solution type) (BioTeke, China). Genome-wide genotyping analysis was performed with Affymetrix AxiomTM Genome-Wide CHB 1 Array Plate (Affymetrix) with 642,832 SNPs at CapitalBio Co. Ltd. Genome-wide genotypes were called by the Axiom TM GT1 algorithm.

10 Quality Control (QC) in GWAS

10.1 QC for Individuals (Including Population Stratification)

A systemic quality control (QC) was performed for individuals and SNPs. The step-wise QC for individuals at the discovery stage (597 cases and 1090 controls) was to remove individuals with:

(1) Affymetrix Dish QC (DQC) <0.82 (following Affymetrix's data analysis guideline) ($n = 0$), (2) per individual call rate <95% ($n = 19$), (3) per individual autosomal heterozygosity >5 S.D. away from mean 2 ($n = 14$), (4) wrong assigned gender ($n = 11$), (5) one of a pair of individuals with cryptic relationship closer than 3rd-degree (proportion identity by descent (IBD) PI_HAT >0.125) or unexpected duplication, i.e., identity >98% (the one with lower call rate was removed in this case) ($n = 39$), and (6) individuals without age information ($n = 12$). After applying these QC steps, 542 cases and 1050 controls remained.

The remaining samples were evaluated for population stratification using principal component analysis (PCA) implemented in EIGENS OFT4.2 3-4 with 209 HapMap subjects as reference panel (60 Utah residents of Northern and Western European descent (CEU), 60 Yoruba from Ibadan (YRI), 45 CHB and 44 JPT) 5. The genotyped autosomal SNPs, which were in low LD (MAF >0.35 and $r^2 < 0.05$ for each pair of SNPs) 6–7 and absent from the five long-range LD regions 8, were included in PCA ($n = 22{,}112$). The format of EIGENSOFT output (eigenvectors, i.e., principal components) was adapted to R 2.8.1 (http://www.r-project.org/) for plotting, and the first two eigenvectors were plotted. Population outlier is defined as individual with >6 S.D. from the mean for any of the top ten eigenvectors.

Results indicated that our samples were of East Asian ancestry and no population outlier was present. Population stratification was also identified by only using the case-control data, and the Tracy–Widom test was employed to detect significant eigenvectors ($p < 0.05$). It turned out that the first eigenvector (eigenvector 1) was significant, which was used as one of the covariates in the subsequent statistical analysis. Summary of the QC-passed individuals is shown in Supplementary Table 1.

10.2 QC for SNPs

After QC for individuals, QC for SNPs was performed to remove SNPs with: (1) per SNP call rate <98% either in cases or controls, (2) Hardy–Weinberg equilibrium (HWE) test p-value < 0.001 for controls, (3) minor allele frequency (MAF) <1% either in cases or controls, (4) significantly different call rates between cases and controls ($p < 1 \times 10^{-5}$, PLINK command—test-missing) 9, and (5) not mapped to autosomes. In total, there were 556,134 SNPs remaining after QC. SNP selection for replication and genotyping in replication stages.

10.3 SNP Selection for Replication I Stage

For SNPs in the non-MHC regions with $p < 5 \times 10^{-5}$, we grouped SNPs based on their LD information and physical location. Any two SNPs with $r^2 > 0.3$ (LD was calculated within a distance of 1 Mb), along with the SNPs located between them, were put into the same group, in such a manner that each group reflect an approximately independent region. For each region, the SNP with the lowest p-value was selected for replication ($n = 23$). For SNPs in MHC region with $p < 5 \times 10^{-8}$, we selected the representative SNPs obtained by conditional logistic regression analysis ($n = 2$).

Besides these SNPs, additional non-MHC SNPs that indicated an association signal with primary Sjögren's syndrome ($p < 5 \times 10^{-2}$) and which had been identified by GWAS (based on NHGRI GWAS Catalog, http://www.genome.gov/gwastudies/) to be associated with rheumatoid arthritis (RA), systemic lupus erythematosus (SLE), systemic sclerosis (SSc), or primary biliary cirrhosis (PBC) were also selected for replication ($n = 8$). We removed the SNPs with problem cluster plots (analyzed by Affymetrix Genotyping Console). In total, 33 SNPs were selected for the replication I stage.

10.4 Genotyping in Replication Stages and Platform Validation

The genotyping at replication stages (I + II) was performed by the iPLEX MassARRAY platform (Sequenom). To validate the concordance between Affymetrix platform and Sequenom platform, we randomly selected 120 QC-passed individuals (60 cases and 60 controls) genotyped at the discovery stage, and re-genotyped the 33 SNPs selected for replication I using MassARRAY. SNPs with per SNP concordance <98% were excluded from further analyses.

10.5 SNP Selection for Replication II Stage

The SNPs which passed platform concordance validation and QC (see below for details) in replication I, as well as with the same association directions in discovery GWAS and replication I stage, were selected for replication II stage (18 out of 33 SNPs).

11 QC in Replication Studies

For quality control (QC) at replication stages I and II, we excluded SNPs with (1) call rate < 95% either in cases or controls, (2) Hardy–Weinberg equilibrium (HWE) test p-value < 0.001 in controls, (3) MAF < 1% either in cases or controls, and (4) significantly different call rates between cases and controls ($p < 1 \times 10^{-5}$, PLINK command—test-missing). After QC on genotype data in replication II, 17 SNPs were left for a combined analysis.

12 Statistical Analysis

The association test of SNPs with primary Sjögren's syndrome risk in the GWAS, replication stages and combined analysis was carried out by PLINK v1.07 11 with an additive model (minor allele count) in logistic regression. At the discovery stage, SNP-trait association p-values were calculated with gender, age, and eigenvector 1 of PCA as covariates. The Manhattan plot of genome-wide $-\log_{10} p$ was drawn by HaploView v4.1 12, and the quantile-quantile (Q-Q) plot of genome-wide $-\log_{10} p$ was generated by R 2.8.1 (http://www.r-project.org/). The genomic control inflation factor λ was calculated by R. Conditional logistic regression (PLINK command—condition) was used to detect independent association signals in the MHC region and residual association in the three non-MHC loci by estimating SNP p-values conditioned upon the most significant SNP. Then all independent SNPs were put into one regression model with adjustment for gender, age, and PCA-based correction to estimate ORs and p-values. The plots of genotype frequency variations between cases and controls of the genome-wide significant SNPs were also drawn by R. Association analyses at the replication stages were implemented with gender and age as covariates, and in combined analysis, besides gender and age, eigenvector 1 and indicator of stage were used as covariates (eigenvector 1 was set to zero for the data of replication I and II).

Meta-analysis was performed by PLINK for the SNPs with genome-wide significance, where Cochran's Q test p-value and heterogeneity index I2 were calculated to test stage heterogeneity. Epistasis (PLINK) and interaction test by partitioning chi-squares were both performed to identify SNPs interacting with the three non-MHC genome-wide significant SNPs (p interaction $<5 \times 10^{-8}$). For pairwise interaction of the three SNPs, all p interaction exceeded 5×10^{-2} after Bonferroni correction for the number of three SNP pairs.

All association test p-values were two-sided and are reported without multiple testing corrections. Genome-wide significance was considered as a p-value $<5 \times 10^{-8}$ as generally used in GWAS.

13 Imputation and Regional Association Plot

We utilized MACH 1.0 to impute non-genotyped SNPs with the CHB data (97 individuals) of the 1000 Genomes Integrated Phase 1 Release 15 as reference panel. Only the imputed SNPs with squared correlation between imputed and true genotypes (Rsq) >0.5 were kept and imputation result (dosage data) was analyzed by mach2dat. Regional results of association were plotted by LocusZoom (stand-alone version).

14 Fine Mapping

We started from the newly identified locus and its downstream gene to detect the exact region for fine mapping. The initial portion and its overlapping LD blocks represent a 220-kb region (74.00–74.22 Mb). To remedy the low coverage of the genome-wide SNP chip in the region, we later extended the 220-kb region to a 560-kb region (73.95–74.51 Mb), which includes all LD blocks possessing SNPs that have an $r^2 > 0.3$ with any of the SNPs in the 220-kb region. Haploview was utilized to select tag SNPs with MAF >0.05, by using Paul de Bakker's Tagger algorithm with threshold $r^2 > 0.9$ based on CHB data of the 1000 Genomes Integrated Phase 1 Release. SNPs with primer design problem were excluded. Genotyping was conducted by the iPLEX MassArray platform (Sequenom). QC for SNPs is similar as that used in replication stages. The results of fine mapping confirmed the association of the new locus and further extended the region of the new locus.

15 Additional Check for Genotyping Quality

Genome-wide significant SNPs were detected through genome wide association studies through a process of three-stage GWAS and by fine mapping. We checked their cluster plots and intensity plots of different stages and detect MAF across plates. The plots are very prominent and the average of MAF through plates is near to the MAF of all plates, with minimal standard deviation.

16 Expression Quantitative Trait Loci (eQTL) Analysis

The eQTL analysis was conducted through Genevar (GENe Expression VARiation) 3.2.0 17.

Spearman's rank correlation coefficient (rho) was explored to detect SNP-gene expression association for the genome-wide significant Single Nucleotide Polymorphisms and their corresponding/nearby (±1 Mb from the SNP) genes based on the gene expression data of lymphoblastoid cell lines from HapMap3 CHB individuals. Accordingly, the corresponding genotypes of these individuals in 1000 Genomes Integrated Phase 1 Release was detected. The p-value was estimated from 10,000 permutation samples and the threshold of the p-value was regarded as 0.001.

References

1. Marees AT, de Kluiver H, Stringer S, Vorspan F, Curis E, Cynthia Marie-Claire C, Derks EM (2018) A tutorial on conducting genome-wide association studies: quality control and statistical analysis. Int J Methods Psychiatr Res 27(2):e1608. https://doi.org/10.1002/mpr.1608
2. Chang M, He L, Cai L (2018) An overview of Genome-Wide Association studies. In: Huang T (ed) Computational systems biology. Methods in molecular biology, vol 1754. Humana Press, New York, NY
3. Tosser-Klopp G, Bardou P, Bouchez O, Cabau C, Crooijmans R et al (2014) Design and characterization of a 52K SNP chip for goats. PLoS One 9(1):e86227. https://doi.org/10.1371/journal.pone.0086227
4. Ramos AM, Crooijmans RPMA, Affara NA, Amaral AJ, Archibald AL, Beever JE et al (2009) Design of a high density SNP genotyping assay in the pig using SNPs identified and characterized by next generation sequencing technology. PLoS ONE 4(8):e6524. https://doi.org/10.1371/journal.pone.0006524
5. Sargolzaei M, Chesnais JP, Schenkel FS (2014) A new approach for efficient genotype imputation using information from relatives. BMC Genomics 15:478. https://doi.org/10.1186/1471-2164-15-478
6. Chang CC, Chow CC, Tellier LC, Vattikuti S, Purcell SM, Lee JJ (2015) Second-generation PLINK: rising to the challenge of larger and richer datasets. Gigascience 4:7. https://doi.org/10.1186/s13742-015-0047-8. PMID: 25722852; PMCID: PMC4342193
7. Tang Y, Liu X (2019) G2P: a genome-wide-association-study simulation tool for genotype simulation, phenotype simulation and power

evaluation. Bioinformatics. 35 (19):3852–3854. https://doi.org/10.1093/bioinformatics/btz126. PMID: 30848784.

8. Henderson CR (1975) Best linear unbiased estimation and prediction under a selection model. Biometrics 31(2):423–447. PMID: 1174616

9. VanRaden PM (2008) Efficient methods to compute genomic predictions. J Dairy Sci 91 (11):4414–4423. https://doi.org/10.3168/jds.2007-0980. PMID: 18946147

10. Aguilar I, Misztal I, Johnson DL, Legarra A, Tsuruta S, Lawlor TJ (2010) Hot topic: 636 a unified approach to utilize phenotypic, full pedigree, and genomic information for genetic 637 evaluation of Holstein final score. J Dairy Sci 93:743–752

11. Liu X, Huang M, Fan B, Buckler ES, Zhang Z (2016) Iterative usage of fixed and random effect models for powerful and efficient genome-wide association studies. PLoS Genet 12(2):e1005767. https://doi.org/10.1371/journal.pgen.1005767

12. Nicodemus KK, Liu W, Chase GA, Tsai YY et al (2005) Comparison of type I error for multiple test corrections in large single-nucleotide polymorphism studies using principal components versus haplotype blocking algorithms. BMC Genet. 6(Suppl 1):S78

13. Tang Z, Xu J, Yin L, Yin D, Zhu M, Yu M, Li X, Zhao S, Liu X (2019) Genome-wide association study reveals candidate genes for growth relevant traits in pigs. Front Genet 10:302. https://doi.org/10.3389/fgene.2019.00302

Chapter 17

Exome Sequencing

Abstract

Exome sequencing, also known as whole exome sequencing (WES or WXS), is a technique for sequencing all the expressed genes in a genome (known as the exome). It has a major advantage over whole genome sequencing since exon or coding region is very less 1–2% of total genome, hence very less sequencing is required and it saves cost, time, and effort of scientists. Outline of the protocol for exome sequencing involves the following steps: DNA samples should be fragmented and biotinylated, oligonucleotide probes (baits) are to be used to selectively hybridize to target regions in the genome. Magnetic streptavidin beads are used to bind to the biotinylated probes. The nontargeted portion of the genome is to be washed away, and the polymerase chain reaction (PCR) is used to amplify the sample, enriching the sample for DNA from the target region. The sample should be sequenced before proceeding to bioinformatic analysis. Array-based methods are similar except that the probes are bound to a high-density microarray. Some commonly employed exome kits were described. A brief idea for data analysis pertaining to exome sequencing has been depicted. Protocol was described with practical examples of case studies.

Keywords Exome, Biotinylated probes, Exome kit, Magnetic streptavidin beads, Molecular inversion probe, Hybrid capture

1 Principle

It is the methodology of first selecting only the subset of coding region DNA which encodes proteins (known as exons) followed by sequencing of DNA using any high throughput DNA sequencing technology.

It has a major benefit over whole genome sequencing since exon or coding region is very less 1–2% of total genome, hence very less sequencing is needed and it saves cost, time, and effort of scientists.

Exome sequencing is also referred to as whole exome sequencing (WES or WXS). It is a technology for sequencing all the expressed genes in a genome (the portion of gene which is expressed is referred to as the exome).

Aruna Pal, *Protocols in Advanced Genomics and Allied Techniques*, Springer Protocols Handbooks,
https://doi.org/10.1007/978-1-0716-1818-9_17,

2 Protocol

2.1 Outline

1. DNA samples should be fragmented and biotinylated, oligonucleotide probes (baits) are to be used to selectively hybridize to target regions in the genome.
2. Magnetic streptavidin beads are employed to bind to the biotinylated probes.
3. The nontargeted portion of the genome is to be washed away, and the polymerase chain reaction (PCR) is employed to amplify the sample, enriching the sample for DNA from the target region.
4. The sample should be sequenced prior to bioinformatic analysis.
5. Array-based methods are similar; the only exception is that the probes are bound to a high-density microarray.

3 Protocol

1. **Target-Enrichment Strategies**: Target-enrichment methods permit one to selectively capture genomic regions of interest from a DNA sample before sequencing.
2. (a) **PCR.**

   Multiplex PCR is employed.

   (b) **Molecular Inversion Probes (MIP).**

   Molecular inversion probe utilizes probes of single-stranded DNA oligonucleotides flanked by target-specific ends. The gaps between the flanking sequences are filled and ligated forming a circular DNA fragment. Probes which did not undergo reaction remain linear and are removed with exonucleases.

   (c) **Hybrid Capture.**

   Microarrays possess single-stranded oligonucleotides with the sequences from the human genome to tile the region of interest fixed to the surface. Genomic DNA is sheared to form double-stranded fragments. The fragments go through end-repair to produce blunt ends and adaptors with universal priming sequences being added. These fragments are hybridized to oligos on the microarray. Unhybridized fragments are then removed through the process and washed away and the desired fragments are eluted. The fragments are then amplified using PCR.

   (d) **In-Solution Capture.**

   In order to capture genomic regions of interest using in-solution capture, a group of custom oligonucleotides (probes) is synthesized followed by hybridization in solution to a fragmented genomic DNA sample. The probes (labeled with beads) selectively hybridize to the genomic regions of interest followed by pulling down of the beads (now including the DNA fragments of interest) and washing to clear excess material. Afterward, the beads are removed and the genomic fragments can be sequenced leading to the selective DNA sequencing of genomic regions (e.g., exons) of interest.

4 Sequencing

Several sequencing platforms are available including the classical Sanger sequencing. Other platforms include the Roche 454 sequencer, the Illumina Genome Analyzer II and the Life Technologies SOLiD & IonTorrent all of which have been used for exome sequencing.

5 Exome Kit Descriptions and Protocol Overview

5.1 Infinium Omni5Exome-4 Kit (Illumina)

The BeadChip array covers >4.3 million whole-genome variants down to 1% minor allele frequency (MAF). It can equally detect novel functional exonic variants.

5.2 Infinium OmniExpressExome-8 Kit

It can discover novel trait and disease associations with optimized tag SNPs and functional exonic content.

5.3 Infinium Exome-24 Kit Arrays

This delivers exceptional coverage of putative functional exonic variants presenting diverse populations and a range of common conditions.

5.4 Infinium Omni2.5Exome-8 Kit

It imparts comprehensive coverage and functional exonic content for next-generation genotyping, GWAS, and CNV analysis.

5.5 Agilent HaloPlex

The HaloPlex Exome Kit consists of 2.5 million probes designed to cover human coding regions. The kit is designed in such a way to target smaller capture regions quickly. The target size is 37 MB. It is designed for 21,522 genes and targets 357,999 exons. The complete work flow requires 1.5 days and need 250 ng of input DNA.

5.5.1 Protocol Overview

1. The procedure starts with a genomic DNA digest with restriction enzymes.
2. Hybridization of HaloPlex probe library to DNA digests. Hybridization aids in genomic DNA fragment circularization and addition of Illumina indexes and flow cell binding motifs.

3. DNA probe hybrids are captured with streptavidin-coated magnetic beads binding to biotinylated probe DNA.
4. Targeted fragments are amplified, leading to the production of a sequence-ready, target-enriched library.

5.6 Agilent SureSelect

The Agilent SureSelect Exome kit consists of an in-solution capture method that utilizes long 120 mer, biotinylated cRNA baits for enriching exome regions from genomic DNA fragments.

5.6.1 Protocol Overview

1. Starting with genomic DNA, samples are sheared leading to small DNA fragments.
2. Libraries are prepared with the aid of Illumina compatible adapters and indices.
3. Biotinylated cRNA baits are incubated with the library for a period of 16 h.
4. Targeted regions are selected with magnetic streptavidin beads.
5. Targeted regions are amplified, leading to the production of a sequence-ready library.

5.7 Agilent SureSelect QXT

The SureSelect QXT kit is a combination of a transposase-based library preparation method with SureSelect's well-characterized target capture system. Enrichment probes are 120 nt long baits which allow the capture of exomes, gene panels, or custom targets. The greatest advantage of QXT is the requirement of less hybridization time, only 90 min. The protocol is similar to SureSelect; the difference lies in requirement of starting material (50 ng) and time (takes less than 1 day to complete).

5.8 IDT xGEN Exome

The IDT xGEN Exome panel comprises of 429,826 biotinylated probes designed to capture 19,396 genes (a 39 Mb region). The probes are synthesized individually and quality controlled as opposed to array synthesized probes, resulting in reduction of truncations in the pools. The protocol needs at least 500 ng of a constructed library. With this technique, hybridization times are relatively short (4 h) compared to that of competitor exome panels. The xGEN panel has the ability to target coding sequences (CDS) in the RefSeq database.

5.8.1 Protocol Overview

1. Blocking oligos are synthesized, combined with a DNA library followed by drying.
2. 5′ biotinylated capture probes are added followed by hybridization to the library for 4 h.
3. Streptavidin beads are prepared followed by addition to the hybridized targets.
4. Unbound DNA is washed off.

5. Remaining fragments are amplified leading to the enrichment of the targeted regions.

5.9 Illumina Nextera Rapid Capture Exome

Nextera exome kits come in two different panels:

1. Nextera Rapid Capture Exome: It hybridizes to 45 Mb of targeted coding sequence.
2. Nextera Rapid Capture Expanded Exome: The expanded panel delivers 62 Mb of exons, untranslated regions (UTRs), and miRNA.

The entire process of library prep and hybridization capture needs only 50 ng of genomic DNA and usually requires 5 h to complete. It makes it a user-friendly exome capture kit.

5.9.1 Protocol Overview

1. Libraries are prepared from small amount as 50 ng of genomic material with Nextera's transposase-based chemistry.
2. DNA libraries are denatured.
3. Denatured single-stranded DNA is hybridized to biotinylated probes designated to regions on the exome.
4. Fragments are enriched with streptavidin beads followed by elution.
5. Fragments are amplified, leading to the production of sequence-ready enriched targets.

5.10 Illumina TruSeq Exome

The TruSeq Exome Enrichment Kits are designed for isolating human exon regions. It is an in-solution sequence capture system. Kits include 340,427,95 mer probes designed against the human NCBI37/hg 19 reference genome. The probe set is designed such as to enrich 201,121 exons spanning 20,794 genes of interest. The kit has the ability to cover 64 Mb of the human genome, with each 95 mer probe targeting libraries that are 300–400 bp. A speciality of the kits is that other than major exon regions, the kit covers non-coding DNA in exon flanking regions, such as UTRs and promoters.

5.10.1 Protocol Overview

1. The workflow starts with the creation of pooled indexed libraries from up to six samples.
2. Sample libraries are denatured into single-stranded DNA followed by hybridization to 95 mer biotin labeled probes.
3. In the next step, pools are enriched with streptavidin magnetic beads and pulled from solution with a magnet.
4. Enriched DNA fragments are eluted from the beads followed by hybridization for a second enrichment reaction.
5. Fragments are amplified leading to the production of sequence-ready enriched targets.

5.11 Roche Nimblegen SeqCap

SeqCap EZ Exome Library Kits has the ability for enrichment of the entire human exome. It utilizes an in-solution-based capture method. The kit has a target of 64 Mb of the human genome with 2.1 million long oligonucleotides.

5.11.1 Protocol Overview

1. The protocol starts with genomic DNA library preparation.
2. The library is hybridized to the SeqCap oligo pool.
3. The sequence library is being hybridized to the oligo pool, then the magnetic beads are employed to pull down the captured genomic DNA fragments.
4. Enriched DNA fragments are eluted from the beads followed by hybridization for a second enrichment reaction.
5. Unbound fragments are washed off and thereby cause the amplification of enriched fragment pool, leading to the production of sequence-ready enriched targets.

5.12 MYcroarray MYbaits

MYbaits are in-solution DNA capture system. The unique feature is that it is fully customizable. It utilizes custom biotinylated RNA baits. Baitsets are priced in modules of 20,000 unique bait sequences, which are particularly affordable for designs targeting hundreds or thousands of loci. The advantage of MYbaits kits is that it is compatible with any type of barcoded NGS library.

5.12.1 Protocol Overview

1. A genomic DNA library is heat denatured followed by hybridization to the RNA baits, usually overnight (or longer for degraded or very rare targets).
2. Followed by hybridization, the biotinylated baits bound to captured genomic DNA are pulled out of solution with streptavidin-coated magnetic beads.
3. Non-specifically bound DNA is washed away followed by capturing of DNA eluted from the RNA baits with heat.
4. Post-captured DNA is amplified, leading to the production of sequence-ready enriched targets.

6 Factors Affecting Capture Efficiency

There are some basic differences in technique of sequencing from general PCR amplification involved in Sanger sequencing. In PCR amplification two primers are needed for annealing near a target in the genome followed by amplification with specificity. But exon capture requires hybridization of a single probe, generally attached to a magnetic bead. After hybridization, we need to rely on the pulldown of that target, followed by amplification. There is significantly more variability with hybridization capture, resulting in on-target rates which are typically no better than 75%. Other factors affecting efficiency of capture are as follows:

1. *GC-rich regions*—UTRs and promoter regions are typically very GC rich. So sometimes it may lower capture efficiency and increases variability between these regions and other more balanced ones.
2. *Quality of DNA*—Poor quality DNA, that with extractions from FFPE, can cause bias as certain regions were observed to be more fragmented than others. In case capture is not balanced, it causes bias and complications during downstream SNP calling and other forms of analysis.
3. *Quantity of DNA*—Low input amounts of DNA generally require more number PCR cycles for getting enough library for the process of hybridization of capture probes to be efficient. Higher PCR cycles lead to a significant amount of PCR duplicates, which make conclusions from data analysis less informative. Usually, 1 ug of a reasonably diverse library was required to capture. Now, newer capture protocols require only 50 ng.
4. *Pseudogenes*—Can decrease evenness of coverage.
5. Fragment or insert size—The inserts necessary from the kits listed above usually vary based on the size of the probes being employed in capture. It is important to note that fragmentation of DNA be analyzed to meet those size requirements for efficient capture.
6. *Repeat elements*—It will decrease the evenness with which reads are distributed across the exome. This results in the need for more sequencing to call de novo SNPs.

It is important to remember that variability in capture efficiency and coverage is inherent in exome sequencing with every kit we discussed. Many of these variables are controllable. In order to increase sequence depth, breath, and evenness of coverage, de novo variants are a lot easier to call.

In subsequent sections, we highlight the calculation of the amount of sequencing necessary for exome-seq project detailing the capture kits commercially available.

7 Calculation of the Amount of Sequencing Needed for Exome Analysis Study: Calculate On-Target Rate or Enrichment Efficiency

Enrichment efficiency (EE) may be calculated as Passing Filter (PF) reads mapped to target divided by Total number of PF reads mapped to reference (TNPF).

$EE = PF/TNPF.$

In case of use of the correct blocking oligos, this rate usually ranges from 0.65 to 0.75.

8 Calculation of Mean Normalized Coverage

Mean normalized coverage means how much sequencing is necessary to yield a given percentage of targeted bases at a particular read depth.

Normalized Coverage (NC) is calculated as Coverage at each base position (CBP) divided by Average Coverage over all base positions (ACBP).

NC = CBP/ACBP.

9 Exome Sequencing Data Requirement

The first step is the identification of the mean sequencing coverage required.

Mean Sequencing Coverage (MSC) is defined as *Desired Coverage* (DC) divided by *Normalized Coverage* (NC).

MSC = (DC)/(NC).

Suppose our desired coverage is 20× and mean normalized coverage is 0.2. The mean sequencing coverage would be: 20×/0.2 = 100×.

10 Determination of the Amount of Sequencing Needed to Meet Coverage Requirement

Required Number of PF Mapped Sequence (RNPFMS) = *Targeted Bases* (TB) × *Mean Sequencing Coverage* (MSC)/*Enrichment Efficiency* (EE).

Let us discuss the case of using the Roche Nimblegen SeqCap EZ Exome v3.0 kit, with a targeted region of 64 Mb. The mean sequencing coverage that we need, calculated above is 100×, and let us assume our on-target rate or enrichment efficiency is ~0.70.

So, 64 Mb × 100/0.70 = 9.1 Gb of passing filter and mapped sequence.

11 Recommended Read Length for Exome Study

While not an absolute necessity, we generally recommend paired-end 2 × 100 read lengths for exome capture sequencing. ~80% of exons are <200 bp in length [9]. Hence a 2 × 100 read should be the ideal for most experiments.

12 Basic Recommendations for Calling Variants and Analyzing Depth of Coverage

12.1 Exome Variant Analysis Recommendations

1. Trim adapter sequences using cut-adapt.
   (a) Input files: FASTA, FASTQ.
   (b) Output files: same as input.
2. Align reads to a human genome build with BWA, post-process data with SAMtools, removal of duplicate reads followed by conversion to a BAM file using Picard.
   (a) Input files: FASTA, FASTQ.
   (b) Output files: BAM.
3. Call SNPs using GATK Haplotype Caller for the production of a VCF file and call germline SNPs and indels.
   (a) Input files: BAM.
   (b) Output: VCF.

12.2 Exome Coverage Analysis Recommendations

1. Use an existing exome BED file or use Bedtools for the creation of intervals in the genomic regions, involved in analysis.
2. Compute read depth of our genomic intervals using GATK Depth of Coverage.

13 Off-Target Reads (Reads Which Are Not Aligning to the Target Region)

Exome sequencing may lead to a large portion of off-target reads. In majority of the cases, as much as 40% of the data produced by capture methods are not of exon regions. The three main types of off-target reads include:

1. Intron and intergenic reads.
2. Viral genome reads.
3. Bacterial genome reads.
4. Mitochondrial reads.

The reasons for such a large portion of exome reads which become off-target are listed below:

1. Poorly designed biotinylated probes (baits).
2. Fragmented input DNA.
3. Small capture regions.
4. Inefficient hybridization conditions.
5. Spurious adapter to adapter annealing.

While calculating the number of reads necessary to fulfill mean on target exon coverage, off-target rates are also taken into consideration.

Off-target reads would involve functionally important genomic regions such as promoters, conserved non-coding sequences, untranslated regions (UTRs), and microRNA. Recently, it has been suggested that data mining and extraction of useful information from non-exonic reads are valuable. Tools for analyzing these off-target reads are also being described.

14 Considerations for Whole Exome Sequencing

14.1 Sequencing Instrument and Read Length for Exome-seq

In order to obtain coverage typically necessary for whole human exome sequencing, we recommend the Illumina HiSeq or Nextseq 500 platforms. For this, a minimum read length of 2 × 100, or 200 bp is necessary. When sequencing at a longer read length gives a slight improvement in single nucleotide variant (SNV) calls, the improvements are observed to be modest at no extra time and cost. Similarly for INDELs, the number of variants identified were up to 2 × 100 bp length sequencing run. Certainly, longer read lengths will lead to an increase of mean coverage. Improved coverage with longer reads increases with increasing sequencing depth. Hence, if we need to obtain 200× coverage, a longer read length needs to be considered. In case we aim for 100× coverage, unless the costs are relatively the same, it is recommended to start with a 2 × 100 bp sequencing run.

14.2 Sequencing Coverage for Exome Sequencing

In order to obtain accurate variant identification, we recommend at least 100× coverage. In most of the cases higher coverage will decrease the chance of a false positive SNV call. Higher coverage will be essential to improve sensitivity and improve the detection of rare variants.

14.3 Sequencing Coverage or Depth Required for Whole Exome Sequencing

Factors to be considered for the calculation of the coverage for exome sequencing include:

1. Average on target rate or probe enrichment efficiency.
2. Mean normalized coverage.
3. Desired coverage vs. mean coverage.

14.4 Choice of Exome Sequencing Capture Kit

In general, we observe the Roche Nimblegen Seq Cap v3.0 + UTR kit is containing the largest designed target and coding regions. Nimblegen SeqCap v3.0 and Agilent SureSelect XT2 v5 + UTR exome capture kits also possess lower off target enrichment in comparison to Illumina Nextera Rapid Capture (expanded exome). The Agilent SureSelect XT2 v5 + UTR exome kits generally have been reported to possess the highest accuracy in SNV detection and best GC-rich region enrichment.

All three kits are commonly used in many facilities leading to most applications. In most of the cases, the choice of kit depends on the amount of DNA. Some available kits can be used with as little as 50 ng of input material, whereas others need at least 1 ug. Most of the exome capture kits can be barcoded and samples are pooled prior to capture. This reduces capture reagent costs. Several older kit versions are available to researchers who need capturing each barcoded library individually.

14.5 Comparison of the Annotation and Exome Capture Design Between Each Kit

Annotated files provide information on genomic regions under coverage by the capture probes and genes included in these regions:

- Agilent SureSelect V5 and V6 BED Files.
- Illumina Nextera Rapid Capture BED Files.
- Nimblegen SeqCap BED Files.

14.6 Choice of Whole Genome Sequencing (WGS) or Whole Exome Sequencing (WES)

There are advantages/disadvantages for both WGS and WES. With the reducing cost of WGS, these are to be carefully considered: Whole Genome Sequencing (WGS) vs. Whole Exome Sequencing (WES).

15 Exome Sequencing Categories

1. **Standard Exome Sequencing includes:**
   (a) 100× sequencing coverage.
   (b) Appropriate for non-cancer-based applications.
   (c) Transposase or ligation-based library preparation.
   (d) Options to choose between IDT xGEN and Illumina Nextera Rapid Captu.
2. **Deep or High Coverage Exome Sequencing Service.**
   (a) 200× sequencing coverage.
   (b) Appropriate for case samples or tumor normal pairs.
   (c) Ligation-based library preparation.
   (d) Options to choose between Agilent SureSelect, NimbleGen SeqCap, IDT, and Illumina Nextera.

   Search for High Depth Whole Exome Sequencing Services.
3. **Clinical Grade Whole Human Exome Sequencing.**
   (a) Performed in a CLIA-certified facility.

16 Let Us Discuss in Details Exome Sequencing Practical Concern

The equipment needed is a computer with the specification of 64 GB RAM with 8 core CPUs in an Ubuntu operating system (14.04 LTS machine).

16.1 Softwares Employed for Data Analysis for Exome Sequencing

The softwares can be downloaded from the locations:

1. FastQC (https://www.bioinformatics.babraham.ac.uk/projects/fastqc/).
2. Bowtie2 (http://bowtie-bio.sourceforge.net/bowtie2/index.shtml).
3. Samtools (http://samtools.sourceforge.net/).
4. Varscan (http://varscan.sourceforge.net/).
5. Bcftools (https://github.co/samtools/bcftools).
6. Vcftools (https://vcftools.github.io/index.html).
7. PANTHER (http://pantherdb.org/).
8. dbSNP (https://www.ncbi.nlm.nih.gov/projects/SNP/).
9. 1000 genomes dataset (http://www.internationalgenome.org/).
10. GeneMania (http://genemania.org/).

16.2 Applications of Exome Sequencing

1. This leads to rare variant mapping in complex disorders.
2. Aids in discovery of Mendelian disorders.
3. Lead to clinical diagnostics.
4. Direct to consumer exome sequencing.

16.3 Application of Exome Sequencing in Animal Science

- The technique is cost-effective and better than whole genome sequencing.
- It has a wide application for Animal Genomic analysis.
- It is very much useful for disease-related traits in animals.
- Very much useful for production-related traits in animals.

Figure 1 gives the comparison of genomic workflow for exome sequencing with DNA microarray, while the schematic outline of exome sequencing is being depicted in Fig. 2.

16.4 Exome Sequencing

- Solution Hybrid Selection: Probes in solution lead to capturing of all exons (exome) for high throughput sequencing.
- 1–2% of whole genome seq.
- Easily multiplex 20 samples in one lane.

16.5 Advantages of Exome Sequencing

- Cost-effective and better than whole genome sequencing.
- Wide application for human genomic analysis.
- Very much useful for disease-related traits in human.

Whole-exome sequencing (WES) is an application of the next-generation technology to detect the variations of all coding regions, or exons, of known genes. Whole-exome sequencing has a coverage

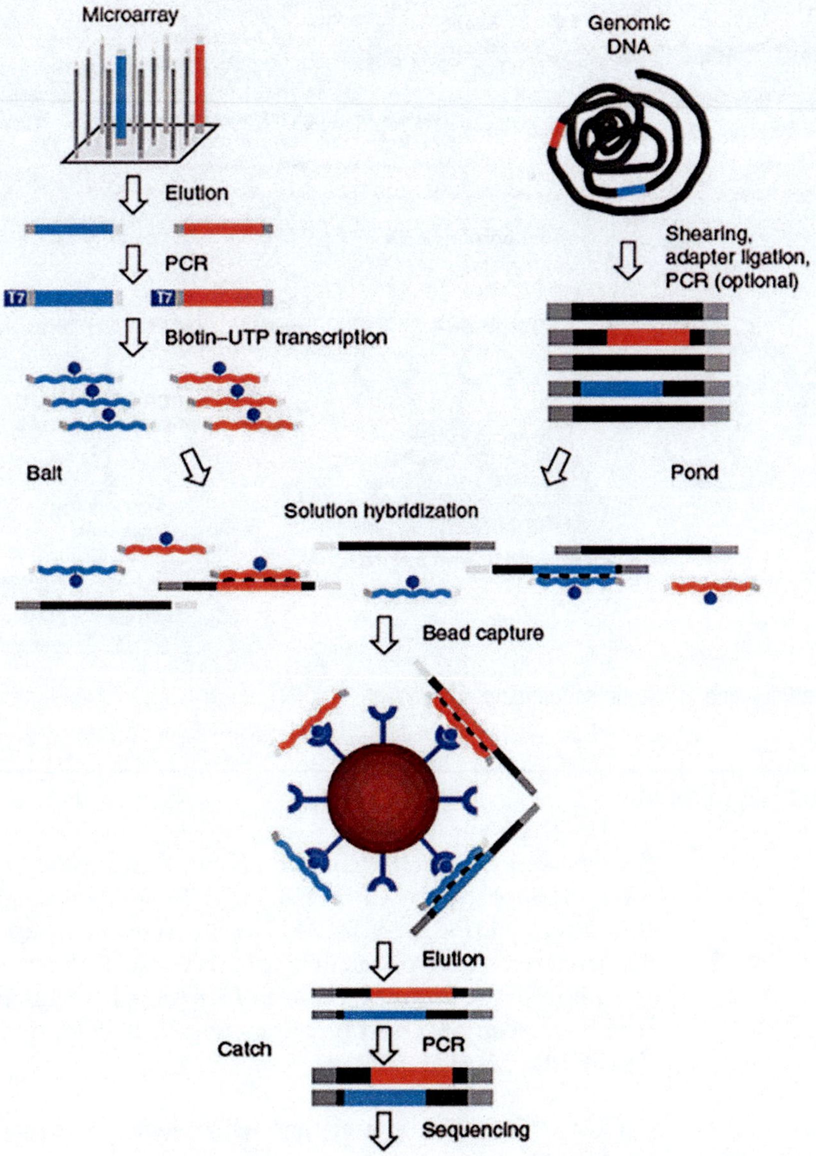

Fig. 1 Comparison of genomic work Flow for exome sequencing with DNA microarray

of more than 95% of the exons. It is expected to possess 85% of disease-causing mutations in case of Mendelian disorders and other disease-predisposing SNPs throughout the genome. The role of more than 150 genes has been identified by means of WES, and this statistics is rapidly growing.

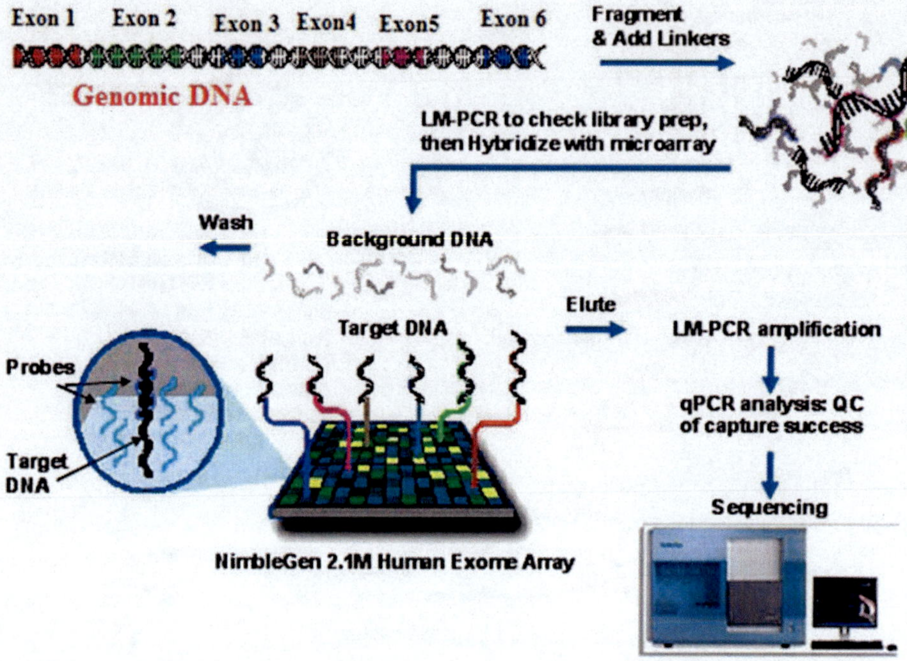

Fig. 2 Schematic outline of exome sequencing

17 Practical Application

The first report which identified an exact diagnosis established by WES was published by Choi et al. in 2009. WES on a patient known as Bartter syndrome (MIM241200), a rare inherited disorder is characterized by hypokalemia, observed that the patient has a novel homozygous mutation in SLC26A3 gene. Mutations in this gene have been observed to cause congenital chloride-losing diarrhea (CLD; MIM214700).

17.1 Exome Sequencing in Human

- Cost-effective and better than whole genome sequencing.
- Wide application for Human Genomic analysis.
- Very much useful for disease-related traits in human.

Figure 3 gives the diagrammatic representation of exome sequencing, while the process of exome sequencing is being described in Fig. 4.

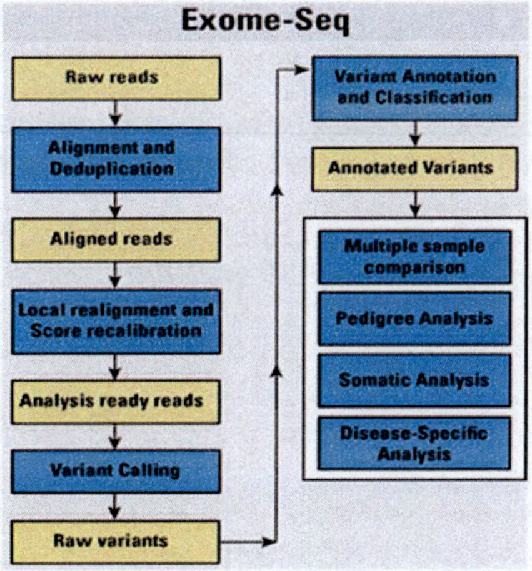

Fig. 3 Layout for exome sequencing

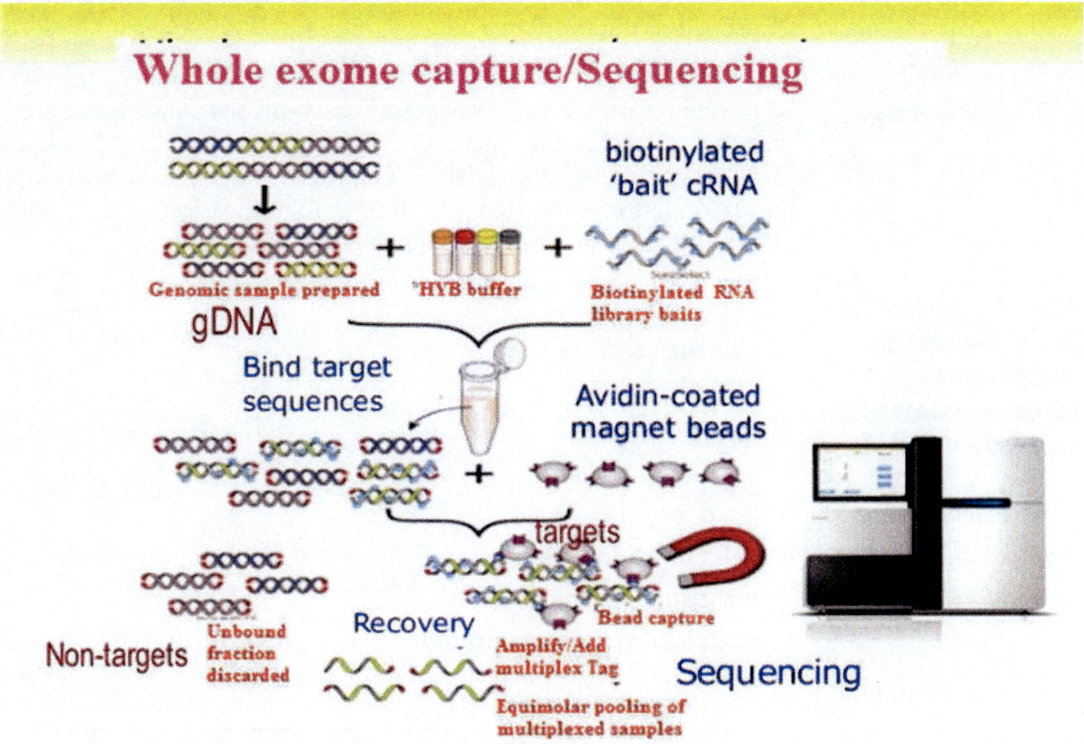

Fig. 4 Diagrammatic representation of the process of exome sequencing

17.2 Comparison for Exome Sequencing with Whole Genome Sequencing

- Detection of somatic mutation between normal/cancer pairs.
- WGS or WES.
- Exome sequencing can lead to more mutation yield leading to better causal gene identification than Mendelian disorders.

Enabling technologies comprise of NGS machines, open-source algorithms, capture reagents, lowering cost, big sample collections.

Exomes more cost-effective: The process involves sequencing patient DNA and filters common SNPs; compares parents child trios; compares paired normal cancer.

17.3 Challenges

- Till date it is not able to interpret many Mendelian disorders.
- Rare variants require large samples sizes.
- Exome might miss region as for example in novel non-coding genes.
- Exome sequencing is unsuccessful at using exome-seq to interpret clinical data.

17.4 Data Analysis

It is one of the most important traits in any next-generation sequencing technique including Exome sequencing. Figure 5 gives an outline of data analysis employed in exome sequencing. Figure 6 identifies the steps involved in identification of mutations pertaining to tumor with respect to healthy subject.

17.5 The Pipeline for Exome Sequencing Includes Three Important Phases as Depicted in Fig. 7

1. Preprocessing.
2. Variant discovery.
3. Prioritization of variants.

17.6 Bioinformatics Issues

- Read mapping.
- SNP, indel detection.
- Heuristic filtering.
- Structural variation detection.

18 Laboratory Protocol: A Case Study—A Protocol for Whole-Exome Sequencing in Newborns with Congenital Deafness—A Prospective Population-Based Cohort

The procedures involve exome sequencing, variant detection, variant filtering followed by interpretation performed in a clinically accredited laboratory as earlier methods applied in other patient cohorts of the Melbourne Genomics Health Alliance.

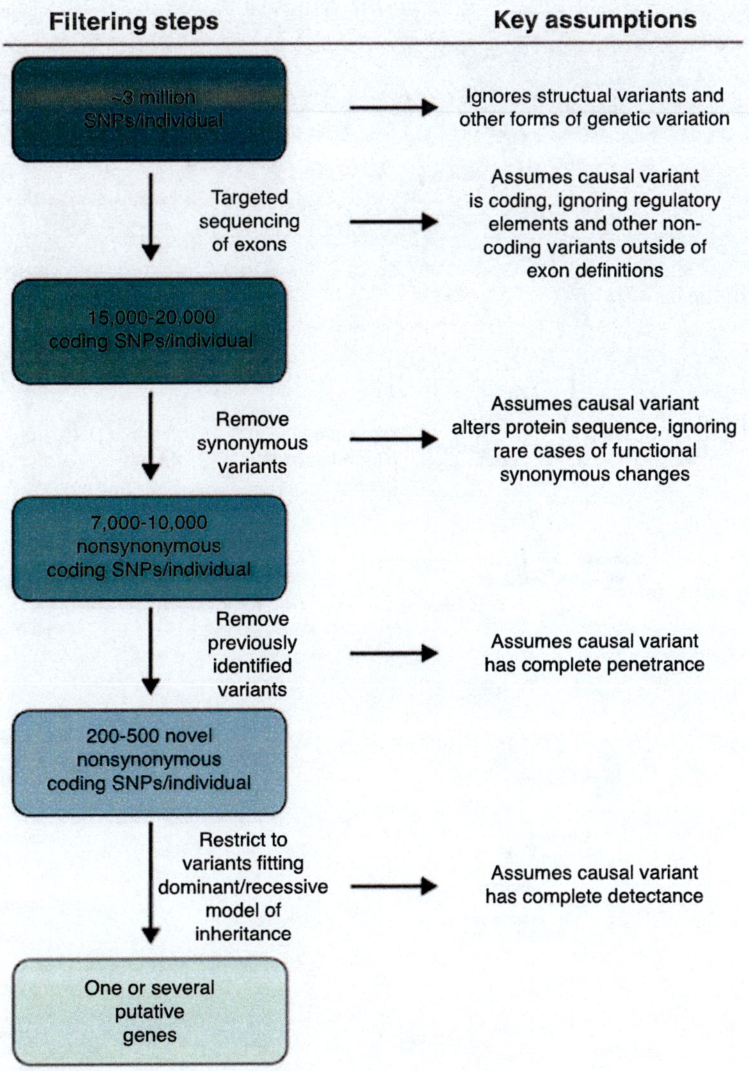

Fig. 5 Outline of data analysis for exome sequencing

18.1 Whole-Exome Sequencing

DNA will be extracted from peripheral blood, and exome sequencing will be conducted with SureSelect Clinical Research Exome V.1 (Agilent) on either a HiSeq4000 or NextSeq500 (Illumina) with a targeted average sequencing depth of 100×.

18.2 Bioinformatic Analysis and Variant Filtering

Data analysis and variant calling will be conducted with Cpipe. Variants will be assessed with LOVD plus. Variants will be prioritized on the basis of the clinician's assessment using information such as likely mode of inheritance. It is assumed to be autosomal recessive unless family history indicates otherwise and any phenotypic information available such as MRI brain findings or

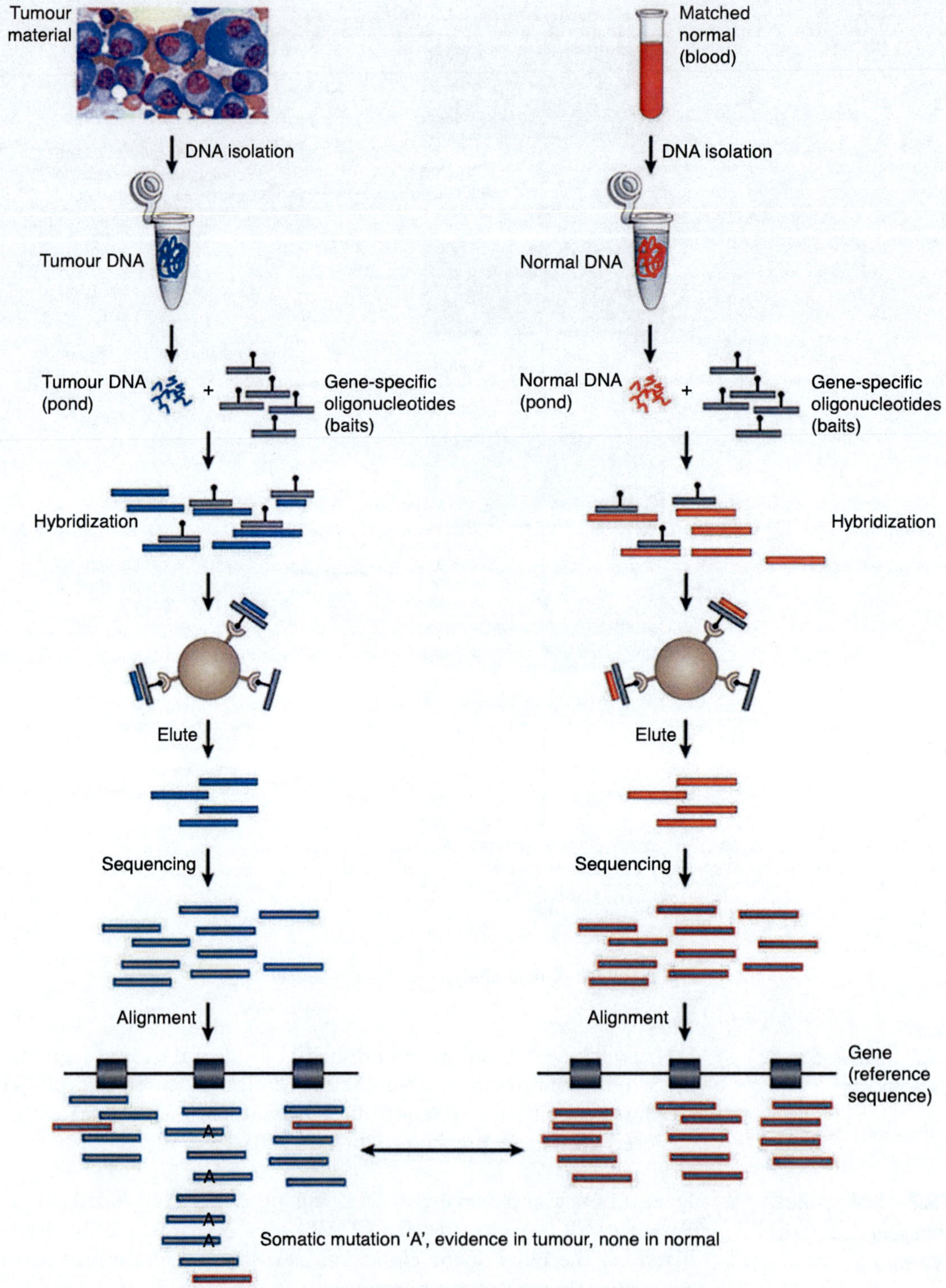

Fig. 6 Identification of mutations pertaining to tumor with respect to healthy subject

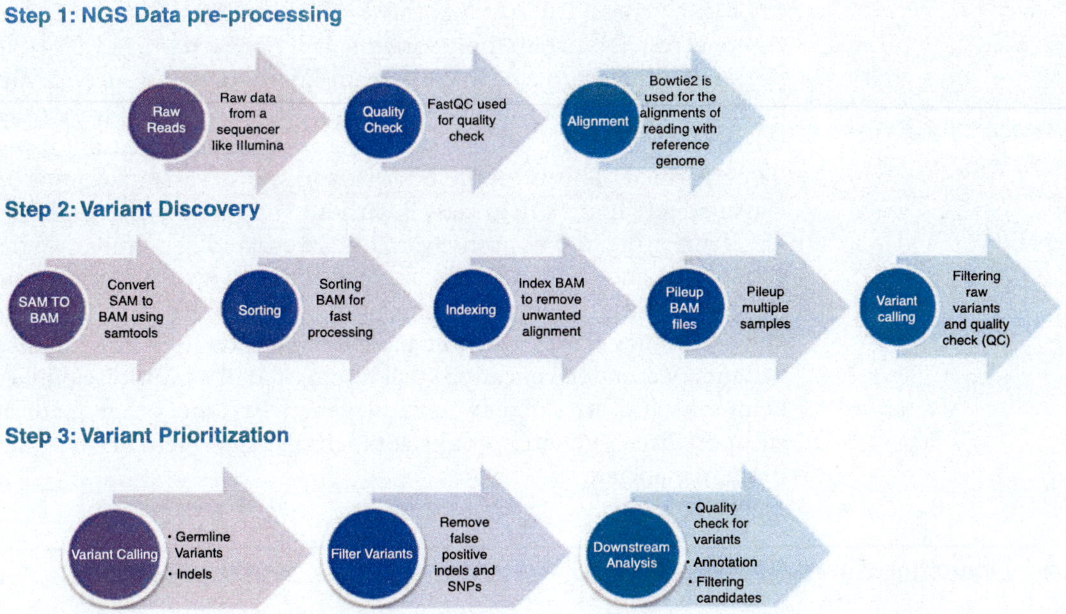

Fig. 7 The pipeline involving three important phases, viz. preprocessing, variant discovery, and prioritization of variants

dysmorphic features on examination. We assume a large proportion of patients with no family history and no additional phenotypic information apart from hearing loss and in this case, all variants detected will be curated.

18.3 Variant Interpretation

The methodology around the interpretation of the additional findings gene lists focuses on the shift in principle from employing WES as a diagnostic tool to use of WES as a screening tool. During diagnosis, a high level of sensitivity is needed from the test to capture all possible causes of a patient's phenotype. In a "screening" or "predictive" analysis, a high level of specificity is necessary so that there is adequate evidence and understanding of the outcome of a gene variant to return results to patients who will often be asymptomatic at that time. The standard diagnostic laboratory variant interpretation protocol was adapted for this cohort. This is due to the novel situation of offering analysis of genes outside of the patient's current phenotype.

Choice A, tier 1 genes will be studied as per the laboratory's National Association of Testing Authorities accredited procedures which are associated with American College of Medical Genetics guidelines, leading to a clinical report.

Choice A, tier 2 genes will be studied by scientists only in patients who do not receive a clinical result from tier 1.

Choice B and C genes will be studied by the same clinically accredited diagnostic laboratory for patients who consented to the analysis, but only variants in class 4 or 5 will be reported to families for these genes. The variants those of uncertain significance, and those were only carriers-only variants, will not be reported. For the purpose of this process to be stream-lined, any missense variant detected in the genes of interest will be searched first. In case the variant is a novel one with no earlier report, no further curation will be performed. This is on the basis that missense variants cannot be classified any higher than class 3 without supporting case reports.

Truncating and nonsense variants are curated in a similar way as variants in choice A. Criteria for classification will be as per the principles outlined in the American College of Medical Genetics and Genomics standards to be used for interpretation of sequence variants. Variant classifications will be checked in a multidisciplinary team meeting attended by clinical geneticists and other medical subspecialists, genetic counselors, molecular geneticists, and bioinformatician.

19 Evaluation Survey

Evaluation surveys have been designed and can be completed online or on paper. Patients will complete the first survey 2 weeks later of their consent for genomic testing and a second survey 2 weeks following their results appointment. The surveys comprise of data captured for Melbourne Genomics across all patient groups as well as data which is specific to children within the Melbourne Genomics Congenital Deafness project. The first survey involves questions about the parents' experience of the consent process involving the decision aid and the genetic counseling appointment. The Decisional Conflict Scale and Decisional Regret Scales are included in survey 1 and 2, respectively, to assess parental decision-making around which gene list they selected for analysis. The first survey involves closed and open questions exploring families' reasons for need to have genomic sequencing for their child. Knowledge questions and queries on the likelihood of the test finding the cause of the hearing loss are involved in gauge understanding. Validated measures to assess potential predictors of uptake of the expanded analysis are involved: tolerance of uncertainty, State-Trait Anxiety scale, medical literacy, and information-seeking preferences. The Perceived Personal Control scale35 is used as a measure of patient empowerment in both surveys. The second survey includes study-specific questions understanding the parents' knowledge of the genomic results, impact on them and their family, and what they have valued about having the test done for their child.

19.1 Analysis of Data

The majority of data analysis will be descriptive in character, because of the small sample size, which is observed to be approximately 100. The primary outcome measures observed will be uptake of testing, the detection rate of causative/pathogenic variants, and detection of carrier status. Detailed sociodemographic characteristics of the participants are essential to study. This includes socioeconomic status, age of parents, language and cultural background, education, and variable aspects of the process and outcomes, followed by comparing groups of interest.

Examples of these include:

1. Information delivery methods (phone call, face-to-face consultation, and written information) and parental responses to the offer of testing and their knowledge related to testing in relation to these various modalities.
2. Psychosocial measures of parents accepting versus decline testing and the scope of results selected by those who participate.
3. Differences in uptake between parents of children who need cochlear implant, those who are systemically unwell or have a broader phenotype (not isolated hearing loss) or whose infant was born prematurely for example. Multivariable statistical analysis can be performed, depending on the sample size pertaining to each group, for example, accept testing versus decline testing and no additional findings versus additional findings.

20 Another Case Study for Exome Sequencing: X-exome Sequencing of 405 Unresolved Families Identifies Seven Novel Intellectual Disability Genes

20.1 Subjects

All index cases possessed a normal karyotype, were negative for FMR1 repeat expansion, and in majority of these large indels had been excluded using array CGH.

20.2 Methods

For each family, DNA from one affected male was employed for constructing a sequencing library with the Illumina Genomic DNA Single End Sample Prep kit (Illumina, San Diego, CA, USA). Enrichment of the X-chromosomal exome was then conducted for each library with the Agilent SureSelect Human X Chromosome Kit (Agilent, Santa Clara, CA, USA). The kit consists of 47,657 RNA baits for 7591 exons of 745 genes of the human X chromosome. Single-end deep sequencing was conducted on the Illumina Genome Analyzer GAIIx (Illumina, San Diego, CA, USA). Read length was observed to be 76 nucleotides. For a subset of families of the second cohort, we conducted droplet-based multiplex PCR (7367 amplicons, 757 genes, 1.54 Mb) in a similar way as the earlier study. Paired-end deep sequencing was conducted on the HiSeq2000 platform (ATLAS, Berlin, Germany).

A scheme outlining the variant discovery workflow is presented in Supplementary Figure 1. Reads were extracted from qseq-files as an output by the Illumina GAII system (Illumina). Reads with ambiguous base calls were not processed for further analysis. The rest of reads were then mapped to the human reference genome (hg18 without random fragments) with RazerS24 (parameters: -mcl 25 -pa -m 1 -dr 0 -i 93 -s 11010111100110001011 1 -t 4 -lm), which tolerates up to 5 bp differences to the reference sequence per read. Only best and unique matches were selected. Rest of the reads and those having indels were subjected to a split mapping procedure of single end reads (SplazerS version 1.0,25 parameters: -m 1 -pa -i 95 –sm 23 -s 111001110011100111 -t 2 -maxG 50000) for the identification of short insertions (≤30 bp) and larger deletions (o50 kb).

For the detection of large insertions/deletions by analyzing changes in depth of coverage along the targeted regions, we employed ExomeCopy. We conducted a quality-based clipping of reads after mapping but before calling variants to reduce the number of false-positive calls. From each end of a read with a sliding window of 10 bp, we processed for the trimming of the read till we find a window with all ten phred base quality values. If there was a variant within 3 bp distance to the clipped region, trimming was expanded up to this potential sequencing error. For both mapping procedures (RazerS + SplazerS), the calling of a variant requires at least three reads with different mapping coordinates with an aim to exclude potential amplification artifacts.

Single-nucleotide polymorphisms (SNPs) and short indels (≤5 bp) were called with snpStore (parameters: -reb 0 -fc 10 -m 1 -mmp -mc 3 -oa -mp 1 -th 0.85 -mmq 10 -hr 0.001 -re -pws 1000), leading to a realignment of the clipped mapped reads, whenever a minimum of three indel-containing reads were found within close proximity. For an indel to be called no more than 75% of the spanning reads were let to contradict it.

For single base variants we employed the Maq consensus statistics integrated into the snpStore code. Larger deletions and small insertions were detected through examination of the split mapping results for potential breakpoint positions. In case of multiple such positions implying varying indel lengths within a 20-bp range such candidate calls were predicted to be unreliable and were so discarded. To identify potential retrocopies, the boundaries of split read mappings were compared with known exon boundaries permitting a tolerance of ±5 bp. In case both of the split ends coincide with exon boundaries these exons were known as being part of a retrocopy event. Completeness of the retrocopy was explained by the highest fraction of exons per transcript for which exon-spanning reads were identified.

In a parallel approach, processing was performed with the sequencing reads with the help of an alternative software, Medical Resequencing Analysis Pipeline (MERAP), for mapping, variant calling, and annotation. In this case, the mapping was performed using SOAP2.2029 leading to about two mismatches. For the calling of single-nucleotide variants (SNVs) and indels, a minimum of four reads and a more stringent Phred-like quality score of ≥20 were necessary. Finally, only those variants called by both approaches were allowed to yield high-confidence candidate variants.

For in silico prioritization of variants, we integrated the following features:

1. Gene/transcript annotations (downloaded from UCSC Genome Browser, hg19).
2. Known sequence variants from the following data sources: dbSNP, 1000 Genomes project, 200 Danish exomes, 30 NHLBI Exome Sequencing Project (ESP6500, version without indels). Base exchanges were considered as "known" (with exception of SNVs detected as only heterozygous in ESP6500 and 1000 Genomes project) if position and type of the nucleotide were identical to entries in the reference databases. We did not utilize a cutoff based on minor allele frequency. For short indels, a tolerance in positional matching was used based on repetitiveness of the deleted/inserted sequence in the SNV flanking sequence.
3. Variants detected in the screen conducted by Tarpey et al. were visible in transcripts derived from ENSEMBL version 54. We explain the amino-acid coordinate shared by most transcripts of a gene as reference. This is sometimes different from the one annotated by Tarpey et al. Conversion of coordinates was successfully obtained for 1647 variants.
4. Evolutionary conserved across 44 vertebrate species.
5. Splice site detection for defining potential cryptic splice sites (software NNSplice; cutoff 0.9 (ref. 32)).
6. Potential functional impact: PolyPhen2,33 SIFT34.
7. Human Gene Mutation Database (HGMD): known variants with Pubmed entries were treated as potentially disease causing in case they were listed in HGMD Professional and annotated in maximally one reference SNV database. Therefore, a prioritization score (PS) was calculated on the basis of basic, computationally tractable criteria like type of variant or evolutionary conservation. Polyphen2/ SIFT gives rise to a categorical output (benign/tolerated, possibly damaging/low confidence, probably damaging/damaging) that were assigned to ordinal variables 1, 2, or 3. Numbers are decreasing with reducing functional impact, missing values are scored nil. Whenever only one of the methods scored 40, the zero score was set to 1 in order to prevent underestimation of the functional impact. PhyloP values were rounded down to decimal numbers, values 45 were set to PHY = 5, for values o2 PHY = 1, for values o0

PHY = 0. Since deletions/insertions are generally not scored by PolyPhen2/SIFT, we defined the following ad hoc weighting scheme: non-sense/frameshift: TYPE = 20 (maximal PS), deletions (450 bp): TYPE = 9 (similar to maximal impact prediction by PolyPhen2 and SIFT), duplications, in-frame deletions, potential splice site variants: TYPE = 3. The score for a change identified in a gene with a role in XLID before this study was set to 3. PS = PP2 * Sift + PHY + TYPE + XLID; if PS420, PS = 20.

We also employed CADD (Combined Annotation-Dependent Depletion) as an additional tool for annotating and interpreting SNVs and small indels (see Supplementary Figure 3 for comparison of the scores).

Analysis of human CLCN4 SNVs in Xenopus oocytes CLCN4 SNVs were introduced into human CLCN4 (NM_001830.3; Gene ID: 118) cDNA cloned into pTLN and pCIneo36 through recombinant PCR. We determined the expression level and stability of wild-type and mutants with p.Gly78Ser, p.Leu221Val, p.Val536-Met or p.Gly731Arg substitutions by western blot analysis of lysates from transiently transfected cells employing standard methods. Xenopus laevis oocytes were injected with 23 ng cRNA that was transcribed with the Message Machine kit (Ambion, Thermo Fisher Scientific Inc., Waltham, MA, USA) from pTLN.

Following 3 days incubation at 17 °C, currents were measured at room temperature with standard two-electrode voltage clamp employing TurboTEC amplifiers (npi electronic, Tamm, Germany) and pClamp10.2 software (Molecular Devices, Sunnyvale, CA, USA). Oocytes were superfused with modified ND96 saline (96 mM NaCl, 2 mM K-gluconate, 1.8 mM Ca-gluconate, 1 mM Mg-gluconate, 5 mM HEPES pH 7.5) and clamped in 20-mV steps to voltages between −100 and +80 mV. The holding potential was −30 mV.

21 Morphological Studies of Mouse Hippocampal Neurons

Mouse embryos were dissected at embryonic Day 16, followed by dissociation of tissue by trypsin as well as by mechanical treatment. Primary cultures of hippocampal neurons were cultured at 37 °C by plating on coated-glass coverslips (poly-L-Lysine and Laminin) at a density of 100,000 per 16 mm Petri dish. Neurons were differentiated for 18 days in vitro (18 DIV) with Neurobasal/B27 medium and antibiotics (Mycozap, Lonza, Basel, Switzerland), replacing half of the media every third day for maintenance as per the standard procedures. Short-hairpin RNA (shRNA) design was prepared by targeting the 3′UTR of each specific gene with Promega shRNA designer tools (Promega BioSciences, San Luis Obispo, CA, USA) or information based on The RNAi Consortium (TCR) shRNA Library.

A control shRNA-producing plasmid was utilized in control experiments as previously described. Three independent shRNA-producing and GFP-expressing plasmids, on the basis of pSystrike vector (Promega), were prepared for each gene and utilized as a pool. Sequences targeted by shRNAs for Cnksr2 and Clcn4–2 (Clcn4) genes and control sequences are GGAGCAGAGG ATGG-CAGTCATTCA, GGTGGGAAGGCTAGCTCTGTTACT, GCGCGGCGTATCAG GGCAAAGCTT, GGGTATGTGG-GAGGGTGTAAATGA, GGGAGAGGCGAGTAC GAAGAT-GAA, GTGGTCTACTCATGGCCATCTCAT, and GCTCACCCTTCCTACTC TC. Full-length murine Cnksr2 (NCBI reference sequence NM_177751.2) and Clcn4 (NCBI reference sequence NM_011334.4) cDNAs were cloned into pFN21A HaloTag® CMV Flexi® vector (Promega). To conduct the rescue experiments, pool of plasmids encoding shRNAs were cotransfected with pFN21A HaloTag® fused to either Cnksr2 or Clcn4. All constructs were sequence verified and plasmids were purified with an endotoxin-free kit (Macherey Nagel, Düren, Germany). Transfections were conducted with Lipofectamine (Invitrogen, Life Technologies, Carlsbad, CA, USA) at 11 DIV and cells were fixed at a later stage of differentiation for analysis (DIV 18). Individual neurons were directly visualized under fluorescence and confocal microscopy (spinning disk microscope, Leica, Leica Microsystems, Wetzlar, Germany) employing GFP labeling as a tracer of morphology. Immunocytochemistry to detect GFP (goat antibody, Abcam, Cambridge, UK) and HaloTag constructs (HaloTag TMR Ligand, Promega) was also analyzed according to standard procedures. Image analysis was performed using Imaris software with "Filament tracer" plugin (Bitplane Scientific Software, Bitplane AG, Zürich, Switzerland) and ImageJ software (Wayne Rasband, Bethesda, NIH).

It leads to quantifying total arborization of neurites (dendrites and axon) for each neuron (total length of neurites, numbers of branches, branching complexity), see Supplementary Figure 4 for details on branching analysis. Quantification was conducted on the basis of three independent experiments with more than 15 cells of each type per experiment analyzed. Mann-Whitney statistical test was employed for the comparison of total neuritic length as well as number of branches, whereas Chi-square test was employed for the evaluation of significance of variations in branching complexity. Primary cultures of hippocampal neurons from wild-type and Clcn4−/− mice were obtained as discussed above with some minor differences.

Animals were dissected at postnatal day 1 (P1) following ethical standards. Papain was utilized for dissociation of tissue. Glass coverslips were coated with poly-L-Lysine and collagen. At DIV

11, neurons were transfected with pEGFP-C1 vector (Clontech, Mountain View, CA, USA) using Lipofectamine-2000 as per the manufacturer's instructions. Cells were fixed and stained at DIV15 or DIV18 as described earlier. Primary antibodies utilized were chicken anti-GFP (Aves Lab, Tigard, Oregon, USA) and mouse anti-microtubule associated protein 2 (Chemicon/Millipore, Merck/Millipore, Darmstadt, Germany) as a neuronal marker. Secondary antibodies conjugated to Alexa Fluor 488 or 546 were obtained from Molecular Probes. Images were undertaken using a LSM510 laser scanning confocal microscope equipped with a ×10 lens (Zeiss). Image analysis was conducted in a blinded manner using ImageJ and its plugin NeuronJ: ns (non-statistically different), *Po0.05, **Po0.01, ***Po0.001 for validation.

22 Other Important Analyses

22.1 Genome Wide Pathway Analysis (GWPA)

GWPA is an approach which harnesses the wealth of information from GWAS or WGS data to detect the additive effects of single variants accumulated in particular gene sets or pathways. The general concept behind pathway analysis of GWAS data can be described as data reduction and aggregation. At first, the data on millions of individual sites around the genome are collected in the form of SNP data. These SNPs are then analyzed for allele frequency differences between case and control groups (as one example) and an assessment of statistical inference is obtained.

The SNPs are placed within genes by genomic position, and the genes are placed into pathways or gene sets on the basis of the information provided by the selected database(s). Thus, individual SNPs lead to collections of SNPs within the framework of a group of genes. Most of the algorithms seek for a preponderance of single-allele *p*-values meeting a specific criterion within the pathway, or gene set, in relation to what one would expect to see by chance.

22.2 Candidate Pathway Analysis

Candidate pathway analysis is a hypothesis-driven approach leading to pathway-level investigation. Pathways, or gene-sets, of interest are preselected on the basis of prior knowledge or investigation, and association testing is conducted on the preselected loci of interest instead of the full genome. It is more helpful as it doesn't require scanning an entire database of pathway information as with GWPA, which can be computationally intensive and requires stringent multiple-testing corrections. Candidate pathway analysis is less demanding and as a result lends itself more readily to large data sets such as imputed and WGS data.

References

1. Meena N, Mathur P, Medicherla KM, Suravajhala P (2018) A bioinformatics pipeline for whole exome sequencing: overview of the processing and steps from raw data to downstream analysis. Bio-protocol Bio101:e2805. https://doi.org/10.21769/BioProtoc.2805
2. Asan et al (2011) Comprehensive comparison of three commercial human whole-exome capture platforms. Genome Biol 12(9):R95
3. Bainbridge MN et al (2010) Whole exome capture in solution with 3Gbp of data. Genome Biol 11:R62
4. Clark MJ et al (2011) Performance comparison of exome DNA sequencing technologies. Nat Biotechnol 29:908. https://doi.org/10.1038/nbt.1975
5. Kingsmore SF, Saunders CJ (2011) Deep sequencing of patient genomes for disease diagnosis: when will it become routine? Sci Transl Med 3(87):1–4. Review of Bainbridge et al and discussion of WGS and targeted or Exome-Seq. They also suggest that an exome costs 5-15 fold less that a WGS
6. Maxmen A (2011) Exome sequencing deciphers rare diseases. Cell 144:635–637. A review of the undiagnosed Diseases Program at NIH. Exome-Seq and high-resolution microarrays for genotyping. They mention the team's first reported discovery of a new disease, which was published in The New England Journal of Medicine
7. Natsoulis G et al (2011) A flexible approach for highly multiplexed candidate gene targeted resequencing. PLoS One 6(6):e21088
8. Parla JS et al (2011) A comparative analysis of exome capture. Genome Biol 12(9):R97
9. Sakharkar MK et al (2004) Distributions of exons and introns in the human genome. In Silico Biol 4(4):387–393
10. Samuels DC (2013) Finding the lost treasures in exome sequencing data. Trends Genet 29:593–599
11. Sulonen AM et al (2011) Comparison of solution-based exome capture methods for next generation sequencing. Genome Biol 12 (9):R94

Chapter 18

RNA Interference

Abstract

RNA interference (RNAi) is a biological process where RNA molecules inhibit gene expression, through a process of the destruction of specific mRNA molecules. Two types of small ribonucleic acid (RNA) molecules, namely microRNA (miRNA) and small interfering RNA (siRNA) or silencing RNA, are essential for RNA interference. A short *hairpin RNA or small hairpin RNA (shRNA/Hairpin Vector)* is an artificial RNA molecule with a tight hairpin turn which can be utilized to silence target gene expression through a process called RNA interference (RNAi). Mechanism of action include cleaving the mRNA strand into two pieces, causing destabilization of the mRNA through shortening of its poly(A) tail, reduced efficiency of translation of the mRNA into proteins by ribosomes. Other form of RNA as ***Piwi-Interacting (pi) RNAs*** are single-stranded 25–31 nt reported to be associated with Piwi protein (a subclass of Argonaute proteins) and the human RecQ1 protein to form a *Piwi-interacting RNA complex (piRC)* for the regulation of the genome within developing sperm cells. Protocol for RNA interference has been described in detail. Certain practical applications were described for RNA interference.

Keywords RNA interference, MicroRNA (miRNA), Small interfering RNA (siRNA), Silencing RNA, Small hairpin RNA, shRNA, Piwi-Interacting (pi) RNAs

1 Basic Principle

RNA interference is a process for RNA-guided regulation of gene expression, when double-stranded ribonucleic acid doesn't allow the expression of genes with complementary nucleotide sequences. The RNAi pathway is conserved in most eukaryotic organisms and is believed to have evolved as a form of innate immunity against viruses. It also plays a major role in the regulation of development and genome maintenance.

RNA interference (RNAi) is a biological process where RNA molecules inhibit gene expression through the destruction of specific mRNA molecules. Two types of small ribonucleic acid (RNA) molecules, namely microRNA (miRNA) and small interfering RNA (siRNA), are essential for RNA interference. RNAs are the direct products of genes. These small RNAs have the ability for binding with other specific messenger RNA (mRNA) molecules, leading to

Aruna Pal, *Protocols in Advanced Genomics and Allied Techniques*, Springer Protocols Handbooks,
https://doi.org/10.1007/978-1-0716-1818-9_18,

either increase or decrease of their activity, for example, the prevention of an mRNA from producing a protein. RNA interference has an essential role in defending cells against parasitic nucleotide sequences, viruses, and transposons. It also affects the process of development.

A short ***hairpin RNA or small hairpin RNA (shRNA/Hairpin Vector)*** is an artificial RNA molecule possessing a tight hairpin turn that can be employed to silence target gene expression by the process of RNA interference (RNAi). Expression of shRNA in cells is accompanied by delivery of plasmids or through viral or bacterial vectors. shRNA is a useful mediator of RNAi which possess a relatively low rate of degradation and turnover. However, it requires use of an expression vector, having the potential to have side effects in medicinal applications.

Small interfering RNA or siRNA (also termed as short interfering RNA or silencing RNA) is a class of double-stranded RNA molecules, which is 20–25 base pairs in length, similar to miRNA. It operates with the RNA interference (RNAi) pathway. It inhibits the expression of specific genes with complementary nucleotide sequences through a process of degradation of mRNA after transcription, accordingly preventing translation.

A **microRNA** (abbreviated **miRNA**) is a small non-coding RNA molecule with about 22 nucleotides, and are observed in plants, animals, and some viruses. It is helpful for RNA silencing and post-transcriptional regulation of gene expression. miRNAs act through base-pairing with complementary sequences within mRNA molecules, leading to silencing of mRNA molecules by one or more of the following processes:

1. Cleaving the mRNA strand into two pieces.
2. Causing destabilization of the mRNA through shortening of its poly(A) tail.
3. Reduced efficiency of translation of the mRNA into proteins by ribosomes.

miRNAs resemble [1, 2] the small interfering RNAs (siRNAs) of the RNA interference (RNAi) pathway, except miRNAs which are derived from regions of RNA transcripts. Such miRNAs fold back on themselves to form short hairpins, whereas siRNAs are derived from longer regions of double-stranded RNA. The human genome may encode over 1900 miRNAs; more recent analysis reports that the number is closer to 600.

miRNAs are common in many mammalian cell types and seem to be about 60% of the genes of humans and other mammals. Many miRNAs are evolutionarily conserved, which signifies to have an important biological function. As for example, 90 families of miRNAs were reported to be conserved as being the common ancestor of mammals and fish. Most of these conserved miRNAs play important functions, as revealed by studies where genes for one or more members of a family have been knocked out in mice.

1.1 Brief History of RNA Interference

Andrew Fire and Craig C. Mello received the 2006 Nobel Prize in Physiology or Medicine for their research on RNA interference in the nematode worm *Caenorhabditis elegans*, published in 1998. The discovery of RNAi and its regulatory potentials has revealed that RNAi can aid in the suppression of desired genes. RNAi is considered as precise, efficient, stable, and better than antisense technology for gene suppression.

1.2 Protocol

A streamlined and simple protocol for the design and implementation of an RNAi experiment in cultured cells is being presented. It aims us to enable those who are inexperienced with RNAi for its application in their research, particularly in the field of neurodegeneration—a case study.

2 Materials

2.1 Making shRNA Constructs

1. Doubly distilled water.
2. Synthetic DNA oligonucleotides encoding shRNA (Integrated DNA Technologies, Standard desalting only).
3. 2× DNA annealing buffer: 20 mM Tris–HCl (pH 7.6), 100 mM NaCl, 2 mM EDTA.
4. shRNA expression plasmid pEGFP-U6.
5. Cloning enzymes (all from New England Biolabs, Ipswich, MA): T4 DNA ligase, T4 polynucleotide kinase, PmeI, EcoRI, calf intestinal alkaline phosphatase.
6. Bacterial culture and transformation reagents: MAX Efficiency DH-5α competent cells, S.O.C. medium, LB broth, LB agar, LB plate with 30 μg/mL Kanamycin.
7. Reagents for colony screen using PCR: forward primer 5′--ATATCCCTTGGAGAAAAGCCTT-3′, reverse primer 5-′-GGCCGCGAACCAATGCA-3′, PCR master mix.
8. 100 bp DNA Ladder.

2.2 Making miRNA Constructs

1. pCAG-RFP-miRint-KX plasmid.
2. Synthetic PCR primers were designed, then dissolved in distilled water at concentration of 10 mM. The sequences are based on the target gene sequence.
3. PCR reagents: Platinum Pfx DNA Polymerase (Invitrogen, Carlsbad, CA) with 10× PCR buffer, 10 mM dNTP mixture (Promega, Madison, Wisconsin), 50 mM $MgSO_4$.
4. Cloning enzymes: Replacement of the two restriction enzymes may be conducted with two new ones: KpnI and XhoI.
5. PCR product purification kit.

2.3 Cell Culture and Transfection

1. Dulbecco's modified Eagle's medium (DMEM) supplemented with 10% fetal bovine serum and antibiotics (100 U of penicillin and 100 μg of streptomycin).
2. 25% Trypsin and 1 mM ethylenediaminetetraacetic acid (EDTA).
3. Lipofectamine 2000.
4. Opti-MEM I.

2.4 Western Blot

1. Phosphate-buffered saline (PBS) (10×): 1.37 M NaCl, 27 mM KCl, 100 mM Na_2HPO_4, 18 mM KH_2PO_4 (adjust to pH 7.4 with HCl if necessary), and autoclave. Store at room temperature.
2. RIPA lysis buffer: 150 mM NaCl, 1.0% IGEPAL® CA-630, 0.5% sodium deoxycholate, 0.1% SDS, and 50 mM Tris–HCl, pH 8.0. Store at 4 °C.
3. Halt Protease Inhibitor Cocktail.
4. BCA Protein Assay Kit.
5. 4× Laemmli's SDS sample buffer: 250 mM Tris–HCl (pH 6.8), 40% glycerol, 8% SDS, 0.02% bromophenol blue, and 8% beta-mercaptoethanol.
6. BioRad Protean II Mini gel system.
7. Separating buffer (4×): 1.5 M Tris–HCl, pH 8.7, 0.4% SDS. Store at room temperature.
8. Stacking buffer (4×): 0.5 M Tris–HCl, pH 6.8, 0.4% SDS. Store at room temperature.
9. 40% acrylamide/bis solution (29:1) and *N,N,N,N*-′-tetramethylethylenediamine (TEMED).
10. Ammonium persulfate: 10% solution in water. Store at 4 °C and make fresh every month.
11. Gel running buffer (10×): 250 mM Tris–HCl, 1.920 M glycine, 1% (w/v) SDS. Store at room temperature.
12. Prestained protein molecular weight markers: Kaleidoscope markers.
13. Protran nitrocellulose membrane (e.g., Whatman, Piscataway, NJ).
14. Transfer buffer (1×): 25 mM Tris–HCl (pH 8.0), 192 mM glycine, and 10–20% methanol.
15. BioRad transblot apparatus.
16. PBST: 0.05% Tween 20 in PBS.
17. Blocking buffer: 5% milk in PBST.

18. Antibodies: primary antibody against the knockdown target protein and secondary antibody conjugated with horseradish peroxidase (HRP).
19. Enhanced chemiluminescent (ECL) reagents.
20. Blot imaging apparatus: Fuji LAS 4000.
21. Stripping buffer.

2.5 Northern Blot

1. Trizol Reagent.
2. 20× MOPS running buffer: 100 mM sodium acetate, 0.4 M MOPS (pH 7.0), 20 mM EDTA (pH 7.0). Store in dark at room temperature.
3. 20× SSC buffer: 3 M NaCl, 0.3 M sodium citrate (pH 7.0). Store at room temperature.
4. 1.5× loading buffer: 75% formamide, 7.5% formaldehyde (37%), 15% 10× MOPS running buffer, 20 μg/mL ethidium bromide. Best to make fresh each time.
5. Loading dye: Bromophenol blue in 35% glycerol and 1× running buffer.
6. Positively charged Nylon membrane (Hybond N+) (e.g., Amersham, Piscataway, NJ).
7. DIG probe kit, consisting of PCR DIG Probe Synthesis Kit, DIG Easy hyb, DIG block and wash buffer, CDP star.
8. UV crosslinker Stratalinker (e.g., Stratagene, LaJolla, CA).
9. HB-1000 Hybridizer (UVP LLC).
10. Blot imaging apparatus: Fuji LAS 4000.

2.6 Reverse Transcription and Quantitative PCR

1. SuperScript® III First-Strand Synthesis System.
2. DNase I.
3. Sybr Green mix.

3 Methods

3.1 Design of siRNA, shRNA, and miRNA

3.1.1 Background

The three types of RNAi triggers, siRNA, shRNA, and miRNA are present. All of them are effective in silencing target genes and they generally can be used interchangeably. siRNA is a simple small RNA duplex. The two strands are chemically synthesized separately followed by annealing to form the siRNA. It can be effectively delivered into many types of cells by standard transfection protocols. shRNA is generated from a gene construct as a single-stranded RNA, which in turn folds for the formation of shRNA. Similarly, miRNA is also developed from a gene construct as a long, single-stranded RNA known as pri-miRNA. The detailed process is being depicted in Fig. 1. Therefore, both shRNA and miRNA need the

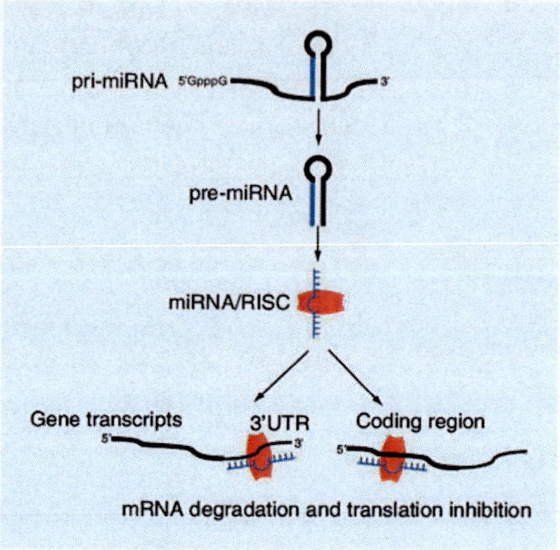

Fig. 1 Diagrammatic representation of the process of miRNA formation leading to mRNA degradation and translation inhibition

construction of plasmids. These plasmids can also be delivered into cells by standard transfection protocols. Generally, the transfection efficiency of siRNA is higher than constructs for shRNA and miRNA. siRNA is able to reach its highest concentration in cells soon after transfection and does not need time to accumulate as shRNA and miRNA constructs do. As a result, siRNA can trigger gene silencing better than shRNA and miRNA constructs. Moreover, the usage of shRNA and miRNA possess some benefits. These gene cassettes may be inserted into viral vectors and can be utilized for making cell lines having the gene silencing construct with proper integration.

Therefore, these constructs may be employed in experiments where long-term, sustained gene silencing is necessary. The purpose of siRNA design is to select siRNAs with the potential knockdown efficiency and minimal off-target effects. This can be acquired based on predefined siRNA sequence features and experiments that test the knockdown efficacy. It is important to note that currently no rational design guarantees a perfect siRNA. Hence, it is essential to design multiple siRNAs followed by testing them. A time-saving approach is to purchase siRNAs and constructs which produce shRNA or miRNA targeting specific genes from commercial sources, most of which offer warranty for a certain degree of knockdown of the target.

For the targets which one cannot find a commercial source, or if the researcher prefers to design and make their own siRNA, we offer the following procedure.

3.1.2 Designing siRNA

The first step in designing siRNA is to detect the target gene sequence. It is preferable to use the RefSeq of the target gene from NCBI, as the RefSeq collection gives nonredundant, curated, and most reliable sequences. The targeted region of the gene should be located within the open reading frame (ORF) between 50 and 100 nt downstream of the start codon and the stop codon.

The UTRs and the beginning of the ORF usually bind to regulatory factors and may be less accessible by the RISC. This need not be regarded as an absolute limit as highly efficient knock-down has been observed with siRNAs directed against the other regions, including 3′UTRs.

At first, the target region is identified, target sequences for RNAi may be selected utilizing one or several siRNA design algorithms which are accessible online (see http://i.cs.hku.hk/~sirna/software/sirna.php). Each siRNA duplex possesses both a sense strand and an antisense strand.

The antisense strand complements the sequence in the target cDNA or mRNA. Each strand of an siRNA duplex is 21 nt long, where the 5′ 19 nucleotides complement each other and match the target gene sequence. The 3′ end two nucleotides of each strand are TT, which are not necessary to match the target sequence and possess the overhang of siRNAs. In order to select the 19 nt target sequence, it is essential to follow some general rules. Sequences are selected so that the GC content lies between 30% and 60%. It is essential to exclude certain cases as follows:

1. Sequence repeats.
2. Sequences with low complexity.
3. Nucleotide polymorphism (SNP) sites.
4. Sequence stretches of four or more bases, such as AAAA, CCCC.
5. Strings of six or more G and C (e.g., GGCGCC) or G and T (e.g., GGTTGT).
6. Sequences containing palindromes, which may form unwanted hairpin structures.

The utilization of RNA folding software, such as RNAfold (http://rna.tbi.univie.ac.at/cgi-bin/RNAfold.cgi) for the exploration of possible secondary structures and exclusion of sequences with undesirable folds is essential. Certain sequences need to be excluded which contain GTCCTTCAA, TGTGT, and CTGAATT sequences. These sequences are known to induce immune response. Sequences are selected with asymmetric thermostability such that the three nucleotides at the 3′ end of the sense strand possess less G and C than its 5′ end. It is important to select two to six target sequences for each target gene.

A negative control siRNA through scrambling targeted siRNA sequence is designed in such a manner that it possesses the same length and nucleotide composition as the target sequence. However, it possesses at least four to five mismatches spread throughout the siRNA. BLAST homology search is conducted to avoid sequences which share a certain degree of homology with nontargeted genes to reduce off-target silencing effects. It is important to exclude a perfect or near perfect homology of the seed region (the six to eight bases at the 5′ end of the two siRNA strands) with any nontarget genes.

The online algorithms follow the general rules stated above and they incorporate supplementary rules derived from their experimentally tested siRNAs. It is necessary to check the rules that are incorporated in the algorithm and check the siRNA against the rules, which are not applied. After the designing of siRNAs, they can be ordered from many commercial sources and utilized directly in experiments. The siRNA sequence may also be inoculated into shRNA or miRNA expression constructs. Moreover, there is no guarantee if a highly efficient siRNA remains as efficient where its sequence is transplanted into an shRNA or miRNA. As a result, it is advisable to make multiple constructs followed by selection of the most efficient ones for use through the process of measuring the target knockdown.

3.1.3 Designing shRNA

Each shRNA possesses a stem and a loop. The stem consists of the 19 nt sense and antisense strands of an siRNA. The antisense strand is complementary to the target mRNA sequence. Pol III promoters, such as U6 and H1, are commonly employed to drive the shRNA synthesis. Both the promoters have defined initiation nucleotide, a G for U6, and an A for H1. There is termination of transcription at a string of four or more Ts. In order to design a shRNA, the 19 nt target siRNA sequence is inserted following the initiation nucleotide. It is necessary to connect the sense and antisense strands with a loop sequence of CAUUCAAGAGAUG and add six Ts at the 3′ end of the antisense strand. Synthesize the two strands of DNA encoding the shRNA with suitable restriction enzyme sites at both ends and clone the sequence behind the Pol III promoter in a DNA plasmid.

3.1.4 Designing miRNA

The expression of miRNAs is generally driven by a Pol II promoter; however, Pol III promoters can also work. One advantage of employing Pol II promoter is that a marker protein can be expressed under the control of the same promoter which drives miRNA expression. The miR30a structural features are generally employed in designing miRNAs. In order to express an miRNA, a pre-miRNA has to be designed and inserted into the expression plasmid. To design premiRNA, replace the 19 nt sequence between the two vertical lines with 19 nt core siRNA sequence. The UC bulge in the sense strand is kept in place. This whole miRNA sequence is transformed to a linear DNA sequence and then it is utilized as the sense strand. The next step involves the addition of the antisense strand.

3.2 Making shRNA Constructs

1. Ordered two strands of synthetic DNA oligonucleotides covering the entire shRNA and the restriction sites PmeI and EcoRI.
2. Annealed the two shRNA oligonucleotide strands: To a 1.5-mL microtube, added 100 μL 2× DNA annealing buffer, 90 μL water, and 10 μL of each oligo (250 μM). Heated the tube with boiling water in a beaker for 5 min. Cooled slowly to room temperature by keeping the beaker on bench top for 1 h. The annealed oligonucleotides will have suitable overhangs for ligation.
3. Added phosphate to the shRNA oligo: To a new 1.5-mL microtube, added 10 μL of the annealed oligo pair, 5 μL of 10× T4 DNA ligase buffer, and 2 μL of T4 polynucleotide kinase. Adjusted the volume to 50 μL using H_2O followed by incubation at 37 °C for half an hour.
4. Digest the expression vector pEGFP-U6 with PmeI and EcoRI:

   To a 1.5-mL microtube, added 15 μL of 10× NEBuffer 4, 1.5 μL of 100 μg/mL bovine serum albumin (BSA), 4 μg of the vector DNA, 2 μL of 10,000 units/mL PmeI, 2 μL of 20,000 units/mL EcoRI, and H_2O to make up the final volume of 150 μL. Incubated the reaction at 37 °C for 1 h. Dephosphorylation of the digested vector is carried out as per the following procedure: To the above digestion reaction tube, added 2 μL of calf intestinal alkaline phosphatase followed by incubation at 37 °C for 1 h. The vector DNA can be purified with the PCR purification kit with a final volume of 80 μL. The concentration of the purified vector is determined by OD measurement.
5. Ligation of the shRNA oligo is conducted into the digested vector as follows: To a 1.5-mL microtube added 2 μL of 10× T4 DNA ligase buffer, 100 ng of vector DNA, 3 μL of the hairpin insert, and 2 μL of T4 DNA ligase. Adjusted the volume to 20 μL using H_2O and incubated at room temperature for 15 min. Transferred the tube on ice.
6. Transformation: Added 50 μL of thawed MAX Efficiency DH-5α competent cells to each prechilled 17 × 100 mm polypropylene tube on ice and 2 μL of the ligation mix. Mixed gently by tapping the tube. Incubated on ice for 30 min. Heat-shocked the cells in a 42 °C water bath for 45 s. Put the cells on ice for 2 min. Added 0.9 mL of SOC medium which has been

warmed to room temperature to the tube and shaken the tube at 225 rpm and 37 °C for 1 h. Spread 20–200 μL of cells on LB plate with 30 μg/mL kanamycin followed by incubation of the plates overnight at 37 °C.

7. PCR screening of colonies: Five to ten colonies were picked up for each shRNA construct. Each PCR reaction contains 1 μL of 10 μM forward and reverse primers and 10 μL PCR master mix. Adjusted the volume to 20 μL by adding H_2O. A micropipette tip may be helpful for picking up cells from single colonies followed by dipping the tip in the PCR tube. PCR reaction is programmed as follows: 5 min at 95 °C, followed by 35 cycles of 95 °C for 30 s, 56 °C for 30 s and 72 °C for 1 min. Checked the PCR products on a 2% agarose gel. Three PCR products are visible: 121 bp from the blank vector control, 91 bp from the self-ligation products, and 148 bp from the vector with shRNA insert (91 bp + 57 bp insert). Picked up two to four positive colonies for each construct, processed for plasmid DNA preparation and finally the DNA sequencing and its verification.

3.3 Making miRNA Constructs

The miRNA constructs can be made with the similar steps as the shRNA constructs as discussed. The major difference is to synthesize the sense and the corresponding antisense DNA strands on the basis of the pre-miRNA sequence, adding overhangs matching the appropriate restriction enzyme cleavage sites at the two ends. Ordered the two synthetic DNA strands followed by annealing as discussed. The annealed DNA sequence can be cloned in the Pol III promoter, overhangs need to be matched with PmeI and EcoRI restriction sites. It may equally be cloned with Pol II promoter in the plasmid pCAG-RFP-miR-int-KX, where the matching is essential for the overhangs with KpnI and XhoI restriction sites. The cloning steps are the same as discussed earlier. We prefer a PCR methodology for the synthesis of the DNA encoding the pre-miRNA and cloning it behind a Pol II promoter in pCAG-RFP-miRint-KX plasmid, as explained below:

1. Designed two strands of synthetic DNA oligonucleotides covering the entire pre-miRNA and the restriction sites KpnI and XhoI.
2. Ordered three strands of oligonucleotide primers, one forward (F1) and two reverses (R1, R2). Primer F1 has a Kpn1 site and primer R2 has an Xho1 site at their 5′ end. Primers F1 and R1 possess 18 complementary nucleotides at their 3′ ends. Primers R1 and R2 possess 18 nt overlapping sequence.
3. Get the full-length pre-miRNA sequence by PCR amplification: Added the following components to an autoclaved microcentrifuge tube on ice: 5 μL 10× PCR buffer, 1 μL $MgSO_4$ (50 mM), 3 μL oligo F1 (10 mM), 1 μL oligo R1 (10 mM),

3 μL oligo R2 (10 mM), 1 μL dNTP mixture (10 mM), 0.5 μL Platinum Pfx DNA Polymerase (2.5 U/μL). Adjusted the volume to 50 μL through the addition of H_2O. PCR using the program: 2 min at 94 °C, followed by 35 cycles of 94 °C for 30 s, 55 °C for 30 s, 72 °C for 30 s. Stored the reaction at 4 °C after cycling or at −20 °C till further use.

4. Inserted the PCR product into the pCAG-RFP-miRint-KX: Digested the PCR product (from **step 3**) and the pCAG-RFP-miRint-KX vector with Kpn1 and Xho1 as per the enzyme manufacturer's instructions.

   Ligated the PCR products into the digested vector as follows: In a 1.5-mL microtube, added 1 μL of 10× T4 DNA ligase buffer, the digested PCR product (the insert) and 100 ng of the digested vector DNA, and 1 μL of T4 DNA ligase. The amount of the insert added in such a way that the molar ratio of the insert to the vector is 3–1. Adjusted the volume to 10 μL using H_2O and incubated at room temperature for 15 min. Kept the tube on ice.

5. Transformation: Added 50 μL of thawed MAX Efficient DH-5α competent cells (Invitrogen) to the 10 μL ligation mixture (from **step 4**). Mixed gently by tapping the tube. Incubated on ice for 5 min. Heat-shock the cells in a 37 °C water bath for 45 s, followed by putting the cells on ice for 5 min. Spread all cells on LB plate with 100 μg/mL ampicillin and incubated the plates overnight at 37 °C.

6. PCR screening of colonies: Picked up five to ten colonies for each miRNA construct. Each PCR reaction possesses 2 μL 10× PCR buffer, 0.8 μL $MgSO_4$ (50 mM), 0.8 μL oligo F1 (10 μM), 0.8 μL reverse primer (10 μM, binding on the vector), 0.5 μL dNTP mixture (10 mM), and 0.25 μL Taq DNA polymerase (2.5 U/μL). Adjusted the volume to 20 μL through the addition of H_2O. A micropipette tip is convenient to pick up cells from single colonies and dipped the tip in the PCR tube. PCR using the program as follows: 2 min at 94 °C, followed by 35 cycles of 94 °C for 30 s, 55 °C for 30 s, 72 °C for 30 s. Checked the PCR products on a 2% agarose gel. Two to four positive colonies for each construct are sufficient for plasmid DNA preparation followed by sequence verification.

3.4 Testing siRNA, shRNA, and miRNA Constructs

Soon after the development of the constructs, they need to be tested to assess which construct works the best and if it meets the knockdown requirement. The most direct way of checking is to transfect our constructs into cultured cells expressing our target gene followed by the measurement of the target gene expression in the transfected cells. We can assess the levels of the target gene expression through Western blot for the protein expression and Northern blot or reverse transcription PCR (RT-PCR) for the mRNA expression. It is essential to utilize cells with high transfection efficiency, assessing the upper limit for the detection of knockdown. We prefer cells which can be transfected with 70% or higher transfection efficiency. It is essential for a good construct to knock down the target by 80% or better. If the transfection efficiency is 70%, an 80% knockdown will lead to the detection of maximally a ~60% reduction in our target expression. For siRNAs, transfection efficiency is normally not a problem as it can be transfected into most cell types with higher efficiency. For shRNA and miRNA constructs, the transfection efficiency would be a problem. This problem can easily be rectified through employing plasmids which express a marker, such as green fluorescent protein (GFP) or red fluorescent protein (RFP). If cells expressing the target gene are not available or if they are available but transfected too poorly to make it of use, alternative approaches are available.

3.4.1 Transfecting Cells

1. Choice of cells: The commonly used cells which are easily transfectable are the best to use. Commonly employed human cell lines include HEK293 and HeLa, and mouse cell lines, such as NF-1, N2a, and NSC-34.

   The protocol with NSC-34 cells is being described since these cells are frequently employed in neurodegeneration research. This is a cell line obtained by fusion of mouse primary motor neurons and a mouse neuroblastoma cell line. The transfection efficiency is less than that of the other cell types as listed above. As a result, the protocols described below can easily be applied to the other cell types.

2. Prepare cells as follows: Warm the culture medium (DMEM plus 10% FBS, 100 U penicillin and 100 μg of streptomycin) in an incubator to 37 °C. Sprayed and wiped all tubes and bottles with 70% EtOH. Thawed one vial of frozen cells in a 37 °C water bath. Added 1 mL medium to the tube of cells. Mixed gently and transferred the cells from the vial with a pipette to a 15-mL tube. Added 5-mL prewarmed medium, mix and centrifuge at 400 × *g* for 3 min. In the next step, removed the medium, followed by addition of 1 mL fresh medium to the cell pellet and resuspended the cells by pipetting up and down till no clumps are visible. Added 10 mL prewarmed medium in addition and mixed properly. Then the entire content is transferred to a T25 flask. Incubated at 37 °C till confluence is observed, which usually takes approximately 3 days.

3. Splitting and plating cells: Aspirated old medium from confluent cell flask and washed cells once with 10 mL warm Dulbecco's Phosphate-Buffered Saline (DPBS) devoid of calcium. The DPBS is aspirated out and added 1 mL 0.5% trypsin to cells. It is then swirled over cells for about 3 min. Added 5 mL of fresh+ medium to the flask followed by resuspension of the cells in the

medium by pipetting up and down until no clumps are visible. Pipetted the cells into a 15-mL tube. Transferred the desired number of cells and proper amount of medium to a new flask (splitting ratio at 1:10) or 6-well plates (1–2 × 10^6 cell per well).

4. Transfected the cells with siRNA, shRNA, or miRNA constructs with Lipofectamine 2000 (Invitrogen) as per the manufacturer's instructions. We normally employ the following procedure: Plate 1–2 × 10^6 cells in 2 mL medium in each of the wells in a 6-well plate. The cells will develop to 90–95% confluency by the following day. Before the transfection, changed medium to DMEM plus 5% FBS devoid of antibiotics and incubated for 1 h. Aspirated the medium and applied to the cells with 500 μL transfection medium consisting of 100 pmol μL siRNA or 4 μg plasmids encoding miRNA or shRNA, and 5 μL (for siRNA) or 10 μL (for plasmids) Lipofectamine 2000 in opti-MEM I. Incubated at 37 °C in a CO_2 incubator for 48–72 h. In the next step, the cells are harvested for protein or RNA extraction to estimate the levels of knockdown.

3.4.2 Determination of Protein Knockdown Using Western Blot

1. For the extraction of the protein, media is removed from the culture plates, followed by twice washing with cold PBS. Any extra PBS is carefully soaked up with the help of a piece of filter paper. Added 50 μL of RIPA lysis buffer with 1× protease inhibitor to each well. It is advisable to use a cell scraper to lift the cells from the bottom of the well. Passed cell lysate through a pipette 20 times to form homogeneous lysate. Transferred lysate to 1.5-mL microcentrifuge tube. Kept samples to stand for 5 min on ice. The next step is centrifugation of the resulting mixture at 14,000 × g for 10 min at 4 °C in order to separate cell debris from protein. The supernatant is transferred to a new tube for use or stored at −80 °C for future use.
2. Take an aliquot of the lysate followed by measuring the protein concentration with BCA Protein Assay Kit (Pierce) as per the manufacturer's instructions.
3. As per the abundance of the protein being identified, mixed 5–100 μg protein with 4× Laemmli's SDS sample buffer followed by boiling for 5 min. Cleared the mixture by a brief (~10 s) centrifugation at the top speed of a microcentrifuge.
4. We routinely employ the Mini gel apparatus for SDS-PAGE. Assemble the gel plates as per the manufacturer's instructions. Prepare the gel mix (for a single 10% separation gel) by mixing 2 mL 40% acrylamide/bisacrylamide (29:1 mixture), 2 mL 4× separation gel buffer, and 3.87 mL water. Prior to pouring the mix, added 12 μL TEMED and 120 μL 10% ammonium

persulfate. The next step is proper mixing and pouring to about 1 cm below the anticipated bottom of the wells. Added about 2 mm thick layer of 75% ethanol on top of the gel to avoid air exposure and accordingly have a smooth top edge of the separation gel. Let the gel polymerize for about 15–30 min.

5. Prepare the 5% stacking gel mix through mixing of 0.5 mL 40% acrylamide/bisacrylamide (29:1 mixture), 1 mL 4× stacking gel buffer, and 2.5 mL water. Following the separation gel is solidified, poured off the ethanol layer, and washed 2× with water. Added 8 μL TEMED and 50 μL 10% ammonium persulfate to the stacking gel mix, immediately mix and pour on top of the separation gel with a pasture pipette. The comb is inserted gently and left for 10 min in order to polymerize. The electrophoresis unit are assembled and immerse it in SDS gel running buffer.
6. The wells are flushed and cleared with gel running buffer. Loaded the protein samples (from **step 3**) and 10 μL kaleidoscope prestained standards to the wells with the help of long gel loading tips. Run through the stacking gel, initially with constant voltage at 60 V, followed by increasing the voltage to 80 V when the bromophenol blue enters the separation gel. Run the gel for about 2 h till the dye has migrated to the bottom of the gel.
7. During running of the gel, a piece of nitrocellulose membrane and two pieces of Waterman filter paper to the same size are cut as per the gel size (5.5 cm × 8.5 cm). Soaked the membrane and sponges in 1× gel transfer buffer for 1 min.
8. Disassembled the gel from the gel running apparatus followed by soaking of the gel and the Waterman filter papers briefly in 1× gel transfer buffer.
9. The transblot apparatus are assembled in the following order (from the negative to the positive pole): sponge, filter paper, gel, membrane, filter paper, sponge. Transferred with a cold pack at 100 V and 4 °C for 1 h.
10. Removed the membrane from the sandwich and washed briefly in PBS. Removed any small pieces of the gel that need to be removed sticking to the membrane.
11. Incubated the membrane in blocking buffer for 1 h at room temperature. Changed to the blocking buffer with the primary antibody and incubated at 4 °C overnight. Washed the membrane thrice with PBST for 10 min each. Incubated with secondary antibody conjugated to HRP in blocking buffer for 1 h at room temperature. Washed the membrane thrice with PBST for 10 min each. All the incubation and washing steps need to be carried out with gentle agitation.

12. Detection of the knockdown target protein:

    Removed the ECL detection reagents from storage at 4 °C. Mixed 1 mL detection solutions A and 1 mL solutions B. The excess wash buffer is drained from the washed membranes and placed the protein side up on a sheet of SaranWrap. The mixed detection reagent is pipetted on to the membrane. It is then incubated for 2 min at room temperature. Drained off the excess detection reagent through tilting the membrane gently with forceps and touching the edge with a piece of filter paper. Placed the blots protein side down on to a fresh piece of SaranWrap. The blots are wrapped up and gently smooth out any air bubbles. Visualize the image of the blot with Fuji LAS 4000 or another imaging machine.

13. In order to confirm the equal loading, stripped the blot if necessary with stripping buffer for 10 min with agitation. Washed the membrane with PBST and blot for tubulin or another housekeeping protein as discussed in earlier steps.

14. The levels of the target protein are quantified using ImageJ or other imaging software and normalized it to the levels of the house keeping protein.

3.4.3 Determining RNA Knockdown Using Northern Blot

1. Prepared probes: Selected a unique segment ranging 200 and 1000 bp in length from the target RNA sequence. A pair of PCR primers are designed to make the DIG labeled DNA probe by PCR utilizing a PCR DIG Probe Synthesis Kit (Roche). Added to an autoclaved microcentrifuge tube on ice: 5 μL 10× PCR buffer with $MgCl_2$ (vial 3), 2 μL forward primer (10 μM), 2 μL reverse primer (10 μM), 5 μL PCR DIG Probe Synthesis Mix (vial 2), 0.75 μL Enzyme mix (vial 1), and template DNA (1–50 ng genomic DNA or 10–100 pg plasmid DNA). Adjusted the volume to 50 μL through addition of water. PCR using the program: 2 min at 94 °C, followed by 35 cycles of 94 °C for 30 s, 60 °C (or the annealing temperature of the primers) for 30 s, 72 °C for 40 s. Used 5 μL of the reaction mixture for checking the results of the reaction on an agarose gel. Stored the reminder of reaction at −20 °C.

2. RNA is extracted from the cells with the Trizol reagent as follows: The media is decanted from 6-well plates followed by washing of the cells in ice-chilled PBS. Excess PBS is soaked with a filter paper. One milliliter of TRIZOL reagent is added to each well and scraped with cell scraper. The cell lysate is passed several times through a pipette.

    Lysate is transferred to 1.5 mL microcentrifuge tube. Leave the samples at room temperature for 5 min. Added 0.2 mL of chloroform per 1 mL of TRIZOL reagent. Sample tubes are capped securely. Vortexed samples vigorously for 15 s and incubated them at room temperature for 2–3 min.

Centrifuged the samples at 12,000 × *g* at 4 °C for 15 min. Transferred upper aqueous phase carefully without disturbing the interphase into new 1.5 mL tube. Added 0.5 mL of isopropyl alcohol per sample. Incubated samples at room temperature for 10 min and centrifuged at 12,000 × *g* at 4 °C for 10 min. Removed the supernatant completely. Washed the RNA pellet with 1 mL 75% ethanol. Mixed the samples by vortexing and centrifuged at no more than 7500 × *g* at 4 °C for 5 min. Repeated the above washing procedure once. Removed all leftover ethanol. Air-dried RNA pellet for 5–10 min and dissolved RNA with 15 μL nuclease free water. Stored the RNA at −80 °C.

3. ***Measurement of RNA concentration***:

   Dilute 1 μL of RNA with 99 μL of nuclease free water (1:100 dilution). OD is measured at 260 and 280 nm to assess sample concentration and purity. The A_{260}/A_{280} ratio need to be above 1.6. Applied the convention that 1 OD at 260 equals 40 μg/mL RNA.

4. ***Prepare and run a denaturing RNA gel:***

   For the preparation of a 200 mL, 0.9% agarose formaldehyde gel, dissolved 1.8 g agarose in 170 mL water in a 65 °C water bath. Added 10 mL formaldehyde (37% stock), and 20 mL 10× MOPS running buffer. Poured the gel to about 0.6 cm thick in a fume hood and allowed it polymerize for ~15 min.

   Mixed 10 μg RNA with 1.5× loading buffer and incubated at 65 °C for 5 min. Cooled on ice, followed by addition of 2 μL loading dye. Prerun gel at 5 V/cm for in 1× MOPS running buffer for 5 min. Loaded samples and run with the same voltage till the bromophenol blue dye reaches the bottom of the gel (2.5–3.5 h for a 10 cm long gel). Soaked the gel thrice for 5 min each in distilled water for the removal of the formaldehyde. Take the photograph of the gel on a UV box.

5. ***Transfer to nitrocellulose membrane:***

   Cut nitrocellulose membrane along with eight pieces of Whatman filter paper to the same size as the gel and soaked in 5× SSC. Assembled transfer stack as follows: Filled a glass tray with 500 mL 5× SSC. Placed a glass plate across the top of the glass tray. Placed a wet filter paper bridge on top of the glass plate, and allowed the two sides drop into the 5× SSC below. Put the layers on top of the paper bridge on the glass plate in the following order: three pieces of wet filter paper, the gel, the wet nitrocellulose membrane, two pieces of wet filter paper, three pieces of dry filter paper, and ~15 pieces of dry paper towel. Place another glass plate and weight (150–200 g) on top of this stack. Allowed to transfer for ~3 h or longer.

6. Cross-link the RNA to the nitrocellulose membrane by placing it on a new piece of filter paper wet with 5× SSC in a UV cross-linker (Stratalinker) and use autocross-link with 1200 J. Removed the membrane and soaked for about 10 s in 1× MOPS buffer for the removal of excess salt and agarose.
7. ***Hybridization and detection of the target RNA:***

   Placed the membrane in a hybridization tube. Prehybridized with 4.5 mL DIG Easy hybridization solution (Roche) for about 1 h at 50 °C with gentle rotation in a hybridization oven (HB-1000 Hybridizer from UVP LLC). Boiled of DIG labeled-DNA probe (250 ng obtained from **step 1**) in DIG Easy hybridization solution (0.5 mL) for 5 min. Chilled quickly in ice-cold water. Added the probe to the hybridization solution in the hybridization tube and allowed to hybridize at 50 °C overnight. Washed the membrane two times with 125 mL 2× SSC plus 0.1% SDS at room temperature for 5 min each. The next step is washing of the membrane twice in 125 mL 0.1× SSC plus 0.1% SDS (preheated to 68 °C) at 68 °C with gentle agitation for 15 min each. Rinsed the membrane in 1× DIG washing buffer for 1 min. Blocked the membrane in 50 mL 1× block buffer (5 mL 10× DIG Maleic acid buffer, 5 mL 10× DIG block buffer, and 40 mL H_2O) with rocking for a period of minimum of 1 h. Added 5 μL Anti-DIG antibody to blocking buffer (1:10,000) and incubated at room temperature with rocking for 30 min. Washed membrane two times in 50 mL DIG washing buffer with rocking for 15 min each. The membrane is kept in 10 mL DIG detection buffer for 5 min for proper equilibration. Placed the membrane with DNA side facing up on a developing folder (or hybridization bag) and applied 1 mL CDP-star working solution. It is essential to cover the membrane with the second sheet of the folder in the next moment and spread the solution evenly. Proper care should be taken to avoid any bubbles on the membrane. Incubated at room temperature for 5 min. The excess liquid is squeezed out and sealed the folder completely. Image the membrane with fuji LAS 4000.
8. Strip the blot as per the necessity through washing the membrane in boiling 0.1% SDS for 5 min, poured off the washing solution, followed washing with 2× SSC at room temperature.
9. Probed the membrane with the help of an 18S or 7S rRNA or any other housekeeping gene probe to be used as a loading control using the same **steps 1–7**. Quantified the bands using imageJ [2].

1. ***Treat RNA samples with DNase:***

   To a microtube, added 2 μL 10× reaction buffer, 10 units of RNase-free DNase, 20 μg of RNA, and H_2O to a final volume up to 20 μL. Incubated at 37 °C for 60 min. Added

3.4.4 Determining RNA Knockdown Using Reverse Transcription and Quantitative PCR (RT-qPCR)

2 μL of DNase stop solution for the termination of the reaction followed by incubation at 65 °C for 10 min to inactivate the DNase.

2. ***cDNA synthesis***:

   To a microtube, added 2.0 μL RNA from **step 1**, 1 μL of Oligo (dT) (50 μM), 1 μL of dNTPs (10 mM each), and 6 μL of RNase-free H_2O. Incubated at 65 °C for 5 min. Moved the tube on ice for at least 1 min. At the same time, two additional two negative controls are constructed as NTC (no template RNA) and NoRT (no reverse transcriptase).

3. Continue the cDNA synthesis:

   To the tubes from **step 2**, added 2 μL 10× RT buffer, 4 μL 25 mM $MgCl_2$, 2 μL 0.1 M DTT 1 μL RNaseOUT (40 U/μL), and 1 μL of SuperScript III RT (200 U/μL). Mixed well followed by a brief spin. Incubated at 50 °C for 50 min followed by 85 °C for 5 min to terminate the reaction. Chilled on ice.

4. Quantitative PCR: The reactions for the target gene and one to two housekeeping genes (as internal controls) would be carried out in parallel. To a microtube, added 2 μL cDNA from **step 3**, 1 μL forward and reverse primer (10 μM each), 15 μL of Sybr Green mix, and 12 μL of H_2O. Set PCR with the program: 3 min at 95 °C, followed by 40 cycles of 15 s at 95 °C, 30 s at 56 °C (2–5 °C below T_m), and 30 s at 72 °C.

5. Quantify knockdown: Derived the Ct values from the PCR data with the threshold set near the middle of the linear range on a log scale. For each sample, calculated the ΔCt by subtracting the Ct value for an internal control. Subtracted the ΔCt from the knockdown samples from the control sample (from cells transfected with nonspecific siRNA, shRNA, or miRNA) for obtaining the ΔΔCt that can be utilized to calculate the percent knockdown [2].

RNA Interference Has a Wide Range of Applications

- Gene knockdown.
- Functional genomics.
- Medicine.
- Antiviral.
- Cancer.
- Safety.

The most common use of RNA interference is the treatment of genetically inherited diseases.

Gene silencing has been proved to an efficient technique for the remedy of clinical disorder [3, 4]. Reports are available for systemically administration of RNAi in humans from siRNA via targeted nanoparticles [5].

Other Forms of RNA Interference

Apart from naturally occurring and manufactured siRNAs, there have been recent publications of alternative forms of both, as follows:

Micro (mi)–RNAs: These are an abundant class of short (19–25 nt) single-stranded RNAs that are expressed in all higher eukaryotes. They are being encoded in the host genome and are thereby processed by Dicer from 70 nt hairpin precursors. Their function is the regulation of endogenous gene expression during development through interfering with mRNA expression by imperfect base pairing and inhibition of protein translation. Current work has detected their specific roles in the regulation of early hematopoiesis and lineage commitment. They are sometimes known as small temporal RNAs as a reflection of their importance in the regulation of developmental timing.

Piwi-Interacting (pi) RNAs

These are single-stranded 25–31 nt RNAs which have recently been detected in mouse, rat, and human testes. They were observed to be associated with Piwi protein (a subclass of Argonaute proteins) and the human RecQ1 protein for the formation of a *Piwi-interacting RNA complex (piRC)*. These complexes are presumed to regulate the genome within developing sperm cells.

Short-Hairpin (sh) RNAs

These are synthetically manufactured and their design is on the basis of a naturally occurring miRNAs. They are able to be expressed by plasmid vectors or endogenously inserted into the genome. The vectors consist of an inverted repeat of 19–29 nucleotides with the desired sequence separated by a short loop of 6–9 base pairs. When the nascent RNA is synthesized, it will develop a stem loop which can then be processed by DICER. These vectors commonly employ pol III promoters, usually utilized by the cell to drive expression of short RNAs, e.g., transfer RNAs.

Small Modulatory (sm) RNAs

These are short, double-stranded RNAs that are observed in the nucleus of neural stem cells of mice. They have an important role in mediating neuronal differentiation through dsRNA/protein interaction.

4 Making RNAi Work: Successful Transfection

Whether in vitro or in vivo, the RNA interference mechanism can be started after the siRNA has been transported into the cytoplasm of the cell. A successful transfection needs that the siRNA molecule carrying a net negative charge under normal physiological conditions, should come into contact, followed by crossing a cell membrane, which also carries a net negative charge. Many methods of siRNA transfection have been developed for use in vitro and in vivo.

4.1 In Vitro Transfection Mechanisms and Protocols Naked Delivery

The challenge of a successful transfection in vitro is to get intracytoplasmic delivery of siRNA to more number of cells as possible, simultaneously maintaining the toxicity associated with transfection or the transfection reagent to a minimum. Optimization of both the method and conditions of transfection is necessary for each experiment and is dependent on the cell line and siRNA being employed.

The most basic approach includes the addition of naked siRNA directly to the cells, not using transfection reagent. This was attempted in neurons. The siRNA was observed to enter the endosomal compartment of the cells while in suspension (identified with fluorescent tagging). However, its release into the cytoplasm when fully functional was not demonstrated. Thus, to ensure functional delivery, the siRNA should either be forced into the cells under high pressure (microinjection) or enter through micropores formed by affecting the charge of the cell membrane with electric pulses (electroporation) (Ambion technotes). High levels of cytotoxicity associated with these methods implies that they are usually reserved for the more difficult to transfect cell lines, e.g., neurons.

4.2 Chemical Transfection

Chemical transfection reagents are the common to use for siRNA transfection vehicle, out of these cationic liposomes (CL) are the most common. Presently, there are more than 30 different commercial varieties of CL formulations that have been constructed specifically for siRNA delivery, e.g., Lipofectamine 2000. They act through surrounding the siRNA with a positively charged lipophilic shield that is able to fuse with the cell membrane and transport the siRNA into the cytoplasm of the cell contained within endosomes. The precise mechanism controlling the timing of their functional release from the endosomal compartment cannot be determined and will affect the timing of silencing onset. Peak mRNA destruction is observed to occur within 24–48 h transfection being transient with an average duration of 2–3 days (Ambion technotes, undated). The exact length of silencing is regulated by the growth rate of the cell line, the level of gene expression, and the half-life of the protein being knocked down.

The development of other transfection reagents is working out simultaneously. These include: (1) Membrane permeant peptides that are short amphipathic peptides translocating across lipid bilayers in an energy independent manner; (2) Atecollagen, a highly purified, pepsin treated type 1 collagen from calf dermis that is positively charged; and (3) Organic-inorganic hybrid nanoparticles that are poly(ethylene glycol)-block-poly (aspartic acid) with calcium phosphate.

All of these have been utilized for successful delivery of siRNA in vitro. There are a number of variables that are known to affect the transfection efficiency and these need to be optimized for each particular cell line.

Factors influencing the transfection rate are as follows:

1. Cell density.
2. Concentrations of transfection reagent and siRNA, the time they are left together before adding to the well.
3. The presence of antibiotics or serum in the cell medium.

As for example, while employing Lipofectamine 2000 (a CL) a cell confluency of 30–50% is advisable. The Lipofectamine and diluted siRNA should be allowed to combine for 30 min for achieving optimal transfection efficiency.

4.3 Vector-Mediated Delivery: Overcoming Transient Silencing

DNA-vector-mediated and virus-vector-mediated mechanisms have been developed for overcoming the limitation of transient silencing. The siRNA expressing vector works as a platform for the production of a large amount of siRNA for a relatively long period. There are two main types of DNA-vector based RNAi systems both of which involve the use of Pol III promotors (U6 or H1).

In the first case, the sense and antisense strands of siRNA are expressed from different cassettes that are aligned in tandem in the same construct, i.e., tandem type. In the subsequent step, they bind and form a short double stranded (si) RNA that is presented directly to RISC and activates the RNAi pathway.

In the second, the sense and antisense strands are expressed as a connected ribonucleic acid with several intermediate bases that form a stem loop structure (short hairpin (sh) RNA). This is later presented to Dicer for further cleavage to siRNAs and RISC presentation.

A particular vector was termed pSUPER, which is used as an RNA polymerase III promotor for expressing short dsRNA in the form of an inverted repeat sequence possessing a hairpin loop. The mammalian cells were transfected with electroporation methods and the targeted protein levels were suppressed for 5–7 days following ten rounds of cell division. Comparison of these two methods has revealed increased efficiency of silencing with shRNA production in lieu of the tandem type.

Retrovirus vectors have also been employed, either through oncoretrovirus or lentivirus vectors. Lentivirus vectors are advantageous, as they can infect both actively dividing and non-dividing cells. They are also able to generate transgenic animals since they are resistant to proviral silencing during development.

4.4 In Vivo Transfection

Once the specificity and efficiency of the siRNA molecule has been validated in vitro, the development of an in vivo application is possible. However, due to the susceptibility of siRNAs to nuclease degradation in serum, and the problem in finding a delivery method which achieves effective systemic distribution to the cell cytoplasm in vivo, this is the area which has proved most elusive.

4.5 Naked siRNA

Initial studies were conducted employing naked siRNA, with the advantage of excluding toxicities associated with transfection reagents. Routes of administration considered included intravascular, intratumoral, intracranial, intraperitoneal, intrasplenic, intramuscular, subretinal, subcutaneous, mucosal, topical application, and oral ingestion. Initial fears regarding the safety of introducing naked siRNAs in mice, with regard to their potential for inducing immune responses, were studied. Through the measurement of levels of plasma interleukins and interferons after intraperitoneal and intravenous injection of siRNAs, an immune response was not observed and systemic siRNAs were also observed to be well tolerated by mice.

To date, the most commonly reported technique of successful administration of naked siRNA in vivo has been the use of intravenous "hydrodynamic delivery."

4.6 Hydrodynamic Delivery of siRNA

This introduces the nucleic acid via the intravascular route at high speed and volume, leading to greater levels of transfection with a more diffuse distribution. Described sites of intravenous delivery include intraportal, intra-saphenous, and the tail vein. Modification of this technique has been successful in the delivery of naked siRNAs silencing both exogenous and endogenous genes in multiple organs in mice. The efficiency of intracellular delivery has been observed to be directly dependent on the volume of fluid given and the speed of injection. In mice this has been standardized at 1 mL/10 g bodyweight injected over a period of 3–5 s. The exact mechanism of action for the transfection of the gene/siRNA into the cells after hydrodynamic delivery is not clearly understood till date.

While delivered via the tail vein, it is presumed that as the injection rate exceeds cardiac output so the introduced fluid aggregates in the superior vena cava. This is later forced out into vessels within organs and then through fenestrae in these vessels into extravascular spaces. In the next step, naked siRNA is brought into contact with the cells of the organ prior to mixing with blood, hence reducing the chance of nuclease degradation.

It was presumed that actual transfection of the hepatocyte by this method is due to "hydroporation." Membrane defects (pores) are formed after the high-pressure delivery and supplies a route for the introduction of the siRNA into the cell. Injection of a membrane permeability marker (Evans blue) reveals that these pores closed up within 10 min, during which the siRNA had entered the cytoplasm and the RNAi mechanism would be underway.

Data regarding exact distribution of siRNA molecules following tail vein injection are scant, though initial experiments reveal a majority of the siRNA distribution is to the liver. This can be due to its proximity to the vena cava and ability to accommodate extra fluid. Due to hydrodynamic injection in mice, certain side effects have been observed. The heart rate reduces from 510 beats per minute to 280 and there are transient irregular rhythms lasting for up to 60 s. A concurrent increase in QRS complex size also indicates increased cardiac chamber size. The liver expands and turns whitish soon after injection.

Clinical side effects involve apnea directly after injection; however, gentle massage of the abdomen improves the condition. This is associated with a transient, right-sided, congestive cardiac failure. Histopathological examination revealed single cell liver necrosis in half the mice after 1 day. The body weight remained increased for 30 min but regains normalcy after 2.5 h due to urination. Despite all these effects, the survival rate was observed to be 100%, and its wide applicability for in vivo siRNA delivery in mice has been observed. Development of this method of delivery is in early stages.

Ethical and animal welfare implications should also be considered as many may regard the intravenous injection of such high volumes of fluids over brief time periods to be unacceptable. Considering the significant side effects, it is uncertain that this technique would be of any use in humans. The use of naked siRNA has been further used through stabilization modifications. Unmodified siRNA molecules are prone to serum nuclease degradation, and without effective modification they will be simultaneously degraded on in vivo administration. A successful stabilizing 2-O-methyl modification was detected, which managed not to affect the efficacy of the siRNA molecule. Stability-modified 2-fluoro pyrimidines siRNAs molecules were employed and observed over 50% to be intact after 24 h of suspension in plasma. This is compared with complete degradation of normal molecules within 4 h. The modifications did not influence their silencing capabilities in vitro or in vivo and acted for their better resistance to nuclease degradation in plasma.

4.7 Chemical Modifications Enhancing Transfection In Vivo

Transfection reagents are a widely experimented technique for in vivo delivery of siRNA. Similar compounds synthesized for in vitro use have been trialed in vivo.

The use of cationic liposomes has been with problem principally due to their poor systemic distribution and toxicity. Preferential deposition of complexes in the lungs and liver, or areas containing proliferating vasculature (e.g., tumors) has been observed to occur after intravenous administration. Their potential toxicity has been observed including the induction of acute inflammation following intraocular, intra-articular, and intratracheal administration and emboli formation from liposome/siRNA complexes and agglutination of red blood cells with respect to their net positive charge. Transient acute toxicities of leukopenia, thrombocytopenia, and increases in alanine transaminase have also been observed. The development of second-generation molecules has helped, thereby increased the stability in serum, improved the target organ distribution and penetrate, and thereby helped the formation of a homogenous population of complexes with siRNA before injection. siRNA molecules have also been stabilized for in vivo administration with cholesterol conjugation. This has been observed to increase resistance to degradation in vivo and possesses good tissue distribution while injecting through the tail vein after 24 h. Utilizing this technique, SiRNA has been identified in the liver, heart, kidney, lung, and adipose tissue. This method was observed to be successful in silencing an endogenous gene (apoB protein) in the liver and jejunum. Plasmid vectors have also been utilized for in vivo siRNA delivery.

However, the relative ease of plasmid regeneration did not compensate for general poor transfection efficiency of the plasmid-based vectors, leading to prompting the construction of viral-/retroviral-based vectors for shRNA delivery. Vector systems that are based on the adenovirus are limited in such a way that they do not integrate into the host genome. In contrast, integration of the shRNA expression cassette into the host genome may happen with lentiviral-based vector systems, e.g., Moloney murine leukemia virus, the Murine stem cell virus, and HIV-1 [6]. As transgenes expressed from these viruses are not silenced during development, these vector systems can also be utilized to generate transgenic animals by the delivery of shRNAs to embryonic stem cells of embryos.

5 Applications of RNA Interference in Pathology

The complete potential of RNAi is not completely recognized, but there exist clear areas for which these are used in both experimental and toxicological pathology.

RNAi in Pharmaceutical Target Validation and Toxicology:

In the pharmaceutical industry, modern scientific techniques have expanded the pool of useful drug targets for the treatment of disease. However, the success rate of new drugs is diminishing. Two significant reasons of failure are unacceptable toxicity and failure of the mechanism for providing significant clinical benefit.

Both of these sources of attrition happen at a late stage in development and are costly both financially and in terms of resources expended. SiRNA has the ability to provide an early, specific, and relatively inexpensive methodology to study a pharmacological mechanism in both preclinical models of efficacy and toxicity. This would lead to a reduction in mechanisms with efficacy and safety issues as progressed into late stages, diverting resource into more productive targets. In particular, the utilization of siRNA in toxicity screening has the ability to decrease the number of animals utilized in toxicity testing [1, 4, 6–8].

5.1 RNAi in Mechanistic Pathology

The understanding of pathogenesis of lesions frequently relies on the testing of generated hypotheses in vivo. This generally involves the exaggerated action or inhibitions of cellular pathways through the overexpression or inhibition of cellular proteins. RNAi has the ability for rapid development of the specific inhibition of protein expression in vivo to either mimic pathological findings or to study the toxicity of administered compounds in animals with lack of pathways suspected of initiating pathological findings.

5.2 RNAi in Animal Models of Disease

Human diseases can be modeled in animals through the inhibition of key regulatory proteins. RNAi has the ability for the production of rapid and cost-effective models of human disease, particularly for proteins whose lacking in the embryo has significant effect on viability, preventing the use of gene knockout technology. This potential has been demonstrated by the use of adenovirus for introduction of siRNA specifically targeting tyrosine hydroxylase mRNA within neurons of the mid brain. This may be regarded as a key enzyme in the production of dopamine, a molecule which is involved in regulating food intake, addiction, and movement control. These animals may act a model of human Parkinson's disease.

5.3 RNAi as a Therapeutic Agent

It is an interesting use of RNAi as a therapeutic agent. Initial research and trial had been undertaken with RNAi as a therapeutic agent in animals. Certain technical hurdles are to be overcome before applying it in man. Initially, the prevention of liver disease was tried through targeting genes related to apoptosis control in the liver in two models of autoimmune hepatitis. SiRNA was directed against caspase and Fas and was delivered hydrodynamically through the tail and portal vein, respectively. In both cases this was able to reduce hepatocyte necrosis and inflammation, and protected the mice from future chronic fibrosis. In another experiment, siRNA targeting Fas in the kidney was employed to reduce ischemia-reperfusion injury. Hydrodynamic and normal volume intrarenal delivery had successfully decreased Fas protein expression by about fourfold.

In addition, the pathology of the renal tissue after the ischemic insult was observed to be much improved in comparison to the control. Following this success, one of the first examples of using RNAi to inhibit viruses in vivo was later attempted. Immunocompromised and immunocompetent mice were hydrodynamically injected with plasmids expressing hepatitis B virus (HBV) and shRNAs targeting HBV. This led to a 99% reduction in HBV detection with antibodies to identify HBV core antigen through immunohistochemical methods, and it was suggested that RNAi could be utilized for the treatment of viral disease in the future.

Later on, a number of experiments with RNA interference to target respiratory viruses have been attempted. At the initial stage, influenza virus was selected due to its significant public health issues, and life long a wholly effective vaccine. The target proteins were selected that are highly conserved across several subtypes of influenza and are essential for viral replication.

It was observed that combined hydrodynamic and intranasal (in a lipid carrier) delivery was most effective for inhibiting virus replication at the site of infection. It also decreased lung viral titers in infected animals and protected animals against lethal challenge. Another experiment with slow I/V delivery of siRNA complexed with a polycation carrier, and its delivery using DNA vectors iv/intra nasally also reflected a dose-dependent decrease in virus production. Intranasal delivery was also observed to be effective while targeting Parainfluenza virus and Respiratory Syncytial virus. Administration of siRNA with and without a vector (Transit TKO) was observed to be both protective and therapeutic. The capability to introduce RNAi across mucosal surfaces is also being studied as a means of treatment of sexually transmitted disease. Intravaginal delivery of RNAi with a target on two viral genes have been observed to protect the mice from the otherwise lethal Herpes simplex virus-2. RNAi has also been employed to alleviate joint inflammation in experimental animals. siRNA targeting tumor necrosis factor alpha was inoculated through injection into the knee joints of mice with collagen induced arthritis (CIA). This was conducted through electroporation. Scoring for the development of arthritis was conducted by assessing the inflammation of joints in the mouse paw, and in mice with CIA, accordingly joint inflammation was successfully inhibited. In the final stage, diminished pain responses have been visualized in rats following intrathecal delivery of an siRNA directed against the pain-related cation channel P2X3.

6 Conclusion

The field of RNAi is expanding at a greater pace leading to new discoveries. SiRNAs are currently being utilized in gene function analysis, target identification and validation, as well as therapeutic agents. Their potentiality for use in the evaluation target toxicity is significant and needs further investigation. Although it is a viable technique for in vitro experimentation, certain constraints include problems with intracellular siRNA delivery and effective gene silencing. Low transfection efficiency and excessive cytotoxicity are commonly encountered. It is important to have the development of an in vitro transfection method with an improved ratio of transfection to efficacy. For example, controlled forced intracellular delivery of siRNA by combining nucleotides with magnetic nanoparticles and application of a pulsed magnetic field would decrease the level of cytotoxicity and increase transfection efficiency more than that fulfilled with electroporation of cells exposed to naked nucleic acids.

Following delivery to the cell, effective knock-down of protein activity in the cell is affected by the half-life of the protein under investigation. As RNAi prevents novel protein production, long half-life proteins can maintain activity for longer periods following effective gene silencing. Knowledge of protein half-life is essential for planning and interpreting RNAi knock-down experiments. Deliveries of RNAi in viral vectors, particularly those that integrate into host genome, cause a prolonged knockdown of protein production. The short-lived nature of inhibition can be advantageous, while using as a therapeutic (to allow for rapid reversal following adverse events).

Demonstration of effective gene silencing depends on the ability to identify both a reduction in mRNA and protein levels of the specified target in tissues—possibly in limited biopsy material. The availability of reagents to accurately measure reduced protein production can lead to a further challenge in the interpretation of experimental results. The transient nature of the inhibition of protein production is important in dividing cells, where cell division results in the loss of RNAi activity. This is of particular concern in vitro, where cell division rates are usually high. Certain in vivo organ systems, for example, the gastrointestinal tract, also possess high endogenous rates of division and gene silencing in these organ systems would be particularly challenging. Virally mediated integration of RNAi sequences into stem cell could overcome this problem, but would result in a permanent knock down of protein expression that may be advantageous for studying in vitro systems, but would carry liability as a therapeutic in case adverse events were encountered. The development of efficient in vivo delivery of RNAi remains problematic. The paucity of publications discussing successful in vivo delivery advises that either this is technologically demanding or else systems are being developed confidentially to ensure commercial exploitation. Efficient and effective in vivo transient and specific protein inhibition with RNAi has proved to have significant potentiality for the generation of many therapeutics with associated financial reward. The search for a transfection method which achieves such efficient delivery of siRNA to a majority of body cell types avoiding significant toxicity is of practical importance. The current "gold standard" of hydrodynamic delivery questions the issues raised related to the adverse events associated with the delivery system. New developments in the field of in vivo delivery are anticipated. Moreover, the cost or production of RNAi molecules is a limitation for the use, particularly the quantities required for in vivo studies. The cost of RNAi molecules has decreased significantly in recent years, and expected to continue. Hopefully, cost should not be the barrier to use this valuable tool. In conclusion, this mechanism of post-transcriptional gene silencing is justified, but experiments should now concentrate on in vivo validation of the technique to continue the path of discovery to continue.

References

1. Kong XC, Barzaghi P, Ruegg MA (2004) Inhibition of synapse assembly in mammalian muscle in vivo by RNA interference. EMBO Rep 5 (2):183–188
2. Song MS, Rossi JJ (2017) Molecular mechanisms of Dicer: endonuclease and enzymatic activity. Biochem J 474(10):1603–1618. https://doi.org/10.1042/BCJ20160759
3. Crunkhorn S (2010) Clinical gene-silencing success. J Nat Rev Drug Discov 9:359. https://doi.org/10.1038/nrd3161
4. Gregory RI, Chendrimada TP, Cooch N, Shiekhattar R (2005) Human RISC couples microRNA biogenesis and posttranscriptional gene silencing. Cell 123:631–640. https://doi.org/10.1016/j.cell.2005.10.022
5. Davis ME et al (2010) Evidence of RNAi in humans from systemically administered siRNA via targeted nanoparticles. Nature 464:1067. https://doi.org/10.1038/nature08956
6. Zou GM, Yoder MC (2005) Application of RNA interference to study stem cell function: current status and future perspectives. Biol Cell 97:211–219
7. Song E, Lee S, Wang J et al (2003) RNA interference targeting Fas protects mice from fulminant hepatitis. Nat Med 9:347–351. https://doi.org/10.1038/nm828
8. Zender L, Hutker S, Liedtke C, Tillmann HL, Zender S, Mundt B, Waltemathe M, Gosling T, Flemming P, Malek NP, Trautwein C, Manns MP, Kuhnel F, Kubicka S (2003) Caspase 8 small interfering RNA prevents acute liver failure in mice. Proc Natl Acad Sci U S A 100 (13):7797–7802. https://doi.org/10.1073/pnas.1330920100
9. Dorn G, Patel S, Wotherspoon G, Hemmings-Mieszczak M, Barclay J, Natt FJ, Martin P, Bevan S, Fox A, Ganju P, Wishart W, Hall J (2004) siRNA relieves chronic neuropathic pain. Nucleic Acids Res 32(5):e49. https://doi.org/10.1038/sj.embor.7400065

Chapter 19

Expressing Cloned Genes for Protein Production, Purification, and Analysis-Biopharming

Abstract

Protein expressed in vitro through expression vector is very important for therapeutic agent, gene therapy, and transgenesis. There are certain *criterions for an expressed protein* to be biologically functional, i.e., the protein should be in proper 3D structure with proper post transcriptional and post-translational modifications. This chapter discusses the protocol for expression and purification of recombinant protein in bacteria and yeast. The first step includes amplification of gene of interest, insertion of the gene of interest into cloning vector, subcloning into expression vector, transformation to protein expressing bacteria or yeast, followed by test for the identification of recombinant protein and large-scale production and isolation and purification. The features for an expression vector are that it should include origin of replication, drug resistance marker, a promoter, transcription terminator, and restriction sites. Practical use of recombinant protein/isolation includes therapeutic use, antibody production, hormone production—e.g., insulin, enzyme production. Future prospect includes regeneration of organ either complete or a part of it which will be very much important for organ transplantation, say kidney transplantation for a patient with renal failure, cornea transplantation for a blind person from his own body. Some practical illustrations have been provided. Mammalian expression has been described, which is the system of choice to study the functional significance of a particular protein in the most physiologically relevant environment, since it allows for the highest level of post-translational processing and functional activity of the protein.

Keywords Gene, Recombinant protein, Expression, Biopharming, Expression vector, Post-translational modification

1 Basic Principle

Protein expressed in vitro through expression vector is very important for therapeutic agent, gene therapy, and transgenesis. In vivo gene expression can be analyzed through various recent biotechniques as real time PCR, DNA microarray, DNA chip, or other NGS studies. The expression profiles of mRNA or cDNA derived from respective mRNA are studied. The studies and production of expressed protein in vitro are of much importance in recombinant protein production for biopharming, therapeutic or other commercial application for recombinant protein. A very good example is therapeutic application for diabetes through recombinant insulin.

Aruna Pal, *Protocols in Advanced Genomics and Allied Techniques*, Springer Protocols Handbooks,
https://doi.org/10.1007/978-1-0716-1818-9_19,

The study of gene expression is directly related to our understanding of proteins. The early work of Christian Anfinsen in the 1950s has depicted that the sequence of amino acids in a protein is responsible for its final three-dimensional structure. Scientists have noticed that protein structure determines where it will act and what it will do. It is an important fact of consideration for the functional aspect of the enzymes, which is determined by the structure. It is a very important breakthrough that the conformational and structural component is the determining factor for gene expression biology. It aids us to establish the link between gene expression and the functional aspect of the cell. There are certain proteins having effect on DNA and RNA sequences, like transcription factors and histones. The study of gene expression also highlights on where the cell expression is modulated. The modulation of gene expression can occur in various subcellular locations, as the nucleus, the cytoplasm, the cell membrane (as there may be impact of proteins on RNA in those cellular subregions).

There are certain ***prime criterions for an expressed protein*** to be biologically functional.

The protein should be in proper 3D structure with proper post-transcriptional and post-translational modifications. It is one of the major reasons that mammalian protein expressed in prokaryotic or yeast vectors is not biologically active. Although the protein is expressed completely, but it lacks proper post-transcriptional and post-translational modifications, hence not biologically active; protein folding is also not proper. There are certain conventional techniques for studying protein: they are mass spectrometry, X-ray crystallography and/or nuclear magnetic resonance spectroscopy (pNMR), SDS-PAGE and double-dimension gels, etc. Certain other protein analysis techniques are ELISA, Western blotting, and immunohistochemistry.

1.1 Study for Protein Shape and Function

Mass spectrometry is the technique which leads the scientists to sequence the amino acids in a protein. Once the sequences are available, scientists are able to compare its amino acid sequence with databases for the identification of any related protein whose function is already known. Usually, similar amino acid sequences have similar functions within a cell. It is the amino acid sequence, which determines the charge of the molecule, its size, and its probable three-dimensional structure. The charge and size can be confirmed afterwards experimentally (via SDS-PAGE and double-dimension gels). To verify the three-dimensional structure, scientists attempt to crystallize the protein for the confirmation of its molecular structure with X-ray crystallography and/or nuclear magnetic resonance spectroscopy (pNMR).

1.2 Study of the Impact of Proteins on Genes or Other Proteins

A good way to cross-check the function of the protein is to determine the changes in cell during the absence of protein. Hence, it is preferable to use model systems (cell culture or whole organisms), where they check the functional aspect of specific proteins or genes through modification or mutation. The expression level of a gene can be estimated through various techniques. Measurement the transcribed mRNA (northern blot) or cDNA detection through real-time PCR, the expressed protein (Western Blot), or by directly staining the protein or mRNA when it is still in the cell (Immunohistochemistry or immunocytochemistry). Are some of the widely employed techniques. Newer techniques have revolutionized the way we studied gene expression—DNA microarrays, serial analysis of gene expression (SAGE), and high-throughput sequencing leads us to larger screens of multiple molecules simultaneously and opened up the possibility of newer and broader kinds of questions in biology. A new discipline, termed as systems biology, has been developed in order to analyze large datasets. It is also involved for checking the networks of molecules which interact and depict the framework for larger and more integrated knowledge of regulatory networks.

It is quite interesting to note that proteins are not the only gene regulators. Other regulatory molecules are RNA, acting on other nucleic acids through certain changes or disruption. As for example, there exists the family of riboswitches. It refers to the ribonucleic acid molecules which form three-dimensional structures which halt or interfere with transcription, provided with the proper external signal. Other illustration of RNA acting on other RNA is the process of RNA interference (RNAi). In this process, double-stranded RNA molecules can degrade mRNA before translation, as a result, it interferes with protein expression very efficiently.

Finally, results from these findings have proved to be very useful in terms of cell differentiation, growth, division, and genetic approaches for treating disease. Reports are available when some human diseases occur due to a defect in a protein's three-dimensional structure. Studies on gene expression and proteins expression allow us to detect the changes at the molecular level.

2 Expression and Purification of Recombinant Protein in Bacteria and Yeast

Recombinant Protein: It may be defined as the protein expressed through the process of introduction of recombinant DNA into a host (microorganism or yeast cell) leading to the production of the gene product.

2.1 Basic Steps to Obtain Recombinant Protein

1. The first step includes amplification of gene of interest (Using PCR).

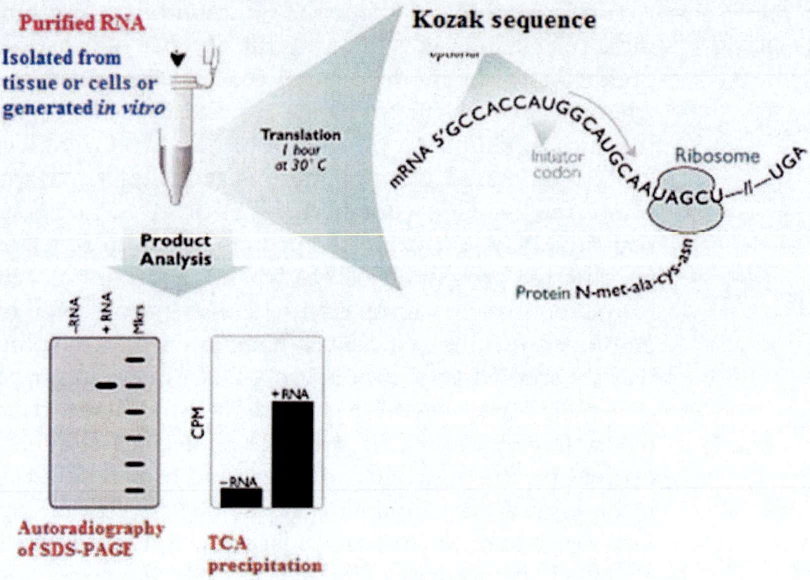

Fig. 1 Schematic representation of Protein expression analysis from RNA

2. Insertion of the gene of interest into cloning vector (Ex: PCR*8).
3. Subcloning into expression vector (Ex: pKK223-3 or PSVK 3).
4. Transformation to protein expressing bacteria (*E. coli*) or yeast.
5. Test for the identification of recombinant protein (Western blot or Fluorescence).
6. Large-scale production (large-scale fermenter).
7. Isolation and purification.

For protein expression, several factors need to be considered, which are expression vector, host system, the site for protein expression, properties of protein (membrane bound, solubility, single or multidomain, size of the protein) and identification, isolation, and purification method. Schematic diagrammatic representation of protein expression analysis from RNA and DNA has been represented in Figs. 1 and 2, respectively. Figure 3 gives an outline of human protein expression analysis.

The features for an expression vector are that it should possess the following criteria:

- Origin of replication.
- Drug resistance marker.
- A promoter.
- Transcription terminator.
- Restriction sites.

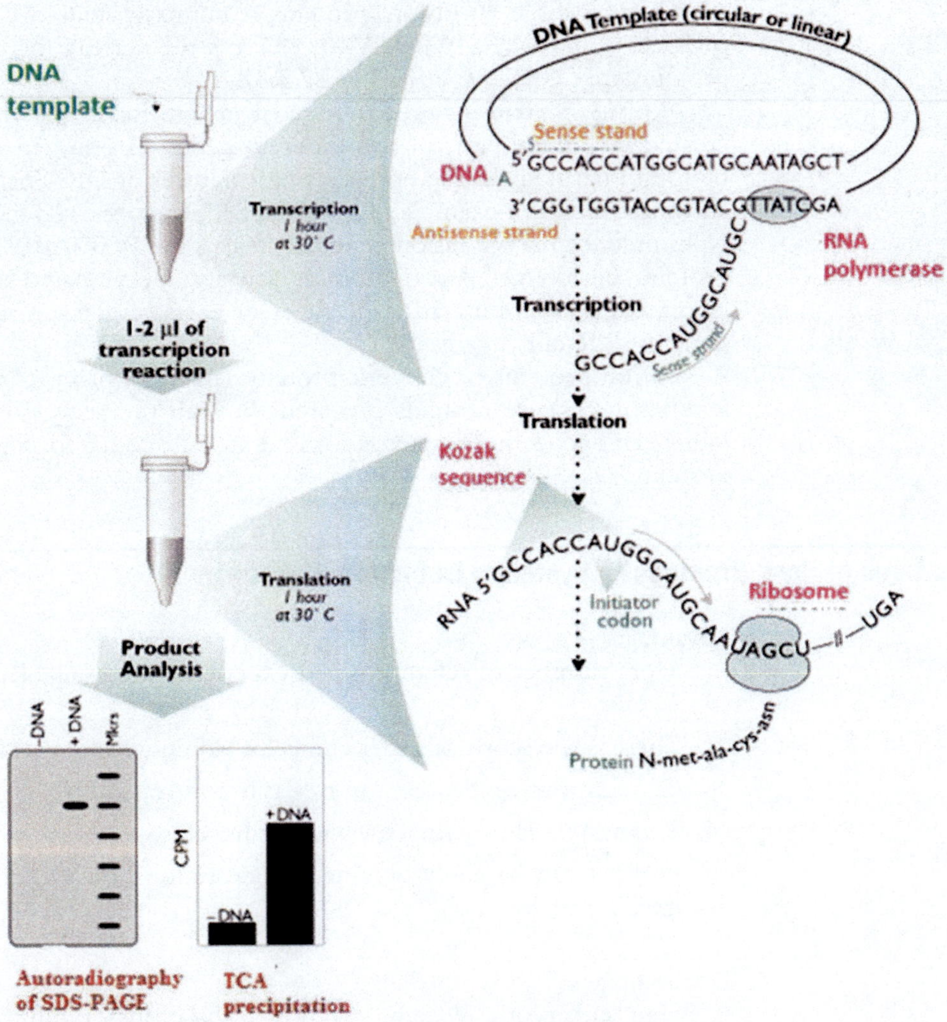

Fig. 2 Schematic representation of Protein expression analysis from DNA

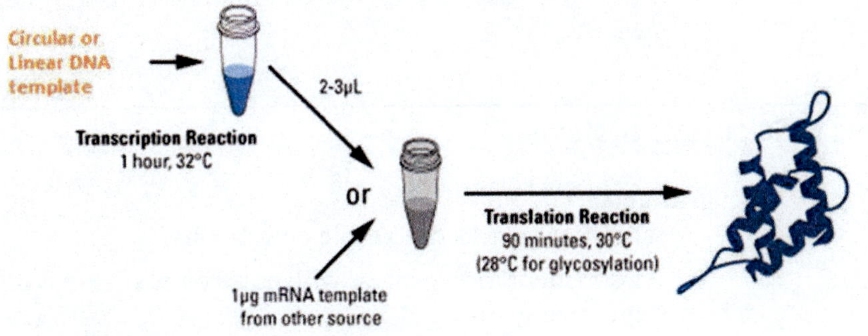

Fig. 3 Outline of Human protein expression analysis

Vectors for non-fusion proteins commonly employed are pKK223-3, pSVK 3, PSVL SV40, whereas the vectors for fusion proteins are pGEX, pQE, pET, pEZZ 18.

A fusion protein is a protein made from a fusion gene, which is created by joining parts of two different genes. Fusion genes may be found naturally in the body through transfer of DNA between chromosomes. For example, the BCR-ABL gene observed in some types of leukemia is a fusion gene that makes the BCR-ABL fusion protein. Fusion genes and proteins can also be prepared in the laboratory by a combination of genes or parts of genes from the same or different organisms.

Fusion proteins or chimeric proteins (made of parts from different sources) are proteins prepared through the process of the joining of two or more genes, which are actually coded for separate proteins.

3 Features of Host: Prokaryotic Systems, Example—*E. coli*

Advantages

1. More number of references and experience are available, e.g., *E. coli*.
2. There is the scope of wider choice of cloning vectors.
3. Gene expression process can be easily controlled.
4. This system is easy to grow with high yields.
5. Product can be easily obtained for secretion into the growth media.

Disadvantages

- Being eukaryotic system, no post-translational modification exists.
- Being eukaryotic system, biological activity and immunogenicity may differ from natural protein.
- There may be the presence of high endotoxin content in gram negative.

4 Features of Eukaryotic Systems

Advantages

- The system lacks detectable endotoxins.
- The fermentation process with eukaryotic system is relatively less expensive.
- The system aids in glycosylation and formation of disulfide bonds.

- Only 0.5% native proteins are being secreted. Hence, the process of isolation of secreted product is comparatively more simplified.
- It involves well-established large-scale production and downstream processing.

Disadvantages

- The process of gene expression is less controlled.
- Post-translational modification is essential for proper functioning of expressed protein. Glycosylation through this system may not be identical to mammalian systems.

There are three **basic steps for isolation of protein**:

1. Disruption of cells (Osmotic, Chemical, or Enzymatic).
2. Mechanical Clarification: Centrifugation Filtration.
3. Concentration: Precipitation or Ultra filtration.

Various Protein purification methods may be summarized as:

1. Charge: IEC/IEF.
2. Size: size exclusion chromatography.
3. Hydrophobicity: Hydrophobic Interaction Chromatography.
4. Ligand specificity: affinity chromatography, nucleotide and coenzymes resins, phosphoprotein resins.
5. Avidin biotin matrices.
6. Carbohydrate binding.
7. Dye resins.
8. Solubility: Precipitation.

5 Applications of Genetic Verification and Detection of Open Reading Frame (ORF) Expression

The most common use of in vitro expression may be to decode a nucleic acid to determine if a gene or ORF is present. In DNA, an ORF comprises of an initiation codon (usually ATG, but can also be GTG). A sequence of nucleotides coding for amino acids and ending with a termination codon (TAA, TAG, or TGA) is also present. For example, a prokaryotic DNA (or eukaryotic cDNA) clone can be assessed for the presence of ORFs through in vitro expression system. In other words, in vitro expression can be utilized for the verification of an ORF predicted by DNA sequencing. In either case, characterization of the resulting proteins is important. The protein size and structure are correlated to the size and sequence of the gene. Specific uses

1. involve analysis of viral RNA genomes in which eukaryotic translation systems are used to determine the number and function of the viral genes, and
2. study of differential protein expression with total cellular mRNA or polysomes from different tissues.

Cloned cDNA Expression: Cloned cDNAs placed behind a phage promoter (e.g., T7, T3 or SP6) are generally utilized for generating gene-specific mRNA useful for programing translation reactions.

Run-off 5′-capped or uncapped mRNA can be obtained in vitro and added to translation extracts. However, these DNA constructs can be utilized directly in eukaryotic coupled transcription/translation reactions.

5.1 Functional Analyses: Enzymatic Activity Analysis

Most of the proteins expressed with in vitro systems are folded in correct manner and processed. Their functional activity is assessed through normal in vivo enzymatic activity. If the extract system is lacking or possessing low level of the enzymatic activity of the expressed protein, the resulting translation reaction can be detected directly without protein purification. Other advantage of in vitro expression is the capability to add exogenous factors to study enzymatic activity, leading to elimination of the need for transfection studies. As for example, expression of adenylate cyclase (ACIV; 110 kDa) in a TNT® System produced a protein with the similar specific enzymatic activity as ACIV produced from a baculovirus expression system.

1. In another example, aromatase synthesized in a TNT® System reaction supplemented with canine pancreatic microsomes and recombinant cytochrome P450 reductase leads to an active enzyme.
2. Mutation Analysis. Followed by the process of cloning a gene, researches have been undertaken to detect the function of a gene product. The first step is to introduce a mutation into the gene for the examination of the effect on the expressed protein.

Methods of mutagenesis include:

1. Serial deletion mutation analysis may be employed by progressive truncation of the 5′- or 3′-ends of the gene or through Bal 31 digestion from an internal single cut restriction site.
2. Site-directed point mutation analysis (e.g., using Promega's GeneEditor™ System(g)).

Both the methods can be employed for the identification of functionally active domains or residues.

5.2 Post-translational Modification Analysis

Post-translational modification of the protein, like proteolytic cleavage or the addition of sugars, lipids, phosphate, or adenyl groups, is prime necessity for functional activity. Each in vitro expression system possesses its own endogenous post-translational modification activities. For example, various phosphorylation, adenylation, myristoylation, farnesylation, isoprenylation, and proteolytic activities have been detected with rabbit reticulocyte lysate. Addition of microsomal membranes leads for studying glycosylation, methylation, and removal of signal sequences. Since all the differences between the various in vitro expression systems are available, it may be essential to try both reticulocyte lysate and wheat germ extract (and in some cases, *E. coli* S30 extract) to detect which system can lead to a functional gene product with the "correct" post-translational modifications. Cellular extracts or different microsomal membrane sources (e.g., Xenopus egg extracts) can be assessed for providing additional modifying activities.

5.3 Molecular Interaction Detection Protein–Protein

Specific protein–protein interactions can be assessed with in vitro expression methods. These interactions can include specific binding (such as antibody-antigen and ligand-receptor binding), macromolecular assembly, and formation of functional transcription complexes. In a general application, one protein partner is expressed in large amounts and purified from *E. coli* as a fusion protein. The other protein partner is expressed in an in vitro expression system as a labeled protein and utilized as a probe for detection of the interaction. This technique is usually useful for the verification of the results from yeast two-hybrid experiments. Various biochemical analysis methods may be employed for the characterization of the expressed proteins.

5.4 Protein-DNA and Protein-RNA

Putative DNA binding proteins, such as transcription factors, can be assessed for their capability to bind to specific sequences on radiolabeled oligonucleotides. The binding is identified by an electrophoretic mobility shift assay (EMSA) where greater retardation of the protein-DNA complex is observed in comparison to unbound DNA.

Normally, the labeled DNA is added directly to the in vitro expression reaction.

5.5 DNA-RNA and RNA-RNA

Antisense DNA oligonucleotides can be utilized for inhibiting expression at both the transcription and translation level. TNT® Systems have been employed for rapid screening of oligonucleotides for those arresting translation.

5.6 Molecular Structure and Localization Analyses: Characterization of Membrane Association

In vitro expression systems are being successfully employed for expressing integral membrane proteins. For example, expression of G-protein-coupled receptors in TNT® Systems supplemented with canine microsomal membranes resulting in the correct folding and insertion of transmembrane domains utilizing the expressed signal anchor and thereby stopping transfer sequences.

5.7 Non-natural Amino Acid Incorporation

Using technology originally developed in 1976, Johnson revolutionized in vitro translation by depicting that non-natural amino acids could be introgressed into polypeptides using epsilon-modified, lysine-charged tRNAs. Further extension of this technology results in the incorporation of photoactivatable crosslinking or fluorescent groups into polypeptides. This is later on monitored for the molecular environment as the labeled peptides pass through the ribosome followed by entry of the endoplasmic reticulum pore. Site-directed incorporation of a photoactivable crosslinker by a non-natural amino acid incorporation was employed to capture protein interacting with different portions of a protein of interest.

Site-specific methods have been developed utilizing an amber suppressor tRNA charged with any number of non-natural amino acids, including fluorescent, spin-label, and isotopic groups. This forms the basis of tRNA-mediated protein engineering (TRAMPE), which is an important field to study.

5.8 Protein Folding and Chaperonin Interactions

In vitro expression is increasingly being utilized to understand the nature of sequential chaperonin interactions needed for protein folding and localization. Workers in the current field have combined the advantages of in vitro expression with the power of instantaneous reporter gene product assays. The folding of polypeptides emerging from ribosomes has been studied with firefly luciferase as a model protein.

5.9 Real-Time Translation/Folding Assays

A novel approach with a wheat germ system has evolved where in the components for the luciferase enzymatic assay is being added directly to the translation reaction and monitored in real time. Luciferase was observed to be fully folded and enzymatically active immediately after release from the ribosome.

Furthermore, luciferase activity was not observed while full-length luciferase remained attached to the ribosome as a peptidyl-tRNA. This may be because the C-terminal portion of the enzyme is masked by the ribosome or ribosome-associated proteins. It has been depicted that the ribosome-bound enzyme attains enzymatic activity when its C-terminus is extended through at least 26 additional amino acid residues. The results depict that the acquisition of the final native conformation by a nascent protein can happen as the protein is being synthesized and that folding does not need release of the protein from the ribosome.

6 Macromolecular Assembly

It is advisable to express numerous gene products in one coupled transcription/translation reaction for the formation of functional transcription factor complexes or viral particles which are identical to those formed in the host.

6.1 Molecular Structure Analysis

The exploration of the function of integral membrane proteins is limited now, since there exists difficulty of producing crystals for use in X-ray diffraction studies. A methodology is being acquired for probing conformational changes in membrane proteins using Fourier transform infrared-difference (FTIR) spectroscopy. In this method, natively folded polypeptides are expressed in vitro with a site-specific insertion of a single isotopic label through amber suppression. This method does not disrupt the protein structure as did earlier site-directed mutagenesis methods and would be applied to a wide range of other proteins. They include those involved in enzyme catalysis, ion transport, and signal transduction.

7 Molecular Diagnostics Protein Truncation Test

Coupled transcription/translation systems is being developed for diagnosis of genetic diseases, a DNA technology-dominated field. The protein truncation test (PTT), otherwise known as in vitro synthesized protein truncation (IVSP) assay, was reported in 1993 for the first time as a rapid method for the detection of translation-terminating mutations in the large gene affecting for Duchenne Muscular Dystrophy and the Adenomatous Polyposis Coli (APC) gene responsible for a type of hereditary colon cancer. In these and other diseases, like hereditary breast cancer, 70–95% of the mutations which cause disease result in a truncated gene product.

In the process of PTT, first step includes purification of genomic DNA or mRNA from the patient's blood or tissue. The next step is either RT-PCR(f) or PCR(f) with the concurrent incorporation of a T7 promoter and optimal translation initiation sequence around the desired start codon.

Usually when the source of mRNA is limiting, a second nested PCR amplification is needed. Large exons are amplified from genomic DNA whereas the smaller exons are amplified together from mRNA. The gene is segmented into overlapping amplified fragments. The amplified DNA is mixed directly to a coupled transcription/translation reaction and translation terminating mutations are identified as faster migrating bands after SDS-PAGE analysis. PTT is able to scan large (2–3 kb) DNA/RNA segments quickly. In addition, PTT detects only disease-causing mutations, thus avoiding evaluation of polymorphisms. The use of the TNT® T7 Quick for PCR DNA System aids in PTT analysis.

7.1 High-Throughput Screening

7.1.1 Screening for Viral-Specific Translation Inhibitory Compounds

Viruses possess a number of different genetic elements utilized for promoting viral expression using host mRNA translation. Several researchers are currently developing screens with in vitro expression of gene constructs consisting of a viral element such as the 5′-UTR which can harbor an IRES (Internal Ribosome Entry Site) followed by a firefly or Renilla luciferase gene. Chemical or antibiotic libraries may be screened for particular translation-inhibiting effects. The viral element can be inserted between the firefly and Renilla luciferase genes with translation of the first gene relying on normal cap-dependent initiation.

The two luciferase genes lead to normalization of the second reporter behind the viral element. The efficacy of compounds may be assessed rapidly. It is now clear that disruption of the folding pathways can result in proteolytic degradation. Researchers are currently utilizing this information to assess the pharmacological activities of benzoquinone ansamycins, such as geldanamycin. These potentially medically important compounds were first detected as interesting due to their ability to inhibit tyrosine kinase activity. This is due to their interaction with Hsp90, which prevents the correct folding of tyrosine kinases leading to their proteolytic degradation. Other important drugs that affect protein folding by inhibition of chaperonin function is detected with this approach.

7.2 Identification of Novel Orphan Receptors

The binding of ligands to in vitro synthesized receptors would be an important aspect for the identification of new receptors. For example, to detect novel "orphan" nuclear receptors and ligands, a novel estrogen receptor was cloned followed by characterization. Saturation ligand binding and ligand-competition assays of the in vitro expressed clone was unique for this novel receptor, as a result, it can be differentiated from a previously cloned receptor.

7.3 Functional Genomics In Vitro Expression Cloning (IVEC)

In this procedure, an oligo(dT)-primed cDNA library is constructed in a high copy expression plasmid possessing a T3, T7, or SP6 promoter. The plasmid library is later on transformed into *E. coli*, and about 105 independent transformants are plated on selective media. The bacterial colonies are developed to a specific size (e.g., 1 mm in diameter), collected and pooled (50–100 clones per pool). Purified plasmid DNA from these pools is directly added to a small-scale (e.g., 10 μL) coupled transcription/ translation reaction, while it is utilized as a template in the presence of [35S] methionine. Based on the number of the full-length cDNA clones in the library, about 30–50 proteins can be produced in a single reaction. Proteins may be assayed for any number of activities, including phosphorylation, proteolysis, or cleavage. Positive pools are further divided until the single cDNA encoding the protein of interest is isolated.

7.4 Ribosomal Display for Cell-Free Protein Evolution

In this procedure, a cell-free system is utilized to transcribe a DNA library, translate the mRNA pools and, with a variety of techniques, the proteins and the encoding mRNAs are retained, that are still attached to the ribosomes. The protein mRNA-ribosome complexes are screened to bind to a target, and the retained mRNA is amplified using RT-PCR and the resulting DNA used for another round of selection. Initially, a prokaryotic *E. coli* coupled transcription/translation system was employed for the generation of large libraries of peptides for receptor ligand screening. Afterwards, improvements lead to the folding of whole proteins into their native structure, when attached to the ribosome.

The first eukaryotic application utilized a coupled rabbit reticulocyte system for studying antibody-ribosome-mRNA (ARM) complexes, resulting in rapid selection and monitoring of antibody combining site evolution.

A cell-free system has been constructed to perform evolution studies, wherein RNA amplification and the coupled reaction can be conducted simultaneously at a given temperature. Following unsuccessful attempts with wheat germ extracts and coupled *E. coli* systems, researchers were able to combine the reactions with a rabbit reticulocyte coupled system.

Through the exertion of selective pressure on functional protein products essential for RNA amplification, this system can be employed for performing laboratory "evolution."

7.5 Preparative Synthesis: Large-Scale Protein Expression and Purification

Cell-free expression systems are usually preferred over in vivo or native systems, since they can be utilized for the expression of toxic, proteolytically sensitive or unstable proteins. Additionally, in vitro systems have the ability to incorporate non-natural amino acids consisting of photoactivatable, fluorescent, or biotin residues. Usually, in vitro systems form nanogram amounts of proteins per 50 μL reaction. Preparative scale methods have been developed that may yield milligram quantities per milliliter of reaction mixture.

Cell-free protein expression system may be developed for in vitro transcription and translation of a particular protein in a quasi-cell environment.

7.6 In Vitro Translation

The in vitro synthesis of proteins in cell-free extracts is regarded as a useful tool for molecular biologists. It has different applications, like the rapid identification of gene products (e.g., proteomics), localization of mutations by synthesis of truncated gene products, aspects on protein folding, and introduction of modified or unnatural amino acids for functional studies.

According to the principle, a cell-free extract for in vitro translation of mRNAs from any type of cells can be prepared. However, practically, only a few cell-free systems are able to develop in vitro protein synthesis. In general, these systems are evolved from cells associated with a high rate of protein synthesis.

8 Cell-Free Expression Systems

The most commonly used cell-free translation systems consist of extracts from rabbit reticulocytes and *Escherichia coli*. All are prepared as crude extracts possessing all the macromolecular components (70S or 80S ribosomes, tRNAs, aminoacyl-tRNA synthetases, initiation, elongation and termination factors, etc.) necessary for translation of exogenous RNA.

For efficient translation, supplementation is to be conducted with amino acids, energy sources (ATP, GTP), energy regenerating systems (e.g., creatine phosphate and creatine phosphokinase for eukaryotic systems). Other requirements may be listed as phosphoenol pyruvate and pyruvate kinase for the *E. coli* lysate, and other co-factors (Mg^{2+}, K^{+}, etc.).

8.1 Two Approaches for In Vitro Protein Synthesis Based on the Starting Genetic Material: RNA or DNA

Standard translation systems, such as reticulocyte lysates and wheat germ extracts, employ RNA as a template. However, "coupled" and "linked" systems start with DNA templates, which are transcribed into RNA then translated.

8.2 Rabbit Reticulocyte Lysate

Rabbit reticulocyte lysate is a greatly efficient in vitro eukaryotic protein synthesis system. It is utilized for the translation of exogenous RNAs (either natural or generated in vitro). In vivo, reticulocytes are highly specialized cells leading to the synthesis of hemoglobin, which represents more than 90% of the protein made in the reticulocyte.

Although these immature red cells have already lost their nuclei, but possess adequate mRNA, as well as complete translation machinery, for extensive globin synthesis. The endogenous globin mRNA can be destroyed through incubation with Ca^{2+}-dependent micrococcal nuclease, that is later inactivated by chelation of the Ca^{2+} by EDTA.

The Kit offers a nuclease-treated reticulocyte lysate. This type of lysate is the most commonly used RNA-dependent cell-free system due of its low background and its efficient use of exogenous RNAs even at low concentrations. Exogenous proteins are synthesized at a rate similar to that seen in intact reticulocyte cells.

Untreated reticulocyte lysate cause translation of endogenous globin mRNA, exogenous RNAs, or both. This type of lysate is employed for studying the translation machinery, e.g., studying the effects of inhibitors on globin translation. Both the untreated and treated rabbit reticulocyte lysates have low nuclease activity and are able to synthesize a large amount of full-length product. Both lysates have the ability to synthesize larger proteins from either capped or uncapped RNAs (eukaryotic or viral).

8.3 *E. coli Cell-Free System*

E. coli cell-free systems consist of a crude extract which is rich in endogenous mRNA. During the process of preparation, the crude extract as described is incubated, to enable this endogenous mRNA to translate and subsequently degrade. Since the levels of endogenous mRNA in the prepared lysate are low, the exogenous product is easily identified.

E. coli extract has a relatively simple translational apparatus as compared to that in eukaryotic system, with less complicated control at the initiation level, making this system to be very efficient in protein synthesis. Bacterial extracts are usually unsuitable for translation of RNA, since exogenous RNA is rapidly being degraded by endogenous nucleases. There are some viral mRNAs which translate efficiently, since they are somewhat resistant to nuclease activity and contain stable secondary structure. However, *E. coli* extracts are suitable for coupled transcription: translation from DNA templates.

8.4 "Linked" and "Coupled" Transcription: Translation Systems

In standard translation reactions, purified RNA is used as a template for translation. In case of "Linked" and "coupled" systems, on the other hand, is used as DNA as a template. RNA is transcribed from the DNA and subsequently translated without any purification. Such systems typically combine a prokaryotic phage RNA polymerase and promoter (T7, T3, or SP6) with eukaryotic or prokaryotic extracts to synthesize proteins from exogenous DNA templates. DNA templates for transcription:translation reactions may be cloned into plasmid vectors or generated by PCR.

8.4.1 Linked Transcription: Translation

The "linked" system is a two-step reaction. The initial step is based on transcription with a bacteriophage polymerase. The next step is translation in the rabbit reticulocyte lysate or wheat germ lysate. Since the transcription and translation reactions are separate, they can be optimized so that both are functioning at their full potential.

On the contrary, many commercially available eukaryotic coupled transcription/translation systems have evolved, compromising one or both reactions so that they can occur in a single tube. Thus, yield is sacrificed for convenience.

8.5 Human In Vitro Protein Expression System

It is a methodology for expressing proteins from DNA or mRNA templates in a cell-free solution comprising of essential components of the cellular translational machinery. Extracts of an immortalized human cell line possess the ribosomes, initiation and elongation factors, tRNAs and other basic components required for protein synthesis. For the synthesis of target proteins from DNA templates up to 6 h, supplementation is essential with proprietary accessory proteins, ATP, and an energy regenerating system. The greatest advantage is that there is no need to remove inhibitory byproducts.

Protein synthesis starts with either an RNA (Fig. 1) or DNA template (Fig. 2). When the starting material is DNA, mRNA is transcribed in 1 h at 32 °C utilizing the included reaction components. The next step follows addition of mRNA to the in vitro translation reaction and incubation for 90 min at 30 °C for general protein expression. Otherwise, the reaction is incubated for 90 min at 28 °C for glycoprotein expression. The diagrammatic view of the outline of Human protein expression analysis is being discussed in Fig. 3.

This bacterial translation system leads to efficient expression of either prokaryotic or eukaryotic gene products within a short period of time. For the highest protein yield and the best initiation fidelity, it is essential that the DNA template has a Shine–Dalgarno ribosome binding. There exist distinct differences in post-transcriptional modification of prokaryotes and eukaryotes as depicted diagrammatically in Fig. 4. Figure 5 provides a schematic representation of protein expression analysis from DNA with coupled transcription/translation.

9 Important Elements for Translation

There are some distinct differences between prokaryotic and eukaryotic mRNA transcripts. Eukaryotic mRNAs are distinguished by two post-transcriptional modifications: a 5′-7 methyl-GTP cap and a 3′ poly(A) tail. Both of these modifications aid in maintaining the stability of the mRNA by preventing degradation. In addition, the 5′ cap structure improves the translation of mRNA by helping to bind the eukaryotic ribosome. It also aids in recognition of the proper AUG initiator codon. This function varies with the translation system and with the specific mRNA being synthesized. The consensus sequence 5′-GCCACCAUGG-3′ (the "Kozak" sequence) is regarded as the strongest ribosomal binding signal in eukaryotic mRNA.

For the initiation of efficient translation process, the key elements are the G residue at the +1 position and the A residue at the −3 position. An mRNA lacking the Kozak consensus sequence can be translated efficiently in eukaryotic cell-free systems if a moderately long 5′-untranslated region (UTR) is present that lacks stable secondary structure (Fig. 6).

In bacteria, the ribosome is guided to the AUG initiation site through a purine-rich region known as the Shine–Dalgarno (SD) sequence. This sequence is complementary to the 3′ end of the 16S rRNA in the 30S ribosomal subunit. Upstream from the initiation AUG codon, the SD region possesses the consensus

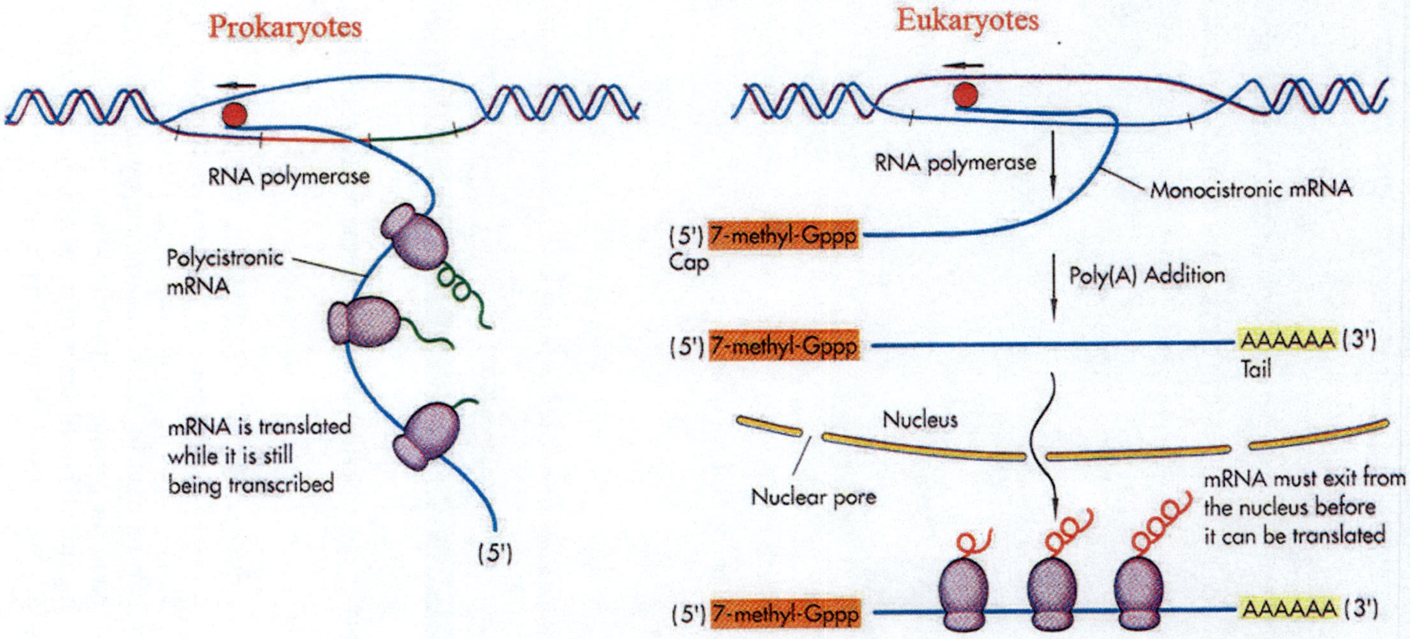

Fig. 4 Diagrammatic illustration of differences in post-transcriptional modification of prokaryotes and eukaryotes

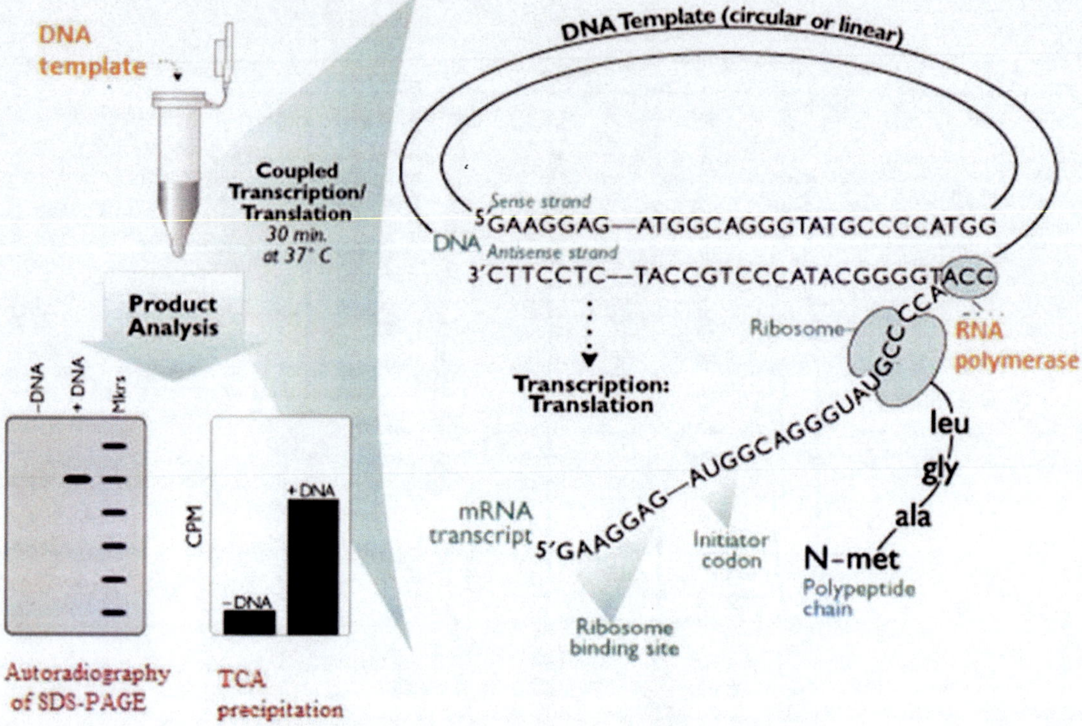

Fig. 5 Schematic representation of Protein expression analysis from DNA with coupled transcription/translation

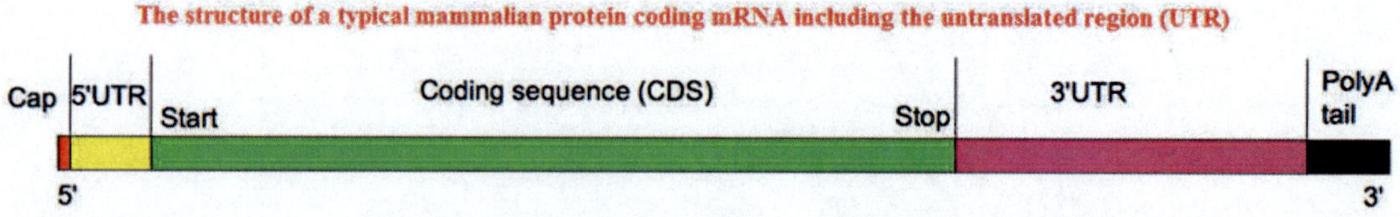

Fig. 6 Schematic representation of a typical human protein coding mRNA including the translated regions (UTR)

sequence 5′-UAAGGAGGUGA-3′. Specific mRNAs have differences in the number of nucleotides complementary to the anti-Shine–Dalgarno sequence of 16S rRNA. It ranges from as few as two to nine or more. The efficiency of translation is dependent upon the position of the ribosome binding site in relation to AUG initiator. The efficiency of translation usually varies from −6 to −10 relative to the A of the initiation site.

10 Ribosomal Binding Site Sequence Requirements

Protein synthesis is regulated by the sequence and structure of the 5′ untranslated region (UTR) of the mRNA transcript. In prokaryotes, the ribosome binding site (RBS), that promotes efficient and accurate translation of mRNA, is called the Shine–Dalgarno sequence. This purine-rich sequence of 5′ UTR is observed to be complementary to the UCCU core sequence of the 3′-end of 16S rRNA. It is located within the 30S small ribosomal subunit. Various Shine–Dalgarno sequences have been observed in prokaryotic mRNAs. These sequences are present about ten nucleotides upstream from the AUG start codon. Activity of a ribosome binding site can be influenced by the length and nucleotide composition of the spacer separating the RBS and the initiator AUG.

In eukaryotes, the Kozak sequence A/GCCACCAUGG, which lies within a short 5′ untranslated region, directs translation of mRNA. An mRNA lacking the Kozak consensus sequence can be translated efficiently in vitro systems, provided, a moderately long 5′ UTR lacking stable secondary structure is present.

Shine–Dalgarno sequence is a characteristic feature for *E. coli* ribosome, whereas eukaryotic ribosomes (such as those found in retic lysate) can efficiently use either the Shine–Dalgarno or the Kozak ribosomal binding sites.

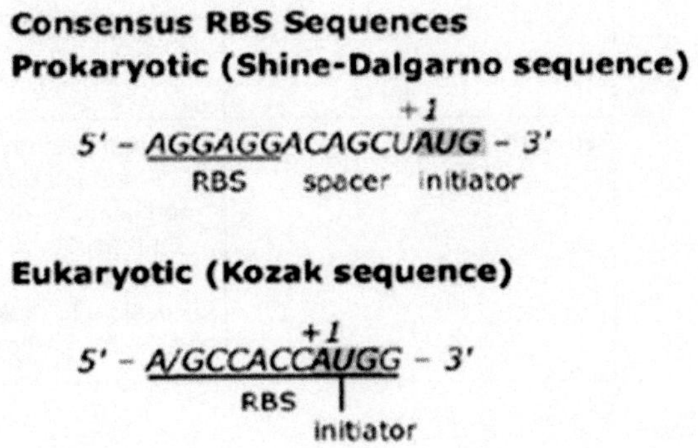

Comparison of in vitro (cell-free) protein expression systems. protein, application, and the template used for protein expression.		
System	**Advantages**	**Disadvantages**
E. coli	•Very high protein yield •Relatively tolerant of additives	•Many eukaryotic proteins insoluble upon expression •Eukaryotic co- and post-translational modifications not possible •Codon usage is different from eukaryotes
Rabbit Reticulocyte (RRL)	•Mammalian system •Cap independent translation	•Sensitive to additives •Protein glycosylation not possible •Co-expression of off-target •proteins
Wheat germ	•Translation of large proteins possible •Devoid of off-target endogenous mammalian proteins •High protein yield	•Mammalian co- and post-translational modifications are not possible •Premature termination of products
Insect	•Translation of large proteins possible •No endogenous mammalian proteins •Certain forms of protein glycosylation possible	•Non-mammalian
Human	•Human system •Co- and post-translational modifications are possible •Synthesis of functional proteins •Possible to make VPLs (virus-like particles)	•Sensitive to additives •Lower yields than E. coli •New system

11 Metagenomics Is the Study of Genetic Material Recovered Directly from Environmental Samples

Metagenomic study is widely applicable for studies of gut bacteria and fungi.

Comparative analyses between metagenomes give rise to added knowledge to the function of complex microbial communities and their role in host health. Pairwise or multiple comparisons between metagenomes can be processed at the level of sequence composition (comparing GC content or genome size), taxonomic diversity, or functional complement. Comparisons of population structure and phylogenetic diversity can be assessed based on 16S and other phylogenetic marker genes. We have already discussed metagenomics in detail in earlier chapters.

11.1 Metatranscriptomics

Through the process of Metagenomics, it is possible to access the functional and metabolic diversity of microbial communities. But it cannot predict the actual active processes. The extraction and analysis of metagenomic mRNA (the metatranscriptome) give an insight on the information concerning regulation and expression profiles of complex communities. Because of the technical difficulties (the short half-life of mRNA, for example) in the collection of environmental RNA there have been relatively few in situ metatranscriptomic studies of microbial communities to date.

Protein detection in any experimental process can be conducted through the following techniques:

1. ELISA.
2. Western blotting.
3. Immunohistochemistry or immunocytochemistry.

Practical use of recombinant protein/isolation:

1. Therapeutic use.
2. Antibody production.
3. Hormone production—e.g., insulin.
4. Enzyme production.

Future prospect includes regeneration of organ either complete or a part of it which will be very much important for organ transplantation, say kidney transplantation for a patient with renal failure, cornea transplantation for a blind person from his own body tissue. Hence there will be no need for immunosuppressive drug administration for the patient.

11.2 Practical Protocols for Production of Very High Yields of Recombinant Proteins Using *Escherichia coli*

11.2.1 Molecular Cloning

The process involves testing of seven different proteins. These are receptor-associated proteins (RAP, 1–210) (RAP, 91–323), truncation mutants of the human apolipoprotein E, apoE (1–183) and apoE (1–214), full-length apoE, a truncation mutant of mouse apolipoprotein AI, apoAI (1–216), and full length human apoAI. The genes of these proteins were subcloned into various expression vectors as follows: RAP (1–210)/pET30a, RAP (91–323)/pET30a, human apoE (1–183)/pET22b, apoE/pET30a-sHT, human apoE (1–214)/pTYB1, mouse apoAI (1–216)/pET30a, and human apoAI/pET30a-sHT. The pET vectors were received from Novagen, WI and the pTYB1 vector was from New England BioLabs, MA. The pET30a vector possesses an introduced Factor Xa site between the long his-tag and the target gene. The pET30a-sHT is also an engineered pET30a vector where the long his-tag was replaced by a six-histidine tag plus a two serine linker. The pET30a and pET30a-sHT are kanamycin-resistant vectors; however, the pET22b and pTYB1 vectors are ampicillin-resistant vectors. The expression vectors were transformed into BL-21 (DE3) bacterial strains [1, 2].

11.2.2 Double-Colony Selection

Initially, LB agar plates were formed either in H_2O or in 70% D_2O (for triple-labeled protein expression). In 70% D_2O plates, instead of autoclaving, the agar medium was microwaved till the agar is dissolved. Three milliliters of agar were poured into a 35 mm × 10 mm petri plate (Corning, NY). Bacterial cells, either from a glycerol stock or 5 μL of a starting culture which has been diluted to an OD_{600} of ~0.05–0.1, were streaked onto the LB agar plates. Several colonies were picked from the plates next morning, followed by the checking of the expression levels of these colonies with the help of the traditional IPTG-induction expression. Glycerol stocks were also developed for each colony. We selected the colony with the highest protein expression and repeated through another round of selection, following the previous procedure described. The colonies selected from the double selection were utilized for the preparation of glycerol stocks and were stored in a −80 °C freezer.

11.2.3 Traditional IPTG-Induction Expression

This expression method is employed to either assess protein expression levels of various colonies during double-colony selection or act as an expression control. For double-colony selection, we employed a small-scale expression with the following procedure: 2 mL of LB media was inoculated with a single colony from a freshly transformed plate to be used as the starting culture and cultured at a temperature of 37 °C. As the OD_{600} of the starting culture becomes 2–3, the culture was gently spun down at 1500 × *g* for 5 min. The part of cell pellets was resuspended in 5 mL of the minimal M9 medium to get an initial O.D between 0.07 and 0.1. Whenever the culture achieved an OD_{600} of ~1.0, it was induced with 0.5 mM IPTG, followed by incubation at 20 °C overnight [Human apoE (1–215), full-length apoE, mouse apoAI, human apoAI, RAP (1–210), RAP (91–323)] or at 28 °C for 16–18 h [Human apoE (1–183)]. Two hundred fifty microliters of cell suspension were collected followed by centrifugation at 3300 × *g* for 5–10 min. The cell pellet was resuspended in 50 μL of 2× SDS gel loading buffer, then heated at 70 °C for 20 min. Cell debris and DNA molecules were pelleted through centrifugation at a maximum speed for a period of 20 min. At the final stage, 10 μL of the supernatant was loaded into the SDS-PAGE for checking the expression level. For the IPTG method as an expression control, we utilized a 50-mL expression with the following procedure: 10 mL of LB media was inoculated with glycerol stock (after double selection) as the starting culture followed by culturing at 37 °C. When the OD_{600} of the starting culture attains 2–3, the culture was gently centrifuged at 1500 × *g* for 5 min. A part of cell pellet was resuspended in 50 mL of the minimal M9 medium to attain an initial O.D between 0.07 and 0.1. When the culture reached an OD_{600} of ~1.0, it was induced with 0.5 mM IPTG and incubated at 20 °C overnight [Human apoE (1–215), full-length apoE, mouse apoAI, human apoAI, RAP (1–210), RAP (91–323)] or at 28 °C for 16–18 h [Human apoE (1–183)]. The cells were harvested through centrifugation and the cell pellet was utilized for protein purification.

11.2.4 Autoinduction Expression

In order to attain autoinduction expressions, we practiced a general procedure: Studier's C-750501 autoinduction minimal medium with 0.75% glycerol was employed for the time course experiments and unlabeled protein production. A modified Studier's C-750501 autoinduction expression medium with 0.4% ^{13}C-glycerol was utilized for triple-labeled protein production. For the time course experiments, the expression medium was prepared either with 99% D_2O or with H_2O, unlabeled glucose, glycerol, and ammonium chloride. For triple-labeled protein expressions, $^{13}C_6$-glucose (0.05%), ^{13}C-labeled glycerol, and $^{15}NH_4Cl$ in D_2O were utilized. The medium was also supplemented with 0.2–2× trace metals as necessary and with vitamins if required. β-Lactose was utilized instead of α-lactose for induction, which contained 30% of α-lactose isomers and provided comparable expression levels. Starting cultures in LB media were prepared with the glycerol stocks from double selection and cultured to an OD_{600} of 1–2 and then centrifuged at 1500 × *g* for 5–10 min.

The cell pellets were resuspended in 25 mL or 50 mL autoinduction expression medium to attain an initial OD_{600} of 0.3. Erlenmeyer flasks of 125 mL and 250 mL were utilized for 25 mL and 50 mL culture volumes, respectively, for better aeration of the cultures. Expressions were allowed at different temperatures for different proteins ranging from 20 to 37 °C.

For the time course experiments, 250 μL culture samples were collected at various time periods. The OD_{600} and pH of the culture at these time points were noted. The cell suspensions were centrifuged, resuspended in 50 μL of 2× SDS gel loading buffer, followed by running on an SDS-PAGE for the comparison of the expression levels at different times. Expression time generally varied from 24 to 64 h. The culture temperature is considered as the factor controlling it.

In case of unlabeled and triple-labeled protein expressions, we first study at different temperatures to obtain the optimal expression time and temperature. For example, the time course of triple-labeled human apoAI reflected that an expression of 36 h at room temperature provides the best protein production. The triple-labeled apoAI expression was then conducted at room temperature with an initial OD_{600} of 0.3. At 36 h, the cells were collected by centrifugation at 3500 × *g* for 15 min. The purifications of the triple-labeled proteins were conducted as described previously.

11.2.5 High-Cell-Density IPTG-Induction Expression

This expression method utilizes rich medium for achieving high cell density. We initiated bacterial expression with a rich medium, such as LB or 2× YT, at 37 °C. As the cell density reaches the middle of the growing phase prior to saturation, we switched the cell culture by gently centrifuging cells and resuspending the pellet into the same volume of minimal medium. We observed that bacterial cells reach saturation at cell densities of OD_{600} of ~10–15 in LB and OD_{600} of ~15–20 in 2× YT. So, bacterial cells are in the middle of the growing phase with a cell density of OD_{600} of 3–5 in LB and OD_{600} of 5–7 in 2× YT. After switching the medium, we cultured bacterial cells for another 1.0–1.5 h devoid of IPTG, at the optimized temperature which is employed for the cell culture after IPTG induction. At this period, the OD_{600} of the cell culture would increase by 1–2 U. IPTG was later added for protein production. The cell culture was incubated at the same temperature for a period which is optimized for different proteins prior to cell harvest. Generally, we found that the OD_{600} value at the end of cell culture increased by 2.0- to 2.5-fold in comparison to the OD_{600} value at IPTG-induction. Thus, prior to cell harvest, the bacterial cells can reach to OD_{600} of 10–15 with a starting medium with LB and OD_{600} of 15–20 with 2× YT. This is about five- to tenfold more OD_{600} than that of the regular IPTG-induction bacterial expression in minimal medium [1, 2].

11.2.6 NMR Spectroscopy

The NMR samples possessed 100 mM phosphate buffer (pH 6.8), 0.01 mM NaN_3, 10 mM EDTA, 50 mM DTT, 5% D_2O, and 0.5–1 mM 40% $^2H/^{15}N/^{13}C$-labeled proteins. The ^1H-^{15}N TROSY-HSQC spectra were received at 30 °C on 600 MHz Varian INOVA spectrometer with a cold probe. Proton chemical shifts were referenced to DSS. Data processing and analysis on an SGI workstation was carried out with nmrPipe and nmrview.

11.3 Selection of Appropriate Expression Vector Is Very Much Essential Because of Post-translational Modifications

One of the most important criteria is the selection of appropriate expression vector. Figure 7 discusses the characteristics of various expression vectors. It is very important to select the appropriate expression system for specific application. Protein solubility, functionality, purification speed, and yield are the critical factors for consideration while selecting an expression system. In addition, each system has its own strengths and challenges, which are important while choosing an expression system. We have six unique expression systems: mammalian, insect, yeast, bacterial, algal, and cell-free systems. As a system is selected, the method of gene delivery will be utilized for consideration for protein expression. The main methodology for gene delivery includes transfection and transduction. Transfection is the process through which nucleic

Expression system	Most common application	Advantages	Challenges
Mammalian 	➡Functional assays ➡Structural analysis ➡Antibody production ➡Expression of complex proteins ➡Virus production	➡Highest level protein processing ➡Can products proteins either transiently or by stable expression ➡Robust optimized transient systems for rapid, ultra yield protein production	➡Gram per liter yields only possible in suspension cultures ➡More demanding culture conditions
Insect 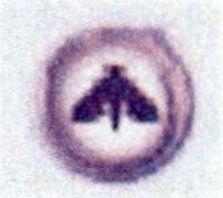	➡Functional assays ➡Structural analysis ➡Expression of intracellular proteins ➡Expression of protein complexes ➡Virus production	➡Similar to mammalian protein processing ➡Can be used in static or suspension culture	➡More demanding culture conditions than prokaryotic systems ➡Production of recombinant baculovirus vectors is time consuming
Yeast 	➡Structural analysis ➡Antibody generation ➡Functional analysis ➡Protein interaction	➡Eukaryotic protein processing ➡Scalable upto fermentation (gm per lit) ➡Simple media requirements	➡Fermentation required for very high yields ➡Growth conditions may require optmization
Bacterial 	➡Structural analysis ➡Antibody generation ➡Functional assays ➡Protein interaction	➡Scalable ➡Low cost ➡Simple culture conditions	➡Protein solubility ➡May require protein specific optimization ➡May be difficult to express some mammalian proteins
Algal 	➡Studying photosynthesis, plant biology, lipid metablism ➡Genetic engineering ➡Biofuel production	➡Genetic modification and expression systems for photosynthetic microalgae ➡Super experimental control for biofuel, nutraceuticals, And specially chemical production ➡Optimized system for robust selection and expression	➡Nascent technology ➡Less developed compared to other host platforms
Cell free 	➡Toxic proteins ➡Incorporation of unnatural label or amino acids ➡Functional assays ➡Protein interactions ➡Translational inhibitor screening	➡Open system: able to add unnatural components ➡Fast expression ➡Simple format	➡Scaling above multimiligram quantities may not be costly

Fig. 7 Characteristics of various recombinant protein expression systems

acids are introduced into mammalian and insect cells. There is wide variation among protocols and techniques vary widely. These include lipid transfection, chemical, and physical methods such as electroporation. To view different transfection methods or our transfection reagent selection guide, go to: thermofisher.com/transfection. For cell types not amenable to lipid-mediated transfection, viral vectors are often employed. Virus-mediated transfection, also known as transduction, is an efficient method to transfect cell types for protein overexpression or knockdown. It is the most commonly used method in clinical research. Adenoviral, oncoretroviral, lentiviral, and baculoviral vectors have been proved to be effective for gene delivery to mammalian cells, both in cell culture and in vivo.

11.4 Cell Lysis and Protein Purification

The next step following protein expression is the isolation and purification of the protein of interest. Protein yield and activity can be improved by selecting the right lysis reagents and appropriate purification resin. Cell lysis formulations have been standardized for specific host systems, such as cultured mammalian, yeast, baculovirus-infected insect, and bacterial cells. Majority of the recombinant proteins are expressed as fusion proteins with shorter affinity tags, like polyhistidine or glutathione S-transferase. This allows for selective purification of the protein of interest. Recombinant His-tagged proteins are purified with the help of immobilized metal affinity chromatography (IMAC) resins. GST-tagged proteins are purified using a reduced glutathione resin.

11.5 Mammalian Cell-Based Protein Expression

A mammalian host system is the preferred expression platform for producing mammalian proteins that have the most native structure and activity. Mammalian expression is the system of choice to study the functional significance of a particular protein in the most physiologically relevant environment, since it allows for the highest level of posttranslational processing and functional activity of the protein. It is commonly employed for the production of antibodies and therapeutic proteins. It is equally helpful for synthesis of proteins helpful for human use in functional cell-based assays.

Gene of interest can be inserted into mammalian cells by transfecting the plasmid DNA using chemical reagents, by electroporation, or by viral transduction using recombinant viruses. The delivery method is so chosen depending upon the mammalian cell type and the experimental effects needed.

Transfection: *is the method of choice for a wide variety of cell types, whereas electroporation using the Invitrogen™ Neon™ Transfection System or viral transduction is the ideal choice for cell types that are difficult to transfect or for nondividing cell types.*

For viral transduction of the gene of interest, lentiviral and adenoviral expression systems may be chosen commercially available.

In case transfection is the delivery method of choice, certain points are taken in consideration:

1. Whether to express the protein transiently or make stable cell lines. Mammalian expression experiments are conducted either transiently or with stable cell lines which express the protein of interest.
2. Transient expression that results in high levels of expression for a few days is most suitable for rapid protein production and quick data generation. Any mammalian expression vector can be utilized for this purpose.
3. Stable expression needs the generation of stable cell lines, where the expression construct is integrated into the host genome. These stable cell lines can be employed over a long experimental time course or used over many experiments. The expression vector possesses a selection marker (such as an antibiotic resistance gene). Hence, the cells with the integrated construct can be easily selected through the addition of a selection agent (e.g., the antibiotic) to the medium. In order to assess the optimal antibiotic concentration to use while establishing a stable cell line, it is recommended to perform a dose-response curve or kill curve. Untransfected cells at 25% confluence and grown in medium containing increasing concentrations of the antibiotic.

For some antibiotics, it is necessary to calculate the amount of active drug to control for lot variation. Replenish the selective medium for every 3–4 days. After 10–12 days, the dishes were examined for viable cells. The cells divide once or twice in the selective medium prior to cell death. It is essential to determine the minimum concentration of antibiotic resulting in complete cell death. This would be considered as the optimal antibiotic concentration for use in stable selection.

12 Protein Expression Protocol and Troubleshooting

With regard to recombinant expression, *E. coli* expression system is the first choice as microbial cell factory. *E. coli* expression system has multiple, significant benefits over other expression systems in terms of cost, ease-of-use, and scale. Now we discuss a general protocol of expression as well as a list of possible solutions when facing the challenge of expressing a new protein in *E. coli*.

12.1 Expression Protocol in E. coli

General protocol of expression process from gene to protein is given below as discussed in Fig. 8.

12.1.1 Phase 1: Codon Optimized Gene Synthesis and Vector Construction

The step involves optimization of codon and mRNA structure. The process is followed by fusion of gene to a protein tag and insertion into an expression vector. In the final step, verification is essential to assess the correctness of construction by sequencing.

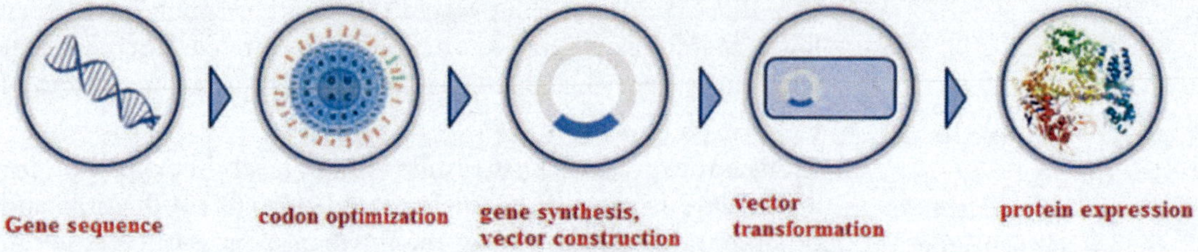

Fig. 8 Diagrammatic representation of protein expression from gene sequence

12.1.2 Phase 2: Transform Expression Vector into E. coli *Competent Cells*

Expression vector is added to thawed competent cells. Heat shocked cells are added to LB broth and shaken. Cell culture is plated onto LB agar plates with appropriate antibiotic.

12.1.3 Phase 3: Starter Culture

Single colony of expression strain is picked up and mixed with 5 mL of LB with appropriate antibiotic. The mixture is shaken at 37 °C for 3–5 h.

12.1.4 Phase 4: Expansion of Starter Culture

The culture is expanded by adding the starter culture to larger volume LB with antibiotic (room temp). It is incubated for 1–4 h until culture density of OD_{600} reaches 0.5–0.6.

Note: Top the flask with cotton or a culture flask cap to avoid culture to environmental contamination.

12.1.5 Phase 5: Induction

Option 1: 37 °C Induction

- Induce expression by adding IPTG to a final concentration of 0.5 mM after culture has reached OD_{600} 0.5–0.6.
- Induce for 3–4 h at 37 °C with shaking.

Note: IPTG is a frozen solution in the −20 °C freezer.

Option 2: Room Temp (20 °C) Induction

- The culture is cooled to room temperature by placing in fridge or iced water bath after it has reached OD_{600} 0.5–0.6.
- Induce expression by adding IPTG to a final concentration of 0.1–1.0 mM.
- Induce overnight (12–18 h) at room (20 °C) temp with shaking.

12.1.6 Phase 6: Cells Collection and Lysis

- Centrifuge the cells at 3500 × *g* for 20 min.
- Resuspend cells in ice cold PBS and re-centrifuge in an appropriate sized tube.
- Remove the supernatant and freeze pellet for later processing.
- Lyse cells using appropriate protocol.

12.2 Expression Troubleshooting

Now, let us discuss certain unavoidable circumstances we encounter for optimizing recombinant protein production in *E. coli*. Possible reasons and solutions in each case are discussed in the following tables.

12.2.1 No/Low Expression

There are certain situations, when the protein of interest is not detectable through a sensitive technique (e.g., Westernblot). In other cases, sometimes protein expression is detectable at a very low level (less than micrograms per liter of culture). These problems may occur due to harmful effect that the heterologous protein exerts on the cell.

	Solutions		
Reasons	**Vector**	**Host strain**	**Growth conditions**
Incorrect vector construction	Confirm vector by sequencing		
Rare codons	Codon optimization	Use strains supplementing rare codons (Rosetta, Codon Plus)	• Lower induction temperature • Grow in poor media
Potein toxicity	Use promoters with tighter regulation	Use pLysS/pLysE bearing strains in T7-based systems	• Start induction at high OD
	Lower plasmid copy number	Use of strains that are better for the expression of toxic proteins (C41 or C43)	• Shorten induction time • Added glucose for the use of expression vectors containing lac-based promoters • Use defined media with glucose as source of carbon

12.2.2 Protein Aggregation

The build ups of protein aggregates are termed as inclusion bodies (IBs). IB formation is due to an unbalanced equilibrium between protein aggregation and solubilization. So, it is possible to get a soluble recombinant protein by methodologies that reduce the factors leading to IB formation.

	Solutions		
Reasons	**Vector**	**Host strain**	**Growth conditions**
Incorrect disulfide bond formation	• Add fusion partners, including thioredoxin, DsbA, DsbC • Clone in a vector containing secretion signal to cell periplasm	Use *E. coli* strains with oxidative cytoplasmic environment	• Lower inducer concentration • Lower induction temperature
Incorrect folding	• Use a solubilizing partner	Use strains with cold-adapted chaperones	• Supplement media with chemical chaperones and cofactors

(continued)

Reasons	Solutions Vector	Host strain	Growth conditions
	• Co-express with molecular chaperones		• Remove inducer and add fresh media • Lower inducer concentration • Lower temperature
Proteins with high hydrophobicity or transmembrane domains	• Add fusion tags, including GST, MBP, SUMO • Generate truncated forms of protein	Use membrane rich strains (C41/C43)	• Lower induction temperature • Shorten induction time • Grow in poor medium • Add heat shock chaperones

12.2.3 Truncated Protein

Sometimes a truncated form of protein is expressed rather than a complete wild protein. Reasons of the phenomenon and possible solutions are given below.

Reasons	Solutions Vector	Host strain	Growth conditions
Rare codon	Codon optimization	Use strains supplementing rare codons (Rosetta, Codon Plus)	• Lower induction temperature • Grow in poor media
Protein degradation	Replace specific protease sites	Use low protease strains	• Induce at high OD • Induce at low temperature • Shorten induction time • Use protease inhibitors when breaking cells
Imbalanced translation process of fusion protein	• Change another fusion protein • Move fusion protein to C-terminal		• Induce at low temperature • Shorten induction time • Change to poor media

12.2.4 Protein Inactivity

Obtaining a nice amount of soluble protein is not the end of the road. The protein may still be of bad quality, i.e., it does not have the activity it should.

	Solutions		
Reasons	**Vector**	**Host strain**	**Growth conditions**
Low solubility of the protein	Fuse desired protein to a solubility enhancer (fusion partners)		Lower temperature
Lack of essential post translational modification	Change expression system		
Incomplete folding	• Use a solubilizing partner • Co-express with molecular chaperones	Use strains with cold-adapted chaperones	• Monitor disulfide bond formation and allow further folding in vitro • Lower temperature
Mutations in cDNA	Sequence plasmid before and after induction	Use a recA− strain to ensure plasmid stability	Transform *E. coli* before each expression round

The target gene of interest may be inserted into mammalian cells through transfection of the plasmid DNA with the help of chemical reagents, by electroporation, or by viral transduction with the help of recombinant viruses. It is very important to determine the best method, which needs to be employed. The choice of delivery method is affected by the mammalian cell type that is being used and the experimental effects desired.

Transfection is applicable for wide varieties of cell types. However, electroporation or viral transduction is applicable for cell types which are difficult to transfect or applicable for nondividing cell types. For viral transduction of the gene of interest, the different expression systems (Invitrogen™ ViraPower™ lentiviral and adenoviral expression systems) were found to be effective.

- When transfection is selected as the delivery method of choice, it is essential to express the protein transiently or make stable cell lines. Mammalian expression experiments are conducted either transiently or with stable cell lines which express the protein of interest.
- Transient expression resulting in high levels of expression for a few days is the best for rapid protein production and quick data generation. Any mammalian expression vector can be used for this purpose.

Table 1
Eukaryotic selection antibiotics

Selection antibiotic	Most common selection usage	Common working concentration	Available powder sizes	Available liquid sizes
Blasticidin S	Eukaryotic and bacterial	1–20 μg/mL	50 mg	10 × 1 mL, 20 mL
Geneticin (G-418)	Eukaryotic	100–200 μg/mL bacteria 200–500 μg/mL mammalian cells	1 g, 5 g, 10 g, 25 g	20 mL, 100 mL
Hygromycin B	Dual-selection experiments and eukaryotic	200–500 μg/mL	–	20 mL
Mycophenolic acid	Mammalian and bacterial	25 μg/mL	500 mg	–
Puromycin	Eukaryotic and bacterial	0.2–5 μg/mL	–	10 × 1 mL, 20 mL
Zeocin	Mammalian, insect, yeast, bacterial, and plants	50–400 μg/mL	–	8 × 1.25 mL, 50 mL

- Stable expression needs the generation of stable cell lines when the expression construct is integrated into the host genome. These stable cell lines can be utilized over a long experimental time course or used over many experiments. Since the expression vector has a selection marker (such as an antibiotic resistance gene), cells with the integrated construct can be selected through the addition of a selection agent (such as the antibiotic) to the medium.

In order to standardize the optimal antibiotic concentration for use during the establishment of a stable cell line, it is recommended to perform a dose-response curve or kill curve. Hence, untransfected cells are plated at 25% confluence and grown in medium consisting of increasing concentrations of the antibiotic. For some antibiotics, we need to calculate the amount of active drug to control for lot variation. We need to replenish the selective medium every 3–4 days. After 10–12 days, examination of the dishes for viable cells need to be practised. Cell division may occur once or twice in the selective medium prior to cell death. Ascertain the minimum concentration of antibiotic that results in complete cell death. This is the optimal antibiotic concentration to use for stable selection (Table 1).

For random integration of the expression construct, Invitrogen™ pcDNA™ vectors may be used. For targeted integration of

Table 2
Mammalian cell-based expression systems compared

System	Basal expression level	Induced expression level	Response time to maximal expression
T-REx system	Low	Highest	24 h
Flp-In T-REx system	Lowest	High	24–48 h
GeneSwitch system	Lowest	High	24 h

the expression construct, we offer the Invitrogen™ Jump-In™ System and Invitrogen™ Flp-In™ System.

It should be taken into consideration if we need inducible or constitutive expression.

- An inducible promoter allows us to control the timing of gene expression. In the absence of the inducer, the gene is not expressed. Addition of the inducer turns on expression. This option is ideal for expressing toxic proteins. For regulated and inducible expression of the gene of interest, we offer the Invitrogen™ T-REx™ Expression System, Invitrogen™ Flp-In™ T-REx™ system, and the Invitrogen™ GeneSwitch™ system (Table 2).

  However, if we are working with a nontoxic gene and the timing of expression is not important, choose an expression vector with a constitutively expressing promoter. For this purpose, the pcDNA vectors are the ideal choice. There are several options to choose from, depending upon the cell type.
  - pcDNA vectors possess certain unique features:
  - The CMV, EF-1, or UbC promoter.
  - Different epitope tags.
  - Standardized detection or purification across a variety of proteins.
  - Several selection markers for creating stable cell lines.
  - Different cloning formats, such as restriction enzyme cloning, TOPO cloning, and Gateway cloning.
- A wide selection of specialized mammalian expression vectors designed for specific applications are necessary, including episomal expression, secreted expression, and intracellular targeting.

  Mammalian transient expression systems allow flexible and rapid production of proteins. They can generate large amounts of protein within 1–2 weeks. Two fully optimized, mammalian transient expression systems are effective for ultrahigh protein yields in CHO and HEK 293 cells. These systems enable

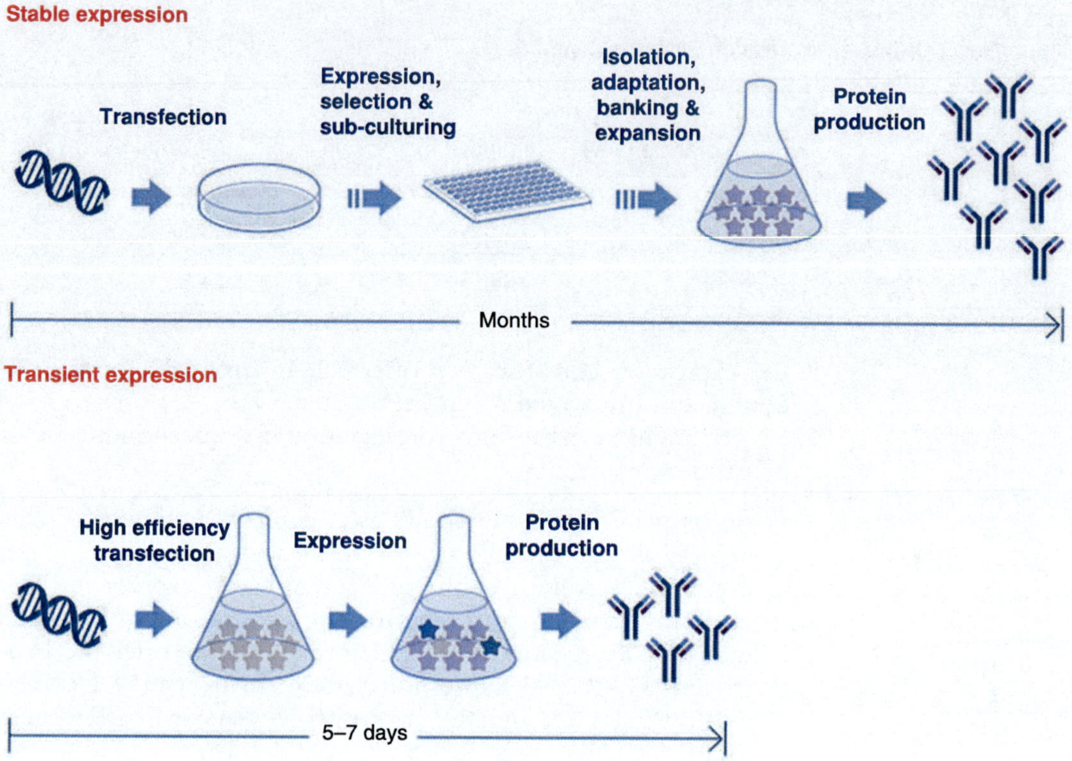

Fig. 9 Comparison of Mammalian cell-based expression systems

expression of human or other mammalian proteins with more native folding and post-translational modifications—including glycosylation—than expression systems based on hosts such as *E. coli*, yeast, or insect cells. Comparison of mammalian cell-based expression system is being diagrammatically represented in Fig. 9.

References

1. Fakruddin M et al (2012) Critical factors affecting the success of cloning, expression, and mass production of enzymes by recombinant *E. coli*. ISRN Biotechnol 2013:590587
2. Francis DM, Page R (2010) Strategies to optimize protein expression in E. coli. Curr Protoc Prot Sci Chapter 5:Unit 5.24.1–Unit 5.2429

Check for updates

Chapter 20

Transgenesis and Biopharming

Abstract

Transgenesis is the process of introducing foreign or exogenous DNA into an animal's genome. The steps include *knockout* or *knockin* of gene. Outline of the protocol of transgenesis process (Transgenic mouse) involves insertion of gene of interest into either the germinal vesicle (nucleus) or the pronucleus of the fertilized egg before fertilization. The next step is implantation into a pseudopregnant mouse followed by checking of expressed gene. The transgenes in many transgenic individuals are also transmitted through the germline to subsequent generations. Protocols were described with different methods of gene transfer in animals, as through (1) Retroviral vectors that infect the cells of an early stage embryo before implantation into a receptive female; (2) microinjection of DNA insert into the enlarged sperm nucleus (the male pronucleus) of a fertilized egg; (3) introduction of genetically engineered embryonic stem cells into an early stage developing embryo before implantation into a receptive female followed by the transfer of diploid somatic nuclei into an enucleated oocyte. Transfection may be carried out through either chemical (Calcium phosphate method, Transfection with polyplexes, Transfection with liposomes and lipoplexes) or physical methods (microinjection, particle bombardment, ultrasound, electroporation). Lastly it discusses the ethical considerations for transgenesis.

Keywords Knock out, Knock in, Retroviral vector, Male pronucleus, Transfection, Electroporation

1 Basic Principle

Transgenesis is the process of introduction of foreign or exogenous DNA into an animal's genome.

The steps include *knockout* or *knock in* of gene. The process of introduction of exogenous donor DNA into a recipient cell is known as Transfection. Intact chromosomes are not taken up in the process. The recipient cell generally gets a part of the DNA. With the aid of the recombinant DNA, it is possible to introduce a particular segment of DNA. Still problems exist for the stability of the new inserts (transient transfectants). DNA technology has been proved to be efficient to introduce genes into animals.

An animal which gains new genetic information with the addition of foreign DNA is assigned as transgenic whereas the introduced DNA is called the transgene.

Aruna Pal, *Protocols in Advanced Genomics and Allied Techniques*, Springer Protocols Handbooks,
https://doi.org/10.1007/978-1-0716-1818-9_20,

2 Outline of the Protocol of Transgenesis Process (Transgenic Mouse)

Step 1:

Plasmids containing the gene of interest are inserted into either the germinal vesicle (nucleus) of the oocyte or into the pronucleus of the fertilized egg before fertilization.

Step 2:

The egg is being implanted into a pseudopregnant mouse.

Step 3:

Soon after the birth, the recipient mouse can be examined to check if it has gained the foreign DNA. The next step is to check if the gene has been expressed. Multiple copies of transgenes are integrated randomly in varying locations in the transgenic individuals' genome.

Step 4:

Reports are available for the transmission of the transgenes in many transgenic individuals through the germline to subsequent generations.

Let us describe pronucleus.

Pronucleus: For a limited time just after fertilization, the male pronucleus and female pronucleus remain separated.

Female pronucleus: During the process of maturation of the ovum before impregnation, a part of the germinal vesicle gets converted to a number of small vesicles. These small vesicles combine into a single clear nucleus that migrates toward the center of the egg and is known as the female pronucleus.

Male pronucleus: During the process of impregnation, the spermatozoon that enters the egg, the tail is lost. The head forms a nucleus, called the male pronucleus, which moves toward the female pronucleus followed by fusion with it, leading to the formation of the first segmentation nucleus. The male pronucleus can be easily visible under a light microscope, since it is larger than the female pronucleus. Figure 1 depicts the microscopic view of fertilization process, where pronuclei, polar body, and embedded sperm are visualized.

3 Transgenesis; Methodology

Transgenic technology has been evolved and practiced in the laboratory mouse. Since 1980s hundreds of different genes have been introgressed into various mouse strains.

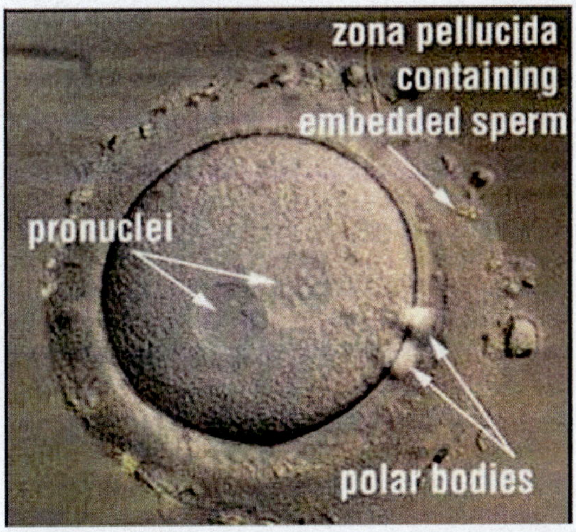

Fig. 1 Microscopic view of fertilization process with presence of pronuclei, polar bodies, and embedded sperm

Methods of Gene Transfer in Animals: For the process of transgenesis, DNA can be introduced into mice genome through one of the following methodologies:

1. Retroviral vectors which infect the cells of an early stage embryo before implantation into a receptive female.
2. Microinjection of DNA insert into the enlarged sperm nucleus (the male pronucleus) of a fertilized egg.
3. Introduction of genetically engineered embryonic stem cells in a developing embryo, which remain at its early stage of development. This is assayed before implantation into a receptive female.
4. Transfer of diploid somatic nuclei into an enucleated oocyte.

- The transgenes are introduced into the pronuclei of fertilized eggs through the above-mentioned processes. The injected embryos are incubated in vitro followed by implantation into the uterus of a pseudopregnant female for further stages of development.

3.1 Retrovirus-Mediated Gene Transfer

The choice of suitable vector is very important for transgenesis. The criteria for useful vectors for the purpose of gene isolation are those that allow themselves for the production of libraries comprising of overlapping fragments of genomic DNA, covering the entire genome several times.

For example, bacteriophage λ genomic library (of 106 viruses, each containing on average 20 Kb of DNA represents 6–7 copies of the entire mouse genome) reported to be useful and the probability of each gene represented is very high.

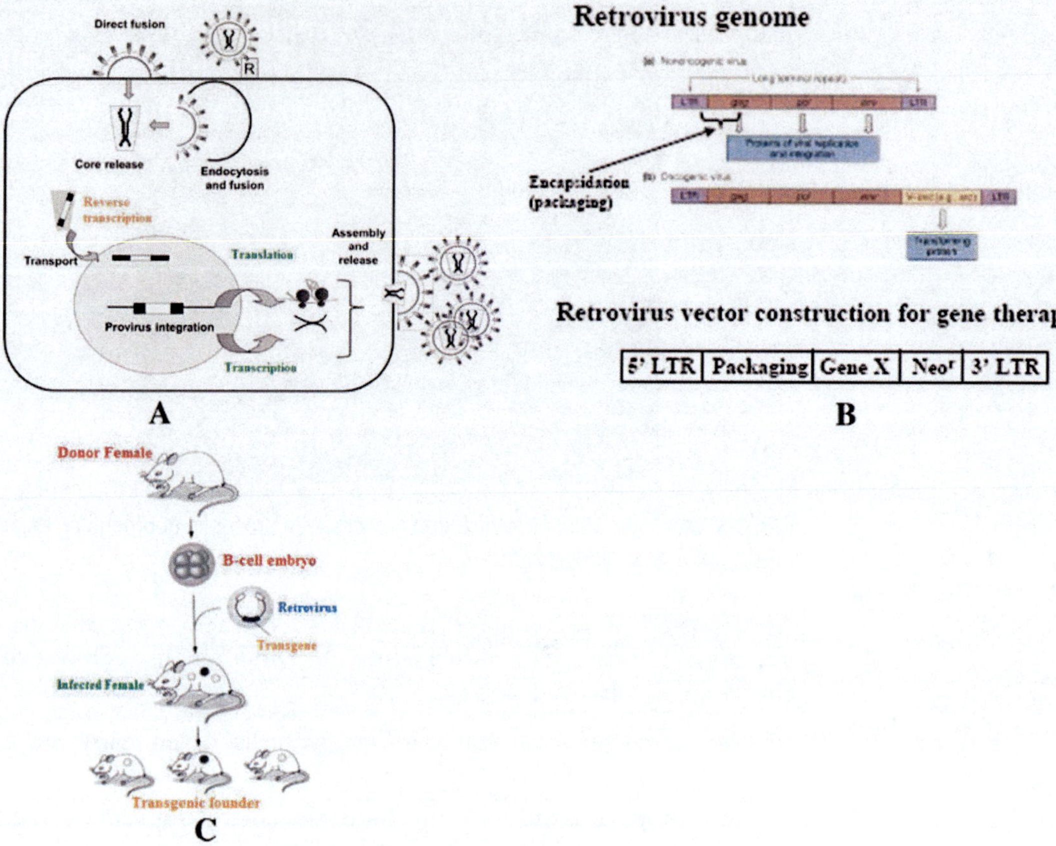

Fig. 2 (**a–c**) Establishing transgenic mice with retroviral vectors

Retroviruses were also found to be useful for the transfer of foreign genes into animal genomes. A retrovirus is a virus whose genetic material is RNA rather than DNA. It acts as vectors for the transfer of genetic material into the host cell, which results in a *chimera*. Chimeras are inbred for about 20 generations until homozygous (carrying the desired transgene in every cell) transgenic offspring are born. Figure 2 diagrammatically describes the process of transgenesis with retroviral vectors with some insight for retrovirus.

Cleavage-Stage embryos (at the eight-cell stage) are infected with a defective retrovirus containing a transgene. Transgenic pups are born from implanted females (foster mothers). Matings are undertaken in order to test which pups have the transgene in their germ line cells. Transgenic lines may be developed from these founder transgenic animals. Diagrammatic flow chart for the development of transgenic cow or chicken has been represented in Fig. 3.

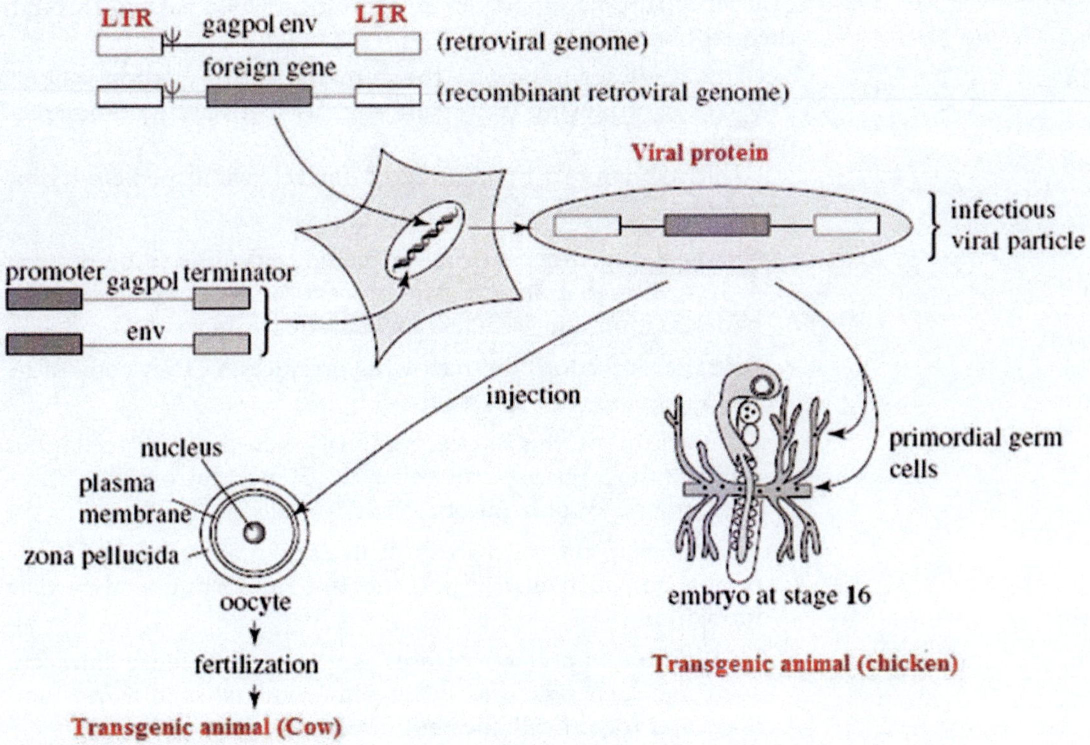

Fig. 3 Utilization of retroviral vectors to develop transgenic animals. The viral genome possessing the foreign gene is introduced into transcomplementing cells. Synthesizing the missing viral proteins. The viral particles are utilized to infect mammal oocytes (cattle, primates), chicken primordial germ cells, or mouse one cell embryos

Advantage: for this technique is that it acts as an effective method for integrating the transgene into the genome of a recipient cell.

Limitations: are that through this technique, transfer of long strand DNA is not possible; only small pieces of the foreign DNA (usually 9–15 kb) can be inserted. It is devoid of essential adjacent sequences for the regulation of the expression of the transgene. The genome of the retroviral strain (helper virus) which is necessary for the creation of large quantities of the vector DNA can be introgressed into the same nucleus as the transgene. It involves low copy number integration. Additional steps are necessary to produce retroviruses. There is every possibility for undesired genetic recombination altering the retrovirus. There exists high frequency of mosaicism. Possible interference by integrated retroviral sequences in transgene expression exists. The genome of the retroviral strain is able to be integrated into the same nucleus as the transgene. This implies that by default the virus itself could be produced by the transgenic organism and is problematic particularly for food animals, i.e., the animal will be utilized for food production. Moreover, the provirus attracts methylation that in conjugation with other mechanisms suppresses its expression while passing through the germ line.

Due to these limitations, the retroviral vector method is rarely utilized for producing transgenic animals that have a commercial potential.

Salient points to be considered for retroviral mediated gene transfer:

- The best stage for retroviral mediated transfer is to be done at 4–16 cell stage embryos. The process can be conducted up to mid-gestation, but with less success rate.
- Just after infection, the retrovirus produces a DNA copy of its RNA genome with reverse transcriptase.
- Completion of this process implies that the host cell goes through the S phase of the cell cycle. Hence, retroviruses effectively transduce only mitotically active cells.
- Alterations to the retrovirus usually consist of removal of structural genes, such as gag, pol, and env, supporting viral particle formation.
- In addition, most retroviruses and complementary lines are ecotropic such that they infect only rodents, such as rats and mice, and rodent cell lines but not humans.
- The DNA copy of the viral genome, or provirus, integrates randomly into the host cell genome, normally without deletions or rearrangements.
- This methodology is not useful for site-directed mutagenesis, as integration is not through the process of homologous recombination.
- Very high rates of gene transfer are attained through the use of retroviruses.

Table 1 lists some of the commonly employed vectors used for retroviral-mediated gene transfer.

3.2 DNA Microinjection Method

Considering the limitations of the retroviral vectors, microinjection of DNA is currently the popular method for producing transgenic mice.

First step is the need of gene of interest in the proper form. A linear transgene construct is prepared, which contains:

- The structural gene of interest, along with introns.
- A strong promoter gene and enhancer from mouse origin to let the gene to be expressed.
- Vector DNA to allow the transgene to be inserted into host DNA.

The immature female mice will be superovulated by sequential administration of FSH/LH and HCG and then mated to fertile

Table 1
Common vectors used for Retroviral-mediated gene transfer

Vector	Origin	Insert size range
Multicopy plasmids	Multicopy plasmids	Up to 20 kb
• Lambda vectors	Bacteriophage λ	Up to 30 kb
• Cosmid	Bacteriophage λ	Up to 40 kb
• P1 artificial chrom	Bacteriophage P1	80–90 kb
• Bacterial artificial chrom.	Large bacteria plasmid	100–300 kb (F factor)
• Yeast chrom. (YAC)	Yeast chromosome	100–1000+ kb

+ means indefinite

males. One-celled embryos are then flushed from the oviducts and put in a drop of medium and visualized by phase-contrast or interference microscopy.

Direct transfer of DNA into the cell without the aid of a carrier is called DNA microinjection (Fig. 4). This can be employed for only a few cells at a time. This technique is useful mainly for large cells such as oocytes, eggs, and the cells of early embryos. The DNA may be directly injected into different tissues, as skin, muscle, or internal organs or it can be directly injected into the blood.

This procedure has the following steps. The number of available fertilized eggs which need to be inoculated is increased by stimulating donor females to superovulate. The process is conducted by giving the mice an initial injection of pregnant mare's serum followed by next injection after 48 h of human chorionic gonadotropin (hCG). Through this procedure, the female produces about 35 eggs on contrary to the normal number of 5–10. These females are then mated and sacrificed. Flushing is advocated for the collection of the fertilized eggs (oocytes) from their oviducts and recovered. Eggs are treated with hyaluronidase to eliminate the adherent follicle cells. Unfertilized eggs are discarded. Figure 5 describes the schematic representation of the production of transgenic animals by microinjection of DNA into fertilized eggs.

The eggs are inoculated immediately with the transgene, following the procedure:

1. An embryo at the pronuclear stage is held in place by suction.
2. A micro needle loaded with a suspension of plasmid DNA (containing DNA insert) will be ready for introgression.
3. It is introgressed through the zona pellucida and plasma membrane into the pronucleus (usually the male).
4. Several hundred molecules of the recombinant DNA are injected in a volume of approximately 1 picoliter (pl) leading to injection of several hundred eggs.

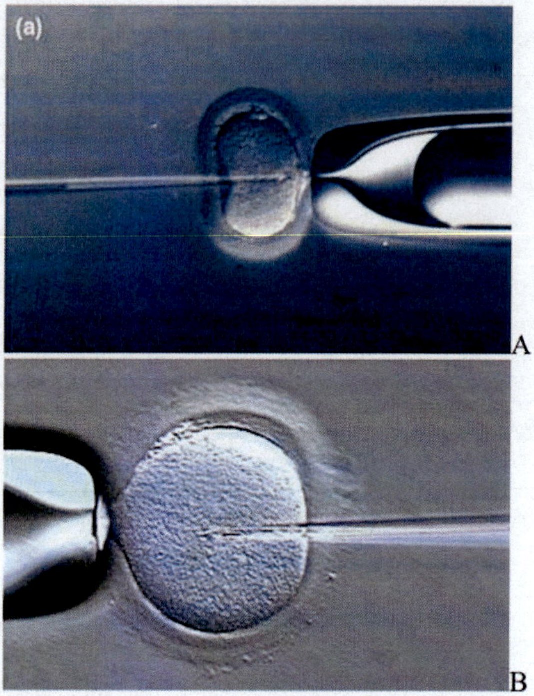

Fig. 4 (**a**, **b**) Microscopic view of DNA microinjection

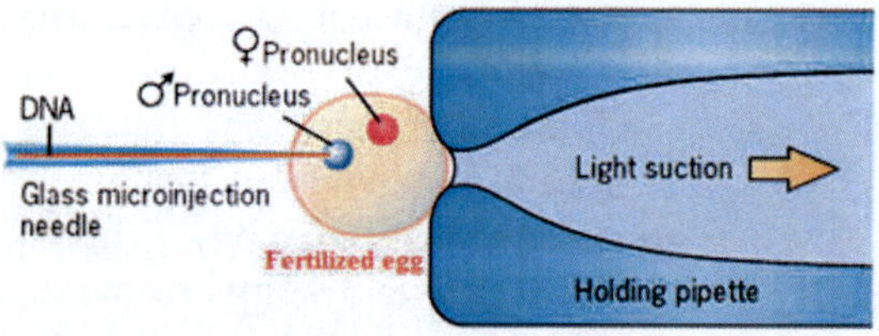

Fig. 5 Schematic representation of the production of transgenic animals by microinjection of DNA into fertilized eggs

5. The male pronucleus can be located with the help of dissecting microscope and the eggs then can be maneuvered, oriented, and held in place followed by the microinjection of DNA.
6. Followed by inoculation, 25–40 eggs are implanted microscopically into a foster mother. Foster mother is the female who has been made pseudo-pregnant through mating to a vasectomized male such that none of the eggs of the foster mother will be fertile. As a result, the foster mother will deliver pups from the implanted fertile eggs 3 weeks after the inoculation.
7. Following birth, the presence of foreign material is detected or verified by DNA hybridization with appropriate probes or PCR.

8. A transgenic mouse can be mated to another to produce homozygous transgenic animal.

3.3 Introduction of Genetically Engineered Embryonic Stem Cells into an Early Stage Developing Embryo Prior to Implantation into a Receptive Female

The method employs cells from the Inner Cell Mass (ICM) of early embryos blastocysts. These cells are able to grow in cell culture and thereby retaining the capability of differentiating into other cell types along with the germ line cells followed by their introduction into another blastocyst embryo. Such cells are called pluripotent (multi) embryonic stem (ES) cells. These cells can be easily manipulated by genetic engineering without changing their pluripotency.

3.3.1 Steps of the Procedure

1. Fertilized eggs (pre-implantation zygotes) are collected from a pregnant mother mouse as described above.
2. Grow zygotes in culture till Day 3.
3. Inner Cell Mass (ICM) is harvested from 3-day-old blastocysts.
4. The Inner Cell Mass (ICM) is cultured on feeder cells to develop Embryonic Stem (ES) Cell lines.
5. Transgenic ES cells are created by microinjection or by introducing cells briefly to an electrical potential. This disrupts the cell membrane and allows the entrance of DNA containing the transgene constructed with the genes of interest. Through this methodology, a functional transgene can be integrated in the place of a dispensable gene in the genome of the ES cell.
6. Injected the transgenic ES cells into the blastocoele of a new 3-day-old host blastocyst. It is to be noted that blastocoele is the fluid-filled cavity of the mass of cells.

   The injected ES cells combine with the host ICM and contribute to the developing embryo. The first-generation offspring are chimaeras. They possess somatic cells composed of both transgenic ES cells and host cells. They also have germ cells composed of both transgenic ES cells and host cells. Normally, a coat color gene is employed in the transgene construct as a visual marker to allow the quick detection of the transgenic (chimeric) pups.
7. The transgene, if in the germ cell lineage, can be successfully transmitted to offspring. As a result, homozygous transgenic lines can be constructed.

Followed by transfection of ES cells in culture with the DNA vector, some cells will have DNA integrated at none-target (spurious) sites, while others will have DNA integrated at target (correct) sites. On the other hand, some cells will not have any DNA integration at all.

Considering the Above Facts, We Can Devise Certain Methods to Enrich DNA Integration at the Specific Sites

This procedure employed ***positive selection***, which refers to the cells that did not accept the DNA inserts and ***negative selection*** refers to the cells who have DNA integrated anywhere in their genome.

In this procedure, a construct will be prepared, which should contain the following:

A Procedure Called Positive/Negative Selection Is Implemented

- Two blocks of DNA sequences (HB1 and HB2) which are homologous to separate regions of the target site.
- The trans gene, TG which will confer a new function on the recipient.
- Neor, a DNA sequence that codes for an enzyme which inactivates neomycin and its relatives such as the drug G418 (lethal to mammalian cells).
- Two different genes for the thymidine kinase (tk1 and tk2), which phosphorylate the nucleoside analogue is called ganciclovir. DNA polymerase is not able to differentiate against the resulting nucleotide and inserts this nonfunctional nucleotide into freshly replicating DNA. Thus, ganciclovir kills cells which possess the tk gene.

Now the arrangement of these sequences is indicative to the positive and negative selection procedure. The schematic representation has been depicted in Fig. 6.

3.3.2 Possible Results

- Most cells are not able to take up the vector. These cells will be killed if exposed to G418 as the neo gene will not be incorporated (positive selection).
- In a few cells, the vector is inserted randomly in the genome. On random insertion, the entire vector, with the tk genes, is inserted into host DNA. These cells were observed to be resistant to G418 but killed by ganciclovir (negative selection).
- In fewer cells, homologous recombination by double crossover at target sites occurs, i.e., stretches of DNA sequence in the vector find the homologous sequences in the host genome. The region between these homologous sequences replaces the equivalent region in the host DNA.
- As a result, tk genes will be excluded and cells survive both G418 and ganciclovir since only the neo and the trans genes are included.

With this methodology, ES cells which carry the target site will be enhanced several thousandfold. This induces better chances of producing a transgenic animal with the desirable characters. Through this methodology, ES cells containing the target are identified and cultured for propagation. Embryonic stem cells possessing an integrated transgene can be cultured and inserted into blastocyst stage embryo, which are then implanted in

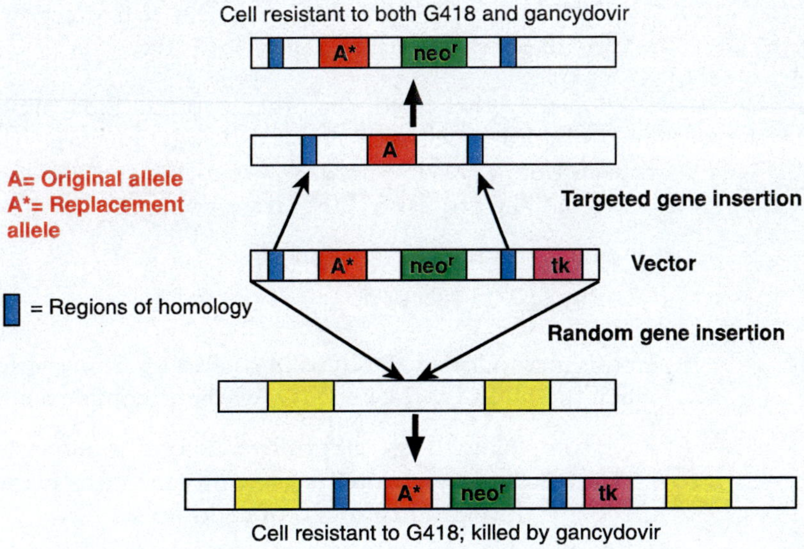

Fig. 6 Schematic representation of genetically engineered embryonic stem cells into an early stage developing embryo

pseudopregnant foster mother. Transgenic lines can then be prepared by initially mating founder transgenic mouse to animals from the same strain and followed by crossing transgenic litter mates to form a homozygous transgenic animal. Unfortunately, pluripotent ES cells comparable to those of mouse were not observed in cattle, sheep, pigs, or chickens.

3.4 Scientific and Medical Applications of the ES Cells Method of Transfection

This route is commonly used to inactivate a gene, alter it, or replace its protein coding region with a reporter (a coding unit whose product is easily assayed. This may be connected to any promoter of interest resulting in expression of the gene employed to assay promoter function.).

The major application of ES cell in transgenic mice lies in medicine and pure science. This includes:

1. Improvement of the understanding of all aspects of healthy animal.
2. Knowledge of therapeutic approaches.
3. Understanding biochemistry and physiology, particularly mammalian development, neurobiology, learning, and memory.

Four major strategies for gene transfer to animal cells are common.

- Two of them are considered biological mechanisms.
  - Through viral (transduction)—the transferred gene represents part of the viral genome.

– Bacteria which invade animal cells (bactofection)—the gene will be transferred as a plasmid.

On the other hand, the other two methods employ chemical and physical methods not involving infection thus termed transfection.

Thus, the process of DNA transfer may be classified as

1. Chemical transfection and
2. Physical transfection.

3.4.1 Chemical Transfection

In this methodology, DNA will be taken up from the surroundings when the DNA is presented as a synthetic complex either as:

- A complex with overall positive charge. It allows it to interact with negatively charged cell membrane, which then promote uptake through the process of endocytosis.
- As lipophilic complex which fuses with the cell membrane, depositing the transgene directly into the cytoplasm.

Calcium Phosphate Method

It employs the step of the formation of a fine DNA/calcium phosphate co-precipitate which initially settles on the cells followed by internalization by endocytosis. The precipitate should be formed freshly at the time of transfection. The DNA gets escaped and reaches the nucleus, which can be then expressed. Since the cells should be coated by the calcium complex, monolayers of cells need to be utilized for maximum efficiency. However, this method possesses only 1–2% transfection efficiency.

Transfection with Polyplexes

Efficiency of this technique is better than that with calcium as it is giving more uniform particle size. Polyplexes are polycationic compounds, which form soluble complexes (polyplexes) by the process of spontaneous electrostatic interaction with DNA. This method is adapted as a plasmid DNA transfer. The efficiency of the method can be improved by exposing cells to osmotic shock or treatment of cells with chloroquine.

A new generation of the polycationic compounds has been evolved. The examples include poly-L-lysine and synthetic poly amines, polyethylenimines and dendrimers. Dendrimers are highly complex molecules arranged in layers from a central initiator like ammonia or ethylenediamine.

Transfection with Liposomes and Lipoplexes

This can be conducted through the packaging of the DNA inside a fusogenic phospholipid vesicle that interacts with the target cell membrane resulting in DNA uptake. In brief, the bacterial cells will be transformed with suitable plasmid vector followed by treatment with chloramphenicol for the amplification of plasmid copy number. The cells will be treated with lysozyme in order to remove the cell wall. The resulting protoplasts will be centrifuged gently onto a monolayer of mammalian cells in order to promote fusion among them with polyethylene glycol.

Use of liposomes is most common for this type of transfection, known as lipofection. The efficiency of this method is better to that of chemical transfection method. With the aid of this method, up to 90% of cells in culture dish can be transected.

3.4.2 Physical Transfection

With help of this methodology, naked DNA is deposited directly into the cell with the aid of a physical force. This includes:

- microinjection.
- particle bombardment.
- ultrasound.
- electroporation.

Whichever method is employed, the result is called transformation. This results in a change in the recipient cell's genome due to the acquired transgene.

1. Electroporation:

   It is a physical transfection technique involved with the creation of transient nanometer size pores in the cell membrane through the exposure of the cells to a brief pulse of electricity. The two most critical parameters are the intensity and duration of electrical pulse.

   Electroporation can be employed for in vivo gene transfer commonly for surface or near surface tissue such as skin, muscle, and certain tumors or even internal tissues like liver. This can be attained through direct application of electrodes to the skin after shaving and mild abrasion with the DNA being injected into the skin prior to electroporation.
2. Ultrasound transfection:

   It involves the exposure of cells to a rapidly oscillating probe like the tip of sonicator. The methodology includes the application of ultrasound waves to a dish or cells or a particular tissue. This results in the formation and collapse of bubbles in the liquid, along with the cell membrane. This process is referred to as cavitation. The transient appearance of such cavities leads the DNA for crossing the membrane into the cytoplasm. The method is applicable for both in vivo or in vitro conditions since the plasmid DNA is left structurally intact. In case of electroporation, the DNA will be injected followed by the ultrasound application.
3. DNA microinjection.
4. Particle bombardment:

It is another direct delivery method evolved for the transformation of plants. The method employs coating small metal particles with DNA. The next step is accelerating them into target tissues with a powerful force like the blast of high-pressure gas or an electric discharge through a water droplet. In animals, this method is greatly applicable for tissues such as skin cells in vivo compared to cultured cells.

5. Vectors for gene transfer:

   It may be of two types: Bacterial or viral.

Bacterial Vector

The exploitation of living bacteria for gene transfer is essential for the genetic manipulation of plants. *Agrobacterium tumefaciens* and its close relatives have been employed for 20 years to generate transgenic plants.

Recently, *A. tumefaciens* has been employed for transfer of DNA to human cells. The protoplast fusion technique can also be considered as highly efficient form of bactofection. This type requires human intervention, while the *A. tumefaciens* does not require human intervention.

Methodology

The bacteria invade the host animal cells and undergo lysis within them, resulting in release of their plasmid DNA. Examples of these bacteria include *Salmonella* species (lysis takes place in the phagocytic vesicle), *Listeria monocytogenes*, and *Shigella flexneri* (lysis takes place for these two species afterwards when they escape from the vesicle).

The plasmid DNA then enters in the nucleus where it gets incorporated with the cell's genome and gets expressed. On the contrary, *A. tumefaciens* attaches itself to the outside surface followed by conjugation. It does not invade the cell.

Viral Vector

Virus particles have a natural ability to absorb to the surface of the cells and gain entry. This can be employed to deliver recombinant DNA into animal cells. Several classes of viruses have been utilized for gene therapy and at least 8 have been used in clinical trials. Transgenes may be integrated into viral vectors either through addition to the whole genome or replacing viral genes. This can be achieved through ligation or by homologous recombination.

If the transgene replaces a non-essential gene the vector is known as ***helper-independent vector***. On the other hand, when it replaces an indispensable gene, then this vector will be ***helper dependent vector***. ***Fully deleted vector or gutted vector or gutless vectors*** are described when all viral coding sequences have been deleted. It is generally recommended to employ this type (fully deleted) of vectors. These vectors possess the cis-acting elements necessary for packaging and for genome replication.

Advantages of using those vectors are high capacity for foreign DNA. Since no viral gene products are prepared, the vector possesses no intrinsic cytotoxic effects.

Examples

1. Adeno virus.

   These are viruses containing a linear, double-stranded genome, of approximately 36 kb.

   (a) Adeno-associated virus (AAV)—Single-stranded and is a member of the parvovirus family.

   (b) Baculovirus vectors (BV)—Have rode-shaped capsid and large double-stranded DNA genomes. They can infect arthropods particularly insects.

- Genotyping Transgenic Mice by PCR to Screen for Potential Founders.

  This is the test method of mice to detect the presence of the transgene by PCR. The sequence of the gene which was inserted into the male pronucleus could be determined if the mouse possesses the transgene of interest through PCR performance.

Tail biopsies: from potentially transgenic mice can be collected 5 weeks after injecting eggs (3 weeks gestation time and 2 weeks of post-natal growth). It can be assessed through DNA extraction from the tail tips and tests for the transgenic by PCR. A set of primers are designed that are taken from the transgene sequence and employing them in a regular PCR for the amplification of the gene of interest if observed. In case of a transgenic mouse, a PCR product corresponding to a known size will appear in the gel. However, non-transgenic mouse can be detected through the absence of PCR product corresponding to that size.

Additionally, for the evaluation of the stability of the insert, some markers need to be checked and the best ones are assayed readily through observing the new phenotype of the progeny.

- Southern Blotting Analysis is yet another efficient technique to determine the number of copies and places.

  Southern blot analysis can be used to determine the number of copies of the transgene integrated along with the number of chromosomal sites the transgene inserted, from samples collected from pup at 6 weeks of age. It is utilized to verify transgenic status and to determine if the transgene is intact (Fig 7). Intensive breeding may be employed with the help of this information of transgenic founders. Transgenic founders are developed in such a manner of a good chance of transmission (at least 5–10 copies) of an intact transgene in a single insertion site.

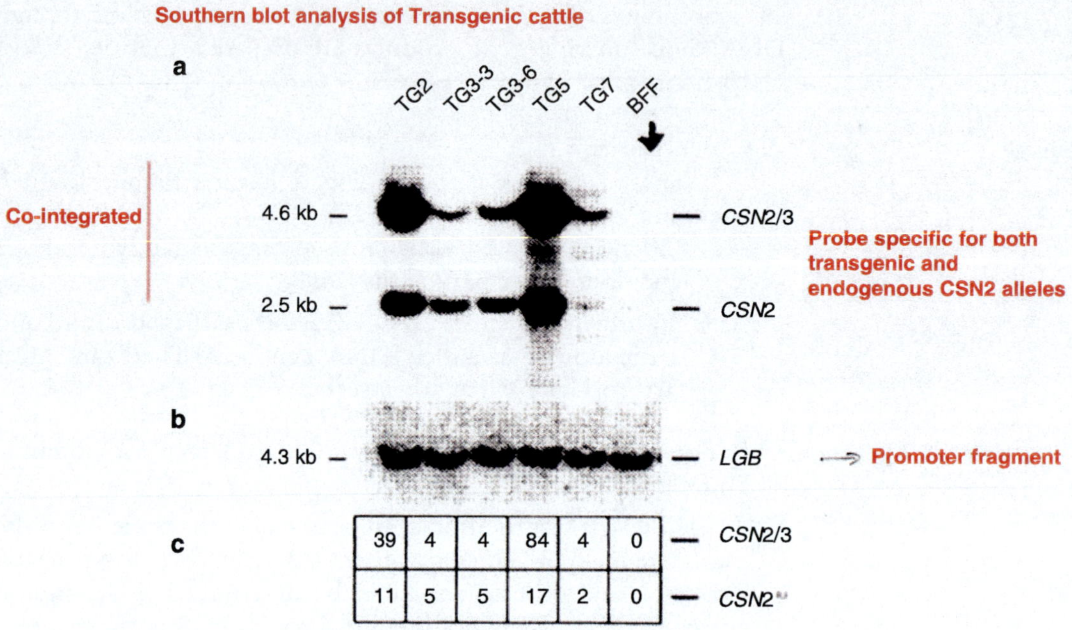

Fig. 7 (**a**–**c**) Southern blot analysis (Actual photograph) of transgenic cattle

One of the problems with transgenesis is based on the fact that after the insertion of DNA (micro-inserted), randomly some genes in the mouse will get replaced, leading to their inactivation. It is dependent on which gene is inactivated and a damage to the progeny may happen.

Nuclear Transfer Method (Nontransgenic Method)

- In this process the sheep Dolly was obtained from an enucleated (nucleus was removed) egg into which the nucleus from a cultured somatic cell of a mature sheep has been introgressed.

- Recovery of Oocytes from animals between 28 and 33 h following injection of gonadotropin releasing hormone.
- Recovered Oocytes are placed in PBS containing 1% FCS and transferred to a new media containing 10% FCS and incubated at 37 °C.
- Removal of the nucleus is conducted manually from an unfertilized oocyte as soon as possible.
- The somatic cell needs to be in a non-dividing stage.
- A non-dividing somatic cell is kept in contact with the oocyte and both are fused together through the application of an electrical potential activating the egg that mimics the process of natural fertilization.
- Hybridization results in an activated oocyte with two chromosome sets (from the diploid somatic cell).

- Generally, the cytoplasm of normal oocyte contains proteins and RNA molecules which are necessary for the early stages of development. However, now the cytoplasm of the somatic cell contains a whole set of genes which are reprogrammed to take control over the developmental program in a similar fashion as the genes of the normal embryo.
- Effect of age: It was observed that cells obtained from fetuses and newborn donors possess better efficiency in nuclear transfer in comparison to the clones derived from adult cells; later reflects more abnormalities.
- It may be due to the fact that somatic cells of adult animals might have more mutations or they are more differentiated compared to that of fetal cells, thus are more likely to fail the full-term development.

Applications of Nuclear Transfer

- Nuclear transfer has potential applications outside the field of transgenesis, for example, the propagation of an animal with a particularly desirable set of genes.
- Propagating the transgenic animals is quite hard to attain and are usually associated with risks.

Hence the nuclear transfer can be employed for propagating a successful transgenic animal making a whole herd of that animal. Successful animals (buffaloes) are present in NDRI herd.

3.4.3 Importance of Transgenesis

1. To improve genetic features of domesticated animals.
2. To act as animal models for study of human diseases.
3. BioPharming: It is the process of using farm animals for production of human pharmaceuticals. For example, mammary glands may be used for production of recombinant product in natural secretion as milk.
4. To study the genes regulation, development of animals.
5. To understand tumor development, e.g., introducing oncogenes and observe the effect.
6. To study immunological specificity, e.g., producing knockout genes which are responsible for some immunological aspects.
7. Other biological interests like examination of the possibility of using transgenic animals in the industrial production of human therapeutic drugs, etc. [1–3].

3.4.4 Prospect for Transgenesis in Livestock Sector

1. Transgenic Cattle:
   Applications.
   (a) The technique is employed for improving the casein content of milk leading to increase of cheese production.

(b) It may be employed for the production of Lactose-free milk (transgene lactase), particularly useful lactose intolerant individuals.

(c) The technique may be employed for the development of disease-resistant cattle, against bacterial or viral diseases.

(d) Transgenesis may be involved with in vivo immunization with specific heavy and light chain genes.

2. Pigs:
   (a) Transgenic pig developed with PST porcine somatotropin (growth hormone) with better performance traits.
   (b) Care should be taken to prevent adverse effects on kidney, stomach, heart, sterility for transgenic pig.
   (c) Transgenic pig may be developed with gene insert for human hemoglobin to replace whole blood transfusions.
3. Sheep:
   (a) Transgenic sheep may have better wool production.
   (b) Transgenic sheep may be developed with keratin promoter.
   (c) Transgenic sheep with better growth factor gene may be developed.
4. Birds:
   (a) Traditional methods cannot be employed, because of avian embryogenesis differences.
   (b) In birds, no ES cells were found.
   (c) ALV-resistant chicken may be developed through transgenesis.
   (d) Transgenic bird may be developed with defective ALV genome.
   (e) Transgenic birds can make viral RNA and protein but blocks assembly of viral particles.
5. Fish:
   (a) Transgenic technique is equally applicable for aquaculture.
   (b) Fishes may be developed with transgene—growth hormone.
   (c) Future prospect includes production of Hilsa fish in fresh water.

3.4.5 Some of the Common Terminologies in Transgenesis

- **Transgenic Mouse**: The term refers to an engineered mouse which has a normal DNA sequence for a gene that has been replaced by an engineered sequence or a sequence from another organism.

- **Knockout Mouse**: A transgenic mouse where the normal gene is missing or engineered leading to loss of transcription or translation. "Knocks out" that gene.
- **Knockin Mouse:** As the term refers to introduction of a foreign gene into the existing genome in contrast to knock out, where a particular gene is expelled from existing native genome.
  - This may include the alteration of the function of the gene (e.g., replace one amino acid with another in a site to detect if that site is essential for the protein's function).
  - Alteration of transcription rate for the overproduction or underproduction of the gene product.
  - Development of a fluorescent gene product to map its distribution in tissue.
- **Conditional Knockout (Knockin) Mouse**: A transgenic mouse wherein the transgene is knocked out (or in) in specific tissues, at a particular developmental stage, or in response to an exogenous substance (e.g., an antibiotic).

4 Ethics and Regulations for Preparation of Transgenic Product

There exist certain rules and regulations for genetically modified organisms.

In case the genetic manipulation of living organisms by humans goes unregulated, it causes exploitation of the organisms along with the harmful impacts on our ecosystem. There is immense necessity for the development of regulatory bodies that regulate the research and process of transgenesis, as biosafety and bioethics are immensely important to regulate the balance in nature and uncontrolled human intervention. Hence transgenesis or other human manipulation of organisms is immensely related to ecological, economic, and bio-safety issues.

4.1 GEAC

In terms of genetically modified organisms (GMOs), the Government of India has developed the **Genetic Engineering Approval Committee (GEAC)**. This organization is bestowed with the responsibility as decision-making authority to take decisions related to the validity of research involving GMOs and regulates the safety of GMOs introduced for public use.

4.2 Patents and Rights

Certain controversial issues are raised when certain companies were granted patents for products and technologies involving genetic material or other resources developed and utilized by people of a specific region over years together.

A recent example is Basmati rice, which was developed by Indian farmers over hundreds of years. An American company, in 1997, got the patent rights for Basmati rice from the US Patent and

Trademark Office. This patent permits this company to sell new variations of Basmati rice and make a profit, at the same time restricts other people from selling it.

Similar cases were attempted to patent Indian herbal medicines also. Hence it is of utmost necessity for the people and countries to be vigilant and counter patent applications exploiting products and techniques native to their land.

4.2.1 Biopiracy

Biopiracy is a term commonly used for the utilization of bio-resources without proper authorization or compensation to the people or country concerned. This is usually committed by multinational companies. Most of the developed, financially rich nations are poor in biodiversity or traditional know-how; on the contrary the developing nations are mostly rich in biodiversity. This causes exploitation of the traditional knowledge for the development of modern, commercial applications thereby saving the maker's time, money, and effort.

People have become more aware of the injustice and inadequacy for the compensation and sharing of benefits between the developing and developed nations. Accordingly, several nations now have developed laws preventing other nations from exploiting their bio-resources and traditional knowledge. As for example, the Indian Parliament cleared the second amendment of the Indian Patents Bill, which considers the issue of biopiracy into consideration. It is a worldwide problem [5–7].

5 Comparison of Transgenesis with Gene Editing

Gene editing is an advanced methodology dealt in Chapter 20. The basic difference between transgenesis and gene editing is that in gene editing the site of the insertion of the "insert" or fragment of interest is at specific site. On the other hand, the site for gene insertion is random in case of transgenesis [2].

References

1. Asfaw AB, Assefa A (2019) Animal transgenesis technology: a review. Cog Food Agric 5:1. https://doi.org/10.1080/23311932.2019.1686802
2. Kalds P, Zhou S, Cai B, Liu J, Wang Y, Petersen B, Sonstegard T, Wang X, Chen Y (2019) Sheep and goat genome engineering: from random transgenesis to the CRISPR era. Front Genet 10:750. https://doi.org/10.3389/fgene.2019.00750
3. Volobueva AS, Orekhov AN, Deykin AV (2019) An update on the tools for creating transgenic animal models of human diseases – focus on atherosclerosis. Braz J Med Biol Res 52(5): e8108. https://doi.org/10.1590/1414-431X20198108
4. Ormandy EH, Dale J, Griffin G (2011) Genetic engineering of animals: ethical issues, including welfare concerns. Can Vet J 52(5):544–550
5. Prakash D, Verma S, Bhatia R, Tiwary BN (2011) Risks and precautions of genetically modified organisms. ISRN Ecol 2011:369573., 13 pages. https://doi.org/10.5402/2011/369573

6. Camara D, Dimitrova Ir, Doynova M, Jachacz L, Kachakova D, Kepka M, Ould Isselmou CB, Vorniere JP, Yungarva Tsv (2008) Transgenic and cloned animals: ethical problems? http://bioethics.agrocampus-ouest.eu/infoglueDeliverLive/digitalAssets/
7. FAO (2001) Ethical issues in food and agriculture. FAO ethics series. Food and Agriculture Organization of the United Nations, Rome

Check for updates

Chapter 21

Rapid Amplification of cDNA Ends (RACE)

Abstract

Rapid amplification of cDNA ends (RACE) or "one-sided" polymerase chain reaction (PCR) or "anchored" PCR is a polymerase chain reaction-based technique that results in the cloning of full-length cDNA sequences in case only a partial cDNA sequence is available. The 5′ RACE system is a set of prequalified reagents utilized for the synthesis of first-strand cDNA, purification of first-strand products, homopolymeric tailing, and preparation of target cDNA for subsequent amplification by PCR. The next step is the amplification of tailed cDNA by PCR with a mixture of three primers: a nested gene-specific primer (GSP2), which anneals 3′ to GSP1, a combination of a complementary homopolymer-containing anchor primer and a corresponding adapter primer, which permit amplification from the homopolymeric tail. Outline of the methodology of the individual steps of 5′ RACE is isolation of RNA, design of 5′ RACE primers, design of 5′ RACE primers, first-strand cDNA synthesis from total RNA, removal of RNA template by RNase mix, purification of first-strand product, homopolymeric tailing of cDNA, amplification of target cDNA, cloning 5′ RACE amplification products. Detailed protocol with illustrations has been discussed along with a troubleshooting guide.

Keywords RACE, Anchored PCR, Nested gene-specific primer, Homopolymeric tailing of cDNA

1 Basic Principle

Rapid amplification of cDNA ends (RACE) is a polymerase chain reaction-based technique that results in the cloning of full-length cDNA sequences in case only a partial cDNA sequence is available. Usually, cDNA clones are obtained from the phage or plasmid library, which are devoid of sequences corresponding to the 5′ or 3′ ends of the mRNA transcripts.

In the RACE technique, PCR is employed to amplify partial cDNAs representing the region between a single point in an mRNA transcript and its 3′ or 5′ end. A prior knowledge of a short internal stretch of the sequence must be known. Through the process of RACE, the missing sequence is to be rebuilt through gene-specific primers oriented in the direction of the missing sequence. Figure 1 depicts the schematic representation of first-strand cDNA 5′RACE and 3′RACE regions.

Aruna Pal, *Protocols in Advanced Genomics and Allied Techniques*, Springer Protocols Handbooks, https://doi.org/10.1007/978-1-0716-1818-9_21,

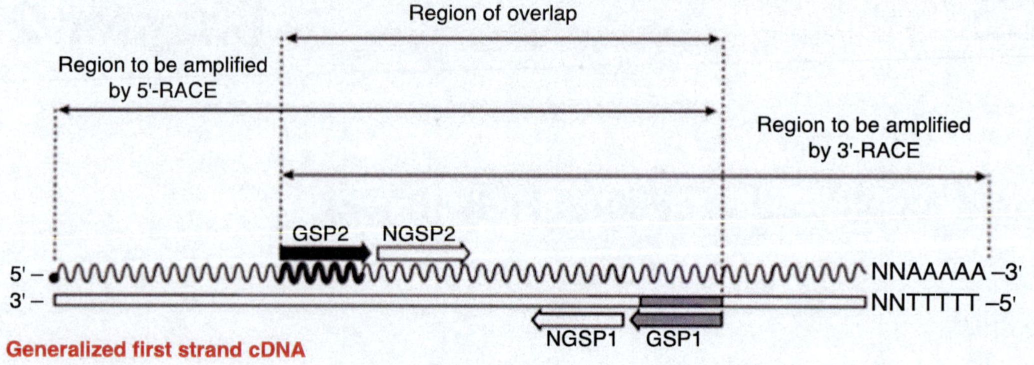

Fig. 1 Schematic representation of first-strand cDNA 5′RACE and 3′RACE regions

2 Practical Protocol

Rapid amplification of cDNA ends (RACE) is a process for amplification of nucleic acid sequences from a messenger RNA template between a defined internal site and unknown sequences at either the 3′ or the 5′-end of the mRNA. This method of amplification with single-sided specificity has been described by others as "one-sided" PCR or "anchored" PCR.

As a general rule, PCR amplification of comparatively fewer target molecules in a complex mixture needs two sequence-specific primers, which flank the region of the sequence to be amplified. However, it is quite difficult to amplify and characterize regions of unknown sequences. 3′ and 5′ RACE methodologies attempt to solve this problem. 3′ RACE employs the advantage of the natural poly(A) tail in mRNA as a generic priming site for PCR amplification (Fig. 2).

The steps involved are the conversion of mRNAs into cDNA using reverse transcriptase (RT) and an oligo-dT adapter primer. PCR-based amplification was employed for specific cDNA with gene-specific primer (GSP), which anneals to a region of known exon sequences and an adapter primer targeting the poly(A) tail region. This allows the capture of unknown 3′-mRNA sequences underlying the exon and the poly(A) tail (Fig. 3). 5′ RACE, or "anchored" PCR, is a technique, which leads to the isolation and characterization of 5′ ends from low-copy messages (Fig. 3).

Although the exact protocol differs among different users, the general strategy remains consistent. First-strand cDNA synthesis is primed with a gene-specific antisense oligonucleotide (GSP1). This allows cDNA conversion of specific mRNA or related families of mRNAs. It increases the potentiality for a complete extension to the 5′-end of the message. After cDNA synthesis, the first strand product is purified from unincorporated dNTPs and GSP1. Terminal deoxynucleotidyl transferase (TdT) is employed for the addition

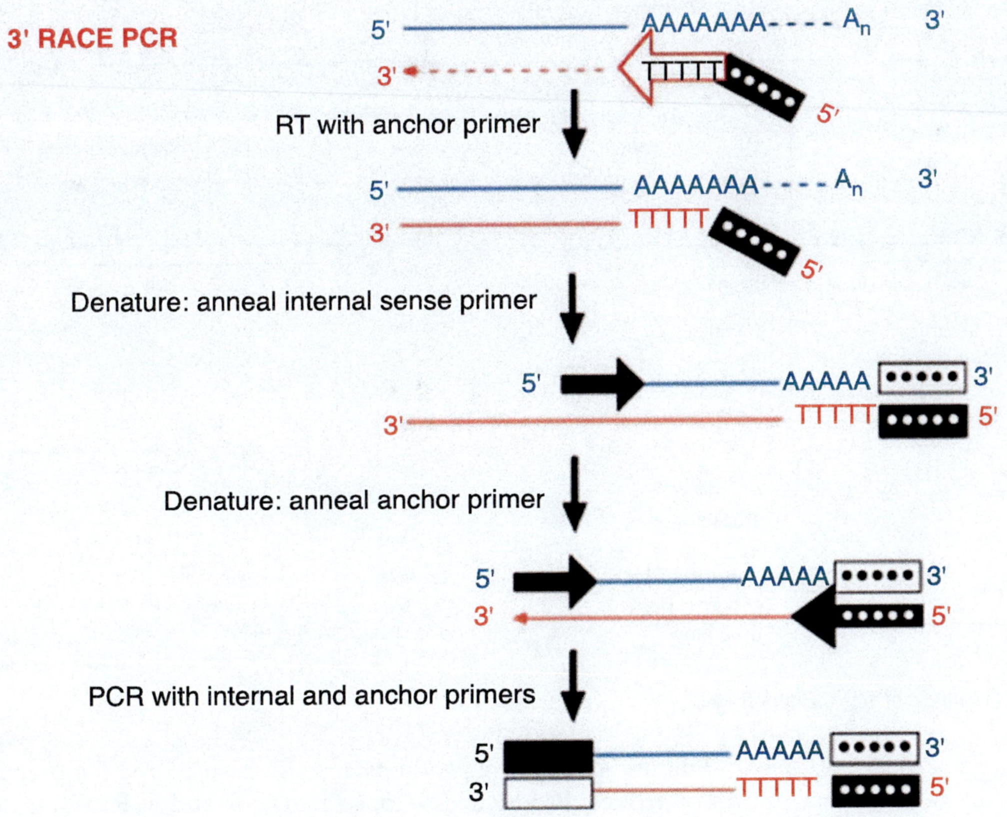

Fig. 2 Outline of 3′RACE PCR

of homopolymeric tails to the 3′ ends of the cDNA. The next step is the amplification of tailed cDNA by PCR with a mixture of three primers:

1. A nested gene-specific primer (GSP2), which anneals 3′ to GSP1.
2. A combination of a complementary homopolymer-containing anchor primer.
3. A corresponding adapter primer, which permits amplification from the homopolymeric tail.

This aids in the amplification of unknown sequences between the GSP2 and the 5′-end of the mRNA. RACE procedures have been utilized for amplification and cloning of rare mRNAs, which may not be detected through conventional cDNA cloning methodologies. RACE is also applicable for the existing cDNA libraries. Random hexamer primed cDNA can also be adapted to 5′ RACE for amplification and utilized for cloning of multiple genes from a single first-strand synthesis reaction. Products of RACE reactions can be sequenced without the need of any cloning steps in between.

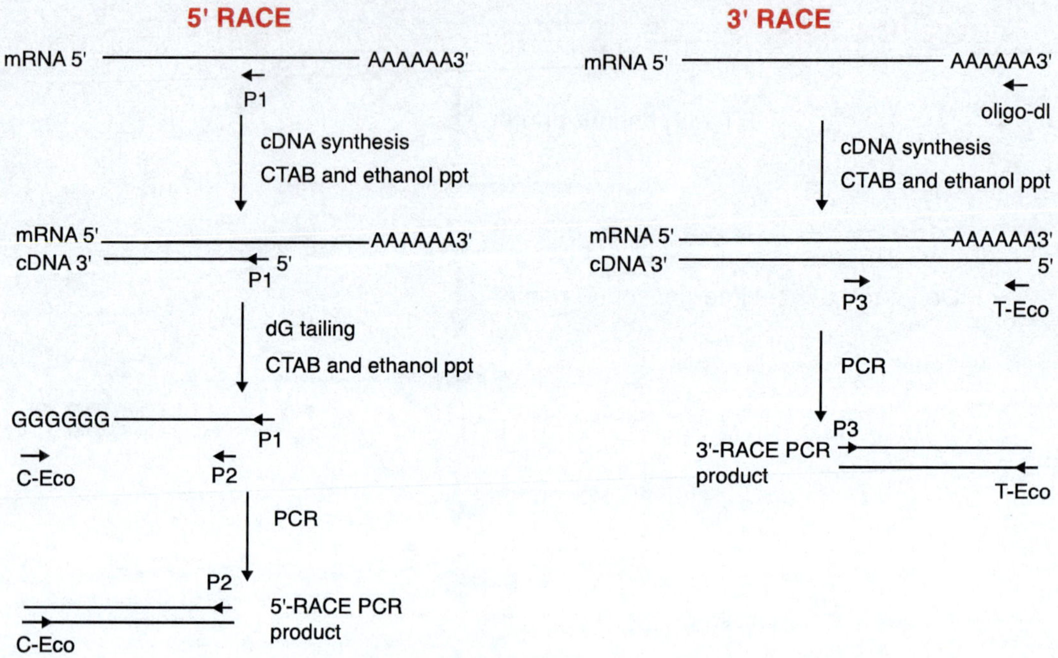

Fig. 3 Outline of 5′RACE and 3′RACE

Simultaneously, the products can also be used for the preparation of probes. Products developed by the 3′ and 5′ RACE procedures may be employed in combination to generate full-length cDNAs. At the end, the RACE procedures can be utilized along with exon trapping methods to allow amplification and subsequent characterization of unknown coding sequences.

3 Summary of the 5′ RACE System

The 5′ RACE system is a set of prequalified reagents used for the synthesis of first-strand cDNA. This is followed by the process of purification of first-strand products, homopolymeric tailing, and preparation of target cDNA for subsequent amplification by PCR. Control RNA, DNA, and primers are essential for monitoring system performance. First-strand cDNA is synthesized from total or poly(A)+ RNA with a gene-specific primer (GSP1) provided by the user and SuperScript™ II reverse transcriptase (RT), a derivative of Moloney murine leukemia virus reverse transcriptase (M-MLV RT) with reduced RNase H activity. Subsequent to first-strand cDNA synthesis, the original mRNA template is removed through treating with RNase mix (mixture of RNase H, which is specific for RNA:DNA heteroduplex molecules, and RNase T1). dNTPs, GSP1, and proteins, which are not incorporated, are separated from cDNA using a S.N.A.P. column.

The addition of a homopolymeric tail to the 3′-end of the cDNA is done with TdT and dCTP.

As the tailing reaction is conducted in a PCR-compatible buffer, the total process of the reaction can be directly amplified by PCR. There is no need of intermediate organic extractions, ethanol precipitations, or dilutions. PCR amplification is carried out using Taq DNA polymerase through a process of nested PCR. Primers used are gene-specific primer (GSP2, designed by the user), which anneals to a site located within the cDNA molecule, and a novel deoxyinosine-containing anchor primer supplied with the system.

After amplification, 5′ RACE products may be cloned into an appropriate vector for characterization processes. The processes

1. include sequencing,
2. restriction mapping,
3. preparation of probes to detect the genomic elements associated with the cDNA of interest, or
4. in vitro RNA synthesis.

The abridged anchor primer (AAP), abridged universal amplification primer (AUAP), anchor primer (AP), and universal amplification primer (UAP) include recognition sequences for Mlu I, Sal I, and Spe I to allow restriction endonuclease cloning of RACE products.

5′ RACE products can also be used for processes, which do not need an intermittent cloning step such as dsDNA cycle sequencing or probe preparation. Other rounds of PCR using the AUAP, or UAP, along with either progressively nested GSPs or size-selected products from the initial PCR, are necessary for imparting sufficient specificity to the process to allow direct characterization of RACE products.

The outline of the methodology of the individual steps of 5′ RACE is discussed below.

1. **Isolation of RNA.**

   The protocol for isolation for RNA has already been discussed in earlier chapters (Chapter 2). The quality of the RNA is very important to detect the sequence information, which can be converted into cDNA. Hence, the optimization of the isolation of RNA, and prevention of the introduction of Rnases, and inhibitors of RT are important criteria. The guanidine isothiocyanate/acid-phenol method is one of the best methods for RNA isolation.

2. **Design of 5′ RACE primers.**

   Good primer design is a prime criterion to determine the sensitivity and the specificity of the first-strand synthesis and the later PCR. At least two antisense gene-specific primers

(GSP) are needed for 5′ RACE and need to be supplied by the user.

GSP1 primes first-strand cDNA synthesis. The primer is designed in such a way so as to anneal at least 300 bp from the mRNA 5′-end. This leads the cDNA for easy purification using a S.N.A.P. column.

A second nested primer, GSP2, annealing to sequences located 3′ (with respect to cDNA **not** mRNA) of GSP1 is needed for PCR. GSP2 primer has the ability to anneal immediately adjacent to GSP1. GSP2 primer can also anneal the sequences located further upstream of GSP1 within the cDNA product. Other sequentially nested GSPs are necessary depending on the efficiency and specificity of the primary PCR.

As a general rule, these primers need to be highly specific for their target sequences and able to form stable duplexes with their target sequences and are free of secondary structure. The primary criteria for primer designing are discussed below.

3. **First-strand cDNA synthesis from total RNA.**

   The capture of mRNA 5′-ends is dependent on complete cDNA synthesis.

4. **Removal of RNA template by RNase mix.**

5. **Purification of first-strand product.**

6. **Homopolymeric tailing of cDNA.**

   (a) A greater proportion of tailed cDNA molecules for efficient amplification of first strand products.

   (b) Homopolymeric tails of sufficient length to permit the primer to anneal.

   (c) Homopolymeric tails of uniform length for the production of a homogeneous amplification product.

   (d) A buffer compatible with the PCR buffer system.

7. **Amplification of target cDNA.**

   There are four times and temperatures that must be considered:

   (a) Preamplification denaturation of the template DNA (PAD).

   (b) Denaturation of product DNA at the start of each cycle.

   (c) Annealing of the primers to the denatured DNA.

   (d) Extension of the primers by the polymerase.

   Steps 2–4 constitute a cycle and are repeated usually 30–35 times followed by a final extension time of 5–10 min and then a holding temperature of 5 °C.

8. **Cloning 5′ RACE amplification products.**

4 Methods

4.1 Components and Storage

The components of the 5′ RACE system are as follows:

Specific cDNA is prepared from 1 to 5 μg of a total RNA or 50–500 ng poly (A)+ RNA preparation for amplification by anchored PCR. The amount of RNA will differ based on the application. Control RNA, DNA, and primers are used to validate the performance of the system. This can be used for experimental RNA preparations in order to assess the efficiency of each step or to troubleshoot potential problems.

Component	Volume	Storage
Reagents		
10× PCR buffer [200 mM Tris–HCl (pH 8.4), 500 mM KCl]	500 μL	−20 °C
25 mM $MgCl_2$	500 μL	−20 °C
10 mM dNTP mix 10 mM each dATP, dCTP, dGTP, dTTP	100 μL	−20 °C
0.1 M DTT	100 μL	−20 °C
SuperScript II reverse transcriptase (200 units/μL)	10 μL	−20 °C
RNase mix	10 μL	−20 °C
5× tailing buffer [50 mM Tris–HCl (pH 8.4), 125 mM KCl, 7.5 mM $MgCl_2$]	500 μL	−20 °C
2 mM dCTP	50 μL	−20 °C
Terminal deoxynucleotidyl transferase	15 μL	−20 °C
5′ RACE abridged anchor primer (AAP, 10 μM)	80 μL	−20 °C
Universal amplification primer (UAP, 10 μM)	40 μL	−20 °C
Abridged universal amplification primer (AUAP, 10 μM)	40 μL	−20 °C
DEPC-treated water	1.25 mL	−20 °C
Control gene-specific primer 1 (GSP1, 1 μM)	25 μL	−20 °C
Control nested gene-specific primer 2 (GSP2, 10 μM)	80 μL	−20 °C
Control PCR primer, gene-specific primer 3 (GSP3, 10 μM)	20 μL	−20 °C
Control DNA (2×10^4 copies/μL; ~0.1 pg/μL)	100 μL	−20 °C
Control RNA (50 ng/μL)	10 μL	−70 °C
DNA purification system		
S.N.A.P. columns	10 columns	4 °C
Collection tubes	10 tubes	4 °C
Binding solution (6 M sodium iodide)	30 mL	4 °C
Wash buffer concentrate	30 mL	4 °C

Certain critical aspects need to be considered. The 10× PCR buffer is without $MgCl_2$. Hence, $MgCl_2$ should be added to the first strand reaction mix. DNA purification system should never be frozen.

4.2 Additional Materials Required

The following items are needed for use with the 5′ RACE system, but are not included.

- sterilized, RNase-free thin-walled 0.2 or 0.5-mL PCR tubes.
- automatic pipettes capable of dispensing 1–20 and 20–200 μL.
- sterilized, RNase-free disposable tips for automatic pipettes.
- disposable latex gloves.
- sterilized, distilled water.
- absolute ethanol.
- GSP1 (cDNA primer, user-defined).
- GSP2 (nested primer for PCR amplification of dC-tailed cDNA, user-defined, appropriately engineered).
- microcentrifuge capable of generating a relative centrifugal force of 13,000 × *g*.
- 37, 42, 65, and 70 °C water baths or heat blocks (or use thermal cycler).
- Taq DNA polymerase.
- programmable thermal cycler; and
- mineral oil (if necessary for your thermal cycler).

4.3 Performance and Limitations of Procedures

The 5′ RACE system is being functionally tested with the control RNA (e.g., with GIBCO BRL Taq DNA polymerase). After PCR, a distinct 711-bp band was observed by agarose gel electrophoresis and ethidium bromide staining. The 5′ RACE system gives a direct and reliable solution for the preparation of tailed cDNA. PCR with single-sided specificity exists as highly challenging. Success with the system is greatly dependent on the efficiency and specificity of the PCR employed for the amplification of tailed cDNA. Care should be taken to employ Taq DNA polymerase and buffer to be used from the same supplier. The PCR protocol is the starting point. The optimal amplification parameters vary as per the gene.

If the 5′ RACE products are longer, i.e., greater than 1 kb, it adds an additional challenge to the procedure. The 5′ RACE system has been employed successfully with a suitable system (e.g., Invitrogen™ eLONGase™ Enzyme Mix), an enzyme system designed for amplification of long templates, to obtain an increased yield of the amplification product. Simultaneously, it also leads to substantially increased length of 5′ RACE products.

The major limitation to long(er) 5′ RACE lies in the specificity and efficiency of PCR. Critical success factors employ primer design, PCR optimization, and a systematic experimental strategy. It includes amplification of primary PCR using nested, gene-specific primers. Truncated products result in informative sequence data that can be applied in additional 5′ RACE experiments as one walks toward the 5′-end.

5 Detail Methodology

5.1 Isolation of Total RNA

It has already been discussed earlier. Full-length cDNA is the isolation of intact RNA is the most important limiting criteria. As a result, it is essential to optimize the isolation of RNA and prevention of the introduction of RNases and inhibitors of RT (reverse transcriptase) such as guanidinium salts, SDS, and EDTA. The effective method of RNA isolation is the guanidine isothiocyanate/acid-phenol method.

The Invitrogen™ TRIzol™ reagent method is an improvement of the original single step that can be used for the preparation of RNA from as little as 10^3 cells. Total RNA isolated with TRIzol reagent is undegraded and essentially free of protein and DNA contamination. To maintain intact RNA, an RNase-free environment is critical.

Note: All temperature incubations need to be carried out in an appropriately programmed thermal cycler. This overcomes the requirements for multiple fixed temperature baths or heat blocks.

Total RNA may contain small amounts of genomic DNA that may be amplified along with the target cDNA. The presence of this double-stranded DNA is not likely to cause problems in 5′ RACE, since it is inefficiently tailed prior to amplification. For better validation, a control experiment needs to be carried out simultaneously without RT. Products developed in the absence of RT are of genomic origin. It is essential to remove all genomic DNA from RNA preparation. Oligo(dT)-selection for poly(A)+ RNA is however not essential although the incorporation of this step may lead to the detection of rare mRNA transcripts.

5.2 Design of the Gene-Specific Primers

Efficient and specific PCR is regulating effective primer design. This is particularly true for RACE applications as the PCR is carried out with only a single GSP. There is not a single foolproof method, which design can guarantee successful amplification. As a result, all primers should be tested in PCR for their effectiveness.

5.3 Primers for PCR (GSP2 and GSP3)

Effective primers form stable, highly specific duplexes with their target sequences, and are free of secondary structure such as hairpin loops and dimers. High stability, i.e., G:C clamps, in the 5′- and central regions of the primer results in hybridization stability with the target sequence. Primers with unstable 3′-ends ($\Delta G > -9$ kCal/mol) give rise to higher specificity, as the potential for mispriming at nontarget sites is minimized. Generally, this condition can be met by including no more than two G or C residues in the last five 3′-bases. In addition, 3′-terminal complementarity needs to be minimized, since primer-dimer artifacts may significantly reduce PCR efficiency. Therefore the nested amplification primer needs to be checked for dimer formation with the anchor primer, also with itself.

Computer algorithms have also been developed, usually leading to this analysis as well as secondary structure analysis.

The next important matter for primers is the T_m (the temperature at which 50% of the primer and its complementary sequence exists in a duplex DNA molecule). The T_m is necessary to establish an annealing temperature for PCR. Optimum annealing temperature should be employed. It needs to be sufficiently low for efficient annealing of the primer to the target, but high enough to minimize nonspecific binding. Since a single GSP is employed in RACE, annealing temperature should be sufficiently stringent in order to reduce the amplification of nonspecific products. As a good starting point, an annealing temperature is chosen a few degrees below the estimated T_m of the primer pair. In practice, primers with T_ms between 60 and 75 °C usually can function effectively in 5′ RACE. Ideally, the T_m of primers utilized for PCR needs to be closely matched. Certain methodologies are present for the calculation of the T_m of primers. They give only estimates. Care should be taken to determine the optimal annealing temperature through experiments. While designing UDG cloning primers, consider dU residues as if they were dT residues for the calculations.

5.4 Primer for First-Strand cDNA Synthesis (GSP1)

As for GSP1, the same rule is observed for PCR primer selection. The T_m should be proper for the relatively low temperature (42 °C) of the cDNA synthesis reaction. A short primer (16–20 bases) is effective for the efficient separation of GSP1 from cDNA product. Efficient recovery of cDNA from the S.N.A.P. column requires a product of at least 200 bases in length. Therefore, we recommend that GSP1 anneal to sequences located at least 300 bases from the mRNA 5′-end. Well-designed primers often fail to function efficiently in priming first-strand synthesis. For example, the secondary structure of the mRNA at the annealing site may prevent efficient annealing or extension of the primer.

5.5 Primers for Subsequent Cloning

The user-defined GSPs need to be compatible with the cloning method. Add the following to the 5′-end of the nested GSP: for

T4 DNA polymerase cloning 5′–CGA–3′ (use with AUAP) for restriction endonuclease cloning, an appropriate adapter region is required. It should be noted that in cases where only limited peptide sequence information is available, degenerate GSPs may be prepared. An alternative to the synthesis of degenerate primers is the substitution of dI residues at wobble-base positions.

5.6 1× Wash Buffer for S.N.A.P. Procedure

Before using the system for the first time, a 1× wash buffer should be prepared from the wash buffer concentrate.

1. Pipetted 1 mL of the wash buffer concentrate and poured into a 50-mL graduated cylinder.
2. Added 18 mL of distilled water and 21 mL of absolute ethanol. Mixed properly.
3. Transferred to an appropriate-sized glass bottle. Put the cCap and stored at 4 °C.

5.7 70% Ethanol Wash for S.N.A.P. Procedure

1. Added 35 mL of absolute ethanol and 15 mL of distilled water and poured to a 50-mL graduated cylinder.
2. Transferred to a suitable-sized glass bottle. Put the cap and stored at 4 °C.

5.8 First-Strand cDNA Synthesis

This procedure is framed to convert specific RNA sequence(s) from 1 to 5 μg of total RNA into first-strand cDNA. Usually, 100–500 ng of total RNA is effective for the amplification of low copy messages by 5′ RACE. Although poly(A)+ RNA can be employed in this procedure for the enrichment for very rare messages, this level of purity is absolutely not essential.

1. Added the following to a 0.5-mL microcentrifuge tube (or thin-walled PCR tube, when thermal cycler is used):

Component	Amount
GSP1	2.5 pmol (~10 to 25 ng)
Sample RNA	1–5 μg
DEPC-treated water (or sterile, distilled water)	Sufficient for a final volume of 15.5 μL

Certain critical aspects need to be considered. Extreme care needs to be taken for avoiding contamination of samples by RNase. While choosing placental RNase inhibitor in the reaction, it needs to be added after the addition of DTT. Appropriate volume adjustments should be made in this step. See precautionary notes regarding the use of RNase inhibitor proteins. While using high GC content mRNA, it is advisable to employ the alternate protocol to help reduce interference by the secondary structure of mRNA.

2. Incubated the mixture 10 min at 70 °C for the denaturation of RNA. Chilled 1 min on ice. Collected the contents of the tube through brief spinning followed by the addition of the following in the order given:

Component	Volume (μL)
10× PCR buffer	2.5
25 mM $MgCl_2$	2.5
10 mM dNTP mix	1
0.1 M DTT	2.5
Final volume	8.5

3. The final volume of **steps 1** and **2** is 24 μL.
4. Mixed gently, and collected the reaction through brief centrifugation. Incubated for 1 min at 42 °C.
5. Added 1 μL of SuperScript™ II RT. Mixed gently and incubated for 50 min at 42 °C.

   Note: 30 min incubation is usually sufficient for short (<4 kb) mRNAs. Longer transcripts require at least 50 min to synthesize enough cDNA for a consistent signal in long PCR.

   Final Composition of the Reaction:
   20 mM Tris–HCl (pH 8.4).
   50 mM KCl.
   2.5 mM $MgCl_2$.
   10 mM DTT.
   100 nM cDNA primer (GSP1).
   400 μM each dATP, dCTP, dGTP, dTTP.
   1–5 μg (~40 ng/μL) RNA.
   200 units SuperScript II RT.

   Note: Keep the enzymes on ice during the procedure to maintain the efficacy. Mixed and quickly centrifuged each component before use.

   Note: More than 5 μg of total RNA, increase reaction volumes and amount of SuperScript II RT proportionately. On the contrary, if there are less than 1 μg of total RNA, no changes to the protocol are essential. Fifty to 500 ng of poly(A) + RNA can be substituted for total RNA in this protocol.
6. Incubated at 70 °C for 15 min for the termination of the reaction.
7. Centrifuged 10–20 s and place the reaction at 37 °C.
8. Added 1 μL of RNase mix, mixed properly, and incubated for 30 min at 37 °C.

9. Collected the reaction by brief centrifugation and placed on ice. The process can be interrupted at this point and the reactions stored at −20 °C.

5.9 S.N.A.P. Column Purification of cDNA

1. Equilibrated the binding solution to room temperature.
2. For each sample to be purified, equilibrated ~100 μL of sterilized, distilled water at 65 °C.
3. Added 120 μL of binding solution (6 M NaI) to the first strand reaction.
4. Transferred the cDNA/NaI solution to a S.N.A.P. column. Centrifuged at 13,000 × *g* for 20 s.
5. The cartridge insert is removed from the tube and transferred the flow through to a microcentrifuge tube. The solution is saved till recovery of the cDNA is confirmed. The cartridge insert is placed back into the tube.
6. 0.4 mL of **cold** (4 °C) 1× wash buffer is added to the spin cartridge. Centrifuged at 13,000 × *g* for 20 s. Discarded the flowthrough. Repeated this wash step **three** additional times.
7. The cartridge is washed twice with 400 μL of **cold** (4 °C) 70% ethanol as explained in the earlier step.
8. Following the removal of the final 70% ethanol wash from the tube, centrifuge at 13,000 × *g* for 1 min.
9. Transferred the spin cartridge insert into a fresh sample recovery tube. Added 50 μL of sterilized, distilled, water (**preheated to 65 °C**) to the spin cartridge. It is then centrifuged at 13,000 × *g* for 20 s for the elution of the cDNA.

   Note: Remove all the ethanol is mandatory for the yield of good quality of DNA.

   Note: The distilled water should be at 65 °C for the highest yield of DNA.

5.10 TdT Tailing of cDNA

Variable quantities of purified cDNA from the S.N.A.P. column purification can be employed in the TdT-tailing reaction. Factors, like the amount of RNA utilized in the first strand reaction and the relative abundance of the desired mRNA, need to be considered. The cDNA pool can be concentrated by lyophilization and the entire content can be employed in the tailing reaction. For the evaluation of the specificity of the subsequent amplification reaction from the oligo-dC tail, the use of a control reaction omitting TdT is recommended.

1. Added the following components to each tube followed by gentle mixing:

Component	Volume (μL)
DEPC-treated water	6.5
5× tailing buffer	5.0
2 mM dCTP	2.5
S.N.A.P.-purified cDNA sample	10.0
Final volume	24.0

2. It is incubated for 2–3 min at 94 °C. Chill for 1 min on ice. Collect the contents of the tube through brief centrifugation and accordingly placed on ice.
3. Added 1 μL TdT, mix gently, and incubate for 10 min at 37 °C.

   Final Composition of the Reaction:

   10 mM Tris–HCl (pH 8.4).

   25 mM KCl.

   1.5 mM $MgCl_2$.

   200 μM dCTP.

   cDNA.

   TdT.

   Note: There is no need to add more amounts of TdT in the tailing reaction may inhibit PCR. $MgCl_2$ as the 5× tailing buffer already contains $MgCl_2$.

   Note: The TdT needs to be optimized for 5′ RACE.
4. Heat inactivate the TdT for 10 min at 65 °C. Collected the contents of the reaction through brief centrifugation and placed on ice.

5.11 PCR of dC-Tailed cDNA

Tailed cDNA obtained from the earlier protocol can be amplified directly by PCR. Amplification of more than 5 μL volumes of the tailing reaction needs appropriate adjustments for a buffer, $MgCl_2$, and dNTP concentrations in the PCR.

1. Equilibrated the thermal cycler block to 94 °C.
2. In the majority of the cases, the "good start" procedure gives rise to specific amplification products. For some target and primer sets, "hot start" has been observed to improve the specificity of the reaction.
3. Added the following to a 0.2 or 0.5-mL thin-wall PCR tube sitting on ice:

Component	Volume (μL)
Sterilized, distilled water	31.5
10× PCR buffer [200 mM Tris–HCl (pH 8.4), 500 mM KCl]	5.0
25 mM $MgCl_2$	3.0
10 mM dNTP mix	1.0

(continued)

Component	Volume (μL)
Nested GSP2 (prepared as 10 μM solution)	2.0
Abridged anchor primer (10 μM)	2.0
dC-tailed cDNA	5.0
Final volume	49.5

Note: It is essential to have the reaction mixture in ice-cold condition in order to avoid nonspecific binding and extension of primers.

4. Added 0.5 μL of Taq DNA polymerase (5 U/μL) prior to mixing.
5. Mixed the contents of the tube (Taq DNA polymerase is mixed immediately before setting the reaction into the thermal cycler) and may be overlaid with 50–100 μL of mineral oil if felt necessary.

   Final Composition of the Reaction:
   20 mM Tris–HCl (pH 8.4).
   50 mM KCl.
   1.5 mM $MgCl_2$.
   400 nM GSP2.
   400 nM abridged anchor primer.
   200 μM each dATP, dCTP, dGTP, dTTP.
   tailed cDNA.
   2.5 units Taq DNA polymerase.
6. Transferred tubes directly from ice to the thermal cycler, which were equilibrated to the initial denaturation temperature (94 °C) earlier.
7. Performed conventionally 30–35 cycles of PCR. A typical cycling protocol for cDNA with less than equal to 1 kb amplified region is:

   PAD: 94 °C for 1–2 min.
   Cycle:
   Denaturation: 94 °C for 0.5–1 min.
   Annealing of primers: 55 °C for 0.5–1 min.
   Primer extension: 72 °C for 1–2 min.
   Followed by:
   Final extension: 72 °C, 5–7 min.
   Indefinite hold: 5 °C, until samples are removed.
8. Analyze 5–20 μL of 5′ RACE products by agarose gel electrophoresis based on the standard protocols, using appropriate size standards. Either TAE or TBE electrophoresis buffer may be employed for the procedure. The volume of the sample employed for analysis will be depended on the volume and

thickness of the sample well. If products will be extracted for reamplification, ultraviolet (UV) visualization of ethidium bromide-stained products should be performed using either a long wavelength (356-nm) UV or 302-nm wavelength source to minimize DNA nicking.

5.12 Nested Amplification

Usually, a single PCR of 25–35 cycles is not sufficient to generate enough specific product to be detectable by ethidium bromide staining. An increase in the number of cycles performed during the PCR beyond 35 cycles may lead to many nonspecific products, not recommended. Accordingly, high target levels may aggravate the amplification of nonspecific products during PCR. It leads to the contribution to the production of a heterologous smear of PCR products. On the other hand, a dilution of the original PCR (0.1–0.05%) may be amplified again with the help of the AUAP or UA and a nested GSP. If there is insufficient sequence information to design a nested GSP, reamplification of gel purified, size-selected PCR products using the UAP, or AUAP, and original GSP is useful for providing specific 5′ RACE products or installation of dUMP-cloning sequences for UDG cloning. Nested PCR is also equally applicable with a plug of agarose from the gel analysis of the initial 5′ RACE reaction.

1. 5 μL aliquot of the primary PCR may be diluted into 495 μL TE buffer [10 mM Tris–HCl (pH 8.0), 1 mM EDTA].
2. Equilibrated the thermal cycler block to 94 °C.
3. Added the following to a 0.2 or 0.5-mL thin-wall PCR tube which is on ice.

Component	Volume (μL)
Sterilized, distilled water	33.5
10× PCR buffer [200 mM Tris–HCl (pH 8.4), 500 mM KCl]	5.0
25 mM $MgCl_2$	3.0
10 mM dNTP mix	1.0
Nested GSP (prepared as 10 μM solution)	1.0
AUAP or UAP (10 μM)	1.0
Dilution of primary PCR product	5.0
Final volume	49.5

Note: It is essential to have the reaction mixture in ice-cold condition for preventing nonspecific binding and extension of primers.

Note: UDG cloning needs amplification with the UAP and properly designed nested GSP. It is preferable to use AUAP while working with an archaebacterial DNA polymerase or long PCR enzyme mixture.

4. Added 0.5 μL of Taq DNA polymerase (5 units/μL) just immediately prior to mixing.
5. Mixed the contents of the tube (Taq DNA polymerase is mixed immediately before putting into the thermal cycler) and overlaid with 50–100 μL of mineral oil (if needed).

   Final Composition of the Reaction:
   20 mM Tris–HCl (pH 8.4).
   50 mM KCl.
   1.5 mM $MgCl_2$.
   200 nM nested GSP.
   200 nM UAP or AUAP.
   200 μM each dATP, dCTP, dGTP, dTTP.
   diluted primary PCR product.
   2.5 units Taq DNA polymerase.
6. Transferred the tubes directly from ice to the thermal cycler pre-equilibrated to the initial denaturation temperature.
7. Performed 30–35 cycles of PCR.
8. Analyzed 5–20 μL of the amplified sample, with agarose gel electrophoresis, ethidium bromide staining, and the appropriate molecular size standards.

   Note: Cycle number is depended on the amount and complexity of the target. About five cycles are sufficient to install dUMP-sequences for UDG cloning.

6 Interpretation of Results

6.1 Analysis of 5′ RACE Results

Following PCR, products may be analyzed through agarose gel electrophoresis (1–2%) followed by staining with ethidium bromide. Band intensity and size distribution of the resulted products rely on the specificity of GSPs utilized for cDNA synthesis and PCR, the complexity and relative abundance of target cDNA, and the PCR conditions employed. Amplification products usually range from a single specific band to multiple discrete products to a broad diffuse smear.

- The complexity of products obtained by 5′ RACE may arise due to the following conditions: Incomplete cDNA synthesis,
- Aberrant priming of GSPs during first-strand synthesis or PCR,
- Mispriming by the anchor primer, as well as primer-dimer and
- other PCR artifacts.

Identification of specific product bands can be complicated by the presence of nonspecific products, which are dependent on both reverse transcription and dC-tailing. If the sequences are present for use as internal probes, it is greatly recommended that Southern blot analysis may be employed for the identification of particular product bands. Specific products may also be detected employing a diagnostic restriction endonuclease digestion, provided the amplified cDNA sequence consists of a known restriction site.

6.2 5′ RACE Controls

Several controls may aid in the interpretation of results. Products resulting from amplification of contaminating genomic DNA can easily be detected from control reactions that exclude RT. An alternative process is the inclusion of control reactions, which employ genomic DNA as a target. The specificity of the anchor primer for the oligo-dC tail needs to be examined by conducting amplification reactions with cDNA subjected to dC-tailing both in the presence and absence of TdT. Additional controls that amplify dC-tailed cDNA with each primer individually (either the abridged anchor primer or GSP2) can be helpful for the identification of nonspecific products resulting from mispriming.

6.3 Nested Amplification

Additional PCR using a nested GSP is a prime requirement for 5′ RACE of rare messages with either the UAP or AUAP. Usually, a dilution of the original PCR is utilized as a target. A nested primer consists of sequences located 3′ to the original primer (GSP2). For 5′ RACE, it would be an antisense primer, which anneals closer to the mRNA 5′-end. Purification of the original PCR product from primers and primer-dimer products is essential for the improvement of the specificity and efficiency of nested amplification processes. Finally, the 5′ RACE procedure would detect a single prominent band on an agarose gel. This may need additional rounds of PCR with successively nested GSPs. The design of a nested primer will be depended on:

1. The data availability of sequence information for the target of interest and
2. The resulting data generated for the original amplification reaction.

While performing 5′ RACE with a nested primer, sequences for downstream cloning manipulations should be designed into the nested GSP [1–4].

7 Troubleshooting Guide

7.1 Testing the 5′ RACE System Using the Control RNA and DNA

While employing the 5′ RACE system for the first time, it is suggested to perform an experiment with the control RNA to familiarize with the 5′ RACE system procedure. It is equally employed for the verification of proper functioning of all components in the protocol, for example, reagents and equipment for PCRs. The control RNA provided with the 5′ RACE System is an 891-bp, which is in vitro transcribed RNA derived from the chloramphenicol acetyltransferase (CAT) gene, which is also engineered to contain a 3′ poly(A) tail. It can be used alone or added to the RNA preparation to test system performance in a background of heterologous nucleic acid. It is necessary to test for the presence of contaminating nucleases. Dilutions of the control RNA can be employed for the determination of the sensitivity of the system or modeling the abundance of the desired mRNA.

The control DNA was prepared by cloning the control 5′ RACE product into pAMP1. Tailed cDNA was amplified with the 5′ RACE anchor primer and control GSP2 consisting of additional UDG cloning sequences. The 4.8-kb pAMP1 5′RACE recombinant possesses the oligo-dC tail sequence and can be used as a PCR positive control for the verification of the performance of the Abridged anchor primer, 5′ RACE anchor primer, AUAP, UAP, or the control GSP3, in addition to the control of GSP2. Otherwise, it may be employed to optimize PCR parameters with the Abridged anchor primer for reaction conditions or thermal cycling device. Two different PCRs are employed for the verification of system performance.

7.2 Conversion of First-Strand cDNA and Recovery of cDNA After S.N.A.P.

Purifications are assayed by a CAT cDNA-specific PCR with the control GSP3 and GSP2. The addition of the oligo-dC tail to purified control cDNA is checked by PCR with the Abridged anchor primer and control GSP2. Sequences for the control primers are presented below.

Note: It will be advantageous to adopt a similar RT-PCR strategy for their message and designing of an appropriate sense gene-specific primer (GSP3). It will be a helpful guide for troubleshooting problems arising during 5′ RACE with their message.

control GSP1 5′-TTG TAA TTC ATT AAG CAT TCT GCC-3′

control GSP2 5′-GAC ATG GAA GCC ATC ACA GAC-3′

control GSP3 5′-CGA CCG TTC AGC TGG ATA TTA C-3′

7.3 Sequences of the Control Primers

Typical results for the procedure employing the control RNA, both alone and in a background of 1 μg HeLa total RNA, are discussed later. A distinct 711-bp 5′ RACE PCR product needs to be visualized by ethidium bromide staining. Other products, normally visible as faint bands or a diffuse smear, may result from spurious priming by GSP2 or the anchor primer, incomplete cDNA synthesis, and primer-dimer artifacts. In case the control RNA is employed in a background of heterologous RNA, nonspecific 5′ RACE products may be observed. These nonspecific 5′ RACE products are dependent on both RT and TdT. This effect is documented by the

310-bp HeLa-derived 5′ RACE product. The presence of these nonspecific but genuine 5′ RACE products principally represents the specificity of the GSPs. It equally urges the need for characterization and enrichment of specific amplification products before cloning or sequencing.

7.4 Control First-Strand cDNA Synthesis

The RNA template employed for this step is control RNA and the primer is control GSP1. For the purpose of testing, the performance in a background of RNA, two first-strand reactions are suggested as follows:

(Control RNA + Control GSP1) and (Control RNA + Control GSP1 in the presence of sample RNA). The volume of DEPC-treated water is adjusted appropriately to make the final volume in **step 1** as 15.5 μL.

Note: Each component should be mixed and centrifuged quickly prior to use.

1. Add the following to a 0.5-mL microcentrifuge tube:

Component	Volume (μL)
Control GSP1 (1 μM)	2.5
Control RNA	1.0
DEPC-treated water	12.0
Final volume	15.5

2. Incubated the mixture 10 min at 70 °C for the denaturation of RNA. Chilled 1 min on ice. Collected the contents of the tube by brief centrifugation and added the following in the order given:

Component	Volume (μL)
10× PCR buffer	2.5
25 mM $MgCl_2$	2.5
10 mM dNTP mix	1.0
0.1 M DTT	2.5
Final volume	8.5

3. The combined total volume of **steps 1** and **2** is 24 μL.
4. Mixed gently and collected the reaction following brief centrifugation. Incubated for 1 min at 42 °C.
5. Added 1 μL of SuperScript II RT. Mixed gently and incubated for 50 min at 42 °C.
6. Incubated at 70 °C for 15 min to terminate the reaction.

7. Centrifuged 10–20 s and placed the reaction at 37 °C.
8. Added 1 μL of RNase Mix, mixed gently, and incubated for 30 min at 37 °C.
9. Collected the reaction by brief centrifugation and placed on ice.
10. Transferred a 2-μL aliquot of the control first-strand cDNA to a 1.5-mL microcentrifuge tube that contains 998 μL of TE buffer [10 mM Tris–HCl (pH 7.5), 1 mM EDTA]. Label tube "A" and retained for later PCR.
11. Purified the remainder of the control DNA using the S.N.A.P. column.

7.5 S.N.A.P. Column Purification of the Control cDNA

1. Equilibration is followed for the binding solution to room temperature.
2. Equilibrated about 100 μL of distilled water at 65 °C (for each sample to be purified) for use in **step 9**.
3. Added 120 μL of binding solution (6 M NaI) to the remainder of the control first-strand reaction.
4. Transferred the cDNA/NaI solution to a S.N.A.P. column. Centrifuged at 13,000 × *g* for 20 s.
5. Removed the cartridge insert from the tube and transferred the flowthrough to a microcentrifuge tube. The solution is saved till recovery of the cDNA is confirmed. The cartridge was inserted back into the tube.
6. Added 0.4 mL of **cold** (4 °C) 1× wash buffer to the spin cartridge. Centrifuged at 13,000 × *g* for 20 s. Discarded the flowthrough. This wash step **is repeated three** additional times.
7. The cartridge is washed twice with 400 μL of **cold** (4 °C) 70% ethanol as described in **step 6**.

    Note: The binding solution should be kept at room temperature for efficient binding of the DNA.

    Note: 1× wash buffer and 70% ethanol must be prepared prior to use and used as cold.
8. Following the removal of the final 70% ethanol wash from the tube, centrifuge at 13,000 × *g* for 1 min.
9. The spin cartridge is transferred and inserted into a fresh sample recovery tube. Added 50 μL of distilled water (preheated to 65 °C) to the spin cartridge. Centrifuged at 13,000 × *g* for 20 s for the elution of the cDNA.
10. 5 μL aliquot of the purified control first-strand cDNA is added to a 0.5-mL microcentrifuge tube containing 495 μL of TE buffer. Labeled the tube "B" and retain for later PCR.

    Note: Care should be taken to remove all the ethanol, otherwise recovery of the DNA will be poor.

Note: It is very important that the distilled water must be at 65 °C in order to maximize recovery.

7.6 TdT Tailing of the Control First-Strand cDNA

A control reaction that omits TdT is based on the amplification of cDNA from this control reaction, which can give important data for troubleshooting RACE. For example, the inclusion of a control reaction that omits TdT will help in the evaluation of the specificity of the amplification reaction from the oligo-dC tail. Labeled two 0.5-mL microcentrifuge tubes "C" and "D," respectively.

1. For the purpose of a more sensitive test, better efficiency of the tailing reaction, prepared a 100-fold dilution of the purified control cDNA. Added 1 μL of the cDNA to a 0.5-mL tube possessing 99 μL of sterilized, distilled water.
2. Added the following components to each tube:

Component	Volume (μL)
DEPC-treated water (or sterile, distilled water)	6.5
5× tailing buffer	5.0
2 mM dCTP	2.5
S.N.A.P.-purified control cDNA (1:100 dilution)	10.0
Final volume	24

3. Incubated for 2–3 min at 94 °C. Chill 1 min on ice. Collected the contents of the tube through brief centrifugation and kept on ice.
4. Added 1 μL TdT to tube(s) "C."
5. Added 1 μL DEPC-treated water to the tube(s) "D."
6. Gently mixed the contents of each reaction and incubated for 10 min at 37 °C.
7. Heat inactivated the TdT for 10 min at 65 °C. Collected the contents of the reaction following brief centrifugation and kept on ice, as such maintaining the cold chain. Retained each reaction for later PCR.

7.7 PCR of cDNA, Tailed cDNA, and Control DNA

The PCR of the different aliquots from each step and the tailed cDNA product(s) are conducted with two sets of primers in two separate PCR mixes. An additional reaction that consists of 5 μL of the control DNA as a target is employed as a positive control for the verification of PCR. One set of primers will be used as control GSP2 and control GSP3. Specific amplification of the control first-strand cDNA product, or control DNA, leads to a prominent 500-bp band after visualizing the products through agarose gel electrophoresis with ethidium bromide staining. The ther set of primers will be control GSP2 and abridged anchor primer that leads to a prominent 711-bp product (and no product from the no TdT control) or both the tailed cDNA and the control DNA. The sample templates to be amplified are:

"A" cDNA was obtained from the first-strand reaction. It checks the efficiency of the cDNA synthesis reaction.

"B" S.N.A.P. eluate sample: (checking the recovery of cDNA from S.N.A.P. column).

"C" tailed cDNA: (It is actually 5′ RACE product; it equally checks the efficiency of tailing).

"D" No TdT control: It evaluates the specificity of amplification of the tailed product, and the presence of inhibitors of PCR.

Control DNA: (checks PCR with both primer sets). There may have more than one tube for each set, depending on the number of first-strand reactions [1].

1. Make two PCR mixes, one for each primer set:

   Mix I = Control GSP2 and GSP3.

   Mix II = Control GSP2 and abridged anchor primer.

   Added the following components on ice to an appropriately sized sterile tube, e.g., 1.5-mL microcentrifuge tube. Prepare enough mix for *n* + 2 reactions, where *n* = number of template samples to be amplified. (Minimum value for *n* is 5: A + B + C + D + Control DNA.) The volumes mentioned are for 50 μL PCRs.

   Mix I (GSP2 and GSP3).

   Note: Added the Taq DNA polymerase prior to the pipetting of the mix into the PCR tubes containing the template.

Component	Volume (μL) per reaction	Volume (μL) 7× mix *n* = 5	Volume (μL) 12× mix *n* = 10
DEPC-treated water	33.5	234.5	402.0
10× reaction buffer	5.0	35.0	60.0
25 mM $MgCl_2$	3.0	21.0	36.0
10 mM dNTP mix	1.0	7.0	12.0
Control GSP2 (10 μM)	1.0	7.0	12.0
Control GSP3 (10 μM)	1.0	7.0	12.0
Taq DNA Polymerase (5 units/μL)	0.5	3.5	6.0
Final volume	45.0	315.0	540.0

Mix II (GSP2 and abridged anchor primer).

Component	Volume (μL) per reaction	Volume (μL) 7× mix $n = 5$	Volume (μL) 12× mix $n = 10$
DEPC-treated water	31.5	220.5	378.0
10× reaction buffer	5.0	35.0	60.0
25 mM $MgCl_2$	3.0	21.0	36.0
10 mM dNTP mix	1.0	7.0	12.0
Control GSP2 (10 μM)	2.0	14.0	24.0
Abridged anchor primer	2.0	14.0	24.0
Taq DNA polymerase 5 units/(μL)	0.5	3.5	6.0
Final volume	45.0	315.0	540.0

2. Pipetted 5 μL of template samples (A, B, C, D, and control DNA) into two sets appropriately labeled thin-walled 0.2 or 0.5 mL PCR tubes. There will be one set for each PCR mix.
3. Added 45 μL of mix I to one set of tubes and add 45 μL of mix II to the second set of tubes.
4. Mixed the contents of the tubes and overlaid with 50–100 μL of mineral oil (if needed).
5. Equilibrated the thermal cycler block to 94 °C.
6. Transferred tubes directly from ice to the hot thermal cycler.
7. Performed 35 cycles of PCR:
   Denaturation: 94 °C for 1 min.
   Annealing: 63 °C for 30 s.
   Extension: 72 °C for 2 min.
8. Incubated the reaction for 10 min at 72 °C following the last cycle of PCR, then maintained reactions at 4 °C.
9. Analyzed 5–20 μL of 5′ RACE products by agarose gel electrophoresis as per the standard protocols, using appropriate size standards. Either TAE or TBE electrophoresis buffer can be used for the procedure. The volume of the sample utilized for analysis will be depended on the volume and thickness of the sample well. If products will be extracted for reamplification, ultraviolet (UV) visualization of ethidium bromide-stained products needs to be performed with either a long wavelength (356 nm) UV or 302 nm wavelength source to minimize DNA nicking.

7.8 General Troubleshooting Guidelines for the 5′ RACE System

Problem	Possible cause	Suggested remedy
No bands after electrophoretic analysis of amplified products	5′ RACE amplification product may be present but in too low a concentration for detection by ethidium bromide staining	Perform Southern blot analysis of amplification products with the internal sequence as a probe. Amplify agarose gel-purified material of expected size range using the AUAP or UAP, and user's GSP2. Perform nested primer amplification from either purified amplification products, dilution of original PCR (0.05%), or agarose gel plug using the UAP, or AUAP, and user's nested GSP.
	Procedural error in first-strand cDNA synthesis, purification of cDNA product, TdT tailing, or PCR	• Use the control RNA to verify the conversion of the first strand product, recovery following purification, and dC tailing. • Design sense GSP from available 5′-mRNA sequence data. Verify first-strand conversion of the desired message by PCR using two gene-specific primers.
	Inhibitors of RT present	• Remove inhibitors by ethanol precipitation of the mRNA preparation before the first-strand reaction. Include a 70% (v/v) ethanol wash of the mRNA pellet. **Note:** Inhibitors of RT include sodium dodecyl sulfate (SDS), EDTA, guanidinium salts, and glycerol. Inhibitors of M-MLV-RT include sodium pyrophosphate and spermidine. SuperScript™ RT is inhibited 50% by 0.0025% SDS, 1 mM EDTA, 15 mM guanidine isothiocyanate, 17% DMSO, 50% glycerol, 5% formamide, 4 μg/mL heparin and 4 mg/mL glycogen

(continued)

Problem	Possible cause	Suggested remedy
		Test for the presence of inhibitors by mixing 1 μg of control RNA with 1 μg of sample RNA and comparing yields of first-strand cDNA or by PCR of control band.
	Target mRNA has a secondary structure that interferes with the annealing of GSP1	Redesign GSP1
	Target mRNA contains strong transcriptional pauses	Maintain an elevated temperature after the annealing step and increase the temperature of the first strand reaction (up to 50 °C).
No bands after electrophoretic analysis of amplified products	RNase contamination	Perform first strand synthesis with the control RNA both alone and added to sample RNA to determine if RNase is present. Assay for control cDNA by PCR using control GSP2 and GSP3.
		Maintain aseptic conditions to prevent RNase contamination.
		Use placental RNase inhibitor during first-strand cDNA synthesis.
		Note: Placental RNase inhibitor requires sulfhydryl reagents for maximal RNase binding activity [5]. Always add RNase inhibitor to reactions after the addition of dithiothreitol. Treatments that denature the protein, such as high-temperature incubation, exposure to oxidizing conditions, or repeated freezing and thawing, can release RNases initially bound by the inhibitor. These RNases may subsequently degrade RNA preparations in downstream procedures.
	Inefficient tailing of cDNA	Perform TdT time course: remove 5-μL aliquots at $t = 2.5$, 5, 10, and 20 min. Amplify each time point using the abridged anchor primer and GSP2.

(continued)

Problem	Possible cause	Suggested remedy
	cDNA did not tail due to strong secondary structure of 3′-end	Denature cDNA prior to tailing by incubation at 94 °C for 2–3 min, then quick chill on ice. Perform the tailing reaction on ice for 1 h to keep 3′-end denatured and available for TdT. Alternatively, the use of cosolvents, e.g., DMSO (≤20%) has been found to increase the efficiency of tailing.
	Inhibition of PCR by TdT	Purify tailed cDNA by ethanol precipitation in the presence of the inert carrier, i.e., glycogen, and 2.5 M NH_4OAc.
	Polysaccharides and small RNAs coprecipitate with mRNA	Ethanol precipitate the RNA preparation; treat the pellet
	Polymerase used in PCR was from an archaebacterium and dUMP primers were used	Use abridged anchor primer, or AUAP, and non-dUMP-containing GSP in PCR using eLONGase® reagents or an archaebacterial polymerase. Use Taq DNA polymerase for PCR with the UAP and dUMP-containing primers.
Agarose gel analysis shows strong primer-dimer product, but no visible gene-specific product	Inefficient removal of GSP1 during S.N.A.P. procedure	Design GSP1 according to the guidelines above. Reduce the concentration of GSP1 used in first-strand synthesis. Remove dC-tailed GSP1 with an additional S.N.A.P. purification after TdT tailing and second-strand synthesis as described above **Note:** dCTP and the 5′ RACE anchor primer may be substituted for dATP and the 3′ RACE system adapter primer described in this protocol. Substitute dU for dT in GSP1. Degrade dU-GSP1 following cDNA synthesis with UDG. Purify cDNA by ethanol precipitation in the presence of the inert carrier, i.e., glycogen, and NH_4OAc.

(continued)

Problem	Possible cause	Suggested remedy
		Note: This technique may be useful for 5′ RACE of short cDNAs that bind poorly to the S.N.A.P. column.
	Primer-dimer product between the anchor primer and GSP2	Minimize nonspecific annealing of primers by the initiation of PCR at an elevated temperature (75–80 °C).
Unexpected bands after electrophoretic analysis of "nested" amplification products	Contamination by genomic DNA spurious priming in the PCR	Pretreat RNA Vary the parameters of the PCR: increase annealing temperature, decrease annealing time, reduce $MgCl_2$ concentration, etc. [6–10]. Minimize nonspecific annealing of primers by the initiation of PCR at an elevated temperature (75–80 °C). Perform PCR control amplifications as discussed above.
Gene-specific 5′ RACE products appear as a smear. Unable to isolate full length 5′ RACE product	RNA preparation is degraded or of poor quality	RNA preparation is degraded or of poor quality
Absence of discrete product bands. 5′ RACE products appear as a heterogeneous smear	Potentially, a normal result	Identify gene-specific products by Southern blot hybridization. Enrich for specific products by reamplification of gel-purified material or nested primer amplification
Absence of discrete product bands. 5′ RACE products appear as a heterogeneous smear	High levels of poly (A)+ RNA (>1 μg) used for first-strand synthesis may contribute to nonspecific products. RNA-primed (GSP1 independent) cDNA may compete for TdT and also for anchor primer during the PCR.	Reduce target level or optimize tailing conditions. Perform TdT time course: remove 5-μL aliquots at $t = 2.5$, 5, 10, and 20 min. Amplify each time point using the abridged anchor primer and GSP2.
5′ RACE product does not correspond to known mRNA sequence. Product is dependent on TdT addition of dC-tail	Target mRNA contains strong transcriptional pauses	Maintain an elevated temperature after the annealing step and increase the temperature of first-strand reaction (up to 50 °C). Use cosolvents, e.g., 5–10% DMSO or 10–20% glycerol compatible with the RT in

(continued)

Problem	Possible cause	Suggested remedy
		first-strand reaction [11] to help eliminate secondary structure while maintaining RT activity.
5′ RACE product does not correspond to full-length mRNA sequence. Product is not dependent on TdT addition of dC-tail	Internal mispriming by the anchor primer at C-rich sequence	Deoxyinosine-containing anchor primer may not be suitable for the amplification of C-rich cDNA. Use alternate 5′ RACE strategy.
5′ RACE product does not correspond to known mRNA sequence.	Aberrant priming of gene-specific primers	Perform control amplification reactions as discussed above, 5′ RACE Controls. Verify the identity of 5′ RACE product by hybridization or diagnostic restriction endonuclease cleavage prior to cloning or sequencing. Increase the specificity of GSP1 by increasing the temperature of first-strand synthesis (up to 50 °C). Optimize PCR conditions to maximize the specificity of GSP2 and anchor primer. Increase annealing temperature, decrease annealing time, decrease $MgCl_2$ concentration, etc. Design new GSP1 and/or GSP2.
Poor cloning efficiency	Inefficient ligation	Increase the incubation time in the ligation reaction; decrease the temperature. Remove dNTPs prior to ligation. • Use a ligation-free cloning method such as the CloneAmp system.
	Poor restriction endonuclease digestion due to residual bound Taq DNA polymerase	Treat PCR products with proteinase K [12].
	Fill-in of overhangs by residual Taq DNA polymerase	Extract PCR products with phenol:chloroform and purify by ethanol precipitation or treat with proteinase K (see above) before restriction endonuclease digestion.
	Clone is unstable in host cells	Try different strains of bacterial cells such as Stbl2™ competent cells.

7.9 Minimizing RNase Contamination

For successful cDNA synthesis, an RNase-free environment is a prime requirement at all times. This normally needs the same level of care utilized to maintain aseptic conditions while working with microorganisms. Several additional guidelines need to be followed:

It will not be wise to assume that anything is RNase-free, except sterilized pipettes, centrifuge tubes, culture tubes, or any similar labware which is explicitly stated to be sterile. Wearing latex gloves is the primary criteria for all manipulations involving RNA.

It is recommended to use a separate set of automatic pipettes for manipulating RNA and the buffers and enzymes used to synthesize cDNA.

It is preferably avoided to use any recycled glassware. Proper treatment needs to be conducted to make it RNase free by rinsing with 0.5 N NaOH followed by sufficient amounts of sterilized, distilled water. Alternatively, bake glassware at 150 °C for 4 h.

Care should be taken to use microcentrifuge tubes preferably from an unopened box, followed by proper sterilization in an autoclave and used for all cDNA work. RNase-free microcentrifuge tubes are commercially available from several suppliers. If it is necessary to decontaminate untreated tubes, soak the tubes overnight in a 0.01% (v/v) aqueous solution of diethylpyrocarbonate (DEPC), rinse them with sterilized, distilled water, and autoclave them. As prepared with RNase-free labware, most of the solutions can be prepared from reagent-grade materials and distilled water, followed by autoclaving. Prepared heat-sensitive solutions with the help of sterilized, distilled water, and filtered them to 0.2 μm with sterilized, disposable filter ware.

Primer	2(AT) + 4(GC)	%GC	NN	NN_{PCR}
GI anchor region	52	72	60.5	66.6
Anchor primer	150	90.3	93.9	90.1
Abridged anchor primer	118	90.0	94.4	92.5
UAP	98	83.8	75.8	75.0
AUAP	66	79.3	67.8	71.5
GSP2	64	75.3	64.5	69
GSP3	66	75.4	66.5	69.2

Most aqueous buffer solutions can be treated with 0.01% (v/v) DEPC and autoclaved. Use of aerosol-resistant pipet tips is recommended.

7.10 T_m Values for 5′ RACE and Control Primers

Oligonucleotide primer T_ms may differ widely, which depends on the method of calculation. A comparative discussion of T_m determinations (°C) for the 5′ RACE anchor primer, abridged anchor primer, UAP, AUAP, control GSP2, and control GSP3 using the 2

(AT) + 4(GC), %GC, and nearest neighbor methods are being explained in the above section. The nearest neighbor analysis method depends not only on the quantity of each base but also on the primary sequence of the DNA. The accuracy of the nearest neighbor method has been reported through a comparative study of experimentally derived T_ms to that predicted through the available estimates for a number of oligonucleotides. Moreover, the precise reaction conditions need to be considered in the calculation. Other methods are either limited by the size of the oligonucleotide for which they are valid or derived for salt and temperature conditions that are different from those for PCR.

References

1. Rashtchian A (1995) Novel methods for cloning and engineering genes using the polymerase chain reaction. Curr Opin Biotechnol 6:30
2. Witcomb JM, Rashtchian A, Hughes SH (1993) A new method for the generation of nested deletions. Nucleic Acids Res 21:4143
3. Carney JP, McKnight C, Van Epps S, Kelly MR (1995) Random rapid amplification of cDNA ends (RRACE) allows for cloning of multiple novel human cDNA fragments containing (CAG)n repeats. Gene 155:289
4. Suggs SV, Hirose T, Miyake T, Kawashima EH, Johnson MJ, Itakura K, Wallace RB (1981). In: Brown DD, Fow CF (eds) Developmental biology using purified genes. Academic Press, New York, NY
5. Sambrook J, Fritsch EF, Maniatis T (1989) Molecular cloning: a laboratory manual. Cold Spring Harbor Laboratory Press, Cold Spring Harbor, NY
6. Delidow BC, Lynch JP, Peluso JJ, White BA (1993) Polymerase chain reaction. In: White BA (ed) PCR protocols. Methods in molecular biology, vol 15. Humana Press Inc, Totowa, NJ, p 1
7. Rychlik W (1993) Selection of primers for polymerase chain reaction. In: White BA (ed) PCR protocols. Methods in molecular biology, vol 15. Humana Press Inc, Totowa, NJ, p 31
8. Innis MA, Gelfand D (1990) Optimization of PCRs. In: Innis MA, Gelfand DH, Sninsky JJ, White TJ (eds) PCR protocols: a guide to methods and applications. Academic Press, San Diego, CA, p 3
9. Saiki RK (1989) The design and optimization of the PCR. In: Erlich HA (ed) PCR technology, principles and applications for DNA amplification. Stockton Press, New York, NY, p 17
10. Saiki RK (1990) Amplification of genomic DNA. In: Innis MA, Gelfand DH, Sninsky JJ, White TJ (eds) PCR protocols: a guide to methods and applications. Academic Press, San Diego, CA, p 13
11. Chomczynski P (1993) A reagent for the single-step simultaneous isolation of RNA, DNA and proteins from cell and tissue samples. Biotechniques 15:532–534. 536-7. PMID: 7692896
12. Breslauer KL, Frank RH, Marky LA (1986) Predicting DNA duplex stability from the base sequence. Proc Natl Acad Sci U S A 83:3746–3750. https://doi.org/10.1073/pnas.83.11.3746. PMID: 3459152; PMCID: PMC323600

Chapter 22

Gene Therapy

Abstract

The gene with the expression vector is inserted in the organ, which needs to be repaired. The gene is expressed locally in the area and recombinant protein will act as a therapeutic agent. It can be broadly classified as somatic gene therapy and germinal gene therapy. There exist four basic approaches for gene therapy: A normal gene inserted in the place of a nonfunctional gene, an abnormal gene that is traded for a normal gene, an abnormal gene that is to be repaired through the process of selective reverse mutation, change the regulation of gene pairs. The approaches for gene therapy are gene augmentation and gene replacement. Protocols were described for specific inhibition of gene expression through antisense therapy, siRNA (small interfering RNA), and ribozymes. Protocols for gene therapy vary as per the mode of insertion. Case studies with suitable examples of gene therapy for the treatment of mastitis have been depicted.

Keywords Somatic gene therapy, Germinal gene therapy, Gene augmentation, Gene replacement, Antisense therapy, siRNA, Ribozymes

1 Principle

The gene with the expression vector is inserted in the organ, which needs to be repaired. The gene is expressed locally in the area and recombinant protein will act as a therapeutic agent.

It can be broadly classified as somatic gene therapy and germinal gene therapy.

Gene therapy is a technique for the remedy of defective genes causing disease.

There exist four basic approaches for gene therapy:

1. A normal gene is inserted in place of a nonfunctional gene.
2. An abnormal gene is traded for a normal gene.
3. An abnormal gene is to be repaired through the process of selective reverse mutation.
4. Alter the regulation of gene pairs [1].

Aruna Pal, *Protocols in Advanced Genomics and Allied Techniques*, Springer Protocols Handbooks, https://doi.org/10.1007/978-1-0716-1818-9_22, © Springer Science+Business Media, LLC, part of Springer Nature 2022

2 History of Gene Therapy

In the 1980s, scientists first started working on gene therapy. Initially, they would insert human genes into a bacterial cell. Then the bacterial cell would transcribe and translate the information into a protein. In the final step, the protein would be introduced into human cells.

The first report of gene therapy was available on September 14th, 1990 when Ashanti DeSilva was subjected to treatment for SCID (severe combined immunodeficiency). Doctors removed white blood cells (WBC) from her body and introduced the missing gene into the WBC, and then replaced back into her bloodstream. This strengthened her immune system. However, it was effective for a few months.

Let us discuss a recent breakthrough achievement in medical Science. Gene therapy has been successful in association with gene editing (CRISPR-Cas9 technique) for the successful treatment of sickle cell anemia and beta-thalassemia [1–9]. This has revolutionized the treatment of two genetic diseases as sickle cell anemia and beta-thalassemia through gene therapy supplemented with gene editing. CRISPR Cas 9 technique has been described in the next chapter in detail. Figure 1 depicts the schematic representation of how gene therapy with CRISPR is helpful for the correction of sickle cell anemia. The US biotech company bluebird bio was approved from the EU conditional market for gene therapy with the ability to replace the need for blood transfusions for people with the blood disorder beta-thalassemia [2]. Instead, gene therapy may be practiced successfully. Figure 2 represents the gene therapy with a healthy gene insert employing retrovirus-mediated gene transfer. However certain ethical concerns still exist in human gene therapy at germline. This is still considered as a punishable offense in the scientific world, as a case reported from China.

3 Protocol

3.1 Mode of Action

1. Delivery of the therapeutic gene into a patient's target cell using suitable vectors.
2. Infection of the target cells through the viral vector.
3. The vector's genetic material is inoculated into the target cell.
4. Functional proteins are synthesized from the therapeutic gene leading the cell to return to a normal state.

There are certain **challenges in gene therapy**:

1. There exists difficulty due to the packaging of the gene.
2. Protect the gene.
3. Targeted delivery of the gene to the nucleus and release in an active form.

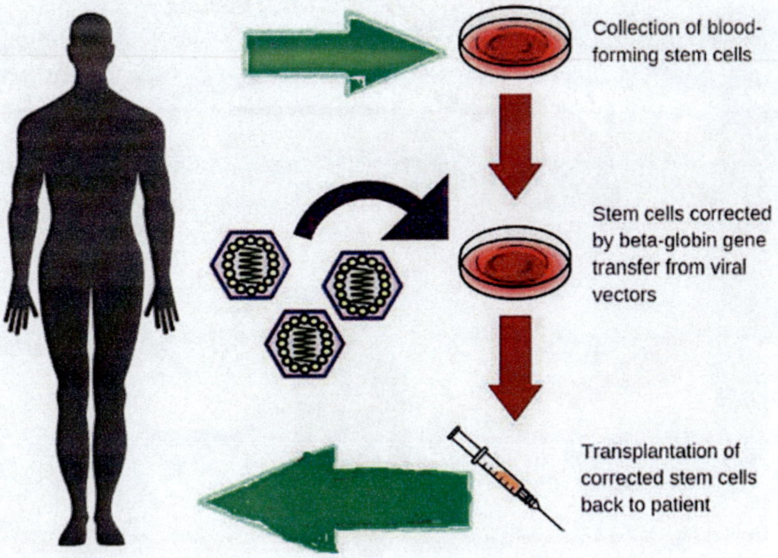

Fig. 1 Schematic representation of the steps involved in conducting gene therapy against sickle cell anemia with CRISPR Cas9 technology

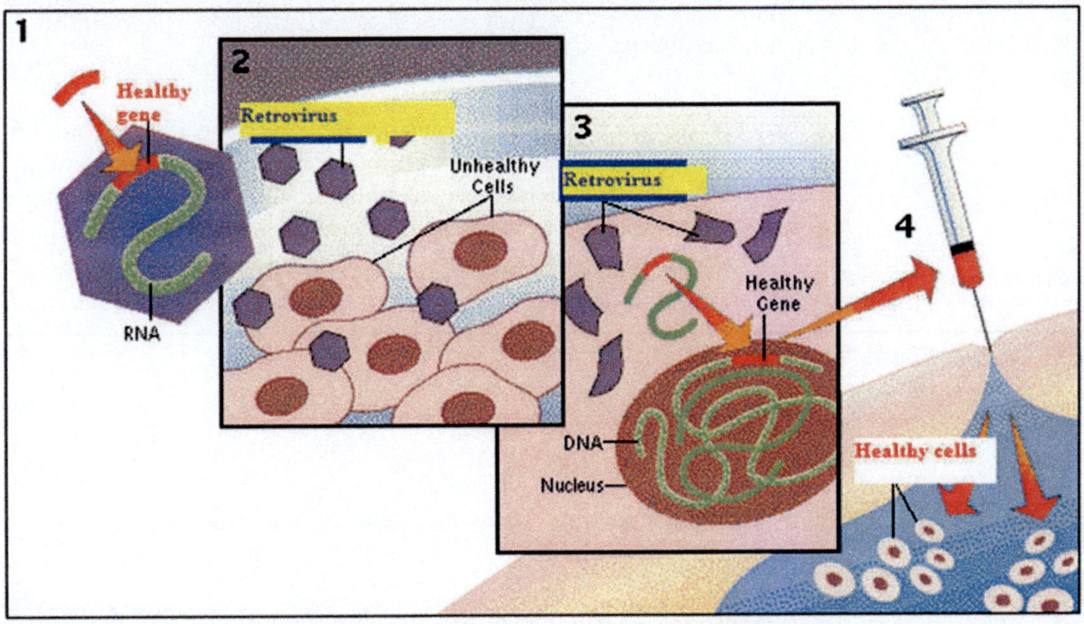

Fig. 2 Gene therapy with healthy gene insert using retrovirus-mediated gene transfer

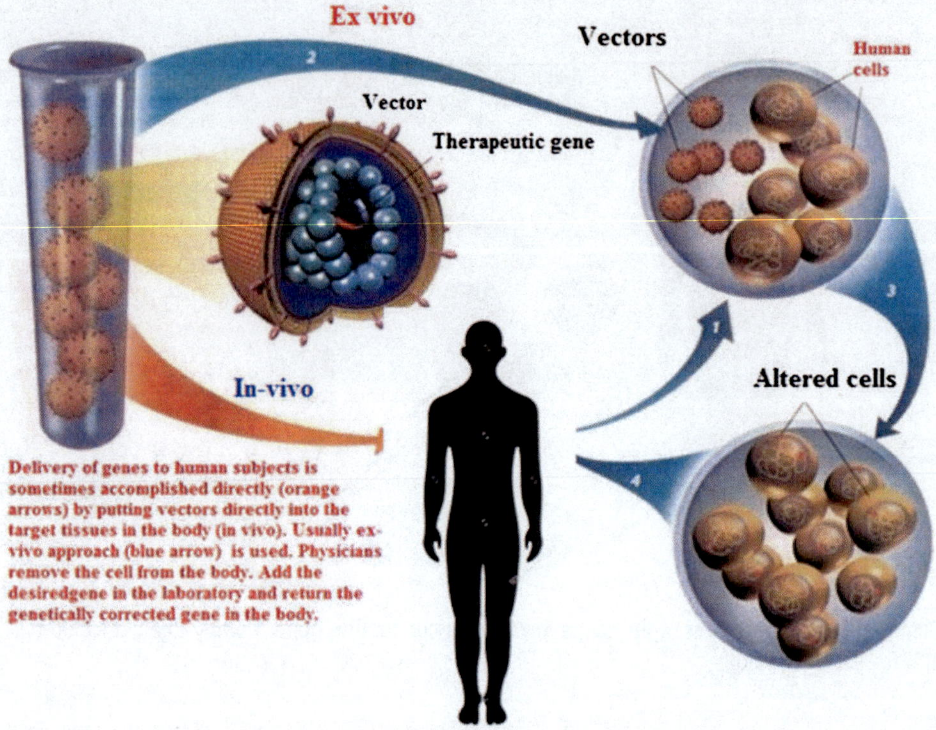

Fig. 3 Delivery of genes in human being through in vivo and ex-vivo techniques

Vectors are the carrier molecules designed specifically to enter cells and deposit therapeutic genes. Vectors can be viral or nonviral. Figure 3 depicts the delivery of genes in human being through in vivo and ex vivo techniques through schematic diagrammatic representation [1, 10, 11].

4 Types of Gene Therapy

1. Germline gene therapy.
2. Somatic cell gene therapy.

It may be further classified as gene augmentation, gene replacement, specific inhibition of gene expression, and targeted cell death.

4.1 Gene Augmentation

It is the process when most therapies simply add a useful gene into a selected cell type to compensate for the missing or flawed version. Useful in treating loss of functional mutations such as tumor genes. Figure 4 depicts the flow diagram of gene augmentation therapy.

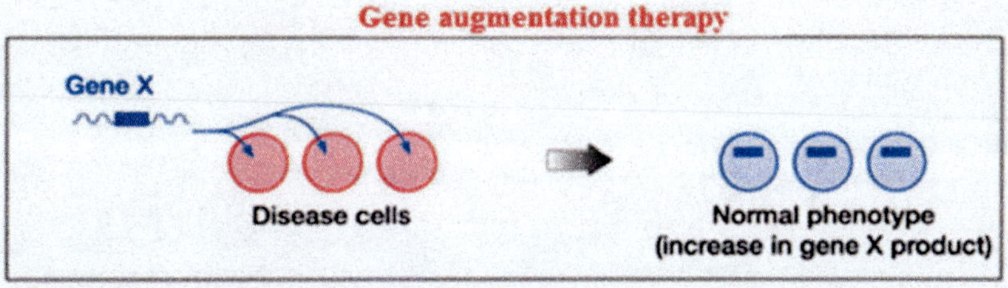

Fig. 4 Diagrammatic representation of gene augmentation therapy

4.2 Gene Replacement

The strategy of this technique is that it replaces the mutant copy with a normal functioning copy in situ. The technique has proved itself useful for gaining functional mutations, for example, oncogenes. Figure 5 represents the flow diagram of gene therapy through targeted gene mutation correction (gene replacement).

5 Specific Inhibition of Gene Expression

It is associated with the silencing of specific genes like activated oncogenes, with the help of the molecules degrading RNA transcripts. Figure 6 depicts a flow diagram of gene therapy through targeted inhibition of gene expression.

The strategies for this technique include:

(a) Antisense therapy.

(b) siRNA (small interfering RNA),

(c) Ribozymes, etc.

Antisense therapy refers to the technique of incorporation of short synthetic ssDNA, which targets the mRNA transcripts of abnormal proteins preventing translation. Figure 7 represents the flow diagram for gene therapy through antisense therapy.

5.1 siRNA Therapy

It is another important technology when small interfering RNAs (short stretches of 21–23 nt of synthetic dsRNA) are utilized. It has a 3′ overhangs of 2 nt incorporating into RISC (RNA induced silencing complex). During the process, target mRNA cleaved in the middle. Figure 8 represents the flow diagram of gene therapy through siRNA.

5.2 Ribozymes

Catalytic RNAs cleave target mRNAs in a sequence-specific manner, e.g., hammerhead ribozymes are engineered to recognize specific sequences and made them resistant to nucleases. Figure 9 represents the diagrammatic view of the catalytic component for gene therapy.

5.3 Targeted Cell Death

It is another useful technology, which causes tissue-specific toxicity due to gene therapy. It is useful in cancer therapy. Even it has a

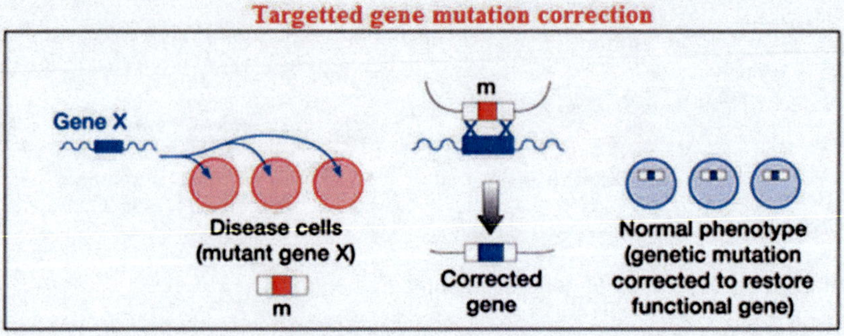

Fig. 5 Diagrammatic representation of gene therapy through targeted gene mutation correction (gene replacement)

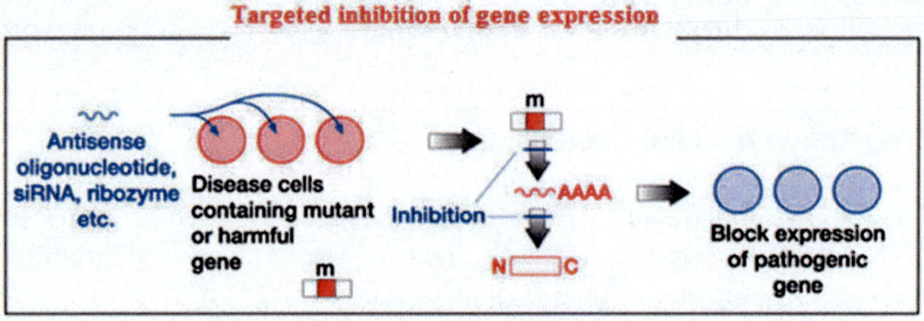

Fig. 6 Gene therapy through targeted inhibition of gene expression

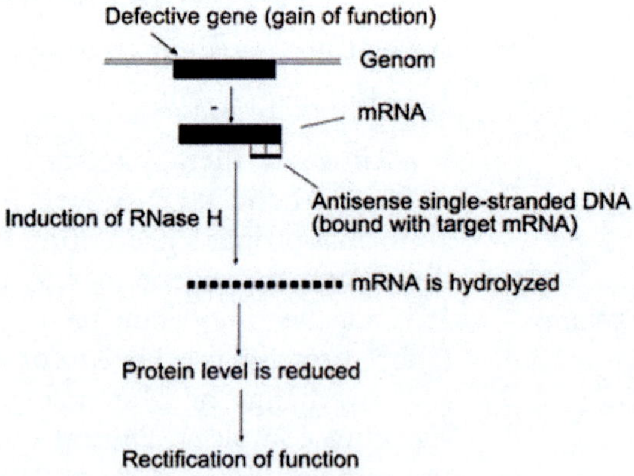

Fig. 7 Gene therapy through antisense therapy

direct approach. It equally has the indirect approach stimulating an immune response against selected cells or eliminating the blood supply. Gene therapy through direct killing of disease cells and the assisted killing of disease cells by immune system cells has been represented diagrammatically in Figs. 10 and 11 respectively.

Somatic gene therapy was observed to be a valuable and effective technique for the prevention and cure in case of mastitis. The basic principle in somatic gene therapy relies on the fact that

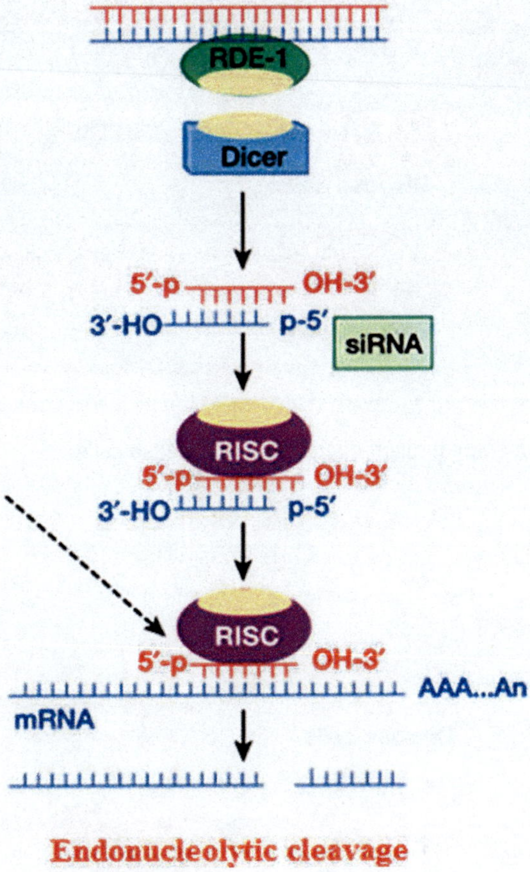

Fig. 8 Gene therapy through siRNA

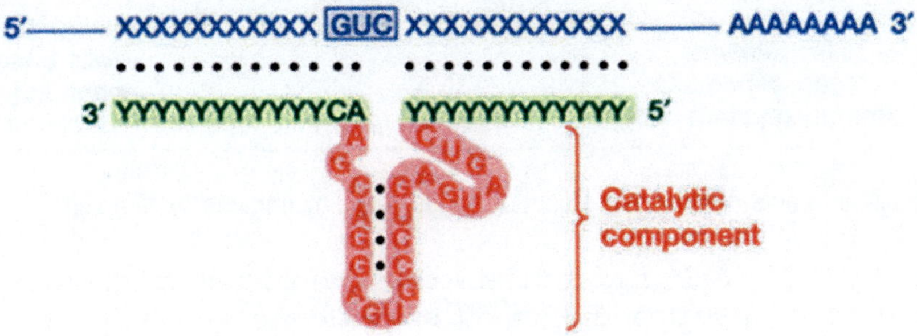

Fig. 9 Catalytic component for gene therapy

appropriate cells are taken from the organism followed by the introduction of a healthy gene. Afterward, the cells are reimplanted into the organism. This kind of therapy leads to the genetic alteration of normal body cells. These cells are expected to restart their normal function with the expression of the "healthy" gene.

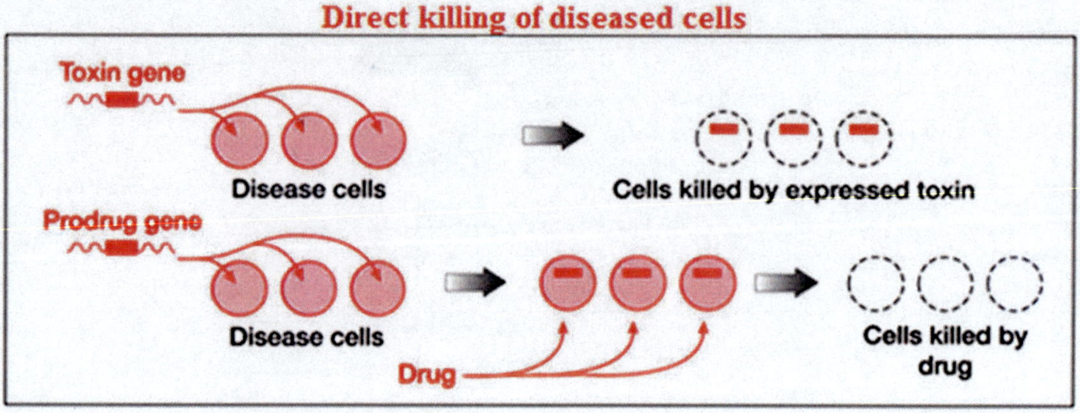

Fig. 10 Gene therapy through direct killing of disease cells

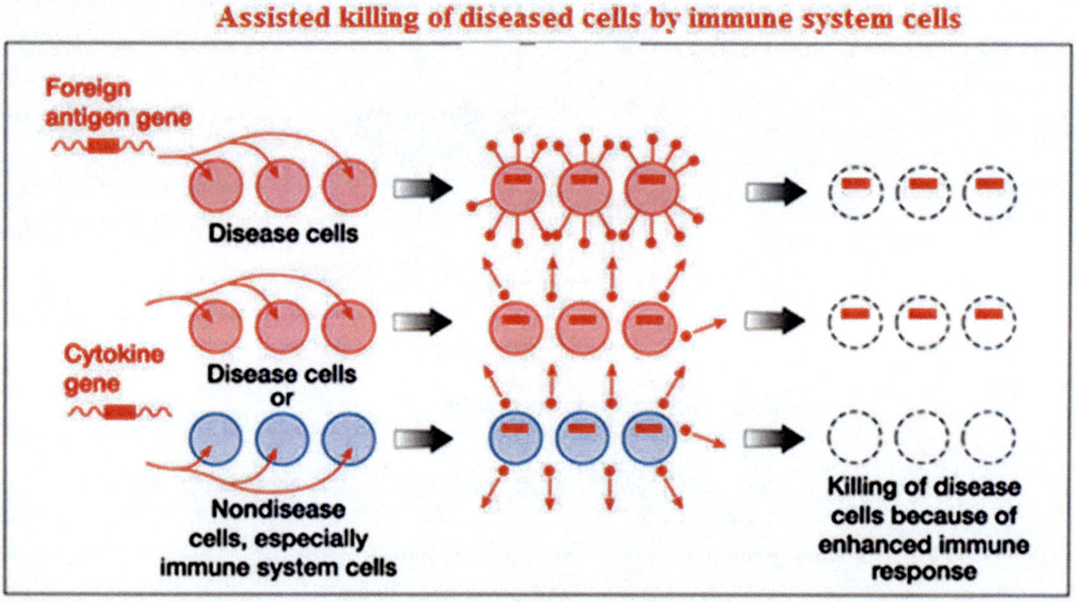

Fig. 11 Gene therapy through the assisted killing of disease cells by immune system cells

In gene therapy, vectors derived from various viruses are being tested for the purpose of transfection of cells since they may easily introduce their genome into host cells. Though it cannot totally be excluded that under somatic gene therapy genes may be introduced in the germline. There are almost no ethical objections against somatic gene therapy. In vivo transfection of the mammary gland for the production of high-value protein against the pathogens responsible for mastitis had been found to be effective. Accordingly, foreign protein production into milk was detected over 60 days after in vivo transfection of the lactating mammary gland of ewes with plasma DNA. Recombinant interferon was synthesized in the milk following mammary gland transfection with recombinant interferon gene construct. Gene therapy strategy has been developed for treating bovine mastitis. A vector specific to mammary gland consisting of human lysozyme (hLYZ) cDNA and kanamycin resistance gene was developed for intramammary expression and clinical studies.

The brief outline of gene therapy may be listed as follows:

1. Insertion of the gene of interest into a benign viral vector. (Choice of
   vector can be important).
2. Use of microinjection to flood a specific area/nucleus of the brain with the viral vector.
3. Some neurons will uptake the virus and incorporate the transgene into the nuclear DNA.
4. Afterward, an assay of the area for mRNA activity is to be done, proteins, etc. to see the effect of the transgene.

Usually, a marker gene is also included in the viral vector to ascertain if the neurons have taken up the vector (e.g., GFP or green fluorescent protein). It is useful to study the overexpression of a gene product. (e.g., overexpression of CREB in the hippocampus) [1, 10, 11].

For the gene therapy, the gene insert (that will express the protein) needs to be inserted in different internal organs. Accordingly, the methods or protocol differ as discussed below:-

1. Methods for gene transfer with DNA adenovirus conjugates.
2. Methods for the construction and propagation of recombinant adeno-associated virus vectors.
3. Generation of high-titer, helper-free retroviruses by transient transfection.
4. Methodology for the construction of retroviral vectors leading to the generation of high titer producers.
5. Development of replication-defective herpes simplex virus vectors.
6. Methods for the use of poliovirus vectors for gene therapy.
7. Methods for the construction of human papillomavirus vectors.
8. Methods for liposome-mediated gene transfer to the arterial wall.
9. Methods for targeted gene transfer to liver with DNA–protein complexes.
10. Methodology for the utilization of retroviral vectors for transfer of the CFTR gene to airway epithelium.

11. Methodology for adenovirus-mediated gene transfer to airway epithelium.
12. Methods for retrovirus-mediated gene transfer to fetal lung.
13. In-situ retrovirus-mediated gene transfer into the liver.
14. Methods for delivery of genes to hepatocytes in vivo using recombinant adenovirus vectors.
15. Methodology for production of high titer, pantropic retroviral vectors for gene transfer into leukemic T cells.
16. Methods for efficient retrovirus-mediated gene transfer to mouse hematopoietic stem cells.
17. Methodology for retrovirus-mediated gene transfer into primary T-lymphocytes.
18. Methods for retrovirus-mediated gene transfer to CD34+ enriched cells.
19. Methods for the use of genetically modified keratinocytes in gene therapy.
20. Methods for particle-mediated gene transfer into skin.
21. Methods for liposome-mediated gene transfer to tumor cells in vivo.
22. Methods for retrovirus-mediated gene transfer to tumor cells.

In the last years, the active immunotherapy approach has been proved to be successful in the treatment of human tumors. This is due to the exciting findings learned from preclinical studies employing genetically altered tumor vaccines. Many clinical protocols are currently being conducted with several different genetic approaches for augmenting antitumor immune responses. These include:

1. The process of ex vivo genetic transfer of cytokine genes into tumor cells.
2. The procedure of ex vivo gene transfer of suicide genes into tumor cells with subsequent activation of the suicide mechanism in vivo.
3. The process of ex vivo cytokine gene transfer to cultured tumor-infiltrating lymphocytes (TIL) to generate more potent T-cells for adoptive transfer.
4. The process of in vivo gene transfer of major histocompatibility complex (MHC) class I molecules by direct injection of the gene into an accessible tumor mass.

6 Cationic Liposome-Mediated Gene Transfer to Tumor Cells In Vitro and In Vivo

We aim at broad applications of gene therapy to human disease. In order to attain our objective, safe and effective technology is essential for delivering functional DNA into cells of an intact organism. Viral and nonviral vectors have their own advantages and limitations. Among the technologies currently studied, the liposomal delivery system has been shown to be much effective. "Cationic liposome-mediated gene transfection (lipofection)," was reported to be highly efficient for transfecting cells in culture. This new technique was first employed by Felgner and co-workers. The liposomes consist of an equimolar mixture of a synthetic cationic lipid *N*-[1-(2,3,-dioleyloxy)propyl]-*N*,*N*,*N*,-trimethylammonium chloride (DOTMA) and a helper lipid dioleoyl-phosphatidylethanolamine (DOPE). The DOTMA/DOPE mixture (Lipofectin) leads to the formation of complexes with DNA by charge interaction upon mixing at room temperature. Other cationic lipids identified are DOTAP, LipofectAMINE, Lipofectam, and DC-chol. The DOTAP is a diester analog of DOTMA and commercially available. LipofectAMINE and Lipofectam are polycationic lipids with a spermine head group that show increased frequency and activity of eukaryotic cell transfection. 3β-[*N*-(*N*′, *N*′-dimethyaminoaminoethane) carbamoyl] cholesterol (DC-chol), a cationic cholesterol derivative, was introduced by Gao and Huang and is routinely used. The DC-chol is both commercially available or can be easily synthesized in a single-step reaction from N,N-dimethylethylenediamine and cholesterol chloroformate. It increases the efficiency of transfection with minimal toxicity. Liposomes thus prepared with DC-chol and DOPE (3:2 molar ratio) were observed to be stable at 4 °C for at least 1 year.

7 Methods for the Use of Cytokine Gene-Modified Tumor Cells in Immunotherapy of Cancer—A Case Study

The importance of gene therapy has become a true perspective. By January 1994, 63 clinical gene therapy protocols (gene therapy or marker studies) have been approved by U.S. control committees after the thorough review process. Out of which, 13 aim at inserting and expressing cytokine genes in tumor cells, with an aim to use such gene-modified cells as vaccines in cancer patients. In experimental mouse models, it was observed that tumor cells transfected with various cytokine genes were rejected in immunocompetent animals. Tumor rejection expressed through the locally produced cytokine results from an inflammatory response at the tumor site. Certain observations are remarkable. The high efficacy of local cytokines was detected to induce an antitumor response in the absence of systemic toxicity. Interestingly, a large number of cytokines possess antitumor activity (IL-1, IL-2, IL-4, IL-6, IL-7, TNF, LT, IFN-γ, MCAF, G-CSF, GM-CSF, IP-10). In several models, cytokine-producing tumors were heavily infiltrated by T-lymphocytes, which cause tumor destruction. This observation is interesting since T-cells can specifically recognize and destroy tumor cells, accordingly, they are involved in tumor immunity. Accordingly, in several tumor models, mice that were immunized by cytokine-gene transfected tumors also rejected the nontransduced parental tumor. They were able to eliminate small preexisting tumor loads in certain cases. Tumor immunity was dependent on T-cells. As a result cytokine gene-modified tumor cell vaccines are believed to trigger a local immune response in cancer patients. This is directed against putative, and some unknown tumor antigens. These are effective against residual (presumably small) tumor loads (e.g., micrometastasis). It is essential to note that these genetic approaches to cancer immunotherapy are the beginning of a developing field. Still there exists certain unresolved problems such as inappropriate presentation of tumor antigens by the tumor cells, tumor-induced immune suppression, tumor heterogeneity. It is essential to ascertain if this technique is better effective than the previous attempts of immunotherapy.

At present, three cytokine gene therapy variations with IL-2, IL-4, TNF, IFN-γ, or GM-CSF genes are tested. Transfected autologous tumor cells, transfected allogeneic tumor cells, and transfected autologous fibroblasts were mixed with tumor cells as vaccine. In the majority of the cases, the cells are irradiated before injection. Results from these initial clinical trials do not exist till date.

8 Methods for Generation of Genetically Modified Fibroblasts for Immunotherapy of Cancer

Many results have indicated that cultured human or rodent tumor cells may be successfully transduced with cytokine genes and selected in the appropriate antibiotic-containing culture media. The selected transductants are generally able to secrete the cytokine code by the transduced gene. In the majority of the cases, substantial levels (e.g., ng quantities) of the cytokine are produced. With the help of retroviral vectors, it is possible to get stably transduced tumor cells with various cytokine genes.

These tumor cells have been utilized for immunotherapy of cancer in various animal models of tumor growth or metastasis, and more recently, in vaccination protocols in patients with cancer. One possible criticism is that genetically modified, and selected tumor cells might have phenotypic characteristics, which are substantially different from those of unmodified tumor cells. Since retroviral vectors are usually used for transduction, it is equally possible that viral antigens expressed on transduced tumor cells leading to the immune response generated due to vaccination. Moreover, primary cultures of human tumor cells are usually difficult to establish and maintain.

9 Methods for Gene Transfer to Synovium

Development of methods for gene transfer to synoviocytes may lead to the idea that gene therapy would be an effective technique employed to treat rheumatoid arthritis (RA) and other joint disorders. The effectiveness is limited because of problems related to inefficient targeting of drugs to the joint, as well as inefficacies of the drugs themselves. The mode of drug delivery to the joint (by traditional oral, iv, and intramuscular routes) depends on passive diffusion of the drug from the synovial vasculature into the joint space. Thus, high systemic concentrations of the drug are essential to achieve therapeutic intra-articular drug levels. In chronic RA, perfusion of the synovium may be compromised, higher doses of drug are required. This is of major concern, as the pharmaceuticals utilized for the treatment of this disease may lead to serious side effects. The problem is further aggravated by the chronic nature of RA, which needs lifelong treatment with high dosages of these drugs.

10 Methods for Adenovirus-Mediated Gene Transfer to Synovium In Vivo

The synovial membrane lining diarthrodial joints consists of a loose layer of cells (1–4 cells in depth). This overlies the surface of articular and periarticular subsynovial tissues, which includes connective tissue, adipose tissue, tendon, bone, and articular cartilage. With the help of light microscopy, the synovial lining tissue shows discontinuous and lacks a clearly defined basement membrane, although the cells seem to be retained by a reticulum. Cells present below this membrane comprise mast cells and endothelial cells (up to 10% of the total cell population). Ultrastructural studies of the synovium have shown that about two major types of cells are present within the synovial membrane.

Type A cells are monocytoid in appearance and possess abundant endoplasmic reticulum with fewer vesicles, vacuoles, and mitochondria.

Type B cells appeared as fibroblastoid containing large granules, numerous filopodia, mitochondria, and intracellular vesicles.

The percentage of Type A and B cells contained within adult synovium varies between species. Along with these, cells with an intermediate ultrastructural appearance and morphology can exist within the adult synovium.

11 Methods for the Use of Stromal Cells for Therapeutic Gene Therapy

The utilization of bone marrow stromal cells in gene therapy indicates the uniqueness of the bone marrow stromal cell as a differentiated cell phenotype, which was observed to be different from the embryo fibroblast, adult connective tissue fibroblast, and cells derived from other sites (e.g. the skin, or connective tissue in muscle). Bone marrow stromal cells consist of the slowly proliferating supportive tissue of the hematopoietic microenvironment. The anatomical localization, biology, physiology, and growth of these cells in tissue culture are investigated intensely by experimental and clinical hematologists for many years. A major controversy in basic hematology raised if bone marrow stromal cells of the hematopoietic microenvironment are transplantable or not. Classic bone marrow transplantation biology indicated that the hematopoietic stem cell and corresponding committed progeny are transplantable in vivo. Techniques for the transplantation of bone marrow stem cells are commonly accepted in both experimental and clinical hematology. On the contrary, transplantation of slowly proliferating stromal cells of the hematopoietic microenvironment is being demonstrated and confirmed by others. The effectiveness of bone marrow stromal cells in somatic cell gene therapy has opened up a new area for clinical therapeutic management.

12 Suppression of the Human Carcinoma Phenotype by an Antioncogene Ribozyme

A number of oncogenes have been identified and are thought to be associated with the progression of neoplasia. These oncogenes are composed of the following: growth factors/receptors, kinases, nuclear proteins (i.e., Fos/Jun), and tumor suppressor genes. Perturbation of one or more of these genes is able to transform normal cells into invasive/metastatic cancer cells. Understanding the role of these oncogenes for specific types of cancer may lead to a more rational basis for tissue-specific targets in cancer gene therapy.

13 Methods for Cancer Gene Therapy Using Tumor Suppressor Genes

Under experimental conditions, a fusion of normal and malignant cells in various different combinations most often results in the suppression of the tumorigenic phenotype of tumor cells. This phenomenon laid down the hypothesis that the normal genome might contain "recessive cancer genes" which, when expressed, would suppress the growth of tumors. The identification and cloning of the retinoblastoma (*Rb*) and *p53* genes have led the study of tumor suppressor genes to progress rapidly. Though tumors generally develop with a series of multiple changes in several genes, the malignant phenotype can be reversed through the introduction of a single chromosome derived from a normal cell. It suggests that single suppressor genes are capable of overcoming the effects of multiple changes related to tumor progression.

14 Intramammary Expression and Therapeutic Effect of a Human Lysozyme-Expressing Vector for Treating Bovine Mastitis: A Case Study

For the purpose to develop a gene therapy strategy for the treatment of bovine mastitis, a new mammary-specific vector was designed. The vector contains human lysozyme (hLYZ) cDNA and the kanamycin resistance gene. It was processed for intramammary expression and clinical studies. Expression was assessed through one-time acupuncture or intracisternal infusion of healthy cows with 400 μg of the p215C3LYZ vector. Over 2.0 μg/mL of rhLYZ could be assessed through enzymatic assay for a period of about 3 weeks in the milk samples.

Western blotting analysis depicted that rhLYZ secreted into milk samples from the vector-injected cows had similar molecular weight as of the natural hLYZ in human colostrums. After 20 days of the primary injection, the quarters were reinjected with the help of the same vector through quarter acupuncture, leading to better concentrations of rhLYZ.

Indirect competitive ELISA of milk samples revealed that the vector injection was not able to induce a detectable humoral immune response against hLYZ. Clinical studies documented better therapeutic effect on clinical and subclinical mastitis with twice acupuncture of quarters with the p215C3LYZ vector showed as treated with antibiotics earlier. It also induces the disappearance of clinical symptoms and relatively high microbiological cure rates. These data give the direction for using the vector to develop gene therapy for treating bovine mastitis.

15 Let us Discuss the Detailed Protocol

15.1 Materials and Methods

15.1.1 Vector Construction

The mammary-specific expression vector p205C3LYZ was constructed as explained by Li et al., [12]. It employs:

1. 5′- and 3′-regulatory elements of goat β-lactoglobulin (blg) gene,
2. the first intron of the β-casein gene of dairy goat to control hLYZ cDNA expression and,
3. bla gene as the selection marker.

In order to construct a new vector better acceptable for clinical use, eukaryotic expression vector pcDNA3 (Invitrogen) was subjected to digestion with restriction enzymes EcoV and PvuII. It is done for the removal of its neo gene and other unnecessary sequences. The bla gene was removed and the kan resistance gene (amplified by PCR from pET-3a vector) was inserted. The hLYZ cDNA expression cassette from the p205C3LYZ vector was inserted into the modified pcDNA3 vector as a SalI/HindIII fragment. The effectiveness of the new mammary-specific vector, designated p215C3LYZ, was subjected to confirmation by restriction digestion and PCR analysis.

As a next step, the hLYZ cDNA was subcloned into eukaryotic expression vector pcDNA3 (controlled by cytomegalovirus or CMT promoter), when the recombinant vector pcDNAKLYZ was employed as the control.

15.1.2 Vector Preparation

The preparation of the vector implies culturing of the components (single colonies of DH5α *E. coli* transformants of p215C3LYZ, the control vector pcDNAKLYZ, and empty vector p215C3) overnight at 37 °C with shaking in Loria broth (LB) with 50 μg/mL kanamycin or ampicillin. The plasmid DNAs were prepared by standard alkaline lysis, purified by cetyltrimethylammonium bromide (CTAB) precipitation, followed by dissolving in 0.15 mol/L sodium phosphate solution at a final concentration of 1 mg/mL while using.

15.1.3 Vector Injection

Expression studies were conducted from the milk samples, which were collected from healthy lactating cows from the Yangzhou University Dairy Farm followed by testing of milk through somatic cell counts (SCC) and the standard California mastitis test (CMT) [13]. Quarters of clinically healthy cows ($n = 24$) with negative CMT score (SCC $\leq$ 100,000 cells/mL) were selected for further study. Prior to intracisternal injection, 25 mL of phosphate-buffered saline (PBS) was infused into each cistern, disposed of with gentle massaging. The next step includes inoculation of 400 μL (400 μg) of the p215C3LYZ vector (5 quarters), control

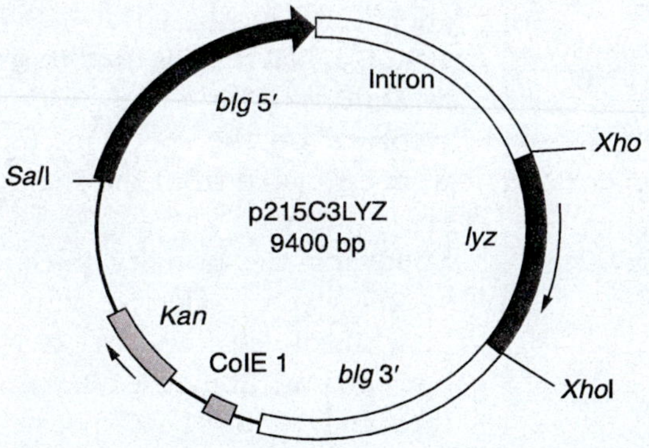

Fig. 12 Schematic representation of the mammary specific expression vector p215C3LYZ

vector pcDNAKLYZ (5 quarters), or empty vector p215C3 (2 quarters) into each quarter cistern through the teat canal. Quarter acupuncture includes 400 μL (400 μg) of the p215C3LYZ vector (5 quarters), pcDNAKLYZ vector (5 quarters), or empty vector (2 quarters), which was injected aseptically through the skin into each quarter using a syringe with a 10-cm needle. For p215C3LYZ-injected (by acupuncture) cows, the secondary injection for vector was performed 20 days after the primary injection employing a similar method. For gene therapy studies, seven quarters with clinical mastitis (characterized by udder redness, swelling, hardness, pain, and grossly abnormal milk) and nine quarters with subclinical mastitis (without visual symptoms but SCC $\geq$ 2,700,000 cells/mL) were chosen based on screened through clinical examination and somatic cell counting of milk samples by CMT. The previous data depicted that there was a history of mastitis for these cows and were subjected to treatment at least once with 750,000 U ampicillin plus 250,000 μg streptomycin per quarter. Following milking, 400 μL (400 μg) of the p215C3LYZ vector (Fig. 12) was injected aseptically into each quarter by acupuncture and the injection was repeated once on the following day.

15.1.4 Enzymatic Assay

About 10 mL of milk samples were aseptically taken from each quarter at successive intervals (0, 2, 4, 6, 8 10, 15, and 20 days) after vector injection. Defattening was an important step and conducted through twice centrifugation at 4000 rpm for 5 min, followed by dilution with one-fold of distilled water. LYZ concentrations were estimated using remazol brilliant blue R (RBB-R)-labeled colorimetric assay [14]. 700 μL of RBB-R-labeled micrococcus lysodeikticus in KH_2PO_3-NaOH solution (pH 7.0) was transferred to each Eppendorf tube, which was prewarmed to 40 °C. Purified hLYZ (100 μL) of different concentrations (for generation of the standard curve) or defatted milk sample was added and incubated at 40 °C for 30 min. Later on, 50 μL of 1 mol/L NaOH was used to stop the reaction. After centrifugation at 1500 rpm for 5 min, OD600 values of the supernatants were recorded on the spectrometer. Averaged concentrations of LYZ were estimated from the standard curve and expressed as μg/mL.

15.1.5 Antibody Assay

Following the primary injection with p215C3LYZ vector, milk samples were collected from three different cows at successive intervals of day 1 and weeks 1, 2, 3, 5 and 7, which were defatted by centrifugation at 4000 rpm for 5 min. The milk samples were then assayed for hLYZ-specific antibody through indirect competitive ELISA [15] using purified human LYZ (coating antigen), rabbit anti-hLYZ serum (the indicator antibody), and HRP-labelled goat anti-rabbit IgG (second antibody).

15.1.6 Western Blotting

After day 4 of vector injection, milk samples were collected from the healthy cow quarters and subjected to centrifugation for 10 min at 4000 rpm. One microliter of the defatted milk sample was initially adjusted to pH 4.3 with 1 mol/L HCl followed by incubation for 20 min at 4 °C for the precipitation of most of the milk proteins. After additional centrifugation for 10 min at 4000 rpm, the supernatant was adjusted to pH 8.0 with 1 mol/L NaOH. 40 μL of the sample was mixed with 10 μL of 5 × SDS-PAGE loading buffer. Normal cow milk, human colostrums with natural hLYZ and mouse milk before and after p215C3LYZ injection were treated similarly and used as the controls. The next step is boiling for 5 min followed by brief centrifugation when 20 μL of each sample was loaded and separated by 12% SDS-PAGE. After the process of transfer onto nitrocellulose membrane, western blotting was conducted using rabbit antihuman lysozyme serum as the first antibody (1:50 in blocking buffer) and IRDyeTM conjugated goat anti-rabbit IgG as the secondary antibody (1:10000 in blocking buffer) as per the manufacturer's instructions. The signals were scanned at 800 nm on Odyssey Infrared Imaging System (LI-COR, Biosciences) according to the manufacturer's instructions.

15.1.7 Bacteriological Test

It is conducted 1 day prior and 10 days postvector injection. Milk samples were collected from the p215C3LYZ injected quarters of the cow udder, 50 μL of each milk sample was spread on blood agar plate fresh laboratory. The plates were incubated for 24 h at 37 °C. Bacteria were primarily identified by colony characteristics. Gram-negative organisms were better documented through a series of biochemical tests, streptococcal organisms by catalase, Gram-stain, esculin, and CAMP reactions, and staphylococcal organisms by the tube-coagulase test and hemolytic patterns.

15.1.8 Cure Definitions

It is important to define what is cure?

The examined data included CMT score, clinical cure, and microbiological cure. A clinical cure may be defined as no visual mastitis symptoms and clots or flakes in the milk 10 days after vector injection. A microbiological cure was defined as no growth of the originally isolated pathogen 10 days after vector injection. CMT score decreased from "+++" or "++" to "+" was defined as improved and to "–" as cured 10 days after vector injection.

16 Result

16.1 Vector Preparation

The purified plasmid DNAs were analyzed by spectrometry with an OD_{260}/OD_{280} ratio of 1.75. Agarose gel electrophoresis revealed that 95% of the plasmid DNAs was supercoiled without visible contamination of bacterial genomic DNA and RNA. Following primary and secondary vector injection, all cows had no visual clinical and milk abnormalities.

16.2 Expression of p215C3LYZ After Primary Injection by Acupuncture

Followed by one-time injection by quarter acupuncture with the mammary-specific vector p215C3LYZ and the control vectors p215C3 and pcDNAKLYZ, milk samples were collected at successive days for enzymatic assay. After the injection with the empty vector p215C3, no increase in enzymatic activity could be detected. On the other hand, following injection with p215C3LYZ vector, concentrations of LYZ in the milk samples started to increase from day 2 and reached the highest level at day 4, followed by a gradual decrease to the prior injection level by day 20 postinjection. Both duration and highest levels of rhLYZ expression after p215C3LYZ injection were similar to that after pcDNAKLYZ injection.

16.3 Expression of p215C3LYZ After Primary Intracisternal Injection

Followed by one-time intracisternal injection with the mammary-specific expression vector p215C3LYZ and the control vectors p215C3 and pcDNAKLYZ, milk samples were collected at successive days for enzymatic assay. Concentrations of LYZ in the milk samples from p215C3LYZ-injected quarters gradually improve from day 2 and reached the highest level at day 4. This is followed by a gradual decrease to the prior injection level by day 20 postinjection. Both duration and the highest level of rhLYZ expression after p215C3LYZ injection were also observed to be similar to those after pcDNAKLYZ injection.

16.4 Dose-Dependent Expression of p215C3LYZ in Mammary Gland

In order to investigate if the expression of p215C3LYZ vector in the mammary gland was dose dependent, 200 or 400 μg of the vector DNA was injected into each quarter through acupuncture and milk samples were collected at successive days after vector injection. Enzymatic assay of milk samples revealed that the concentrations of rhLYZ secreted into milk after injection with 200 μg of p215C3LYZ vector were slightly lower than those with 400 μg of the vector.

16.5 Expression of p215C3LYZ Vector After Secondary Injection

The primary vector injection leads to a relatively high level of rhLYZ expression in milk, but hLYZ was observed to be different from bovine LYZ at the amino acid sequence level. Hence, primary vector injection could induce immune responses with an inhibitory effect on rhLYZ expression following secondary vector injection. In order to overcome it, 20 days after primary vector injection, the quarters were reinjected by acupuncture with the same dose of the p215C3LYZ vector and milk samples were collected for enzymatic assay. After the secondary vector injection, the concentrations of rhLYZ secreted into milk were even higher than those after the primary injection.

16.6 Detection of hLYZ by Western Blotting

In order to test the identity of the rhLYZ, milk samples were procured on day 4 from p215C3LYZ-injected cow quarters followed by analysis with Western blotting using rabbit antihuman lysozyme serum as the probe. A specific protein band was observed in human colostrums and milk from p215C3LYZ-injected cow or mouse, however, it was absent in normal cow and mouse milk. The molecular weight (about 14,000) of the rhLYZ expressed in cow or mouse mammary gland was found to be the same as the natural hLYZ in human colostrums.

16.7 Antibody Response After p215C3LYZ Injection

To identify the immunogenicity of p215C3LYZ vector, milk samples were collected at successive intervals after primary and secondary vector injection for antibody assay by indirect competitive ELISA. In comparison to the milk samples prior to vector injection, differences were absent in OD490 value in milk samples from p215C3LYZ-injected cows 3 weeks after primary vector injection and two additional weeks after secondary vector injection.

16.8 Therapeutic Effect of Vector Injection on Bovine Mastitis

Clinical examination revealed that all cows (7/7) had an absence of visual mastitis symptoms and milk abnormalities 10 days after twice quarter acupuncture with p215C3LYZ vector. In comparison to prior vector injection (86% (6/7) and 78% (7/9) of milk samples (quarters) from cows with clinical and subclinical mastitis, respectively, were having lowered CMT scores and 57% (4/7) of quarters with clinical mastitis had a normal SCC. Microbiological tests revealed that most milk samples from cows with the clinical disease had dose-dependent expression of p215C3LYZ vector in mammary gland. However it is to be noted that the cows already had a primary injection of LYZ gene containing vector. Higher concentration of lysozyme was also detected in the injected quarters [16, 17].

Certain attempt of gene therapy had been tried in treatment of other diseases, such as muscular dystrophy [18] and X-linked severe combined immunodeficiency [19].

References

1. Burdette WJ (2001) The basis for gene therapy. Charles C Thomas, Springfield
2. Smith J (2019). First Gene Therapy for Beta Thalassemia Approved in Europe. https://www.labiotech.eu/medical/bluebird-bio-gene-therapy-thalassemia/
3. Annon (2019). NIH researchers create new viral vector for improved gene therapy in sickle cell disease. Oct 2, 2019. https://www.nih.gov/news-events/news-releases/
4. Belomlinsky A (2019). Gene therapy vehicle for sickle cell disease shifts forward. Oct 3 2019. https://www.genengnews.com/news/gene-therapy-vehicle-for-sickle-cell-disease-shifts-forward/
5. Cavazzana (2018). New gene therapy success in beta thalassemia: 22 patients treated in France, United States, Thailand and Australia. https://presse.inserm.fr/en/
6. Harrison C (2019) First gene therapy for β-thalassemia approved. Nat Biotechnol 37:1102–1103. https://doi.org/10.1038/d41587-019-00026-3
7. Karponi G, Zogas N (2019) 2019. Gene therapy for Beta-thalassemia: updated perspectives. Appl Clin Genet 12:167–180. https://doi.org/10.2147/TACG.S178546
8. Nowogrodzki A (2018) Gene therapy targets sickle-cell disease. Nature 564:S12–S13. https://doi.org/10.1038/d41586-018-07646-w
9. Stephenson J (2019). Sickle cell patient's recovery after gene therapy heightens hopes for a cure. NHLBI IN THE PRESS. National Heart, Lungs and Blood Institute. March 12, 2019. https://www.nhlbi.nih.gov/news/2019/
10. Lander RJ, Winters MA, Meacle FJ, Buckland BC, Lee AL (2002) Fractional precipitation of plasmid DNA from lysate by CTAB. Biotechnol Bioeng 79(7):776–784
11. Peel D (2006). Virus vectors & gene therapy: problems, promises & prospects. Virus Vectors & Gene Therapy. 1998. Department of Microbiology & Immunology, University of Leicester. http://www.tulane.edu/~dmsander/WWW/335/peel/peel2.html
12. Li GC, Sun HC, Sun Q, Chen G, Zhang Q, Shi WQ, Li HD (2004) Transgenic mice expressing human lysozyme gene in milk. Chin J Microbiol Immunol. 24(2):142–145
13. Middleton JR, Hardin D, Steevens B, Randle R, Tyler JW (2004) Use of somatic cell counts and California mastitis test results from individual quarter milk samples to detect subclinical intramammary infection in dairy cattle from a herd with a high bulk tank somatic cell count. J Am Vet Med Assoc. 224(3):419–423. https://doi.org/10.2460/ja
14. Ito Y, Yamada H, Imoto T (1992) Colorimetric assay for lysozyme using Micrococcus luteus labeled with a blue dye, Remazol brilliant blue R, as a substrate. Chem Pharm Bull 40 (6):1523–1526. https://doi.org/10.1248/cpb.40.1523
15. Hurley IP, Coleman RC, Ireland HE, Williams JH (2004) Measurement of bovine IgG by indirect competitive ELISA as a means of detecting milk adulteration. J Dairy Sci. 87 (3):543–549. https://doi.org/10.3168/jds.S0022-0302(04)73195-1
16. Düchler M, Pengg M, Schüller S, Pfneisl F, Bugingo C, Brem G, Wagner E, Schellander K, Müller M (2002) Somatic gene transfer into the lactating ovine mammary gland. J Gene Med 4(3):282–291
17. Sun HC, Xue FM, Qian K, Fang HX, Qiu HL, Zhang XY, Yin ZH (2006) Intramammary expression and therapeutic effect of a human lysozyme-expressing vector for treating bovine mastitis. J Zhejiang Univ Sci B 7(4):324–330. https://doi.org/10.1631/jzus.2006.B0324
18. Crayton S. 2006 "First Clinical Trial Of Gene Therapy For Muscular Dystrophy Now Under Way." Medical News Today. 1 April 2006. University of North Carolina at Chapel Hill. Nov 11 2006. www.medicalnewstoday.com
19. McCormack MP (2002) Activation of the T-cell oncogene LMO2 after gene therapy for X-linked severe combined immunodeficiency. N Engl J Med 346:1185–1193. http://content.nejm.org

Chapter 23

Nutrigenomics

Abstract

Nutrigenomics is an emerging issue that combines a certain area of nutrition and molecular tools to study the level of responses of different nutrients in the body of a population. It is a field of research related to the analysis and understanding of the molecular level of interactions between dietary bioactive components with gene expressions at different levels. The exact amount of nutrient taken by the body from food or exogenous source is controlled by genetics. The entire process of digestion including uptake, absorption, and assimilation of nutraceuticals preventing diseases is also controlled by nutrigenomics. This branch of science mostly works through the process of gene–diet interaction or epigenetics. The onset of various diseases (as diabetes, cardiovascular diseases, osteoporosis, aging, cancer, immunological diseases, and neurological disorders) directly depends on the nutrient imbalances, but the susceptibility varies from individual to individual with a similar kind of nutrition level, which is entirely due to nutrigenomics. The study is equally applicable to humans and animals. To study the efficacy of newer probiotics or vitamins or minerals as nutraceuticals in basic biological processes, as growth, reproduction, or immunity, nutrigenomics can be very well studied with real-time polymerase chain reaction (PCR) or DNA microarray or through next-generation sequencing approach as transcriptomics. Differential mRNA expression profile of genes involved with various levels of nutrients supplementation as probiotics or micronutrients provides us the most useful information for determination of efficacy and appropriate dosage of probiotics or micronutrients and prevention of toxicity induced by excess administration. Such studies are in progress in our laboratory. Ongoing research project on the production of antibiotic-free chicken through probiotics as a replacer to antibiotic growth promoters with an aim to prevent the risk of antibiotic resistance in humans upon consumption is in progress through nutrigenomics study. Vitamins and minerals are helpful for the vital function and metabolism of the body and they are involved in various metabolic pathways, regulated by various enzymes. These enzymes are mostly protein and are controlled by respective genes and nutrigenomics. The concept of the biomarker in nutrigenomics has evolved as an important avenue to avert disease and slow down the process of aging, leading to the importance of the pharmaceutical capabilities of the food. Nutrigenomics focuses on nutritional interventions with an effect to prevent, delay, and treat diseases such as asthma, obesity, Type 2 diabetes, cardiovascular disease, and prostate cancer. So, it is the genomics that regulates the nutrition level of the body.

Keywords Nutrigenomics, Gene–diet interaction, Epigenetics, Disease, Probiotics, Real-time PCR

1 Basic Principles

Nutraceuticals are the food constituents that are of therapeutic importance as probiotics, vitamins, and minerals required in less

Aruna Pal, *Protocols in Advanced Genomics and Allied Techniques*, Springer Protocols Handbooks,
https://doi.org/10.1007/978-1-0716-1818-9_23,

quantity. Basic nutrients are carbohydrates, protein, fat, vitamins, and minerals. Although we eat similar kinds of food, its absorption and assimilation depend upon the host genetics. Nutrigenomics is a field of research focusing on the molecular level of interactions between dietary bioactive components with gene expressions at different levels.

Nutrigenomics is the branch of science dealing with the research of the effects of food and food constituents on gene expressions. Nutrient absorption, metabolism, elimination, or biological effects are regulated by the genetic variation existing due to single nucleotide polymorphisms (SNPs), CNV, SNV leading to altered gene expression. Diet–gene interaction and epigenetics are the main factors regulating the concept of nutrigenomics. The aim of nutrigenomics is to understand the body's reaction to various techniques including genomics, transcriptomics, proteomics, and metabolomics. Nutrigenomics aims to optimize rational means for nutrition with respect to the subject's genotype. The basic principle of nutrigenomics first includes diet as a serious risk factor for a number of diseases for certain individuals under particular circumstances. Second, the components of the food intake can influence directly or indirectly on the human genome to alter the expression or even the genetic structure. Third, the individual genetic makeup or genotype influences the balance between health and disease, and finally the genes which are regulated by dietary factors can play a role in the onset, incidence, progression, and severity of chronic diseases. Figure 1 represents the schematic representation of the possible mode of nutrigenomics.

Nutrients Effects Gene in the Following Ways

1. It can as an integral part of the enzyme.
2. It can act as a cofactor in metabolism.
3. It acts through hormonal regulation.
4. It may act as a transcription factor.

2 Protocol

We discuss the protocol with suitable examples.

As discussed, in a simple way nutrigenomics may be explained as the epigenetic regulation of gene through nutrient incorporation. Hence the basic protocol to study nutrigenomics is similar to that of epigenetics. Genes affected by diet regulation may be identified through differential mRNA expression profile, DNA microarray, transcriptomics, GWAS, exome sequencing various biotechnological techniques.

In this current section, the possible mode of epigenetic modifications due to food consumption influencing an impact on

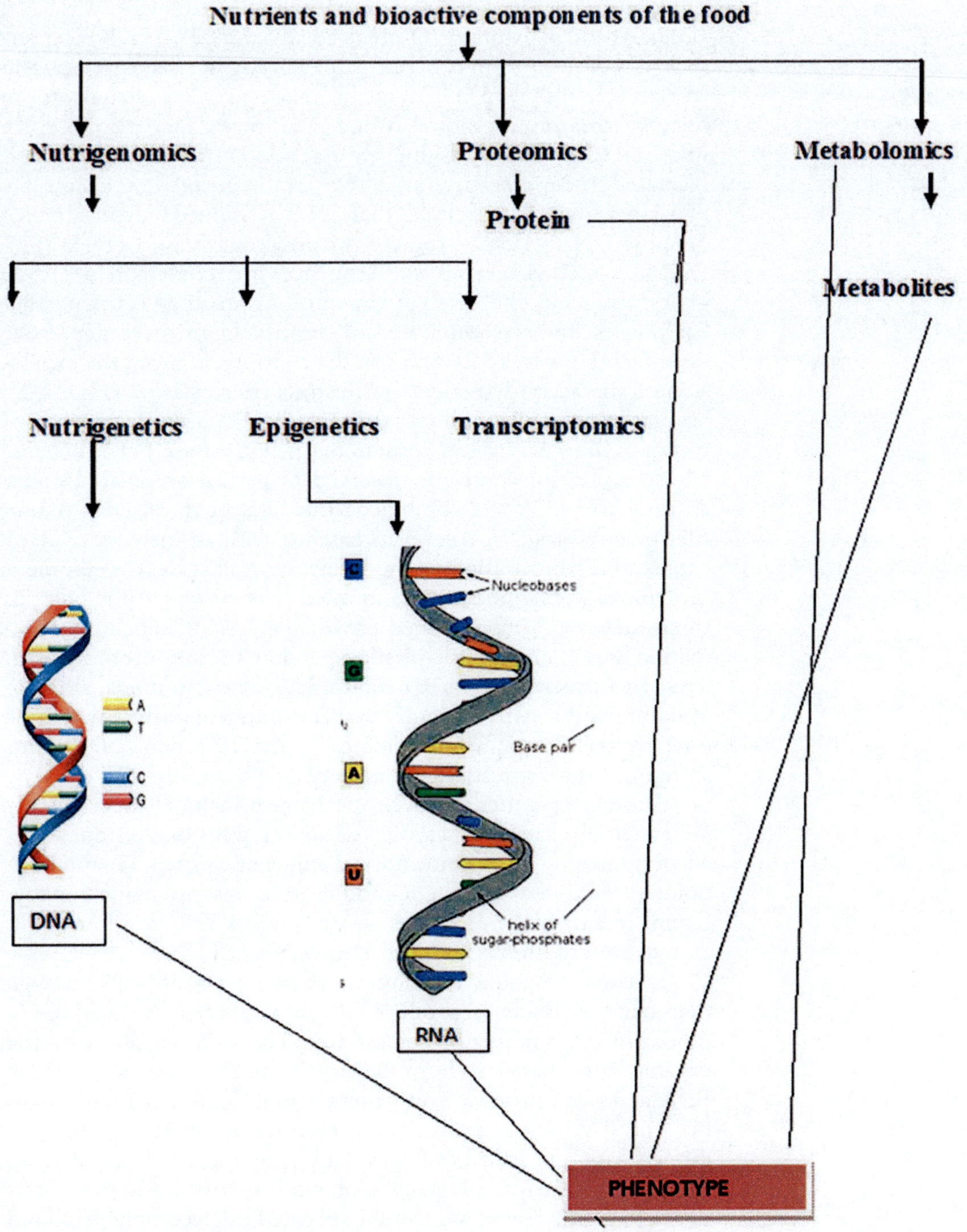

Fig. 1 Schematic representation of the possible mode of nutrigenomics

individual health would be discussed. One possibility is through directly affecting catalytic activities of the enzymes responsible for "writing" or "erasing" the epigenetic modifications.

Wang et al. identified two phytochemicals—dihydrocaffeic acid (DHCA) and malvidin-3′-O-glucoside (Mal-gluc), metabolic intermediates derived from concord grape juice, grape seed extract, and transresveratrol—which attenuate depression-like behaviors in mice. With a mouse model, Wang et al. reveal that treating mice with DHCA and Mal-gluc, when added to drinking water, increased their resilience to stress and reduced depression-like behaviors. It was observed that DHCA reduced expression of "methyl-DNA writer"—the DNA methyltransferase 1 (DNMT1). DNMT1 methylates intronic sequences of interleukin 6 (IL-6) genes, reducing the level of this proinflammatory cytokine that has previously been implicated in the development of depressive disorders. However Mal-gluc, on the contrary, reduced the expression of an "acetyl-lysine eraser," histone deacetylase 2 (HDAC2), leading to a significant increase in histone H3 acetylation on promotor of the RAS-related botulinum toxin substrate 1 (Rac1) gene, whose expression level was observed to be downregulated under chronic stress. When combined, the treatment of DHCA and Mal-gluc reduced IL-6 level to baseline with an increase of Rac1 expression, contributing to the resilience against the development of depression-like phenotypes in mice. The current study gives us the evidence of grape-derived phytochemicals influencing inflammation and brain synaptic plasticity, and as a result, promote resilience against stress in mice. Accordingly, the compounds, alone or in combination with currently available antidepressants, would be effective as therapeutic candidates in the treatment of patients suffering from depression and anxiety.

Another epigenetics-driven mechanism includes metabolically generated intermediates capable of hijacking histone residues usually modified by more conventional epigenetic marks. Huang et al. pointed out a new epigenetic modification, histone lysine benzoylation, identified by HPLC-MS/MS analysis first in human cells and later on in mouse liver and Drosophila cells.

In vitro metabolic labeling assays performed in cells revealed that benzoyl lysine is produced from a cofactor benzoyl CoA. Benzoyl CoA is in turn generated from chemical reaction with the organic compound sodium benzoate. Sodium benzoate is an FDA-approved preservative, generally employed as an antimicrobial and flavoring agent for the prevention of mold formation of canned fruit and fruit beverages. ChIP-seq and RNA-seq analyses revealed that the lysine benzoylation mark is enriched on promoters of active genes. Genes are those implicated in glycerophospholipid metabolism, phospholipase D signaling, ovarian steroidogenesis, serotonergic synapse, and insulin secretion. Furthermore, it was observed that on the contrary to other known acyl lysine modifications which are "erased" by several HDACs, the lysine benzoylation mark can be removed only by SIRT2, a NAD+-dependent protein deacetylase. It is presumed that benzoyl lysine level is dynamic and independent of histone acetylation.

If there is effect of dietary sodium benzoate can lead to epigenetic changes in vivo is yet to be explored further and confirmed, the study indicates a possible effect of the compound on lysine benzoylation in cultured cells. This is a very good example of epigenetic mechanism, when the metabolic products (sodium benzoate-induced histone lysine benzoylation) affect gene expression. Thus, it is a good example for nutrigenomics too.

Malnutrition in early life is directly and indirectly related to a number of disorders in adulthood, which give rise to the hypothesis of epigenetic memory. It has been reported that lactation in mice causes epigenetic changes influencing the possibility of the development of obesity in later life. Milk lipids have been reported to activate the nuclear receptor, PPARα (peroxisome proliferator-activated receptor), which is an important transcriptional regulator of hepatic liver metabolism.

2.1 Using a Genome-Wide Analysis of DNA Methylation

A number of PPARα target genes, including the fibroblast growth factor-21 (Fgf21), undergo PPARα-dependent DNA demethylation upon pharmacological activation of PPARα in mice. Fgf21 is a liver hormone effective in a number of metabolic functions, such as body weight maintenance and regulation of energy homeostasis. It has been reported that once established in early life, the DNA (de)methylation status of Fgf21 does not change in adulthood, and that the reduced DNA methylation of Fgf21 correlates with a reduction in diet-induced obesity in older animals. It has been further documented that Fgf21 demethylation is stimulated during the lactation period of newborn mice. The results of this study are interesting leading to a new avenue in understanding the mechanisms contributing to obesity. The information correlates to a relationship between breastfeeding and weight gain suppression and also the mechanism of specific epigenetic modifications built in early life produces a long-lasting effect. Other factors were reported to markedly contribute to changes in epigenetic programs and responses. It was observed that gut microbes and their metabolites affect the host chromatin state, increase histone polyacetylation leading to the production of short-chain fatty acids (SCFAs).

A "Western-type" diet rich in processed foods and high sugar drinks was observed to limit microbial SCFAs production, cause prevention of occurrence of many of the microbiota-dependent events, leading to alterations in hepatic gene expression. Other examples include the effect of a low-carb ketogenic diet was found to rescue hippocampal memory defects in a mouse model of Kabuki syndrome, which was characterized by loss of site-specific histone methylation and deficiency in chromatin opening. This diet augments the formation of β-hydroxybutyrate, an HDAC inhibitor, leading to changes in H3ac and H3K4me3 in the hippocampus. This causes rescue of the neurogenesis and memory phenotypes of these mice models. These facts lead to a profound knowledge of biological, mechanistic, and physiological aspects of epigenetics. The studies attempt to highlight the potential roles of the environmental epigenetic changes, which have to be put into the equation while drawing a comprehensive map of epigenetic networking. The influence of dietary components on epigenetic imprinting, detection of alternative nutrition-based therapeutic approaches, and designing tools for personalized diet result in improvement of health and life expectancy in the recent past.

3 Impact and Application of Nutrigenomics

3.1 Food Industry

Dietary reference intakes include vitamins, macronutrients, and micronutrients. Genomic technologies are affecting food processing, food safety, and quality assurance. The DNA sequence of the human genome varies by only 0.1% between individuals, called as polymorphisms. Some polymorphisms have effect on the functioning of the proteins encoded by the genes. Single nucleotide polymorphism (SNPs) is one of the most common types of nucleotide variation affecting the metabolism of food. SNPs lead to altered peptide synthesis, leading to change in the domains of the protein, change in function of the protein. The SNPs are the key enabler and the most important marker for the study of nutrigenomics [1]. Nutrigenomics played an important role in food processing also. This includes the discovery of markers and to guide the industrial process or improvement of supply chain management, safety evaluation of food components, detection and identification of microorganisms responsible for food spoilage, authentication of plants, animals, and packaged foods are now determined by DNA fingerprinting within the food industries [2].

3.2 Nutrient Uptake and Assimilation

Nutrients are involved in various processes of nutrient uptake by the body. Nutrients are absorbed from the alimentary tract via the small intestine. Threonine, an essential amino acid is responsible for the genetic control of mucin from the small intestine, thus aiding in the nutrient absorption process. Hemoglobin present in RBC is responsible for the distribution and assimilation of nutrients. Iron is an integral part of hemoglobin and iron absorption is genetically controlled.

3.3 Nutrigenomics in Diseases

Several diseases are influenced by the presence or absence, or varying levels of different nutritional ingredients. Certain diseases or disorders were reported to be controlled by genetic and nutritional components. This opens the scope for nutrigenomic studies exploring the role of dietary alterations affecting the outcome of the disease [3].

Alterations in genotype are responsible for causing gastrointestinal cancers, inflammatory diseases, and osteoporosis. Nutrient imbalances cause diseases like aging, cancer, cardiovascular disease (CVD), chronic fatigue, diabetes, immune disorders, neurological disorders, and Parkinson's disease. Genetic mutations in genes can cause susceptibility to cancer. Gene–diet interaction too has an important role in developing cancer. High consumption of red meat may cause cancer risks, whereas dietary fibers having a protective effect against bowel cancer.

Nowadays, DNA kits are available for better management of human health. Kits detect if the person is genetically susceptible to a particular disease that is influenced by nutrients, for example, heart disease, bone health, and insulin resistance (Tables 1, 2, 3, 4) [1, 4–8].

3.4 Impact and Applications of Nutrigenomics in Animal Sector

Cost involvement for the feeding of animals is more than 70%. Hence optimization of the cost of feeding is of utmost essential to get economic benefit from the livestock sector. Nutrigenomics is very much useful for studying nonruminants as the pork and poultry industry.

Application of nutrigenomics to resolve the molecular markers is important in nutrition research. Very limited information is available about the effect of diet on the expression of genes regulating production and reproductive traits in livestock. There is utmost necessity to understand how gene expression is being regulated by nutrients in livestock.

Microarray study demonstrated that a glutamine supplemented diet is better for piglets with modulating expression of genes, which are crucial for intestinal metabolism and function. The influence of dietary protein/carbohydrate ratio on fat deposition and gene expression (heart fatty acid-binding protein or H-FABP, peroxisome proliferator-activated receptor Y or PPARY) was investigated in the mussel of pigs. This results that low protein content in the diet enhances the intramuscular fat by regulation and the above-mentioned gene enhances the quality of pig meat. Dietary proteins are also important in maintain pig health. It was reported that the hepatic gene expression of cytochrome 8B1(Cy8B1), aldehyde dehydrogenase 2(Aldh2) and thiosulphate transferase(TST) in pigs fed with *Camelina* sativa. Nutrition-related genes in pigs are very important to optimize feed efficiency.

Broiler chicken is in high demand nowadays, and feed optimization in broiler chicken is nearly impossible without nutrigenomics. It was stated that nutrigenomics play a vital role in broiler production through different factors like the correlation of nutrition and genetics in breeding programs, epigenetic, delivering of better health, and increasing the meat quality.

There has been a recent advancement of the poultry industry with a high feed conversion ratio. Hence each and every nutrient is crucial and critical [4–8]. Recently use of antibiotics as growth

Table 1
List of genes and SNPs for cardiac diseases, which are influenced by level of nutrition

Genes analyzed	Roles of the gene in heart health	Genetic variation screened for variations found in your gene
MTHFR	Use of folic acid for DNA synthesis and repair	C677T A1298C
MS-MTRR	Metabolism of vitamin B12	A66G
MTR	Removal of homocysteine	A2756G
CBS	Metabolism of vitamin B6 & removal of homocysteine	C699T
MnSOD SOD3	Antioxidant defense	C(-28)T T175C C760G
IL-6 TNFα	Inflammatory response	G(-174)C G(-308)A
APOC3	Triglyceride metabolism	C3175G
CETP LPL	Cholesterol metabolism	G279A C1595G
eNOS ACE	Blood flow	G894T DEL

Table 2
List of genes and SNPs for bone health, which are influenced by the level of nutrition

Gene analyzed	Role of the gene in bone health	Genetic variation screened for variations found in your gene
VDR VDR VDR	Regulates calcium and vitamin D roles in bone formation	CTaqlT TBsmlC Fokl
COL1A1	Collagen role in bone formation	GSp1T
IL-6 IL-6 TNF-α	Old Bone Elimination New Bone Creation cycle	G(-174)C G(-634)C G(-308)A

promoter has been banned by a regulation of Govt. of India. It is due to the fact that antibiotic residue in chicken meat possesses health hazards to the end consumer. Hence it is of utmost importance to find out the alternative to antibiotics in chicken. We have undertaken a project on probiotics to be used as an alternative to

Table 3
List of genes and SNPs for insulin resistance, which are influenced by level of nutrition

Gene analyzed	Role of die gene in insulin resistance	Genetic variation screened for variations found in your gene
VDR VDR	Mechanism of insulin secretion	CTaqlT TBsmlC
IL-6 TNF-α	Inflammatory response; response of cells to insulin	G(-174)C G(-308)A
PPARy	Glucose and lipid metabolism	Prol2A
ACE	Blood pressure regulation	II/DD

Table 4
List of genes and SNPs for genetic assessment for antioxidant and detoxification, which are influenced by level of nutrition

Gene analyzed	Role of the gene in antioxidant/ detoxification activity	Genetic variation screened for variations found in your gene
eNOS	Vascular Function	G894T
GSTM1 GSTP1 GSTT1	Detoxification	(DEL) A313G A341G (DEL)
MnSOD SOD3	Destroys free radicals	C(-28)T T175C C760G

antibiotic growth promoters entitled "Effect of GALLIPRO (probiotics) Supplemented feed on Growth Performances of Broiler Chicken—A Challenge Study" funded by Evonik (SEA) Pvt. Limited, Singapore. Probiotics are supplied at a different level (dose) and check growth and immune status of control and challenge through differential mRNA expression profile of growth and immune response genes (innate and humoral). Vitamin E-containing diet that can enhance the immune protection against bacterial lipopolysaccharides associated infection in chickens. Chicken growth and development can be accelerated with a proper low protein diet. Data analysis by nutrigenomics showed that an algae-based diet is important to reduce stress profile in chickens. Minerals like zinc can actively regulate the transporter gene in the intestine of broiler chicken and the expression pattern is directly related to the amount of zinc in the feed. It was reported that selenium had a major role in gender-specific gene regulation in chicken.

Therapeutic effect of different amino acids is also of utmost importance to regulate the genes involved in nutrient absorption. We have undertaken another project in our university entitled "Influence of Threonine on mucin biology for better gut health in commercial broiler chicken." Threonine supplied at the different level (dose) and check mRNA expression profile of mucin genes. Mucin encoded by mucin genes aids in the proper absorption of nutrients.

4 Future Scope of Nutrigenomics

The objective of nutrigenomics is to affect the genomics of individuals through nutritional interventions for the prevention, delay, and treatment of diseases such as asthma, obesity, Type 2 diabetes, cardiovascular disease, and prostate cancer. But, the current benefits from nutrigenomics are restricted. Encouraging people to become more health conscious, acquiring a personalized diet, and discovering genetic susceptibilities as a strong motivator for making dietary and lifestyle changes are the basic beneficiaries. Nutrigenomics approach is divided into: a) gene switching; b) emphasize in gene protein relation; c) influence of food. To evaluate the interaction between diets and genes, DNA microarray technique, real-time polymerase chain reaction can be applied. 2 D gel electrophoresis may act as an important tool to explore the effect of individual amino acid on protein composition.

Moreover, the potential future benefits from nutrigenomics are tremendous:

1. The limits for intake of essential macronutrients—proteins, carbohydrates, and fats—and micronutrients (vitamins and minerals) will be better explored and understood.
2. Diseases that are genetically controlled and influenced by nutrients may be avoided or abolished.
3. Wastage of vitamins and other dietary supplements that is administered in excess can be avoided.
4. People whose health is relatively not affected by diet can continue to eat foods that they enjoy after they are genetically screened that they do not contain any SNP or deleterious mutation for diseases.
5. Span of life may be increased.

Further development of nutrigenomics promises us the better development of functional food and usefulness in studying of different human diseases. Researchers prove its potentiality in food and animal sectors, growth, and health, and safer life of animals and human beings. Genes that we inherit cannot be altered but we can definitely change their response and expressions to the environment through Nutrigenomics. In the future, food products are tailored to promote the health and well-being of groups in the population detected on the basis of their individual genomes.

References

1. Doo M, Yangha K (2015) Obesity interactions of genome and nutrients intake. Prev Nutr Food Sci 20(1):1–7
2. Brown L, van der Ouderaa F (2007) Nutritional genomics: From farm to fork. Br J Nutr 6:1027–1035
3. Bhatt SN, Sharma AD (2011) Nutrigenomics: a nonconventional therapy. Int J. Pharm Sci Rev Res 8:100105
4. Brennan KM, Pierce JL, Cantor AH, Pescatore AJ, Power RF (2012) Source of selenium supplementation influences testis selenium content and gene expression profiles in single comb white Leghorn roosters. Biol Trace Elem Res 145:330337
5. Guo X, Tang R, Wang W, Liu D, Wang K (2011) Effects of dietary protein/carbohydrate ratio on fat deposition and gene expression of peroxisome proliferator activated receptor γ and heart fatty acid binding protein of finishing pigs. Livestock Sci 140:111116
6. Meadus WJ, Duff P, McDonald T, Caine WR (2014) Pigs fed camelina meal increase hepatic gene expression of cytochrome 8b1, aldehyde dehydrogenase and thiosulfate transferase. J Anim Sci Biotechnol 5:1
7. Naji TAA, Amadou I, Zhao RY, Tang X, Shi YH, Le GW (2014) Effects of phytosterol in feed on growth and related gene expression in muscles of broiler chickens. Trop J Pharm Res 13:9–16
8. Xiao R, Power RF, Mallonee D, Routt K, Spangler L et al (2012) Effects of yeast cell wall derived Mannan oligosaccharides on jejunal gene expression in young broiler chickens. Poult Sci 91:16601669

Check for updates

Chapter 24

Gene Editing

Abstract

Gene editing is a technique of genetic engineering when a particular fragment of DNA is either inserted, replaced, or removed from a genome using artificially engineered nucleases or known as "Molecular scissors." It is also popularly known as genome editing with engineered nucleases (GEEN). Three available genome editing tools are zinc finger technology (ZFN), TALEN (transcription activator-like effector nuclease technology), CRISPR-Cas9 (clustered regularly interspaced short palindromic repeats (CRISPR) technology, which is the better and cost-effective technique. These nucleases create a specific cut or break (DSB or double-strand breaks) at desired locations in the genome through two processes—homologous recombination (HR), nonhomologous end joining (NHEJ). The outline of the protocol for CRISPR technology may be explained as initial preparation of CRISPR components (Cas9/sgRNA[CrRNA tracRNA duplex]), RNP complex, and ssDNA homology directed repair (HDR) templates of gene of interest. The next step is microinjection of CRISPR components into one-day-old chicken zygotes, transferred to pseudopregnant foster mothers/embryonated egg of chicken; sperm-guided gene transfer may also be attempted. After 7 weeks of microinjection, biopsy samples were collected from pups/ microinjected zygotes, leading to windowing and ex vivo culture to hatch chicks, check gene expression for the desired insert.

Keywords Genome editing with engineered nucleases (GEEN), Zinc finger technology (ZFN), TALEN (transcription activator-like effector nuclease technology), CRISPR-Cas9 (clustered regularly interspaced short palindromic repeats,CRISPR), sgRNA, CrRNA, tracRNA

1 Basic Technique

It is a technique of genetic engineering when a particular fragment of DNA is either inserted, replaced, or removed from a genome with the help of artificially engineered nucleases or known as "Molecular scissors." It is also popularly known as genome editing with engineered nucleases (GEEN).

These nucleases create specific cut or break (DSB or double-strand breaks) at desired locations in the genome (Fig. 1) and aid the cell's endogenous mechanisms for repairing the induced break by two different processes:

1. Homologous recombination (HR).
2. Nonhomologous end joining (NHEJ).

Aruna Pal, *Protocols in Advanced Genomics and Allied Techniques*, Springer Protocols Handbooks,
https://doi.org/10.1007/978-1-0716-1818-9_24,

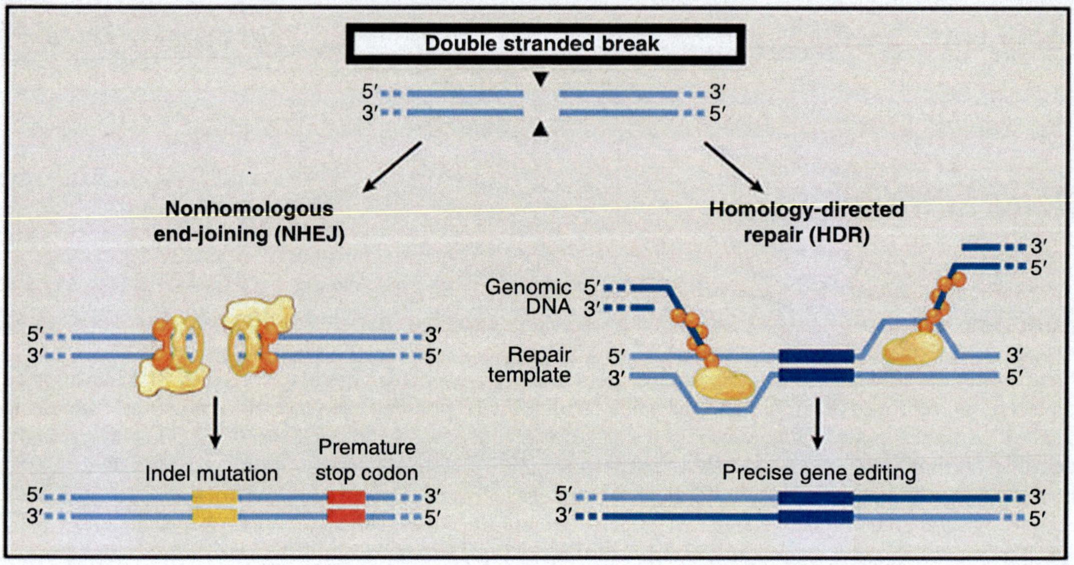

Fig. 1 Gene editing through double-stranded break of DNA

2 Advantages of Gene Editing over Transgenesis

Gene editing is a newer technology over transgenesis. There are limitations as well as ethical concerns for transgenesis. These are overcome with the technologies of gene editing where the desired gene can be inserted at a specific and desired site. Hence the chances for deleterious effects may be nullified as in the case of transgenesis.

Available tools till date available for genome editing are as follows:

2.1 Zinc Finger Technology (ZFN)

Zinc finger technology is a genome editing technology involving zinc finger nucleases. These are artificial restriction enzymes created by the fusion of a zinc finger DNA-binding domain to a DNA-cleavage domain. Thus, zinc finger domains can be designed to target specific desired DNA sequences. This allows zinc finger nucleases to target unique sequences within complex genomes. With the help of endogenous DNA repair machinery, these are useful to precisely change the genomes of higher organisms. Along with Cas9 and TALEN proteins, ZFN is gaining importance as an efficient tool in the field of genome editing. Figure 2 depicts a schematic representation of zinc finger technology.

A zinc finger nuclease is a form of endonuclease, which is highly site specific. It is designed with an aim to bind and cleave DNA at specific positions. It contains two protein domains. The first domain is the DNA-binding domain, containing eukaryotic transcription factors and the zinc finger. The second domain is the

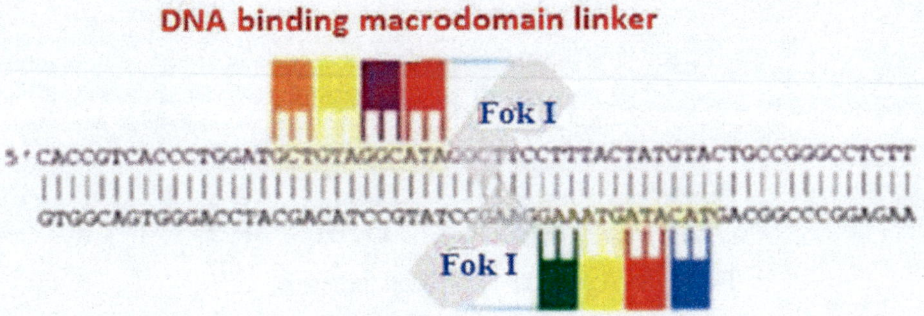

Fig. 2 Gene editing through zinc finger technology

nuclease domain, which is consisting of the FokI restriction enzyme and leads to the catalytic cleavage of DNA.

2.1.1 DNA-Binding Domain

The DNA-binding domains of individual ZFNs contain three and six individual zinc finger repeats, which can each recognize between 9 and 18 base pairs, respectively. In case, the zinc finger domains are specific for their intended target site, a pair of three-finger ZFNs that recognize a total of 18 base pairs. The best method to create new zinc-finger arrays is to combine smaller zinc-finger "module-s"of known specificity. The most common modular assembly process combines three separate zinc fingers. These zinc fingers recognize a three base pair DNA sequence to give rise to a three-finger array responsible for recognizing a nine base pair target site.

2.1.2 DNA-Cleavage Domain

FokI is commonly used as the cleavage domain in ZFNs. FolkI is a cleavage domain of type II restriction endonuclease. This cleavage domain dimerizes in order to cleave DNA. A pair of ZFNs is necessary to target nonpalindromic DNA sites. C-terminus of each zinc finger domain fuses with the cleavage domain of standard zinc fingers. The two individual zinc fingers bind the opposite strands of DNA located at a certain distance with the C-terminus. This aids in the dimerization of two cleavage domains and cleavage of DNA.

FokI nuclease acts as the DNA-cleavage domain and binds DNA by engineered Cys2His2 zinc fingers. Specific zinc fingers aid in the recognition of different nucleotide triplets and dimerization of the FokI nuclease (induce DSB).

2.2 TALEN Technology

Transcription activator-like effector nuclease (TALEN) systems are a fusion of TALEs, which are derived from the *Xanthomonas* spp. lead to a restriction endonuclease FokI. By altering the amino acid repeats in the TALEs, the researcher can customize TALEN systems to specifically bind target DNA and induce DNA breaks (Fig. 3).

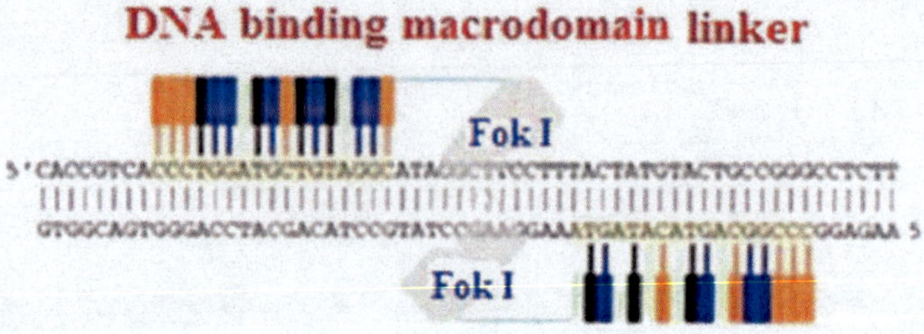

Fig. 3 Gene editing through TALEN technology

2.3 CRISPR-Cas Technology

RNA-guided endonucleases (RGEN) use a short guide RNA (gRNA) to recognize DNA, bind an endonuclease, leading to the site-specific cleavage. Derived from the clustered regularly interspaced short palindromic repeats (CRISPR) found in bacteria that serve to identify/destroy foreign DNA (Fig 4a, b).

CRISPR is a better cost-effective technique than TALENS across the board. CRISPR is a comparatively cheap genome-editing solution than TALENs, there are certain limitations. CRISPR technology may lead to off-target effects, which means it may cause the unintentional modification of nontargeted genes [1, 2].

3 History of CRISPR-Cas Technology

1987: There is the first report of CRISPR clustered repeat.

2000: It was recognized that CRISPR families are present throughout prokaryotes.

2002: The name of CRISPR was coined. The signature of cas gene was defined.

2005: The foreign origin of spacers was proposed to have a role in adaptive immune functions.

2007: The first experimental evidence of CRISPR adaptive immunity was established.

2008: CRSPR acts upon DNA targets.

2009: Type III B Cmr CRISPR complexes cleave RNA.

2010: Cas 9 is guided by spacer sequences and cleaves target DNA via DSBS.

2011: tracrRNA forms a duplex structure with crRNA in association with Cas9.

Type II CRISPR systems are modular and can be heterologously expressed in other organisms.

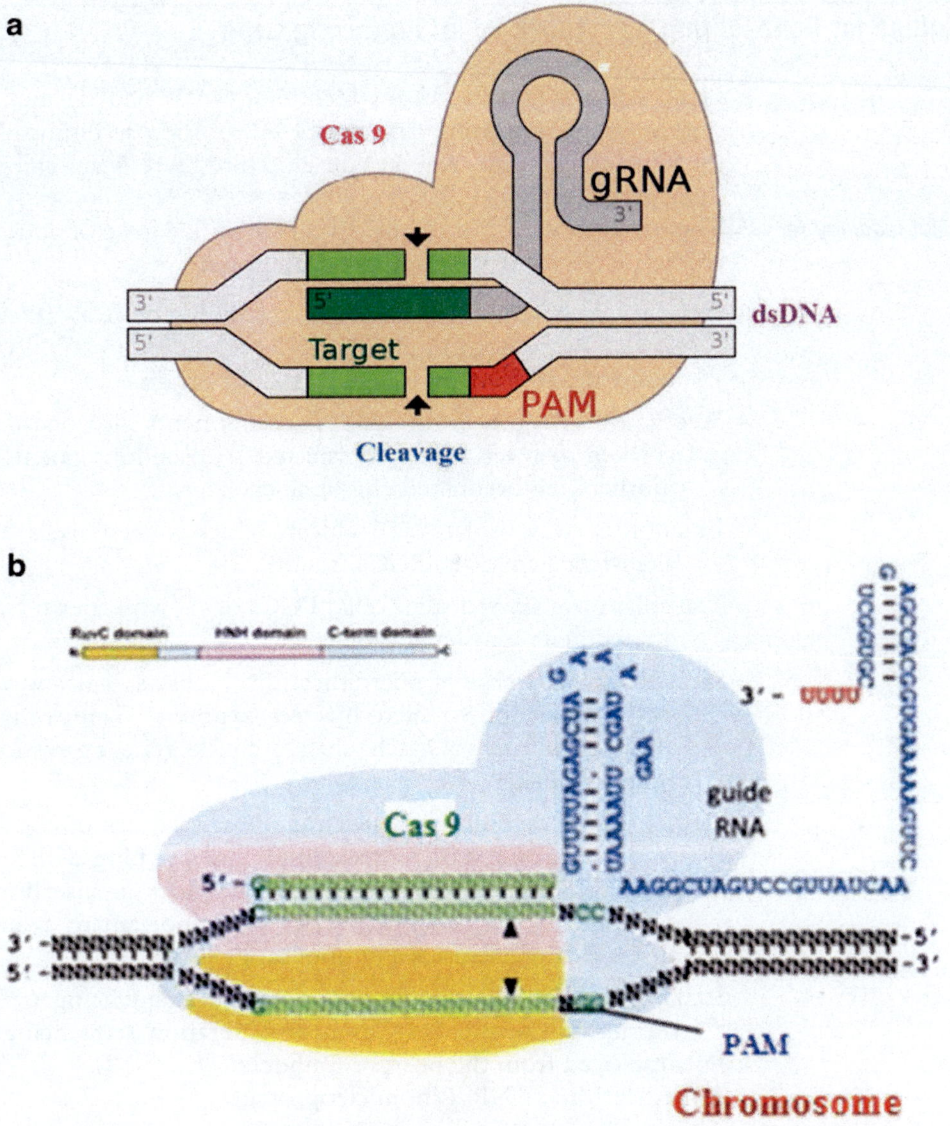

Fig. 4 (**a**) Gene editing through **CRISPR-Cas technology**. (**b**) Gene editing through CRISPR-Cas technology

2012: In vitro characterization of DNA targeting by Cas 9 was established.

2013: First demonstration of Cas 9 genome engineering in eukaryotic cells.

2014: Genome-wide functional screening in Cas 9 was assessed. Crystal structure of apo Cas 9 was predicted. The crystal structure of Cas 9 is deciphered in complex with guide RNA and target DNA.

4 Outline for Gene Editing for "Knock in" of Promising Gene

Gene editing is a successful technology. We had developed an outline for gene editing for "knock in" of the promising gene of duck in chicken with an aim to develop bird flu resistant chicken, as discussed below:

4.1 Methodology for Step 1

The basic steps for "Knock in" of the identified gene of duck in the chicken embryo may be outlined as follows:

Step 1: Preparation of CRISPR components (Cas 9/sgRNA [CrRNA tracRNA duplex]), RNP complex, and ssDNA HDR templates of RIGI of indigenous duck origin.

Step 2: Microinjection of CRISPR components into one-day-old chicken zygotes. Then transferred to pseudopregnant foster mothers/embryonated egg of chicken.

In another way, isolate and culture chicken stem cells/PGCs. Transfect stem cells/PGCs/sperm.

Transplant transfected stem cells/PGCs or AI with sperm (sperm-guided gene transfer).

Step 3: After 7 weeks of microinjection, biopsy samples were collected from pups/microinjected zygotes. Windowing and ex vivo culture to hatch chicks, check gene expression for RIGI of duck.

Step 4: For the confirmation of the insert, a series of checks are employed. PCR-based check employ genotyping with specific primers designed for confirmation if knock-in insertions are complete. The checking is extremely important to assure the absence of truncations and duplications.

Step 5: For the final confirmation, Sanger sequencing (or high throughput sequencing, as next-generation sequencing) was employed from the positive founders.

Abbreviations: RNP: Ribonucleoprotein

CrRNA tracRNA duplex: Cas9-guide RNA

HDR: homology directed repair, ssDNA: single stranded DNA.

4.2 Methodology for Step 2

Analyzing the functionality of the "Knockin" chicken developed through a series of immunological tests pertaining to viral infections.

5 Protocol for Gene Editing

I. Let us describe the protocol with a suitable example of gene knockout for Stra 8 gene.

5.1 Gene Cloning and gRNA Design

The primers used for amplification for Stra 8 gene obtained from the NCBI database (ID: XM_416179): forward, 5′

CCCAAGCTTATGCAAGAATGTGAAAAAC3'; reverse, 5′ CCGGAATTCTTATAAATCTTCATCATCA3'. The coding sequence region was detected based on the protospacer adjacent motif (NGG or GGN) with the initial 19–21 bp for gRNA design. In order to overcome off-target effects, the entire genome was searched for possible sites off-target. The gRNA sequence or nucleotide sequence of gRNA to the target site is being described.

The following oligo primers were used as per the selected gRNA, as 5′AAACACCG-gRNA-1, 5′CTCTAAAAC-gRNA-1-anti; 5′AAACACCG-gRNA-2, 5′CTCTAAAAC-gRNA-2-anti; 5′AAACACCG-gRNA-3, 5′ CTCTAAAAC-gRNA-3-anti.

Step 1: Double-stranded oligo was formed through annealing of sense strand and antisense strand. The annealing system comprises of 1 μL of each sense strand and antisense strand, 1 μL solution, 3 μL H_2O.

Annealing procedures involve 95 °C for 5 min, natural cooling to 16 °C, 16 °C for 10 min.

Step 2: The next step comprises of insertion of oligo dipolymer into vector.

The reaction system consists of 1 μL Cas9/gRNA vector, 2 μL oligo dipolymer, 7 μL H_2O and kept at 25 °C standing for 5 min. The 5–10 μL final product obtained in step 2 was added into 50 μL DH5α competent cells, which were thawed freshly, fully mixed, put on ice for 30 min. The cells were subjected to 42 °C heat shock on ice for 90 s, standing for 2 min, and cultured in LB solid medium supplemented with 0.1% ampicillin for 12–16 h. Positive clones were identified picked out and cultured in LB liquid medium supplemented with 0.1% ampicillin for 12–16 h.

The recombinant plasmids were extracted with the appropriate technique. Plasmid containing the right insertion was then sequenced. The sequencing primer was TGAGCGTCGATTTTTGTGATGCTCGTCAG.

5.2 sCell Culture and Transfection

DF-1 cells (chicken embryo fibroblast cell line) were cultured in DMEM (Dulbecco's modified Eagle's medium, high glucose, Sigma). The culture media was supplemented with 15% (v/v) fetal bovine serum (Invitrogen), 100 IU/mL penicillin, and 100 μg/mL streptomycin. All cultures were maintained in a 5% CO_2 humidified atmosphere at 37 °C and passaged for every 2–3 days. Cells were grown to about 60% confluency for transfection purposes. Fugene was employed as the transfection reagent. The proportion of Fugene to plasmid was maintained as 3:1. Flow cytometry and cell sorting were assessed after 48 h of transfection to collect the positive cells.

5.3 Luciferase-SSA Assay, T7 Endonuclease I Assay (T7E1) and TA Clone Sequence

The next step involves the construction of the single-strand annealing (SSA) luciferase reporter. With this aim, a terminator sequence followed by the gRNA target sequence was inserted into a plasmid encoding the luciferase gene. The SSA luciferase reporter was cotransfected with the CRISPR/gRNA vector and a Renilla luciferase reporter, which acts as an internal control. An empty plasmid was taken as a negative control. After 48 h of transfection, luciferase SSA activity was detected by Synergy™ 2 Multi-Mode Microplate Reader (Biotech, Vermont, USA). To assess luciferase signal, terminator and target site was inserted into the Luciferase vector to coinfect the CRISPR/gRNA vector, the newly constructed luciferase reporter gene, and the internal control Renilla plasmid. Transfected void vector was used as control. The target site is localized after the terminator. 48 h post-transfection, GFP positive cells were screened out with flow cytometry from the secondary generation cells. Genomic DNA of positive GFP cells was extracted for the initial step, followed by PCR with the primers: F: 5′ GTATTTCCTTTGCTTCCTATGC 3′ and R5′ TGATACCACACTGGGATTCCAT 3′. The fragment obtained was of 511 bp fragment along with the target site, which was cloned and confirmed through T7E1 enzyme. TA clones ($n = 30$) were screened and selected for sequencing and the efficiency of CRISPR/Cas9 gene knockout vector was estimated. Gene knock out efficiency (KOE) was calculated as the counts of the mutated bacterial community(CMBC) divided by the total sequenced bacterial community(TSBC), multiplied by a hundred.

$$\text{Gene KOE} = \text{CMBC}/\text{TSBC} \times 100\%.$$

5.4 In Vitro Transcription of gRNA and Cleavage Activity

The fragment of gRNA was set as a template, T7-gRNA was employed for amplification to get PCR product of 120 bp. PCR product was purified and utilized as DNA template for the in vitro transcription. The target gRNA (PAM sequences) was inserted into the PCR products through the process of bridging PCR. In vitro transcription processes were depicted in vitro transcription kit.

PCR reaction conditions employed are:

gRNA plasmid 10 ng.

T7-gRNA-FPg (10 μM) 1.5 μL.

gRNA-RP (μM) 1.5 μL.

2 × Pfu Mix 2 μL.

Added double distilled water up to 50 μL.

PCR amplification program followed.

Step1: 95 °C 3 min; Step 2: 94 °C 30 s, 58 °C 30 s, 72 °C 30 s for 35 cycles; Step 3: 72 °C 10 min; 16 °C 10 min.

Meanwhile, PCR products of standard gRNA1 (g1) and gRNA2 (g2) were prepared. The primer sequence of T7-gRNA-F was: TAATACGACTCACTATAG-gRNA-GTTTTAGAGCTA-GAAATAGC and gRNA-R: AGCACCGACTCGGTGC-CACTT. In vitro enzyme digestion conditions were shown below.

Table . In vitro enzyme digestion conditions.

	1	2	3
Cas9	1U	1U	1U
10X Cas9 buffer	2 μL	2 μL	2 μL
gRNA	50 ng (gRNA)	50 ng (Positive)	50 ng (Negative)
ddH_2O	X μL	X μL	X μL
dsDNA	50 ng	50 ng	50 ng
Total	20 μL	20 μL	20 μL

5.5 *In Vitro Induction Experiment*

5.5.1 *Isolation and Culture of Chicken ESCs*

Freshly fertilized eggs were collected and disinfected with 1% benzalkonium bromide and 75% alcohol. Initially, the blunt end of egg was broken by the tweezer and egg white was removed. In the next step, blastoderm cells at stage X were collected by spoon method in tissue culture dishes followed by rinsing with Ca^{2+} and Mg^{2+} free phosphate-buffered saline (PBS-CMF) to remove the yolks and vitelline membrane. ESCs washed with PBS were transferred into fresh tissue culture dishes with PBS supplemented with 0.25% trypsin and 0.04% ethylenediaminetetraacetic acid (EDTA). The process was followed by digestion at 37.0 °C for 5–8 min. Collection of the dissociated cells was done through centrifugation at 1000 rpm for 8–10 min and suspended. ESCs were then maintained in a 5% CO_2 humidified atmosphere at 37.0 °C with cytokine culture medium (DMEM (Sodium pyruvate, L-glutamine), 10% FBS, 1% nonessential amino acid, 5.5×10^{-5} mol/L β-ethyl alcohol, 2% chicken serum, 10 ng/mL bFKF, 0.1 ng/mL LIF and 5 ng/mL SCF).

5.5.2 *In Vitro Induction*

Twenty-four well plate with a density of 10^5 cells per well was employed for culturing the third-generation ESCs followed by transfection of the overexpression or CRISP/Cas9 mediated knockout vector of Stra8 gene. The corresponding control vector was transfected and designated as the reference.

The medium used was DMEM (sodium pyruvate, L-glutamine) along with 10% FBS+RA (10^{-5} mol/L) after 24 h transfection. The cell culture media were replaced every 2 days, the morphological changes of the cells were detected by fluorescence inverted microscope. The samples were collected every 2 days during the process of induction. Western blotting and qRT-PCR were evaluated for the confirmation of the expression of NANOG, Sox2, C-kit, Cvh, integrinα6, and integrinβ1. Indirect immunofluorescence assay was employed for the detection of the expression of germ cell-specific genes following induction [3].

5.6 Microinjection of the Cas9/gRNA Plasmid into Chicken Embryos

CRISPR/Cas9 vector employed was capsulated with polyethylenimine (PEI), followed by injection into chicken embryos. It was then sealed with paraffin, followed by incubation at 38.5 °C for 4.5 days in order to avoid contamination. DNA isolation was conducted for analysis using the T7EI assay.

The validation of the technique was conducted through the following techniques:

5.6.1 Quantitative Real-Time PCR (qRT-PCR)

The cells were collected at successive intervals of 0 day, 4 days (PGC-like cell appear in vitro), 10 days (SSC-like cell appear in vitro) and RNA was extracted using the RNeasy kit and reverse transcribed to cDNA. The primers utilized for qRT-PCR were listed as follows:

Table . Primer information for qRT-PCR.

Gene	Primers for qRT-PCR	Tm(°C)	Size(bp)
Nanog	F: TGGTTTCAGAACCAACGAATGAAG	64	180
	R: TGCACTGGTCACAGCCTGAAG		
Sox2	F: GAAGATGCACAACTCGGAGATCAG	64	100
	R: GAGCCGTTTGGCTTCGTCA		
C-kit	F: GCGAACTTCACCTTACCCGATTA	64	150
	R: TGTCATTGCCGAGCATATCCA		
Cvh	F: TGTCTTGAAGGCCTCGTTTG	61	138
	R: CATATCCTTGGCAGGTTGTTGA		
integrinα6	F: TGTTTGTGGGGACCAGATTG	53	120
	R: CCAGGTGACATTTCCCATCA		
integrinβ1	F: GAAACCCGGGATATCATTGG	53	140
	R: CAGCAACACCTTGCTGACAG		

Analysis was performed with the SuperReal PreMix Color (SYBR Green) and performed on an ABI two-step RT-PCR system (Applied Biosystems 7500, USA) with diluted first-strand cDNA as per the manufacturer's instructions. For each reaction, 0.1 ng of total RNA was employed as input. The qRT-PCR reaction mixtures consist of 1 μL of cDNA, 10 μL of 2× SuperReal Color PreMix, 0.6 μL of the forward and reverse primers (10 μmol/L), 0.4 μL of 50× ROX Reference Dye, and RNase-free ddH_2O in a total reaction mixture volume of 20 μL. The mixtures were poured into a clear tube (0.2 mL thin wall, Axygen, USA). The reactions involve 40 cycles of the following program: 95 °C for 15 min, followed 30 °C for 10 s, 60 °C for 30 s, and 72 °C for 32 s. The β-actin gene (house-keeping genes) was employed as an internal reference gene, and all reactions were conducted in triplicate. Gene expression levels were calculated using the $2^{-\Delta\Delta CT}$ method.

5.6.2 Immunofluorescence

Cell suspensions were incubated on poly-L-lysine-coated slides for adherence followed by fixation for 10 min in 4% paraformaldehyde. Following 1 h of blocking in 10% bovine serum albumin (BSA) in PBS, antibodies against Integrin α6 were added at a 1:200 dilution in 10% BSA in PBS, followed by incubation for overnight at 4 °C.

Following the process of washing thrice with 0.05% Tween-20 in PBS, FITC-labeled secondary antibodies (Biosynthesis Biotechnology Co., Ltd.) were added at a 1:100 dilution for 2 h. The step was repeated for washing again with 10% BSA-PBS, and the cells were visualized under an inverted fluorescence microscope.

5.6.3 FACS

Cell suspension induced of 10 days were collected in a 1.5 mL centrifuge tube, 1000 × *g* centrifugal for 5 min. Discard the supernatant and add the integrinα6 (1:200) and integrin β1 antibody. The above suspension was incubated at 4 °C for 1–2 h followed by centrifugation. The supernatant was discarded and cells were washed two times with precooling PBS. Precooled PBS @500 μL was added for FACS analysis.

5.7 Data Analysis

t-test analysis was carried out ($P < 0.05$ for significant differences, $P < 0.01$ for highly significant differences) (SPSS19.0 software). EXCEL2003 was used to generate figures.

6 Protocol for CRISPR/Cas9-Based Knock-In Using the VIKING Method

CRISPR/Cas9-based genome editing has been proved to be a very efficient technique for genome engineering.

Here we attempt to describe the versatile nonhomologous end joining-based knock-in module for genome editing (VIKING) method1. The uniqueness of this method is that cells are simultaneously transfected with three vectors:

1. Donor vector,
2. Donor cleavage vector, and
3. Target cleavage vector.

The VIKING method utilizes the "VKG1 sequence" which do not require the customization of the donor vector. This method has the following three advantages:

1. There is no need to modify donor and donor cleavage vectors if they contain the VKG1 sequence. Certain vectors (such as pUC19, pcDNA3, pENTR/D-TOPO) contain this sequence. On the contrary, other techniques involve the design and construction of new donor vectors, which is complicated.
2. In this technique, vector transfection for nonhomologous end-joining repair-dependent knock-in is being optimized. Whereas in other technologies, the process becomes hampered by random integration.

In this technique, DNA sequences longer than 10 kbp can be inserted efficiently, which is usually not possible with general methods. A longer sequence can be inserted using this method, rather than utilizing a viral vector. Let us describe the knock-in/knock-out and knock-in procedures using the VIKING method. Figure 5 represents workflow of VIKING method and schematic image of electrophoresis shows typical results of genotyping (Protocol I,), whereas Fig. 6 represents a schematic diagram of the protocol for knock-in in cultured cells using the VIKING method. (Protocol II). Figure 7 describes Knock-in of a VDR variant (Q259P) in VDR-knock-out HaCaT cells using the VIKING method.

6.1 Reagents

6.1.1 Cell Culture Reagents

Dulbecco's modified Eagle's medium (DMEM) [4].

Fetal bovine serum, Irish Origin.

Penicillin-streptomycin-amphotericin B suspension (×100).

Phosphate-buffered saline.

Trypsin.

6.1.2 Transfection

Opti-MEM™ I Reduced Serum Medium, no phenol red (11058-021; Life Technologies, California, USA).

TurboFect transfection reagent (R0531; Thermo Fisher Scientific, Waltham, MA).

Puromycin dihydrochloride (P8833; Sigma-Aldrich Japan, Tokyo, Japan).

Blasticidin S, Hydrochloride (KK-400; Kaken Pharmaceutical, Tokyo, Japan).

6.1.3 Plasmids

Target genome cleavage vector (we use pX330*; https://www.addgene.org/42230/).

AAVS1 locus-specific cleaving vector (Fig. 6 blue).

Donor vector* (pVKG1-PURO, Puromycin resistance, or pVKG1-BSD, Blasticidin resistance, which are distributed by Addgene (108,670 and 108,680, respectively)) FLAG-tagged VDR variant (Gln259 > Pro) expression vector (pcDNA3-FLAGhVDRQ259P****) Donor cleavage vector* (VKG1-gRNA-pX330, which is distributed by Addgene (108671)) *A human codon-optimized SpCas9 and chimeric guide RNA expression plasmid. ****This pCDNA3 backbone contains a VKG1 sequence for cleavage by Cas9-gRNA. *The donor cleavage and donor vectors are known as the VIKING module and need not to be modified according to the target sequence.

6.1.4 DNA Extraction

Dulbecco's phosphate-buffered saline.

50 mM NaOH.

1 M Tris–HCl, pH 8.0.

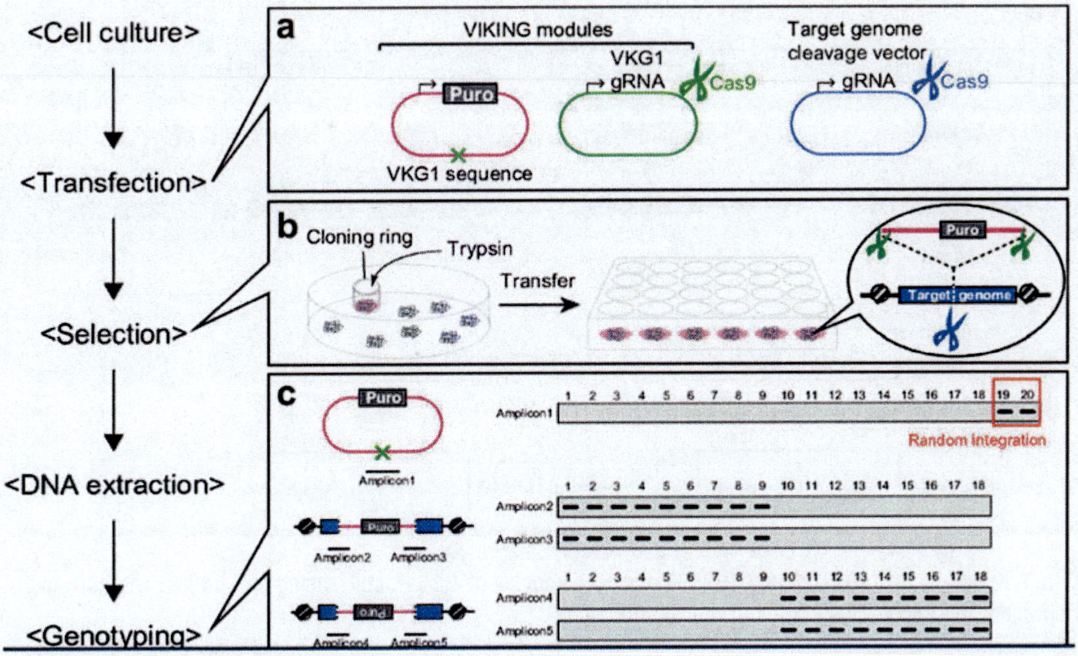

Fig. 5 Workflow of VIKING method. Schematic image of electrophoresis shows typical results of genotyping (Protocol I). (**a**, **b**, **c**) Represent the successive stages of the methodology

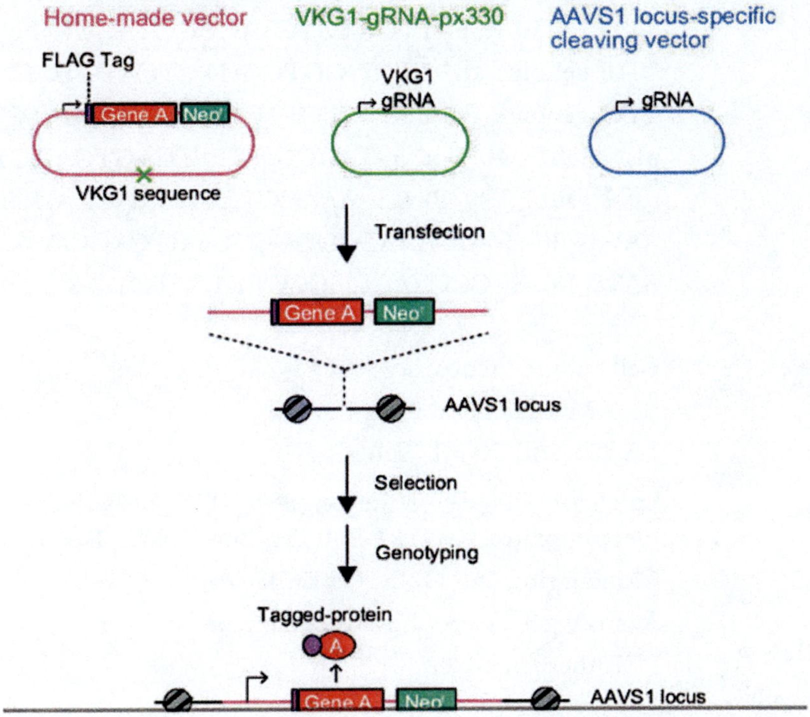

Fig. 6 Schematic diagram of the protocol for knock-in in cultured cells using the VIKING method. (Protocol II)

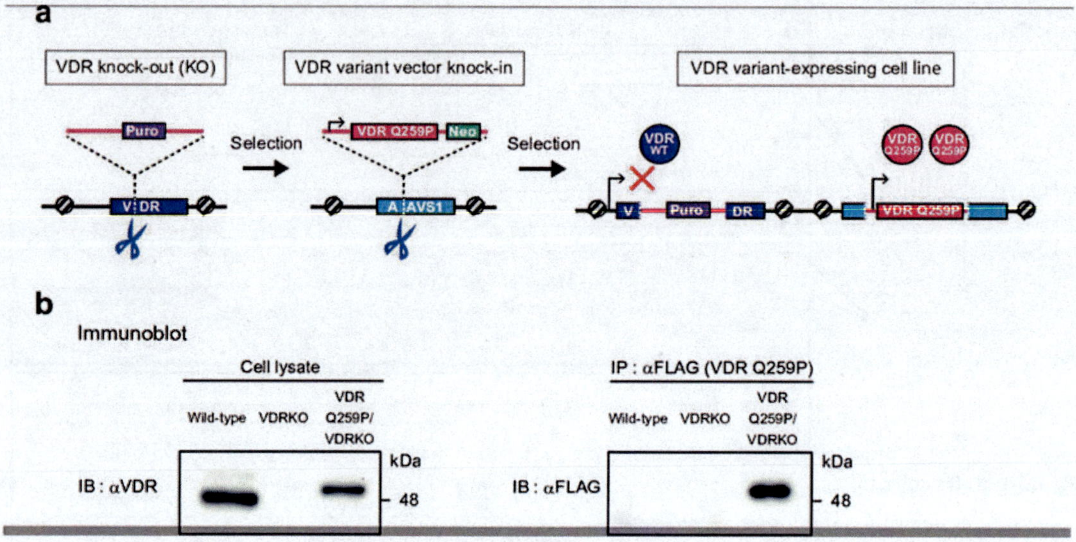

Fig. 7 Knock-in of a VDR variant (Q259P) in VDR-knock-out HaCaT cells using the VIKING method. (**a**, **b**) Represent the successive stages of the methodology

6.1.5 Genotyping

GoTaq Polymerase (M3001; Promega, Wisconsin, USA).

Primers (change according to the target locus for knock-in).

VKG 1_4561 F, 5′-GCCTATGGAAAAACGCCAGC-3′.

VKG 1_250 R, 5′-TTCCTGTCTAGCGGTACGCG-3′.

VDR genome_del 3 F, 5′-GGTGGGCCTCATGTCTTCTG-3′.

VDR genome_del 1 R, 5′-CCTTCATCATGCCGATGTCC-3′.

pEN_ObLi_40_For, 5′-TACCGCCTTTGAGTGAGCTG-3′.

pEN_ObLi_190_Rev, 5′-AAACCTGTCGTGCCAGCTGC-3′.

AAVS _ F5, 5′-GGAAATGGGGGTGTGTCACC-3′.

AAVS_R6, 5′-CCCTACCCCCCTTACCTCTC-3′.

6.1.6 Equipment

Cell culture incubator (5% CO_2/37 °C).

BSL-2 biosafety cabinet.

1.5, 15, and 50 mL Tubes.

Cell culture plasticware (e.g., dishes, plates, and pipettes).

Electroporator (CUY21 EDITII; Bex, Tokyo, Japan).

Cloning ring (AS ONE, Osaka, Japan).

Agarose gel electrophoresis equipment.

PCR thermal cycler.

7 Procedure

7.1 Protocol I: Procedure for Knock-in/Knock-out in Cultured Cells Using the VIKING Method

7.1.1 Day 1: Cell Culture

HaCaT or human embryonic kidney 293F (HEK293F) cells were incubated in Dulbecco's modified Eagle's medium (DMEM) consisting of 10% fetal bovine serum and 1 U/mL penicillin-streptomycin (D10 medium).

7.1.2 Day 2: Transfection (Fig. 1a)

Transfection is being carried out with electroporation method (HaCaT cells). Cells are suspended in 400 μL Opti-MEM. Cells are transfected with the VIKING module. The target genome cleavage vector employs a total of 15 μg DNA (molar ratio of donor cleavage vector (VKG1-gRNA-pX330):donor vector (pVKG1-PURO):target genome cleavage vector (pX330) = 1:1:17) with an electroporator. Cells are transfected to 100 mm dishes and culture in D10 medium lacking antibiotics for 24 h.

Lipofection method (HEK293F cells): Cells are suspended in 400 μL Opti-MEM. Cells are transfected with the VIKING module and the target genome cleavage vector with a total of 15 μg DNA (molar ratio of donor cleavage vector (VKG1-gRNA-pX330): donor vector (pVKG1-PURO):target genome cleavage vector (pX330) = 1:1:17) with TurboFect transfection reagent. Cells are transferred to 100 mm dishes and culture in D10 medium lacking antibiotics for a period of 24 h.

7.1.3 Day 4: Selection

D10 medium is being refreshed. Puromycin (0.3 μg/mL) is added to the D10 medium and culture cells for 1–2 weeks.

7.1.4 Day 14: Clone Picking (Fig. 1b)

Cells are washed with phosphate-buffered saline. The cloning ring is placed on a single colony. At least 20 colonies are picked up. Limiting dilution is also possible. 20 μL trypsin is added to each ring. Cells are incubated at 37 °C for 2 min in case of HEK293F cells or 10 min in the case of HaCaT cells, followed by transfer of cells are to a 24well or 48-well plate. Cells are incubated at 37 °C for several days. Cells are transferred to a 12-well plate and incubated. A portion of the cells are being isolated for DNA extraction.

7.1.5 Day 16: DNA Extraction

Added 180 μL of 50 mM NaOH and transferred to a 1.5 μL tube. Incubated at 95 °C for 30 min. Added 20 μL of 1 M Tris–HCl, pH 8.0. Centrifuged at 12,000 rpm for 10 min. Transferred the supernatant to a new tube.

7.1.6 Day 17: Genotyping (Confirmation of Knock-in by PCR)

PCR of the extracted DNA as a template is conducted to confirm knock-in and random insertion. Special polymerase is not needed, GoTaq polymerase is used in this protocol. Primers are designed to amplify the VKG1 sequence.

Example:

Primer 1 (VKG 1_4561 F), 5′-GCCTATGGAAAACGCCAGC-3′.

Primer 2 (VKG 1_250 R), 5′-TTCCTGTCTAGCGGTACGCG-3′.

Primers are designed that specifically bind to the target genome sequence.

Example:

Primer 3 (VDR genome_del 3 F), 5′-GGTGGGCCTCATGTCTT CTG-3′.

Primer 4 (VDR genome_del 1 R), 5′-CCTTCATCATGCCGATG TCC-3′.

7.2 Genotyping by PCR

PCR sample.

2× GoTaq GreenMaster Mix, 12.5 μL.

10 μM Primer A, 2.5 μL.

10 μM Primer B, 2.5 μL.

Nuclease-free water, 6.5 μL.

Genomic DNA, 1 μL.

Total, 25 μL.

Thermocycler reaction conditions:

95 °C for 2 min.

30 cycles of 95 °C for 30 s, 55 °C for 1 min, and 72 °C for 2 min.

72 °C for 4 min.

12 °C hold.

Step 1. Confirm random insertion.

PCR is conducted with Primer 1 and Primer 2 for Amplicon1.

Random insertion is confirmed by amplification of Amplicon1.

Step 2. Confirm knock-in at the target locus.

Forward direction.

PCR is conducted with Primer 1 and Primer 3 for Amplicon2, or Primer 2 and Primer 4 for Amplicon3.

Knock-in in the forward direction is confirmed by amplification of Amplicon2 and Amplicon3.

Reverse direction.

PCR is conducted with Primer 2 and Primer 3 for Amplicon4, or Primer 1 and Primer 4 for Amplicon5. Knock-in in the reverse direction is confirmed by amplification of Amplicon4 and Amplicon5 [4].

8 Protocol II: Procedure for Knock-in in Cultured Cells Using the VIKING Method (Fig. 2)

8.1 Day 1–16

Same as Protocol I.

8.2 Day 17: Genotyping

Design primers to amplify the VKG1 sequence.

Same as Protocol I.

Design primers to amplify the VKG1 sequence.

Example:

Primer 1 (pEN_ObLi_40_For), 5′-TACCGCCTTTGAGTGAG CTG-3′.

Primer 2 (pEN_ObLi_190_Rev), 5′-AAACCTGTCGTGCCAGC TGC-3′.

Design primers which specifically bind to the target genome sequence.

Example:

Primer 3 (AAVS_F5), 5′-GGAAATGGGGGTGTGTCACC-3′.

Primer 4 (AAVS_R6), 5′-CCCTACCCCCCTTACCTCTC-3′.

8.3 Genotyping by PCR

Same as Protocol I.

8.4 Remarks

1. This vector is selected as per the target locus for knock-in.
2. Other than, HaCaT and HEK293F cells, the VIKING method was observed to be useful and validated in the C4-2 (human prostate cancer), UMR-106 (rat osteogenic), and MC3T3-E1 (mouse preosteoblastic) cell lines.

3. These two transfection methods are examples and need to be changed according to the cell lines to attain a high transfection efficiency.
4. The backbone of the donor vector is integrated into the genome in principle.
5. Tandem insertions of the donor vectors occur often [4].

9 Easi-CRISPR Protocol for Creating Knock-in and Conditional Knockout Mouse Models Using Long ssDNA Donors

CRISPR/Cas9-based genome editing has the ability to easily generate knockout mouse models by disrupting the gene sequence. Its efficiency for the creation of models was observed to be poor which needs either insertion of exogenous DNA (knock-in) or replacement of genomic segments is very poor. The majority of mouse models utilized in research are knock-in (reporters or recombinases). Gene-replacement (for example, conditional knockout alleles containing *LoxP* sites flanked exons) is also a common methodology. The common methodology for the creation of such models has been documented with double-stranded DNA as donors. Their efficiency was observed to be usually low (1–10%) and hence not suitable for routine use.

Recently it has been observed that long single-stranded DNAs (ssDNA) act as very efficient donors, both for insertion and for gene replacement. This method is referred to as *Easi*-CRISPR (*E*fficient *a*dditions with *s*sDNA *i*nserts-CRISPR), a highly efficient technology (30–60%, or up to 100% in some cases). The protocol takes about 2 months to generate the founder mice.

This protocol is based on the principle of generating Cas9-mediated fluorescent protein knock-ins with a self-exciting selection cassette (SEC). It was established by Dan Dickinson in 2015 [1]. Choice of Cas9 target site is very important.

1. Identified a 100–200 bp region wherein the Cas9 target site needs to be located. We generally employ a 200 bp window centered on the start codon (for N-terminal tags) or stop codon (for C-terminal tags).
2. Submitted this genomic sequence to the Zhang lab's CRISPR design tool at http://crispr.mit.edu. Make sure *Caenorhabditis elegans* is being selected as the genome for checking specificity.
3. The design tool lists a set of potential targeting sequences, ranking is conducted in order of predicted specificity. The target sites are chosen in such a way with a specificity score > 95, and in most of the cases the site should score 98 or 99 (100 indicates perfect specificity). In case, there are several candidate

sites with high specificity, the site is so chosen, which is closest to the desired insertion site (start or stop codon).

It makes our work easier if we list out potential off-target sites, copied, and pasted into a Word document or Excel sheet, and saved this list for future reference.

Added the target sequence to the Cas9–sgRNA construct.

1. The design tool includes target sites of the form 5′N20-NGG-3′, where N is any base. Insertion of the N20 sequence into the Cas9–sgRNA construct (pDD162, Addgene #47549) is the prime step. NEB's Q5 Site-Directed Mutagenesis Kit is employed for this purpose. Forward primer 5-′-N20GTTTTAGAGCTAGAAATAGCAAGT-3′ (where N20 is our 20 bp targeting sequence from the design tool), and reverse primer 5′-CAAGACATCTCGCAATAGG-3′ was employed for this purpose.
2. PAM (NGG motif) should not be included in primers for the Cas9-sgRNA construct. Care should be taken to include NGG motif in the target DNA, although it is not part of the sgRNA.
3. Sequencing primer 5′-GGTGTGAAATACCGCACAGA-3′ is employed to verify the correct insertion of the targeting sequence.

Primers are designed to add homology arms to an FP–SEC vector. Our strategy is for cloning homology arms into FP–SEC vectors. Homology arms are evolved through PCR and inserted in the place of the ccdB negative selection markers, that are flanked by restriction sites.

Particular restriction sites are so selected that no residual sequence is left after the homology arms are added. ccdB negative selection leads to a robust and efficient cloning strategy. In a case study, out of eight repair templates generated, seven yielded more than 80% correct clones. Four primers need to be developed, two for each homology arm. The homology arms will be amplified and added sequence overlaps for Gibson assembly to the ends of each arm.

In case FP::SEC insertion will not disrupt the Cas9 target site, primers will play the role. Primers will be responsible for introducing silent mutations, thereby preventing Cas9 from cutting the repair template. We can choose either of the two possible pairs of restriction enzymes to digest the FP–SEC vector: AvrII+SpeI or ClaI+SpeI.

In case the former combination (AvrII and SpeI) is used, the repair template will consist of a flexible linker between the 5′ homology arm and FP. This is essential for generating C-terminal tags. If ClaI and SpeI are employed, the 5′ homology arm will be fused directly to the FP, with no added sequence. This is essential for N-terminal tags, or when no flexible linker is desired [5].

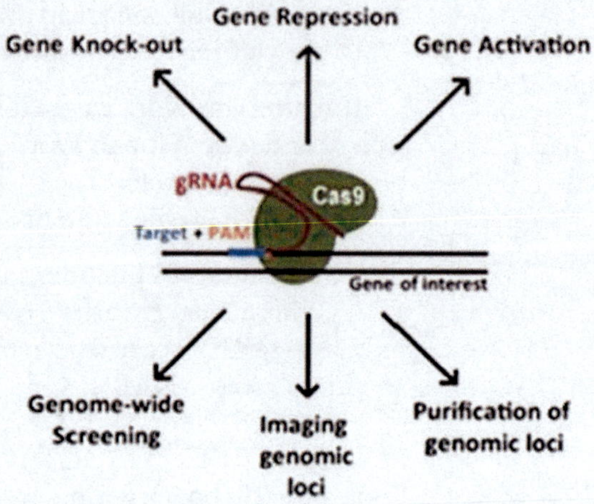

Fig. 8 Outcomes of CRISPR Cas9 technology using guide RNA guided gene editing

9.1 Application for CRISPR Technology

CRISPR technology has a widespread avenue as figured out in Fig. 8.

9.2 Applications of CRISPR Genome Editing

Applications of CRISPR-Cas9 technology have been indicated in the fields of basic and clinical research, drug development, therapeutics, the environment, and agriculture. Clinical research has proved the utilization of CRISPR in such diseases as cancer, AIDS, Huntington's disease, Duchenne muscular dystrophy, and others.

The applications include the creation of genetically modified cell lines and animal models. In addition to create gene knockouts and more specific modifications, CRISPR technology to modulate gene expression via interference (CRISPRi) or activation (CRISPRa), without altering the genomic sequence are other equally important techniques (view table).

9.2.1 Types of CRISPR-Enabled Genome Edits

Genomic manipulation	application	Required components
Knock-out	Disruption of function for a specific gene can result from various different modifications	Cas9 (or Cas9 nickase); single (or dual) gRNA introducing disruptive edits
Edit	Execution of a specific sequence modification at a specific genomic location	Cas9 (or Cas9 nickase); single (or dual) gRNA targeting the region where the edit should be made; repairing template
Interference (CRISPRi), or activation (CRISPRa)	Reduction or increase in expression of a gene without permanent modification of the genomic sequence	dCas9 or dCas9-repressor or dCas9-activator; gRNA(s) targeting promoter elements of target gene

9.2.2 Methods to Check Genome Editing On-Target Efficiency

Type of on-target assessment	Type of identified variation	Results	Information content	Scale
Cleavage assay	Single mismatches and indels	Semiquantitative	Low	Low to high throughput
PCR and RFLP	HDR insertions	Quantitative	Low	Low to high throughput
PCR	Deletions	Quantitative	Low	Low to high throughput
Sanger	All edits	Quantitative + qualitative	High	Low throughput
NGS	All edits, genome-wide off-target effects	Quantitative + qualitative	High	Low to high throughput

Both CRISPR genome editing and next-generation sequencing (NGS) have proved to be advanced technologies providing additional information. A synergistic combination of these two technologies may result in better relaxation of the biological systems they are modifying. NGS technologies may be employed at various stages of a genome editing workflow. Thus, it leads to the confirmation of CRISPR knockouts for the analysis of off-target effects and assessment of the functional impact of gene edits.

9.2.3 Confirming CRISPR Knockouts and Other Edits

CRISPR genome editing experiments lead to mixed cell populations, with only a small subset carrying the desired edit. It becomes necessary to detect which cells have the desired CRISPR knockout or targeted mutation. Currently, we employ the methodologies to identify the edits involved through cleavage assays, PCR, Sanger sequencing, and NGS.

NGS is the assay providing both qualitative and quantitative information at high resolution across the full range of modifications. It can cater to the needs of any throughput and can be utilized for the monitoring of off-target effects. NGS-based targeted sequencing gives rise to a cost-effective solution to confirm CRISPR-induced edits by focusing on regions targeted for modification.

9.3 CRISPR off-Target Analysis with NGS

Successful implementation of CRISP/Cas9 technology needs to include strategies for the identification and reduction of off-target effects, or unintended modifications at sites other than the intended target. Computational methods are employed for the evaluation of RNA specificity and prediction of off-target sites during genome editing experiments.

Online tools and web-based algorithms are present publicly (view table). However, genome-wide analyses such as NGS-based whole-genome sequencing (WGS) are essential for the discovery of off-target sites which may escape prediction algorithms.

9.3.1 Publicly Available Analysis Tools to Predict Off-Target Effects

Off-target analysis tools	Website
Benchling	https://benchling.com
Cas-OFFinder	http://www.rgenome.net/cas-offinder
MIT CRISPR design tool	http://crispr.mit.edu
Deskgen	https://www.deskgen.com/landing
DNA 2.0 CRISPR gRNA design tool	https://www.dna20.com/products/crispr#4
E-CRISP	http://www.e-crisp.org/E-CRISP
EuPaGDT	http://grna.ctegd.uga.edu
GenScript gRNA design tool	http://www.genscript.com/gRNA-design-tool.html
ZiFiT	http://zifit.partners.org/ZiFiT
Broad GPP portal	http://portals.broadinstitute.org/gpp/public
CROP-IT	http://cheetah.bioch.virginia.edu/AdliLab/CROP-IT/homepage.html

9.3.2 Unbiased Methods to Analyze Off-Target Effects

Method	Type	Description	Advantage	Limitation
IDLV capture (integrase-defective lentiviral vector capture)	Cell-based (live cells)	Integration of IDLVs occurs with a selectable marker into DSBs sites. This is followed by the recovery of integration sites through LAM-PCR, high-throughput sequencing	It can identify DSBs in living cells	Relatively insensitive
GUIDE-Seq (genome-wide unbiased identification of DSBs enabled by sequencing)	Cell-based (live cells)	DSBs are tagged with double-stranded oligodeoxynucleotides (dsODNs), tag-specific amplification, high throughput sequencing	Simple, efficient, and precise, straightforward protocol, and availability of open-source software for data analysis	The efficient delivery dsODN may be potentially harmful to cells and it has not been tested in vivo
HGTST (high-throughput genome-wide translocation sequencing (HGTST)	Cell-based (live cells)	Generation of prey and bait DSB through the expression of two nucleases, use of a biotinylated primer against the bait DSB junction, linear amplification mediated PCR to recover translocations between prey and bait, streptavidin-based enrichment, high throughput sequencing	Requires the delivery of only the editing complex and can potentially be used in vivo	Nuclease-induced translocations are rare events and tend to occur between sites on the same chromosome
BLESS (breaks labeling, enrichment on streptavidin and next-generation sequencing)	Cell-based (fixed cells)	Isolation and fixation of treated cells, isolation and permeabilization of intact nuclei, in situ ligations of adapters to transient nuclease induced DSBs, enrichment, high throughput sequencing	It has been used in tissues where Cas9 had been delivered in vivo, independent from endogenous DNA repair machinery	Technically challenging, only allows to identify DSBs present at a specific moment, cannot identify DSBs before permeabilization
Digenome-Seq (digested genome sequencing)	In vitro	Extraction of genomic DNA from in vitro treated cells, followed by ligation of sequencing adapters, finally whole-genome sequencing	No limitations related to cell-based factors	Sequencing-inefficient, high background noise

(continued)

Method	Type	Description	Advantage	Limitation
CIRCLE-seq (circularization for in vitro reporting of cleavage effects by sequencing)	In vitro	Genomic DNA is sheared and circularized through intramolecular ligation. Linear DNA molecules, which escape degradation earlier, are degraded away now. Circular DNA molecules with cleavage sites are later on relinearized with the original CRISPR complex. This ultimately releases newly cleaved DNA ends for adaptor ligation, PCR amplification, and high throughput sequencing	No limitations related to cell-based factors	

9.3.3 More Ways to Use NGS in Genome Editing

NGS can be incorporated at various stages of a CRISPR genome editing workflow. It is useful for high-resolution on- and off-target assessment, functional verification, and evaluation of CRISPR knockouts and edits.

During the initial design phase, resequencing of a locus or genome (for species that lack a reference genome) is helpful in RNA selection. During the process of cloning, CRISPR-Cas9/ guide RNA constructs, resequencing of the resulting plasmids can lead to rapid and high-confidence verification of the CRISPR delivery vectors. It is helpful in particular for high-throughput experiments with large plasmid libraries.

In the following discussion, we will discuss about use of NGS into a CRISPR genome editing. Figure 9 explains the tentative work flow.

Genome editing is helpful for both gene knockout or gene knock-in or knockdown.

9.3.4 CRISPR-Cas9 Knockdown

The technique for obtaining knockdowns with CRISPR-Cas9 involves the use of proteins known as CRISPR-associated genes. These genes can be programmed to insert exogenous DNA fragments into a CRISPR repeat locus. This methodology leads to the production of small RNAs, which can serve as a template for Cas proteins to silence that same sequence in a cell. CRISPR is usually employed by many researchers for obtaining knockdowns in various mouse models. This is essential for the purpose of need for their genetic research. The main objective is usually to find the precise mechanism through which the silenced genes work, and the consequences when their expression is diminished or stopped.

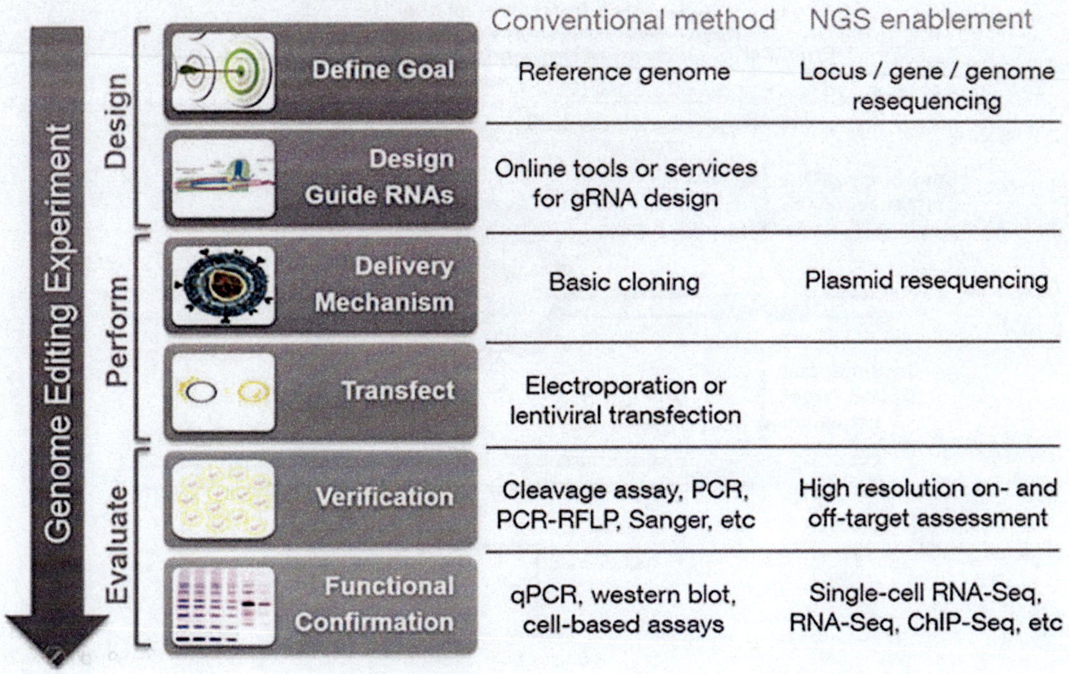

Fig. 9 Workflow for genome editing through CRISPR

9.3.5 Advantages of CRISPR-Cas9 Knock Down

Knockdown genes can be achieved either through genetic manipulation, through CRISPR, or with reagents such as RNA oligonucleotide or a short DNA. While comparing to other acute knockdown methods, CRISPR-Cas9-mediated knockdown is either more efficient and complete (e.g., RNA-interference based approaches) or faster (generation of conditional knockout mice and subsequent transfection with Cre), leading to the ideal candidate for the identification of neuronal function. Although not capable of disrupting gene function completely, hypomorphic knockdown may offer many advantages over knockouts. Gene knockdown via RNAi or CRISPRi is independent of the frameshift mutations or ploidy.

The partial gene knockdown technique is helpful for researchers to study the functional significance of certain necessary genes whose knockout would otherwise be cell-lethal. In addition, applications like drug target discovery would be advantageous from phenotypic hypomorphs, since an RNAi or CRISPRi knockdown rather than a knockout may mimic the effects of an inhibitive drug more closely. In certain cases, a full knockout may show an idealized loss-of-function phenotype for the most potent inhibitive drugs. However, gene knockdowns are reversible, a feature that has proven very useful in the past [1, 2].

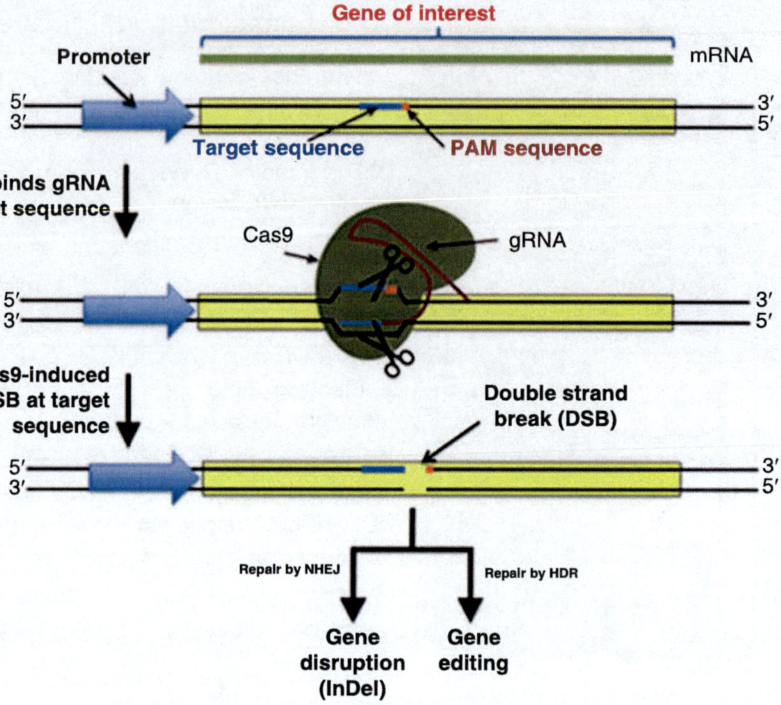

Fig. 10 Principle for gene editing with CRISPR Cas9 technology

9.3.6 CRISPR-Cas9

The principle of CRISPR-Cas9 has been explained in Fig. 10.

We list out certain abbreviations commonly used for gene editing.

DSB: Double-strand break, a break in both strands of DNA. There is Cut, two proximal, opposite strand nicks can be treated like a DSB.

NHEJ: It is the abbreviated form for nonhomologous end joining, a DNA repair mechanism that often introduces InDels.

HDR: Homology directed repair, a DNA repair mechanism that uses a template to repair nicks or DSBs.

Figure 11 represents the adaptive bacterial immunity based on CRISPR to protect against viral infection.

Cas: CRISPR associated protein, the Cas9 nuclease is the active enzyme employed for the Type II CRISPR system.

CRISPR: It refers to clustered regularly interspaced short palendromic repeat, which is actually a region in bacterial genomes used in pathogen defense.

crRNA: It refers to the endogenous bacterial RNA conferring target specificity, it needs tracrRNA to bind to Cas9.

gRNA: guide RNA, which is a fused product of the crRNA and tracrRNA. It provides both targeting specificity and

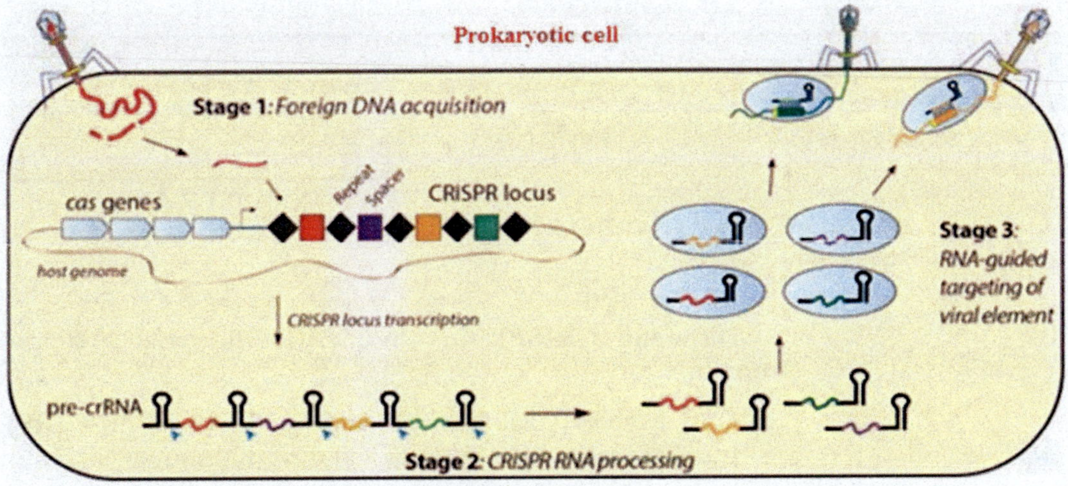

Fig. 11 Diagrammatic stage of CRISPR Cas9 technology

scaffolding/binding ability for Cas9 nuclease. It is to be kept in mind that it does not exist in nature.

PAM: It is the abbreviated form for protospacer adjacent motif, a required sequence that must immediately follow the gRNA recognition sequence but is NOT in the gRNA.

Target sequence: This refers to the 20 nucleotides, which are incorporated into the gRNA plus the PAM sequence. The target sequence is from the genomic DNA.

tracrRNA: The endogenous bacterial RNA which links the crRNA to the Cas9 nuclease, can bind any crRNA.

In the following section, we can summarize the CRISPRCas9 technology with suitable diagrammatic illustrations:

CRISPR-Cas9 System.

These are the parts of the bacterial immune system which detects and recognize the foreign DNA and cleaves it.

1. CRISPR loci: The abbreviation stands for clustered regularly interspaced short palindromic repeat loci (Fig. 12).
2. Cas (CRISPR associated protein): It can target and cleave DNA in a sequence-specific manner.

   A CRISPR array consists of a series of repeats interspaced by spacer sequences acquired from invading genomes.

Components of CRISPR:

1. Protospacer adjacent motif (PAM).
2. CRISPR RNA (crRNA).
3. Transactivating crRNA (tracrRNA).

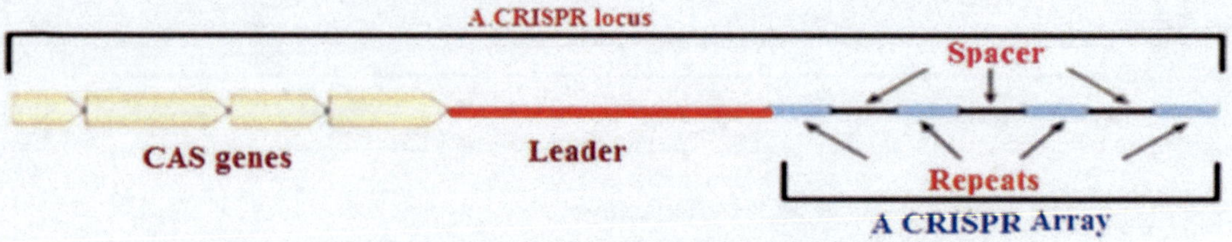

Fig. 12 A CRISPR locus

Different CRISPR-Cas System in Bacterial Adaptive Immunity

1. Class 1 type 1(CRISPR-Cas3) and type III (CRISPR-Cas10): It utilizes several Cas proteins and the crRNA.
2. Class 2 type II (CRISP-Cas9) and type V (CRISPR-Cpf1): It employs a large single component Cas9 protein in conjunction with crRNA and tracrRNA (Fig. 13).

Action of CRISPR in bacteria:

The CRISPR immune system acts to protect bacteria from repeated viral attack through three basic steps:

1. Adaptation.
2. Production of cr RNA.
3. Targeting.

We have attempted to summarize some basic aspects of CRISPR based on diagrammatic illustrations. These include the mechanism of action of CRISPR- Cas9 (Fig. 14). Structure of both crRNA (Fig. 15) and Cas 9 protein (Fig. 16). Figure 17 includes the diagrammatic representation of how Cas 9 complex is activated in response to guide RNA. Fig. 18 depicts step wise representation of the action of CRISPR Cas9. Schematic representation of the mechanism of action of CRISPR Cas9 (Fig. 19) and the Versatile nature of CRISPR technology (Fig. 20) has been described diagrammatically.

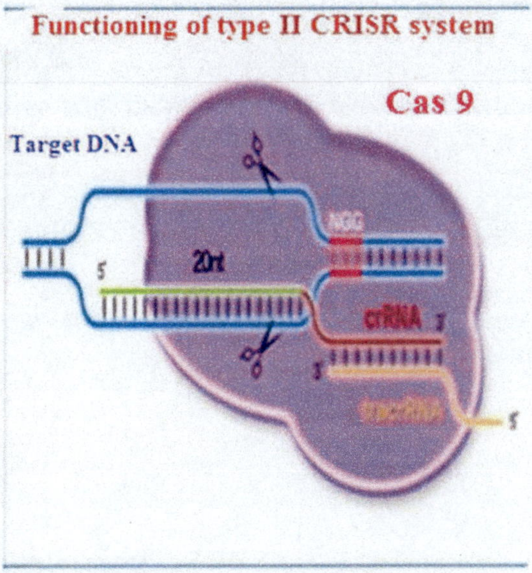

Fig. 13 Functioning of type II CRISPR system

Different Cas proteins and their function

Protein	Distribution	Process	Function
Cas1	Universal	Spacer acquisition	DNAse, not sequence specfic, can bind RNA; present in all Types
Cas2	Universal	Spacer acquisition	specific to U-rich regions; present in all Types
Cas3	Type I signature	Target interference	DNA helicase, endonuclease
Cas4	Type I, II	Spacer acquisition	RecB-like nuclease with exonuclease activity homologous to RecB
Cas5	Type I	crRNA expression	RAMP protein, endoribonuclease involved in crRNA biogenesis; part of CASCADE
Cas6	Type I, III	crRNA expression	RAMP protein, endoribonuclease involved in crRNA biogenesis; part of CASCADE
Cas7	Type I	crRNA expression	RAMP protein, endoribonuclease involved in crRNA biogenesis; part of CASCADE
Cas8	Type I	crRNA expression	Large protein with McrA/HNH-nuclease domain and RuvC-like nuclease; part of CASCADE
Cas9	Type II signature	Target interference	Large multidomain protein with McrA-HNH nuclease domain and RuvC-like nuclease domain; necessary for interference and target cleavage
Cas10	Type III signature	crRNA expression and interference	HD nuclease domain, palm domain, Zn ribbon; some homologies with CASCADE elements

12

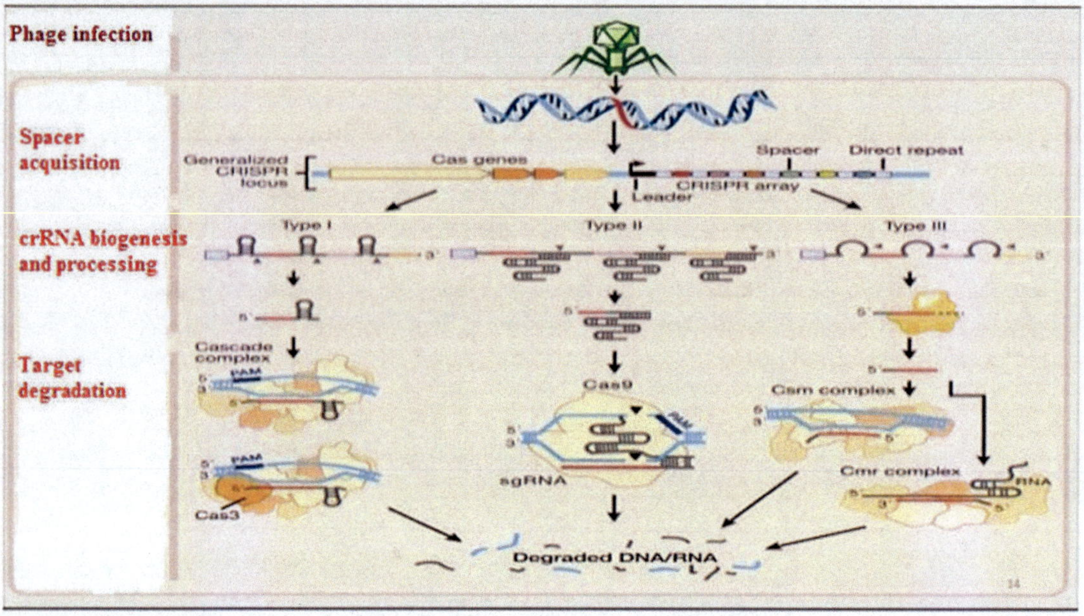

Fig. 14 Mechanism of action of CRISPR- Cas9

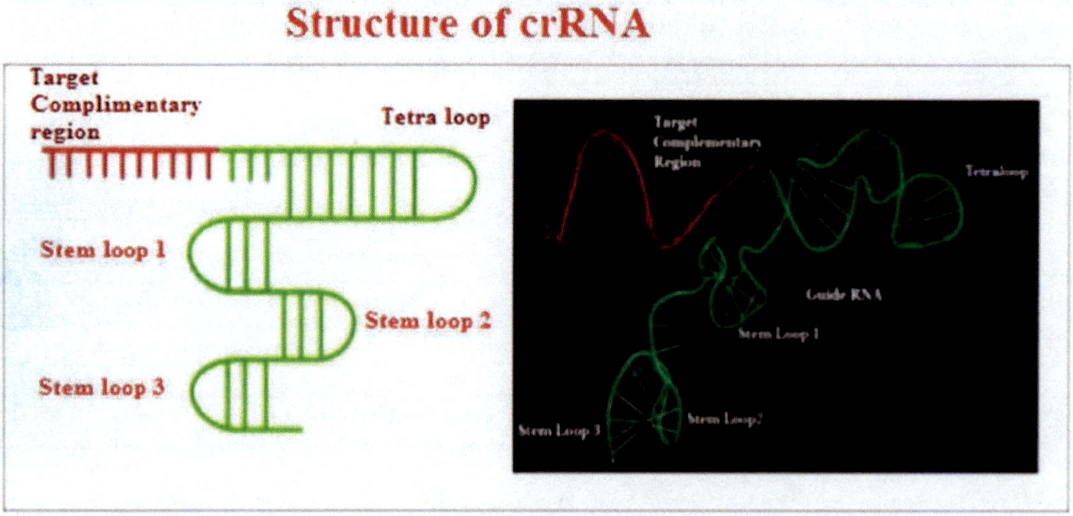

Fig. 15 Structure of crRNA

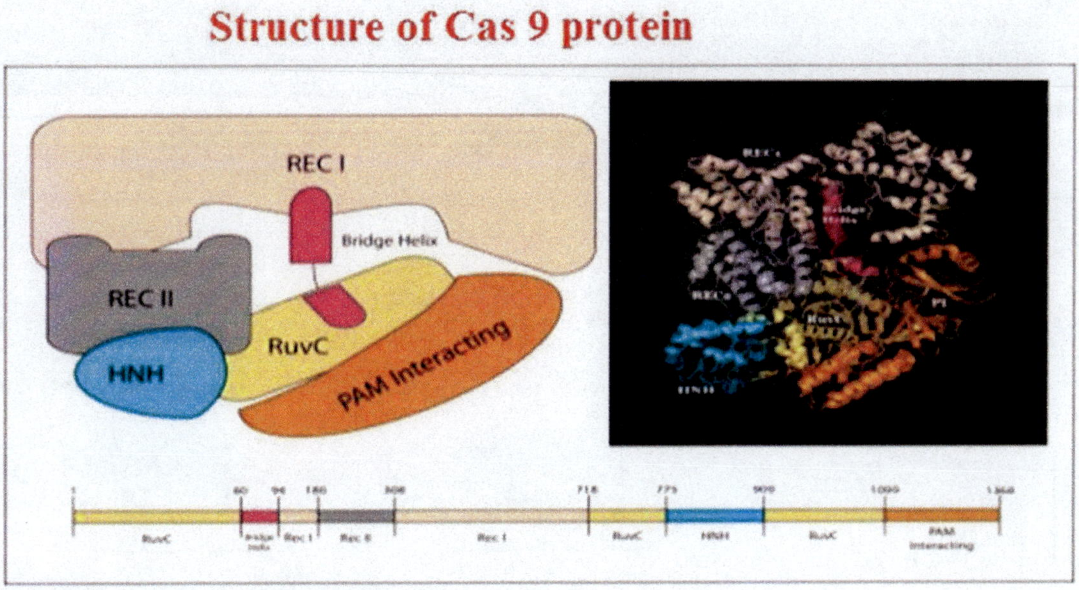

Fig. 16 Structure of Cas 9 protein

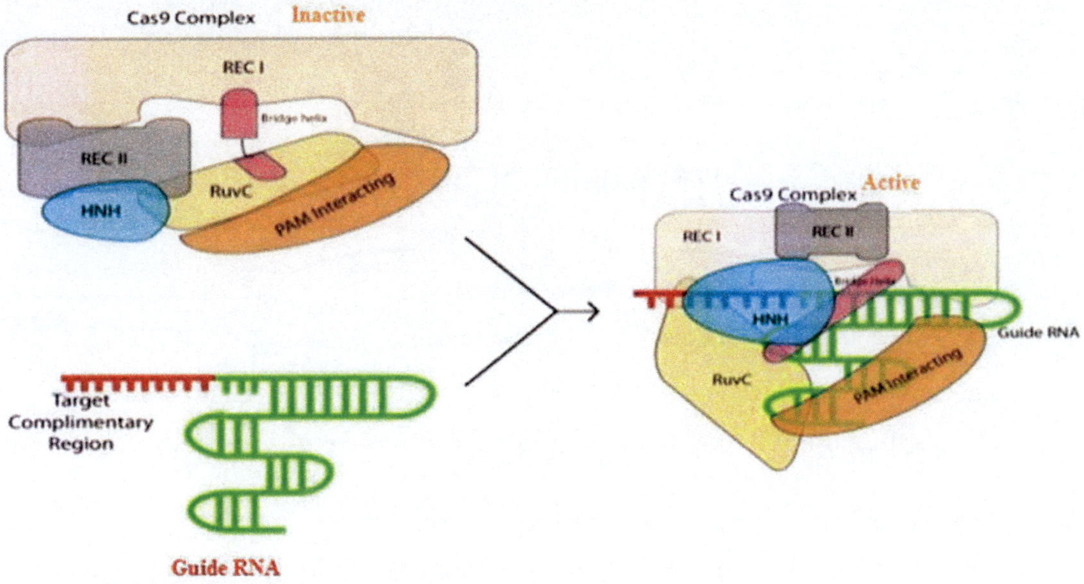

Fig. 17 Diagrammatic representation of how Cas 9 complex is activated in response to Guide RNA

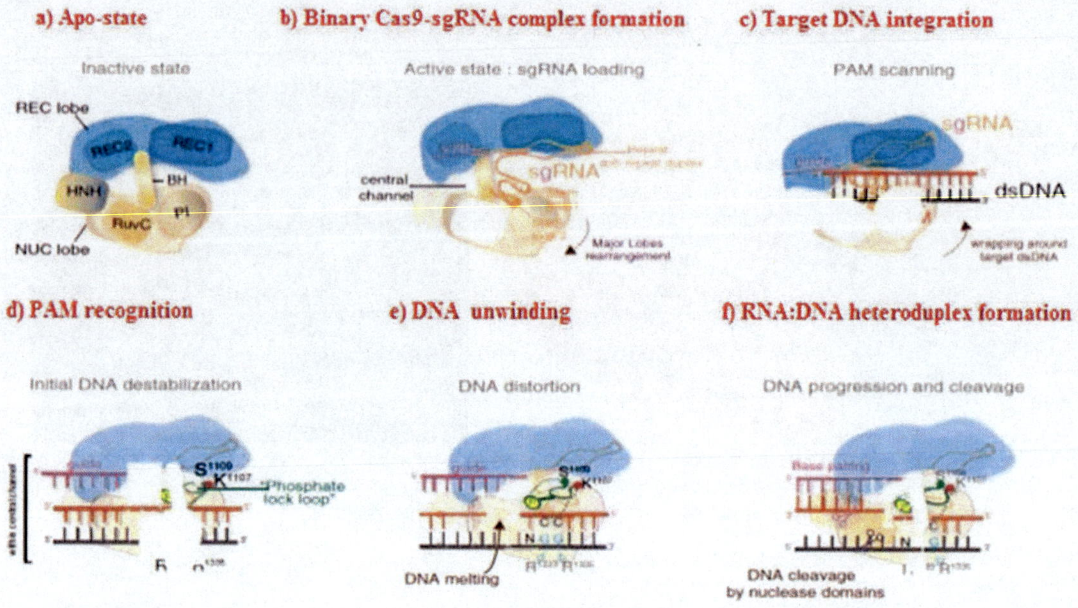

Fig. 18 Stepwise representation of action of CRISPR Cas9. (**a–f**) Represent the transition or progress for successive stages

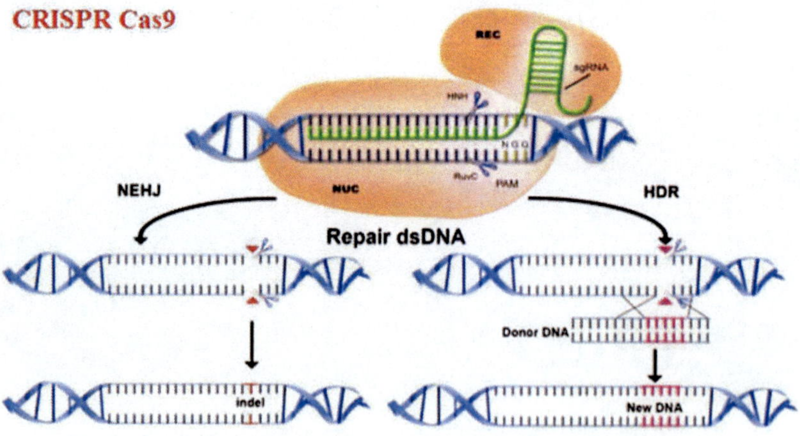

Fig. 19 Mechanism of action of CRISPR Cas9 (schematic representation)

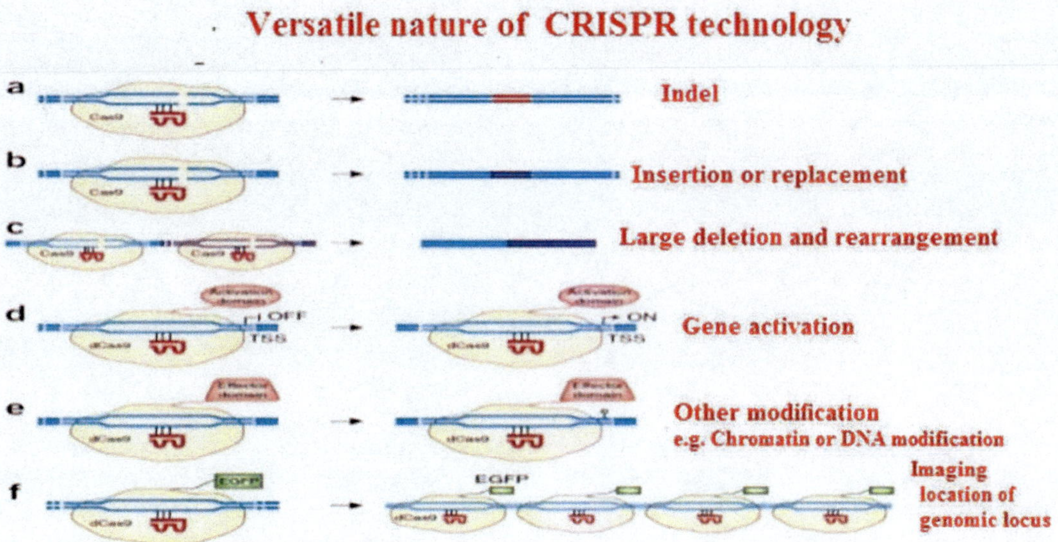

Fig. 20 Versatile nature of CRISPR technology. (**a–f**) Represent the successive stages of progress of the process

References

1. Dickinson DJ, Pani AM, Heppert JK, Higgins CD, Goldstein B (2015) Streamlined genome engineering with a self-excising drug selection cassette. Genetics 200(4):1035–1049. https://doi.org/10.1534/genetics.115.178335
2. Han R, Li L, Ugalde AP, Tal A, Manber Z et al (2018) Functional CRISPR screen identifies AP1- associated enhancer regulating FOXF1 to modulate oncogene-induced senescence. Genome Biol 19:118. https://doi.org/10.1186/s13059-018-1494-1
3. Zhang Y, Wang Y, Zuo Q, Li D, Zhang W, Wang F et al (2017) CRISPR/Cas9 mediated chicken Stra8 gene knockout and inhibition of male germ cell differentiation. PLoS One 12(2): e0172207. https://doi.org/10.1371/journal.pone.0172207
4. Sugano SS, Sawatsubashi S, Joko Y Protocol for CRISPR/Cas9-based knock-in using the VIKING method, 24 January 2019, PROTOCOL (Version 1) available at Protocol Exchange. https://protocolexchange.researchsquare.com/, https://doi.org/10.1038/protex.2019.010
5. Miura H, Quadros RM, Gurumurthy CB, Ohtsuka M (2018) Easi-CRISPR for creating knock-in and conditional knockout mouse models using long ssDNA donors. Nat Protoc 13 (1):195–215. https://doi.org/10.1038/nprot.2017.153

Chapter 25

Introduction to Stem Cell Biology

Abstract

Stem cell is a type of cell that has the capacity to continuously divide and differentiate (develop) into various other kind(s) of cells/tissues. The basic characteristics for stem cell are that these are known as unspecialized cells, capable of dividing, renewing, and proliferating; have the potentiality to give raise to particular cell types (differentiation). Three types of stem cells are totipotent, pluripotent, and multipotent. Molecular markers for stem cells are being discussed. Basic pluripotent stem cell culture protocols have been discussed in the current chapter. Other methods for neural stem cell and mesenchymal stem cells are described. Certain case studies are discussed with suitable examples.

Keywords Totipotent, Pluripotent, Multipotent, Molecular markers, Neural stem cell, Mesenchymal stem cells

1 Basic Principle

Stem cell: It is a type of cell that has the capacity to continuously divide and differentiate (develop) into various other kind(s) of cells/tissues.

2 History of Stem Cell Biology

1998—Stem cells were first extracted from human embryos for the first time.

1999—Insulin-making cells were successfully transplanted in human from cadavers.

2001—Federal funding for embryonic stem-cell research was restricted by President Bush.

2002—Juvenile Diabetes Research Foundation International contributed to $20 million fund-raising attempt to support stem-cell research.

2002—California stem cell research.

2004—Private funding was used by Harvard researchers to grow stem cells from embryos.

Aruna Pal, *Protocols in Advanced Genomics and Allied Techniques*, Springer Protocols Handbooks,
https://doi.org/10.1007/978-1-0716-1818-9_25,

There are three basic characteristics for stem cell:

1. These are blank cells or known as ***unspecialized cells.***
2. These cells are ***capable of dividing, renewing, and proliferating.***
3. These cells ***have the potentiality to give raise to particular cell types (differentiation).***

Other important points to be measured for stem cell research are as follows:

1. A cell is said to be ***self-renewable*** that has the ability to divide continuously.
2. A cell is said to be ***pleuripotent*** when it has the capacity to develop into several different types of cells or tissues.
3. ***Repair*** is referred to as the ***capacity to return purpose*** to damaged cells in the living organism.

Stem cells are of the following types:

1. Totipotent: Every cell can expand into a new individual. Example: Cells from early (1–3 days) embryos.
2. Pluripotent: Cells can develop to any (over 200) cell types. Example: Some cells of blastocyst (5–14 days).
3. Multipotent: Cells are differentiated, but can shape a lot of tissues. Example: Fatal tissue, cord blood, and adult stem cells.

These may be clearly described in tabular form as follows [1, 2]:

Stem cell type	Description	Examples
Totipotent	Each cell has the ability develop into a new Individual	Cells from early (1–3 days) embryos
Pluripotent	Cells has the ability to form any (over 200) cell types	Some cells of blastocyst (5–14 days)
Multipotent	Cells differentiated, but can form a lot of tissues	Fetal tissue, cord blood, and adult stem cells

3 Molecular Marker for Stem Cell

Classical and new candidate molecular markers which have been described to demonstrate a molecular profile for human embryonic stem cells (hESCs), mesenchymal stem cells (MSCs), and hematopoietic stem cells (HSCs) are being discussed. Usually cited markers for embryonic ESCs are Nanog, Oct-4, Sox-2, Rex-1, Dnmt3b, Lin-28, Tdgfl, FoxD3, Tert, Utf-1, Gal, Cx43, Gdf3, Gtcml,

Terfl, Terf2, Lefty A, and Lefty B. MSCs are primarily recognized by the expression of CD13, CD29, CD44, CD49e, CD54, CD71, CD73, CD90, CD105, CD106, CD166, and HLA-ABC and lack CD14, CD31, CD34, CD45, CD62E, CD62L, CD62P, and HLA-DR expression. HSCs are mainly isolated based on the look of CD34, but the marker with CD133 and CD90, jointly with a lack of CD38 and other lineage markers, gives rise to the most homogeneous pool of SCs.

4 Basic Pluripotent Stem Cell Culture Protocol

4.1 Outline of the Protocol

1. Stem cell culture considerations:
   (a) Successful stem cell culture.
   (b) Quality control of cell cultures.
2. Reagent preparation
   (a) Preparation of 0.1% gelatin solution.
   (b) Preparation of iMEF culture medium.
   (c) Preparation of basic fibroblast growth factor (b-FGF) stock solution
   (d) Preparation of 0.1% bovine serum albumin (BSA) solution.
   (e) Preparation of 10 μg/mL b-FGF stock solution.
   (f) Aliquots of b-FGF (10 μg/mL) stock solution.
   (g) Preparation of pluripotent stem cell culture medium.
   (h) Preparation of culture medium (CM) preparation.
   (i) Preparation of collagenase solution.
3. Cell culture.
   (a) Gelatin coating of culture plates.
   (b) Coat culture plates using 0.1% gelatin solution.
   (c) Thawing and seeding of frozen inactivated mouse embryonic fibroblasts (iMEFs).
   (d) Prepare gelatin plate(s) in the biosafety cabinet.
   (e) Thaw iMEFs, Seed iMEFs in 6-well plate, Post seeding.
   (f) Thawing and seeding, and replacement of medium for pluripotent stem cells.
   (g) Monitor iMEFs under microscope (Fig. 1).
   (h) Thaw pluripotent stem cells.
   (i) Seed pluripotent stem cells in 6-well plates.
   (j) Replacement of pluripotent stem cell culture medium.
   (k) Monitor hPSC growth. Change medium for hPSC culture.

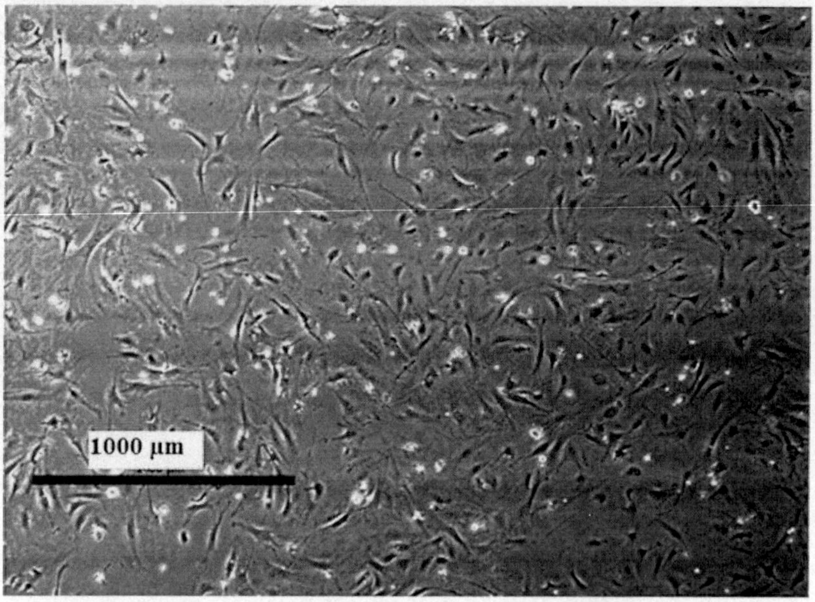

Fig. 1 Microscopic view of MEF cells at meeting of 0.1–0.13 × 10^6 iMEFs/mL. Bar represents 1000 µm

(l) Passaging of pluripotent stem cells on fresh iMEF plates.

(m) To assess how to passage (or split) pluripotent stem cells.

(n) Calculation of medium and solution needed.

(o) Evaluate & prepare fresh iMEF plates.

(p) Procedure-pick to discard-enzyme Harvest cells (no more than two plates at a time).

(q) Spin and disrupt colonies.

(r) Added cells to iMEF plates (processing no more than three plates at a time).

(s) Procedure-pick to keep sterile bent Pasteur glass.

(t) Transferred areas of undifferentiated cells to new iMEF plates (Pick to keep).

(u) Harvesting and cryopreservation of pluripotent stem cells.
- Determine when to harvest cells.
- Prepared cryovials.
- Calculation of the amount of culture medium and collagenase solution required for this process.
- Preparation of 2× cryopreservation media (or freezing medium, FM) in the biosafety cabinet.
- Incubation of cells with collagenase solution (prepare no more than two plates at a time):

- Harvesting of cells.
- Freezing of cells.

Stem cell research is a quickly growing field with the possibility for the development of therapeutic agents for the treatment of diseases as well as analyzing disease development from early stages. The culture of human pluripotent stem cells shares a lot of the similar protocols as benchmark for mammalian cell culture. However, the successful culture and maintenance of human pluripotent stem cells (hPSCs) in an undifferentiated state needs extra considerations to ensure that cells maintain their key type of self-renewal and pluripotency. The protocols in this document symbolize a subset of the standard operating procedures employed to maintain and culture stem cells at the Massachusetts Human Stem Cell Bank, and have been carefully tested subject to verification [3–5].

5 A Stem Cell Culture Considerations

Stem cell research is a quickly increasing field with the hidden for the development of therapeutic agents for the treatment of diseases as well as study disease development from early stages. However, to fulfill this promise, researchers require to have access to standardized protocols for the development, upholding, and differentiation of these unique cells. Such "best practices" will authorize comparisons of different studies and enhance the refinement of these techniques. Such stability can be driven by capital such as StemBook and by stem cell banks. Stem cell banks also serve as expensive capital of properly identified, quality inhibited and characterized cell lines along with the help to find the way the legal and IP issues which are common to working with stem cells.

The protocols in this text stand for a subset of the benchmark operating events employed to maintain and culture stem cells at the Massachusetts Human Stem Cell Bank, and have been carefully tested and verified.

5.1 A1 Successful Stem Cell Culture

The culture of human pluripotent stem cells resembles a lot of the same protocols as standard mammalian cell culture. Moreover, the successful culture and maintaining human pluripotent stem cells (hPSCs) in an undifferentiated state need extra considerations to guarantee that cells retain their key characteristics of self-renewal and pluripotency.

Successful hPSC culture needs the recreation of the in vivo stem cell microenvironment, or "niche," that includes growth factors, cell-to-cell communications, and cell-to-matrix adhesions. Unlike numerous cell types, hPSCs are grown in aggregates, or colonies, helping them to make this niche. Standard culture of hPSCs includes the experience to media enriched with growth factors found in fetal bovine serum (FBS) or defined serum replacements. In addition, standard hPSC culture systems used to hold cells such as an inactivated mouse embryonic fibroblast (MEF) feeder layer to maintain growth and avoid differentiation. These cells give necessary intercellular communications, extracellular scaffolding, and factors making a robust and stable hPSC culture environment.

There are a few fundamental protocols essential for the culturing of mammalian cells, including thawing frozen stocks, plating cells in culture vessels, changing media, passaging, and cryopreservation. Passaging means the barring of cells from their current culture vessel and culturing them to one or more new culture vessels. Passaging is essential to decrease the harmful effects of overcrowding and for expansion of the culture. Protocols helped in this manual explain the standard culture of hPSCs. These events have been shown to be dependable and make reproducible experimental data.

Although the actual culture process described here use a diversity of animal products, any clinical use of hPSCs will need removal of these products as they pose a risk of experience to retroviruses and other pathogens from the culture environment.

5.2 A2 Quality Control of Cell Cultures

It is fundamental that researchers confirm the sterility, authenticity, and genetic stability of cell lines utilized in their job in order to publish and provide reproducible and useful experimental data. After receiving a new cell line, it is highly optional that cells assayed for the criteria outlined below and monitored at normal intervals in order to confirm these characteristics.

5.2.1 Cell Line Sterility

Cells in continuous culture are normally weak to microbial contamination. Bacterial and fungal contamination can reason cell death and eventual loss of complete cultures. Human stem cell cultures are mainly vulnerable as they are commonly cultured in enriched media without antibiotics. It is greatly advisable that cells received from any provider need to be tested for microbial contamination before early use and at usual intervals during routine culture in the laboratory. Due to abundance of microbes in the atmosphere of the culture environment, aseptic technique needs to be practiced to prevent culture contamination.

In addition to bacteria and fungi, another general contaminant is mycoplasma. These intracellular microorganisms are usually lesser than bacteria and may influence cell growth, normally at high levels of contamination. However, stable infections of cells can lead to genetic and phenotypic changes. Frequent sources of Mycoplasma involve contaminated materials of animal origin such as serum, trypsin, and primary feeder cell cultures. Thus, it is vital to check for the presence of mycoplasma on a normal basis and remove contaminated cultures. Common process employed for the detection of mycoplasma includes enzymatic assays, polymerase chain reaction (PCR), culture in selective media, and DNA staining of test cells to observe mycoplasma, which grow in close organization with the cell membrane. Although the least sensitive, DNA staining with 4′6-Diamidino-2-phenylindole (DAPI), a fluorescent stain which binds strongly to DNA, is a simple process to be employed in most laboratories to identify mycoplasma. Commercially existing kits are existing by a number of reagent companies.

Viruses are an added form of contamination. Viruses can change a cell line's characteristics to varying degrees through multiple activities. These comprise use of host cellular resources for viral replication and integration into the host genome. Viral infection may affect experimental data and could result in false interpretations. In addition, cultures which are contaminated with blood borne viruses capable of human infection pose a serious health risk. Common sources of viral contamination comprise animal products and preparations, for example, bovine serum, antibodies, and mouse embryonic fibroblasts. Although testing for viral contamination is performed by stem cell banks and repositories on a regular basis, this testing is not common practice for individual research laboratories. It is optional that laboratories test cells for viral infection prior to their distribution and that recipient laboratories request documentation of this testing. Several companies propose a wide range of viral testing services; costing depends on the breadth and types of tests performed.

5.2.2 Cell Line Authenticity

Cells in culture are at high danger for cross-contamination because of the fact that most laboratories routinely culture multiple cell lines simultaneously. Studies have shown that up to 30% of cell lines donated to public repositories are contaminated by quickly growing cell types as HeLa cells, which can outgrow and replace the original cell lines. This cross-contamination may be more, leading to false data and misleading conclusions. This is of particular interest in regulated environments when authenticity or identity is the prime component of quality controls systems for clinical applications.

Several approaches were used to confirm cell lines. Recent DNA typing uses PCR-based techniques to analyze similar hypervariable satellite DNA sequences and single nucleotide polymorphisms. Multiple companies propose genotyping services for a small fee.

5.2.3 Cell Line Stability

Cells developed in culture usually have accumulated genetic and/or phenotypic changes. Studies have revealed that long-term culture of hPSCs results in genetic abnormalities including changes at the chromosomal level that can be identified through karyotyping. The karyotypes of human pluripotent stem cell cultures need to be regularly tested by a certified cytogenetics laboratory.

Several techniques are used to make sure that pluripotent stem cell lines retain their stem cell phenotype, and include expression of stem cell markers, both molecularly and as expressed in the intracellular level. It is highly suggested that several methods be used to monitor cell line stability. Teratoma formation, embryoid body formation, RT-PCR, immunocytochemistry staining, and flow cytometry are some of the more common process employed. While many of these methods are common laboratory techniques, it is shrewd to consider periodic secondary testing through an independent laboratory or core facility, in order to confirm.

As previously stated, the protocols accessible here have been thoroughly tested and verified. However, as stem cell research field works to evolve, new practices and processes are emerging like, while feeder-free culture was once a new technology, nowadays many standardized feeder-free protocols have been observed. As new techniques arise for more competent and effective hPSC culture, series of evaluation and adoption after rigorous testing and standardization are necessary [6–10].

6 B Reagent Preparation

Several reagents must be organized prior to maintaining human pluripotent stem cells (hPSCs) in culture. It is vital to have a firm grasp of the characteristics of each reagent, including components, appropriate preparation, storage, and shelf life. This chapter deals with the preparation of the following reagents necessary for hPSC culture:

- *0.1% Gelatin Solution*
- *Inactivated MEF Medium.*
- *Basic Fibroblast Growth Factor (bFGF) Solution.*
- *hPSC Pluripotent Culture Medium,*
- *Collagenase Solution.*

Many reagents, especially Knockout Serum Replacer (KOSR), bFGF, and collagenase, can have major variation between lots. When working with a new lot, it is vital to compare it directly to the lot that is now in use, to ensure that the current culture quality and viability are maintained.

6.1 B1 Preparation of 0.1% Gelatin Solution

6.1.1 Supplies

- 500 mL sterile glass bottle
- Weigh boat.

6.1.2 Reagents

- Gelatin, Porcine.
- Endotoxin-free.

6.1.3 Procedure

Prepare 0.1% Gelatin Solution

Note: Adjust the volume of water and gelatin powder proportionally if a volume other than 500 mL of gelatin is made.

- Addition of 0.5 g of gelatin powder to a clean 500 mL pyrex bottle.
- Addition of 500 mL endotoxin-free (e.g., MilliQ) water to the bottle.
- Swirl to mix. (At this stage, the gelatin is not soluble.)
- Autoclave for 30 min within 2 h after mixing.
- Cool down the 0.1% gelatin solution to normal temperature and store at 4–8 °C until use. Use this solution within 2 month of preparation.

6.2 B2 Preparation of iMEF Culture Medium

6.2.1 Supplies

- 5 mL sterile serological pipettes.
- 10 mL sterile serological pipettes.
- 25 mL sterile serological pipettes.
- 500 mL bottle connected with a 0. 22 μm Stericup.

6.2.2 Reagents

- DMEM-liquid.
- MEM Non-Essential Amino Acid Solution (NEAA).
- Heat Inactivated Fetal Bovine Serum (HI- FBS).
- 70% ethyl alcohol (Diluted from 95% ethanol).

6.2.3 Procedure

Preparation of iMEF Culture Media

- In the hood, open a 500 mL filter and bottle unit and mark the bottle:
  - iMEF and date of preparation,
  - Finishing date (mm/dd/yyy, 14 days after media preparation). For example: Exp: 07/10/2008.
  - Your initials.

- Add the suitable amount of ingredients to the 500 mL filter cup as exposed in the table below:

Ingredient	100 ml total	250 ml total	500 ml total
DMEM	90 ml	225 ml	445 ml
Heat inactivated FBS	10 ml	25 ml	50 ml
NEAA	1 ml	2.5 ml	5 ml

Note 1: The DMEM may be calculated by pouring directly into the graduated filter cup of the filter/bottle unit.

Note 2: Scale up or down proportionally if one more quantity of medium is required.

- Filter the media using the Stericup filter into the attached bottle.
- Store the medium bottle at 4 °C, and utilize the medium within 14 days.

6.3 B3 preparation of Basic Fibroblast Growth Factor (B-FGF) Stock Solution

6.3.1 Supplies

- Sterile microcentrifuge tubes.
- 1000 μL sterile pipette tips
- 10 mL sterile serological pipettes.
- 25 mL sterile serological pipettes.
- 50 mL sterile centrifuge tubes.
- 70% ethanol spray.
- Forceps.

6.3.2 Reagents

- Basic Fibroblast Growth Factor (b-FGF).
- PBS with 0.01% $CaCl_2$ and 0.01% $MgCl_2$.
- 30% BSA (Bovine Serum Albumin).

6.3.3 Procedure

Making of 0.1% Bovine Serum Albumin (BSA) Solution

- 0.1 mL of the 30% BSA is added to 30 mL of PBS.
- Close the tube and mix the solution by inverting 3–4 times. It is now a 0.1% BSA solution.

Preparation of 10 Ug/mL b-FGF Stock Solution

- Place one vial of 100 μg lyophilized b-FGF in a 50 mL tube and temporarily centrifuge the b-FGF vial at 200 × *g* for 5 min for bringing the lyophilized b-FGF to the base of the vial.
- A pair of forceps will be useful for removal of the vial from the 50 mL tube with greatest care. In a biosafety cabinet, cautiously remove the cap gently. Care should be taken to prevent contact with any b-FGF powder that has stuck to the cap.
- Add 500 μL of 0.1% BSA solution to the b-FGF vial. (Do not touch FGF pellet at the bottom.) Carefully replace the cap. Carefully remove the cap and gently pipette up and down a

few times. Transfer the b-FGF solution to the 50 mL tube. Rinse the b-FGF vial with 500 μL solution from the "FGF" tube to collect any residual b-FGF protein in the vial. Repeat this 2–3 times.

Aliquots of b-FGF (10 μg/mL) Stock Solution

In the hood, keep a set of 50 sterile 1.5 mL or 2 mL vials in a cryo-vial holder.

- Aliquot the solution.
- Store at −80 °C.

6.4 B4 Making of Pluripotent Stem Cell Culture Medium

6.4.1 Supplies

- 1000 μL sterile pipette tips.
- 20 μL sterile pipette tips.
- 5 mL sterile serological pipettes.
- 10 mL sterile serological pipettes.
- 25 mL sterile serological pipettes.
- 50 mL sterile centrifuge tubes.
- 250 mL Stericup Filter.
- 500 mL Stericup Filter.
- 70% ethanol spray.

6.4.2 Reagents

- DMEM-F12 media (Invitrogen 11330-032).
- Knockout Serum Replacer (KOSR) (Invitrogen 10828-028).
- L-glutamine, non-animal, cell culture tested (Sigma G-8540).
- MEM Non-Essential Amino acid solution (Invitrogen 11140-050).
- (b-FGF); Stock 10 μg/mL.
- beta-Mercaptoethanol (Sigma M7522).

6.4.3 Procedure

Preparation of Culture Medium (CM) Preparation

Note: Please note that a lot of iPS culture media formulas suggest the inclusion of antibiotics. When working with any cell line, check with the cell line provider for specific media instructions.

Note: If the final total volume of medium is dissimilar from the one listed in the table below, adjust the volume of ingredients proportionally.

- In a biosafety cabinet, opened a 250 mL or 500 mL filter and bottle unit and labeled the bottle of the filter unit with the Culture Medium and date of preparation, expiration date (14 days after media preparation), initials.

- According to the final total volume to be made, added appropriate amounts of ingredients to the 250 mL or 500 mL filter unit cup as shown in the table below.

Note: The DMEM-F12 may be calculated by pouring directly into the graduated filter cup. Other reagents should be additional with sterile serological pipettes.

Note: When thawing cell lines which have been previously cultured under dissimilar conditions, in majority of the cases, it is possible to adjust the cells to the culture medium explained here through the process of thawing of the cells into this formulation. Thawing hPSCs is completely described in the methods; *C3 Thawing and Seeding, and replacement of medium for Pluripotent Stem Cells.*

		hESC Media		iPSC Media	
Ingredient	**Stock concentration**	**Final concentration**	**100 ml**	**Final concentration**	**100 ml**
DMEM-F12	100%	NA	80 ml	NA	80 ml
K/O serum replacer	100%	20%	20 ml	20%	20 ml
100 mM L-Glutamine	200 mM	2 mM	1 ml	2 mM	1 ml
Nonessential amino acids	10 mM	0.1 mM	1 ml	0.1 mM	1 ml
beta-mercaptoethanol	14.3 M	0.1 mM	0.7 μl	0.1 mM	0.7 μl
b-FGF*	10 μg/ml	4 ng/ml	0.04 ml	10 ng/ml	0.1 ml

- Filter the media through the 0.22 μM filter.
- Keep the medium bottle at 2–8 °C, and the medium to be utilized within 14 days.

6.5 B5 Preparation of Collagenase Solution

6.5.1 Supplies

- Weigh boat.
- 25 mL sterile serological pipettes.
- 150 mL Stericup filter.
- 70% ethanol spray.

6.5.2 Reagents

- DMEM-F12 media.
- Collagenase Type IV.

6.5.3 Procedure

Note: The following procedure is to arrange a collagenase solution at a concentration of 1 mg/mL.

- Weight 100 mg of Collagenase Type IV powder.
- Using a 25 mL pipette, transport 20 mL of DMEM/F12 medium to the weigh boat containing the collagenase.

- Pipette DMEM/F12 medium up and down in the boat to suspend the collagenase powder. The collagenase should suspend almost instantly.
- Transfer the solution to a 50 mL tube after dissolving of the collagenase.
- Rinse the residual collagenase in the boat with an extra 20 mL of DMEM/F12 medium and add this to the 50 mL tube containing the collagenase solution, tighten cap.
- Spray the 50 mL tube containing the collagenase solution with 70% ethanol, and keep the tube in the biosafety cabinet.
- Pour 60 mL of DMEM/F12 medium to a 150 mL 0.22 μm filter unit. Add 40 mL of collagenase solution from the 50 mL tube.
- Filter sterilize the collagenase solution (1 mg/mL).
- Place the solution at 4 °C, and use within 14 days.

7 C Cell Culture

The protocols provided here explain the standard culture techniques employed for pluripotent stem cells.

7.1 C1 Gelatin Coating of Culture Plates

7.1.1 Supplies

- 6-well tissue culture plates.
- 5 mL sterile serological pipettes.
- 10 mL sterile serological pipettes.

7.1.2 Reagents

- 0.1% gelatin solution.

7.1.3 Procedure

Warm to room temperature a suitable amount of gelatin solution. For 6-well plates, apply 2 mL of 0.1% gelatin solution for each well.

Coat Culture Plates Using 0.1% Gelatin Solution

- Put the plates which are to be coated in the biosafety cabinet.
- Tag the cover of the plate (not over the wells) with: "G" for gelatin, date, initials.
- Added 2 mL of 0.1% gelatin solution to every well.
- Tilted or swirled the plates in several directions to enable the liquid covering the total surface area.
- Placed the plates in a 37 °C incubator.

Note: The plates will be ready for use in 4 hours and can be utilized for up to 7 days.

7.2 C2 Thawing and Seeding of Freezing Inactivated Mouse Embryonic Fibroblasts (iMEFs)

7.2.1 Supplies

- 5 mL sterile serological pipettes.
- 10 mL sterile serological pipettes.
- 25 mL sterile serological pipettes.
- 15 mL sterile centrifuge tubes.
- 50 mL sterile centrifuge tubes.
- 70% ethanol spray.

7.2.2 Reagents

- iMEF culture medium.
- Gelatin-coated plates.
- Frozen iMEFs.

7.2.3 Procedure

Prepare Gelatin Plate(S) in the Biosafety Cabinet

- Eliminate gelatin plate(s) from the incubator and place inside the biosafety cabinet.
- Aspirate the gelatin solution from the plate(s) entirely with a Pasteur pipette.
- Keep the plate(s) to the incubator for later use.

Thaw iMEFs

Note 1: To seed 6 wells in a 6-well plate, 1.2–1.5 × 10^6 iMEFs are necessary (assuming a 90% recovery of iMEFs from one freezing and thawing cycle).

Note 2: Thaw no more than two vials at a time.

Note 3: Use eye protective safety glasses and insulated gloves when removing cryovials from the liquid nitrogen freezer.

- Bring a suitable amount of MEF medium (iMEF CM) to 37 °C (assuming 2 mL per well +9 mL for thawing).
- Place the warmed iMEF CM in the biosafety cabinet after carefully cleaning the outside of the container with 70% ethanol.
- Tag a sterile tube (both on the cap and side) in the biosafety cabinet as "iMEF."
- Transport 9 mL iMEF CM into the "iMEF" tube.
- Unwrap the LN_2 freezer and take out the exact vial(s) of iMEFs.
- Warm the vial(s) slightly between gloved hands as walking to the water bath.
- In the 37 °C water bath, immerse the vial in the water without submerging the cap. After about 30 s, check often to see if the frozen solution is beginning to melt. This procedure may take up to 1 min.
- When there is just a small piece of ice floating in the vial, eliminate the cryovials from the water bath.
- Put the vial(s) in the biosafety cabinet after thoroughly cleaning the outside of the cryovials with 70% ethanol.

- In the biosafety cabinet, pipette the cells up and down softly a couple times, and then transfer the cell suspension dropwise to the tube labeled "iMEF" while gently swirling the tube. This will help to decrease osmotic shock to the cells.
- Centrifuge it for 5 min at 200 × *g*.
- Return the 50 mL centrifuge tube to the biosafety cabinet after spreading it with 70% ethanol.
- Be careful while aspirate the supernatant from the tube as not to touch the cell pellet.
- Softly flick the bottom of the tube to loosen the cell pellet.
- Add 10 mL of iMEF CM to resuspend and mix the cells by pipetting softly up and down a few times.
- If necessary, count iMEFs.
- Add iMEF CM to the iMEF tube for a last cell concentration of 0.1–0.13 × 10^6 iMEFs/mL. Image shows right concentration of MEFs (Fig. 1).

Seed iMEFs in 6-Well Plate

- Take the gelatin plate(s) from the incubator to the biosafety cabinet and label the plates with "iMEF," date, and initials.
- Resuspend the iMEFs totally by gently pipetting up and down a few times and add 2 mL/well to gelatin coated wells.
- When all plates in the biosafety cabinet are seeded, slide the plates back and forth and side to side (cross-motion) three to five times within the biosafety cabinet to constantly distribute cells and make another 3–5 cross-motions with the plates when transferred to the incubator.

Post Seeding

- Do not upset the freshly seeded plates for at least the next 12 h.
- Check each plate of cells under the microscope the next day for MEF quality and for signs of contamination.
- Record observations.
- Repeat the above each day until the plates are utilized for pluripotent stem cell culture.

Note: Remove unused or contaminated iMEF plates after 5 days.

7.3 C3 Thawing and Seeding, and Substitute of Medium for Pluripotent Stem Cells

7.3.1 Supplies

- 5 mL sterile serological pipettes.
- 10 mL sterile serological pipettes.
- 25 mL sterile serological pipettes.
- 15 mL sterile centrifuge tubes.
- 50 mL sterile centrifuge tubes.
- 70% ethanol spray.

7.3.2 Reagents

- DMEM-F12.
- hES Cell Culture Medium.
- iMEF plate containing iMEFs.

Monitor iMEFs under Microscope

- Observe the number of plates essential for this protocol one under a microscope.
- Return the plates to the incubator if there is no contamination.

Thaw Pluripotent Stem Cells

Note 1: Use eye protective safety glasses and insulated gloves when removing cryovials from the liquid nitrogen freezer.

Note 2: Never thaw cells simultaneously from diverse stem cell lines. If they are from the same line and same lot, thaw no more than two vials at a time in order to reduce prolonged exposure of thawed cells to cryopreservation medium.

- Bring a suitable amount of DMEM-F12 and CM to room temperature. Keep it in the biosafety cabinet after cleaning with 70% ethanol.
- Place the hot DMEM-F12 and CM aliquots in the biosafety cabinet after carefully cleaning with 70% ethanol.
- Mark a sterile tube (both on the cap and side) in the biosafety cabinet as “cells.”
- Move 9 mL CM per cryovial of hPSCs into the “cells” tube.
- Eliminate the cell vial(s) from the LN_2 freezer and verify the label on the cryovial(s).
- Heating the vial(s) slightly between gloved hands while walking to the water bath.
- In the 37 °C water bath, immerse the vial(s) in the water without submerging the cap. Swirl the vial gently.
- After 30 s, check regularly to see whether frozen medium is beginning to melt. This process may take up to 1 min.
- When there is only a small piece of ice floating in the cryovial(s), eliminate the vials(s) from the water bath.
- Place the vial(s) in the cryovial holder (if using) after systematically cleaning with 70% ethanol.

- Pipette the cells lightly up and down a few times and transfer the cell suspension drop-wise to the CM in the labeled "cell" tube while gently swirling the medium to reduce osmotic shock to the cells.
- Centrifuge for 5 min at 200 × *g*. And take the standard iMEF plates from the incubator and keep it in the biosafety cabinet:
- Mark the plate(s) in the biosafety cabinet (the date, the words "thaw" and the cell line ID, the passage number plus one, and initials).

After completion of centrifugation is complete, the centrifuge tube is returned to the biosafety cabinet after thorough cleaning with 70% ethanol.

- Remove the supernatant from the tube (be careful not to touch the cell pellet).
- Add the 2.5 mL of CM to the tube and pipette cells softly a several times to mix.

Seed Pluripotent Stem Cells in 6-Well Plates

Note: To avoid mislabeling cells and cross-contamination, seed cells from diverse lines in different iMEF plates.

- Remove the iMEF medium completely from the iMEF plates with a Pasteur pipette.
- Wash and aspirate the wells with 2 mL/well of DMEM-F12.
- Mix the cells in the tube totally by pipetting softly up and down a few times, and dropwise add cells at 1 vial per well.
- When all plates are seeded, slide the plates in back and forth and side to side (cross-motion) three to five times inside the biosafety cabinet.

Post Seeding

- Do not shift the freshly seeded plates for at least the next 12 h.

Note: When thawing cell lines that have been previously cultured under different conditions, in most cases, it is possible to adjust the cells to the culture medium described in these protocols by directly thawing the cells into it. The complete formulation for culture medium is helped in the protocol.

B4 Preparation of Pluripotent Stem Cell Culture Medium.

Figure 2 depicts the microscopic view of hPSCs at 2 days post thaw, while microscopic view of healthy and differentiated colonies of hPSCs is being represented in Fig. 3.

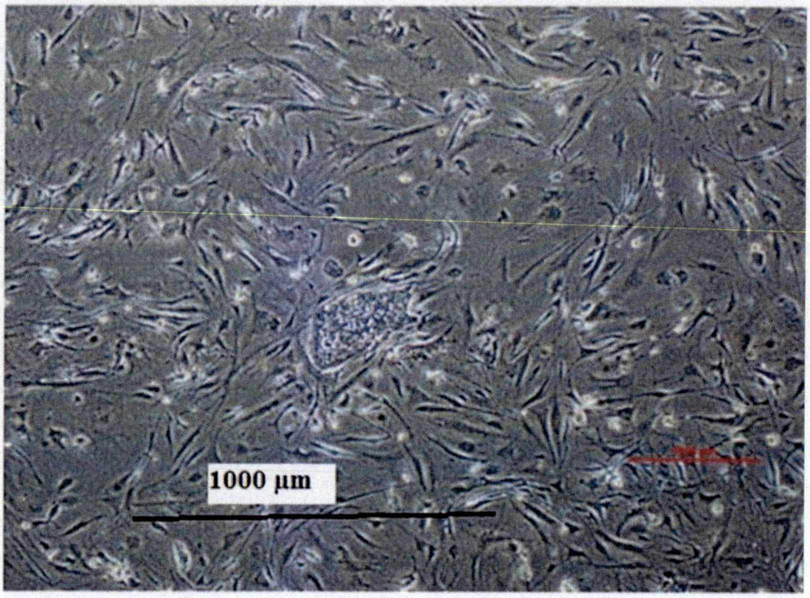

Fig. 2 Microscopic view of hPSCs at 2 days post thaw. Bar represents 1000 μm

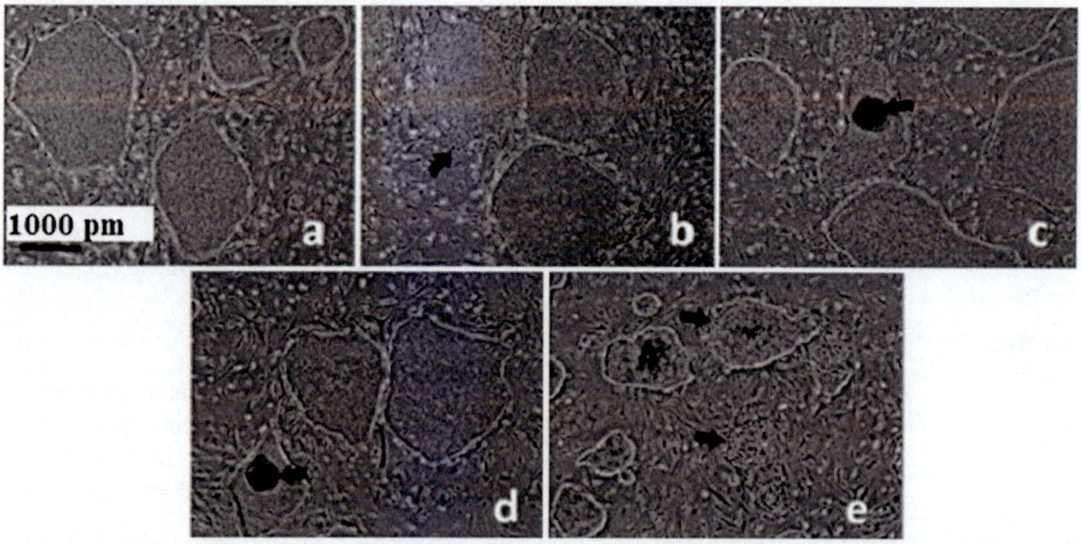

Fig. 3 Microscopic view of Healthy and differentiated colonies: (**a**) Healthy, undifferentiated hPSC colonies usually have well-defined borders; (**b**) Spontaneously differentiated cells may resemble cobblestones and differentiated cells may reside at the colony perimeter or in the center of the colony; (**c**, **d**) different culture conditions can make diverse types of differentiated cells (**e**). Bar represents 1000 μm

7.4 C4 Substitute of Pluripotent Stem Cell Culture Medium

7.4.1 Supplies

- 5 mL sterile serological pipettes.
- 10 mL sterile serological pipettes.
- 70% ethanol.

7.4.2 Reagents

- hPSC culture medium.

7.4.3 Procedure

Monitor hPSC Growth

- Take the hPSC plates from the incubator one by one and assess cell growth under the microscope. Return the culture plates to the incubator. If no further culturing will be done with the plate, abandon all contaminated culture wells/plates with at least 250 μL bleach (per well) for at least 30 min. If uncontaminated wells in a plate with contamination are to be cultured further, then perform the following:
- Move the plate to a biosafety cabinet and disinfect every contaminated well with 250 μL bleach for 5 min. After removing the disinfected content from the well completely, change medium for other wells and place the plate in a specifically elected incubator for observation during the next few days. Once the culture is free of contamination for 5 days, the plate can be returned to usual incubator for further culture.

Change Medium for hPSC Culture

- Warm a proper amount of CM to room temperature (For a 6-well plate, 2.5 mL of CM is needed per well.)
- Place the warm "CM" tube or bottle in the biosafety cabinet after carefully cleaning with 70% ethanol.
- Take the hPSC plates from the incubator and place them in the biosafety cabinet.
- Aspirate the spent medium totally using a Pasteur pipette.
- Add 2.5 mL clean CM to each well in a 6-well plate.
- Return the hPSC plates to the incubator.

Note: Repeat the above process daily until the cells are ready for passage or splitting to new iMEF plates.

7.5 C5 Passaging of Pluripotent Stem Cells on New iMEF Plates

7.5.1 Supplies

- 5 mL sterile serological pipettes.
- 5 mL sterile glass pipettes.
- 10 mL sterile serological pipettes.
- 15 mL centrifuge tubes.
- 50 mL centrifuge tubes.
- Colony marker.

7.5.2 Reagents

- DMEM-F12.
- Collagenase Solution.
- hPSC Culture Medium.
- 6-well plates with iMEFs.

7.5.3 Preparation

Establish when and how to Passage (or Split) Pluripotent Stem Cells

Note: Different cell lines have different growth kinetics so, different splitting schedules may be used for different cell lines. However, it is normally true that hPSCs grow slowly during the first couple of weeks after being thawed, then faster until the growth rate reaches a plateau. The cell growth rate will then stay in that plateau for many passages if cells are cultured accurately. It is necessary to observe cell growth daily to become familiar with the growth kinetics of the cell line being cultured.

Under a microscope, carefully estimate cell growth and quality. Explanation will be used to determine when and how often the cells are passaged.

Healthy, undifferentiated hPSC colonies generally have well-defined homogeneous borders and the individual cells within the colony appear to be similar. The exact colony morphology will vary with different cell lines and culture conditions (i.e., whether iMEFs, mTeSR1, or conditioned medium with Matrigel are used for hPSC culture).

Simultaneously, differentiated cells may resemble cobblestones or trophoblast-like cells. The differentiated cells may exist in at the colony perimeter or in the center of the colony. There can also be differentiated colonies containing great clumps of swirled tissue, or little balls of cells, like embryoid bodies. Different culture conditions yield diverse types of differentiated cells.

- Generally, passage cells when any of the following occur:
  - iMEF feeder layer is 2 weeks old.
  - 50% of the colonies are (greater than 2000 μm).
  - Colonies are also dense (at approximately 70% confluence).
  - Colonies exhibit increased differentiation [11–17].

Determine How to Channel (or split) Pluripotent Stem Cells

The process of choice depends on the differentiation level in the cultures and how many colonies to be passaged.

% Differentiation	Method to use	Split ratio
0–20%	Remove differentiated areas, Pick to Discard then enzymatically passage remaining hPSCs (i.e., PTD-Enzyme)	1:1 to 1:3
>70%	Pick only undifferentiated areas of cells (i.e., Pick to Keep; PTK)	1:1
20–70%	Either above method is fine	1:1 to 1:3

- If discrimination level is low (less than 20%), the method of PTD-Enzyme is recommended.
- If discrimination level is high (more than 70%), the method of PTK is recommended.
- If differentiation level is intermediate (20–70%), either process can be chosen depending on how many good colonies are available and how many colonies are to be passaged:
  - If there are many undifferentiated colonies existing and only limited colonies need to be passaged, PTK is recommended.
  - If there are many undifferentiated colonies available and as many good colonies are possibly required to be passaged, PTD-Enzyme is recommended.
  - If there are a limited number of undifferentiated colonies available and as many good colonies as possible need to be passaged, either method is suitable.
- Verify the splitting ratio to be used.
- Observe the records from the previous passages (up to three passages). Evaluate the cell splitting ratios and the colony confluence levels at the harvest days between sequential passages.
  - If cells grew extensively faster than the previous passage, increase the splitting ratio.
  - If the growth rate is parallel to the previous passage, use the same cell splitting ratio. If cells grew drastically slower than the previous passage, split cells at a lower ratio than the previous one. Figure 4 represents the screening criteria to determine if hPSC colonies can be passaged or not.

Note: Regardless of the process used, it is prudent to set aside one well of cells from the same culture used for passaging for an additional 1–2 days in case of errors or contamination during passaging.

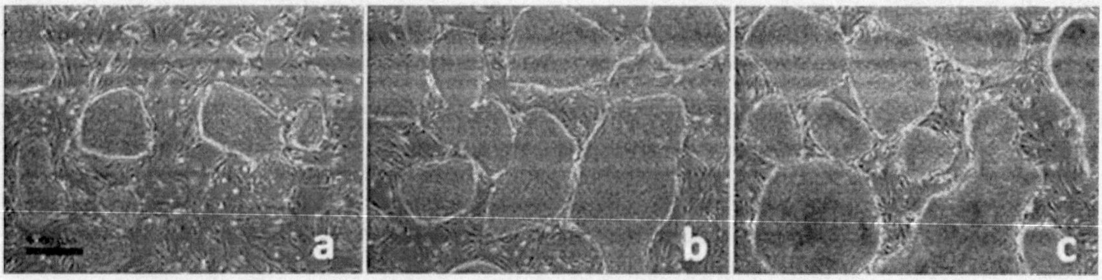

Fig. 4 Microscopic view of the culture confluence determines when hPSC colonies should be passaged: (**a**) colonies are not ready to be passaged at 50% confluence; (**b**) Cultures should be passaged at around 70% confluence; (**c**) before most of the colonies contains darkened areas. Bar represents 1000 μm

Determine How Much Medium and Solution Are Needed for this Process

- For harvesting and plating hPSCs on iMEFs, the next are required:
  - DMEM-F12 (2 mL for each well for each iMEF Plate).
  - Collagenase solution (1 mL for each well to be harvested).
  - Culture medium (2.5 mL for each well for harvest; 2.5 mL per well for seeding + extra).
- Analyze and record the amount of the above materials required for this process.

Evaluate and Prepare Fresh iMEF Plates

- Take the freshly seeded iMEF cell plates from the incubator one at a time and estimate cell culture under microscope.
- Return the plates to the incubator if there is no observable contamination.
- Take the standard iMEF plates and place them in the biosafety cabinet.
- Mark the plates with the cell line name, new passage number, the date, and initials.
- Totally aspirate the medium from the iMEF plate(s) with a Pasteur pipette.
- Add 2 mL/well of fresh DMEM-F12 medium.
- Softly swirl the plate to rinse the well walls.
- Aspirate the DMEM-F12 medium entirely from the plate (s) with a Pasteur pipette.
- Add 1.5 mL of fresh culture medium/well.
- Return the plate(s) to the incubator.

Procedure-Pick to Ruin-Enzyme Harvest Cells (No More than Two Plates at a Time)

- Mark a sterile tube as "Cells" for collecting hPSCs later.
- Take the hPSC plate(s) from the incubator and estimate under a microscope. If using a colony marker, mark the differentiated

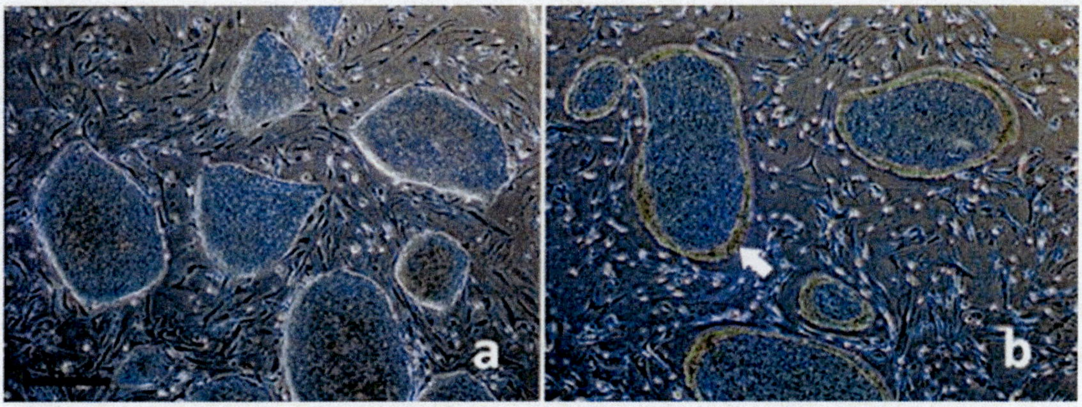

Fig. 5 Colonies before (**a**) and after collagenase treatment (**b**). Colonies are ready to be harvested if edges of colonies are obviously curled and lifted off feeder layer (arrow, **b**). Bar represents 1000 μm

colonies on the base of the plate and place in the biosafety cabinet.

- Using a Pasteur pipette, aspirate all marked (differentiated) cells in the plate by moving the colonies with the tip of the pipette so the differentiated areas will be sucked away leaving the undifferentiated colonies in the plate.
- Aspirate the remaining medium along with the floating differentiated cells or cell debris using a straight Pasteur pipette and set the timer at 3 min and 10 min, respectively.
- Add 1 mL collagenase per well and return the plate(s) to the incubator. Start the timer.
- After 3 min of incubation, check the plate(s) under a microscope to decide whether the cell colonies are ready to be harvested. If they are ready, the perimeter of the colony should appear folded back, unraveling from the iMEFs. If they are not ready, return the plate back to the 37 °C incubator for one more 3–5 min of incubation. Figure 5 represents the microscopic view of the colonies before and after collagenase treatment.

Note: Incubation time-span depends on the freshness of the collagenase solution. The older it is, the longer the incubation point will be, do not exceed 15 min of incubation time.

- When collagenase incubation is complete, aspirate collagenase from the wells in the biosafety cabinet. Be careful not to upset the cell/colony layer.
- Add 2 mL culture medium to every well in the plate(s).
- Utilizing a sterile 5 mL glass pipette, take up generally of the medium from one well. Use the pipette to scrape the cells/colonies in the well while gradually releasing the medium into the well to wash the cells/colonies off the surface. Pipette up

and down softly to minimize bubbles. Repeat these steps until most of the cells/colonies are disconnected from the surface.

- When cells in all wells of one plate are detached, move the cell/colony solution from all wells into the labeled sterile tube. Wash all the wells in the plate with another 3 mL of CM. Transfer the cell/colony solution to the same tube.

Spin and Disrupt Colonies

- Centrifuge cells at 200 × *g* for 5 min and aspirate the supernatant.
- Resuspend the cell pellet in an suitable volume of culture medium (1 mL/well to be seeded, plus 0.5 mL extra for pipetting error).
- Using a 5 mL glass pipette, disrupt the cell clusters by pipetting cells up and down quietly a few times. Pipette cautiously to avoid bubbles by keeping the pipette tip in the solution at all times.

Note: In the above step, pipette the cell suspension up and down moderately fast, initially, to disturb big cell clusters. Check regularly to see if the big cell clusters are disrupted. When most of the colonies are little enough to remain suspended in solution, the cells are ready to be plated. When mixing, pipette gradually as not to break down clusters any more than needed.

Addition of Cells to iMEF Plates (Processing No More than Three Plates at a Time)

- Returning to the iMEF plates (containing 1.5 mL/well culture medium) to the biosafety cabinet.
- Using a 5 mL glass pipette, quietly pipette cells up and down a few times. Then take 3 mL of the cell suspension and drop-wise (in a circular motion) add 1 mL/well to each of the three wells.
- Repeat the above two steps until cells are circulated to all wells in the iMEF plate.
- When complete, slide the plates in a cross-motion along the within surface of the biosafety cabinet.

Note: Do not shift the freshly seeded plates for at least the next 12 hours. Make sure cells under microscope the next day.

Procedure-Pick to Keep Prepare Sterile Bent Pasteur Glass

Note: Bent Pasteur pipettes are necessary to remove undifferentiated hPSCs for transfer to new iMEF plates.

- Take a 9-in. Pasteur glass pipette and heat the area about 2 in. from the tip temporarily over the open flame of an alcohol burner (about 3–5 s).
- Once the heated area is pliable, remove the pipette from the flame and quietly pull the tip at a 45° angle until it separates from the base of the pipette. Place the tip temporarily over the open

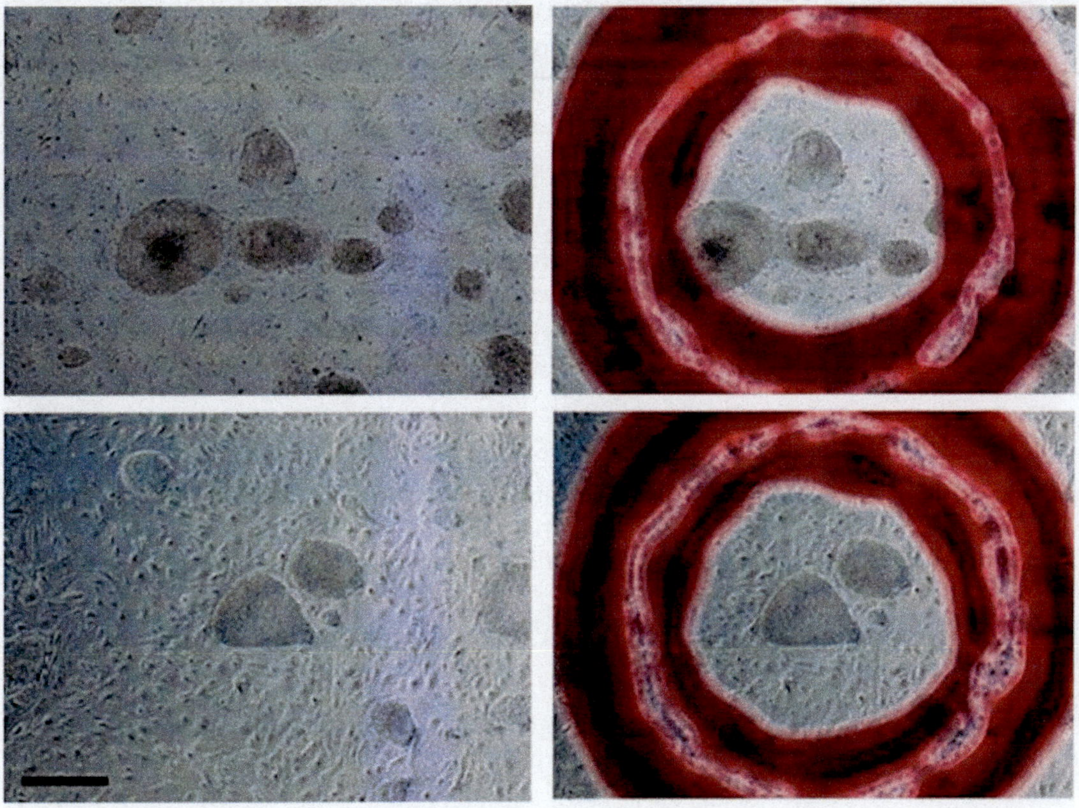

Fig. 6 Mark plates: Top row; mark undifferentiated areas with the colony marker if PTK is to be used. Base Row; mark differentiated areas if PTD is to be used. Bar represents 500 μm

flame to form a small ball. Ideally, the curved pipette should now have a small 0.5-in. hook that ends with a small ball at the tip.

- Repeat steps above to make additional bent Pasteur pipettes and put them in a Pasteur pipette container.
- Autoclave bent Pasteur glass pipettes prior to use.

Transfer Areas of Undifferentiated Cells to Fresh iMEF Plates (Pick to Keep)

Note: This division describes PTK at a ratio of 1:1 (1 well to 1 well). If there are a few viable colonies, a ratio of 2:1 or higher may be needed. The detail pictorial view is being represented in Fig. 6.

- Remove a hPSC plate from incubator and place it in a biosafety cabinet.
- Aspirate the spent medium from the wells using a sterile straight Pasteur pipette and add 1.5 mL fresh culture medium to each well containing cells.
- Spot the undifferentiated area in the wells on the plate bottom using a colony marker attached to an inverted microscope.

- Transfer the plate to the biosafety cabinet. Remove the undifferentiated cells in the well with a modified/bent pipette. Pipette the cell clusters a few times to break them into aggregates.
- Transfer the freshly equipped iMEF plate containing 1 mL/well culture medium from the incubator and place it in the biosafety cabinet.
- From the stem cell plate, gently move the medium (about 1.5 mL) along with the floating colony pieces into the freshly prepared iMEF well at 1:1 ratio (one well to one well).
- Return the seeded plate to the incubator. Slide the plate in a cross-way on the shelf of the incubator 3–5 times.
- Add 2.5 mL of medium to the well of the old plate and return it to the incubator for 1–2 days of additional culture.

Note 1: If the need arises to open the incubator while the cells are attaching to the surface (during the next 12 h), open and close the incubator door gently.

Note 2: Replace culture medium daily [13–17].

8 Stem cell culture-Requirements and protocol

8.1 Supplies

- 5 mL sterile glass pipettes.
- 5 mL sterile serological pipettes.
- 10 mL sterile serological pipettes.
- 15 mL sterile centrifuge tubes.
- 50 mL sterile centrifuge tubes.
- Cryovials (Nalgene 5000-0020).
- Mr. Frosty™ (Nalgene® 5100-0001).

8.1.1 Reagents

- Heat inactivated defined FBS.
- DMSO (Diethyl sulfoxide).
- Isopropanol.
- Collagenase solution 1 mg per mL.
- hPSC Culture Medium.

8.1.2 Procedure

Note: Harvest cells when they are in the log phase of growth, which is usually between 4 and 6 days after plating.

Determine when to Harvest Cells

- Using a microscope, evaluate cell growth and quality carefully.
- All of the following observations will be used to decide when cells are ready to be harvested. Harvest cells when:

- There is a smaller amount than 10% differentiation in the previous passage.
- Colonies of cells reach 70% confluence.
- About 50% of colonies are greater than 2000 μm.

Prepare Cryovials

- Make sure there are enough freezing containers on the bench near the biosafety cabinet for freezing cells. Make sure all freezing containers have an enough amount of fresh (made on the day of harvest) isopropanol in them. Each freezing container can grasp 18 cryovials.
- Mark each cryovial (Provider's cell line name and passage number, date, initials) and expose to UV light for at least 30 min.

Determine how Much Culture Medium and Collagenase Solution Is Needed for this Process

- For harvesting hPSC, the following are needed:
  - Collagenase solution (1 mL/ well to be harvested).
  - Culture medium (for freezing media preparation).

Note: Do not place collagenase in water bath to warm.

Set up 2× Cryopreservation Media (or Freezing Medium, FM) in the Biosafety Cabinet

Note: Make 2 × FM media just before use (fresh).

- Decide the amount of cryopreservation media required:
  - The media required = 0.5 mL × the number of cryovials +4 mL extra for pipette and filter error.
- According to the volume designed above, label one sterile 50 mL conical tube or a larger size sterile bottle as 2 × FM (on both the cap and side).
- Add the following, proportionately, to the labeled container:
- Keep the above solution on ice or in a 4–8 °C refrigerator and utilize it within the day.

Note: This 2× FM will be added to cell suspension at a 1:1 volume ratio, resulting in 1× freezing medium with a final concentration of 10% DMSO and 30% FBS, respectively.

Incubate Cells with Collagenase Solution (Make No More than Two Plates at a Time)

- Place the warm culture medium aliquot in the biosafety cabinet after carefully cleaning with 70% alcohol.
- Mark a sterile container as "Cells" for collecting hPSC for 3 min.
- Take out one or two plate(s) from the incubator and place it/them in the biosafety cabinet.
- Aspirate the medium entirely from the plate(s) using a Pasteur pipette.
- Add 1 mL collagenase/well.
- Return the plate(s) to the incubator.

- Start the timer for 3 min.
- After 3 min of incubation, check the plate(s) under a microscope to decide whether the cell colonies are ready to be harvested. When they are ready, the perimeter of the colony should emerge folded back, separating from the iMEFs.

Note: Incubation time varies on the freshness of the collagenase solution. The older it is, the longer the incubation will take.

- When collagenase incubation is complete, keep the plate in the biosafety cabinet.
- Aspirate collagenase from the wells. Be careful not to disturb the cell/colony layer.

Harvest Cells

Note 1: HPSCs recover improved from freezing and thawing when frozen in large aggregates.

Note 2: Run off the contents in the well until all wells are scraped and add 2 mL culture medium to each well.

- Using a 5 mL glass pipette, take up the majority of the medium from one well. Scrape the cells/colonies in the well using the pipette while gradually releasing the medium into the well to wash the cells/colonies off the surface. Softly pipette the medium up and down during the wash to minimize bubbles and to disrupt cell clumps. Repeat these steps until most of the cells/colonies are separated from the surface.
- Separate cells/colonies in the other wells by repeating the above steps.
- When all wells are complete, combine the cell/colony solution from all wells into the base of the marked sterile "Cells" container. Rinse all of the wells in one plate with another 3 mL of culture medium, transfer the cell/colony solution to the same sterile "Cells" container.

Freeze Cells

Note 1: Once in DMSO freezing medium, cells should be frozen as rapidly as possible (within 5 min).

Note 2: When freezing a large lot, different parts of the freezing method may be completed by different personnel in order to reduce the time that cells are in the freezing medium.

- Centrifuge cells at 200 × *g* for 5 min.
- During the centrifugation phase, loosen all cryovial caps in the biosafety cabinet without taking them off.
- When centrifugation is complete, revisit "Cells" container to the biosafety cabinet after cleaning with 70% ethanol.
- Aspirate the supernatant off the cell pellet.

- Resuspend the cell pellet gently in a proper volume of culture medium (0.5 mL/well plus 0.5 mL extra).
- After washing with 70% ethanol, place the cold 2 × freezing medium (FM) in the biosafety cabinet.
- Dropwise, add the same volume of 2 × FM to the cell suspension.
- After pipetting softly to mix the cells, transfer cells to cryovials at 1 mL/vial.
- Transfer cryovials to Mr. Frosty freezing container and place it in −80 °C freezer overnight.

  Note: Cells require to cryopreserved in LN_2 freezers.

- The next day, rapidly move the cryovials into the LN_2 freezer on dry ice.

Other methods for neural stem cell and mesenchymal stem cells are described here in brief.

Neural stem cell.

8.2 Protocols

- Cell viability assays for neural stem cells.
- Characterizing neural cells by qPCR.
- Cryopreservation and healing of mature differentiated neural cells.
- Cryopreserving neural stem cells.
- Culturing human neural stem cells.
- Culturing rat fetal neural stem cells.
- Determination and culture of cortical astrocytes.
- Determination and culture of dopaminergic neurons (from midbrains of rodents).
- Determination of dopaminergic neurons (from human embry onic stem cells).
- Distinguishing glial precursor cells into astrocytes and oligodendrocytes.
- Distinguishing neural stem cells into neurons and glial cells.
- Expansion of neural stem cells utilizing StemPro NSC SFM.
- Induction of neural stem cells from human pluripotent stem cells utilizing Gibco PSC Neural Induction Medium.
- Purification and transfection of mouse primary neurons.
- Purification, culture, and characterization of cortical and hippocampal neurons.
- Determining neuronal cell health through viability and neurite outgrowth.

- Quick reference: TaqMan® Array Plates.
- Thawing the rat glial precursor cells (GPCs).
- User Protocols: Human Neural Stem Cell Immunocytochemis try Kit.
- Xeno-free culture and distinguishing of neural stem cells into neurons.

Mesenchymal stem cell.

9 Culturing Mesenchymal Stem Cells (MSCs) in 2% Reduced Serum Medium—a Case Study

MSCs contain a mere 0.01–0.0001% of total bone marrow nucleated cells and need in vitro cell culture expansion in order to get enough numbers for clinical applications. MesenPRO RS™ Medium is an abridged serum (2%) medium specifically formulated to support the growth of human mesenchymal stem cells (MSCs) in culture. MesenPRO RS™ Medium helps improve MSC expansion compared with conventional serum-supplemented media while maintaining a similar gene expression profile.

Decreasing the concentration of FBS to 2% in a highly consistent formula produced under cGMP conditions helps decrease the variability typically introduced by the 10–20% FBS used in classical media. There are also many unfamiliar elements in FBS, such as signaling molecules, apoptotic factors, and nutrients. The uneven concentrations of these components can cause lot-to-lot variation, which means that some FBS lots do not support MSC culture. When used to recover cryopreserved MSCs, our MSC-Qualified FBS [18, 19] minimizes the require to test multiple FBS lots to recognize the optimal one for your use.

MesenPRO RS™ Medium [18, 20–23] (Eibes et al., 2010) provides improved expansion of MSCs compared with classical media (DMEM +10% FBS) and steady trilineage mesoderm differentiation potential as depicted in Figs. 7 and 8.

9.1 Materials Needed

- MesenPRO RS™ Medium (Cat. no. 12746-012) composed of MesenPRO RS™ Basal Medium and MesenPRO RS™ Growth Supplement. The medium is also existing as MesenPRO RS™ (New Zealand origin) kit under Cat. no. E07-1000. (PDF).
- L-Glutamine-200 mM (100×), Liquid (Cat. nos. 25030-149, 25030-164) or CTS™ GlutaMAX™-I Supplement (Cat. no. A12860-01) or human ADSCs (StemPro®Human Adipose-Derived Stem Cells (Cat. no. R7788-115).
- CTS™ (Cell Therapy Systems) DPBS, lacking calcium chloride, without magnesium chloride (Cat. no. A12856-01).
- Countess® Automated Cell Counter (Cat. no. C10227).

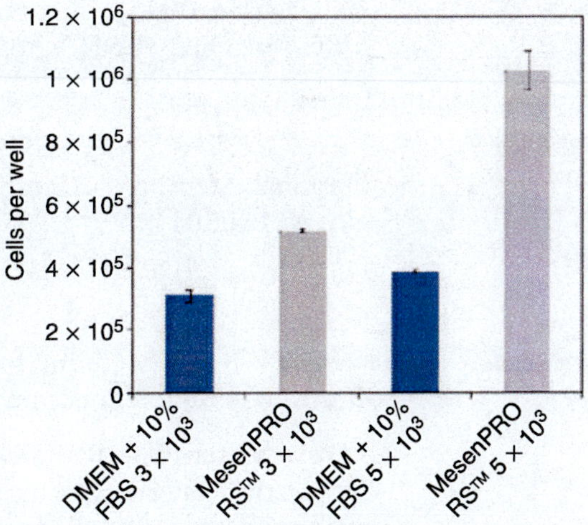

Fig. 7 MesenPRO RS™ Medium helps a 69–169% development in expansion over 6 days, compared to classical media. MSCs were isolated from normal human bone marrow mononuclear cells by standard protocols. Early-passage cells were plated into 6-well plates at either 3×10^3 or 5×10^3 cells/cm^2 in DMEM (low glucose) having 4 mM L-glutamine, 5 μg/mL gentamicin, and 10% MSC-mixed FBS; or MesenPRO RS™ Medium containing 2 mM L-glutamine and 5 μg/mL gentamicin. Cells were incubated at 37 °C with 5% CO_2 in humidified air and fed on day 3. Sixth day, cells were harvested using TrypLE™ reagent and counted. Data symbolize cell count averages from duplicate wells ($p \leq 0.007$ and $p \leq 0.002$, respectively, by Student's t-test)

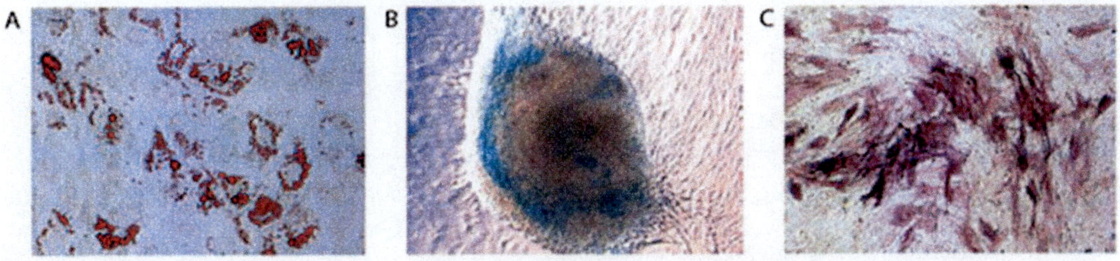

Fig. 8 Human MSCs cultured in MesenPRO RS™ Medium keep trilineage differentiation potential. Human MSCs cultured in MesenPRO RS™ Medium were seeded into adipogenic, chondrogenic, or osteogenic differentiation medium for 14 days, useful (**a**) adipocytes (oil red O lipid stain), (**b**) chondrocytes (alcian blue glycosaminoglycan stain), and (**c**) osteoblasts (alkaline phosphatase stain)

- 37 °C water bath.
- Appropriate tissue culture vessels and supplies.

9.1.1 For Recovery Only of Cryopreserved MSCs

- Dulbecco's Modified Eagle Medium (DMEM) [DMEM/F-12, GlutaMAX™ (Cat. nos. 10565-018, 10565-042) or DMEM, little glucose, pyruvate, no glutamine, no phenol red (Cat.

no. 11054-001)] including 10% Fetal Bovine Serum, MSC-Qualified (USDA Approved) (Cat. no. 12662-002).

9.1.2 For passaging and cryopreservation of cells

- CTS™ TrypLE™ Select Enzyme (Cat. no. A12859-01).
- Optional: StemPro® Accutase® Cell Dissociation Reagent (Cat. No. A1110501).
- Dimethyl sulfoxide (DMSO) (Cat. no. D12345).

9.2 Preparing Media and Materials

MesenPRO RS™ Medium for improving and Culturing MSCs (500 mL of complete medium)

1. Thaw MesenPRO RS™ Growth Supplement at 2–8 °C 24 h, correctly thawed. Do not thaw at 37 °C. Avoid frequent freeze/thaw cycles of MesenPRO RS™ Growth Supplement.
2. To make 500 mL of complete MesenPRO RS™ Medium, aseptically mix the following components:

MesenPRO RS™ Basal Medium	485 mL
MesenPRO RS™ Growth Supplement	10 mL
200 mM L-glutamine or GlutaMAX™-I	5 mL (2 mM final concentration)

3. Complete, supplemented MesenPRO RS™ Medium confined from light can be stored at 2–8 °C for up to 2 weeks.

Note: Before use, warm total medium required for that day at room temperature until it is no longer cool to the touch. Do not warm the medium at 37 °C.

9.3 Recovery of Cryopreserved MSCs

Note: We advise that you recover MSCs into complete MesenPRO RS™ Medium from thaw*.

*You can use conventional medium for recovery of MSCs. To make 500 mL of complete serum-supplemented DMEM medium, thaw Fetal Bovine Serum, MSC-Qualified (USDA Approved) at 2–8 °C.

1. Quickly thaw (<1 min) frozen cells in a 37 °C water bath.
2. Pipette the whole contents of the cryovial into a sterile 50-mL conical tube.
3. Cautiously, by dropwise addition (2–3 drops per 10 s), add 5–7 mL of prewarmed complete MesenPRO RS™ Medium while gently swirling the tube.
4. Add a further 5 mL prewarmed complete MesenPRO RS™ Medium while quietly swirling the tube.
5. Transfer all the contents of the conical tube into a suitable tissue culture flask.
6. Incubate at 37 °C in a humidified ambience of 5% CO_2 in air.

7. Switch over spent medium with fresh prewarmed complete MesenPRO RS™ Medium 24 h post-thaw.

   Note: For recovery of MSCs, it is optional to seed cells at $\geq 7 \times 10^3$ cells/cm^2 for the initial recovery passage.

9.4 Subculturing MSCs

MesenPRO RS™ has been improved for the culturing of human mesenchymal stem cells that have been started utilizing standard adherent isolation and growth circumstances (e.g., DMEM +10% FBS) and has not been tested for the early expansion of MSCs directly from primary tissue sources. MesenPRO RS™ has been improved for culturing MSCs at greater than clonal densities (*see* **step 9**).

It is optional to subculture human mesenchymal stem cells directly into MesenPRO RS™ complete medium. It is vital that cell confluence be 60–80%, cell viability be at least 90%, and the growth rate be in mid-logarithmic phase prior to sub culturing.

Note: The following events apply to adherent cultures in a T-75 culture flask (75 cm^2). Volumes should be used to accordingly for desired vessel size.

1. Watch culture flask on inverted microscope and confirm that the cells are ready to be subcultured (60–80% confluent).
2. Aspirate medium and floating cells starting a confluent monolayer and discard.
3. Clean cells with 5–10 mL Dulbecco's Phosphate Buffered Saline (DPBS), without calcium, magnesium, or phenol red.
4. Take away DPBS and add 5–7 mL of prewarmed CTS™ TrypLE™. Select Enzyme to the culture flask and incubate at 37 °C until cells have fully detached (approximately 3–5 min).
5. Observe cell monolayer using an inverted microscope to make sure complete detachment from the surface of the flask.
6. Stop cell dissociation by adding 10 mL of prewarmed absolute MesenPRO RS™ Medium to flask and gently pipette up and down to disperse cells into a single-cell suspension.
7. Remove cell suspension into a sterile conical tube. Wash flask with an additional 5 mL complete MesenPRO RS™ Medium and unite into the conical tube.
8. Centrifuge cell suspension at $100 \times g$ for 5–10 min.
9. Aspirate supernatant and resuspend the pellet in a suitable volume of prewarmed complete MesenPRO RS™ Medium.
10. Determine whole viable cell density using a Countess® Automated Cell Counter (alternative automated or manual methods may be used).
11. Inoculate flask at 3–5 $\times 10^3$ viable cells/cm^2 and return to incubator.

Note: For best performance and cell growth, cultures should be re-fed every 3–4 days with fresh complete medium.

9.5 Cryopreservation of MSCs

1. Organize a 2× cryopreservation solution on the day of use:
   (a) Make 10 mL of 2× cryopreservation solution by supplementing MesenPRO RS™ complete medium with 20% Dimethyl Sulfoxide (DMSO). Store at 4 °C until use.

MesenPRO RS™ complete medium	8 mL
DMSO	2 mL (20% final concentration)

2. Harvest cells for cryopreservation using the MesenPRO RS™ total medium according to **steps 1–9** of Subheading 9.4.
3. Reconstitute the harvested cell pellet to twice the desired final cell concentration (i.e., 2 × 10^6 cells/mL) in 0.5 mL of pre-warmed MesenPRO RS™ total medium.
4. Slowly add 0.5 mL of the 2× cryopreservation solution to the cell suspension, and quietly mix to ensure even cell distribution.
5. Immediately add the needed volume of cell suspension (i.e., 1 mL) to pre-chilled (2–8 °C) cryovials.
6. Keep the cryovials at −70 °C in a cryogenic freezing container (e.g., Mr. Frosty® (1 °C) Freezing Container).
7. After 24 h, transfer the frozen cells to liquid nitrogen (vapor phase); storage at −200 °C to 125 °C is suggested [1, 2, 13, 14].

10 Culturing Human Mesenchymal Stem Cells

(MSCs) under Serum-Free and Xeno-Free Conditions.

Serum-Free, Xeno-Free Medium.

StemPro® MSC SFM XenoFree (Chase et al., 2013) has been improved for the growth and expansion of human mesenchymal stem cells (MSCs) and Adipose-derived Stem Cells (ADSCs) under totally serum-free and xeno-free parameters.

Due to the low frequency of human MSCs in primary tissue, expansion of this stem cell population is critical and provides enable basic biological studies and clinical research. In addition, human MSCs can only be propagated a limited number of times, thereafter exhibiting reduced proliferation and differentiation potential. Development of human MSCs and adipose-derived stem cells (ADSCs) (Lindroos et al., 2009) [20] in StemPro® MSC SFM XenoFree is comparable to classical

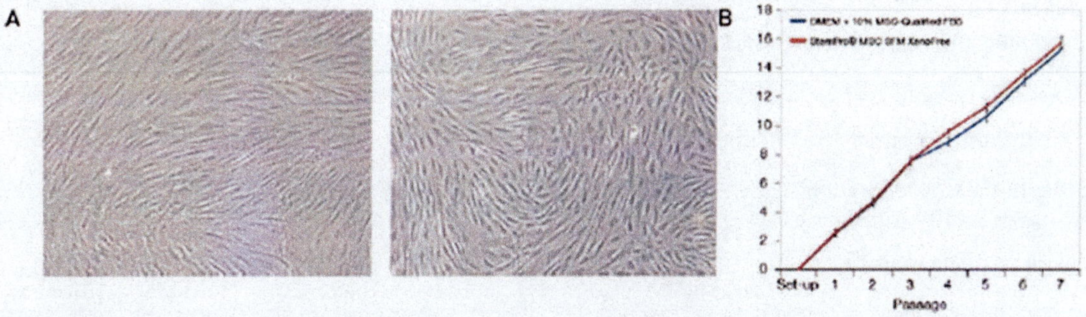

Fig. 9 Human MSC expansion under xeno-free conditions. Human bone marrow–derived MSCs stretched in DMEM (low glucose) + 10% MSC-Qualified FBS, or in StemPro® MSC SFM XenoFree + CTS™ CELLstart™ substrate-coated plates, revealed a similar expansion rate. (**a**) Morphology of extended (passage 3) human MSCs (10× objective). (**b**) Net expansion of human MSCs. Input human MSCs = passage 5, 4-donor pool. Passage frequency = every 4–6 days. Seed density = 5–7 × 10^3 cells/cm^2. Harvest enzyme = TrypLE™ Express. Counting method = Countess® Automated Cell Counter

medium (DMEM +10% MSC-Qualified FBS) in conditions of morphology and growth features as represented in Fig. 9 .

StemPro® MSC SFM XenoFree offers a completely xeno-free system when used in conjunction with CTS™ CELLstart™ substrate; thus, cells are grown in a physiological environment that allows for clinically applicable results.

Serum-Free Medium:

CTS™ StemPro® MSC SFM is a serum-free medium particularly formulated for the growth and expansion of human mesenchymal stem cells (MSC) or adipose-derived stem cells (ADSC) (Table 1).

StemPro® MSC SFM (Chase et al., 2010, Agata et al., 2009) [18] helped greater effectiveness of human MSC expansion at high cell densities, requiring less medium, surface area, and time compared with traditional medium (DMEM (low glucose) + 10% FBS). Human MSCs grown in classical medium have a compacted cell morphology and reach confluence at 1.0–3.0 × 10^4 cells/cm^2. Human MSCs cultured in StemPro® MSC SFM with a much smaller, spindle-shaped morphology and can attain densities of >1.0 × 10^5 cells/cm^2 as depicted in Fig. 10.

10.1 Materials Needed

10.1.1 For Either Culture and Isolation from Bone Marrow

- L-Glutamine-200 mM (100×), Liquid (Cat. nos. 25030-149, 25030-164) or CTS™ GlutaMAX™-I Supplement (Cat. no. A12860-01).
- 37 °C water bath.
- Appropriate tissue culture vessels and supplies.
- Cryogenic storage containers and supplies.

Table 1
Selecting the right MSC culture system for your research needs

MSC Culture system	Reduced-serum*	Serum-free*	Xeno-free, serum-free*
Supports MSC derivation from primary tissue	✓	✓	✓**
Cell or tissue source	Bone marrow, adipose, cord blood, placenta	Bone marrow, adipose, cord blood, placenta	Bone marrow, adipose, cord blood, placenta
Category	Human, mouse, rat, sheep, goat, pig, horse	Human	Human
Maintains MSC phenotype	✓	✓	✓
Supports trilineage differentiation	✓	✓	✓
Offers enhanced chondrogenesis	✓	✓	✓
Produced under cGMP-manufacturing standards	✓	✓	✓
Offers lot-to-lot consistency	✓	✓	✓
Free of animal components			✓

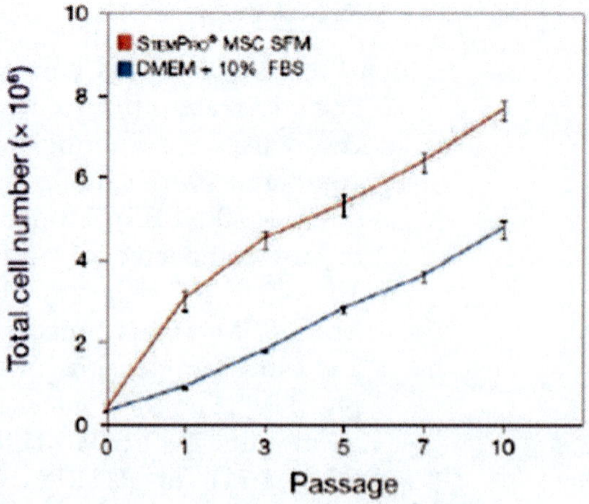

Fig. 10 hMSCs grown on CTS™ CELLstart™ substrate-coated dishes in StemPro® MSC SFM exhibit a 166% improvement in expansion over 10 passages, compared to classical medium. Average net whole cell number per T25 flask was designed for human MSCs growing in StemPro® MSC SFM or classical medium ($n = 3$). The culture had a seed density of 1×10^4 cells/cm^2, a split frequency of 3 days, and a medium alter every 2 days

- CTS™ CELLstart™ Substrate (Cat. no. A10142-01).
- Dimethyl sulfoxide (DMSO) (Cat. no. D12345).
- CTS™ DPBS, calcium, magnesium (Cat. no. A12858-01).
- CTS™ (Cell Therapy Systems) DPBS, devoid of calcium chloride, without magnesium chloride (Cat. no A12856-01).
- Gentamicin Reagent Solution (10 mg/mL) (Cat. no. 15710-072).
- CTS™ TrypLE™ Select Enzyme (Cat. no. A12859-01).
- CTS™ Synth-a-Freeze® Medium (Cat. No. A13713-01).

10.1.2 For Serum-Free, Xeno-Free Cultures

- StemPro® MSC SFM XenoFree (Cat. no. A10675-01) composes of StemPro® MSC SFM Basal Medium and
- StemPro® MSC SFM XenoFree Supplement.
- Gentamicin (50 mg/mL), liquid (Cat. no. 15750-060).

10.1.3 For Serum-Free Cultures

- CTS™ StemPro® MSC SFM (Cat. no. A10332-01) composes of CTS™ StemPro® MSC SFM Basal Medium and CTS™ StemPro® MSC SFM Supplement.

10.1.4 For Isolating MSCs from Bone Marrow

- Human bone marrow aspirate.
- HBSS, calcium, magnesium, no phenol red (Cat. no. 14025-134).
- Heat-inactivated and pooled AB-Human Serum (Cat. no. 34005-100).
- ACK Lysing Buffer (Cat. no. A10492-01).
- Trypan Blue Solution, 0.4% (Cat. no. 15250-061).

10.2 Preparing Media and Materials

10.2.1 StemPro® MSC SFM XenoFree entire Medium for Serum-Free, Xeno-Free Cultures (500 mL of complete medium)

1. Thaw StemPro® MSC SFM XenoFree Supplement at 2–8 °C overnight (thawed supplement will have a little cloudy appearance). Do not thaw at 37 °C. Use thawed material directly or aliquot (i.e., 1 mL) unused material and store at −20 °C to −5 °C. Avoid additional freeze-thaw cycles.

   Note: When stored frozen at −20 °C to −5 °C, the StemPro® MSC SFM XenoFree Supplement is secure for 12 months.
2. To prepare 500 mL of complete StemPro® MSC SFM XenoFree Medium, aseptically mix the following components:

StemPro® MSC SFM Basal Medium	490 mL
Additional: Gentamicin (50 mg/mL), liquid	50 μL (5 μg/mL final concentration)

Complete StemPro® MSC SFM XenoFree Medium can be stored in the dark at 2–8 °C for up to 2 weeks.

Note: Before utilize it, warm total medium necessary for that day at room temperature until it is no longer cool to the touch. **Do not warm the medium at 37 °C.**

10.2.2 Primary Isolation Medium: StemPro® MSC SFM XenoFree whole Medium include 2.5% human AB Serum for Isolation of MSCs from Bone Marrow

1. Thaw StemPro® MSC SFM XenoFree Supplement at 2–8 °C overnight (thawed supplement will have a slightly cloudy appearance). Do not thaw at 37 °C. Use thawed material urgently or aliquot (i.e., 1 mL) unused material and store at −20 °C to −5 °C. Avoid additional freeze-thaw cycles.

   Note: When preserved frozen at −20 °C to −5 °C, the StemPro® MSC SFM XenoFree Supplement is stable for 12 months.

2. To prepare 500 mL of complete StemPro® MSC SFM Xeno-Free Medium, aseptically mix the following components:

StemPro® MSC SFM Basal Medium	477.5 mL
StemPro® MSC SFM XenoFree supplement	5 mL
AB-human serum	12.5 mL (2.5% final concentration)
Additional: Gentamicin (50 mg/mL), liquid	50 μL (5 μg/mL final concentration)

Note: Filtering the AB-Human Serum through a 0.22 μm filter is recommended before use.

3. Complete StemPro® MSC SFM XenoFree Medium with 2.5% human AB serum can be stored in the dark at 2–8 °C for up to 2 weeks.

   Note: Before use, warm complete medium need for that day at normal temperature until it is no longer cool to the touch. **Do not warm the medium at 37 °C.**

10.2.3 StemPro® MSC SFM CTS™ Complete Medium Lacking Serum Cultures (500 mL of Complete Medium)

1. Thaw CTS™ StemPro® MSC SFM Supplement at 2–8 °C overnight, do not thaw at 37 °C. Use thawed material instantly or aliquot (i.e., 15 mL) unused material and store at −20 °C to −5 °C protected from light. Avoid additional freeze-thaw cycles.

   Note: When stored frozen at −20 °C to −5 °C in the dark, the CTS™ StemPro® MSC SFM Supplement is steady for 12 months.

2. To prepare 500 mL of full StemPro® MSC SFM XenoFree Medium, aseptically mix the following components:

CTS™ StemPro® MSC SFM Basic Medium	420 mL
CTS™ StemPro® MSC SFM supplement	75 mL
Additional: 10 mg/mL gentamicin reagent solution	50 μL (5 μg/mL final concentration)

CTS™ StemPro® MSC SFM Medium can be stored in the dark at 2–8 °C for up to 4 weeks, within the expiration date of all three components.

10.2.4 Coating Culture Vessels with CTS™ CELLstart™ Substrate

1. Reduce CTS™ CELLstart™ concentration in a ratio of 1:100 in DPBS with calcium and magnesium (i.e., 100 μL CTS™ CELLstart™ Substrate into 10 mL of CTS™ DPBS) or in prime isolation medium if you want to isolate MSCs from bone marrow (i.e., 100 μL CTS™ CELLstart™ components into 10 mL of primary isolation medium) and mix by moderately pipetting or inversion, do not vortex. Add 10 mL of the CTS™ CELLstart™ Substrate solution to every T-75 flask and make sure of complete surface coverage.
2. Incubate at 36–38 °C in a humidified atmosphere of 4–6% CO_2 in air for 60–120 min. After incubation, eliminate flasks from the incubator and provisionally place them in a laminar flow hood until use. Instantly before use, remove all CTS™ CELLstart™ Substrate solution and replace with complete medium. Do not rinse coated plates before use. **Note**: If CTS™ CELLstart™ Substrate coated plates (with coating solution in the plate) are wrapped tightly with Parafilm® sealing film, they can be stored at 2–8 °C for 2 weeks. **DO NOT REMOVE COATING SOLUTION PRIOR TO STORAGE.**

10.2.5 Recovering Cryopreserved Human MSCs

Note: These commands can be used for either serum-free and xeno-free cultures or for just serum-free cultures by adding the appropriate complete medium in **steps 3, 5, 6,** and **8**

1. Rapidly thaw (<1 min) a frozen vial of human MSCs in a 37 °C water bath until a small amount of ice remains.
2. Pipette the entire contents of the cryovial into a sterile 50-mL conical tube.
3. Carefully add 5–10 mL of pre-warmed (37 °C) complete medium to the conical tube at a rate of approximately 1 drop every 2 s and gently swirl after every addition to ensure homogeneity of the cell suspension.
4. Centrifuge the tubes at 100–200 × *g* for 5 min at room temperature. Aspirate and discard supernatant, being careful not to disturb the cell pellet.

5. Resuspend the cell pellet in a minimum volume of pre-warmed (37 °C) complete medium for cell counting. Determine total viable cell density with a Countess® Automated Cell Counter (alternative automated or manual procedures may be used). Calculate the volume of cell suspension required to seed cells at a density of $\geq 5 \times 10^3$ cells/cm^2.
6. Add the cell suspension to an appropriate CTS™ CELLstart™ substrate-coated flask containing 10–15 mL pre-warmed complete medium (*see* Subheading 10.2.4).
7. Incubate at 36–38 °C in a humidified atmosphere of 4–6% CO_2 in air.
8. Replace the medium in the flasks every 2–3 days with fresh pre-warmed complete medium.

10.2.6 Isolating MSCs From Bone Marrow

1. Mix human bone marrow aspirate 1:2 with HBSS (i.e. 20 mL aspirate +20 mL HBSS).
2. Pipette 15 mL gradient density media into a 50-mL centrifuge tube.
3. Slowly layer 20 mL of dilute bone marrow aspirate on top of the gradient density media, taking care not to mix cells and the media.
4. Centrifuge (with brake off) tubes at 400 × *g* for 35 min at 20 °C.
5. Carefully remove the mononuclear cell (MNC) layer with a sterile transfer pipette or 1-mL serological pipette. Add 30 mL HBSS to the tube containing harvested MNCs and pipette to dissociate the cell pellet.
6. Centrifuge at 400–500 × *g* for 10 min. Remove supernatant.
7. Repeat **step 7**.
8. Resuspend cells in a minimal volume of primary isolation medium (StemPro® MSC SFM XenoFree Complete Medium with 2.5% human AB serum).
9. Transfer 25 μL of the cell solution to a microfuge tube and add 50 μL of ACK Lysing Buffer for RBC lysis.
10. Pipette cell solution to mix and incubate at room temperature for 5 min.
11. Dilute cell lysis solution 1:2 with Trypan Blue Solution.
12. Count the cell number using the Countess® automated cell counter or other preferred method (e.g., a hemacytometer).
13. Seed viable MNCs at a density of 2.67×10^5 cells/cm^2 on CTS™ CELLstart substrate-coated flasks, using primary isolation medium to dilute the substrate.

14. After 3–4 days, harvest the supernatant and centrifuge at 300 × *g* for 5 min.
15. Add pre-warmed primary isolation medium to the flask.
16. Change complete primary isolation medium every 2 days.
17. Upon reaching 70–80% confluence, propagate cells under serum-free conditions as noted below.

10.2.7 Subculturing MSCs

General Guidelines

For xeno-free, serum-free cultures:

- StemPro® MSC has been designed for the multi-passage expansion of human bone marrow-derived MSCs and Adipose-derived Stem Cells (ADSCs) at larger than clonal densities ($\geq 5 \times 10^3$ cells/cm^2). Decreased seeding densities may result in suboptimal cell expansion, although optimal growth conditions must be determined for each application.
- When subculturing human MSC input cell confluence should be 60–90%, cell viability should be at least 90%, and the growth rate should be in mid-logarithmic phase prior to subculture which initiating cultures under suboptimal conditions may affect product performance.
- For the best possible performance, re-feed the cultures every 2 days with StemPro® MSC SFM XenoFree complete medium.

For serum-free cultures:

- CTS™ StemPro® MSC SFM has been developed for the primary isolation and multi-passage expansion of human bone marrow-derived MSC and ADSC at better than clonal densities, with optimal cell expansion analyzed at $\geq 5 \times 10^3$ cells/cm^2.
- Reduced seeding densities may result in suboptimal cell expansion where optimal growth conditions must be determined for each application.
- For the best possible performance and cell growth, cultures should be re-fed every 2 days with fresh complete CTS™ StemPro® MSC SFM.

Note: Protocols detailed below are for cultures maintained in a T-75 culture flask (75 cm^2). Adjust the volumes accordingly for required vessel size (*see* Table 2)

10.3 Propagating MSCs

1. Study the stock culture flask (~60–90% confluent for xeno-free, serum-free medium and ~60–80% confluent for serum-free medium).
2. Pre-warm CTS™ TrypLE™ Select reagent (Cat. no. 12859) and the appropriate complete medium before use.
3. Add 10 mL of pre-warmed complete medium to a 50 mL conical tube for each flask being harvested.

Table 2
Reagent volumes

	Area of vessel	Surface area (cm^2)	CTS™ TrypLE reagent (mL)	Complete medium (mL)	CTS™ CELLstart™ solution (1:100) (mL)
Dishes	35-mm	9.62	0.5–1	2–3	0.5–1
	60-mm	28.27	1–2	3–5	2–3
	100-mm	78.54	2–3	10–15	7–10
	150-mm	176.71	3–5	15–20	15–20
Plates	6-well	9.62	0.5–1	2–3	2.0
	12-well	4.01	0.2–0.5	1–2	1
	24-well	2.00	0.1–0.2	0.5–1	0.5
Flasks	T-25	25.00	1–2	4–5	2–3
	T-75	75.00	2–3	12–15	7–10
	T-175	175.00	3–5	35–40	15–20
	T-225	225.00	5–7	40–45	20–25

4. Aspirate spent medium from the T-75 flask and discard.
5. Clean the cell monolayer surface with 5–10 mL of CTS™ DPBS.
6. Add 3–5 mL of CTS™ TrypLE™ reagent to the T-75 flask, tilt the flask in all directions to ensure complete coverage of cell monolayer and incubate the cells in CTS™ TrypLE™ reagent for 2–10 min at 37 °C.
7. After incubation, check the flask with an inverted microscope for cell detachment. Tightly tap the flask as necessary to help complete cell detachment.
8. Upon detachment, add 5 mL of pre-warmed StemPro® MSC SFM XenoFree (for xeno-free, serum-free cultures) or 5 mL pre-warmed CTS™ DPBS with calcium and magnesium (for serum-free cultures) to each flask to completely cover the surface area.
9. Centrifuge the tubes at 100–200 × *g* for 5 min at room temperature and aspirate medium and discard being careful not to disturb cell pellet.
10. Resuspend the cell pellet in a minimal volume of pre-warmed complete medium for cell counting, using a preferred counting method (e.g., Countess® Automated Cell Counter).
11. Remove the CTS™ CELLstart™ substrate solution from each coated flask and add 15 mL of complete medium.

12. Add enough cell suspension to each flask to provide $\geq 5 \times 10^3$ cells/cm^2 (i.e., 3.75×10^5 viable cells per T-75 flask). Softly mix or swirl the cell suspension to make sure of even distribution.
13. Place the culture flask(s) in the incubator at 37 °C with a humidified atmosphere of 4–6% CO_2.

Restore the spent culture medium **every 2–3 days** with pre-warmed complete medium for the best possible performance and cell growth.

10.4 Cryopreservation of MSCs.

1. Prepare a cryopreservation solution on the day of utilize according to the type of culture or use CTS™ Synth-a-Freeze® Medium [13]:
   (a) **Used for xeno-free, serum-free cultures**: Prepare 10 mL of XFM cryopreservation solution (2×) by supplementing StemPro® MSC SFM XenoFree complete medium with 20% Dimethyl Sulfoxide (DMSO). Store at 4 °C until use.

StemPro® MSC SFM Basal Medium	7.9 mL
StemPro® MSC SFM XenoFree supplement	0.1 mL (1% ultimate concentration)
DMSO	2.0 mL (20% final concentration)

   (b) **Used serum-free cultures:** Prepare 10 mL of SFM cryopreservation solution on day of use by supplementing CTS™ StemPro® MSC SFM Basal Medium with 25% CTS™ StemPro® MSC SFM Supplement and 10% Dimethyl Sulfoxide (DMSO). Store at 4 °C until use.
2. Harvest cells for cryopreservation using the appropriate complete medium according to **steps 1–8** of Subheading 10.3.
3. Reconstitute the harvested cell pellet to twice the desired final cell concentration (i.e., 2×10^6 cells/mL) in 0.5 mL of pre-warmed complete medium. Gradually add 0.5 mL of cold XFM or SFM cryopreservation solution (2×) to the cell suspension, and softly mix to certify even cell distribution. If you use CTS™ Synth-a-Freeze® Medium, resuspend the cell pellet to the required ultimate concentration (1×10^6 cells/mL) and chase **steps 4–6** below.
4. Immediately add the desired volume of cell suspension (i.e., 1 mL) to cryovials.
5. Use an automated or manually controlled rate freezing apparatus following standard procedures (1 °C decrease per minute) and place the cryovials at −70 °C in a cryogenic freezing container (e.g., Mr. Frosty® (1 °C) Freezing Container).
6. After 24 h, transfer the frozen cells to liquid nitrogen (vapor phase); storage at −200 °C to −125 °C is recommended.

Characterization of human MSCs grown under xeno-free conditions has been depicted in Fig. 11, whereas hMSCs cultured in StemPro® MSC SFM retain trilineage differentiation possible through long-term passaging have been represented in Fig. 12.

Expected Results

Maintaining Trilineage Potential Under Serum-Free and Xeno-Free Conditions

Using CTS™ StemPro® MSC SFM, human MSC or ADSC can be extended for multiple passages while maintaining their trilineage mesoderm possible (i.e., ability to differentiate into osteogenic, chondrogenic, and adipogenic lineages).

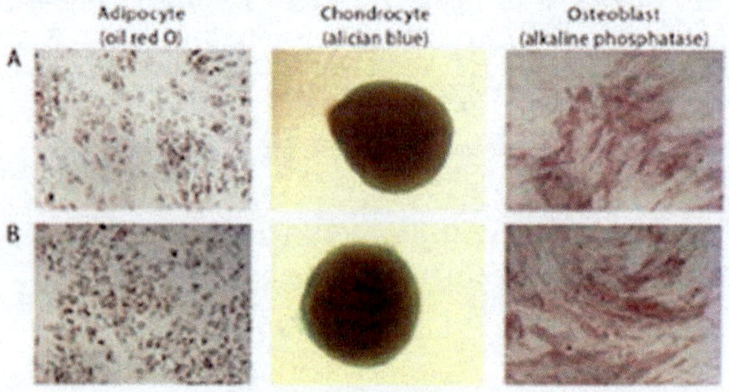

C

Marker	% Positive
DMEM + 10% MSC-Qualified FBS	
CD73+/NEG-	99.7
CD90+/NEG-	99.6
CD105+/NEG-	100.0
CD34+	0.4
StemPro® MSC SFM XenoFree	
Marker	% Positive
CD73+/NEG-	98.2
CD90+/NEG-	99.6
CD105+/NEG-	100.0
CD34+	0.2
NEG = multiplex analysis of CD14, CD19, CD45 and HLA-DR	

Fig. 11 Characterization of human MSCs grown under xeno-free conditions. Human bone marrow–derived MSCs extended in (**a**) DMEM (low glucose) + 10% MSC-Qualified FBS, or (**b**) StemPro® MSC SFM XenoFree + CTS™ CELLstart™ substrate–coated plates, exhibited trilineage mesoderm differentiation potential as revealed through oil red O staining (adipocyte), alcian blue staining (chondrocyte), and alkaline phosphatase staining (osteoblast)

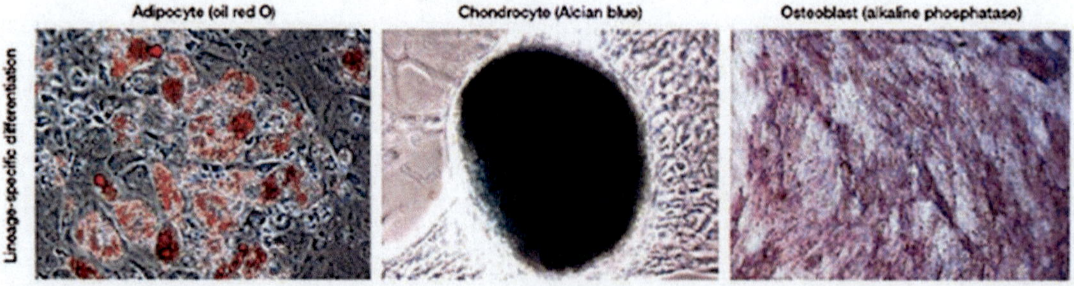

Fig. 12 hMSCs cultured in StemPro® MSC SFM retain trilineage differentiation possible through long-term passaging. hMSCs cultured and were seeded into adipogenic, chondrogenic, or osteogenic differentiation medium for 14 days

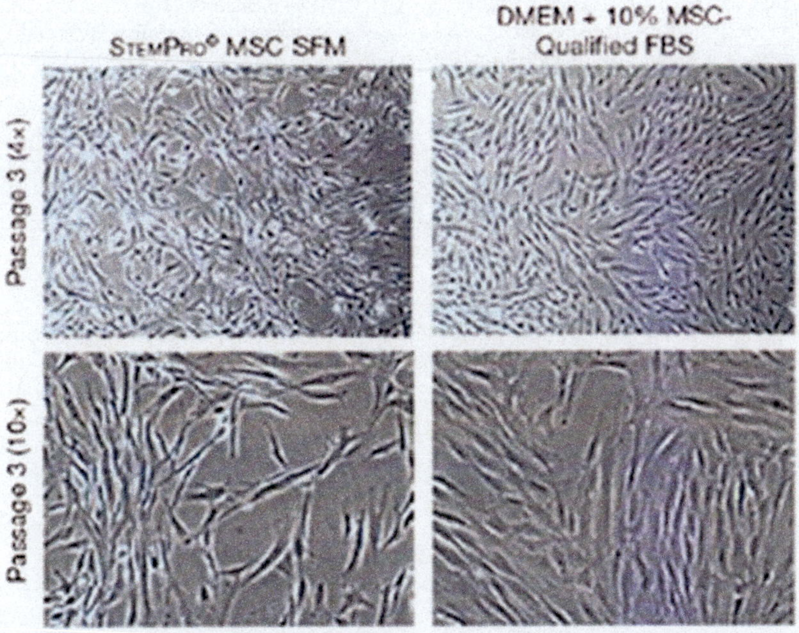

Fig. 13 hMSCs developed in StemPro® MSC SFM exhibit a less flattened, spindle-shaped morphology. Human MSCs prolonged in StemPro® MSC SFM or classical medium are shown

Morphology of hMSCs Developed Under Serum-Free and Xeno-Free Environments

Human MSCs developed in classical medium reached cell confluence at 1.0–3.0 × 10^4 cells/cm^2, human MSCs developed in StemPro® MSC SFM have a much lesser, spindle-shaped structures and can reach densities of >1.0 × 10^5 cells/cm^2 (Fig. 13).

References

1. WiCell Research Institute. Introduction to Human Embryonic Stem Cell Culture Methods, September, 2007. http://www.wicell.org/index.php?option=com_content&task=category&sectionid=7&id=246&Itemid=248
2. Gonzalez R, Wesselschmidt RL, Schwartz PH, Loring J (2007) Human Embryonic Stem Cell Culture. In: Loring J, Wesselschmidt RL, Schwartz PH (eds) *Human Stem Cell Manual: A Laboratory Guide*. Academic Press, New York, pp 5–7
3. Yao S, Chen S, Clark J, Hao EB, Hayek GM, Ding A, S. (2006) Long-term self-renewal and directed differentiation of human embryonic stem cells in chemically defined conditions. Proc Natl Acad Sci U S A 103:6907–6912
4. McDevitt TC, Palecek SP (2008) Innovation in the culture and derivation of pluripotent human stem cells. Curr Opin Biotechnol 19:527–533
5. Chase LG, Firpo MT (2007) Development of serum-free culture systems for human embryonic stem cells. Curr Opin Chem Biol 11:367–372
6. Nicklas W, Kraft V, Meyer B (1993) Contamination of transplantable tumors, cell lines, and monoclonal antibodies with rodent viruses. Lab Anim Sci 43:296–300
7. Stacey GN, Cobo F, Nieto A, Talavera P, Healy L, Concha A (2006) The development of 'feeder' cells for the preparation of clinical grade hES cell lines: challenges and solutions. J Biotechnol 125:583–588
8. Nelson-Rees WA, Daniels DW, Flandermeyer RR (1981) Cross-contamination of cells in culture. Science 212:446–452

9. MacLeod RA, Dirks WG, Matsuo Y et al (1999) Widespread intraspecies cross-contamination of human tumor cell lines arising at source. Int J Cancer 83:555–563
10. Freshney IR (2005) Culture of Animal Cells: A Manual of Basic Technique, 5th edn. John Wiley & Sons, Inc., Hoboken, New Jersey, pp 192–195
11. Upoff C, Drexler H (2011) Elimination of mycoplasmas from infected cell lines using antibiotics. Methods Mol Biol 731:105–114
12. Massachusetts Human Stem Cell Bank Standard Operating Procedures, 2009. https://www.umassmed.edu/mhscb
13. Katkov II, Kim MS, Bajpai R, Altman YS, Mercola M, Loring JF, Terskikh AV, Snyder EY, Levine F (2006) Cryopreservation by slow cooling with DMSO diminished production of Oct-4 pluripotency marker in human embryonic stem cells. Cryobiology 53:194–205
14. Ware CB, Nelson AM, Blau CA (2005) Controlled-rate freezing of human ES cells. BioTechniques 38(879–880):882–873
15. Borowski M, Giovino-Doherty M, Ji L, Shi MJ, Smith KP, Laning J. Basic pluripotent stem cell culture protocols (June 10, 2012), StemBook, ed. The Stem Cell Research Community, StemBook. https://doi.org/10.3824/stembook.1.63.1. https://www.stembook.org
16. Freshney RI (2005) Culture of animal cells: a manual of basic technique, 5th edn. John Wiley & Sons, New Jersey, pp 73–85
17. Massachusetts Human Stem Cell Bank (2009). Standard Operating Procedures. Available at. http://www.umassmed.edu/mhscb
18. Ng F et al (2008) PDGF, TGF-beta, and FGF signaling is important for differentiation and growth of mesenchymal stem cells (MSCs): transcriptional profiling can identify markers and signaling pathways important in differentiation of MSCs into adipogenic, chondrogenic, and osteogenic lineages. Blood 112 (2):295–307
19. Kuçi S et al (2010) CD271 antigen defines a subset of multipotent stromal cells with immunosuppressive and lymphohematopoietic engraftment-promoting properties. Haematologica 95(4):651–659
20. Sugii S et al (2010) Human and mouse adipose-derived cells support feeder-independent induction of pluripotent stem cells. Proc Natl Acad Sci U S A 107(8):3558–3563
21. Tsigkou O et al (2010) Engineered vascularized bone grafts. Proc Natl Acad Sci U S A 107 (8):3311–3316
22. Schraufstatter IU et al (2009) C3a and C5a are chemotactic factors for human mesenchymal stem cells, which cause prolonged ERK1/2 phosphorylation. J Immunol 182 (6):3827–3836
23. Carreras A et al (2009) Obstructive apneas induce early release of mesenchymal stem cells into circulating blood. Sleep 32(1):117–119

Chapter 26

Transcriptomics

Abstract

The **transcriptome** is the set of all RNA molecules in one cell or a population of cells. Transcriptome analysis leads to the identification of genes that are expressed differentially in distinct cell populations with respect to different treatments. Two methodologies for studying transcriptomics are DNA microarray and RNAseq. The current chapter discusses the methodologies for SAGE, CAGE, and DNA microarray. RNAseq as high-throughput technique is discussed in other chapter. Microarrays for transcriptomics typically fall into one of two broad categories: low-density spotted arrays or high-density short probe arrays. The application for transcriptomics includes diagnostics and disease profiling for human and animals and production traits for animals exclusively.

Keywords Transcriptome, DNA microarray, RNAseq, SAGE, CAGE

1 Basic Principle

The **transcriptome** may be defined as the set of all RNA molecules either in one cell or in a population of cells. Transcriptome analysis leads to the identification of genes that are expressed differentially in distinct cell populations with respect to different treatments. These are studied through high-throughput techniques as microarray analysis or RNAseq. mRNA acts as a transient intermediary molecule in the information network, whereas noncoding RNAs perform additional diverse functions. A transcriptome reflects the total transcripts present in a cell. Transcriptomics technologies gives a broad account of which cellular processes are active and dormant. One of the major challenges in molecular biology lies in analyzing how the same genome evolves to different cell types and how gene expression is regulated.

The first attempt for studying whole transcriptomes began in the early 1990s. Later on, technological advances since the late 1990s have repeatedly revolutionized the field. Transcriptomics have evolved as a widespread discipline in biological sciences.

There are two key contemporary techniques in the field:

Aruna Pal, *Protocols in Advanced Genomics and Allied Techniques*, Springer Protocols Handbooks,
https://doi.org/10.1007/978-1-0716-1818-9_26,

1. Microarrays that quantify a set of predetermined sequences, and.
2. RNAseq that uses high-throughput sequencing to record all transcripts.

   With the advent of newer technologies, the volume of data generated by each transcriptome experiment is enhanced. Accordingly, data analysis methods have steadily been adapted, leading to more accurate and efficient analysis with increasingly large volumes of data, referred to as Big data.

   Transcriptome databases have increased in volume and utility as more transcriptomes are collected and shared by researchers. It would be difficult for the interpretation of the information contained in a transcriptome without the reference of earlier experiments.

Measurement of the expression of an organism's genes in different tissues or conditions, or at different times, provides necessary information on the mechanism of gene regulation and other details of an organism's biology. It can also be employed to infer the functions of earlier unannotated genes. Transcriptome analysis has allowed the study of gene expression changes in different organisms and to understand human disease.

2 Protocol

In this chapter, we shall discuss the methodology of transcriptomics. In this chapter, we shall restrict with methodologies for DNA microarray, SAGE, and CAGE. RNAseq is discussed in separate chapter.

The primary step involves:

3 Isolation of RNA

All transcriptomic methods need RNA to first be isolated from the experimental organism for the recording of transcripts. RNA extrac tion techniques involve a sequence of steps—including mechanical disruption of cells or tissues membrane, followed by disruption of RNase with chaotropic salts and disruption of macromolecules and nucleotide complexes. It is followed by separation of RNA from undesired biomolecules as DNA and concentration of the RNA through precipitation from solution or elution from a solid matrix. Isolated RNA needs to be treated with DNase for digestion of any traces of DNA. It is essential to enrich messenger RNA since total RNA extracts are almost 98% ribosomal RNA. This can be aided through poly-A affinity methods or rRNA depletion using sequence-specific probes. Degraded RNA may affect downstream results; for example, mRNA enrichment from degraded samples will lead to the depletion of 5′ mRNA ends and an uneven signal across the length of a transcript.

4 Expressed Sequence Tags

An expressed sequence tag (EST) is a short nucleotide sequence derived from a single RNA transcript. The first step involves copy of RNA as complementary DNA (cDNA) with the help of a reverse transcriptase enzyme, followed by sequencing of the resultant cDNA. Since, ESTs can be collected without prior information of the organism from which they come, they can be derived from mixtures of organisms or environmental samples. Although higher-throughput methods are commonly used in current days, EST libraries usually provide sequence information for early micro-array designs. As for example, a barley microarray was designed from 350,000 previously sequenced ESTs.

5 Serial and Cap Analysis of Gene Expression (SAGE/CAGE)

5.1 Summary of SAGE

Within the organisms, mature mRNA transcripts (red) are synthesized from genes are transcribed and spliced (in eukaryotes). The mRNA is extracted from the organism, and subjected to reverse transcription, in order to copy the mRNA into stable double-stranded–cDNA (ds-cDNA; blue). In SAGE, the ds-cDNA is digested by restriction enzymes (at location 'X' and 'X' + 11) for the production of 11-nucleotide "tag" fragments. These tags are concatenated and sequenced with long-read Sanger sequencing (different shades of blue indicate tags from different genes). The sequences are deconvoluted to detect the frequency of each tag. The tag frequency can be employed to report on transcription of the gene that the tag came from.

Serial analysis of gene expression (SAGE) evolved as EST methodology to enhance the throughput of the tags generated, allowing some quantitation of transcript abundance. cDNA is developed from the RNA but is then digested into 11 bp "tag" fragments using restriction enzymes that cut DNA at a specific sequence, and 11 base pairs along from that sequence. The next process is the joining of cDNA tags head-to-tail into long strands (>500 bp) followed by sequencing using low-throughput, but long read-length methods such as Sanger sequencing.

The next step is the joining of these cDNA tags. In the next step, the sequences are divided back into their original 11 bp tags using computer software, the process is referred to as deconvolution. In case a reference genome is available, these tags can be matched to their corresponding gene in the genome. During

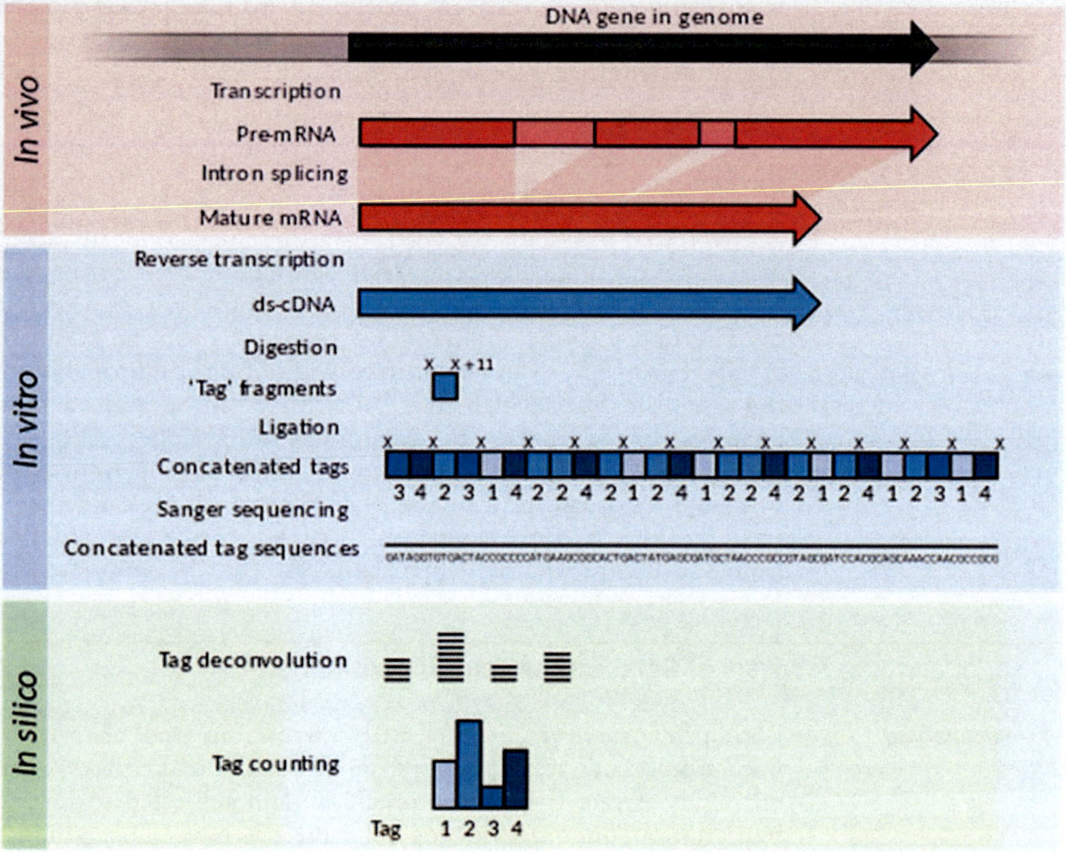

Fig. 1 Schematic representation of Transcriptomics with SAGE/CAGE

the absence of a reference genome, the tags can be directly employed as diagnostic markers if found to be differentially expressed in a disease state.

The cap analysis gene expression (CAGE) method is a variant of SAGE, where the sequences tags from the 5′ end of an mRNA transcript only. As a result, the transcriptional start site of genes can be identified in case the tags are aligned to a reference genome. Identifying gene start sites is employed for promoter analysis and for the cloning of full-length cDNAs.

SAGE and CAGE methods employ information on more genes as was possible when sequencing single ESTs. However, sample preparation and data analysis are more labor-intensive in SAGE and CAGE. Figure 1 represents schematically the Transcriptomics with SAGE/CAGE.

6 Microarrays

6.1 Summary of DNA Microarrays

In the process of DNA microarray, within the organisms, genes are transcribed and spliced (in eukaryotes) to generate mature mRNA transcripts (red). The mRNA is extracted from the organism and subjected to reverse transcription, leading to copying of the mRNA into stable ds-cDNA (blue). In the process of microarrays, the ds-cDNA is fragmented followed by fluorescent labelling (orange). Figure 2 schematically represents Transcriptomics with DNA microarray.

The labelled fragments bind to an ordered array of complementary oligonucleotides. Measurement of fluorescent intensity across the array determines the abundance of a predetermined set of sequences. These sequences are specifically chosen for reporting on genes of interest within the organism's genome.

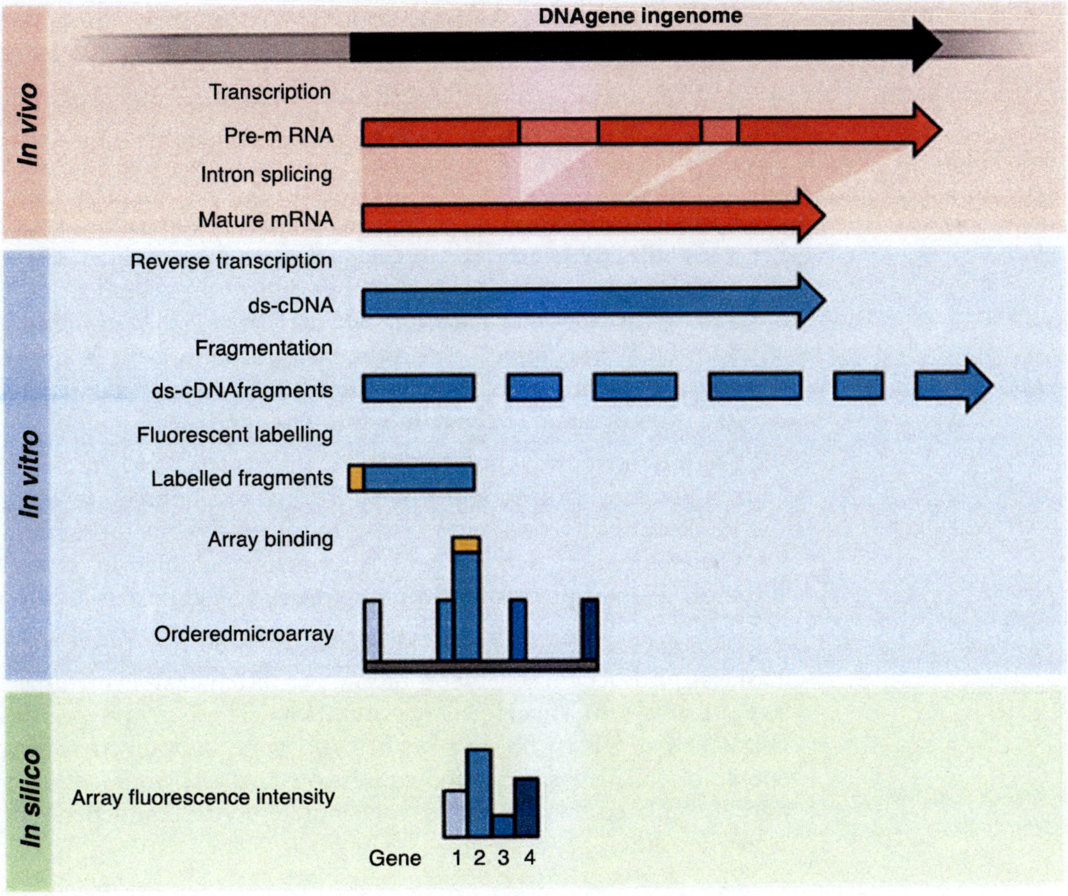

Fig. 2 Schematic representation of Transcriptomics with DNA microarray

7 Principles and Advances

Microarrays comprises of short nucleotide oligomers, referred to as "probes ", that are arrayed in a grid on a glass slide. Transcript abundance is controlled by hybridization of fluorescently labelled transcripts to these probes. The fluorescence intensity at each probe location on the array determines the transcript abundance for that probe sequence.

Microarrays need some knowledge for the genomics of the organism of interest. As for example, in the form of an annotated genome sequence, or a library of ESTs that can be employed to generate the probes for the array.

8 Methods

Microarrays for transcriptomics typically fall into one of two broad categories:

1. Low-density spotted arrays or.
2. High-density short probe arrays.

Transcript abundance is retrieved from the intensity of fluorescence received from fluorophore-tagged transcripts that bind to the array.

1. Spotted low-density arrays indicate picolitre drops of a range of purified cDNAs arrayed on the surface of a glass slide. These probes are longer than those of high-density arrays and lack the ability for the identification of alternative splicing events. Spotted arrays employ two different fluorophores to label the test and control samples, and the ratio of fluorescence is utilized to calculate a relative measure of abundance.
2. High-density arrays employ a single fluorescent label, and each sample is hybridized and thus individual detection is possible. High-density arrays were commonly used by the Affymetrix GeneChip array, when each transcript is quantified by several short 25-mer probes that together assay one gene.

NimbleGen arrays were a high-density array developed by a maskless-photochemistry method, that permitted flexible manufacture of arrays in small or large numbers. These arrays possess 100,000 s of 45 to 85-mer probes and were hybridized with a one-color labelled sample for expression analysis. Some designs were formulated that are able to incorporate up to 12 independent arrays per slide.

8.1 RNA-Seq

8.1.1 Summary of RNAseq

Within the organisms, genes are transcribed and spliced (in eukaryotes) resulting in the production of mature mRNA transcripts (red). The mRNA is extracted from the organism followed by fragmentation and copied into stable ds-cDNA (blue). The ds-cDNA is sequenced with high-throughput, short-read sequencing methods. These sequences are to be alignedlater to a reference genome sequence to reconstruct genome regions that were being transcribed. This data can be employed to annotate where expressed genes are, their relative expression levels, and presence of any alternative splice variants. Figure 3 represents transcriptomics with RNAseq schematically.

A large number of reads are necessary to ensure better coverage of the trancriptome, allowing detection of low abundance transcripts. Experimental design is further complicated by sequencing technologies that were subject to limited output range, the variability in the efficiency of sequence creation, and variability in sequence quality. It is to be kept in mind that each species has a different number of genes and as a result need a customized sequence yield for an effective transcriptome. Studies determined suitable thresholds empirically, but with the maturation of the technology, suitable coverage was predicted by transcriptome saturation with the help of computational methodologies. It is better to add more biological replicates rather than adding more reads in order to improve detection of differential expression in low expression genes. The Encyclopedia of DNA Elements (ENCODE) had recommended for 70-fold exome coverage for standard RNAseq. It has equally recommended for up to 500-fold exome coverage in order to detect rare transcripts and isoforms.

8.2 Data Analysis

Transcriptomics methods were observed to be highly parallel and need certain computations for the production of meaningful data for both microarray and RNAseq experiments. Microarray data is documented as high-resolution images, with arrangements for fea ture detection and spectral analysis. Raw image files are each about 750 MB in size, whereas the processed intensities are around 60 MB in size in a microarray. Multiple short probes are employed, which match a single transcript revealing details about the intron-exon structure. Suitable statistical models are needed for the determination of the authenticity of the resulting signal. RNAseq studies lead to the produce billions of short DNA sequences, which need to be aligned to reference genomes comprised of millions to billions of base pairs. *De novo* assembly of reads within a dataset needs constructing highly complex sequence graphs. RNAseq operations are highly repetitive and benefitted from parallelized computation. But modern-day algorithms implies that consumer computing hardware is enough for simple transcriptomics experiments that do not need de novo assembly of reads. A human transcriptome could be accurately captured with RNAseq with 30 million

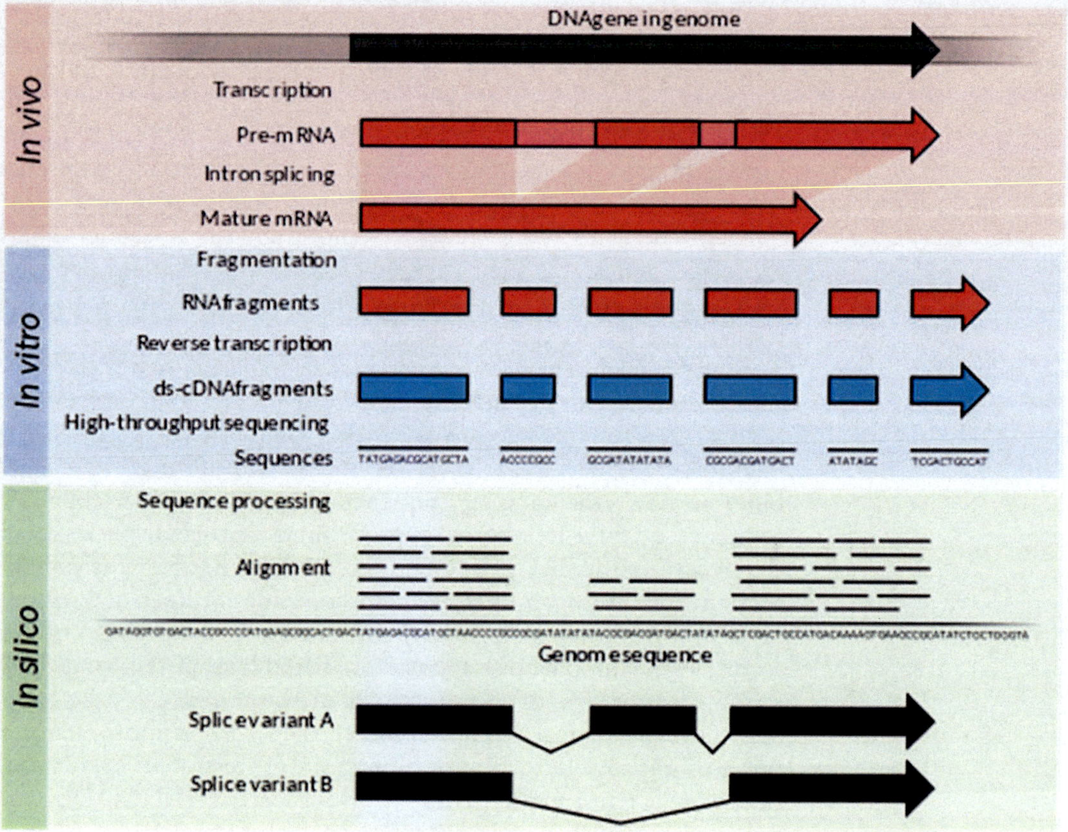

Fig. 3 Schematic representation of transcriptomics with RNAseq

100 bp sequences per sample. Approximately 1.8 gigabytes of disk space per sample is the pime necessity, when stored in a compressed fastq format. Processed count data for each gene would be much less, equivalent to processed microarray intensities. It is essential to store sequence data in public repositories, such as the Sequence Read Archive (SRA). RNAseq datasets can be uploaded via the Gene Expression Omnibus.

8.3 Image Processing

8.3.1 Microarray and Sequencing Flow Cell

Microarrays and RNAseq depend on image analysis in various ways (Fig. 4). In a microarray chip, each spot on a chip is a designated oligonucleotide probe, and fluorescence intensity directly detects the presence and quantity of a specific sequence (Affymetrix). In a high-throughput sequencing flow cell, spots are sequenced one nucleotide at a time. The color at each round indicates the next nucleotide in the sequence (Illumina Hiseq). Other variations of these techniques use more or fewer color channels.

Microarray image processing should correctly identify the regular grid of features within an image and deal with the quantification of the fluorescence intensity for each feature independently. Image

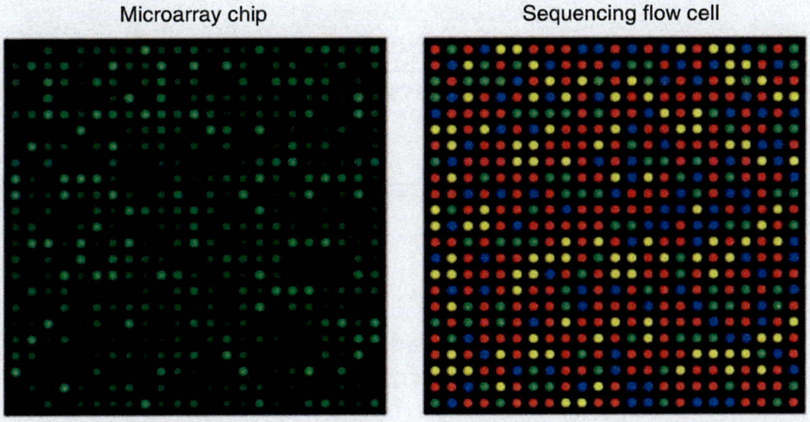

Fig. 4 Visualization of DNA microarray chip and RNAseq

artefacts need to be additionally identified and subsequently removed from the complete analysis. The detection and amount of each sequence is assessed through fluorescence intensities directly. It is possible since the sequence of each probe on the array is previously known.

The first steps of RNAseq also include similar image processing; however, conversion of images to sequence data is typically handled automatically by the instrument software. The Illumina sequencing-by-synthesis method results in an array of clusters distributed over the surface of a flow cell. The flow cell is imaged up to four times during each sequencing cycle, with tens to hundreds of cycles in total. Flow cell clusters are analogous to microarray spots and must be correctly identified during the early stages of the sequencing process. In Roche's pyrosequencing method, the intensity of emitted light determines the number of consecutive nucleotides in a homopolymer repeat. There are many variants on these methods, each with a different error profile for the resulting data.

9 Quality Control

Since the sequence reads are not totally perfect, the accuracy of each base in the sequence should be estimated for downstream analyses. It is essential to examine the raw data for the following points to be highlighted:

1. Quality scores for base calls are high,
2. The GC content matches the expected distribution,
3. Short sequence motifs (k-mers) are not over-represented,
4. The read duplication rate is acceptably low.

Several software options are present for sequence quality analysis, like FastQC and FaQCs. Abnormalities need to be removed through trimming or sequence need to be tagged for special treatment during later processes.

10 Alignment

The step involves linking of sequence read abundance to the expression of a specific gene. Two situations may arise. Firstly, when reference sequence is available for the species, transcript sequences are aligned to a reference genome. Secondly, when no reference sequence is available, *de novo* alignment to one another is recommended. The major challenges for alignment software involve:

1. sufficient speed to permit billions of short sequences need to be aligned in a meaningful timeframe,
2. flexibility to recognize and deal with intron splicing of eukaryotic mRNA, and,
3. correct assignment of reads that map to multiple locations.

Recent advanced softwares are available to deal with these issues, and they lead to increases in sequencing read length thereby reducing the chance of ambiguous read alignments. Various currently available high-throughput sequence aligners are maintained by the EBI.

Specialized handling of intron sequences are the primary requirement, as intron is absent in mature mRNA and it is conducted through alignment of primary transcript mRNA sequences derived from eukaryotes. Short read aligners undergo an additional round of alignments specially designed for the identification of splice junctions, with the information by canonical splice site sequences and known intron splice site information. Detection of intron splice junctions do not allow reads from being misaligned across splice junctions or erroneously discarded, which in turn allows more reads to be aligned to the reference genome and thereby accelerate the accuracy of gene expression estimates. Due to the fact that gene regulation can occur at the mRNA isoform level, splice-aware alignments also allow detection of isoform abundance changes that would otherwise be lost in a bulked analysis.

De novo assembly is applicable to align reads to one another to construct full-length transcript sequences without use of a reference genome. Challenges to de novo assembly involve larger computational requirements in comparison to a reference-based transcriptome, additional validation of gene variants or fragments, and additional annotation of assembled transcripts. The first metrics employed to detail the transcriptome assemblies, like N50, was observed to be misleading and currently better evaluation methods are available. Annotation-based metrics are of improved assessments of assembly completeness, as for example, contig reciprocal best hit count. The de novo assembly can be employed as a reference for subsequent sequence alignment methods and quantitative gene expression analysis.

11 Quantification

Quantification involves Heatmap identification of gene co-expression patterns across different samples as depicted in Fig. 5.

Each column represents the measurements for gene expression change for a single sample. Relative gene expression is denoted by color: high-expression (red), median-expression (white), and low-expression (blue). Expression profiles for the genes and samples that resemble can be automatically grouped (left and top trees). Samples can be various individuals, tissues, environments, or health conditions. In this example, expression of gene set 1 is high and expression of gene set 2 is low in samples 1, 2, and 3.

Quantification of sequence alignments can be conducted at different levels, particularly at genomic, exonic, or transcriptomic level. Typical outputs comprise a table of read counts for each feature supplied to the software; i.e., for genes in a general feature format file. Gene and exon read counts can be calculated efficiently with HTSeq. Quantitation at the transcript level is further complicated requiring probabilistic methods for the estimation of transcript isoform abundance from short read information. Cufflinks software is a good example. Reads which align equally well to multiple locations need to be identified. They may either be removed, aligned to one of the possible locations, or aligned to the most probable location.

Some quantification methods need an exact alignment of a read to a reference sequence. The kallisto software method is based on a combination of pseudoalignment and quantification into a single step that runs 2 orders of magnitude faster than contemporary methods employed as those of tophat/cufflinks software, with less computational burden.

12 Validation

Transcriptomic analyses can be validated with an independent technique. Quantitative PCR (qPCR) is a valid and effective technique, which is also statistically assessable. Gene expression is assessed against defined standards both for the gene of interest and control genes. The assessment by qPCR resembles to that obtained by RNAseq, wherein a value can be measured for the concentration of a target region in a given sample. qPCR is applicable to

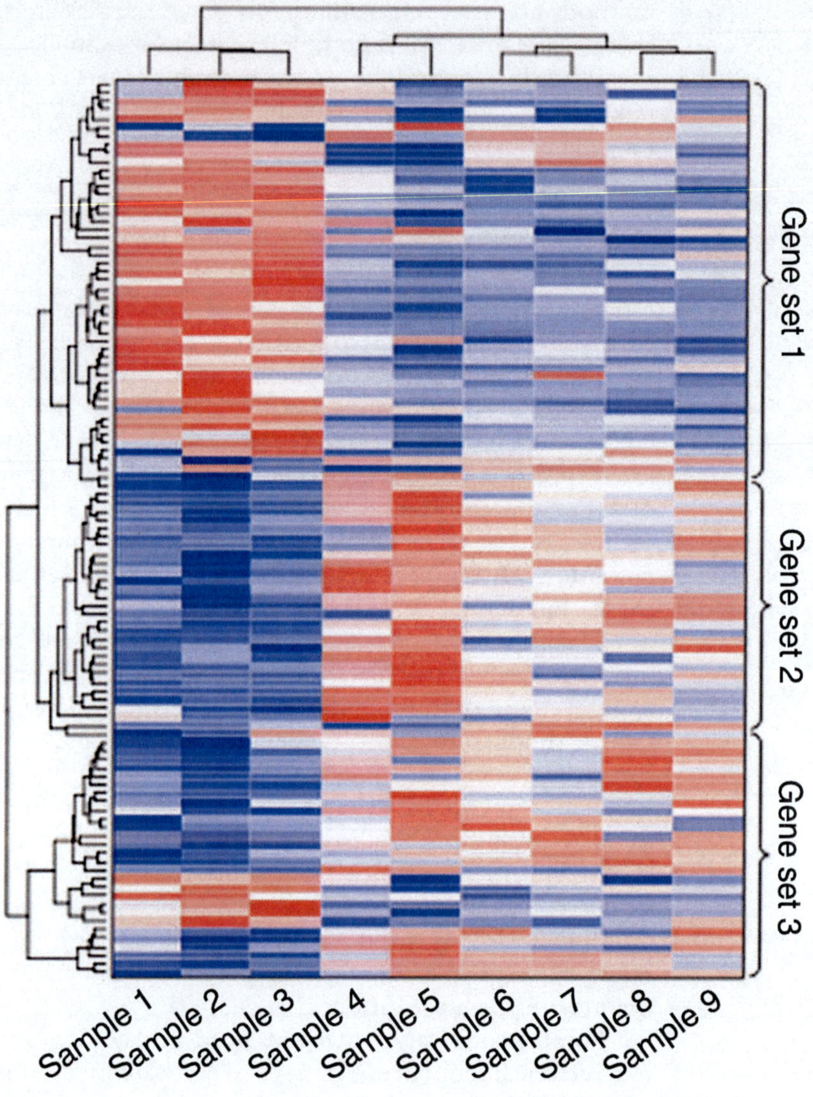

Fig. 5 Heatmap identification of gene co-expression patterns across different samples

amplicons less than 300 bp, generally towards the 3′ end of the coding region, avoiding the 3′UTR. In case validation of transcript isoforms is needed, an inspection of RNAseq read alignments depict where qPCR primers might be placed for maximum discrimination. The measurement of multiple control genes together with the genes of interest gives rise to a stable reference within a biological context. qPCR validation of RNAseq data has revealed that various RNAseq methods are highly correlated.

It is one of the most important phenomena to have functional validation of key genes for post transcriptome planning. Observed gene expression patterns can be functionally linked to a phenotype through an independent knock-down /rescue study in the organism of interest.

13 Applications

13.1 Diagnostics and Disease Profiling

Transcriptomic strategies have observed broad application across different areas of biomedical research, including disease diagnosis and profiling. RNAseq approaches have led to large-scale identification of

1. transcriptional start sites,
2. uncovered alternative promoter usage, and,
3. novel splicing alterations.

   These regulatory elements are of practical importance in human disease and, as a result, defining such variants is crucial to the interpretation of disease-association studies. RNAseq can also identify disease-associated single nucleotide polymorphisms (SNPs), allele-specific expression, and gene fusions , contributing to the understanding of disease causal variants [1, 2].

Retrotransposons are transposable elements that proliferate within eukaryotic genomes by a process of reverse transcription. RNAseq can inform about the transcription of endogenous retrotransposons influencing the transcription of neighboring genes by various epigenetic mechanisms that lead to disease. Accordingly, the importance using RNAseq to understand immune-related dis ease is increasing at a rapid rate because of the ability to dissect immune cell populations and to sequence T cell and B cell receptor repertoires from patients.

References

1. Hong SP, Chan TE, Lombardo Y et al (2019) Single-cell transcriptomics reveals multi-step adaptations to endocrine therapy. Nat Commun 10:3840. https://doi.org/10.1038/s41467-019-11721-9
2. Das Mahapatra K, Pasquali L, Søndergaard JN et al (2020) A comprehensive analysis of coding and non-coding transcriptomic changes in cutaneous squamous cell carcinoma. Sci Rep 10:3637. https://doi.org/10.1038/s41598-020-59660-6

Check for updates

Chapter 27

Development of Biomarker

Abstract

Biomarkers (or biological marker) are usually referred to as the naturally occurring molecule, gene, or characteristic by which a particular pathological or physiological process, disease, etc., can be identified. In this chapter, two approaches for biomarkers have been discussed—clinical biomarker for human and biomarker for animal in terms of health and production. Different types of commonly used molecular markers include RFLP, SSCP, AFLP, VNTR, and DNA sequencing. Different biomarkers for different diseases in human have been tabulated. Apart from these, different protocols were discussed for the diagnosis of human diseases (metabolic disease, tumor, neurodegenerative diseases, prion diseases, viral infection) with exosomes as biomarker. Molecular biomarkers were discussed for animals as marker-assisted selection or genomic selection.

Keywords Clinical biomarker, Molecular biomarker, QTL, Genomic selection, Marker-assisted selection, Exosomes

1 Basic Principle

Biomarkers are usually known as the naturally occurring molecule, gene, or characteristic through which a particular pathological or physiological process, disease, etc., can be detected.

Alternatively, biomarkers may also be defined as a biological characteristic, which can be measured, evaluated, and act as an indicator of normal physiological or pathological process. Biomarkers are equally useful for analyzing the efficacy of therapeutic agents or drugs. Biomarkers may include patterns of gene expression, levels of protein in body fluids, and alterations in electrical activity in the brain.

2 There Are Two Approaches for Biomarkers

In human, biomarkers are used mostly for the clinical diagnosis of diseases. In animal science, apart from health, production is also equally important. Hence, in this current chapter, we shall discuss about biomarkers for both human and livestock. Biomarker is referred as the short form for biological marker.

Aruna Pal, *Protocols in Advanced Genomics and Allied Techniques*, Springer Protocols Handbooks,
https://doi.org/10.1007/978-1-0716-1818-9_27,

2.1 Definition of a Biomarker

A biomarker has to be something that can be measured: accurately and reproducibly. There are several definitions of a biomarker. As per the World Health Organisation, biomarker is referred to as *any measurement reflecting* an interaction between a biological system and a potential hazard, which may be chemical, physical, or biological. The response can be measured at functional, physiological, biochemical, at the cellular level or at a molecular interaction level.

Biomarkers are not merely the medical symptoms, for example, say high temperature. However, an increased heart rate due to physical exertion may be referred to as biomarker. The increased heart rate is a physiological response to the exercise. Till date medical science utilizes biomarkers for assessing functionality of different body organs, say liver, kidney, and heart. We had already discussed in our earlier published article.

In this current chapter, we attempt to explore the biomarkers at molecular level.

2.2 Diagnostic Use of Biomarkers

A simple example of a diagnostic use of biomarkers is the estimation of biomarkers in blood to ascertain heart attack. Measurement of the levels of enzymes, hormones, and proteins in blood helps a doctor to assess the severity of heart attack and to assess the extent of damage of heart. Analysis of the blood sample can be conducted and the levels of marker chemicals can be estimated. In the case of a heart attack, a doctor assesses the blood level of two important chemicals as cardiac troponin and creatinine kinase. These may be regarded as the biomarker for heart attack or cardiac arrest. *Cardiac troponin* is a protein that enters the bloodstream soon after a heart attack and the level is maintained till other biomarkers return to normal levels. *Creatinine kinase* is an enzyme whose concentration rises after a heart attack. A certain type of CK can be estimated to assess the damage to the heart. It is to be kept in mind that the levels of biomarkers may alter over time and appear or disappear at different stages after an attack. Preventative biomarkers give an indication of the risk of a person getting a particular disease; for example, a genetic mutation is currently utilized to indicate the increased risk of getting breast cancer or heart disease.

3 Marker

Types of markers in animals:

1. Morphological marker: In this case, animals are selected based on appearance, for example, pigmentation. The main drawback is lack of polymorphism. Unless there is sufficient variability, selection becomes highly ineffective.

2. Biochemical marker: Here, the animals are selected based on biochemical characteristics, for example, variations in hemoglobin, amylase, and blood groups.
3. Chromosomal marker: Animals are selected based on structural and numerical variations in chromosome. Structural variations include deletion, insertion, and substitution. Numerical variations include trisomy and monosomy.

   Disadvantages are reduced polymorphism.
4. Molecular marker: These are the markers that reveal variation or polymorphism at the molecular level. The main advantages of molecular marker are that they are (i) highly polymorphic, (ii) co-dominant in expression, (iii) complete penetrance, (iv) non-destructive assay, (v) random distribution throughout genome, (vi) can be detected at day old stage, (vii) can be detected from any body tissue, (viii) available in both sexes, (ix) highly applicable for sex linked and sex-limited traits.

4 Markers

Diagnostic markers are biological parameters helpful for the diagnosis of diseases. These markers may consist of the presence of particular genetic variants and the levels of particular proteins in body fluids.

Molecular markers are a small fraction of the DNA sequence that usually makes up genes, while the major portion of the DNA represents non-coding sequences (97%), the exact function of which is not properly understood. Molecular markers may be present in the coding region (exon of gene), and may be in the non-coding region (intron within the gene or junk DNA). Most of the DNA (97%) remains outside gene, referred as junk DNA. These contain QTL markers, known as quantitative trait loci (QTL). Other markers may be RAPD, microsatellite, VNTR, etc. Linkage disequilibrium is an important criterion to be taken into consideration. If QTL is to be used for a marker, it should be closely linked with the genes of economic importance.

Different kinds of molecular markers exist. The efficacy of the marker differs with the kind of marker. Different markers (based on the method of detection) are as follows:

1. RFLP.
2. SSCP.
3. AFLP.
4. VNTR.
5. DNA sequencing.

Molecular markers as biomarkers are useful for both human and animals. But as per the need, their outcome is also different. In human, molecular marker or biomarkers are useful for disease diagnosis, whereas in animals, these are used for the selection of economically beneficial livestock, which are used mostly for food production. Molecular markers are equally applicable for animal Health. These facts are being discussed in details.

1. *Biomarkers for human- used for diagnosis.*
2. *Biomarkers for animals- to be used for marker-assisted selection or genomic selection.*

Let us discuss this topic in detail.

5 Biomarkers for Human Used for Diagnosis Purpose

In this section, molecular biomarkers are being described. In Human, molecular biomarker is commonly used for disease detection or diagnosis. Some of the particular diseases related to endocrine system, nervous system, or so on can be easily detected through biomarkers. For example, recently, it had been reported that cardiac troponin may act as biomarker for cardiac disease. Other recently included biomarkers include AX1N1 gene polymorphism for ovarian cancer. Intracellular coenzymes act as natural biomarkers for metabolic activities and mitochondrial anomalies. Neutrophil gelatinase-associated lipocalin act as promising bio marker for human acute kidney injury. Table 1 describes certain biomarkers pertaining to certain diseases [1–3].

Further list of molecular biomarkers with a potential role in heart health, insulin resistance, bone health, and genetic assessment of insulin and detoxification includes Tables 2, 3, 4, and 5.

6 Summary for Exosomes as Biomarker for Various Diseases

Protocol for molecular biomarker differs from case to case. Exosomes were observed to be potential biomarker for metabolic disease, tumor, cancer, neurodegenerative diseases, prion diseases, viral disease, and kidney related diseases. An attempt has been done to discuss these facts.

6.1 Exosome as a Biomarker for Metabolic Diseases—A Case Study

Exosomes are extracellular vesicles possessing the molecules that regulate the metabolic functions of adjacent or remote cells. Exosomes are released from various cells types, which have been documented through in vitro, in vivo, and clinical studies. It was observed that exosomes have role in metabolic disorders including type 2 diabetes. Hence, the proteins, mRNA, microRNA, and lipids in exosomes isolated from biological fluids were utilized as

Table 1
Some of the recently identified biomarkers for human diseases

Biomarkers	Disease
AX1N1 gene polymorphism	Ovarian cancer
Intracellular coenzymes	Metabolic activities and mitochondrial anomalies
Neutrophil gelatinase-associated lipocalin	Human acute kidney injury
CD44 epithelial isoform	Colorectal cancer
Cardiac troponins	Cardiac disease
hsa_circRNA_103636	Major depressive disorder
α-Synuclein as biofluid marker	Neurodegenerative diseases
Biomarkers of oxidative damage and inflammation	Alzheimer's disease
Folate metabolism abnormalities	Autism
Mitochondrial DNA deletions	Endometriosis in symptomatic females of child-bearing age
Exosome	Diagnosis and prognosis of malignant tumors
CD163+ immune cell infiltrates, presence of CD54+ microvessels	Diagnosis of patients with embryonal rhabdomyosarcoma

Table 2
Molecular biomarkers for identified genes responsible for heart function [1, 3–14]

Genes analyzed	Genetic variation screened for variations found in your gene
MTHFR	C677T, A1298C
MS-MTRR	A66G
MTR	A2756G
CBS	C699T
MnSOD	C(-28)T, T175C
SOD3	C760G
IL-6	G(-174)C
TNFα	G(-308)A
APOC3	C3175G
CETP	G279A
LPL	C1595G
eNOS	G894T
ACE	DEL

Table 3
Molecular biomarkers for identified genes responsible for insulin resistance

Gene	Role in insulin resistance	Polymorphism or variation detected at molecular level
VDR	Mechanism of insulin secretion	CTaqIT
VDR	Mechanism of insulin secretion	TBsmIC
IL-6 TNF-α	Inflammatory response; response of cells to insulin	G(−174)C G(−308)A
PPARγ	Glucose and lipid metabolism	Pro12A
ACE	Blood pressure regulation	II/DD

Table 4
Molecular biomarkers for identified genes responsible for bone health

Gene	Role of the gene in Bone Health	Genetic variation screened for variations found at genomic level
VDR	Regulates absorption and metabolism of vitamin D and calcium Roles in bone formation	CTaqIT
VDR		TBsmIC
VDR		FokI
COL1A1	Collagen role in bone formation	GSp1T
IL6	Old bone elimination and new bone creation cycle	G(-174)C
IL6		G(-164)C
TNFα		G(-308)A

biomarkers in metabolic disorders. Till date, certain problems with the development of clinically applicable biomarkers have not yet been worked out completely. The biologic functions of exosomes are briefly introduced in this chapter, discussing the technical and practical advantages and limitations of different methods of exosome isolation for the detection of exosomal biomarkers of metabolic disorders. Useful exosomal biomarkers have the potentiality to provide enough information about the pathogenic mechanisms and complications of metabolic syndrome. This needs isolation and characterization of high-purity exosomes.

Exosomes are nano vesicles (50–100 nm diameter), which are released from most viable cells. Recent studies depict that exosomes correspond to the intraluminal vesicles of endosomal multivesicular bodies (MVBs). These are released in the extracellular environment following MVB fusion with the plasma membrane. They develop

Table 5
Molecular biomarkers for identified genes responsible for genetic assessment for antioxidant and detoxification

Gene	Role of gene in antioxidant/detoxification	Polymorphism/variation detected
eNOS	Vascular function	G894T
GSTM1	Detoxification	(DELETION)
GSTP1		A313G
GSTT1		A341G
MnSOD	Destroys free radical	C(-28)T
		T175C
SOD3		C760G

Study (year)	Biosample	Pathology	Levels	Exosomes proteins	Exosomes nucleic acids
Smalley et al. (2008)	Urine	Bladder cancer	Up	EGF receptor pathway proteins, retinoic acid-induced protein 3, α-subunit of GsGTP-binding protein and resistin	ND
Nilsson et al. (2009)	Urine	Prostate cancer	Up	PCA-3, TMPRSS2:ERG and PSA	ND
Zhou et al. (2006)	Urine	Acute kidney injury	Up	Fetuin-a	ND
Sonoda et al. (2009)	Urine	Renal ischemia reperfusion	Down	Aquaporin-1	ND
Rood et al. (2010); Wittmann and Jäck (2010)	Urine urine	Nephrotic syndrome	Up	Neprilysin, aquaporin-2 and podocalyxin	ND
Wittmann and Jäck (2010)	Plasma	Prostate cancer	Up	ND	miR-16, miR-34b, miR-92a, miR-92b, miR-103, miR-107, miR-197, miR-328, miR-485-3p, miR-486-5p, miR-574-3p, miR-636, miR-640, miR-766 and miR-885-5p
Corcoran et al. (2011)	Plasma	Breast cancer	Up	ND	miR-141 and miR195
Keller et al. (2009); Taylor and Gercel-Taylor (2008); Li et al. (2009)	Plasma	Ovarian cancer	Up	ADAM 10, CD24, L1CAM, EMMPRIMN and claudin	miR-21, miR-141,miR-203, miR-205, miR-214, miR-200a, miR-200b and miR-200c
Bala et al. (2012)	Plasma	Alcoholic liver disease	Up	ND	miR-122 and miR-155

ND: No data

through membrane invagination of late endosomes that lead to vesicles containing cytosolic components and expose the extracellular domain of some plasma membrane receptors at their surface. The concept of other MVB-independent mechanisms to the exosome formation process may also exist. Exosomes possess specific protein markers of the endosomal pathway, such as tetraspanins (CD63, CD9, and CD81), heat shock proteins (HSP70), and proteins from the Rab family, Tsg101 and Alix, that are absent in other types of vesicles of similar size. Ectosomes are synthesized upon plasma membrane shedding and possess markers such as integrins and P-selectin that are unrelated to the endosomal/exosomal pathway and are CD63 negative. Likewise, apoptotic particles may be differentiated as they selectively possess histones. Exosomes were the first to be detected in the harvested media of reticulocyte cell cultures as microvesicles that contain membrane proteins, as the transferrin receptor. Later on, several cell types have been revealed to release exosomes into the extracellular environment. A group of hematopoietic cells such as B cells, T cells, dendritic cells, mast cells and platelets, intestinal epithelial cells, Schwann cells, neuronal cells, adipocytes, fibroblasts (NIH3T3), and tumor cells were detected to secrete MVB internal vesicles.

Platelets secrete exosomes that are able to regulate the coagulation response; B cells and dendritic cells act as Immune system cells. They aid in mediating MHC-dependent immune responses on nanovesicle secretion. Hence, vesicles possess specific adhesion molecules for narrow targeting of recipient cells. Dendritic cell-derived exosomes, which were collected from a variety of different tumors, eradicate and suppress the growth of murine tumors in a T-cell-dependent manner. It is interesting to note that immunogenicity is more in an antigen-presenting vesicles, exosomes, than that of soluble molecules and accordingly; currently, investigation is focused on the production of safe vaccines with exosomes.

It has been observed that exosomes secreted by glioblastoma and mast cells may carry mRNAs encoding for up to 1300 genes that are being delivered and properly translated into target cells. The vesicle-associated RNA is known as exosomal shuttle RNA including small RNA species, such as miRNAs. Exosomal miRNAs are functional in target cells producing effective gene silencing.

It has been suggested that exosomes are cell "cleaners" as well as cause cell shuttles. They can act a crucial role in cell-to-cell communication. They act as highly specialized functional extensions of cells, increasing their area of action. They are usually secreted in biological fluids, as for example, plasma, urine, cerebrospinal fluid, epididymal fluid, amniotic fluid, malignant and pleural effusions of ascites, bronchoalveolar lavage fluid, synovial fluids and breast milk. This suggests a mechanism of the exchange of information across distant body compartments.

The molecular contents of the exosomes that are secreted in body fluids have shown to be highly specific and a precious biomedical tool. It has been suggested that exosome purification may be a more informative diagnostic tool than whole body fluid. Cells for the drainage system, as for example, kidney tubular cells, provide exosomal markers that cannot be detected in whole urine.

Urine exosomal preparations from prostate cancer patients secrete two products resulting from an *ERG* fusion chromosomal rearrangement, such as mRNA of PCA3 and TMPRSS2. In a similar way, certain urine exosomes may act as diagnostic and prognostic markers of renal ischemia/reperfusion injury. Antidiuretic hormone action, such as aquaporin-1 and -2, has been also been detected during renal ischemia/reperfusion injury. However, these molecules are not easily detected in whole urine.

The power of nanovesicles as biomarkers are based on the enrichment of highly selected markers during exosomal sorting. It consists of a little proportion (less than 0.01%) of the total proteome of body fluids. The enhancement of diagnostic markers at the exosomal source, aids in the detection of relatively lowly expressed biomarkers that would otherwise go undetected. In healthy human-circulating exosomes, approximately 66 proteins were detected, out of which majority were involved in vesicular trafficking pathways. The exosome biomarker strategies were considered important, as minute quantities of proteins and RNAs give very insightful information which is specifically associated with pathological conditions.

Exosomes are the small vesicles that are secreted into the extracellular environment through the process of endosome fusion with the plasma membrane. The molecular content of exosomes secreted into the body fluid have been regarded as a precious diagnostic tool [15, 16].

6.2 Exosome-Based Strategies for the Diagnosis of Tumors—A Case Study

An adapted ELISA test can detect, characterize and quantify exosomes. It has been detected that tumor patients possess significantly higher plasmatic levels of exosomes expressing *CAV1* in comparison to the plasma of healthy donors. It has been detected that plasmatic exosomes deliver lung tumor-associated miRNAs. This implies that circulating exosomal miRNAs represent to those expressed in the tumors can detect disease earlier. Earlier studies have depicted that salivary exosome protein and miRNA content may be advantageous to detect disease conditions, like Sjogren's syndrome, and exosomes from amniotic fluids can be useful for prenatal diagnosis.

It is a unique idea that exosomal biomarker may be responsible for more than one condition. Plasma exosomes are released by various cell sources, from both blood cells and tissues that are in turn filtered by endothelial cells and organs like the placenta and the liver. Thus, isolation of cell-specific exosomes through immunoprecipitation would be a successful methodology to detect differential changes, which as a result discriminate the cell lineages that produce them. Plasma exosomes of hepatic origin can be isolated with antibodies that are specific to the liver marker carbamoyl phosphate synthetase 1. As a result, body fluid exosomes comprise a source of multiple markers that gives clinically useful information.

An initial screening in conditioned media from cultured cells is useful for building a predictive model of deregulated pathways depicted in exosomes. Models can be translated in vivo into biological fluid sampling from a large patient cohort as the next step.

Publications depict the association of exosomal markers with several cancer types. Some exosome-based diagnostic assays are presently been reported for clinical use and are in vogue.

An exosome-based tumor diagnostic assay called ExoTEST™ (HansaBioMed, Estonia), which is a sandwich ELISA has evolved for capturing and quantification of exosomes in human plasma and other body fluids, also in cell culture supernatants. The basic principle involves that proteins of exosomes such as CD63 and Rab-5b, and a tumor-associated marker (e.g., CAV1) are useful for immunoisolation of nanovesicles. Western blot and flow cytometry analysis were employed for estimate the accuracy of the method.

In clinical studies, it was realized that use of ExoTEST, $CD63^+$ and $CAV1^+$ plasma exosome levels were significantly more in melanoma patients in comparison to healthy donors. ExoTEST was proved to have a sensitivity of 43 and 68% to detect $CD63^+$ and $CAV1^+$ plasma exosomes, respectively. In melanoma patients, plasma exosomes that express $CAV1^+$ were shown to be significantly greater in comparison to plasma exosomes expressing $CD63^+$. The ExoTEST could detect and quantify exosomes purified from human tumor cell culture supernatant and plasma from severe combined immunodeficiency mice engrafted with human melanoma. It was observed that plasma levels of exosomes in melanoma-engrafted severe combined immunodeficiency mice were correlated to tumor size.

The greatest limitation of use of exosome for diagnostic purpose is the lack of throughput of plasma/biological fluids nanovesicle isolation. In spite of a number of commercially available assays, this methodology for exosome isolation provides fast protocols. Ultracentrifugation protocols of large volumes will lead to a suitable yield. This is the limitation for high-throughput implementation of exosome-based diagnostic approaches. Additionally, the specificity and sensitivity in plasma and urine can be diminished by immunoglobulin interference [15, 16].

Majority of the misfolded proteins linked to neurodegeneration are shuttled through exosomes. Superoxide dismutase, α-synuclein,

6.3 Exosomes: Novel Biomarkers for Neurodegenerative Diseases

amyloid, and tau are associated with amyotrophic lateral sclerosis, Parkinson's disease, Alzheimer's disease (AD), and tauopathies, respectively. These are all processed through the MVB pathway and secreted along with exosomes by cultured cells. The presence of exosome-associated tau is also detected in human cerebrospinal fluid samples of early AD patients.

It has been suggested that misfolded proteins linked with the neurodegenerative diseases may spread the "misfolded status" to unaffected brain areas through a prion-like mechanism. The exosomal shuttling system is responsible for the spread of the newly formed toxic proteins within the CNS. Exosomes secreted from α-synuclein overexpressing SH-SY5Y cells possess α-synuclein. These exosomes can effectively transfer the protein to normal SH-SY5Y cells. Exosome-associated α-synuclein oligomers are to be taken up by recipient cells, leading to the induction of more toxicity compared with free oligomers. Neuronal cell also secretes another toxic form of synuclein, γ-synuclein, into conditioned medium as exosomes.

It is crucial to validate this hypothesis in the future with large-scale in vivo studies and ascertain the potential of exosomal miRNAs in the diagnosis of prion diseases.

Majority of the misfolded proteins associated with neurodegeneration act through exosomes. As per the "prion-like" theory, neurotoxic misfolded proteins are commonly transmissible. Their spread within the CNS probably occurs through the exosomal shuttling system. miRNAs are deregulated in the CNS through neurodegeneration. Although it is not properly known if they are released in exosome body fluids. Further research is necessary to assess the diagnostic value of exosomes in neurodegenerative diseases.

6.4 Exosome and Other Human Pathological Conditions

It was very interesting to reveal that exosomes possess functional nucleic acids. It suggests that exosomes are similar to viral-like particles. It is remarkable to note that many retroviruses, including HIV, have been revealed to use the MVB machinery for propagation. Other type of virions such as Epstein–Barr virus (EBV) employ the exosomal pathway. As for example, EBV miRNAs are secreted by EBV-infected B cells by exosomes. These exosomal miRNAs can be internalized by dendritic cells when there is silencing of specific EBV target immunoregulatory genes, leading to the spread of the virus. Human intestinal epithelial cells infected with rotavirus (RV) secrete exosomes containing both viral proteins (VP6), and immunomodulators, for example, heat shock proteins and TGF-β1.

Thus, exosomes lay a role in the spread of pathogens including some types of viruses like Epstein–Barr virus and rotavirus. They have been regarded as the potential biomarkers for kidney-related diseases.

6.5 Diagnosis of Prion Diseases

Diagnostic test for transmissible spongiform encephalopathies is currently unavailable at an early stage, and blood carriers of prions are unknown. Prions are shuttled through exosomes in various in vitro cell models; however, its occurrence in body fluids is still not confirmed. Some miRNAs are particularly deregulated in the brain of numerous transmissible spongiform encephalopathy models. After prion infection, these miRNAs are secreted in the extracellular environment in exosomes with cultured cells. It suggests that they might also be secreted in exosomes into body fluids in vivo.

7 Biomarkers for Animals to Be Used for Marker-Assisted Selection or Genomic Selection

It is difficult to explain the protocol for biomarker as such, since there are numerous ways of studying it. Hence protocol differs. We shall describe some case studies.

Selection is the process of differential rate of reproduction where some parents with better phenotype were allowed to reproduce whereas others are not. We select the food animals reared in farm for their production. Cattle and buffalo are selected for milk production. Hence our aim would be to select those cattle yielding more quantity of milk than others. Similarly, goat and sheep are reared for meat production. Sheeps are also reared for wool production. Poultry are reared for both egg and meat production. Conventional selection has its own limitations with genetic gain up to a limited extent. Certain traits are difficult to measure in the lifetime of the animal, say for example the meat traits, which are obtained only after slaughter. It is equally difficult to select for sex limited traits.

Marker-assisted selection (MAS): It is the process of indirect selection aided by molecular markers. The addition of genomic information to phenotypic information in order to improve the selection response to the traditional method is known as marker-assisted selection (MAS). It is the method when marker genes are used to detect the presence of desirable genes is called as marker-assisted selection.

MAS is based on the concept to refer to the presence of a gene from the presence of a marker, which will be tightly linked to the gene of interest. The marker and the gene are should be tightly linked, then there is possibility of their transmission together to the progeny individuals.

Marker-assisted selection (MAS) is indirect selection process when a trait of interest is selected based on a marker linked to the trait, but not based on the trait itself. The idea is to combine all genetic information at markers and QTL with the phenotypic information in order to improve genetic evaluation and selection. Marker-assisted selection is also known as marker-assisted breeding, marker-assisted introgression, or marker-aided selection. It differs from gene assisted selection (GAS) that refers to the selection which is based on QTLs.

8 History of MAS

The concept of marker-assisted selection (MAS) using the information of polymorphic loci act as an aid to selection, which was introduced as early as in 1900. From 1970 onwards, technology was developed for identification and genotyping of large number of genetic markers. Researches were undertaken to demonstrate the identification of genomic regions that control variation in quantitative traits. The work was conducted on QTL to enhance selection, it is expected to have potential application of gene or marker-assisted selection in livestock.

Relative advantage of MAS/GAS over traditional selection is higher in case of

1. traits with low heritability.
2. the QTL is of larger effect.
3. the favorable allele is initially rare.
4. QTL and marker are closely linked.
5. mode of gene action is nonadditive.

Gene markers are most effective for traits that are difficult to improve under traditional selection. They include.

1. Slaughter and carcass traits, for example, pH, Tenderness, colour of meat.
2. Traits measured on one sex only.
3. Sex limited trait as milk production.
4. Traits that are measured late in life.
5. Lifetime fecundity.
6. Traits that are difficult or expensive to measure.
7. Disease resistance traits.

8.1 Methodological Advantage of MAS

The advantages include that DNA can be isolated from any tissue, for example, blood and hair. DNA can be isolated at any stage even during fetal life. DNA has longer shelf-life readily exchangeable b/w labs. Analysis of DNA can be carried out at early age/ even at the embryonic stage. DNA can be isolated for all the sexes.

Successful MAS breeding programs require advances in following areas:

1. Gene mapping: Detection and mapping of genes and genetic polymorphisms.
2. Marker genotyping: Genotyping of large numbers of individuals for numerous markers at a justified cost for QTL detection with routine application for MAS.
3. QTL detection: It is the process of identification and measurement of associations of identified genes and genetic markers with economic traits.
4. Genetic evaluation: It is the process of integration of phenotypic and genotypic data in statistical methods in order to estimate breeding values of individuals in a breeding population.
5. MAS: It is the process of development of breeding strategies and programs for utilizing the molecular genetic information in selection and mating programs.

The major Steps/ Methodology involved in MAS are as follows:

1. Validation of molecular markers:

   DNA is isolated from test individuals and association with marker and the trait is established.
2. The next step involves DNA extraction from the breeding population and adoption of Marker-assisted selection. Accordingly, selection of individuals on the basis of presence of desired molecular markers for the concerned trait.
3. From markers to MAS:

It is necessary to construct high density DNA marker map for economically important farm animals, which involves many markers of known location, distributed throughout the genome and interspersed at relatively short intervals. It provides a framework for further application for MAS. The next step is the detection of putative genes affecting particular traits of interest through the process of testing for associations between marker variants and any trait of interest.

8.2 Limitations of MAS

1. The initial cost involvement is high.
2. It requires high cost at initial stages.
3. It requires expert technical skill.
4. To maximize the benefit, it requires certain automated techniques.
5. Genotype environment interaction plays a major role in affecting the phenotypic traits. Hence, it is essential to conduct phenotyping in multiple environments.

6. DNA marker need to be validated separately for each the breeding population. Assumptions for the validity of markers need to be considered strictly.
7. Large population size is utmost essential for its validation of MAS.

9 Current Application of MAS in Livestock

The genome map that was reported for the first-time livestock was for the chicken in 1992, which was subsequently followed by publication of maps for cattle, pigs and sheep. Thereafter, the search for useful markers has continued with the addition of further species, including the goat, horse, rabbit and turkey (*see* http://www.thearkdb.org). Microsatellite markers have been one of the most promising marker. Markers have been detected for almost all farm animal species, along with that for the milk production in dairy cattle, buffalo, growth and carcass traits of beef cattle, chicken, and goat.

There are reports of methodology for MAS, genomic selection strategies, use of molecular technologies various application of biomakers in animal science. These include the advancement of animal breeding, the efficiency of MAS and genomic selection, QTLs and epistatic effects, types of selection model, genome-wide screening for markers, MAS in dairy breeding, relationship between MAS and linkage analysis, relationship between MAS and inbreeding and selection for sex limited characters. ISAG–FAO recommended certain microsatellite markers for different species, as cattle, buffalo, sheep, goat, horse, donkey, camelid, pig and chicken.

Holstein Association, USA has established genomic testing technologies with a wide array of tests (Holstein Association USA, 2011). Illumina designed cost-effective chips and the services to evaluate the genetic merits of cattle are open. An Animal QTL database (Animal QTLdb) collects all publicly available trait mapping data, for example, QTL (phenotype/expression, eQTL), candidate gene and association data (GWAS). Copy number variations (CNV) mapped to livestock animal genomes was developed for the location and comparison of discoveries within and between species (http://www.animalgenome.org/cgi-bin/QTLdb/index).

The practical application of MAS with Booroola Fecundity gene has a major effect for improving prolificacy, which needs to be implemented in future breeding policy. MAS is also very much effective for qualitative traits [17–25].

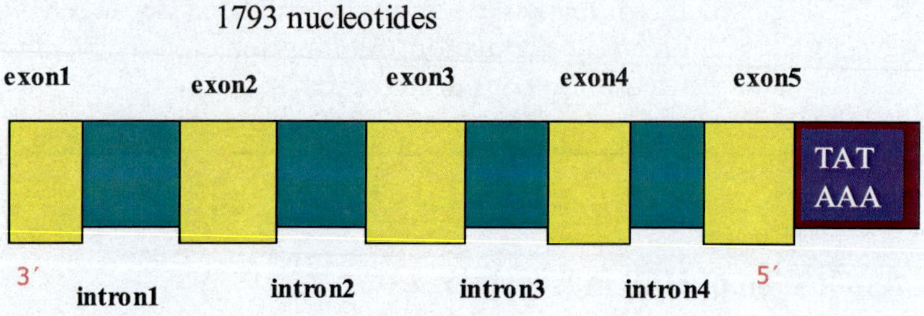

Fig. 1 Schematic representation of Growth hormone gene

10 A Simple Example of Molecular Marker (RFLP)

In the fifth exon of growth hormone gene, CTG codes for leucine, while GTG codes for valine (Fig. 1). Through phenotypic association study, it was observed that leucine variant codes for better phenotype in terms of growth, reproduction, and production. Hence, it may be considered as a marker for above-described phenotypic traits. It has to be observed if the marker is segregated or not and it is transmitted from generation to generation.

Application of Molecular markers:

1. Gene mapping.
2. Pre and postnatal diagnosis of diseases.
3. Anthropological and molecular evolution studies.

11 Conventional Techniques for Detection of Polymorphism

RFLP, SSCP, RAPD, AFLP, Sequencing (Sanger's sequencing).

In earlier chapters, we have already discussed the protocol for conventional techniques (as RFLP, SSCP).

12 Conclusion

The application of MAS in breeding programs is dependent on the information to breeders about variable marker information from animal to animal and the different effects on multiple traits. It is also dependent on the genotypic information that is helpful to improve their commercial breeding activities. MAS can equally evolve an apparently possible methodology for the selection for genetic disease resistance animals and detection of possibility of diseases at an earlier age in case of human. In the future to make MAS effective in large breeding populations or cohorts, the availability of large-scale genotyping methods and infrastructure that allows the generation of hundreds of thousands of molecular data at a reasonable cost will be necessary. Apart from MAS, molecular data are very much useful in any conservation program or evolutionary or demographic studies for the assessment of biodiversity through heterozygosity index, polymorphism information index (PIC) and for studying ethnic population in human race.

References

1. Heikal AA (2019) Intracellular coenzymes as natural biomarkers for metabolic activities andmitochondrial anomalies. Biomark Med 4 (2):241–263. https://doi.org/10.2217/bmm.10.1
2. Creed J, Maggrah A, Reguly B, Harbottle A (2019) Mitochondrial DNA deletions accurately detect endometriosis in symptomatic females of child-bearing age. Biomark Med 13 (4):291–306. https://doi.org/10.2217/bmm-2018-0419
3. Devarajan P (2010) Neutrophil gelatinase-associated lipocalin: a promising biomarker for human acute kidney injury. Biomark Med 4(2):265–280. https://doi.org/10.2217/bmm.13.63. Review
4. Simonsen AH, Kuiperij B, El-Agnaf OMA, Engelborghs S, Herukka S-K, Parnetti L, Rektorova I, Vanmechelen E, Kapaki E, Verbeek M, Mollenhauer B (2015) The utility of α-synuclein as biofluid marker in neurodegenerative diseases: a systematic review of the literature. Biomark Med 10(1):19–34. https://doi.org/10.2217/BMM.14.105
5. Sadeghi A, Roudi R, Mirzaei A, Mirzaei AZ, Madjd Z, Abolhasani M (2019) CD44 epithelial isoform inversely associates with invasive characteristics of colorectal cancer. Biomark Med 13(6):419–426. https://doi.org/10.2217/bmm-2018-0337
6. Passino C, Aimo A, Masotti S, Musetti V, Prontera C, Emdin M, Clerico A (2019) Cardiac troponins as biomarkers for cardiac disease. Biomark Med 13(5):325–330. https://doi.org/10.2217/bmm-2019-0039
7. Cui X, Niu W, Kong L, He M, Kong L, He M, Jiang K, Chen S, Zhong A, Li W, Lu J, Zhang L (2016) hsa_circRNA_103636: potential novel diagnostic and therapeutic biomarker in major depressive disorder. Biomark Med 10 (9):943–952. https://doi.org/10.2217/bmm-2016-0130
8. Galasko D, Montine TJ (2010) Biomarkers of oxidative damage and inflammation in Alzheimer's disease.Theme: Alzheimer's disease: the biomarker revolution-review. Biomark Med 4 (1):27–36. https://doi.org/10.2217/bmm.09.89
9. Kather JN, Hörner C, Simon-Keller K (2019) CD163+ immune cell infiltrates and presence of CD54+ microvessels are prognostic markers for patients with embryonal rhabdomyosarcoma. Sci Rep 9:9211
10. Wojsiat J, Laskowska-Kaszub K, Mietelska-Porowska A, Wojda U (2017) Search for Alzheimer's disease biomarkers in blood cells: hypotheses-driven approach. Biomark Med 11 (10):917–931. https://doi.org/10.2217/bmm-2017-0041
11. Ryan NS, Rossor MN (2010) Correlating familial Alzheimer's disease gene mutations with clinical phenotype. Biomark Med 4 (1):99–112. Theme: Alzheimer's disease: The Biomarker revolution. 2010. https://doi.org/10.2217/bmm.09.92
12. Frye RE, Slattery JC, Quadros EV (2017) Folate metabolism abnormalities in autism: potential biomarkers. Biomark Med 11 (8):687–699. https://doi.org/10.2217/bmm-2017-0109. Review
13. Wan Y, Zheng SY (2019) Enhanced detection of tumor secreted vesicles. Nat Biomed Eng 3:421–422
14. Zhang Y, Wang Y, Huang X, Li Q, Li K, You D, Song Y, Min S, Zhou B, Wang W (2019) Association between *AXIN1* gene polymorphisms and epithelial ovarian cancer in Chinese population. Biomark Med 13(6):445–455. https://doi.org/10.2217/bmm.10.12
15. Properzi F, Logozzi M, Fais S (2013) Exosomes: the future of biomarkers in medicine. Biomark Med 7(5):769–778. https://doi.org/10.2217/bmm.10.12
16. Salehi M, Sharifi M (2018) 2018. Exosomal miRNAs as novel cancer biomarkers: challenges and opportunities. J Cell Physiol 233 (9):6370–6380. https://doi.org/10.1002/jcp.26481
17. JCM D (2004) Commercial application of marker- and gene-assisted selection in livestock- strategies and lessons. J Anim Sci 82:313–328
18. Pal A, Chakravarty AK, Chatterjee PN (2014) Polymorphism of growth hormone gene and its association with seminal and sexual

behavioral traits in crossbred cattle. Theriogenology 81:474–480

19. Pal A, Chakravarty AK, Bhattacharya TK, Sharma A (2005) Polymorphism of growth hormone gene and its association with expected milk production traits in dairy bulls. J Appl Animal Res 27(9):29–33
20. Pal A, Chakravarty AK, Bhattacharya TK, Joshi BK, Sharma A (2004) Detection of polymorphism of growth hormone gene for the analysis of relationship between allele type and growth traits in Karan fries cattle. Asian Aust J Animal Sci 17(9):1334–1337
21. Pal A, Chatterjee PN, Das S, Batobyal S, Biswas P, Sharma A (2017) Biodiversity among sheep and goat reared under different agroclimatic regions of West Bengal, India. Indian J Animal Sci 87(1):80–86
22. Ruane J, Colleau JJ (1996) Marker assisted selection for a sex-limited character in a nucleus breeding population. J Dairy Sci 79:1666–1678
23. Thomasen JR, Danner CE, Willam A, Sorensen C (2013) Genomic selection strategies in a small dairy cattle population evaluated for genetic gain and profit. J Dairy Sci 97 (1):458–470. https://doi.org/10.3168/jds.2013-6599
24. Uleberg E, Meuwissen THE (2007) Fine mapping of multiple QTL using combined linkage and linkage disequilibrium mapping - a comparison of single QTL and multi QTL methods. Genet Sel Evol 39:285–299
25. der Werf JHJ V (2007) Marker assisted selection in sheep and goats. In: Marker assisted selection (MAS) in crops, livestock, forestry and fish: current status and the way forward, FAO Invited Book Chapter 13

Chapter 28

Single-Cell Genomics

Abstract

Single-cell sequencing is the process for examination of the sequence information from individual cells with the help of appropriate next-generation sequencing (NGS) technologies. The advantages of single-cell sequencing are that it provides a better resolution of cellular differences and a detail understanding of the function of an individual cell with respect to its microenvironment. Certain limitations associated with single-cell genomics involves degradation, sample loss and contamination, uneven coverage, noise, and erroneous quantification of sequencing data due to picogram quantities of samples used. It is effective in studying heterogeneous samples, rare cell types, analysis of microorganisms that cannot be cultured, cell lineage relationships, evolution of diseases, relationship among cell lineages, and mosaicism of somatic tissue. The current chapter discusses the single-cell sequencing technologies for single-cell RNA sequencing (scRNA seq), single-cell genome (DNA) sequencing, and single-cell DNA methylome sequencing- epigenetic study. Application for genomics include cancer, metagenomics, stem cells, developmental biology, immunology, neurobiology, drug discovery, reproductive health, microbial ecology and evolution, plant biology, and forensics and allele-specific gene expression. Various case studies of single-cell genomics for early maternal and fetal interface, for evolutionary and taxonomic studies, and for studying cancer genomics have been included.

Keywords Single-cell sequencing, Single-cell RNA sequencing (scRNA seq), Single-cell genome (DNA) sequencing, Single-cell DNA methylome sequencing

1 Basic Principles

Single-cell genomics includes studying various biotechniques at single-cell level.

Single-cell sequencing is the methodology for examination of the sequence information from individual cells using appropriate next-generation sequencing (NGS) technologies. The advantages of single-cell sequencing is that it provides a better resolution of cellular differences and a detail understanding of the function of an individual cell with respect to its microenvironment. Sequencing the DNA of individual cells provides information about mutations possessed by small populations of cells, as in cancer. On the other hand, sequencing the RNAs expressed by individual cells can give

Aruna Pal, Protocols in Advanced Genomics and Allied Techniques, Springer Protocols Handbooks,
https://doi.org/10.1007/978-1-0716-1818-9_28,

information about the existence and behavior of different cell types, as in development.

It is more difficult to conduct single-cell sequencing as compared with sequencing from cells in bulk. The small amount of starting materials from a single cell may cause degradation, sample loss, and contamination, which in turn exert pronounced effects on quality of sequencing data. Since very small amount of nucleic acid (in pictogram quantity) is used, usually heavy amplification is necessary for sample preparation of single-cell sequencing. This results into uneven coverage, noise, and erroneous quantification of sequencing data. Recent technical advancements build single-cell sequencing a promising tool for approaching a set of seemingly inaccessible problems. Single-cell genomics have multifactorial effects. It is effective in studying heterogeneous samples, rare cell types, analysis of microorganisms that cannot be cultured, cell lineage relationships, evolution of diseases, relationship among cell lineages, and mosaicism of somatic tissue. Considering the importance of the technique, single-cell sequencing was regarded as "the method of the year 2013 by Nature Publishing Group."

2 Introduction to Single-Cell RNA Sequencing

Complex biological systems are governed by the coordinated functions of individual cells. Conventional techniques involving bulk genome or transcriptome data are not able to explore the cellular heterogeneity leading to the complexity. Single-cell sequencing is a next-generation sequencing (NGS) method that analyzes the genomes or transcriptomes of individual cells, which provides a high-resolution view of cell-to-cell variation.

Highly sensitive ultra-low-input and single-cell RNA sequencing (RNA-Seq) methods are able to explore the distinct biology of individual cells in complex tissues leading to the understanding of cellular subpopulation responses to environmental cues. These improve the study of cell function and heterogeneity in time-dependent processes such as differentiation, proliferation, and tumorigenesis [1, 2].

As regards the protocol, it is different for each and every different genomic techniques that are dealt separately in brief. It is considered as one of the most latest technologies in genomics developed so far. A schematic representation of single-cell gene expression workflow has been depicted in Fig. 1.

The basic protocol for single-cell genomics involves the following steps:

1. Biological input.
2. Library preparation: It is conducted with chromium controller single-cell kits.

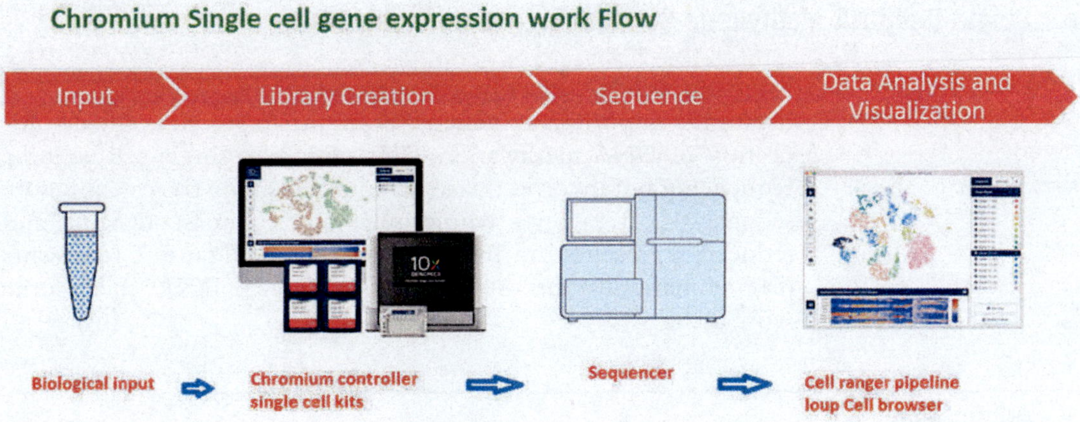

Fig. 1 A schematic representation of single-cell gene expression workflow

3. Sequencing.
4. Data analysis and visualization.

Some of the techniques are being described in brief:

3 Single-Cell Genome (DNA) Sequencing

The entire technique involves a series of processes:

1. Isolation of a single cell.
2. Whole-genome-amplification (WGA).
3. Construction of sequencing libraries.
4. Sequencing the DNA using a next-generation sequencer (e.g., Illumina and Ion Torrent).

A genome constructed in this way is generally referred to as a single amplified genome or SAG. It is equally applicable in microbiome studies for obtaining genomic data from uncultured microorganisms. It can also be united with high-throughput cell sorting of microorganisms and cancer.

One of the common methods employed for single-cell genome sequencing is multiple displacement amplificationleading to research into various avenues such as microbial genetics, ecology, and infectious diseases. Data retrieved from microorganisms might establish processes for culturing in the future. Examples of the genome assembly tools that can be employed in single-cell genome sequencing include SPAdes, IDBA-UD, Cortex, and HyDA.

4 Single-Cell DNA Methylome Sequencing—Epigenetic Study

"Single-cell DNA methylome sequencing" has been regarded as one of the important epigenetic techniques involved in the quantification of DNA methylation. This resembles single-cell genome sequencing but involves the addition of a bisulfite treatment before sequencing. It involves whole genome bisulfite sequencing and reduced representation bisulfite sequencing. Figure 2 represents diagrammatically the outline of single-cell DNA methylome sequencing.

5 Applications

Single-cell genomics has a wide range of applications as described as below diagrammatically (Fig 5):

scRNA-Seq is gaining popularity across different biological disciplines such as development, neurology, oncology, autoimmune disease, and infectious disease. scRNA-Seq has provided considerable insight into the development of embryos and organisms. The first vertebrate animals to be mapped through scRNAseq were Zebrafish and Xenopus laevis. In such cases, multiple stages of the embryo were studied, allowing the entire process of development to be mapped on a cell-by-cell basis.

6 Single-Cell RNA Sequencing (scRNA-Seq)

The standard procedure for RNA-seq analysis such as microarrays and standard bulk RNA-seq analysis involves the expression of RNAs from large populations of cells. However, with this technique, it is not possible to identify the differences between individual cells within these populations.

Single-cell RNA sequencing (scRNA-seq) provides opportunity to study the expression profiles of individual cells. Gene clustering analysis is employed to study the pattern of gene expression. Since small amount of material is available, it is quite difficult to get complete information on every RNA expressed by each cell.

This can reveal the existence of rare cell types within a cell population that may never have been seen before, which are the novel findings. Rare specialized cells in the lung called pulmonary ionocytes have been identified, which are expressed in case of cystic fibrosis with this technique. Transmembrane conductance regulator was identified in 2018 by two groups performing scRNA-Seq on lung airway epithelia, which is also a novel finding.

Protocol for single-cell RNA sequencing:

Current scRNA-seq protocols involve the following steps:

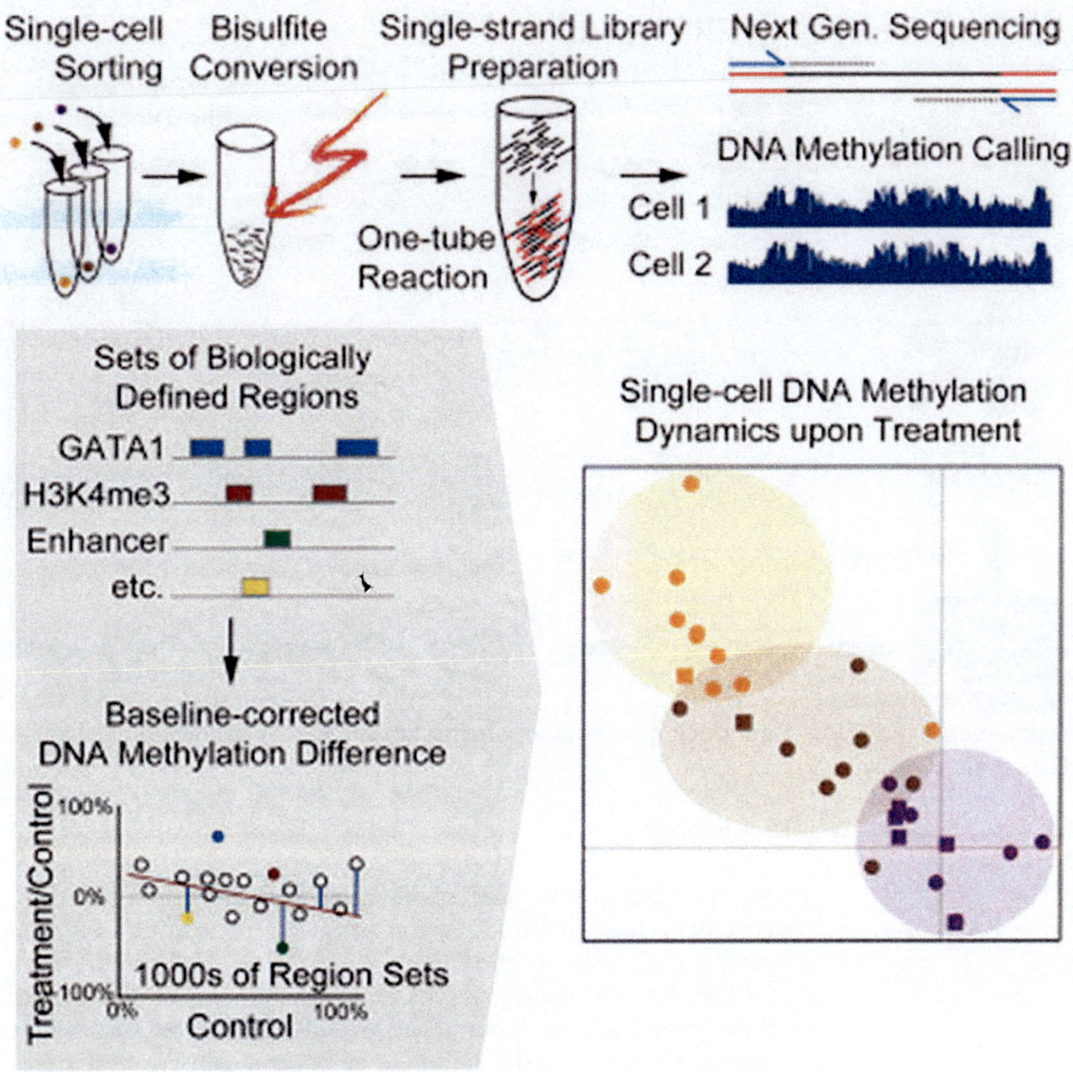

Fig. 2 Outline of single-cell DNA methylome sequencing

1. Isolation of single cell and RNA.
2. Reverse transcription (RT).
3. Amplification.
4. Library generation and sequencing.

Earlier, individual cells were separated into separate wells. The recent methodology includes the encapsulation of individual cells in droplets in a microfluidic device. The reverse transcription reaction is being performed, converting RNAs to cDNAs within each droplet. Each droplet is labeled with a DNA "barcode" containing cDNAs derived from a single cell. After the process of reverse transcription, the cDNAs from many cells can be pooled together

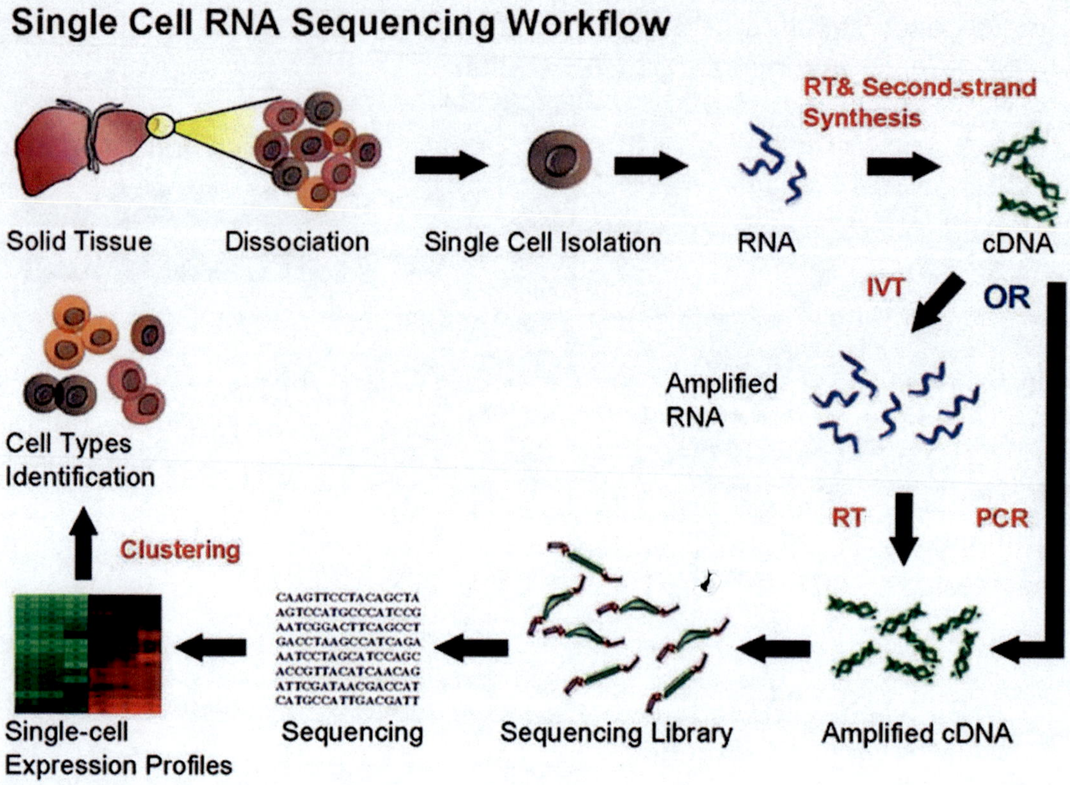

Fig. 3 Single-cell RNA sequencing workflow

for sequencing. In the process of sequencing, the transcripts from a particular cell are identified by the unique barcode. Figures 3 and 4 describe the outline of the steps involved in single-cell RNA-seq.

The critical stage for scRNA-Seq is the process of preservation of the initial relative abundance of mRNA in a cell as well as identification of rare transcripts. The reverse transcription step is crucial as the efficiency of the RT reaction is determined by the amount of RNA that will be analyzed by the sequencer. Reverse transcriptases and the priming conditions have a positive effect on full-length cDNA production and the generation of libraries biased toward 3′ or 5′ end of genes.

In order to amplify cDNA, currently two techniques are in use: PCR or in vitro transcription. The advantage of PCR-based method is that it can produce full-length cDNA.

But the major limitation is that with PCR amplification, particular sequences (for instance, GC content and snapback structure) can also be exponentially amplified, leading to libraries with uneven coverage. On the other hand, libraries generated by IVT can overcome such PCR-induced sequence bias. As a result, specific sequences can be transcribed inefficiently, which leads to sequence drop-out or production of incomplete sequences. Several scRNA-

Fig. 4 Diagrammatic representation of scRNA-Seq

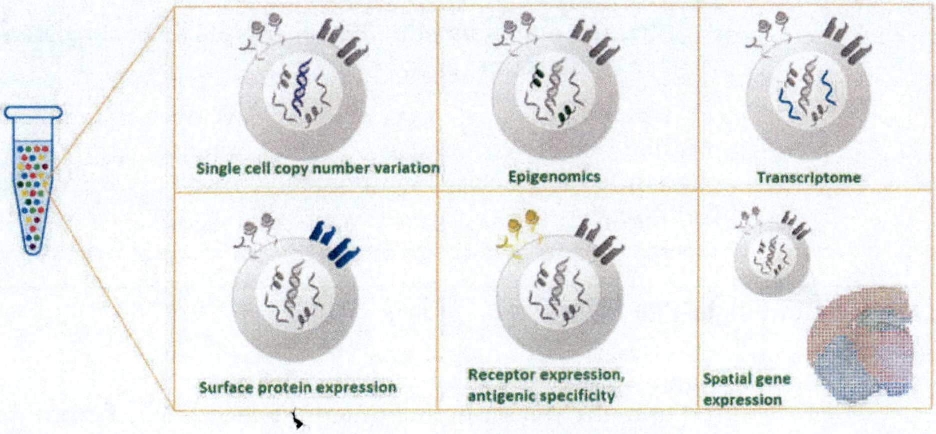

Fig. 5 Diagrammatic representation of applications for single-cell genomics

seq protocols are available, for example, STRT, SMART-seq, CEL--seq, RAGE-seq, Quartz-seq., and C1-CAGE. Each of these protocols is different in terms of processes for reverse transcription, cDNA synthesis, and amplification. Differences also exist in the possibility to include sequence-specific barcodes (i.e., UMIs) or the ability to process pooled samples.

scRNA-Seq gives rise to deeper insight to the multi-tiered complexity of different cells within the same tissue type. scRNA-Seq is now applicable for other methods in the single-cell RNA sequencing field. The methodology involves manual isolation of single cells under a microscope and lysed accordingly. In the next process, mRNAs are purified and primed with a poly(T) primer for reverse transcription. Unreactive primers are eliminated by exonuclease I digestion. The subsequent steps include addition of Poly (A) tails to the first strand cDNA at the 3′ end and annealing of poly (T) primers for second-strand cDNA generation. Ultimately, the cDNAs are PCR-amplified, sheared, and prepared into sequencing libraries.

There are certain pros and cons for scRNA seq, as described below:

Pros:

- It leads to single-cell resolution transcriptomic analysis.
- It can detect unknown splice junctions.
- In this technique, various RNAPII-specific antibodies can be employed to improve targeting accuracy.

Cons:

- It has very low throughput.
- In this technique, there is no scope to use molecular barcodes.
- It can be expensive to scale up.
- It has the requirement of TruSeq RNA Library Preparation Kit v2, a cumbersome process.

Later in 2017, two processes were developed to simultaneously measure single-cell mRNA and protein expression with oligonucleotide-labeled antibodies known as REAP-seq and CITE-seq.

7 Advantages of Single-Cell RNA-Seq

Single-cell and ultra-low-input RNA-Seq are powerful techniques to study the transcriptome in an unbiased manner from minimal input.

1. The technique can lead to robust transcriptome analysis down to single-cell input levels for high-quality samples.
2. Integrated protocol of scRNA seq involves directly from whole cells and preserves sample integrity.
3. High-resolution analysis leads to the discovery of cellular differences, which are generally masked by bulk sampling methods.

8 The Chromium Single-Cell Gene Expression Solution

The Chromium Single-Cell Gene Expression Solution is one of the most advanced technologies, which provides a solution for cell characterization and gene expression profiling of hundreds to tens of thousands of cells. The technique enables the detection of even more unique transcripts per cell. Addition of Feature Barcoding technology, a more complete molecular readout cell by cell is possible. It aids in identification of cell-specific CRISPR-mediated perturbations, measurement of gene, and cell surface protein expression in the same cell. It has the ability with unlimited possibilities in near future.

It has the ability for further gene expression analysis to characterize cell populations, cell types, cell states, and more on a cell-by-cell basis. Chromium single-cell gene expression solution can have an added advantage of assessing tumor heterogeneity, stem cell composition, and dissecting neuronal populations. It acts with

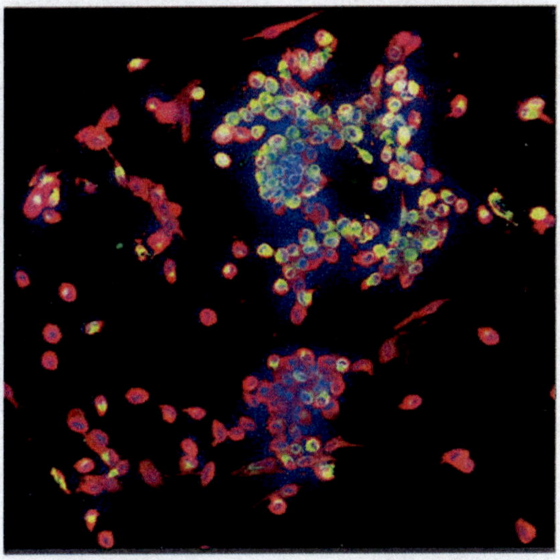

Fig. 6 Overview of single-cell visualization genomics

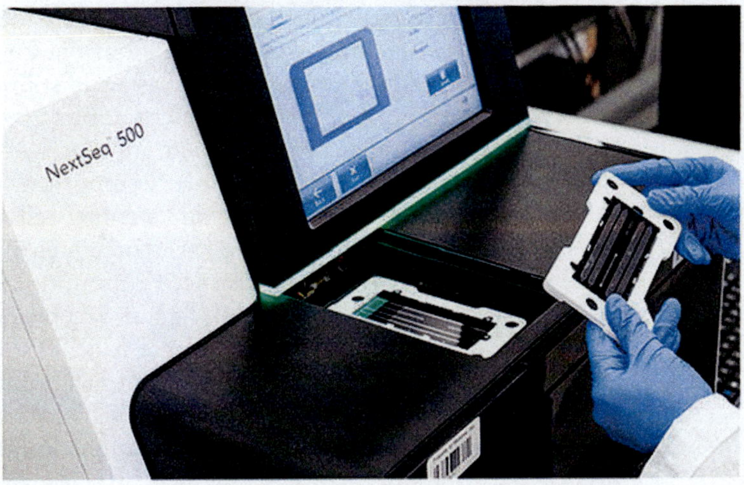

Fig. 7 Photograph for Next Seq 500 applicable for single-cell genome studies

turnkey software tools. This technology permits the creation of highly complex libraries from single cells to maximize insight from any sample type. Figure 6 describes the overview of single-cell visualization genomics. Figure 7 depicts the Photograph for Next Seq 500 applicable for single-cell genome studies.

The latest functions of Chromium Single-Cell Gene Expression Solution are listed as follows:

1. Identification and characterization of rare cell types.
2. Analyzing and understanding cellular heterogeneity and its contribution to biological system.

3. Performing cellular phenotyping with single-cell RNA-seq for the identification of novel targets, biomarkers, and cell types and states without the requirement for preselected targets.
4. Evaluation of mRNA and cell surface protein expression profiles within the same cell.
5. Performing high-throughput and high-resolution functional genetic screens in tens of thousands of cells at a time.
6. Assessing comprehensive gene expression phenotypes for individual CRISPR perturbations.

9 A Case Study: Single-Cell Reconstruction of the Early Maternal–Fetal Interface in Humans

Here, we discuss a case study when single-cell genomics is applicable for studying reproductive biology in human. During early human pregnancy, the uterine mucosa is modified into the decidua, where the fetal placenta implants. Placental trophoblast cells then intermingle and communicate with maternal cells. Trophoblast–decidual interactions are responsible for common diseases of pregnancy, including pre-eclampsia and stillbirth. It will be helpful if we profile the transcriptomes of about 70,000 single cells from first-trimester placentas, which are matched with maternal blood and decidual cells. The cellular composition of human decidua reflects subsets of perivascular and stromal cells that are located in distinct decidual layers. There exist three major subsets of decidual natural killer cells having distinctive immunomodulatory and chemokine profiles. A repository of ligand–receptor complexes were developed and analyzed with statistical tool for the prediction the cell-type specificity of cell–cell communication through these molecular interactions. The data identify many regulatory interactions preventing harmful innate or adaptive immune responses in this environment. Thus, single-cell atlas of the maternal–fetal interface details about the cellular organization of the decidua and placenta. It also reflects the interactions that are critical for placentation and reproductive success [3, 4].

10 A Case Study: Molecular Classification and Comparative Taxonomics of Foveal and Peripheral Cells in Primate Retina

In primates, including human, high acuity vision in primates is controlled by a small central retinal region known as the fovea. Since fovea is absent in most of the accessible organisms, its specialized function and dysfunction in ocular diseases are not properly understood. In the current study, 165,000 single-cell RNA-seq profiles were generated to study the complete picture of cellular taxonomies of macaque fovea and peripheral retina. The results reveal that more than 80% of >60 cell types match between the two regions. However, differences were observed in proportions and gene expression, which are related to functional differences. Comparisons were made for retinal types of macaques with those of mice. It was observed that interneuron types are tightly conserved. On the contrary, diversity was observed between projection neuron types and programs, although conserved transcription factor codes were reported. Key macaque types are conserved in humans, leading to the mapping of cell-type and region-specific expression of >190 genes responsible for seven human retinal diseases. The present work gives rise to a framework for comparative single-cell analysis across tissue regions and species [5].

11 Acquired Cancer Resistance to Combination Immunotherapy from Transcriptional Loss of Class I HLA: A Case Study

Resistance to cancer immunotherapy is critical in case of positive outcome. It becomes extremely essential to understand the mechanism of such resistance. Cellular immunotherapy trials serve as a means to probe complex tumor–immune interfaces with the help of defined T-cell/antigen interactions. Two patients were treated possessing metastatic Merkel cell carcinoma with autologous Merkel cell polyomavirus specific CD8+ T cells and immune-checkpoint inhibitors. In either case, dramatic recovery was due to dense infiltration of activated CD8+s into the regressing tumors. Moreover, late relapses occur at 22 and 18 months, respectively.

In the current study, single-cell RNA sequencing with identified dynamic transcriptional suppression of the specific HLA genes was reported. HLA genes represent the targeted viral epitope in the resistant tumor as a result of intense CD8-mediated immunologic pressure. This observation is different from genetic HLA-loss by its reversibility with drugs. Transcriptional suppression of Class I loci may be responsible for resistance to other immunotherapies, including checkpoint inhibitors. These have positive implications for the design of improved immunotherapy treatments [6].

12 Applications for Single-Cell Genomics

Single-cell genomics is an effective and advanced technology with varieties of applications. These include cancer, metagenomics, stem cells, developmental biology, immunology, neurobiology, drug discovery, reproductive health, microbial ecology and evolution, plant biology, forensics, and allele-specific gene expression.

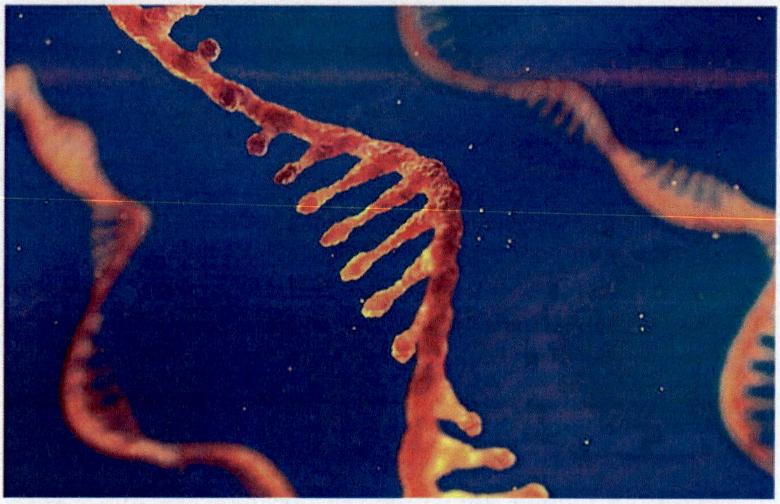

Fig. 8 Concept of single cell genomics

13 High- and Low-Throughput Methods

Single-cell sequencing methods can be differentiated based on cell throughput. Low-throughput methods involve the process of mechanical manipulation or cell sorting/partitioning technologies. Low-throughput single-cell sequencing has the ability to process dozens to a few hundred cells per experiment.

Recent methodologies of high-throughput single cell profiling leads to the advances in microfluidic technologies. This allowed the researchers to examine hundreds to thousands of cells per experiment in a manner, which is cost-effective.

Illumina sequencing methodologies are used for both the high- and low-throughput methods. It is one of the most widely adopted NGS technologies, which generates approximately 90% of sequencing data worldwide.

References

1. Nojima T, Gomes T, Carmo-Fonseca M, Proudfoot NJ (2016) Mammalian NET-seq analysis defines nascent RNA profiles and associated RNA processing genome-wide. Nat Protoc 11:413–428
2. Nojima T, Gomes T, Grosso AR et al (2015) Mammalian NET-Seq reveals genome-wide nascent transcription coupled to RNA processing. Cell 161:526–540
3. Vento-Tormo R et al (2018b) Single-cell reconstruction of the early maternal–fetal interface in humans. Nature 563(7731):347–353. https://doi.org/10.1038/s41586-018-0698-6
4. Vento-Tormo R, Efremova M, Botting RA, Turco MY, Vento-Tormo M, Meyer KB, Park JE, Stephenson E, PolaÅski K, Goncalves A, Gardner L, Holmqvist S, Henriksson J, Zou A, Sharkey AM, Millar B, Innes B, Wood L, Wilbrey-Clark A, Payne RP, Ivarsson MA, Lisgo S, Filby A, Rowitch DH, Bulmer JN, Wright GJ, Stubbington MJT, Haniffa M, Moffett A, Teichmann SA (2018a) Single-cell reconstruction of

the early maternal-fetal interface in humans. Nature 563:347–353

5. Peng YR et al (2019) Molecular classification and comparative Taxonomics of foveal and peripheral cells in primate retina. Cell 176 (5):1222–1237.e22. https://doi.org/10.1016/j.cell.2019.01.004

6. Paulson KG et al (2018) Acquired cancer resistance to combination immunotherapy from transcriptional loss of class I HLA. Nat Commun 9(1):3868. https://doi.org/10.1038/s41467-018-06300-3

Check for updates

Chapter 29

Big Data Analysis

Abstract

Big data is a term applicable to data that are too large or complex for traditional data-processing application software to adequately deal with. The importance of big data relies with the advent of high-throughput sequencing techniques, known as next-generation sequencing techniques. Big data analysis includes capturing data, data storage, data analysis, search, sharing, querying, updating, information privacy, transfer, visualization, and data source. The possible application of Big data relies on detection and estimation of QTLs, molecular marker–based classification, genotypes analysis of molecular variance (AMOVA), statistical issues in whole genome association (WGA) analysis, high dimensional data analysis by R, biological data handling—bioconductor, classification, prediction, and data reduction. Data mining, commonly used technique, is an interdisciplinary field of computer sciences, which is involved in application of statistical methods, artificial intelligence, and database management. Analysis was described with suitable examples, say using PLINK software. It also explains Data simulation using HapMap data.

Keywords Big data, Next-generation sequencing, AMOVA, Whole genome association, R package, Data mining, PLINK software, HapMap data

In this chapter, we will be dealing briefly with the concept with very large data and its analysis, popularly referred to as Big data. We commonly deal with the genomic data arising from high-throughput sequencing, known as next-generation sequencing. Big data may be defined as the large or complex data applicable for data-processing application software. Large data with large number of cases (rows) need a robust statistical analysis. However, highly complex data may result into a great FDR (false discovery rate). It equally refers to Genomic data analysis.

Big Data Analysis Involves a Number of Steps as Follows

1. Capturing data.
2. Data storage.
3. Data analysis.
4. Search.
5. Sharing.

Aruna Pal, *Protocols in Advanced Genomics and Allied Techniques*, Springer Protocols Handbooks,
https://doi.org/10.1007/978-1-0716-1818-9_29,

6. Querying.
7. Updating.
8. Information privacy.
9. Transfer.
10. Visualization.
11. Data source.

There are three basic concepts associated with Big data:

- Volume.
- Variety.
- Velocity.

Another basic concept associated with big data is veracity. Veracity refers to the noise in the data and its value.

Analysis of Big data involves strategic intervention of analyzing large data size. This big data may be generated from a wide variety of sources. The ultimate objective of analyzing the data is to reveal the underlying science, which were otherwise remain unexplored and give rise to newer information.

Nowadays, big data analytics was proved to be a valuable tool for analyzing the next-generation sequencing (NGS) data developed through high-throughput sequencing technologies.

Figure 1 diagrammatically gives an idea for a big data.

Some of the practical applications for the use of Big data analysis:

1. Quantitative trait loci (QTLs) detection and estimation.
2. Molecular marker–assisted selection.
3. Analysis of molecular variance (AMOVA) of genotypes.
4. Statistical analysis of whole genome association.
5. Whole genome analysis (WGA) analysis by R.
6. Bioconductor-based handling of data.
7. Data handling includes classification, prediction, and reduction of data.

1 The Advantages of Analysis for Big Data

- **Reduction of Cost**. Technologies for big data (e.g., Hadoop and cloud-based analytics) were observed to be cost effective for the storage of large data size.
- **Fast Decision Making**. Technologies (e.g., Hadoop and in-memory analytics, associated with the capability for the analysis of newer data sources. Big data analysis supports us for

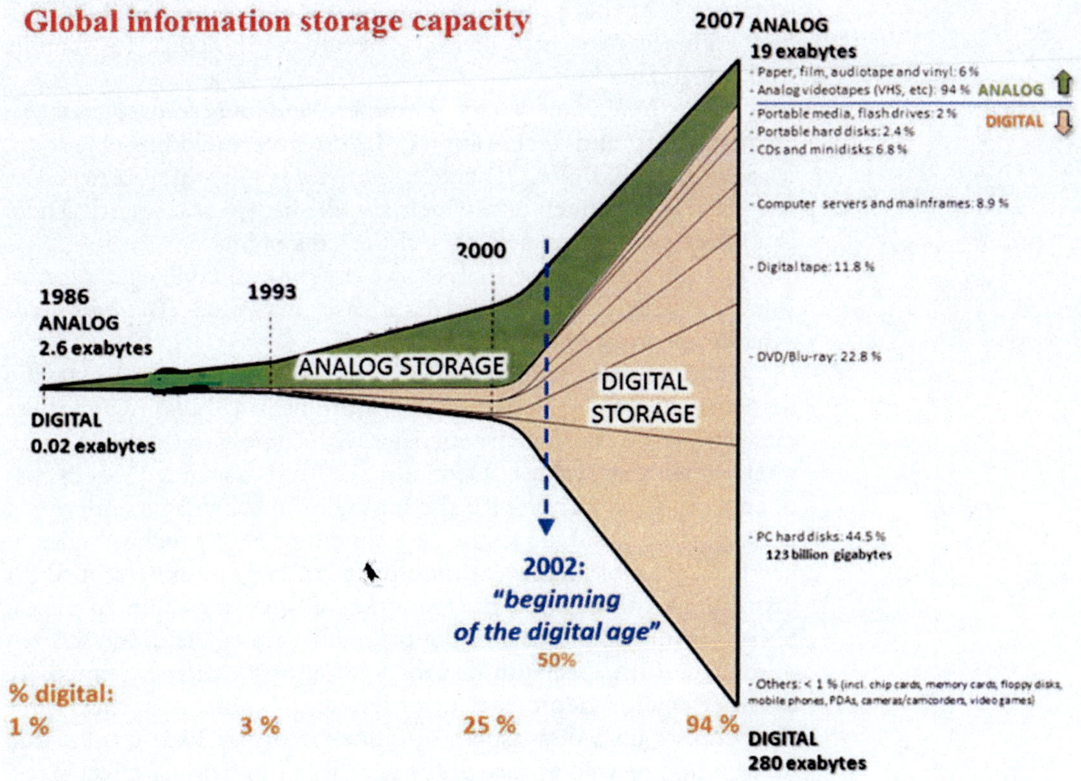

Fig. 1 Big data

instant analysis of information for effective decision-making process.

- **Newer Products and Services**. This technology can monitor the need of the customer and attain satisfaction through effective analytics. It can cater the personal need of the customers.

2 Data Mining

Data mining is a field of computer science, which is interdisciplinary in nature. It involves use of statistical methods, artificial intelligence, and database management.

The process involves exploration of new patterns from large data sets [1, 2]. It involves a process of machine learning technique detecting earlier unknown patterns and learning known patterns.

Big data analytics is a process allowing scientists involved with data analysis (data scientists) for the evaluation of greater volumes of data sources that were not able to deal with traditional systems. Traditional systems are not able to analyze large data structures. Sensitive software programs are employed for Big data analysis. The unstructured data to be utilized in Big data analysis may not be suitable for further processing. Big data requires processing at high end. Hence, occasionally, it becomes difficult to handle traditional data.

Thus, there is necessity for newer and bigger data analytics environments and technologies. It involves including Hadoop, MapReduce, and NoSQL databases. As a result, there is necessity for a software framework, which should be open accessed. These can process large data sets over clustered systems.

For the subsequent analysis, we describe suitable examples for Big data analysis. Let us discuss a case of ***GWAS (Genome wide association study) with Plink software***.

Genome-wide association studies (GWAS) is aimed at the detection of single-nucleotide polymorphisms (SNPs), when there exist differences in allele frequencies with respect to the function of phenotypic trait values. Detection of trait-associated SNPs has opened up new avenue into the biological mechanisms underlying these phenotypes. Due to the evolvement of newer technologies, it has been possible to reveal the impact of large numbers of SNPs distributed throughout the genome. GWAS was able to reveal SNPs contributing to the risk of psychiatric traits, including schizophrenia, autism spectrum disorders, attention deficit hyperactivity disorder, major depressive disorder, and bipolar disorder. This study opens up a new aspect of influence of psychiatric traits due to common as well as rare SNPs with small individual effect sizes. GWAS is based on in-depth knowledge of the genetic architecture of the human genome.

It is necessary to mention two important research initiatives, namely, the International HapMap Project and the 1000 Genomes project related to this aspect. The International HapMap Project (http://hapmap.ncbi.nlm.nih.gov/) explains the patterns of common SNPs in the human DNA sequence. On the other hand, the 1000 Genomes (1KG) project (http://www.1000genomes.org/) deals with map of both common and rare SNPs. As GWAS results explore that effect sizes of individual SNPs are small, scientists in the psychiatric have gained an interest in methodology aggregating the effect of SNPs.

Now, it becomes essential to explain polygenic risk score (PRS) analysis. This is considered as the most relevant method due to certain advantages:

1. It is relatively easy to conduct.
2. It can be applied to target samples with relatively modest sample sizes.

PRS deals to the effect sizes of multiple SNPs into a single aggregated score that can be utilized to predict disease risk. The PRS is an individual-level score calculated based on the number of risk variants carried by a person. It is weighted by SNP effect sizes that are derived from an independent large-scaled discovery GWAS. The score is an indication of the total genetic risk of a specific individual pertaining to a particular trait, utilized for clinical prediction or screening (e.g., breast cancer). In case of psychiatric traits, PRS is also significantly associated with case–control status. But, its discriminative accuracy may not be sufficient for clinical applications. PRS has led to the exploration of the genetic architecture of psychiatric traits through its ability to predict disease status. It has further been employed for the exploration of genetic effect sizes obtained from a GWAS of a particular phenotype of interest, which in turn may be employed for the prediction of the risk of another phenotype. Although recent GWAS have a major achievement to the identification of significantly associated SNPs for many phenotypic traits, social scientists and clinicians can explore newer researches to the genetics field, improving our understanding of the precise nature of the specific behavioral, cognitive, or neural correlates of detected risk SNPs.

But the analysis of genetic data needs several quality checks and statistical analyses to avoid error due to several potential sources of confounding (e.g., ethnic stratification). At least a fair knowledge of genetic power calculation is essential in order to avoid performing underpowered studies. Another tutorial gives a guideline to researchers who intend to incorporate genetics into their studies but do not possess a formal background in the particular topic.

1. First step is the rigorous quality control (QC) procedures on genotype before conducting GWAS. It includes the application of appropriate methodology to take into account the ethnic heterogeneity.
2. Second, the commonly used tests of association between SNPs and phenotypic traits of interest are discussed, while controlling for potential confounders.
3. Third, to explain the process to conduct PRS analysis.

Example R and Unix scripts are provided at https://github.com/MareesAT/GWA_tutorial/. It is not essential to elaborate on SNP imputation of unmeasured SNPs based on reference data. Imputation is a commonly used method in GWAS; it is out of the scope of this article. Open-source whole-genome association analysis toolset PLINK version 1.07 [3] that can be easily downloaded from http://zzz.bwh.harvard.edu/plink/. The PLINK 1.9 beta version possesses the same options but be much faster https://www.cog-genomics.org/plink/1.9/. As PLINK 1.9 is currently a beta version, it is advisable to employ official PLINK version in this tutorial. However, it is equally possible to complete all tutorials using PLINK 1.9. However, some of the steps explained in this article could be analyzed in conventional statistical packages such as R, it is recommended to use a software package particularly used for the analysis of genetic data. Other than PLINK, many other software are also available for the analysis of SNP data such as Genabel and SNPTEST. Other methods that allow for testing association in family-based GWAS have also evolved. We advocate GNU/Linux-based computer resources; however, many of the options are also available through the windows version of PLINK. A basic user guideline to shells and command lines can be found at http://www.ee.surrey.ac.uk/Teaching/Unix/. The graphs by the GitHub example scripts will be available using the free, open-source programming language R (https://www.r-project.org/).

3 Data Format

PLINK software uses the data available in either read text-format files or binary files. Since reading large text files can be time-consuming, it is advisable to use binary files. Text PLINK data comprise two files: the first contains information on the individuals and their genotypes (*.ped), and the other possesses the information on the genetic markers. On the contrary, binary PLINK data comprise three files, a binary file containing individual identifiers (IDs) and genotypes (*.bed), and two text files that contain information on the individuals (*.fam) and on the genetic markers (*.bim).

For example, in a study of bipolar disorder, the *.bed file would possess the genotyping results of all patients and healthy controls; the *.fam files have the subject-related data (family relationship with other participants in the study, sex, and clinical diagnosis). On the other hand, the *.bim file would show the information on the physical position of the SNPs. Analysis using covariates usually need a fourth file, containing the values of these covariates for each individual.

4 Basic PLINK Command

PLINK is a command line program; hence, its usage requires an active shell waiting for commands. This can be recognized by its prompt ($ or >) just before the cursor. Generally, the path of the current directory will be visualized before the prompt. The present directory is a central notion for PLINK usage, since by default, PLINK will load data files from and save result files in this directory. The current directory can be altered to any directory with the help of conventional Unix commands, typically cd. Following the prompt, the use of PLINK is determined by typing the plink keyword. In case, PLINK is not installed in a standard directory, the path to the directory where PLINK is installed needs to be typed in front of the command, for example, /usr/local/bin/plink. After the plink keyword, other options controlling PLINK's workflow will follow, separated by spaces. These options all begin with two dashes (--). One of the first options to provide is the format and the name of the data files: use --file {your_file} for text files and --bfile {your_file} for binary files. After that, all other required options can be added, for instance, the --assoc option to perform an association analysis; this specific option will tell PLINK to perform an X2 test for each SNP to the phenotype of interest. Multiple options can be combined within a single command line. Within PLINK, a default order is implemented that works irrespective of the order of the commands within the command line. A useful, and sometimes mandatory, option is --out {outfile}, which gives rise to a name to output files (suffixes will be added as needed by PLINK). Care should be taken that PLINK will delete without notice any existent file with the same name. Please note that the options within PLINK extend beyond what is being discussed in the current article; for the full set of options, *see* http://zzz.bwh.harvard.edu/plink/. Figure 2 gives an overview of various commonly used PLINK files, whereas Fig. 3 gives the structure of the PLINK command line [4].

5 QC of Genetic Data

One of the most important steps that is employed as a part of any GWAS is the use of appropriate QC (Quality control). Extensive QC is vital for GWAS to generate reliable results since raw genotype data are inherently imperfect. Errors in the data can give rise due to a number of reasons as listed below and others:

1. due to poor quality of DNA samples,
2. poor DNA hybridization to the array,
3. poorly performing genotype probes, and,
4. sample mix-ups or contamination.

6 Data Simulation Using HapMap Data

In order to be able to illustrate all analysis steps using realistic genetic data, we simulated a dataset ($N = 207$) with a binary outcome measure utilizing the publicly available data from the International HapMap Project (http://hapmap.ncbi.nlm.nih.gov/downloads/genotypes/2010-05_phaseIII/ plink_format/) [5]. For this tutorial, for the creation of an ethnically homogenous dataset, only Utah residents with ancestry from Northern and Western Europe (CEU) were included. Due to comparative small

*.ped

FID	IID	PID	MID	Sex	P	rs1	rs2	rs3
1	1	0	0	2	1	CT	AG	AA
2	2	0	0	1	0	CC	AA	AC
3	3	0	0	1	1	CC	AA	AC

*.map

Chr	SNP	GD	BPP
1	rs1	0	870000
1	rs2	0	880000
1	rs3	0	890000

*.fam

FID	IID	PID	MID	Sex	P
1	1	0	0	2	1
2	2	0	0	1	0
3	3	0	0	1	1

*.bed

Contains binary version of the SNP info of the *.ped file. (not in a format readable for humans)

*.bim

Chr	SNP	GD	BPP	Allele 1	Allele 2
1	rs1	0	870000	C	T
1	rs2	0	880000	A	G
1	rs3	0	890000	A	C

Covariate file

FID	IID	C1	C2	C3
1	1	0.00812835	0.00606235	–0.000871105
2	2	–0.0600943	0.0318994	–0.0827743
3	3	–0.0431903	0.0133068	–0.000276131

Legend			
FID	Family ID	rs{x}	Alleles per subject per SNP
IID	Individual ID	Chr	Chromosome
PID	Paternal ID	SNP	SNP name
MID	Maternal ID	GD	Genetic distance (morgans)
Sex	Sex of subject	BPP	Base-pair position (bp units)
P	Phenotype	C{x}	Covariates (e.g., Multidimensional Scaling (MDS) components)

Fig. 2 Overview of various commonly used PLINK files; SNP, single-nucleotide polymorphism

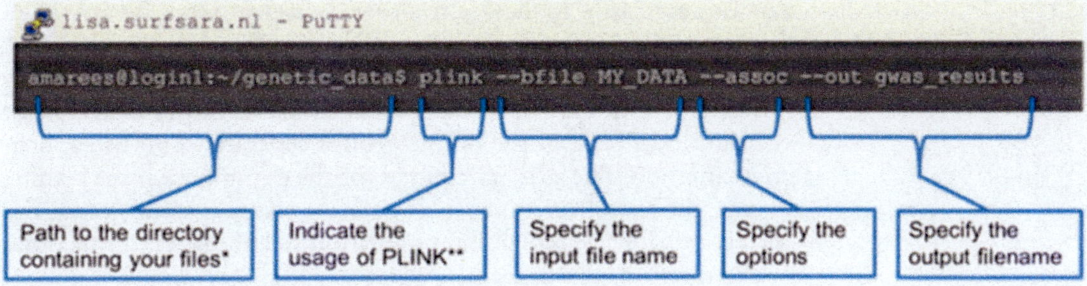

Fig. 3 Structure of the PLINK command line

sample size of the HapMap data, genetic effect sizes in these simulations were set at values larger than that realized in genetic studies of complex traits. Larger sample size (e.g., at least in the order of thousands but likely even tens or hundreds of thousands) is a prime requirement for the detection of genetic risk factors of complex traits. The HapMap data with a simulated phenotypic trait can be observed at https://github.com/MareesAT/GWA_tutorial/ (1_QC_GWAS.zip).

7 Overview of QC Steps

Due to the challenges faced while characterizing GWAS, it is essential to illustrate essential QC steps and to provide example scripts. The following points describe the summary of seven QC steps, which include recommendations regarding the specific thresholds. However, thresholds may vary as per the study specific characteristics.

The seven QC steps are based on the filtering out of SNPs and individuals based on the following:

1. individual and SNP missingness,
2. inconsistencies in assigned and genetic sex of subjects,
3. minor allele frequency (MAF),
4. deviations from Hardy–Weinberg equilibrium (HWE),
5. heterozygosity rate,
6. relatedness, and.
7. ethnic outliers.

Hands-on Experience with the performance of QC **Steps 1–7** is essential through following all steps outlined in the online tutorial at https://github.com/MareesAT/GWA_tutorial/ (1_QC_GWAS. zip +2_Population_stratification.zip). It leads to the scripts for data QC and visualization of potential sources of bias. These scripts analyze QC on the CEU group of the HapMap data but are easily applicable to other datasets, except family-based datasets and datasets involving multiple distinct ethnic groups. As a general rule, when a sample includes multiple ethnic groups (e.g., Africans, Asians, and Europeans), it is advisable to perform tests of association in each of the ethnic groups separately and use appropriate methods like as meta-analysis for the combination the results. In case, the sample includes subjects from a single ethnic group, the corrections for the remaining population stratification can be done by the methods discussed below.

8 Controlling for Population Stratification

A significant source of systematic bias in GWAS is population stratification. It has been observed that even subtle degrees of population stratification within a single ethnic population may exist. Hence, an essential quality control step is the testing and controlling for the presence of population stratification. Several methods are in vogue to correct for population stratification.

In this tutorial, the method will be discussed which is incorporated in PLINK: the multidimensional scaling (MDS) approach. This method is based on the calculation of the genome-wide average proportion of alleles shared between any pair of individuals within the sample in order to obtain quantitative indices (components) of the genetic variation for each individual. The individual component scores can be plotted for the exploration of groups of individuals that are genetically more similar to each other than expected. For example, in a genetic study involving subjects from Asia and Europe, MDS analysis would explore that Asians are genetically more similar to each other than to Europeans. For the investigation of the individuals with the generated component scores, which deviate from the samples target population, plotting of the scores of the sample under investigation will be helpful. A population of known ethnic structure (e.g., HapMap/1KG data) is helpful in this regard. This step is known as anchoring. This allows the researcher to get ethnic information on their sample and to assess possible ethnic outliers. A script is present at https://github.com/MareesAT/GWA_tutorial/ (2_Population_stratification.zip) for performing MDS on own data anchored by data of the 1KG project (http://www.1000genomes.org/). Individuals identified as outliers based [on the MDS analysis need to be removed from further analyses.] Followed by the exclusion of these individuals, a new MDS analysis need to be conducted, and its main components are to be utilized as covariates in the association tests. It is done to correct any remaining population stratification within the population. The number of components need to be included is dependent upon the population structure and the sample size. However, the inclusion of up to 10 components is normally accepted.

Note: Please check that it is always advisable to use Linux-based system for big data analysis.

References

1. Big data Analytics. https://www.sas.com/en_in/insights/analytics/big-data-analytics.html
2. Kirshners A, Parshutin S, Leja M (2012) Research on application of data mining methods to diagnosing gastric cancer. In: Perner P (ed) Advances in data mining. Applications and theoretical aspects. ICDM 2012. Lecture notes in computer science, vol 7377. Springer, Berlin, Heidelberg
3. Purcell S, Neale B, Todd-Brown K, Thomas L, Ferreira MA, Bender D, Maller J, Sklar P, de Bakker PI, Daly MJ, Sham PC (2007) PLINK: a tool set for whole-genome association and population-based linkage analyses. Am J Hum Genet 81(3):559–575. https://doi.org/10.1086/519795
4. PLINK 1.9. 2014. http://zzz.bwh.harvard.edu/plink/
5. The International HapMap Consortium (2003) The International HapMap Project. Nature 426:789–796. https://doi.org/10.1038/nature02168

INDEX

A

B

Aruna Pal, *Protocols in Advanced Genomics and Allied Techniques*, Springer Protocols Handbooks,
https://doi.org/10.1007/978-1-0716-1818-9,

D

E

F

G

H

M

N

S

T